AF835287

ANTARCTIC ENVIRONMENTS AND RESOURCES

A geographical perspective

Antarctic Environments and Resources

A Geographical Perspective

JAMES D. HANSOM & JOHN E. GORDON

Routledge
Taylor & Francis Group
LONDON AND NEW YORK

First published 1998 by Addison Wesley Longman Limited

Published 2013 by Routledge
2 Park Square, Milton Park, Abingdon, Oxfordshire OX14 4RN
711 Third Avenue, New York, NY 10017, USA

First issued in hardback 2017

Routledge is an imprint of the Taylor & Francis Group, an informa business

Copyright © 1998, Taylor & Francis.

The right of James D. Hansom and John E. Gordon to be identified
as authors of this Work has been asserted by
them in accordance with the Copyright,
Designs and Patents Act 1988.

All rights reserved. No part of this book may be reprinted or reproduced or utilised in any form or by any electronic, mechanical, or other means, now known or hereafter invented, including photocopying and recording, or in any information storage or retrieval system, without permission in writing from the publishers.

Notices
Knowledge and best practice in this field are constantly changing. As new research and experience broaden our understanding, changes in research methods, professional practices, or medical treatment may become necessary.

Practitioners and researchers must always rely on their own experience and knowledge in evaluating and using any information, methods, compounds, or experiments described herein. In using such information or methods they should be mindful of their own safety and the safety of others, including parties for whom they have a professional responsibility.

To the fullest extent of the law, neither the Publisher nor the authors, contributors, or editors, assume any liability for any injury and/or damage to persons or property as a matter of products liability, negligence or otherwise, or from any use or operation of any methods, products, instructions, or ideas contained in the material herein.

ISBN 13: 978-1-138-18051-2 (hbk)
ISBN 13: 978-0-582-08127-7 (pbk)

British Library Cataloguing in Publication Data
A catalogue record for this book is available from the British Library.

Library of Congress Cataloging-in-Publication Data
A catalog entry for this book is available from the Library of Congress.

Set by 30 in Times NR MT

Contents

Preface

This book has been written over a period of great change in Antarctic affairs. In the last decade or so, Antarctica has emerged as one of the centrepieces of a global environmental agenda, a modern symbol of the last great wilderness on Earth. Several catalysts have sparked this emergence: following the overexploitation of the great whales, much concern was expressed about the impact of commercial fisheries on the marine ecosystem; there have been concerns about the impact of human activities on global atmospheric processes and the resultant pollution and climate change impacting on Antarctica and its ice cover; there has been the dramatic appearance of the ozone 'hole'; and there has been deep concern surrounding speculations on the exploitation of any mineral wealth that the continent might have. The emergence of these concerns has been paralled by the recognition of the role played by Antarctica in the control and stabilisation of many fundamental global climate and oceanographic processes. Its importance as an archive of climate history that helps us to understand how climate change may affect the global environment is unsurpassed. This image of Antarctica at the centre of global climatic and environmental processes, as well as environmental concerns, is very different from the picture of an insular and remote Antarctica that existed as recently as the 1960s. The growth of mass communications and tourism has further opened Antarctica to the world and it is no longer the exclusive preserve of a few explorers or scientists; it is a region whose processes influence the whole world and potentially affect the whole of society.

It is our belief that the geography of Antarctica is best understood through the interactions of the natural environment with human activities. It is also our belief that the geography of Antarctica is rooted in the events of the past as well as the present. These events span not only the discovery and exploration of the continent, as well as the exploitation of its resources and scientific interests and the resulting impact on the environment, but also the development of a unique international political framework for the management of the region. The book therefore highlights the legacy of past developments in these areas, which provides the background for understanding the new environmental agenda and the forces that will shape the future geography of the region. A key turning point was the shelving of the minerals convention, CRAMRA, and the rapid development of the Environmental Protocol to the Antarctic Treaty in 1991.

Against this background, this book was conceived with a view to attempt an integration of a wide range of environmental information within the context of human use of Antarctic resources and to provide an overview of the material necessary for an understanding of Antarctic environments and resources, as well as of their exploitation and management. In presenting what should amount to a 'new geography' of Antarctica, we are very much aware that oversimplification is the mother of misconception and have attempted to construct a balanced view of a wide range of material. This has not been an easy task in the rapidly changing areas of scientific research and international policy developments, and such an attempt risks simply reflecting the state of play at the time of writing. To help to circumvent this, we have attempted to give a flavour of the types of outcome that the future might hold for Antarctica and to suggest ways forward. Nevertheless, some areas, particularly those dealing with some aspects of terrestrial and marine biology, are touched on only in passing and we are only too aware that a fuller treatment would be an advantage. However, given such a general book and the need to be both brief and succinct, it is inevitable that there will be areas that are either omitted or given only limited treatment. For the topics that are covered, an extensive bibliography is included for those wishing to pursue issues of interest in greater depth than is possible in a general text such as this. A small amount of recommended reading, chosen to reflect breadth, balance, and integration of topics rather than publication date, is also included at the close of each chapter.

We have also turned to the experts in many fields for advice and guidance on a variety of aspects of the text and have been grateful for their willingness to point out our errors, misconceptions and often outdated ideas. Various aspects of the text and illustrations were expertly reviewed by P.J. Beck, A. Clarke, P.B. Davis, I. Everson, R.K. Headland, J. Priddle, M.G. Richardson, J.R. Shears, R.I.L. Smith, B.E. Storey, D.E. Sugden, M.B. Usher, D.W.H. Walton and R. Willan. Our thanks go to all of these Antarctic scientists. Our great thanks also go to P.D. Clarkson, C. Eriksson, R. Gambell, R. Galbraith, R.K. Headland, I. Reddish, J.R. Shears and B. Stonehouse for providing advice and information on a range of human impacts and environmental issues. Thanks are due to W.H. Mills and S. Sawtree at Scott Polar Research Institute Library for help in tracing material and to Willie Greenfield of Christian Salvesen plc for tracing and providing historical whaling pictures. We are grateful to Andy Alsop, an experienced Antarctic pilot, for allowing us to use his excellent aerial photographs. Donna Barfield of Quark Expeditions could not have been more helpful in providing up-to-date tourist information and illustrations. We are also grateful to J. Bonner, I.B. Campbell, T. Chinn, E.Stump, B. Kirk, D. Sugden, Discovery Point, Dundee, Canterbury Museum, K.-H. Kock and C.W.M. Swithinbank for providing photographs. The British Antarctic Survey was instrumental in providing access to a range of both illustrations and Antarctic experts who were willing to provide critical advice. Many of the photographic illustrations were expertly copied and handled by Ian Gerard and Les Hill, with computer image enhancement of the *Scotia* photography carried out by Norman Tait, all of the University of Glasgow. Yvonne Wilson of the Department of Geography and Topographic Science at the University of Glasgow transformed our diagrammatic offerings into elegant and stylish line drawings with a commendable degree of patience. Dick Birrie, Gordon Thorn and Roger Timmis were excellent Antarctic field companions. All at Addison Wesley Longman are also remembered. Finally, both of us owe a debt of gratitude to David Sugden and Chalmers Clapperton for their inspiring undergraduate and postgraduate teaching and guidance and for encouraging our lifelong fascination with, and concern for, the Antarctic environment. However, in spite of all the above expert help, any failings in the following text are solely the responsibility of the authors.

Acknowledgements

We would like to thank the following for permission to reproduce copyright material;

Charles Swithinbank for the photograph on the front cover, and figures 8.3 and 9.6; SCAR for figure 1.2 redrawn from a figure in *The Role of Antarctica in Global Change* (1993) and for figure 6.6 redrawn from a figure in *SCAR Bulletin*, **127** (October 1997); US Geological Survey for figures 1.3 and 9.2a; Cambridge University Press for figures 1.4 and 2.15 redrawn from two figures in JR Dudeny 'The Antarctic atmosphere' and figures 5.3, 5.6 and 5.7 redrawn from three figures in I Everson 'Life in a cold environment' in *Antarctic Science* (1987) edited by DWH Walton; Ed Stump for figures 2.3 and 8.4; C Bridge for figure 2.5 redrawn from a figure in RCR Willan et al 'The mineral resource potential of Antarctica: geological realities' in *The Future of Antarctica. Exploitation versus Preservation* (1990) edited by G Cook; PF Barker and Oxford University Press for figure 2.7d from *The Geology of Antarctica* (1991) edited by PF Barker, IWD Dalziel and BC Storey; Cambridge University Press and PD Rowley for figure 2.8 redrawn from a figure in PD Rowley et al. 'Metallogenic provinces of Antarctica' in *Antarctic Earth Science* (1983) edited by RL Oliver, PR James and JB Jago; DH Elliot and Oceanus for figure 2.9; The American Geographical Society for figure 2.10 redrawn from a map in HG Goodell 'The Sediments' in *Marine Sediments of the Southern Ocean*; University of Chicago Press for figure 2.13 from WD Sellars *Physical Climatology* (1965); Methuen & Co for figure 2.14 redrawn from T. Hatherton (ed.) *Antarctica* (1965); The Royal Meteorological Society for figure 2.17 redrawn from a figure in P Stark (1994) 'Climatic warming in the central Antarctic Peninsula area' in *Weather*, **49**(5); Macmillan Magazines Ltd for figure 2.19 - reprinted with permission from *Nature*, AE Jones and JD Shanklin 'Continued decline of total ozone over Halley, Antarctica since 1985' (1995). Copyright 1995 Macmillan Magazines Limited; The Scott Polar Research Institute for figure 2.20 redrawn from a figure in KB Mather and GS Miller (1967) 'The problem of the katabatic winds on the coast of Terre Adelie' in *Polar Record*, and for figures 8.5a and b redrawn from two figures in CM Harris (1991) 'Environmental effects of human activities on King George Island, South Shetland Islands, Antarctica' in *Polar Record*; DJ Drewry and Belhaven Press for figure 2.21; Trevor Chinn for figures 3.3 and 9.2b; Trevor Chinn and SIR Publishing for figure 3.36 which appeared as fig. 10.4 in TJH Chinn 'The Dry Valleys' in *Antarctica, the Ross Sea Region* (1990) edited by T Hatherton; DE Sugden or figures 3.4, 3.6 and 3.14, 3.10, 5.17, 6.1c,

6.7; Hermann Englehardt for figure 3.7; Macmillan Magazines Ltd and DG Vaughan for figure 3.8 – reprinted with permission from *Nature*, DG Vaughan and CSM Doake 'Recent atmosphere warming and retreat of ice shelves on the Antarctic Peninsula' (1996). Copyright 1996 Macmillan Magazines Limited; Macmillan Press Ltd for figure 3.10 from Clark, Gregory and Gunnell (1988) *Horizons in Physical Geography*; JR Petit for figure 3.11; Academic Press Inc. and J Jouzel for figure 3.12 redrawn from J Jouzel *et al* (1989) 'A Comparison of deep Antarctic ice cores and their implication for climate . . .' *Quaternary Research*, **31** (135–50); Macmillan Magazines and J Jouzel for figure 3.13 - reprinted with permission from *Nature*, J Jouzel et al 'Extending the Vostok ice-core record of palaeoclimate to the penultimate glacial period' (1993). Copyright 1993 Macmillan Magazines Limited; Blackwell Science Ltd for figure 3.16 redrawn from fig. 7 in PJ Barrett and MJ Hambrey (1992) 'Plio-Pleistocene sedimentation in Ferrar Fiord' in *Sedimentology*, and Blackwell Science Ltd and Ed Murphy for figures 4.19a and b redrawn from fig. 2a and b in EJ Murphy (1995) 'Spatial structure of the Southern Ocean ecosystem...' in *Journal of Animal Ecology* Julian Priddle for figures 3.18 and 5.21; James G Bockheim and John Wiley & Sons Ltd for figure 3.20 redrawn from a figure in JG Bockheim (1995) 'Permafrost distribution in the Southern Circumpolar Region and its relation to the environment...' in *Permafrost and Periglacial Processes*; S.W. Byers for figure 3.23c; Cambridge University Press for figure 3.24 redrawn from a figure in *The Biology of Polar Bryophytes* (1988) by RE Longton; RE Longton for figure 3.25; Andrew Kennedy and the Institute of Arctic and Alpine Research at the University of Colorado for figure 3.26 redrawn from fig. 2 in AD Kennedy (1993)'Water as a limiting factor in Antarctica' in *Arctic and Alpine Research*, **25**(4); H Janetschek for figure 3.27; The British Antarctic Survey for figures 3.28, 3.34, 4.14, 5.2, 5.8, 6.4, 6.5, 7.12 and 8.10 from its archive material and figure 8.9 redrawn from a figure in JR Shears (1993) *The British Antarctic Survey Waste Management Handbook*; Elsevier Science. RI Lewis-Smith for figure 3.29 reprinted from Van der Maarel, *Dry Coastal Ecosystems 2A: Polar Regions and Europe* (1993) pp 73–93 and pp 51–71, figure 7.3 and 6.3 with permission from Elsevier Science, RI Lewis-Smith 9.3, 9.4; Andrew Clarke for figure 3.35; DJ Webb for figure 4.3; HJ Zwally for figure 4.4; D Kelletat for figure 4.5; P Wadhams and Springer-Verlag GmbH & Co. KG for figure 4.7 redrawn from fig.5 in P Wadhams 'The Antarctic Sea Ice Cover' in *Antarctic Science: Global Concerns* (1994) edited by G Hempel; The Open University for figure 4.9 adapted from figure 6.19 in *Ocean Circulation* (1993); Fishing News (Books) Ltd and Blackwell Science Ltd for figure 4.10 redrawn from a figure in *The Stocks of Whales* (1965) by NA Mackintosh; Cambridge University Press and GA Knox for figures 4.11, 4.23 and 4.24 redrawn from three figures in *The Biology of the Southern Ocean* (1994) by GA Knox; Cornelius W Sullivan for figure 4.12a; Kluwer Academic Publishers M Spindler for figure 4.12b from Bleil, *Geologic History of the Polar Ocean* (1990); Paul Treguer and Springer-Verlag GmbH & Co. KG for figure 4.13 redrawn from fig.1 in P Treguer 'The Southern Ocean: Biogeochemical Cycles and Climate Change' in *Antarctic Science: Global Concerns* (1994) edited by G Hempel; RM Laws and American Scientist for figure 4.15 redrawn from fig. 6 in RM Laws (1985) 'The Ecology of the Southern Ocean' in *American Scientist*, **73**; RM Laws for figures 4.21 and 4.25; JP Croxall for figure 4.18; Scientific American for figure 5.1 from 'Antarctic Fishes' (November 1986) by Joseph T. Eastman and Arthur L. De Vries; Academic Press Ltd and RM Laws for figure 5.12 redrawn after a figure in RM Laws 'Seals' in *Antarctic Ecology*, Vol. 2 (1984) edited by RM Laws. By permission of the publisher Academic Press Limited, London; The American Meteorological Society for figure 5.15 from S Mannabe, MJ Spelman and RJ Stouffer 'Transient responses of a coupled ocean-atmosphere model...' in the *Journal of Climate*, Vol 5(2), February 1992; Macmillan Magazines Ltd for figure 5.16 - reprinted with permission from *Nature*, RJ Stouffer, S Manabe and K Bryan 'Interhemispheric asymmetry in climate response to a gradual increase of atmospheric CO2' (1989). Copyright 1989 Macmillan Magazines Limited; Canterbury Museum archives in Christchurch, New Zealand for figures 6.2 and 9.6a; The Public Record Office for figure 6.5; Jennifer Bonner for figure 7.3; W Greenfield and Christian Salvesen plc for figures 7.6 and 7.7; Cambridge University Press and K-H Kock for figures 7.8 and 7.10 redrawn from two figures in *Antarctic Fish and Fisheries* (1992) by K-H Kock; K-H Kock for figure 7.9; Springer-Verlag GmbH & Co. KG for figures 5.13 and 7.13 which appeared as fig. 6 and fig. 5 respectively in *Antarctic Science: Global Concerns* (1994) edited by G Hempel, figure 5.21 which appeared in *Antarctic Ocean and Resources Variability* (1988) edited by D Sahrhage, and figure 8.7 from *Antarctic Ecosystems. Ecological Change and Conservation* (1990) edited by KR Kerry and G Hempel; IB Campbell for figure 8.2; Academic Press Ltd for figure 8.6 reprinted from *Journal of Environmental Management*, **36**, JS Burgess AP Spate and FI Norman 'Environmental impact of station development in the Larsemann Hills...', pp 287–289, (1992) by permission of the publisher Academic Press Limited; Quark Expeditions Inc. for figures 8.11 and 8.14; Stephano Nicolni and Quark Expeditions Inc for figure 8.13; Elsevier Science for figure 9.1 reprinted from a figure in CM Harris and J Meadows ' Environmental Management in Antarctica . . .', *Marine Pollution Bulletin*, **25**: 239–49 (1992); SCAR for table 7.2 which appeared as table 1 in Rutford, 1986, and table 8.1 which appeared as table 5 in Benninghoff and Bonner, 1985; Cambridge University Press for Table 7.1 which appeared as table 5 in I Everson 'Antarctic Fisheries' (1978) in *Polar Record*, **19**(120), and for Table 8.6 extracted from table 10.4 in N Leader-Williams (1988) *Reindeer on South Georgia. The Ecology of an Introduced Population.*

Whilst every effort has been made to trace copyright holders, in a few cases this has proved impossible and we would like to take this opportunity to apologise to any copyright holders whose rights we may have unwittingly infringed.

List of acronyms and abbreviations

AAT	Australian Antarctic Territory
AGCM	Atmosphere General Circulation Model
AMCAFF	Agreed Measures for the Conservation of Antarctic Fauna and Flora
ASMA	Antarctic Specially Managed Area
ASOC	Antarctic and Southern Ocean Coalition
ASPA	Antarctic Specially Protected Area
ASTI	Area of Special Tourist Interest
ATCM	Antarctic Treaty Consultative Meeting
ATCPs	Antarctic Treaty Consultative Parties
ATS	Antarctic Treaty System
AVHRR	Advanced Very High Resolution Radiometer
BAS	British Antarctic Survey
BIOMASS	Biological Investigations of Marine Antarctic Systems and Stocks
BIOTAS	Biological Investigation of Terrestrial Antarctic Systems
BWU	Blue Whale Unit
CCAMLR	Convention on the Conservation of Antarctic Marine Living Resources
CCAS	Convention for the Conservation of Antarctic Seals
CEE	Comprehensive Environmental Evaluation
CEMP	Committee for the Ecosystem Monitoring Programme
CEP	Committee for Environmental Protection
CFCs	Chlorofluorocarbons
CLIMAP	Climate, Long-Range Investigation, Mapping and Prediction
COMNAP	Council of Managers of National Antarctic Programs
CRAMRA	Convention on the Regulation of Antarctic Mineral Resource Activities
DSDP	Deep Sea Drilling Program
EIA	Environmental Impact Assessment
ENSO	El Niño–Southern Oscillation
ENVISAT	Environmental Satellite
ERS	Earth Resources Satellite
ESMR	Electronically Scanning Microwave Radiometer
FAO	Food and Agricultural Organisation (United Nations)
GLOCHANT	Global Change and the Antarctic
IAATO	International Association of Antarctic Tour Operators
IEE	Initial Environmental Evaluation
IGY	International Geophysical Year
IPCC	Intergovernmental Panel on Climate Change
IUCN	[International Union for the Conservation of Nature and Natural Resources] – The World Conservation Union
IWC	International Whaling Commission
MARPOL	International Convention for the Prevention of Pollution from Ships
MIZ	Marginal Sea Ice Zone

MPA	Multiple-Use Planning Area
MSY	Maximum Sustainable Yield
NGOs	Non-Governmental Organisations
NSF	National Science Foundation
PAHs	Polycyclic Aromatic Hydrocarbons
PCBs	Polychlorinated Biphenyls
PEP	Protocol on Environmental Protection
RADARSAT	Radar Satellite
SCAR	Scientific Committee on Antarctic Research
SMMR	Scanning Multichannel Microwave Radiometer
SOI	Southern Oscillation Index
SPA	Specially Protected Area
SRA	Specially Reserved Area
SSSI	Site of Special Scientific Interest
THIR	Temperature Humidity Infrared Radiometer
TOMS	Total Ozone Mapping Spectrometer
UN	United Nations
UNCLOS	United Nations Convention on the Law of the Sea
UNEP	United Nations Environmental Programme

CHAPTER 1 Introduction

Objectives

Key themes and questions covered in this chapter:

- images of Antarctica
- why study the Antarctic?
- scale and global effects of Antarctic physical systems
- significance of Antarctic science
- changing perceptions of Antarctic resources and environments
- Antarctica at centre stage of world environmental awareness
- limits of the Antarctic and the Southern Ocean

1.1 Introduction and aims

1.1.1 'Antarctic'

In literal terms, the word 'Antarctic' has its origins in the Greek name for the Arctic regions. The ancient Greeks named the polar constellation above the North Pole *arktos*, the bear, and the region which lies opposite *arktos* became known as the anti-Arctic, or Antarctic. Yet there are other meanings and connotations attached to the word 'Antarctic'. There are few locations on Earth whose very name conjures up such evocative and stark images in the popular imagination as does the name 'Antarctic'. The word stands for remoteness and desolation, a place of pristine cold crystallised and set amidst a frozen ocean. It lies somewhere on the very frontier of the popular perception, remote and obscure and cold:

> . . . the vast stillness of the polar wilderness, the plateau of lifelessness where men come and go like passing shadows and some are gone for ever, entombed in the white shroud; this is a place of the most terrible beauty, of the white night and darkness at noon. This is Antarctica . . . spreading away from us in giant white waves, waves without beginning, without ending, spreading far away across all her countless horizons
> (Mickleburgh, 1987).

This perception of a marginal, peripheral and inhospitable continent has been perpetuated in literature: Coleridge's most famous work 'The Rime of the Ancient Mariner' is full of Antarctic imagery of loneliness, isolation, icebergs and the albatross, a solitary bird that haunts the remote and cold southern seas. All of these images are returned to time and again in accounts of the experiences of Antarctic explorers (Simpson-Housley, 1992). The central themes underlying such a perception of Antarctica spring essentially from the unique geography, history and cartography of the continent. It is isolated geographically from its nearest neighbour by 1100 km of windswept and ice-infested ocean. Only the bravest and most skilled of the early navigators dared risk their small wooden sailing vessels in the waters guarding the great southern continent, or 'Terra Australis Incognita' as it came to be known. This isolation resulted in the relatively tardy discovery of the coastline early last century and, although more recent exploration during the present century has greatly expanded our knowledge of the geography of the interior, the popular image of a remote and impenetrable wilderness has been perpetuated.

Cartography has consolidated this image: after an uncertain start, when the hypothetical continent was shown on Greek maps, it was subsequently lost or omitted from most mediaeval maps. Parts of the coast and interior then appeared intermittently on maps in the eighteenth and nineteenth centuries; even modern world maps often show only a thin coastal strip stretched across the southern extremities or omit the continent altogether (Figure 1.1). The evolving cartography has

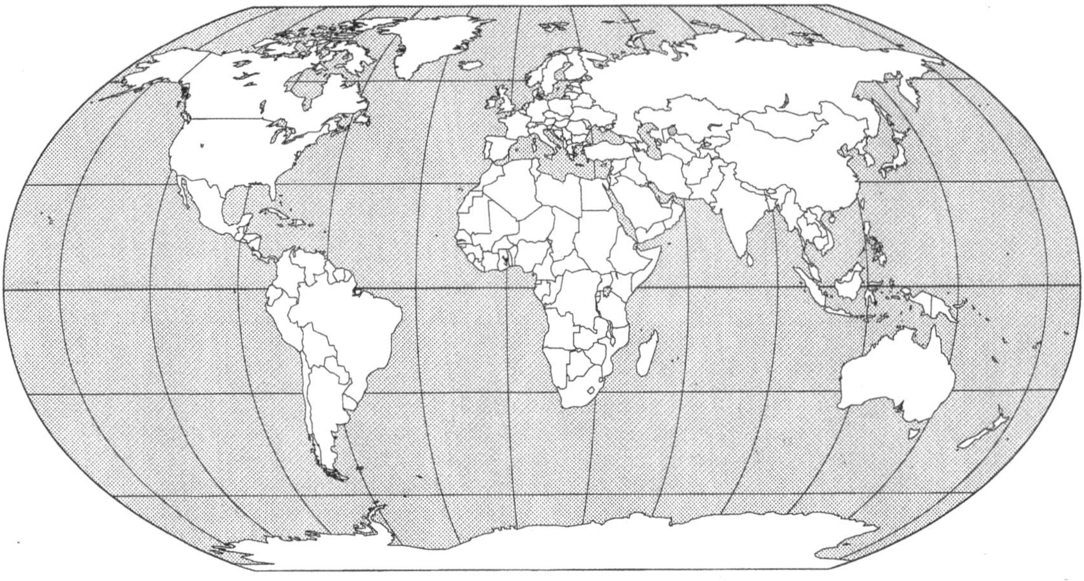

Fig 1.1 Modern map of the world showing Antarctica as a peripheral region.

been a reflection of changing interest in Antarctica: a theoretical possibility for Greek geographers; a source of economic wealth for the sealing and whaling industries; its unknown coasts and interior a compelling attraction for explorers; its physical and biological systems of immense interest for science; and, more recently, a concern for its environment in a global context. Clearly there has been a fundamental shift in the geographical perception of the Antarctic, and its status as a 'pole apart' has now been replaced with a vision of a continent at the heart of many of the vital atmospheric and oceanic systems that sustain the Earth in its present condition. As a result, most of the maps contained within this book reveal an altogether different perspective to that of Figure 1.1 and show Antarctica at the core of global processes (Figure 1.2).

Yet it is perhaps this image of Antarctica at the core of the Southern Hemisphere's oceanic and atmospheric systems that raises concern for its future. This concern springs from a relatively new, radically different and perhaps misguided perspective of the Antarctic where the continent is viewed as a mine of untapped mineral resources awaiting exploitation and the surrounding ocean as the last great open-access fishery. If the continent is truly at the core of many global processes then any future mineral exploitation or overfishing of species at the heart of the marine ecosystem raises serious environmental issues. Equally, there is a growing concern about the environmental impacts that may arise from climate change and stratospheric ozone depletion that could produce far-reaching effects both in Antarctica and elsewhere. The pristine environment of Antarctica provides essential data for a global baseline against which to monitor climate-related changes that may affect global processes.

1.1.2 Aims of this book

Over the past 15 years or so, unprecedented interest has focused on Antarctica arising from a number of major developments concerning interest in the minerals resources of the region and the negotiations for a minerals regime; the discovery of the Antarctic ozone 'hole'; the effects of global warming on the stability of the ice sheet and on sea-level rise; the effects of human activities on the terrestrial and marine ecosystems; the growth and environmental impacts of Antarctic tourism; the impact and management of Antarctic fisheries (particularly for krill); the high-profile campaigns by environmental non-governmental organisations (NGOs) to have Antarctica declared a 'World Park'; the possible review of the Antarctic Treaty in 1991; the emergence of Antarctica as a political issue in international forums such as the UN; and advances in scientific understanding, notably the role of Antarctica in global environment change. Following this period of rapid change in

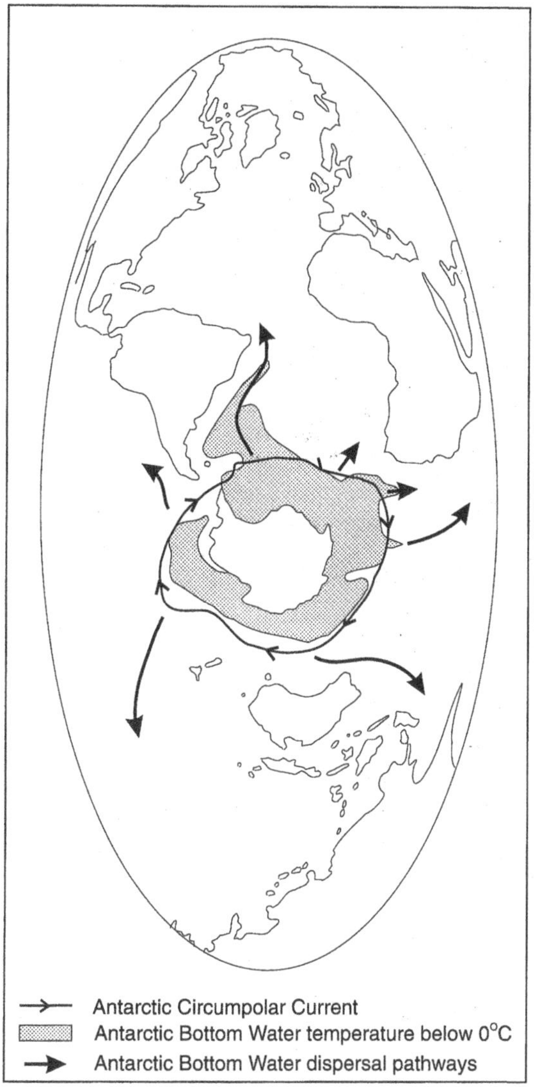

Fig 1.2 Antarctica at the core of global systems: Antarctic Bottom Water transfer into the world's oceans via deep pathways and beneath the Antarctic Circumpolar Current (abyssal isotherms in °C) (source: redrawn after Pudsey in SCAR, 1993a).

Antarctic affairs and growth in scientific knowledge, there is now a need for a 'new' geography of Antarctica to integrate the legacies of past exploitation and human activities with the new environmental agenda that now underscores all human use and management of the continent and its oceans. Antarctica, its environment and resources are now more firmly on the world stage than at any time in the past, and its future as a 'natural reserve devoted to peace and science' depends on the success of international treaties and agreements, namely the Antarctic Treaty and its associated instruments, including the Protocol on Environmental Protection. The time appears right for the emergence of a 'new' geography of Antarctica and so the main aim of this book is to present a geographical perspective on the nature, functioning and interrelationships of Antarctic environments, together with the past and present patterns of human interaction and management of resources. The way in which the resources of the Antarctic have been exploited in the past provides lessons for future environmental management. However, in view of the relatively rapid changes occurring in the amount, type and accessibility of information on a wide range of Antarctic issues, a secondary aim is to provide, in one book, an up-to-date source of information that focuses on the key issues and questions related to the Antarctic environment and the management of its resources. With such broad aims the book very much represents an overview of the way in which the physical and biological systems operate geographically on different scales in space and time, and of how humans have interacted with these systems.

The principal systems examined include the atmospheric and climatic systems of the Antarctic region, the geological, glacial, periglacial and biological systems of the Antarctic continent, and the physical and biological systems of the Southern Ocean. The book examines the ways in which these systems interact and influence one another and attempts to place these interactions within a wider schema of the potential changes contingent upon any future global warming. The intrusion of humans into the previously pristine natural environment is examined against the background of past resource exploitation and impacts, and in the context of the present and future management regime emplaced under the Antarctic Treaty System to regulate the use of the continent and its surrounding ocean. Conservation and environmental management must cope with not only human interactions with the environment but also wider political issues. This is true of conservation anywhere in the world but is brought into sharp focus in a continent where sovereignty claims have been suspended for a number of years and where the method of enforcing international agreements is far from clear. Hence, an awareness of the political background is essential to an understanding of the pressures on the Antarctic environment and the regulatory approaches and measures adopted.

It is in the nature of an overview that some areas of study and interest are given only limited or no attention, especially in view of the huge amount of published material that has emerged, particularly over the past few decades. Substantial recent literature spans areas as diverse as glaciology, ice sheet modelling, invertebrate physiology, ozone chemistry, minerals potential, political sovereignty, marine biology and international law. Of course, this selection is just the tip of the iceberg not only of Antarctic science but also of issues of more general concern. However, the amount of literature concerned with Antarctica is fairly restricted in comparison with that for the other continents, and it is the only continent and ocean with a virtually complete bibliography available on both paper and CD-ROM! Rather than examine a few issues in great depth, the approach adopted has been to provide enough of the background information to highlight the changes in, and interaction between, these disparate areas of study. Such an approach clearly places itself at risk of superficiality, yet many of the problems facing polar regions stem from a lack of appreciation of the broad interrelationships between systems, resulting from past compartmentalisation of fields of study (Sugden, 1982). This book aims to fill this gap in the Antarctic literature and to provide an overview and synthesis of the nature of, and issues relating to, Antarctic environments and resources, and their management.

1.2 Why study the Antarctic?

1.2.1 Antarctic statistics

The Antarctic is unique amongst continents: the 98% that is ice-covered contains almost 90% of the world's fresh water in an ice sheet that reaches just over 4.7 km thick in East Antarctica (NERC, 1989, 1993) (Figure 1.3). It is therefore an important part of the global hydrological cycle, the changes in the relative amount of ice contained within its mass being intimately related to changes in sea level over time. At 13.6 million km^2 (NERC, 1993), the area of the continent with all its islands and ice shelves is about 30% larger than the area of the USA and accounts for 10% and 30% of the land surface of the globe and the Southern Hemisphere, respectively. It is the world's windiest and coldest continent yet, in spite of the volume of frozen water that it contains, it is also the largest desert on Earth, with most of the polar plateau having an annual water-equivalent snowfall of less than 5 cm (Dudeney, 1987). At some locations the wind is channelled by topography to blow uninterruptedly for weeks on end: a mean annual wind speed of 80 km h^{-1} and gusts in excess of 320 km h^{-1} were recorded in 1912 by the Australian explorer Douglas Mawson. Typical temperatures in the interior reach lows of between –40 °C and –50 °C but Vostok, the highest and most remote station on the polar plateau, holds the record as the coldest place on Earth at –89.5 °C.

Above all, Antarctica is a continent of extremes: not only is the climate extreme; a mean elevation of 3 km above sea level makes it one of the highest of continents; its ice sheets are the biggest and thickest on Earth; and its annual girdle of sea ice seasonally doubles the ice-covered area of the region. The existence of such extremes means that the influence of the Antarctic region on both climate and ocean extends not only to its immediate area but also into mid-latitude global systems. This strongly suggests that the Antarctic is worthy of scientific study in its own right, particularly considering the present incomplete knowledge both of some parts of the interior and of the detailed functioning of its physical systems. The global significance of Antarctic systems is further emphasised by noting the ways in which it is fundamentally different from its closest comparison, the Arctic.

1.2.2 Arctic and Antarctic contrasts

The Arctic and Antarctic regions are similar in that they are the major heat sinks of the Earth, yet they are also polar opposites in more than a purely cartographic sense. The fundamental difference lies in the geographies of the Arctic and Antarctic: the Arctic is in essence a polar ocean basin capped by a thin skin of sea ice and surrounded by land; the Antarctic consists of a pole-centred, high and ice-covered continent surrounded by a deep ocean. Temperatures decline with altitude and since 50% of the Antarctic continent is higher than 2 km above sea level and 25% is higher than 3 km above sea level, the temperatures are up to 40 °C lower than they would be for a continent at sea level (Benson, 1962). Apart from in Greenland, much of the Arctic basin is at sea level and is thus warmer than the Antarctic. Furthermore, in the south there are few land barriers between 40° and 60° S to interrupt the flow of westerlies and the

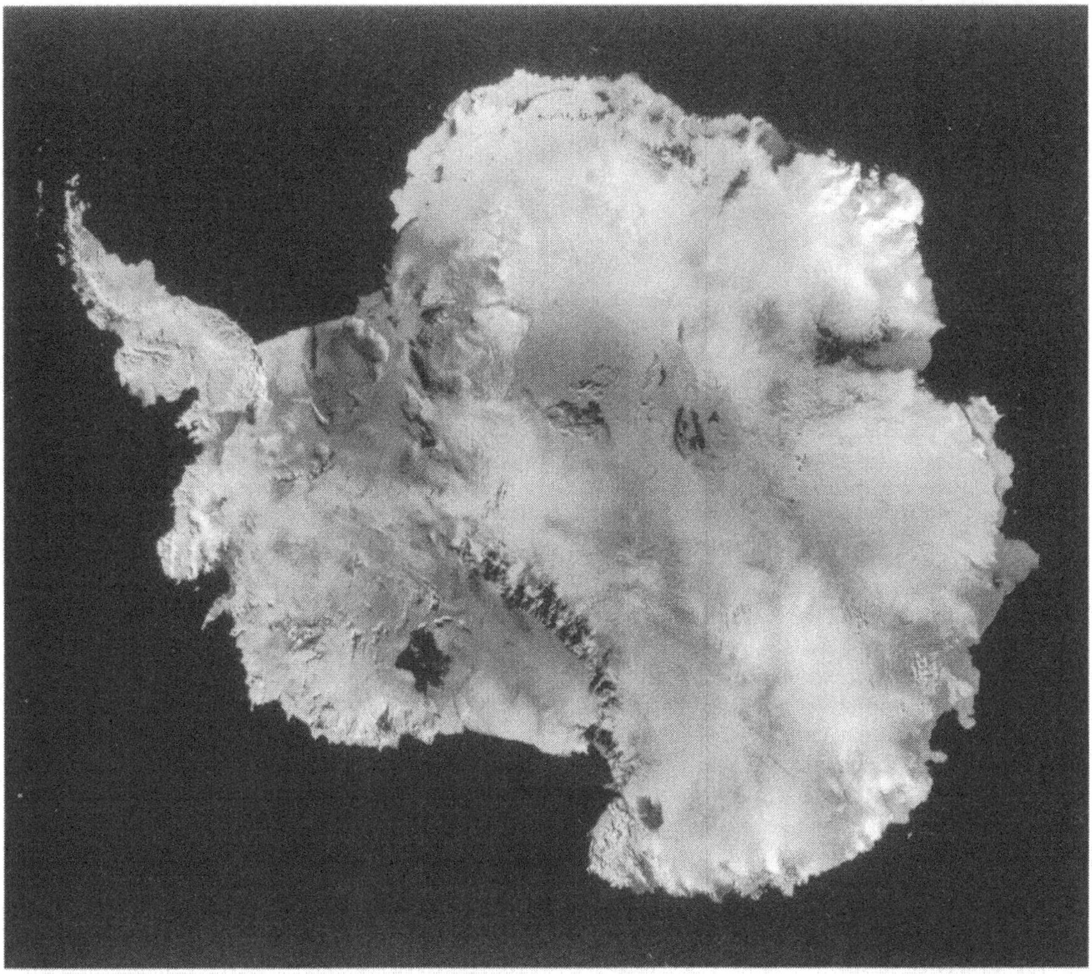

Fig 1.3 Antarctica from space: a spectacular mosaic image of the frozen continent produced from AVHRR scenes collected by NOAA, 1980–1987 (data available from US Geological Survey, EROS Data Center, Sioux Falls, South Dakota, USA).

ocean currents that they drive. This is in sharp contrast to the Arctic, where continents interrupt both air and ocean flows. Averaged over the year for all latitudes, the southern atmospheric circulation pattern possesses 50% more momentum than does its northern equivalent (Dudeney, 1987). As a result, there is a marked hemispherical asymmetry in surface pressure between north and south and a correspondingly strong westerly circumpolar circulation in the Antarctic that inhibits the transfer of heat from lower latitudes (Figure 1.4).

The strong westerly winds produce a westerly oceanic circulation that is also strong enough to prevent the penetration of warm lower-latitude water into the ocean surrounding Antarctica. In contrast, the Arctic Ocean is fed by the warm waters of the North Atlantic Drift, which flow between Greenland and Norway and result in a relatively warm ocean at –2 °C (Dudeney, 1987). These differences ensure that less heat is transferred from lower latitudes to the Antarctic than is the case to the Arctic. Heat loss to the atmosphere is also much less from the ice-covered Antarctic continent than from the relatively warm Arctic Ocean in spite of its cover of sea ice. The contrasting geographies of north and south and the sheer size of the Antarctic heat sink have resulted in a long-term mean temperature in the southern hemisphere that is 1.6 °C lower than in the northern hemisphere; a thermal equator that is 10° north of the geographical equator; glaciers that reach sea level 10°

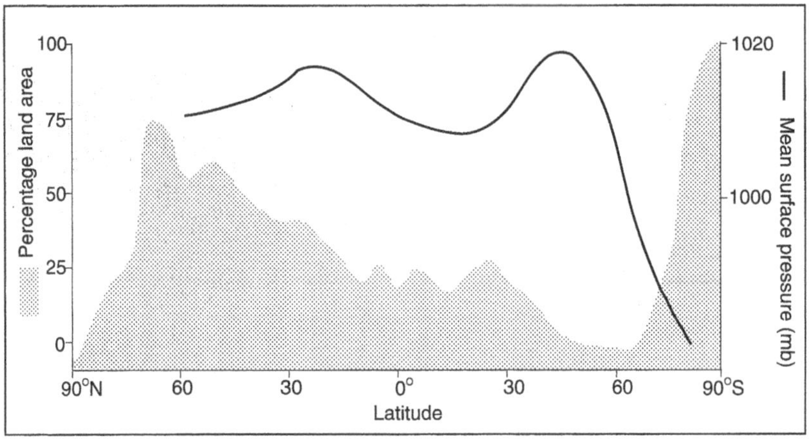

Fig 1.4 Contrasts in land area between the Northern and Southern Hemisphere. Mean surface pressures are lowest in the Southern Hemisphere, and this is also the area of strong westerly winds (source: redrawn after Dudeney, 1987).

closer to the equator in the south than in the north; and coastal temperature lows that are 20 °C lower in the Antarctic than in the Arctic.

There are other fundamental differences that make the Antarctic a profitable place for study. These include: a unique and isolated flora and fauna distributed along its perimeter, in contrast to a pan-Arctic biology of numerous European, American and Asiatic elements; no indigenous human population or settlements, whereas the Arctic has several tribal peoples; and economic activity that is limited to fishing and tourism at present with very limited environmental disruption on land, in stark contrast to the numerous areas of oil, gas and minerals extraction, hydro-electric power generation, reindeer herding and fishing in the Arctic. There are also political contrasts such as the demilitarised status of the Antarctic compared with the high military interest in the Arctic, and the suspension of sovereignty claims in the Antarctic by virtue of the Antarctic Treaty, as opposed to the ownership of Arctic land by individual nations (although dispute remains over some boundaries of continental shelves and economic zones).

1.2.3 The scale and global effects of Antarctic systems

A grasp of the scale and nature of the physical background of Antarctica is essential to an understanding of the natural systems that have developed on and around the periphery of this vast and cold continent (see Figure 1.3). Four major physical systems underpin the natural environment. The atmospheric system over Antarctic regions plays a key role in the heat balance of the world, and its climate affects areas well beyond the immediate environs of the continent. The breakup of the supercontinent of Gondwanaland provides the global tectonic and geological context of the continent and its Southern Hemisphere neighbours. The great mass of ice on both the land and the surrounding ocean surface comprises a cryosphere that has no equal on Earth. Such ice as breaks off or melts at the periphery feeds into the cold Antarctic ocean. The currents in these waters radiate outwards and serve to exchange nutrients with the lower-latitude oceans. They are a vital link in the productivity and heat balance not only of the Southern Ocean ecosystem but also of the world's oceans as a whole. The ecosystem of the Southern Ocean, although not as productive as was once thought, still supports such numbers of animals as to make the adjacent terrestrial ecosystems of the continent itself look meagre in comparison. For what is popularly portrayed as a peripheral and forgotten continent, the Antarctic plays a remarkably central role in the global environmental system in terms of climate, global heat balance, oceanic circulation and marine nutrient cycling. As such it is worthwhile to summarise some of the ways in which the geography of its systems have global significance. Spatial and temporal variations in these systems are crucially important not only to human understanding of how the planet currently functions but also to prediction of how it may change in the future.

The role of the Antarctic in global change is now recognised widely (Hempel, 1994) and in particular by the Scientific Committee on Antarctic Research (SCAR), a committee of the prestigious and influential International Council of Scientific Unions (SCAR, 1993). Following a report on the role of the Antarctic in global change, SCAR (1993a) was responsible for the creation of an advisory group of specialists on Global Change and the Antarctic (GLOCHANT), whose remit was to plan and implement a regional research programme of global-change research in the Antarctic and to provide the linkage and liaison of this programme to other global-change programmes. This group has been operational since 1992. The six core projects of the SCAR (1993a) plan are:

1 The Antarctic sea ice zone: interactions and feedbacks within the global geosphere–biosphere system.
2 Global palaeoenvironmental records from the Antarctic ice sheet and marine and land sediments.
3 The mass balance of the Antarctic ice sheet and sea level.
4 Antarctic stratospheric ozone, tropospheric chemistry, and the effects of UV radiation on the biosphere.
5 The role of the Antarctic in biogeochemical cycles and exchanges: atmosphere and ocean.
6 Environmental monitoring and detection of global change in Antarctica.

Antarctic atmospheres, past and present

As a result of its geographical position, physical characteristics and relative isolation from direct human activity and pollution, a fundamental attribute of the Antarctic for science is the pristine quality of its environment. Because it is the world's biggest pollution-free zone it is an ideal location for measuring not only the present levels of pollution in the Earth's atmosphere and oceans but also the past levels. The Antarctic ice sheet presents unparalleled scope for reconstructing past climates by means of the proxy record contained within the layers of ice in cores taken from deep within the ice sheet (Petit et al., 1997). The ratios of oxygen and hydrogen isotopes from these layers give information about past temperatures, while acids, particles, solutes and gases within the cores also trace the composition of ancient atmospheres. They also show the more recent inexorable rise of carbon dioxide (CO_2), methane and oxides of nitrogen that has resulted from industrial activity elsewhere. Ice-core chemistry is thus able to reveal the nature of past atmospheres and allow monitoring of the changes in present global levels of chemicals that have been transported to the Antarctic by atmospheric processes (SCAR, 1993a). Such work also provides a vital link between past atmospheric and climate change and the result of such change on the continent's ice sheets and peripheral environments.

The global importance of atmospheric research in Antarctica was dramatically demonstrated by the discovery of a springtime hole in the stratospheric ozone layer in 1984 (Farman *et al.*, 1985). The thinning of the ozone layer has detrimental effects on biota via enhanced receipt of harmful UV-B radiation and has since affected adjacent continents such as South America, New Zealand and Australia. In a very public acknowledgement of the global interconnection of atmospheric processes, governments moved with unprecedented haste via the Montreal Protocol (1987) to limit the source of the pollution causing ozone depletion. This was one of several factors that resulted in Antarctica being thrust onto, and subsequently remaining on, the world stage. Climate modelling of increases in CO_2 and resulting global warming predicts that the polar regions are likely to show the greatest changes in climate, albeit with a range of possible feedback effects from ice sheet and sea ice changes (Cattle *et al.*, 1997). Antarctica plays a key role in the modelling of present global climate, which depends increasingly on more detailed data from the polar regions. If the Earth's climate is to be understood, then we need to understand the heat balance over Antarctica and the nature of the ocean–atmosphere–sea ice coupling, which determines the heat exchange over the Southern Ocean (Fifield, 1987; Hanna, 1996).

In the study of geospace (the ionosphere, magnetosphere and upper atmosphere) the Antarctic has no rival and has been called a unique 'window on space'. The geomagnetic and geographic poles are a full 23^0 of latitude apart in the Antarctic and this eccentricity produces a variety of atmospheric phenomena, which make it an excellent research platform (Walton and Morris, 1990). Plasma, a hot electrified gas ejected from the Sun, flows along the magnetic field that connects the north and south magnetic poles, and any disturbances in the Earth's magnetosphere are propagated along the field by plasma flow. This is a global effect that is of increasing importance to humans because disturbances can severely disrupt communications between ground stations using high-frequency radio links or via geostationary satellites (NERC, 1989). Accurate prediction of the effects of these disturbances is of immense practical value for two reasons: the level of the atmosphere under study is the same as that at which satellites and space stations orbit and disturbances can damage on-board solar cells and electronic equipment as well as distorting remotely sensed images of Earth; and disturbances can induce, over a wide range of latitudes, electrical currents in power transmission lines, which lead to increased corrosion and power cuts (Fifield, 1987).

Studies of past and present Antarctic climates comprise two core projects of the SCAR (1993a) plan: the environmental monitoring and detection of global change in Antarctica and the effects of Antarctic ozone on the biosphere. International research programmes related to these projects include the use of the Total Ozone Mapping Spectrometer (TOMS) and Solar Backscattered Ultraviolet instruments to measure changes in the ozone amounts; the Network for the Detection of Stratospheric Change, a network of at least five sites worldwide aimed at identifying latitudinal ozone changes; and the Airborne Antarctic Ozone Experiment, aimed at the simultaneous measurement of trace gases from airborne campaigns, balloon flights and, in the future, pilot-less aircraft. Biological effects programmes include the Biological Investigations of Terrestrial Antarctic Systems (BIOTAS), an international programme to assess the effects of ultraviolet radiation on both terrestrial organisms and phytoplankton.

The geological evolution of Antarctica

The geology of Antarctica is fundamental to the understanding of how the continents of the Southern Hemisphere developed from the fragments of the supercontinent Gondwanaland, which broke up about 180 Ma ago. Because of the distinct palaeogeography of Gondwanaland, with Antarctica at its centre and the other southern continents distributed around it, the correct reconstruction and chronology of the break-up is important to the understanding of any possible mineral resources and the development of circumpolar seaways that allowed palaeocirculation systems to develop. This has fundamentally affected the subsequent course of global biogeographical and climatic change and, as a result, is responsible for the present atmospheric and oceanic circulation, which so influences the present global climate. In addition, the intervening areas between the smaller Gondwanaland fragments of West Antarctica, along with several areas offshore of East Antarctica where deep sedimentary basins and rift systems are filled with thick piles of undisturbed sediment, are speculated to be likely locations for the accumulation of hydrocarbons.

West Antarctica also contains areas where processes related to tectonic subduction are well-displayed over Cenozoic time. For example, in the Antarctic Peninsula area there is a geological record of almost continuous subduction of oceanic lithosphere for over 200 Ma (Walton and Morris, 1990). There is also evidence in Antarctic geology of significant sublithospheric thermal anomalies that may document the causes of the break-up of Gondwanaland itself. Such evidence is of clear global significance as it provides clues to the general interpretation of destructive plate margins and in developing models for the mineralisation of magmatic arc systems.

Global palaeoenvironmental records from the Antarctic ice sheet and marine and land sediments

In order to detect and understand changes in global climates and environments, and to predict the magnitude and direction of future changes, it is necessary to have comparative data from the past. In mid-latitudes, a major drawback for palaeoenvironmental studies has been the lack of a terrestrial record of change that is both reliable and long enough to compare with the long and unbroken records from deep-sea cores. The ice sheets of Antarctica, like that of Greenland, provide a solution to this dilemma because ice cores provide a terrestrial record of comparable quality, and often better resolution, than cores taken from the adjacent floor of the Southern Ocean: the deep ice cores from Vostok, East Antarctica, yield data with better than 100 years resolution and extend to c.400 Ka ago, covering four glacial/interglacial cycles (Jouzel *et al.*, 1987; Petit *et al.*, 1997). Such continuous cores are an ideal link between the short modern instrumental records of climate change, the longer record of continuous deep-sea cores, and the intermittent record provided by the terrestrial and continental shelf glacial sediments that give independent validation of the deep-sea record (SCAR, 1993a). They are also a link between terrestrial events such as glaciation and the response of the oceans to climatic change. The most recent coring project is the European Project for Ice Coring in Antarctica (EPICA), where two deep cores in East Antarctica will allow study of the climatic shifts that have characterised the past several glacial–interglacial cycles (Peel, 1997). A 3.5 km core at Dome Concorde, south of the Indian Ocean, will allow the continental ice core record to be set in a global context, whilst a shorter core in Queen Maud Land, south of the Atlantic Ocean, will allow the influence of the Atlantic on snowfall and ice fluctuations to be assessed.

The key role of the Antarctic ice sheet and land and marine sediments in studies of global change and palaeoenvironmental reconstruction is recognised internationally, as shown by its adoption as a core project area in the SCAR plan (SCAR, 1993a). Around 70 cores have been drilled in sediments on the Antarctic margin and deep ocean since 1972 under a range of international and national programmes including the Deep Sea Drilling Program (DSDP), the Offshore Drilling Program, the Dry Valleys Drilling Program, and the Cenozoic Investigations of the Western Ross Sea. These have been joined recently by the Antarctic Offshore Stratigraphy Project, coordinated by a SCAR Group of Specialists on Cenozoic Palaeoenvironments.

The Antarctic ice sheet and world sea level

The fluctuations of the Antarctic ice sheet volume and sea ice extent provide important modulating effects on radiation budgets and atmospheric circulation at a variety of timescales. Since prediction of the behaviour of the Antarctic ice sheet and sea ice extent in the future must necessarily depend on an understanding of how they have reacted to changes in the past, such work is of direct global relevance to a world that is likely to get warmer. Melting of all the Antarctic ice would lead to an estimated sea-level rise of up to 80 m (Drewry, 1991) and force some drastic changes to the world's densely populated coastal lands. However, such scenarios tend to ignore the phased nature of warming and the opportunity of feedback effects to modulate the process. For example, the low temperatures of the Antarctic continent mean that the most direct and immediate effect of warming would be a more limited sea ice extent. However, since warm air contains more moisture than cold air a possible feedback effect in a warmer world, with a more restricted sea ice cover, might be more open water and an increase in coastal precipitation. This would lead to ice sheet expansion rather than ice sheet contraction. Exactly how a reduction in the extent of sea ice might affect the stability of the floating ice shelves, which themselves contain the drainage of parts of the continental ice sheets, is an area of further uncertainty. The linkage mechanisms are as yet unclear, but either way the indications are clearly that the future behaviour of global sea level depends to a great extent on the stability of the Antarctic ice sheet.

The global significance of the mass balance of the Antarctic ice sheet on sea level has been recognised in the SCAR plan (SCAR, 1993a). Existing international programmes include the United States West Antarctic Ice Sheet Initiative, which aims to produce a numerical model for the West Antarctic ice sheet; the German, Norwegian, British and Russian cooperation on the Filchner–Ronne Ice Shelf Programme, to develop numerical simulations of ice shelf and ocean current behaviour; the use of the European Space Agency's ERS-1 satellite for Ice-Sheet Research, to detail the altimetry and thermal characteristics of the ice sheet; and the European Ice-Sheet Modelling Initiative, aimed at intercomparison and validation of models (*ibid.*).

The global role of the Southern Ocean: the importance of the marginal sea ice zone; and the global role of biogeochemical cycles

Antarctic sea ice is a major element in the global climate system. The sea ice region is responsible for the formation of cold and dense Antarctic Bottom Water: the salt rejection from the freezing of sea ice, in addition to cooling beneath ice shelves, produces a density-driven deep-water mixing and thermohaline convection. The world's oceans are cooled both by this deep water as it is transported north and by other Antarctic surface water masses also flowing north (see Figure 1.2). Replaced by equal volumes of warmer water flowing south, these Antarctic currents are vital cogs in the world's heat balance. The extent and enhanced albedo of the winter sea ice region and the restricted summer ice extent strongly affect temperature and precipitation variations over several degrees of latitude. This time-varying state of the marginal sea ice zone (MIZ) fundamentally affects ocean–atmosphere exchanges in ways that are poorly understood and remain unaccounted for in the numerical models that are used to predict global changes.

The MIZ is also a key habitat for marine biota, and its seasonal variation produces a strong annual cycle in the productivity of the marine plants and microscopic animals that form the basis of the marine food web. The presence of sea ice limits photosynthetic production in the surface layers of the ocean, but it also provides a solid platform that organisms use as a permanent or temporary habitat. Of particular importance is the way in which sea ice influences the size and duration of the springtime phytoplankton blooms that provide food for the zooplankton species. This includes the Antarctic krill, the single most important food source of many Antarctic birds and marine mammals. The MIZ may also play a critical role as a habitat, shelter and foraging area in the life cycles of krill, other zooplankton, and seabirds and marine mammals. There is general agreement on the biological importance of the MIZ and that climate-driven variations in its timing and cycle will impact on marine biota (SCAR, 1993a; Spindler and Dieckmann, 1994). However, much of the detail is unclear: for example, on the extent to which the MIZ algae are able to seed phytoplankton blooms in the Southern Ocean, or the extent to which the MIZ affects the biomass, diversity, turnover and organisation of pelagic species as well as of seabirds and mammals (SCAR, 1993a).

The Southern Ocean also plays an important role in the global biogeochemical cycling and exchange of gases between the atmosphere and the ocean. The cold Antarctic waters have a high capacity to absorb CO_2 and other gases from the atmosphere, so the areas close to the continent where these cold waters are near the surface are important sinks for CO_2. Phytoplankton fix inorganic carbon in the surface layers of the ocean, so photosynthesis is the principal route for CO_2 to move between atmosphere and ocean. Two areas of upwelling and mixing are of paramount importance (Treguer, 1994): at the Polar Frontal Zone the cold waters sink and move north, exporting CO_2 and other nutrients to the world's oceans; at the Antarctic Divergence the upwelling of warm water enriched in CO_2 and other nutrients from the mid-latitudes results in a net flux of CO_2 to the atmosphere and fertilisation of the surface layers of the ocean. Such biogeochemical cycling is of critical importance to the productivity and functioning of the ecosystems of both the Southern Ocean and of adjacent mid-latitude oceans. It is also of vital importance to global climate change via the global flux and balance of CO_2.

When assessing the role of the Antarctic in global change, the SCAR plan identified the importance of the Southern Ocean in the MIZ and biogeochemical cycling as two core areas of concern (SCAR, 1993a). There are numerous national and international research programmes related to the physical and biological structure of the Southern Ocean, including the 40-nation World Ocean Circulation Experiment; the Programme for International Polar Oceans Research (PIPOR), which uses the ERS-1 satellite to study sea ice dynamics; the British Fine Resolution Antarctic Model (FRAM), which has successfully modelled detailed circulation and flow patterns such as eddies (Webb *et al.*, 1991); the Ecology of the Antarctic Sea Ice Zone (EASIZ), which aims to determine the role of the MIZ in Antarctic marine systems and biogeochemical cycling; the Coastal and Shelf Zone Ecology of the Antarctic Sea Ice Zone (CS-EASIZ), which aims to improve understanding of the Antarctic coastal and shelf marine ecosystem (SCAR, 1994); and the Southern Ocean Joint Global Ocean Flux Study (SO-JGOFS), which aims to analyse and quantify the physical and biogeochemical processes relating to CO_2 flux in the oceans. These programmes have recently been joined by the Southern Ocean Global Ocean Ecosystems Dynamics Research (SO-GLOBEC), which aims to emulate the 1976–1993 success of the international Biological Investigations of Marine Antarctic Systems and Stocks (BIOMASS). The new programme aims to improve understanding of both the variability of the ocean and sea ice habitat and the population characteristics and dynamics of the zooplankton, benthos and top predators.

The value of the Antarctic as a model for international resource management

By any standards, the level of scientific activity in the Antarctic is impressive and is above all an index of the global significance attached by international science to the processes operating there. However, there is also great value in viewing the Antarctic as a model for studies of international regime development and resource management. The Antarctic Treaty System (ATS) is a unique international agreement encompassing a wide range of issues. Conservation and environmental management of Antarctic resources within the ATS have had to address not only the problems of human/environment interactions, including resource exploitation, but also the wider political issues that underlie international agreement in Antarctica. Hence, the political framework is an important backdrop in understanding the pressures on Antarctic environmental systems and the regulatory approaches and measures adopted. Overall, there has been an impressive level of international agreement in the way that the Antarctic has been managed under the Antarctic Treaty, and this has implications for the management of common resources and environments elsewhere.

1.3 Structure of this book and definitions

1.3.1 Structure of this book

The book is divided into three main parts. Part 1 provides an outline of the natural environment of the Antarctic and summarises the fundamental physical and biological systems and their interrelationships. Part 2 of the book introduces human involvement in the Antarctic via exploration, exploitation of resources and the resulting environmental impacts. Part 3 examines the management framework in place under the Antarctic Treaty System and assesses the performance of the various measures adopted to regulate resource exploitation and to mitigate the human impact on the environment. It also outlines possible developments in environmental management and future concerns.

In Part 1, Chapter 2 examines the morphology, geology and climate of the continent and provides the backdrop to a review of the physical systems of the terrestrial environment. Chapter 3 comprises a summary of the nature and functioning of the Antarctic ice sheets and ice shelves and of the ice-free areas, which support a range of glacial, periglacial, coastal and freshwater processes. These physical systems provide the basis for the habitats of the plants and animals that make up the Antarctic terrestrial, coastal and freshwater ecosystems. The establishment and distribution of biota is partly controlled by geography via the availability of ice-free ground and the development of rudimentary soils following deglaciation and partly by the climatic variables of light, temperature, seasonality and water availability, which are important to growth and productivity.

The other major systems of the Antarctic are the physical and biological systems of the Southern Ocean (Chapter 4). The geography of the ocean floor owes much to its tectonic past, and this has also influenced the timing and development of a circumpolar oceanic circulation. Driven by the circumpolar westerly wind systems, the ocean circulation influences the transfer of cold water to lower-latitude oceans and is influenced in turn by the return of warmer waters; both transfers involve water masses laden with different types of nutrients. The resultant nutrient mix provides the ideal conditions for a seasonal burst of phytoplanktonic productivity that is related to the spring break-up of the sea ice. The zooplankton that graze the phytoplankton form the basis of a food web that supports all higher forms of life in the Southern Ocean, from penguins and other seabirds to seals and whales.

All of the terrestrial and oceanic systems are subject to constraints imposed by the geographic location and climate of the Antarctic. Chapter 5 summarises the ways in which the biota interact with the harsh physical systems that affect life south of the Polar Frontal Zone. The flora and fauna do so via an impressive range of adaptations and adjustments that allow them to mitigate the harmful effects of cold, desiccation and extreme seasonality of light and heat, but some animals manage to avoid them altogether, for example by migration. However, both natural and human-induced climate change in the future may also induce as yet undetermined impacts. These include alterations in temperature and precipitation and changes in the extent of sea ice, all of which will affect biota to varying degrees. For example, cold-adapted species of low mobility are at a disadvantage in warmer conditions. The types of likely changes to both physical and biological systems are also summarised in Chapter 5.

The various resources and environments of the Antarctic have provided the stimulus for past human exploration and exploitation, resulting in a distinct human impact on both the terrestrial environment and the marine ecosystem. This is addressed in the second part of the book. Chapter 6 reviews the changing motives for human interest in the Antarctic, from the early sealing and whaling expeditions through the heroic era of exploration to modern developments in scientific studies, exploitation of marine resources and tourism. The modern period has also seen the intrusion of geopolitics through sovereignty claims and the use of science to maintain a strategic presence on the continent in support of these claims and possible future benefits from mineral resource exploitation. Thus, the geography of human activity, already influenced by the geography of the physical environment, became influenced by wider political interests. Growing political tensions led to the development of the Antarctic Treaty System, an international agreement to maintain the continent in the interests of peace, science and protection of the environment. Human activity in the Antarctic has also had an ecological impact, and Chapter 7 deals with the results of the exploitation of the marine resources represented by seals, whales and finfish as well as the present exploitation of finfish, cephalopods and krill. In each case the development, impact and regulation of the respective industries is analysed. After many years of speculation concerning the extent and value of any minerals on the continent and the hydrocarbon potential of the Antarctic seabed, a moratorium on minerals activities for 50 years has been agreed under the Protocol on Environmental Protection.

A variety of other, 'non-extractive', impacts on the Antarctic environment arising directly from the human presence on the continent or indirectly from more distant sources are assessed in Chapter 8. The causes and impacts are summarised, together with the regulations and controls aimed at reducing their effects. By world standards the overall level of Antarctic marine pollution is low, but in spite of this there have been a few well-publicised pollution incidents associated with the loss of support ships. Antarctic atmospheric pollution has come from two sources: local sources from generator exhausts and waste incineration at scientific bases (now discontinued); and the more widespread effects of global atmospheric pollution and springtime ozone depletion. The localised effects and impacts of scientific stations are related to a wide range of activities such as construction work, often involving heavy machinery, waste disposal, fuel spills, trampling of sensitive habitats and disruption of wildlife, often at critical stages in their breeding cycles. The pristine and untouched image of the Antarctic has also proved to be a strong selling point for the burgeoning Antarctic tourist industry, yet, paradoxically, the popularity of this destination has increased the impacts at sites important for wildlife. The development, impact, regulation and future of tourism is also assessed in Chapter 8. Not only have humans locally modified the Antarctic environment, particularly in the ice-free areas, but they have also deliberately introduced species exotic to the Antarctic and sub-Antarctic. The horses and sled dogs first introduced by the early explorers have had no lasting natural impact on the continent, but on the sub-Antarctic islands introduced species such as reindeer, cats and rats have generally been successful immigrants, resulting in the local elimination or reduction of indigenous species (Leader-Williams, 1985).

Past and present environmental management in the Antarctic forms the basis for Part 3 of the book. The principal way in which environmental management has been achieved is via the Antarctic Treaty and its various instruments and measures (Chapter 9). Many of the treaty provisions have been very successful in achieving their aims, but the success of others has been compromised by a need for political consensus among the treaty nations. The reasons behind the failure to implement the Convention on the Regulation of Mineral Resource Activities and the resulting emergence of the Protocol on Environmental Protection to the Antarctic Treaty are examined in Chapter 9, together with a review of the contributions of NGOs to the formulation of environmental policy and its effective enforcement. Some alternative approaches to Antarctic environmental management are also considered, as well as an assessment of the provisions for the management of the sub-Antarctic islands.

Chapter 10 draws together the past experiences of human interaction with Antarctic environments and resources and assesses whether the objectives of the Antarctic Treaty have been achieved. This forms the basis for examination of what the future might hold, both for the Antarctic environment and for the exploitation of resources from both scientific and commercial perspectives. The chapter starts by retrospectively evaluating the impacts of past and present resource issues relating to living and minerals resources and the environmental management questions that they have raised. Political issues related to the efficacy of the Antarctic Treaty in the face of pressures on resources and sensitivities over sovereignty are examined, together with an evaluation of the success of the Antarctic Treaty itself. The book ends by assessing prospects for the future across a range of issues: the future development of the Antarctic Treaty and the implementation of the Environmental Protocol; the future role of science in providing high-quality data to help to understand Antarctic systems and their global connections; and the possible future impacts of global climate change and ozone depletion on the Antarctic environment.

1.3.2 The limits of the Antarctic

In a book that attempts to deal with the interrelationships of the major physical and biological systems and the importance and connection of these to lower-latitude systems, it seems unnecessary and possibly misleading to define boundaries, with all the rigidity that these imply. Boundaries may exist on maps for human convenience but the reality is that the natural characteristics of many locations result from their position along a continuum or gradient of environmental variables that may describe very wide and mobile zones. These flexible boundaries may also be time-transgressive, depending on changes in space or in time of, for example, weather patterns, temperature or sea ice extent. Notwithstanding this position, it seems helpful to sketch some broad limits to the area that this book covers and to ask where the limits of the region known as the Antarctic might actually lie.

The two terms 'Antarctic' and 'Antarctica' are more often than not used interchangeably, but in general 'Antarctica' is taken to refer to the continent itself and the term Antarctic is used to denote the region that includes both ocean and continent. The boundary that separates the Antarctic from lower latitudes lies in mid-ocean along the line where the cold and dense Antarctic waters meet and sink beneath the warmer and less dense waters of the Atlantic, Pacific and Indian Oceans (Figure 1.5). This oceanographic boundary is popularly known as the Antarctic Convergence (now technically referred to as the Polar Frontal Zone). It extends in an unbroken circumpolar line that roughly parallels the mean

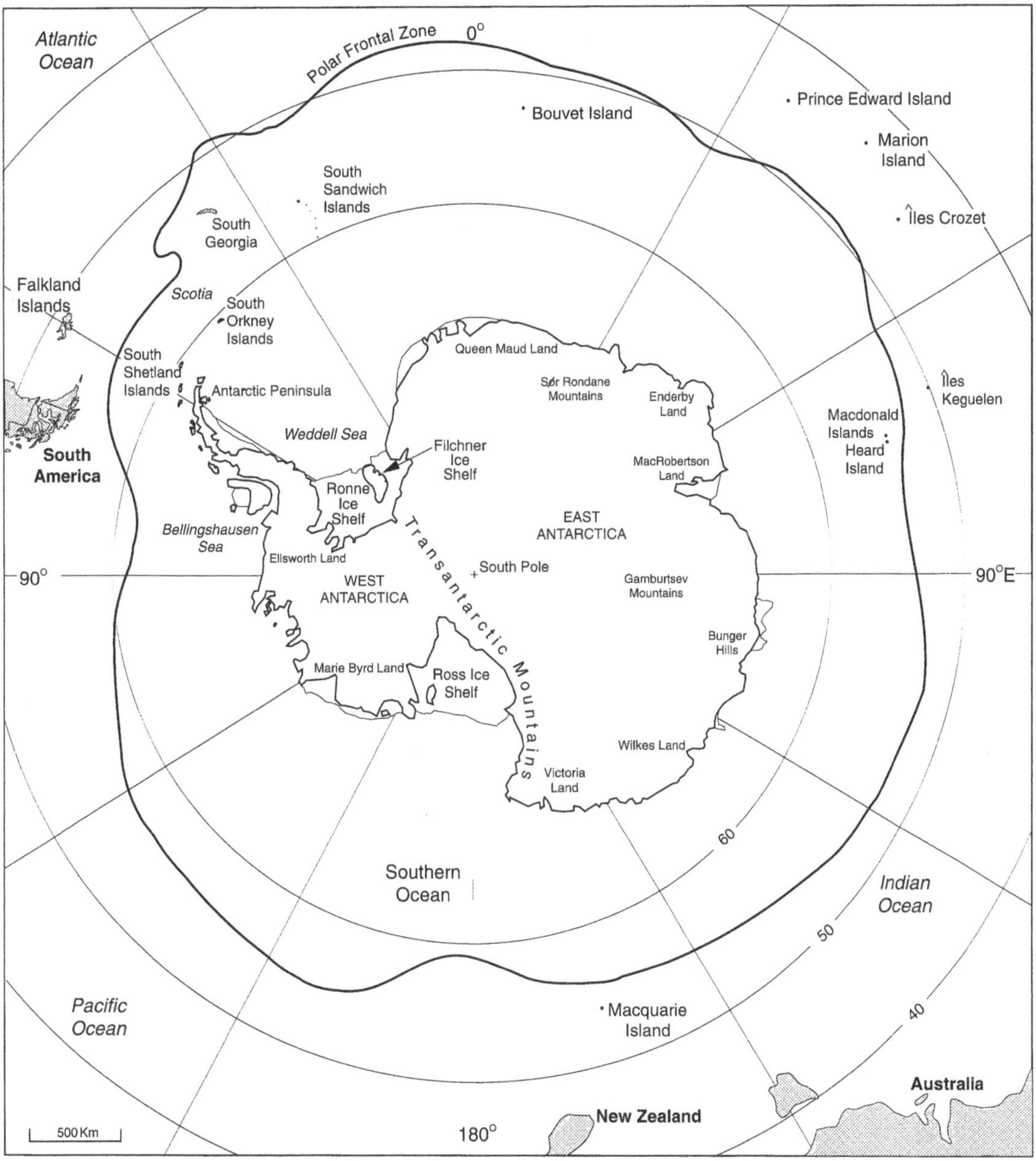

Fig 1.5 Antarctic limits and locations of main places cited in the text.

February isotherm (10 °C) and thus is also of climatic significance. The Polar Frontal Zone shifts with time and longitude but on average at about 58 ° S, it lies well to the north of the Antarctic Circle. Above all, the Polar Frontal Zone represents the biogeographical limits of the Antarctic region: for example, both air and sea-surface temperatures fall sharply as it is crossed and abrupt changes in the composition of plankton and seabirds occur (Stonehouse, 1972; Sugden, 1982).

Much of the continental land mass is covered by large ice sheets, but enough of the subglacial topography is known to indicate that there are two fundamental units: East Antarctica (or Greater Antarctica) and West Antarctica (or Lesser Antarctica), separated by the Transantarctic Mountains, which extend across the continent from the Weddell Sea to the Ross Sea (see Figure 1.5). Terms such as *east* and *west* are not particularly helpful descriptors for a continent that straddles the pole but unfortunately they are now commonly used. East Antarctica comprises the bulk of the land that lies mainly to the east of longitude 315 °E to 165 °E and includes the Transantarctic Mountains and the large area between the mountains and the Indian Ocean. West Antarctica lies mainly to the west of longitude 165 °E to 315 °E and includes the Antarctic Peninsula, Marie Byrd Land and Ellsworth Land and those areas that lie between the Transantarctic Mountains and the Atlantic and Pacific Oceans. Impressed by the uniformity of climate, winds, currents and biological activity of the Antarctic seas, Captain Cook used the name Southern Ocean to describe the vast unbroken ring of ocean south of 50 ° S that surrounds the continent, and this name is widely used today. The term Antarctic Ocean is also used, but mainly for the sectors south of the Polar Frontal Zone (Deacon, 1984). The term sub-Antarctic is used for the eight island groups that lie close to but mostly north of the Polar Frontal Zone and have comparable climates and vegetation (Selkirk, 1992). Îles Kerguelen lie on the mean position of the Polar Frontal Zone; South Georgia, the South Sandwich Islands, Bouvet Island, Heard Island and the Macdonald Islands lie to its south, while Macquarie Island, Îles Crozet, Marion Island and Prince Edward Island lie to its north (see Figure 1.5). Finally, the Antarctic Treaty area refers to all land and ice shelves south of 60 ° S, but not the high seas. Sovereignty claims exist for parts of this area but are deferred under the Antarctic Treaty (see Chapter 6). All of the sub-Antarctic islands are under national sovereignty although the ownership of some, e.g. South Georgia and the South Sandwich Islands, is disputed.

Further reading

Chaturvedi, S. (1996) *The Polar Regions. A Political Geography*. John Wiley, Chichester. 806pp.

Lovering, J.F. and Prescott, J.R.V. (1979) *Last of Lands . . . Antarctica*. Melbourne University Press, Carlton, Victoria. 212pp.

SCAR (1993) *The Role of Antarctica in Global Change*. Scott Polar Research Institute, Cambridge. 54pp.

Simpson-Housley, P. (1992) *Antarctica, Exploration, Perception and Metaphor*. Routledge, London. 131pp.

Sugden, D.E. (1982) *Arctic and Antarctic: A Modern Geographical Synthesis*. Blackwell, Oxford. 472pp.

Walton, D.W.H. (ed.) (1987) *Antarctic Science*. Cambridge University Press, Cambridge. 280pp.

PART 1

The natural environment of the Antarctic

CHAPTER 2

Morphology, geology and climate of Antarctica

Objectives

Key themes and questions covered in this chapter:

- morphology of Antarctica
- geological and tectonic framework of Antarctica
- possible metallogenic and hydrocarbon potential of Antarctica
- climate of Antarctica
- temperature patterns and trends
- atmospheric circulation, regional and local winds
- precipitation patterns and trends

2.1 Introduction

The aim of the four chapters that comprise this section of the book is to describe and explain the principal controls on the functioning and spatial variation in the major physical systems of the Antarctic.

2.2 The morphology of Antarctica

Viewed from space (see Figure 1.3), the continent of Antarctica describes a shapely comma composed of two distinct, ice-sheet-covered areas. In the east, the approximately circular bulk of the East Antarctic ice sheet rises to altitudes of 4200 m above the rocks that comprise the foundation of a continent that lies mostly at altitudes close to or below sea level. Bounded on the seaward and northern side by the Indian and Atlantic Oceans, and on the landward side by the Transantarctic and Pensacola Mountains, only isolated mountain chains break this pattern: for example, the 3000 m high subglacial Gamburtsev Mountains in the centre of the continent, the 3200–4200 m high nunataks of the Sør Rondane Mountains and adjacent ranges of Queen Maud Land (see Figure 1.5).

West Antarctica, the tail of the Antarctic comma, straggles 1200 km north as a bedrock archipelago with the ice sheet submerging the lower parts of the continental sections. West Antarctica encompasses the islands of the Scotia Sea, the Antarctic Peninsula, Ellsworth Land and Marie Byrd Land. At an average elevation of 2000 m, the West Antarctic ice sheet is generally lower than its eastern neighbour, but it is also the home of the 4897 m high Vinson Massif, Antarctica's highest peak, together with numerous summits over 3000 m (BAS, 1993). In spite of this, much of the West Antarctic ice sheet between the Ross and Weddell Seas rests on bedrock that is, for the most part, more than 800 m below sea level (Drewry, 1983). In the Antarctic Peninsula, spectacular mountains tower above an icing-sugar landscape of steep valley walls flanked by tidewater glaciers that calve into an iceberg-studded system of deep fjords (Figure 2.1). The boundary separating West Antarctica from East Antarctica is dramatically marked by the 3500 km long Transantarctic Mountains, a vast range of impressive peaks that crosses the continent from Cape Adare in Victoria Land to the Pensacola Mountains, south of the Filchner Ice Shelf in the Weddell Sea. The highest point of the Transantarctic range lies at 4500 m at Mount Wade in the Queen Maud Range, close to the South Pole.

The Antarctic coastline generally rises steeply from rocky margins in mountainous areas such as those adjacent to the Ross Sea and Antarctic Peninsula or as a steep ramp or cliff of ice along much of the coast of East Antarctica. Elsewhere, the coastline

Fig 2.1 The tidewater glaciers and fjords of the Antarctic Peninsula (source: aerial photograph courtesy of A. Alsop).

comprises vast floating ice shelves where glaciers have flowed from the continent into the sea, their seaward edge marked by vertical ice cliffs and backed by almost flat monotonous upper surfaces of snow and ice (Figure 2.2). On all sides the continent is flanked by continental shelf, a large proportion of which is covered by ice shelves. At its widest, the Antarctic continental shelf reaches 1000 km wide in the Bellingshausen, Weddell and Ross Seas, but its mean width and depth is 200 km and 500 m, respectively (J.B. Anderson, 1991). As such it is both wider and deeper than other continental shelves. In many areas the inner part of the shelf slopes towards the continent, not offshore, and this, together with the great depth of the shelf, is probably the combined result of isostatic depression and glacial erosion. Theoretical models suggest that, as a result of the weight of an ice sheet like that of East Antarctica, a structurally uniform shelf will be depressed to form a proglacial isostatic depression nearly 300 m deep near the ice front, becoming shallower over a distance of approximately 180 km seaward (Walcott, 1970). However, the Antarctic shelf is characterised by major differences in structure, sediment thickness and basement rock type and may not conform everywhere to predicted configurations (J.B. Anderson, 1991). There is also a great deal of evidence to suggest that the shelf has generally been subjected to extensive glacial erosion, resulting in a very rugged surface topography, the removal of unknown amounts of bedrock, and a lowering of the shelf surface, which probably occurred during the relatively recent Pliocene–Pleistocene glacial advances (Anderson, 1991; Anderson *et al.*, 1991).

From the outer part of the shelf, the continental slope falls relatively steeply to ocean basins, which lie at depths of 3000–6000 m. At distances roughly midway between Antarctica and the adjacent continents to the north, the ocean floor is traversed by mountainous submarine ridges whose origins are related to the tectonic development of Antarctica and that all but encircle the continent (Heezen *et al,* 1972). The one exception is the Scotia arc, the name given to the eastward-closing recurved loop of mountains, arcuate submarine ridges and rugged islands that connects the Antarctic Peninsula to the Andes of South America (Barker *et al.*, 1991). Although composed of different types and ages of rocks, the mountains, mid-ocean ridges and islands of the Scotia arc result from a common tectonic history that links the evolution of West and East Antarctica to the break-up of the supercontinent of Gondwanaland.

Fig 2.2 The edge of the Brunt Ice Shelf in the Weddell Sea. The visible part of the ice wall of ice shelves may be 30 m above the water level, but the true thickness of many ice shelves lies in the range 100–500 m (Swithinbank, 1988). Such ice shelves are the floating seaward edge of large glaciers that descend from higher ground inland (source: aerial photograph courtesy of A. Alsop).

2.3 The geology and tectonics of Antarctica

2.3.1 Introduction

Understanding the geology of Antarctica is made difficult by the lack of rock outcrops, which are estimated to account for less than 2% of the 14×10^6 km^2 continental area (Walton and Morris, 1990). However, the amount of Antarctic rock actually exposed may be greater than the amount of rock exposed in, for example, Great Britain. These rock outcrops occur mostly in the Transantarctic Mountains (Figure 2.3) or along a narrow coastal strip usually less than 300 km wide. In contrast, the interior of East Antarctica is largely devoid of outcrops, which, together with extremely severe climatic conditions, makes for great difficulty in both the gathering of geological evidence in the field and its subsequent correlation between isolated rock units across vast intervening ice-covered areas. In spite of this, the broad outline of Antarctic geology and its evolution is becoming increasingly clear (Thomson *et al.*, 1991; Tingey, 1991).

2.3.2 Outline of Antarctic geology

Geologically, East Antarctica is a stable ancient shield of Precambrian (Table 2.1) metamorphic and igneous rocks, analogous to the Precambrian shields of South America, South Africa, India and West Australia (James and Tingey, 1983; Tingey, 1991). The Antarctic shield is composed of a series of ancient and stable cratons separated by younger mobile belts and flanked on its Pacific side by younger orogenic rocks (*ibid.*) (Figure 2.4). The distribution of most of the rocks of Antarctica and its neighbouring continents is related to the amalgamation and break-up of previous continental arrangements including Gondwanaland, the most recent supercontinent (Figure 2.4). The supposed nature of the links to other continents and the timing and mode of break-up are discussed below, but first the rocks outcropping in Antarctica are described.

The rocks of East Antarctica formed mostly over 500 Ma ago, when a period of widespread metamorphism, tectonism and plutonism occurred, but many date from a similar but less well-defined event 1000 Ma ago, and a few are amongst the oldest rocks in the world at 3800 Ma old (Harley, 1989). These ancient rocks are exposed in various places, such as

Fig 2.3 This panorama of the Transantarctic Mountains northwest of Mount Griffith (3095 m) shows in the foreground the plutonic rocks of the Ross Orogen, with the dark cliffs in the distance capped by a mid-Palaeozoic erosion surface formed after the Ross Orogeny. On the skyline, the Rawson Plateau contains the sedimentary rocks of the Beacon Supergroup, which overlie the unconformity (source: photograph by kind permission of Ed Stump).

the coastal mountains of Enderby Land and Queen Maud Land, where the exposures are excellent in quality, commonly comprising large cliff sections of glacially cleaned, fresh rocks (Tingey, 1991). Younger folded sediments of Upper Proterozoic–Lower Palaeozoic age are exposed in the fold belts that border the Precambrian metamorphic shield: mostly in, or adjacent to, the Transantarctic Mountains (Laird, 1991) (see Figure 2.4). These rocks are mainly turbidites, quartz sands and mudstones and appear to have been deposited within a series of basins on the Pacific margins of the shield: the oldest being in the region of the Pensacola Mountains at over 1200 Ma old, and the remainder in the Transantarctic Mountains at 800–600 Ma old (*ibid.*). Between 660 and 580 Ma ago the subsidence within the basins was terminated by uplift, deformation and metamorphism during the Beardmore Orogeny. Deposition was resumed on parts of the eroded remnants of the Beardmore orogen, with marine carbonates and muds being common over much of the Middle Cambrian. Some contemporary vulcanism is also known from the Ellsworth Mountains. The period of deposition along the site of the Transantarctic Mountains was halted about 500 Ma ago by the Ross Orogeny, a major tectonic event that resulted in widespread uplift, deformation, plutonism and metamorphism involving the Lower Palaeozoic and earlier strata (Stump, 1992). However, such are the uncertainties over dating the absolute age of the Cambrian–Precambrian boundary, and thus the duration of the Cambrian, that Stump (1992) suggested that the concept of the Beardmore Orogeny as distinct from the Ross Orogeny was now blurred.

Following the uplift associated with the Ross Orogeny, a period of relatively widespread erosion appears to have led to the development of a distinct mid-Palaeozoic erosion surface that separates the Proterozoic–early Palaeozoic orogenic belt (Ross orogen) from the flat or gently tilted Devonian to Triassic sedimentary rocks (Beacon Supergroup) and Jurassic continental rocks (Ferrar Supergroup) that rest unconformably above (*ibid.*) (Figure 2.3). Reaching thicknesses of up to 2.5 km, the rocks of the Beacon Supergroup are best exposed in the Transantarctic Mountains, but strata similar in lithology and age are exposed elsewhere, such as in Queen Maud Land and the Prince Charles Mountains. The Ellsworth Mountains also have similar rocks, but these are both folded and thicker (Barrett, 1991b). The

Table 2.1 Geological timescale (redrawn after Palmer, 1983)

CENOZOIC (AGE (Ma): 5, 10, 15, 20, 25, 30, 35, 40, 45, 50, 55, 60, 65)

PERIOD		EPOCH	
QUATERNARY		HOLOCENE	
QUATERNARY		PLEISTOCENE	
TERTIARY	NEOGENE	PLIOCENE	LATE
TERTIARY	NEOGENE	PLIOCENE	EARLY
TERTIARY	NEOGENE	MIOCENE	LATE
TERTIARY	NEOGENE	MIOCENE	MIDDLE
TERTIARY	NEOGENE	MIOCENE	EARLY
TERTIARY	PALEOGENE	OLIGOCENE	LATE
TERTIARY	PALEOGENE	OLIGOCENE	EARLY
TERTIARY	PALEOGENE	EOCENE	LATE
TERTIARY	PALEOGENE	EOCENE	MIDDLE
TERTIARY	PALEOGENE	EOCENE	EARLY
TERTIARY	PALEOGENE	PALEOCENE	LATE
TERTIARY	PALEOGENE	PALEOCENE	EARLY

MESOZOIC (AGE (Ma): 70, 80, 90, 100, 110, 120, 130, 140, 150, 160, 170, 180, 190, 200, 210, 220, 230, 240)

PERIOD	EPOCH
CRETACEOUS	LATE
CRETACEOUS	EARLY
JURASSIC	LATE
JURASSIC	MIDDLE
JURASSIC	EARLY
TRIASSIC	LATE
TRIASSIC	MIDDLE
TRIASSIC	EARLY

PALEOZOIC (AGE (Ma): 260, 280, 300, 320, 340, 360, 380, 400, 420, 440, 460, 480, 500, 520, 540, 560)

PERIOD	EPOCH
PERMIAN	LATE
PERMIAN	EARLY
CARBONIFEROUS	LATE
CARBONIFEROUS	EARLY
DEVONIAN	LATE
DEVONIAN	MIDDLE
DEVONIAN	EARLY
SILURIAN	LATE
SILURIAN	EARLY
ORDOVICIAN	LATE
ORDOVICIAN	MIDDLE
ORDOVICIAN	EARLY
CAMBRIAN	LATE
CAMBRIAN	MIDDLE
CAMBRIAN	EARLY

PRECAMBRIAN (AGE (Ma): 750, 1000, 1250, 1500, 1750, 2000, 2250, 2500, 2750, 3000, 3250, 3500, 3750)

EON	ERA
PROTEROZOIC	LATE
PROTEROZOIC	MIDDLE
PROTEROZOIC	EARLY
ARCHEAN	LATE
ARCHEAN	MIDDLE
ARCHEAN	EARLY

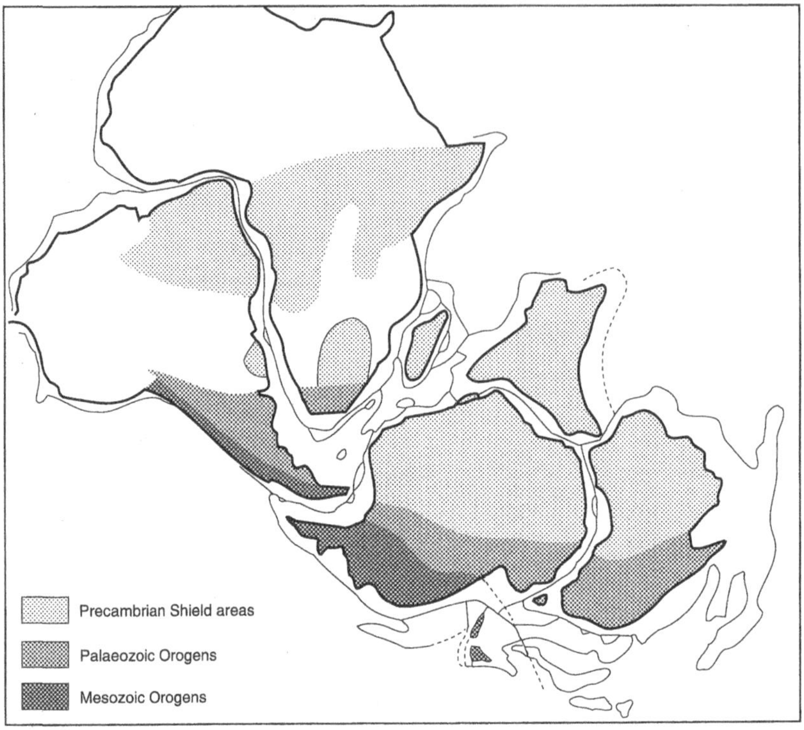

Fig 2.4 The Gondwanaland supercontinent about 180 Ma ago reveals the original geography of the shield and orogenic belts of Antarctica. Reconstructions such as these and those in Figure 2.5 provide the basis for speculation concerning mineral occurrences in Antarctica (source: redrawn from Lovering and Prescott, 1979).

Beacon sediments range in age from late Palaeozoic to mid-Mesozoic (400–200 Ma) and contain sandstones, shales and conglomerates together with coal-bearing Permian strata. The Beacon sediments are characterised by fish and plant fossils (Truswell, 1991; G.C. Young, 1991) as well as by palynomorph assemblages that allow them to be correlated with those of eastern Australia and other parts of Gondwanaland (Barrett, 1991). The Beacon sequence is capped by basalt flows and intruded by dolerite dykes and sills of mid-Jurassic age (Ford and Himmelberg, 1991; Tingey, 1991). The Ferrar Supergroup dolerites constitute one of the most striking geological features of Antarctica and provide a pre-break-up link between the broadly similar rocks found in the other fragments of Gondwanaland: the Tasmanian and Western dolerites of Australia; and the Karoo dolerites of South Africa (*ibid.*). The largest single occurrence of Ferrar rocks in Antarctica is the Dufek intrusion in the Pensacola Mountains, the youngest rocks of which are Permian in age, with the dominant phase of emplacement at 175 Ma ago (Elliot *et al.*, 1985). The total volume of Ferrar magma has been estimated at 500 000 km^3 (Kyle *et al.*, 1981), but since recent modelling suggests that the Dufek intrusion is a 6600 km^2 dyke-like body, the total volume of Ferrar rocks may have been overestimated (Ferris *et al.*, 1997). In the context of Gondwanaland, the Ferrar magma may have stretched in a belt 4500 km across Antarctica from Queen Maud Land to northern Victoria Land and beyond into southern Africa and Australia (Ford and Himmelberg, 1991) (Figures 2.4, 2.5). The uplift of the Beacon and Ferrar rocks to their present topographic position occurred much later during a period of rapid Cenozoic uplift that began about 50 Ma ago and led to the uplift of the Transantarctic Mountains (Gleadow and Fitzgerald, 1987).

The geology of West Antarctica is dominated by Mesozoic and Cenozoic rocks (Barrett *et al.*, 1991), but there are some outcrops of older rocks. One important outcrop of Precambrian rocks occurs on the northern part of the Scotia arc region at Cape Meredith, on the southwestern tip of the Falkland Islands. This provides a firm tie with the stable

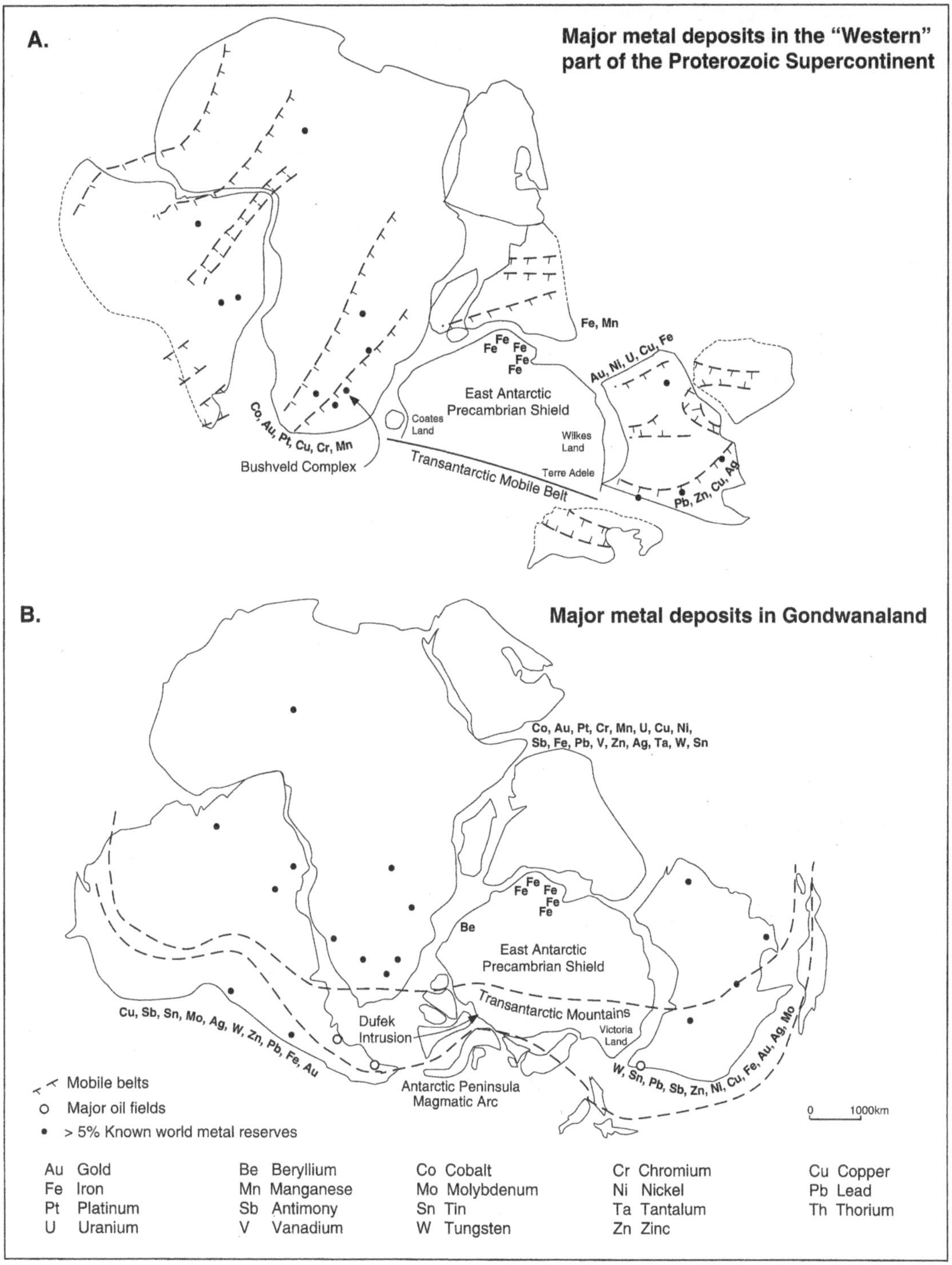

Fig 2.5 Continental reconstructions for (A) 2500–600 Ma ago showing the western part of Proterozoic supercontinent. The position of Laurentia, which abutted the 'Pacific' side of Antarctica, is not shown; and (B) the period up to 180 Ma ago showing the former relationship of Antarctica to the other Gondwanaland continents (source: after Willan *et al.*, 1990). It is important to recognise that such reconstructions may mislead, since the mineral potential of one continental terrane to another has significance only when the two parts were in similar environments, formed by the same process, at the same time.

craton of Gondwanaland and is important for understanding the evolution of the Scotia arc region (*ibid.*). Proterozoic gneisses and schists and thick sequences of Palaeozoic sediments with some igneous and metamorphic rocks occur in Ellsworth Land and Marie Byrd Land (Laird, 1991). Sequences of Devonian–Permian Beacon sediments occur, mainly in the Ellsworth Mountains, but also in the Ohio Range, and are capped in both areas by the Ferrar basalts of the Jurassic (Tingey, 1991).

Elsewhere, the geology of West Antarctica is intimately linked to the events that surrounded the accretion and break-up of Gondwanaland (Storey, 1993). These events are discussed below but broadly involve a long period in the Mesozoic and Cenozoic characterised by phases of crustal extension in some places and compression in others. During this period, extensive volcanic and plutonic rocks, which make up most of the Antarctic Peninsula and the southernmost Andes of South America, were emplaced (Thomson *et al.*, 1983; Barrett *et al.*, 1991). These rocks extend south through Marie Byrd Land, which was the site of extensive basaltic vulcanism in the Cenozoic (Elliot, 1991; Pankhurst *et al.*, 1991). Much of the magmatism, folding and uplift of the rocks of the Antarctic Peninsula relates to the Mesozoic and Cenozoic Andean Orogeny (Elliot, 1975), but advances in the dating of rocks have now blurred the distinction between the Andean events and earlier events associated with Gondwanaland (Storey and Garrett, 1985).

2.3.3 Plate tectonics and Antarctica

Although some of the details are still imperfectly known, it is generally accepted that Antarctica was once the central keystone of a vast supercontinent called Gondwanaland, which contained South America, Africa, India, Australia and New Zealand (Lawver *et al.*, 1991) (see Figures 2.4, 2.5). Ever since Wegener proposed the theory of continental drift in 1915, geologists have recognised many geological similarities between the five continents (Wilson, 1976). For example, seeds from the fossil fern *Glossopteris* have been found in Antarctica, Africa, Australia, South America and India (Truswell, 1991). The remains of an Upper Cretaceous ankylosaur found in the Antarctic Peninsula may demonstrate a land link to South America at that time (Olivero *et al.*, 1991), and Triassic reptiles and amphibians bearing a strong resemblance to those of the other southern continents have been found in the Transantarctic Mountains (Gee, 1989). In the same way, many other species of plants and animals had a common home in these continents until the mid-Jurassic Period, some 180 Ma ago. However, at this point the commonality between many of the species disappears and new and distinct forms emerge on each of the continents, producing a very different biogeography than had existed before (Hallam, 1976; Truswell, 1991). A similar pattern exists in the matching of rock types and ancient fracture zone geometry between the once contiguous parts of the continents (Lawver *et al.*, 1991). The original interpretations linking the continents together used tenuous land bridges that later disappeared. Such constructions are not required if the continents themselves have drifted over time to both amalgamate and break up.

The Gondwanaland supercontinent, which first amalgamated about 550 Ma ago and started to break up at about 180 Ma ago, was only the most recent of the supercontinents in which the Antarctic crust has played a part (Stump, 1992; Storey, 1993). Based on comparative geology, palaeomagnetic records and the trends of Precambrian orogenic belts, the constituent parts of Gondwanaland are thought to have been part of a supercontinent in the Precambrian that included Laurentia (North America and Greenland) and Eurasia (Siberia, Arabia and Baltica) (Piper, 1982), the western part of which is shown in Figure 2.5. Although the exact continental palaeogeographies are imprecise, recent hypotheses suggest that the margins of the southwest United States and East Antarctica (the SWEAT hypothesis) were conjugate prior to the break-up of Laurentia from Antarctica (Stump, 1992; Storey, 1993). The basement rocks of the Ross orogen in East Antarctica match the age and type of rocks in the southwest USA and are associated in both locations with the later deposition of turbidites coincident with volcanic rocks. However, the widespread plutonism that commenced by 550 Ma ago in the Ross orogen was not matched on the western margin of Laurentia, which remained passive (Stump, 1992). Thus the separation is thought to have occurred prior to the Ross event at about 750 Ma ago (Storey, 1993). Hoffman (1991) proposed that the separation of Laurentia in a clockwise rotation led to the fan-like convergence of what then became the constituent cratons of Gondwanaland and led to the opening of the Pacific Ocean. The proposed re-assembly 'turned the constituent cratons of Gondwanaland inside out' (*ibid.*), so that rift zones that had formed in the interior of the continent later became the external plate boundaries.

The break-up of the Gondwanaland supercontinent led to a number of smaller fragments drifting apart and the extrusion of flood-basalts, which mainly formed the ocean floors, in between. In time the newly created Antarctic fragments became progressively surrounded by an ever-expanding basin of new rock as neighbouring fragments drifted north (Bradshaw, 1991; Lawver *et al.*, 1991). The nature and timing of this break-up is partly recorded in the geomagnetic signatures of the new lavas extruded on the seabed and partly by dating the rocks that were left behind (Lawver *et al.*, 1991). The geomagnetic records show that the oldest parts of the ocean floor are of Jurassic age and located close to the continental margins, with progressively younger rocks arranged in parallel bands towards the mid-ocean ridges, where the suture or fracture lines occur. Storey (1995) identified three main episodes in the disintegration of the Gondwanaland continent (see Figure 2.4). The initial rifting stage took place in the early Jurassic (180 Ma ago) and led to a seaway forming between West (South America and Africa) and East Gondwanaland (Antarctica, Australia, India and New Zealand). The second stage commenced in the early Cretaceous (130 Ma ago), when South America separated from the African–Indian plate and the African–Indian plate from Antarctica. Finally, at the begining of the late Cretaceous (100–90 Ma ago), Australia and New Zealand separated from Antarctica. Before and during the initial rifting stage, the Pacific margin of the Gondwanaland continent was the site of active subduction. At the same time, a large magmatic province developed along a linear belt that stretched from the Karoo province of southern Africa, through the Antarctic Ferrar province to the Tasman province in Australia (*ibid.*).

In the late Jurassic (150 Ma ago), the extensive Chon-Aike volcanic province in South America was split by the formation of a rift basin. The rocks that infilled the basin were subsequently uplifted and are now represented by the Rocas Verdes in South America and the Larsen Harbour complex in South Georgia. The separation of Antarctica and Australia from India was also preceded by rifting and extensive magmatism in eastern India, the Kerguelen Plateau and Western Australia. At the beginning of the late Cretaceous (100 Ma), sea-floor spreading occurred between Australia and Antarctica but, unlike the earlier break-up events, was not linked to vulcanism and was essentially passive. It may have taken some time to accomplish, for the passive margin was characterised by initially slow rifting and spreading rates: palaeomagnetic anomalies (Weisel and Hayes, 1972) and dates from deep-sea cores (Kennett, 1978) indicate the creation of new sea floor about 55 Ma ago in this part of the Southern Ocean (Figure 2.6). However, by 45 Ma ago the rate of sea-floor spreading between Antarctica and Australia had begun to speed up markedly (Storey, 1995). The final break-up phase in which Antarctica played a part was the rifting of New Zealand from the Marie Byrd sector of Antarctica, which occurred 84 Ma ago. In Marie Byrd Land, this break was preceded 100 Ma ago by a shift from arc- to rift-related vulcanism as subduction was replaced by spreading along the New Zealand margin (*ibid.*).

In a similar fashion, in the South America–Africa–Antarctic region, the initial fracture between South America and Africa probably occurred 135 Ma ago, but the link between South America and the Antarctic Peninsula may have persisted well into the Tertiary as a thin sliver of continental material (Barker and Burrell, 1977). Forests of the southern beech, *Nothofagus*, and even the remains of marsupials are known from the early Tertiary of Seymour Island, Antarctic Peninsula (Francis, 1991; Truswell, 1991), the annual growth rings showing a cold but uniform climate. In contrast to the relatively simple mode of break-up of other areas of the Antarctic continent from Gondwanaland, most reconstructions produce unacceptable overlaps in the position of the Antarctic Peninsula and southern South America (Dalziel, 1983; Storey, 1991). The anomalies are removed if West Antarctica is regarded as a mosaic of small crustal blocks or micro-continents that have moved relative to each other as Gondwanaland finally fragmented (*ibid.*).

The post-break-up development of the Antarctic Peninsula and Scotia arc region has been reviewed by Barrett *et al.* (1991), and the following account is largely drawn from their interpretation. Prior to the mid-Mesozoic break-up of Gondwanaland, there was a period of subduction of the proto-Pacific Ocean floor along the Pacific edge of the supercontinent, and major magmatic arcs became established along the entire length of the Antarctic Peninsula and southernmost Andes during the Mesozoic and Cenozoic (Barrett *et al.*, 1991). Initially, these were associated with the widespread extension that accompanied the opening of the South Atlantic Ocean and the Weddell Sea basin so that by the late Jurassic, the Antarctic Peninsula was probably a narrow magmatic

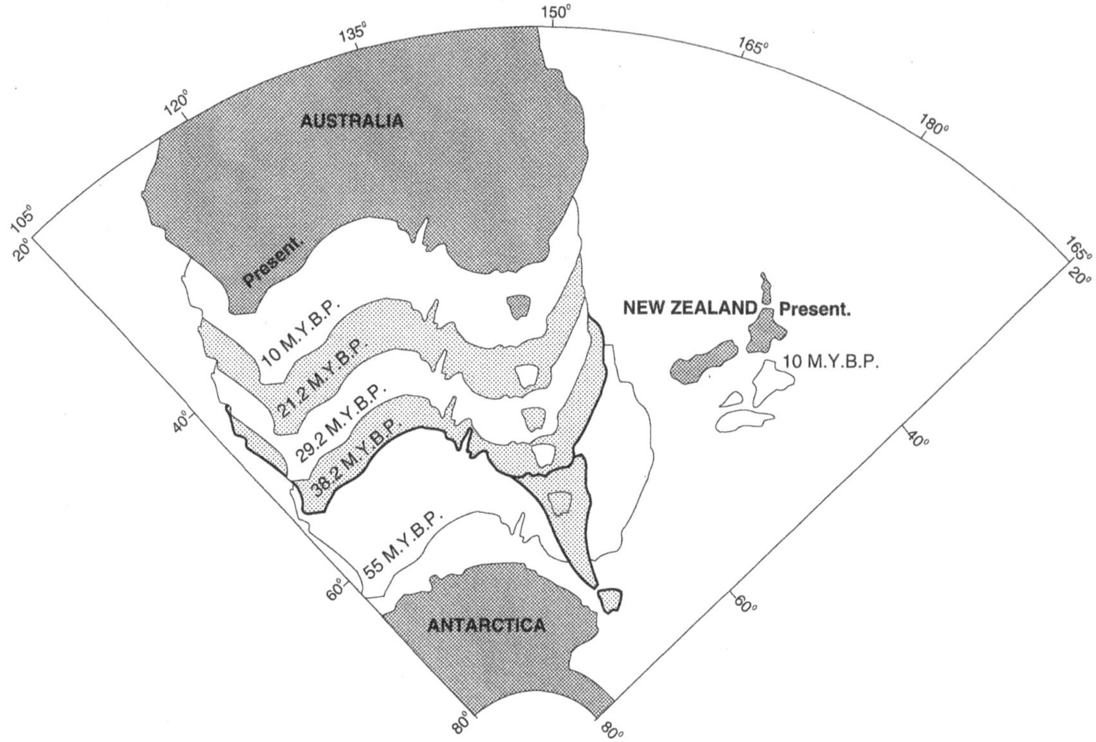

Fig 2.6 The drift of Australia away from Antarctica was in progress about 55 Ma ago, but full ocean circulation in the region developed later, about 38.2 Ma ago, when the area of the South Tasman Rise opened (source: modified from Weissel and Hayes, 1972)

arc terrane that was breaking northeastwards off the Pacific margin of Gondwanaland, with the Weddell Sea opening behind. The rocks that formed the ocean floor in the marginal back-arc basin at this time (late Jurassic to early Cretaceous) were part of the 'Rocas Verdes' basin, the uplifted and folded remnants of which are now found in the southernmost Andes and the island of South Georgia. Much of this basin was subsequently destroyed by a major episode of compressional deformation in the mid-Cretaceous. Because this deformation affected South Georgia and parts of the Antarctic Peninsula as well as initiating the tectonic uplift of the Andes, it was probably widespread and related to ongoing subduction along the entire Pacific margin of South America and the Antarctic Peninsula (*ibid.*).

The subsequent evolution of the Scotia Sea region arose from a complication of the boundary between the Antarctic plate (ANT) and the South American plate (SAM) (*ibid.*). After about 137 Ma ago and the opening of the South Atlantic, SAM–ANT relative movement became more complex: the Antarctic Peninsula moved first independently, then with East Antarctica, about a rotational centre in the southeast Pacific, and this extended and flexed the narrow link with the Rocas Verdes so that a broken cusp had formed by the early Cenozoic (*c.* 65 Ma ago) (Figure 2.7). At the start of the Scotia Sea opening, the micro-continental fragments that are now widely dispersed within the region formed the fragments of this compact cuspate connection between South America and the Antarctic Peninsula (see Figure 2.7). As the Antarctic plate rotated, the direction of motion along its boundaries changed so that the northwestern part of the boundary became convergent, the ensuing subduction proceeding by tearing along an east–west line as well as by an eastward movement of the subduction zone (*c.* 45 Ma ago). This led to back-arc extension and the development of the Scotia Sea, the eventual fracturing and opening of the Drake Passage probably taking place during the late Oligocene between 30 and 22 Ma ago (Craddock,

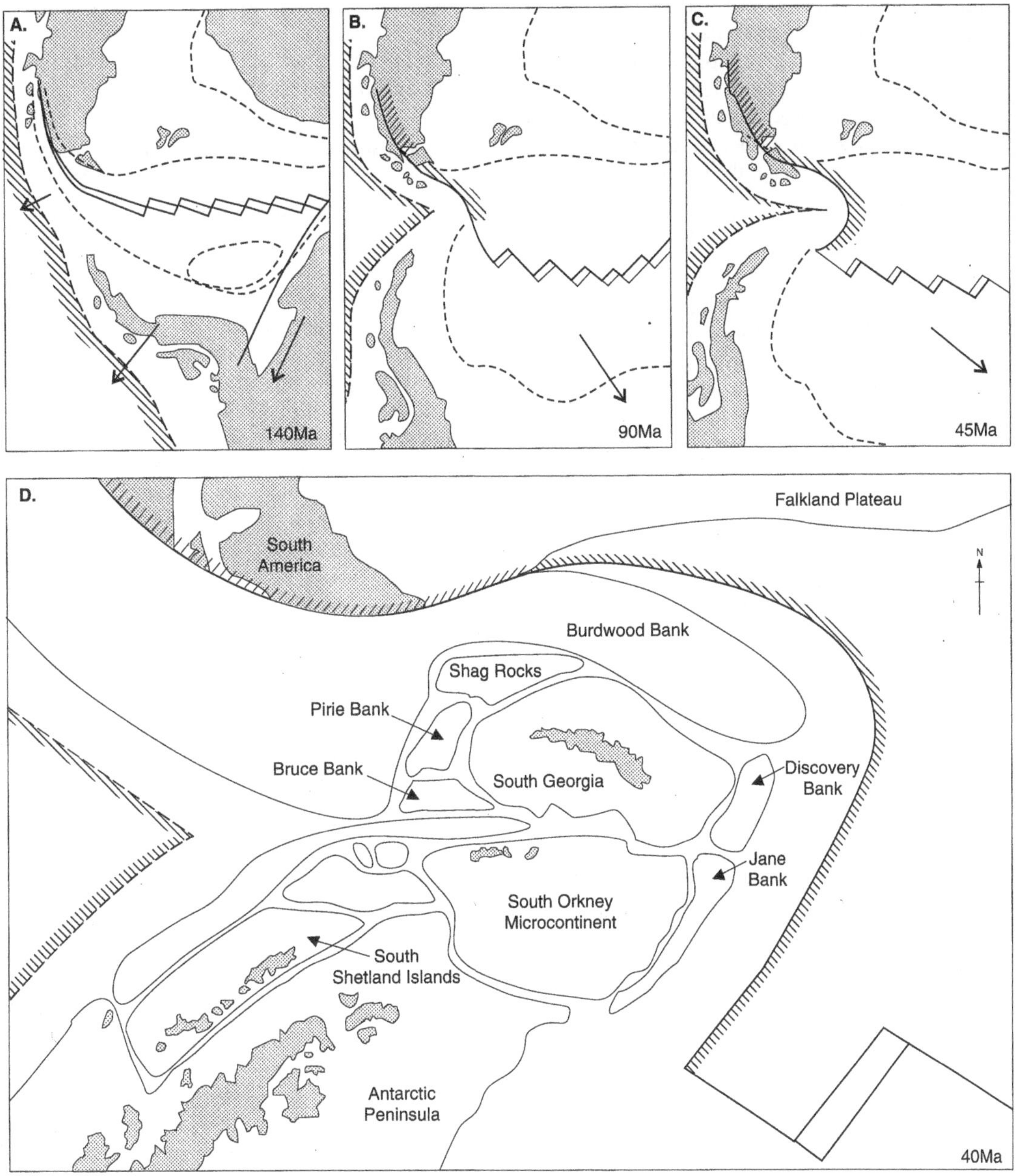

Fig 2.7 Schematic reconstruction of the Scotia Sea region in (A) late Jurassic–early Cretaceous, (B) mid–late Cretaceous, (C) early Tertiary and (D) components of the Scotia Arc region in the mid-Tertiary (sources: modified after King and Barker, 1988; Barker et al., 1991).

1982). The fragments of the thin strip of continental material that once connected South America with the Antarctic Peninsula (Barker and Griffiths, 1972) have thus moved eastwards, mainly on the northern (Burdwood Bank, Shag Rocks and South Georgia) and southern (South Orkney micro-continent, Bruce, Jane and Discovery Banks) limbs of the extending arc whose current subduction zone lies along the

South Sandwich Trench (Dalziel, 1983). The opening of the Scotia Sea and Drake Passage signalled the final event in the break-up of Gondwanaland and the isolation of the Antarctic continent.

Subduction in the eastern Scotia Sea and extensional activity along the northern end of the Antarctic Peninsula continued through the late Tertiary and into the Recent Period. Eight of the eleven islands of the volcanic arc of the South Sandwich Islands show signs of volcanic activity in historic times and none is older than 4 Ma (Barker and Hill, 1981). At the northern end of the Antarctic Peninsula, recent volcanic activity on Deception Island and Bridgeman Island suggests that the Bransfield Strait is still opening northwest–southeast. Volcanic activity elsewhere in Antarctica is associated with major rifting of the West Antarctic during the Cenozoic, probably within the past 25–30 Ma (LeMasurier and Rex, 1991). The rift system is bounded on its poleward side by the Pacific flank of the Transantarctic Mountains and in West Antarctica by a rift running from the western Ross Sea to Ellsworth Land. Volcanic eruptions associated with the rift have been used to infer the timing and extent of continental glaciation in this part of Antarctica. At Mount Murphy, on the Marie Byrd coast, alternate layers of subaerial rocks and subglacial rocks, erupted from beneath the West Antarctic ice sheet, indicate a fluctuating ice cover between 24 and 3.5 Ma ago (LeMasurier *et al.*, 1994). Within the Marie Byrd Land volcanic province, almost all the nunataks are chains of Cenozoic volcanoes or basement rock overlain in some cases by volcanic rocks. On the other side of the rift, the topographical relief of the Transantarctic Mountains is a product of rapid uplift that began about 50 Ma ago. Still active in the western Ross Sea, the Terror Rift extends north–south between the active volcanoes Mount Erebus and Mount Melbourne. The implications of the volcanic evidence for our understanding of the development of the Antarctic ice sheets are discussed further in Chapter 3.

The above account of tectonics, together with the requirements of geometry, suggest that if volumes of new oceanic crust are added at mid-ocean ridges, which encircle the Antarctic continent, then these ridge systems must all have been migrating northwards for the later part of the period of the break-up of Gondwanaland. It follows that for the past 100 Ma Antarctica has been situated, and then isolated, in an approximately polar location (Walton and Morris, 1990; Lawver *et al.*, 1991). Part of the importance of the above tectonic reconstructions lies in their influence on world climate and palaeo-circulation patterns of ocean and atmosphere as well as on the possible location of any geological resources that may occur in Antarctica as a result of its former Gondwanaland connections.

2.3.4 Antarctica and geological resources

The pattern of plate tectonics outlined above has been used to speculate that the parts of Antarctica that were once juxtaposed at the same time with mineral-rich parts of other continents may contain similar mineral resources (see Figure 2.5) (de Wit, 1985). Similar arguments have been used to speculate on the locations of hydrocarbons. The important word here is 'speculation', since most of the claims of potential mineral wealth have been made on the basis of superficial continental comparisons. There have been no exploratory studies for commercial mineral occurrences by resource geologists. Perhaps more important than continental refit and age is the need to correlate geological environments and processes, yet such comparisons of Antarctic formations and terranes with the rest of the world have not been done: continental comparisons have to be specific about the instant in time to which the continental refit refers (Willan, personal communication, 1997). Correlation with Gondwanaland is valid only for rocks of Gondwanaland age (500–180 Ma) and is irrelevant to older terranes. Most of East Antarctica is formed of fragments amalgamated from older supercontinents that are now widely dispersed as parts of other continents, for example North America (see the SWEAT hypothesis, page 24). Similarly, the refit is irrelevant to formations and terranes that were to form subsequently such as the Antarctic Peninsula. A telling point is made by Willan *et al.* (1990), who note that the three continents *not* used in the Gondwanaland refit (North America, Greenland and Eurasia) are also mineral-rich and contain two-thirds of the world's mineral resources. It follows that whatever continental refits are used, Antarctica will be surrounded by mineral-rich continents and comparisons with Gondwanaland state the obvious: that Antarctica is probably as mineralised as the other continents (*ibid.*).

It is crucial to define the terms used to define mineral resources. Mineral occurrences are localities where the rocks contain anomalous mineralogy and geochemistry (a few centimetres to hundreds of metres across). They may be localities recorded by the geologist in passing, or they may be intensively explored localities where the location, structure or size make it unprofitable to exploit in the foreseeable

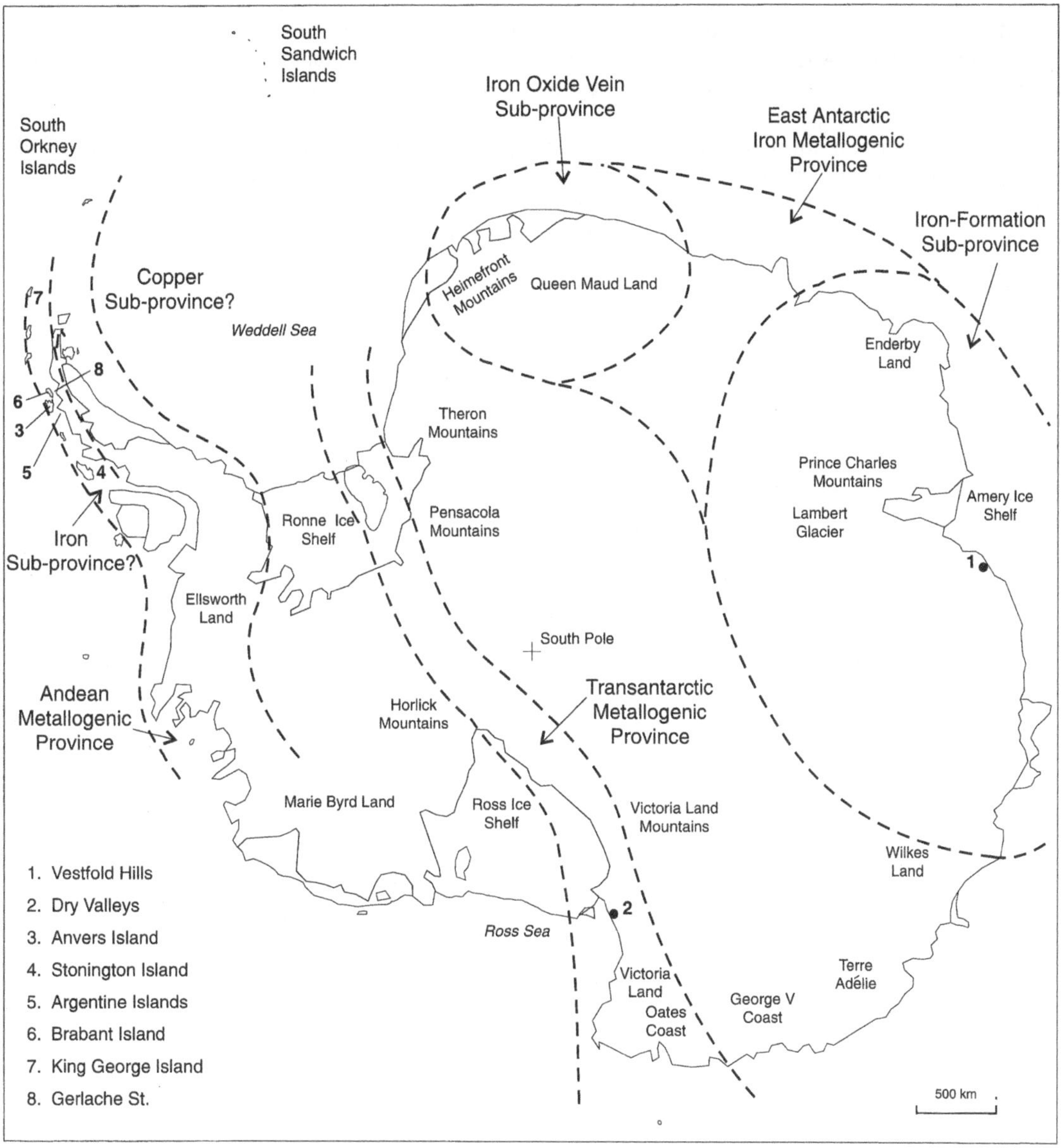

Fig 2.8 Speculated 'metallogenic' provinces of Antarctica. It should be emphasised that maps such as this are based on extremely limited field evidence of mineral occurrences rather than on formal resource assessments by economic geologists. Antarctica has no proven mineral resources (source: redrawn after Rowley *et al.*, 1983).

future. Mineral deposits are substantial volumes of rock whose geology, mineralogy and geochemistry allow profitable exploitation of either metals, non-metals, oil, gas, or rocks used in their entirety (bulk minerals, coal). Mineral resources may be defined, speculative or hypothetical, and the mineral resource potential is a qualitative measure of the possibility of rocks occurring that may be of potential use to society (*ibid.*). Economic deposits are volumes of rock that can be extracted and utilised at a profit within the economic and political conditions prevailing. Reserves are that portion of a mineral resource that could be mined and from which valuable or useful minerals could be recovered economically under conditions realistically assumed at the time of reporting (Miskelly, 1994). No reserves exist in Antarctica (Rowley *et al.*, 1991). Antarctica is bound to have large mineral occurrences that would be exploited if they were located in the inhabited continents, but their locations are unknown so they fall into the 'speculative' class of mineral resource. In addition, any discussion of Antarctic mineral resources should ideally be placed in a global context, since the significance of an Antarctic 'mineral occurrence' can only be judged in the light of the global mineral situation as outlined, for example, by Evans (1993, 1995).

There have been several attempts to summarise the mineral resource potential of Antarctica, most notably by Rowley *et al.* (1983), who recognised three main 'metallogenic' provinces: in East Antarctica; the Transantarctic Mountains; and the Andean Province (most of West Antarctica) (Figure 2.8). In reality, since many of the localities used are poorly defined and in some cases may well be part of the background variation rather than anomalies, use of the term 'metallogenic' is misleading (Willan, personal communication, 1997). For example, in the Andean Province, subsequent work has shown these early predictions of distinct provinces to be premature (Rowley *et al.*, 1991). Nevertheless, the provinces that Rowley *et al.* define are broad lithotectonic provinces that can be used to examine the mineral resource potential of East Antarctica, the Pacific margin of Gondwanaland, and the development of post-Gondwanaland magmatic arcs.

Supercontinental reconstructions and resources: East Antarctica

Speculation about the possible mineral wealth of Antarctica has been based on comparisons with the mineral resources of the fragments of the continents that may formerly have been adjacent to Antarctica. However, continental reconstructions for pre-600 Ma ago are uncertain and recent work suggests that East Antarctica, given its size, may be a collage of Archean and Proterozoic fragments that are unrelated to each other. One of these reconstructions for the Proterozoic speculates that the silver, lead, copper and zinc deposits of central Australia and the gold, nickel, uranium, copper and iron formations of Western Australia may continue into East Antarctica in Terre Adélie and Wilkes Land, respectively (Willan *et al.*, 1990) (see Figure 2.5A). The Indian manganese and iron province appears to continue into Queen Maud Land and Enderby Land, but it is possible that the South African cobalt, gold, platinum, copper, chromium and manganese province (Wilsher and de Wit, 1990) may not extend into Queen Maud Land (Willan *et al.*, 1990).

In East Antarctica, the so-called 'metallogenic' province of Rowley and Pride (1983) can be subdivided into two sub-provinces (see Figure 2.8). The iron formation sub-province extends from Wilkes Land to west Enderby Land and contains 400 m thick exposures of banded iron formation (jaspilite) rocks at Mount Ruker in the Prince Charles Mountains. Aeromagnetic anomalies that extend 180 km west of Mount Ruker beneath glacial ice indicate that the banded iron formation extends over a large area comparable in size with banded iron formations found in other parts of the world (Rowley *et al.,* 1983, 1991). The Antarctic rocks so far discovered contain an average iron content of 33.5% (Tingey, 1990). The 34% iron content of rocks at Neuman Nunataks in Enderby Land and 32% at Mount Ruker are comparable with the 36% iron content of the Lake Superior ores that are mined in the United States but are well below the 64% iron content of the high-grade ores of Western Australia (Lovering and Prescott, 1979). In the Vestfold Hills, abundant erratics of banded iron testify to subglacial source rocks (Ravich *et al.*, 1982; Tingey, 1990). Clearly the full extent of the banded iron formation has yet to be established. The iron oxide vein sub-province identified by Rowley *et al.* (1983) occurs in Queen Maud Land, where iron and lesser amounts of copper- and lead-bearing rocks occur (Neethling, 1969).

Because it was constructed from an amalgam of earlier fragments, each with their own geological inheritance, the Gondwanaland refit between 500 and 180 Ma ago adds little to the above assessment of potential resources (see Figure 2.5B). However, it does alter their present-day spatial arrangement, since the fragments have certainly been re-assembled from an earlier collage (Hoffman, 1991). In addition, the Gondwanaland refit for 180 Ma ago shows the situation after the rocks of the Transantarctic Mountains had been emplaced between southern South Africa and eastern Australia and also shows the developing Pacific margin with fragments of the Antarctic Peninsula in place (Willan *et al.*, 1991). These two developments resulted in additional mineralisation.

The Pacific margin of Gondwanaland: Transantarctic Mountains

The Transantarctic Mountains are in large part underlain by Proterozoic and Palaeozoic rocks that were both folded and intruded into in the late Proterozoic and early Palaeozoic (Laird, 1991). The folding associated with the Ross deformational belt is thought to extend through the Transantarctic Mountains to central Australia in one direction and possibly as far as Mozambique in the other (see Figure 2.5B). (Craddock, 1982). Mineralising processes or events within this belt have resulted in widespread occurrences of tin, lead, zinc, copper, gold, silver, barium, manganese and other metals. The Flinders Ranges in Australia, which may correlate with parts of the Transantarctic Mountains, contain many occurences of the above metals. Parts of the eastern Tasman belt, which geologically resemble Victoria Land, contain gold, copper, silver, arsenic, lead, zinc, molybdenum, bismuth, tin, tungsten and antimony (Rowley *et al.*, 1991).

Unconformably overlying the rocks of the Transantarctic Mountains are the flat-lying sedimentary and igneous rocks of the Beacon Supergroup and Ferrar Supergroup. The largest and best known of the Ferrar rocks are the layered gabbroic intrusions of the Dufek Massif in the Pensacola Mountains. The Dufek intrusion was once thought to cover an area of 50 000 km^2 but is now modelled as a much smaller 6600 km^2 dyke-like body (Ferris *et al.*, 1998). In spite of this recent re-evaluation in size, the intrusion remains an important layered igneous complex, similar although smaller than the mineral-rich intrusions of the Bushveld, Stillwater and Sudbury complexes of South Africa, the USA and Canada, respectively (Ford and Himmelberg, 1991). Ford (1990) describes layers of iron–titanium oxides and copper and iron sulphides from the exposed top layers and speculates on the existence of platinum group (PG) minerals in the unexposed middle part. At present, PG minerals are the most likely targets for potential exploration (*ibid.*). Other Dufek-type intrusions are reported from the dry valleys of Southern Victoria Land and throughout the Transantarctic Mountains, but they have not been studied in any detail (Hamilton, 1964).

The development of post-Gondwanaland magmatic arcs: West Antarctica

West Antarctica consists of the Mesozoic and Cenozoic rocks of several small lithospheric plates that have moved independently of each other (see Figure 2.7) (Barker *et al.*, 1991), but there are outcrops of older rocks such as the folded Palaeozoic clastic rocks of Ellsworth Land (Rowley *et al.*, 1991). The Antarctic Peninsula forms part of a magmatic arc that developed on continental lithosphere and has a similar geological and tectonic setting to the Andean belt of South America. The Andean orogen extends from New Zealand and the western Pacific, through West Antarctica and the Antarctic Peninsula, to the Andes of South America (see Figure 2.4). The Andes have long been known to host some of the richest and largest metal deposits on Earth, and the apparent continuity of the West Antarctic magmatic arc with the Andes has led to speculation that it may host similar metal deposits (see Figure 2.5B) (Rona, 1976). Destructive plate boundaries are prime sites for the accumulation of metals melted from both the descending plate and the older continental basement and carried through the overlying crustal layers. The central Andes contain large deposits of copper, antimony, tin, molybdenum, silver, tungsten, zinc, lead, iron and gold (Rowley *et al.*, 1991), but the distribution of major deposits is not continuous along-strike, because the southern Andes have had a very different Cenozoic evolution (Rowley and Pride, 1982). Subduction is still active in the northern Andes, whereas in the southern Andes and Antarctic Peninsula it has been much slower and stopped variously in the Tertiary. Nevertheless, Wright and Williams (1974) and Rowley *et al.* (1991) consider the Antarctic Peninsula to represent one of the most likely places in Antarctica for the discovery of mineral resources.

One area of copper mineralisation occurs in the extreme northwest of the Antarctic Peninsula but in the absence of any exploration, geotechnical or economic studies in the region, the potential is unknown (see Figure 2.8) (Rowley and Pride, 1982). Other copper occurrences in Antarctica are on the islands west of the Antarctic Peninsula (Rowley *et al.*, 1991), and Pride *et al.* (1990) identify sites of copper and related base metal mineralisation on King George Island in the South Shetland Islands and at the Gerlache Strait and Anvers Island in the northern Antarctic Peninsula, but they emphasise that the resource potential of these localities is unknown. Occurrences of magmatic iron oxides are common in the Antarctic Peninsula, but magnetite-rich layers are confined to the western flank and offshore islands of the peninsula, where they occur as veinlets in lava flows and gabbroic rocks on Brabant Island and in the Argentine Islands, respectively (Rowley *et al.*, 1991).

The conditions under which such mineral deposits might become exploitable in the future, together with any likely environmental impacts, are reviewed later, and it remains here to reiterate the caution that mineral exploration has not yet been carried out and hence no commercial mineral deposits have yet been identified anywhere in Antarctica. Nor is there likely to be exploration in Antarctica in the foreseeable future, given the current environmental regime, and the vigour and success with which abundant resources are being discovered elsewhere in the world. Of the larger mineral occurrences that exist, only iron in the Prince Charles Mountains, the Dufek sills for PG minerals, and the Antarctic Peninsula for copper, gold and silver might have been explored if located on another continent (*ibid.*).

Other possible geological resources

Amongst the other possible geological resources are the occurrences of bulk minerals, aggregates, coal, hydrocarbons, manganese nodules and icebergs. Of these, there has been most interest in the resource value of the coal, hydrocarbons and manganese nodules. The resource potential of icebergs is covered in Chapter 7.

Coal occurs in Permian to Triassic strata in Antarctica and the largest known quantities occur in Permian sandstones in the Transantarctic Mountains (Rose and McElroy, 1987). Thick beds of largely unfolded coal are especially prominent in the Beacon Supergroup of the Transantarctic Mountains and in the Prince Charles Mountains (Coates *et al.*, 1990; Rowley *et al.*, 1991). Also, widely distributed from the Theron Mountains, on the Weddell Sea coast, to central Victoria Land on the Ross Sea coast, the coal has a high ash and low sulphur content and its quality ranges from low-volatile bituminous to anthracite (Splettstoesser, 1985; Coates *et al.*, 1990). Coates *et al.* suggest that 153×10^{12}t of coal may exist in three basins in the Transantarctic Mountains along the coast of the Ross Sea. However, since most coal occurs in inaccessible and mountainous areas it is unlikely to be mined for shipment off the continent (Rowley *et al.*, 1991).

There are no petroleum resources in Antarctica. However, it has long been speculated that extensive hydrocarbon deposits occur in the sedimentary basins in and around Antarctica (Behrendt, 1983, 1990, 1991). This speculation is based on the relationships of hydrocarbon occurrence in sedimentary basins and undisturbed tectonic plate edges elsewhere in the world (Anderson, 1990) (Figure 2.9). Since most of the sedimentary basins of the Antarctic continent are either disturbed by metamorphosis, deformation or igneous intrusion, or are buried beneath ice in excess of 2 km thick, the more likely hydrocarbon occurrences lie almost entirely within the younger sedimentary fringes of the Antarctic continental shelf. However, the development of glaciers in the Eocene and Oligocene led to glaciomarine conditions on the shelves surrounding the continent and this serves to limit further the areas of potential for hydrocarbon development to those shelves with pre-late Oligocene deposits (*ibid.*).

On this basis, the most promising sites appear to be along the coast of West Antarctica, where rapidly deposited wedges of sediments may have trapped hydrocarbons (Behrendt, 1991). The geophysical data and sparse drilling information suggest that the most likely places are the deep sedimentary basins of the Ross Sea continental shelf (14 km thick), the Weddell Sea continental shelf (14–15 km thick) (*ibid.*), the Bellingshausen Basin (greater than 3 km thick), the Wilkes Land coast (greater than 2 km thick), and the Prydz Bay–Amery Basin and Lambert Graben (5–12 km thick) (Holdgate and Tinker, 1979; Office of Technology Assessment, 1989; Anderson, 1991) (see Figure 2.9). In the Ross Sea basin, the 1973 Deep Sea Drilling Program showed flows of methane in association with the petroleum gases of ethane and ethylene that are no different from drill holes elsewhere (Behrendt, 1991). In other words, like on-land mineral occurrences, they are indistinguishable from 'background' occurrences. However, asphaltic residues reported from the CIROS-1 drill hole in the Ross Sea indicated that petroleum had been present there (*ibid.*). In spite of this, there is no drilling or seismic evidence for hydrocarbon presence and the most recent study of the Larsen Basin (Macdonald *et al.*, 1988) identified only a modest hydrocarbon potential. Antarctic hydrocarbon resources are speculative, but it is clear that anywhere else in the world deep sedimentary basins and structural rift systems containing undisturbed sediment up to 14 km thick would be regarded as targets for hydrocarbon exploration.

Manganese nodules and encrustations are known from many areas of the Southern Ocean and elsewhere. These consist of a nucleus of a volcanic shard, shell, or iceberg drop-stone covered with thick layers of manganese and iron-rich hydrated oxides precipitated from the surrounding ocean waters. Goodell (1973) showed that they contain an array of exploitable metals including iron, manganese, tin, nickel and copper. Nodules are related to areas of low rates of sedimentation and strong bottom currents (Frakes and Moreton, 1990). Figure 2.10 shows that the most abundant areas lie in the Pacific sector of the

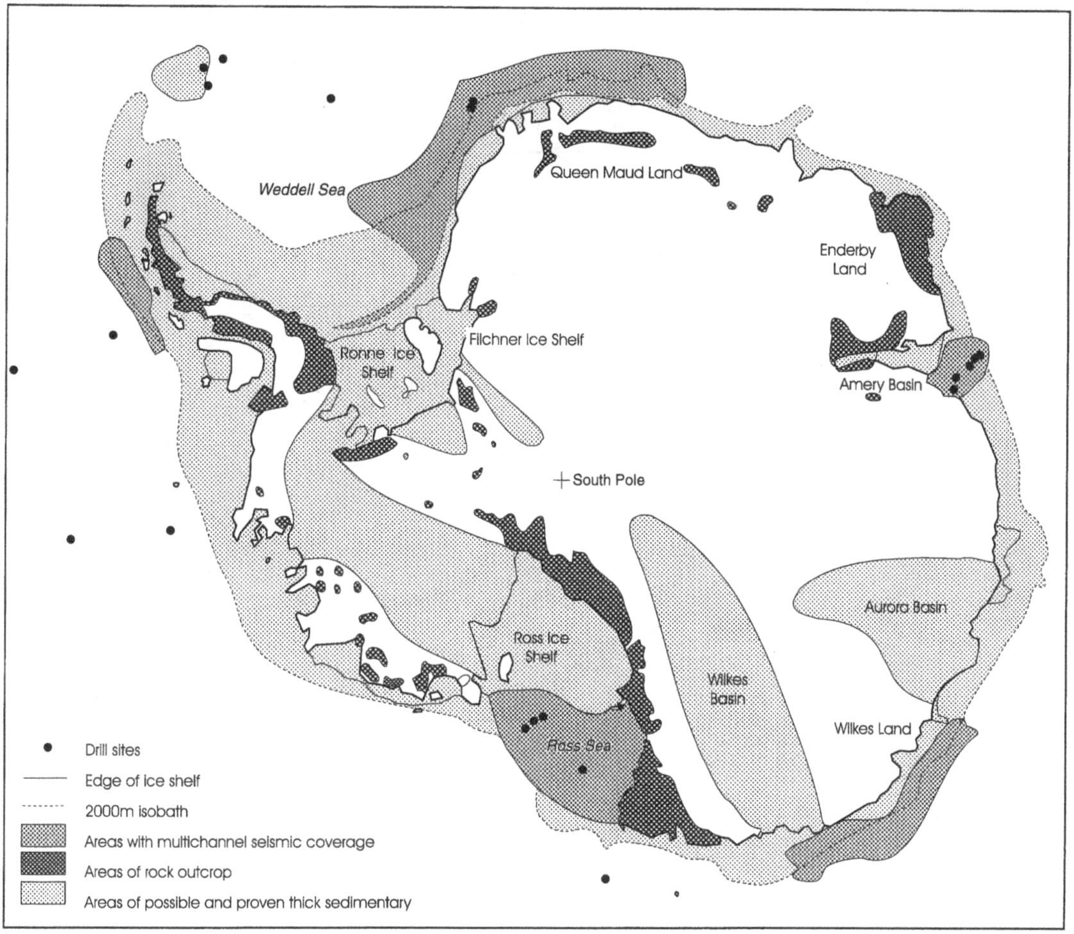

Fig 2.9 Potential hydrocarbon exploration areas, sedimentary sequences, rock outcrops and seismic coverage. Antarctica has no proven hydrocarbon resources (source: redrawn after Elliot, 1988).

Southern Ocean close to the location of the Polar Frontal Zone. According to Frakes and Moreton, the most striking feature of Antarctic manganese nodules is the enormous quantity that is known to exist around the continent at sea-floor coverages of up to 60% and concentrations of greater than 20 kg m^{-2}. However, manganese nodules have attracted little economic interest even in the accessible oceans of the world, and since no one nation has sovereignty over the mineral resources of the deep ocean, uncertainty over ownership represents a barrier to exploitation.

2.3.5 Summary

Much remains to be understood about the geology of Antarctica and, in spite of great advances in geophysical techniques that now allow sub-ice topography to be surveyed, the depth and extent of the ice cover will ensure that progress is slow and sampling sparse. For example, only the broadest of magnetic and gravity surveys have been done and sub-ice seismic surveys for structures are in their infancy. However, the broad outline of the continental structure seems fairly clearly related to the amalgamation and break-up of Gondwanaland and earlier continents. The present-day continent comprises a complex of Precambrian cratons constituting the stable shield of East Antarctica. These rocks are overlain unconformably by the widespread and flat-lying sedimentary Beacon Supergroup, which itself is intruded by the basaltic rocks of the Ferrar Group. These are well exposed in the uplifted Transantarctic Mountains. West

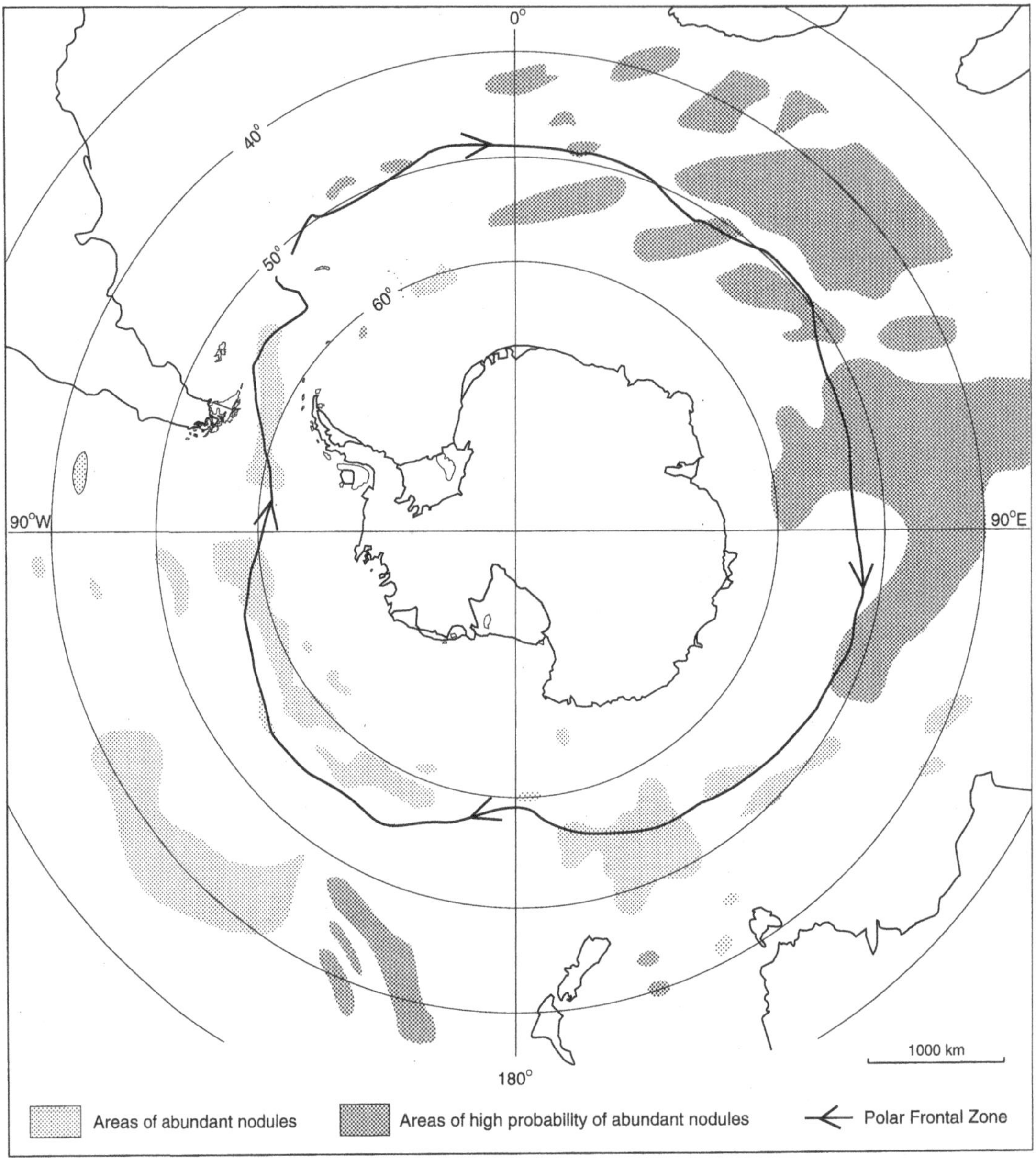

Fig 2.10 The estimated geographical distribution of seabed manganese nodules in the Southern Ocean (source: redrawn after Goodell, 1973 adaptation, by permission of the American Geographical Society

Antarctica consists of several microplates that share a common history with South America.

There has been much speculation over the potential of Antarctic rocks to yield mineral resources on a scale similar to those found in the rocks of formerly contiguous continents. This speculation has extended to the seabed, where hydrocarbons may exist and manganese nodules are known to exist. Whether minerals exist in exploitable quantities in Antarctica remains as yet unproven, irrespective of the physical problems of extraction and transport and the issues of environmental protection, mineral rights and sovereignty that would require to be addressed before any extraction could begin.

2.4 The climate of Antarctica

2.4.1 Introduction

As one of the world's greatest heat sinks, Antarctica plays a central role in the global climate system (Phillpot, 1985). Its principal effect as a regulator of heat derives from an average receipt of radiation at the poles that is about 40% less than at the equator (Figure 2.11), and the transfer of heat between the poles and the tropics is the driving force of the global meteorological system. Low polar temperatures occur as a result of both the curvature of the Earth and the 23.3° inclination of the Earth's axis of rotation in relation to the plane of its orbital axis round the Sun (Figure 2.12). There is also increasing evidence that global climatic variability arises from a structured, globally interconnected system (Yuan *et al.*, 1996), a striking example of which is the Antarctic Circumpolar Wave (White and Peterson, 1996). This is a disturbance in Antarctic sea ice extent, sea-surface temperature, surface wind speed and atmospheric pressure at sea level that may affect equator to pole

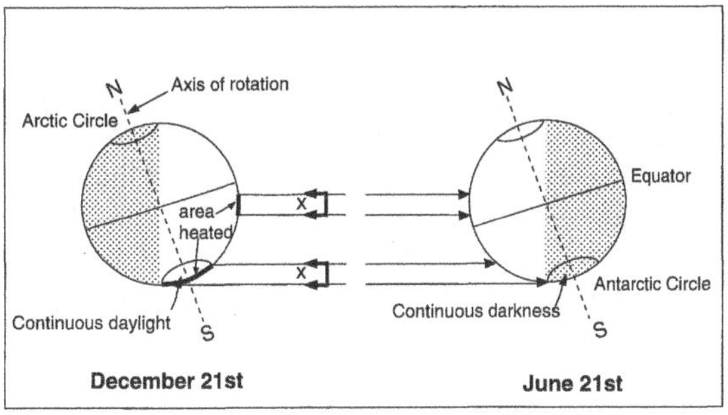

Fig 2.12 The influence of polar location and curvature on radiation receipt and duration at mid-winter and mid-summer.

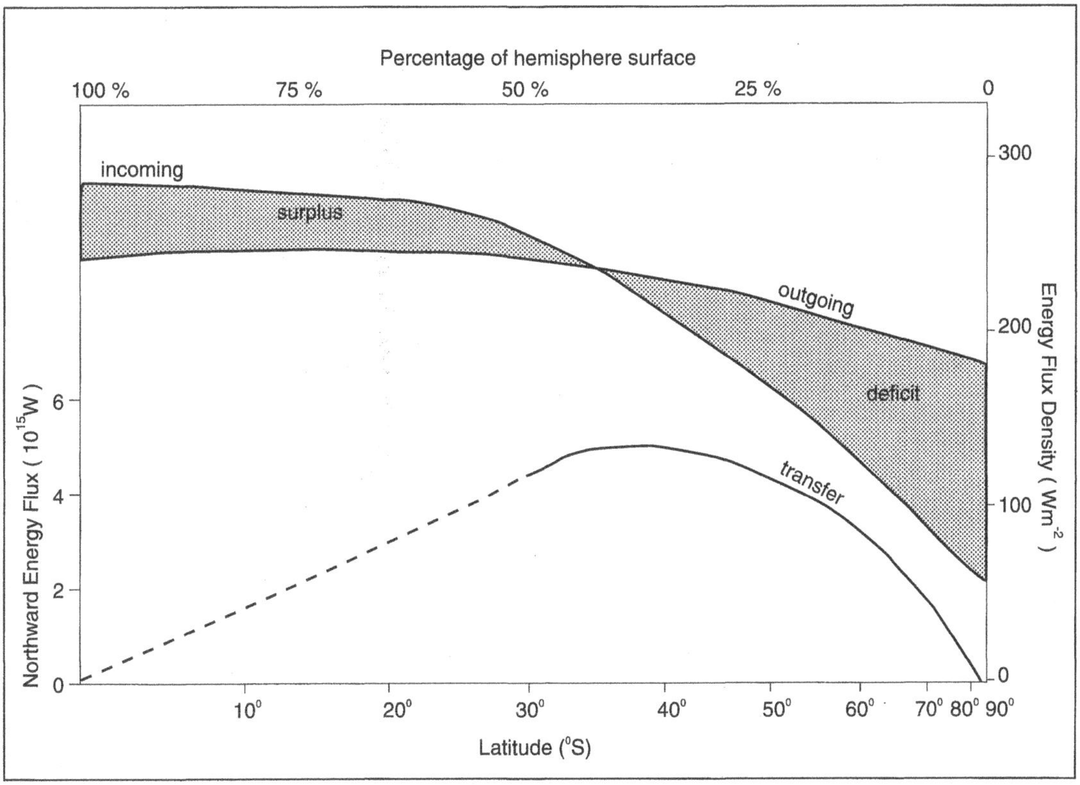

Fig 2.11 Incoming solar radiation and outgoing radiation from Earth and atmosphere create zones of surplus and deficit that are balanced by poleward energy transfers (source: data from Gabites, Houghton and Newell, redrawn with permission from Barry and Chorley, 1992).

temperature gradients and drive the dynamics of the atmosphere and oceans (Yuan, 1996).

The climate of Antarctica is dominated by three main features: the cold waters of the Southern Ocean; the variable sea ice cover; and the high continental ice sheet. The Southern Ocean is characterised by two wind-driven currents, which result in the circulation of cold water around the continent. The zone of mixing of the cold water with warmer sub-Antarctic water at the Polar Frontal Zone produces a zone of great temperature and salinity gradients in the ocean (see Figure 1.5). The winter advance of the sea ice effectively doubles the ice-covered area of the Antarctic region and further depresses temperatures by increased albedo effects; 98% of the 14 million km^2 of the Antarctic continent is covered by snow and ice and is high, cold, windy and extremely dry.

Most of the climate stations in the Antarctic are far apart and coastal in location, having been chosen for ease of access rather than for meteorological reasons (Phillpot, 1985). Few have records that extend over more than 30 years and this imposes real limitations on both the interpretation of Antarctic climate and on the identification of trends in climate (Ellis, 1991). In spite of the spacing and paucity of stations, particularly in the interior, the relative uniformity of the continental plateau has allowed realistic climatic assessments of large parts of the interior, but the steep coastal slope zone presents more complex problems (Phillpot, 1985). The ice-free areas that have been exposed by recession of the ice, and the 'oases' are of local rather than general significance for climate.

2.4.2 Radiation and temperature regime

Radiation

The amount of incident radiation fluctuates enormously between the lowest values recorded on Earth during the darkness of the polar winter to the highest values recorded on Earth during the continuous daylight of the polar summer (see Figure 2.12). However, much of this incident radiation is lost to the Antarctic heat budget for a variety of reasons. Fresh snow and ice surfaces reflect 80–90% of incoming solar radiation (Oke, 1978), so that rather than being absorbed as in lower latitudes, this potential summer warmth is simply reflected by the extensive coverage of ice and snow on the continent (Figure 2.13). Solar radiation peaks in summer at a time when much of the Antarctic sea ice is still present and this further contributes to a high degree of radiation reflectance (albedo). Since polar air holds ten times less moisture than temperate air, the amount of solar radiation absorbed by moisture in the atmosphere reaches a global minimum over the Antarctic continent (Barry and Chorley, 1992) (see Figure 2.13). The amount of radiation received is also affected by cloud cover. Mean cloudiness over the oceans is high, is lower over the continental coastline and falls to a minimum over the interior. West Antarctica is a more cloudy region than East Antarctica (Phillpot, 1985). A distinct lack of atmospheric dust also characterises Antarctica (Sugden, 1982), and the combined effect of low moisture and dust levels allows long-wave radiation to escape from the atmosphere without providing the heating so characteristic of lower latitudes. However, because of atmospheric clarity Antarctica is ideally suited to studying atmospheric processes. Radiational losses on the scale experienced in Antarctica (together with the huge heat surfeits of the tropics) are compensated for within the global energy budget by the poleward flow of sensible heat in warm air from the tropical regions.

Temperature

With the above radiation patterns it is clear that temperatures will be low over much of the Antarctic continent. However, the geography of Antarctica also plays a vital part in depressing temperatures. Unlike the Arctic, the polar continental land mass of Antarctica is surrounded by ocean. Much of the continental interior is therefore remote from the moderating influence of the ocean, an effect that is accentuated during winter, when expansion of sea ice leads to a doubling of the summer ice-covered area and that affects not only temperature but also precipitation (Rubin and Weyant, 1965). The mean annual temperature on the polar plateau is a chilly –50 °C (Figure 2.14). This ranges between –70 °C in July and –30 °C in January (Schwerdtfeger, 1970), with an extreme of –89.5 °C, the world's lowest recorded temperature, being measured in July 1983 at Vostok on the polar plateau. In coastal locations, the mean annual temperature is about –10 °C, with a range between –25 °C in July and –4 °C in January. The great variation between the polar plateau and the coast reflects not only the unequal receipt of radiation at the surface but also the effect of altitude, radiational cooling and distance from the ocean (see Figure 2.14).

Antarctica is a high continent. Schwerdtfeger (1970) estimates that 55% of its area lies at an elevation of more than 2000 m, and around 25% is more

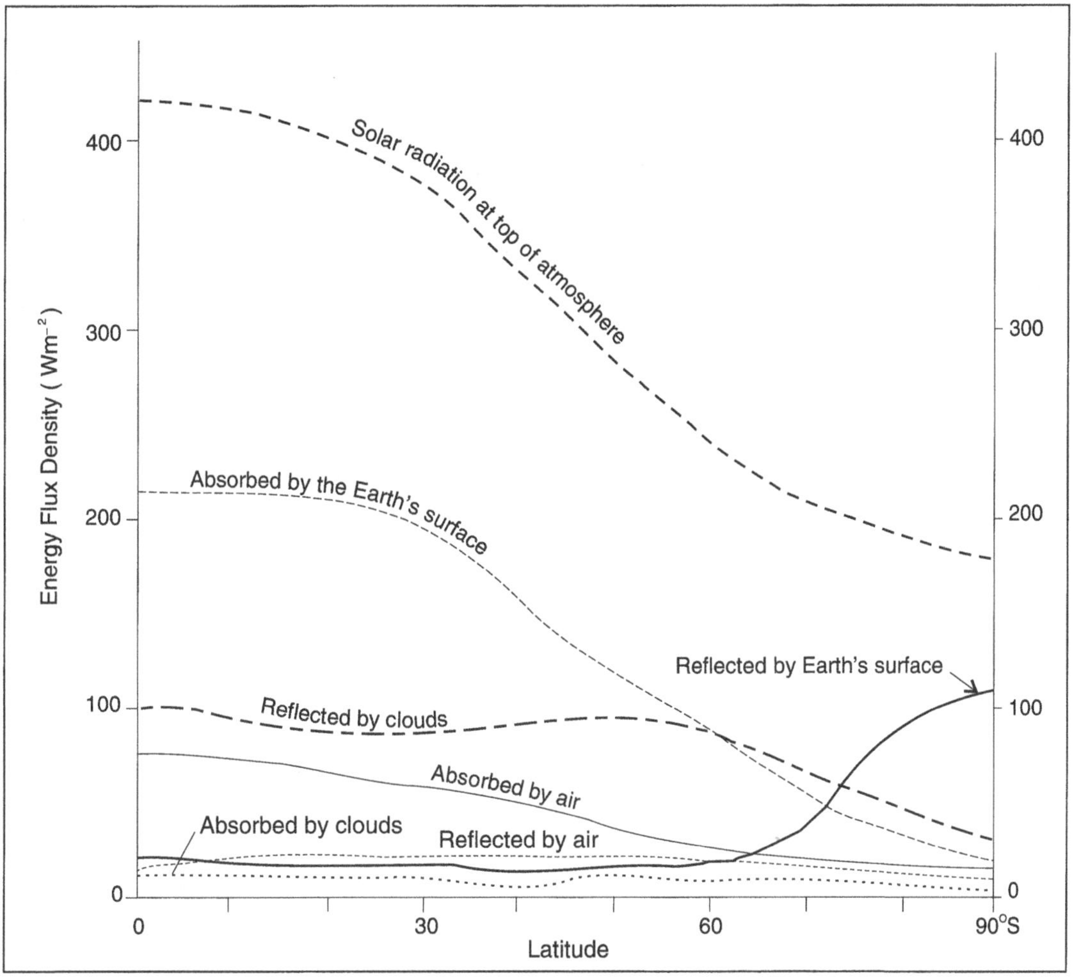

Fig 2.13 The average latitudinal variation of radiation receipt, reflection and absorption (source: after Sellars, 1965).

than 3000 m above sea level. Since temperature declines by 1 °C for every 100 m altitude above ice sheets (Benson, 1962), it follows that the mean annual temperature over 55% of the Antarctic continent should be 20 °C lower than its sea-level equivalent. On the polar plateau, there is a very marked asymmetry between the durations of summer and winter (Schwerdtfeger, 1984). Summer is very short, and only from mid-December to mid-January are temperatures above –30°C. The dome-shaped temperature graph has been called the 'pointed summer' and is rapidly followed by a sharp fall in temperature so that, when the Sun sets by the end of March, temperatures vary by only a few degrees during the 'coreless winter' (Figure 2.15). This asymmetry may have played a part in the tragic last journey of Captain Robert Scott, the British polar explorer. Prior reports by another British explorer, Ernest Shackleton, in 1909 had indicated that mean temperatures of –29 °C in early January could be expected on the polar plateau. However, by late January 1912, Scott had experienced the unexpectedly low temperatures that eventually exacted a fatal toll on his party in March 1912. Bechervaise (1962) provides a graphic insight into the implications of existence at the extremely low temperatures that have been recorded in Antarctica:

> if you try to burn a candle the flame becomes obscured by a cylindrical hood of wax, if you drop a steel bar it is likely to shatter like glass, tin disintegrates into loose granules,

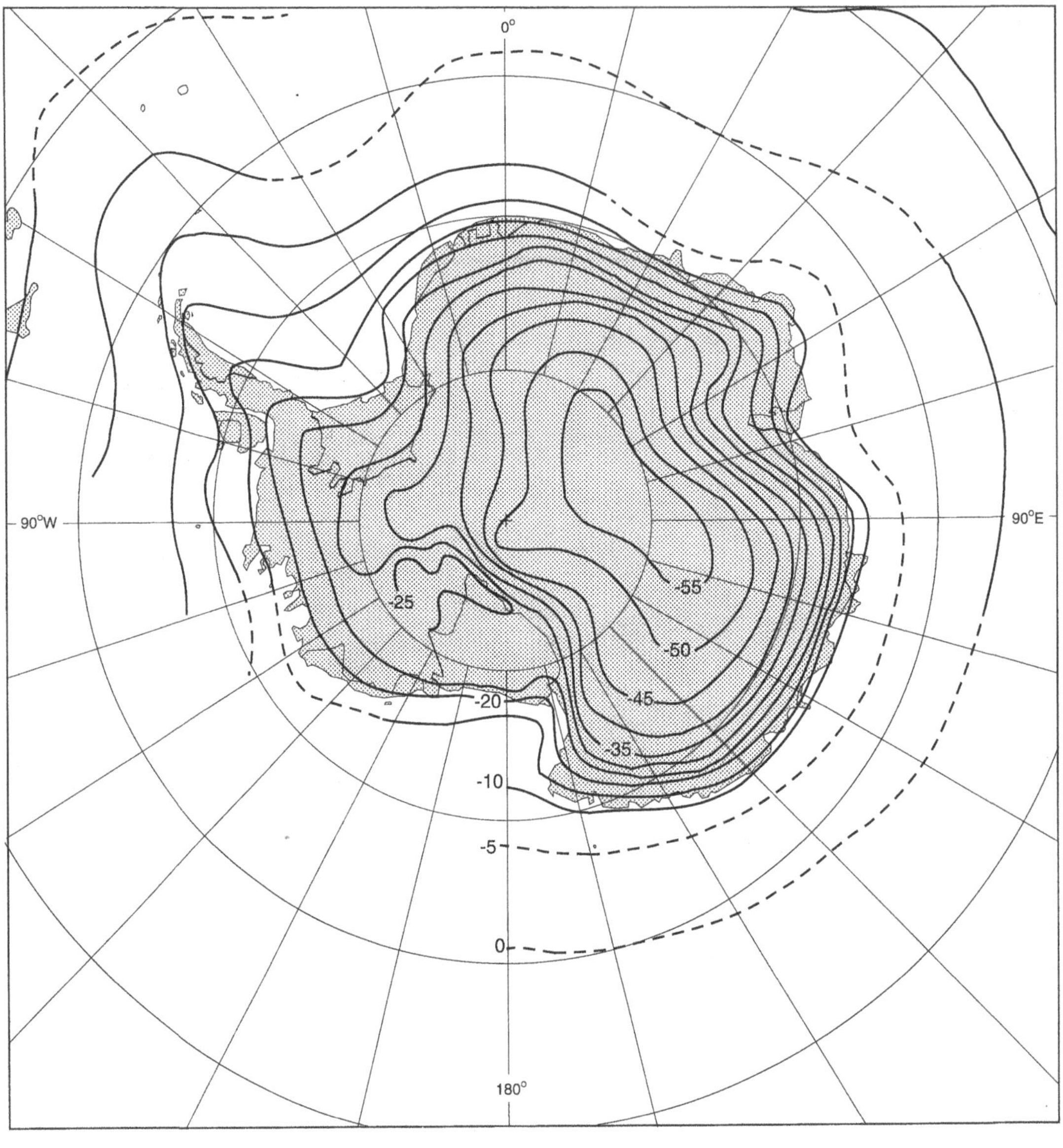

Fig 2.14 Mean annual surface temperatures over Antarctica (°C) (source: data from Rubin and Weyant, 1965).

> mercury freezes into a solid metal, and if you haul up a fish through a hole in the ice within five seconds it is frozen so solid that it has to be cut with a saw.

The temperature regime in the Antarctic Peninsula and the sub-Antarctic islands is more benign than that of East Antarctica, reflecting latitude, more variable sea ice conditions and the moderating influence of the Southern Ocean. The mean annual temperature at Adelaide Island, which lies just south of the Antarctic Circle on the west of the Antarctic Peninsula, is about –1.8 °C, with the summer mean reaching 0.9 °C and a winter mean of –11.3 °C (Morrison, 1990). However, winter temperatures for the Antarctic Peninsula are highly variable depending on the source of the air: northerly winds bringing mid-latitude air can elevate temperatures above freezing; southerly winds from the polar plateau can send the temperature down to –40 °C. The extent of the sea ice in winter also exerts a strong

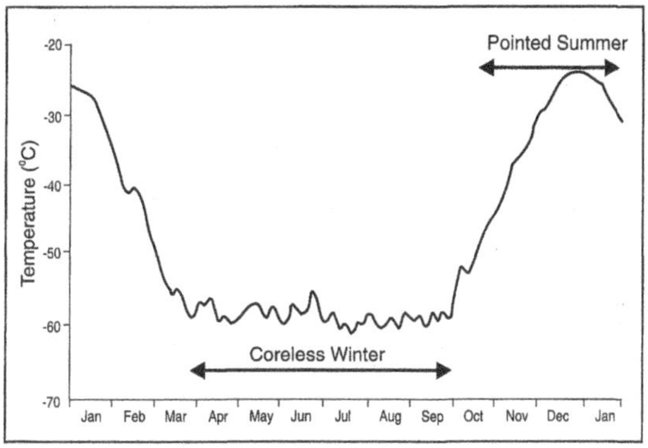

Fig 2.15 Mean monthly temperatures from the Polar Plateau. Note the rapid onset of the long polar winter and the sharp and pointed summer (source: modified from Dudeney, 1987).

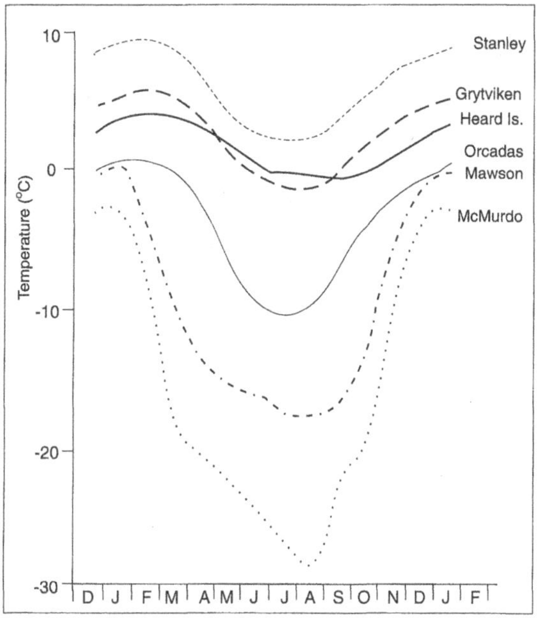

Fig 2.16 Mean monthly temperatures at some sub-Antarctic island and coastal stations around Antarctica. Locations affected by sea ice show enhanced temperature differences between summer and winter (source: after Budd, 1982).

influence on coastal temperatures (Figure 2.16), and coastal continental stations such as McMurdo and Mawson experience lower than expected winter temperatures as a result. With the sea ice gone, the summer temperature difference between sub-Antarctic and coastal continental stations is much less. In the sub-Antarctic, South Georgia records mean annual air temperatures at sea level of almost 2 °C, with a seasonal variation of only 7 °C (Headland, 1984). However, under some conditions, temperatures can exceed 20 °C, and 15 °C is not uncommon on still days in summer (*ibid.*). Temperature maxima of 12.4 °C and 14.4 °C have been recorded for Macquarie Island and Heard Island, respectively (Phillpot, 1985).

Low temperatures apart, one of the most remarkable features of the Antarctic temperature regime is the development of strong radiational cooling over the snow and ice surfaces, resulting in well-defined surface inversions (Phillpot and Zillman, 1970). They are a common feature at Antarctic coastal continental stations but are highly variable in intensity. With the exception of the two summer months, the inversion layer is ever-present over the inland plateau and increases in strength to reach a maximum in winter (Phillpot, 1985). Temperature inversions are mainly associated with the calm anticyclonic conditions that occur in winter and are disrupted by cyclonic activity and strong winds. Sugden (1982) reports the thickness of the surface inversion layer in polar environments to be 10–100 m, above which temperatures sometimes climb to 30 °C higher than at the surface, although Schwerdtfeger (1970) suggests that the top of some Antarctic inversions can reach 1000 m. At the 3642 m high Plateau Station in Queen Maud Land, observations using a 32 m mast showed the temperature at 24 m altitude to be 20 °C higher than at the surface (*ibid.*). Any vertical mixing tends to destroy the inversion, so Antarctic blizzards are usually accompanied by a rise in temperature. This is of little benefit to polar humans, since the enhanced wind-chill more than offsets any rise in temperature!

Temperature trends

Whether there exist any long-term temperature changes in the Antarctic is of paramount interest at a time when global temperatures are generally thought to be rising. Temperature measurements from early expeditions are available at some sites over the last 100 years, but systematic measurements for contintental stations date from only 30 years ago (Ellis, 1991). Best estimates from comparisons of the temperatures recorded at the beginning of this century suggest that Antarctic air temperatures are about 1 °C higher at present, 0.57 °C of this occurring between 1957 and 1994 (Jones, 1990, 1995). However, Sansom (1989) found no significant trends over the past 30 years in data from four stations: Faraday, Scott, Amundsen–Scott, and Mirny. Part of the problem may lie in the predominantly coastal distribution of stations, with only very few long-term data from the polar plateau. There is also variability in the statistical significance values with season as well as variations in the decadal temperature values, which are probably associated with the distribution of sea ice (King, 1994). In spite of uncertainty over the trend for Antarctica as a whole, there is some evidence of regional trends in temperature. Analysis of the average temperatures over 25 years (1957–1981) for eight stations in each of West and East Antarctica allowed Mayes (1981) to identify steady warming up to 1974 followed by cooling to 1981 for West Antarctica, but only very slight warming for East Antarctica. The cooling in West Antarctica was then reversed to give a warming of summer and autumn temperatures of 0.05 and 0.1 °C a^{-1}, respectively, over the period 1958–1989, although no spring or winter increases have occurred (Zwally, 1991).

In the Antarctic Peninsula at Faraday, Stark (1994) found a statistically significant warming trend of 2.7 °C over the 44 years to 1990 (Figure 2.17), reported by many to be a warming trend in the region of 2.5 °C over 50 years (Doake and Vaughan, 1991; Cattle *et al.*, 1997; Morris *et al.*, 1997); 1989 was the warmest year on record in the Antarctic Peninsula and together with the warming trend may be responsible for the break-up of the northernmost ice shelves (Morrison, 1990; Vaughan and Lachlan-Cope, 1995; Vaughan and Doake, 1996). One of the longest temperature records in the Antarctic comes from Signy Island in the South Orkney Islands, where Smith (1990) shows a predominantly warming though variable trend in mean summer air temperatures over the past 40 years of approximately 1 °C; this may have contributed to a 35% reduction in ice cover. However, Stark (1994) showed that over the same period there was no statistically significant increase in annual temperature at Signy Island. In the sub-Antarctic, mean annual air temperatures at South Georgia show a 0.6 °C rise between 1906 and 1982 (BAS, 1987).

Extrapolation of regional warming trends from the Antarctica Peninsula to more southern parts may be unwarranted since no significant trends in temperature are detectable from the continent itself (Ellis, 1991; Zwally, 1991). However, according to Stark (1994), the geography of the Antarctic Peninsula has an important influence in reducing continental influences on the climate: Faraday shows a warming trend, while Signy, in the South Orkney Islands, shows no significant trend because it is affected by the climate of the Weddell Sea. This suggests that the mid-Antarctic Peninsula is an area of high climatic sensitivity and is uniquely placed to observe any global warming trend (*ibid.*). There is also evidence of a warming trend from the sub-Antarctic islands. Frenot *et al.* (1995) link the recession of glaciers on Îles Kerguelen to a 1.27 °C increase in mean annual air temperature between 1964 and 1982. Mean annual air temperatures have also risen by approximately 1 °C at Marion Island over the past 40 years, reflecting a more widespread trend that also includes Macquarie Island and Heard Island (Allison and Keage, 1986; Chown and Smith, 1993). However, while there is limited evidence for an increase in temperature at some Antarctic continental stations (Jacka and Budd, 1991), the distribution of stations remains too sparse, and the record too short and

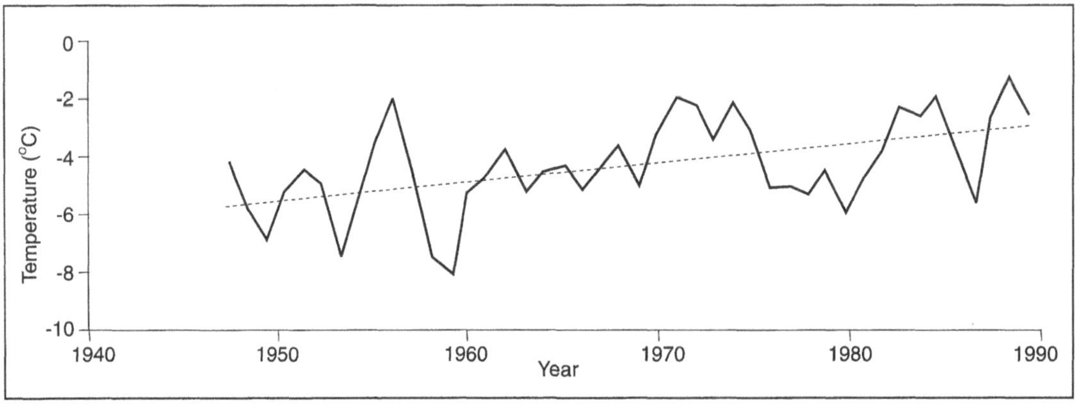

Fig 2.17 Mean annual temperatures at Faraday Base, central Antarctic Peninsula. The gradient of the regression line indicates a statistically significant warming of 2.67 °C over 44 years (source: data from Stark, 1994).

variable, to show any statistically significant trends for the Antarctic as a whole (Allison, 1992).

2.4.3 Atmospheric circulation and wind patterns

Atmospheric circulation

The atmospheric circulation over Antarctica is as close to the circular theoretical model for a polar location as could be hoped for (Figure 2.18A). The cold air over the South Pole is denser than warmer air in lower latitudes and as it progressively subsides along a gradient towards the pole it is deflected by the rotation of the Earth to form a clockwise vortex over the South Pole, a pattern that is mirrored by an anticlockwise vortex over the North Pole. On the surface of a uniform geoid the polar vortex is predicted to be perfectly symmetrical and indeed, at the 500 mb altitude, a striking symmetry exists in the observed vortex over the continent (see Figure 2.18A). Even the surface pressure pattern is convincingly symmetrical, although the influence of the coast and the mountains of the Antarctic Peninsula is apparent (Figure 2.18B). During the polar winter, the circulation over Antarctica is dominated by steep temperature gradients between the warmer air to the north and the very cold air of the vortex (Solomon, 1990). These act as a barrier to inflowing air (Drake, 1995) and are accompanied by a rapid circumpolar flow of strong westerly winds and an extremely isolated polar vortex (Schroeberl and Hartmann, 1991). In summer, the polar vortex disappears quite suddenly in late November or early December as the atmosphere warms, meridional exchange becomes possible and a rapid influx of air from lower latitudes occurs. Winter is therefore dominated by stable pressure systems, intense cold and persistent westerly winds, while summer pressure and winds are more variable.

The polar vortex and ozone

The nature of Antarctic meteorology and the isolated air mass of the winter polar vortex plays an important part in the destruction of the stratospheric gas ozone (O_3): 95% of the world's O_3 occurs between 15 and 40 km altitude in the stratosphere, where it shields the Earth and lower atmosphere from the harmful effects of solar ultraviolet (UV-B) radiation (Solomon, 1990). Ozone is produced when solar radiation causes normal oxygen molecules (O_2) to split and recombine with other oxygen molecules. It is created at the equator and then transported polewards via atmospheric circulation, where it is destroyed (Drake, 1995). Over the Antarctic winter and polar night, the absence of sunlight allows temperatures in the vortex to fall to −183 °C and polar stratospheric clouds composed of ice crystals are formed. The

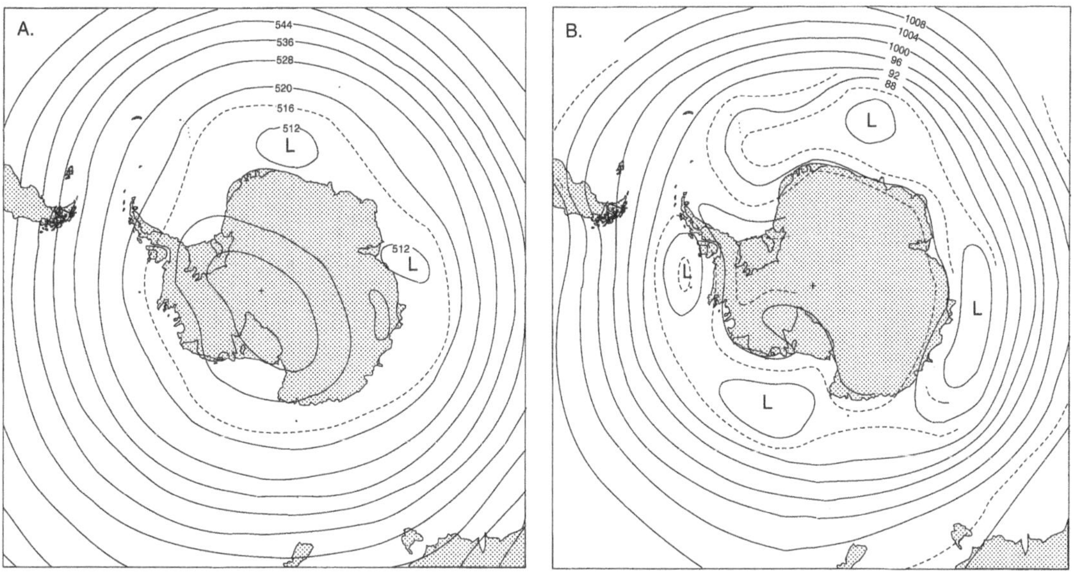

Fig 2.18 Mean January pressure at (A) 500 mb altitude and (B) sea level. Note the symmetrical 500 mb pattern and the sea level/surface pattern disrupted by the influence of topography (source: after Schwerdtfeger, 1970).

springtime appearance of sunlight converts the chlorine compounds that adhere to the ice crystals into reactive species, mainly chlorine monoxide (ClO), which begin to destroy O_3 (Tolbert *et al.*, 1987). The initial loss of O_3 decreases the absorption of solar energy, leading to atmospheric cooling, further ice cloud production and loss of O_3: a positive feedback cycle (Voytek, 1989). In late spring, solar radiative warming removes two factors central to the destruction of O_3: the ice crystals that are host to ClO evaporate and the polar vortex itself begins to weaken, allowing meridional exchange of air. This results in rising O_3 levels due to natural regeneration over Antarctica and an influx from lower latitudes (Solomon, 1990).

Under normal circumstances, the production and destruction of O_3 is thought to be broadly self-regulating, but since 1978 there has been an alarming decrease in springtime values over Antarctica (Farman *et al.*, 1985; Jones and Shanklin, 1995) (Figure 2.19). Low O_3 values are typical of the Antarctic spring, but the appearance of a 30–40% decrease in O_3 in the spring of 1985 was a bombshell and heralded the beginning of an international awareness of the so-called Antarctic 'ozone hole' (Farman *et al.*, 1985; Stolarski *et al.*, 1986). Both ground data and satellite data from the Total Ozone Mapping Spectrometer (TOMS) have now confirmed statistically significant falls in the amount of O_3 over the past decade and more (Drake, *et al.*, 1995). The most depleted levels over Antarctica were recorded in 1993, when the monthly mean total was reduced to about 35% of the historical value (Gardiner *et al.*, 1997) (see Figure 2.19). The O_3 hole has been observed to be continent-wide up to 60° S and is not static but an elliptical feature that changes its position and shape over a few days. The addition of large amounts of sulphate aerosols into the stratosphere following the eruption of Mount Pinatubo in 1991 has exacerbated the problem in the short term, since chemical reactions similar to those associated with O_3 also occur on the surface of sulphate aerosols. The potential impact on both physical and natural systems of O_3 reduction is reviewed in Chapter 5.

Wind patterns

The surface pressure fields (Figure 2.18B) control the main features and wind speeds of the surface winds: a strong oceanic gradient from about 35° S to 60° S;

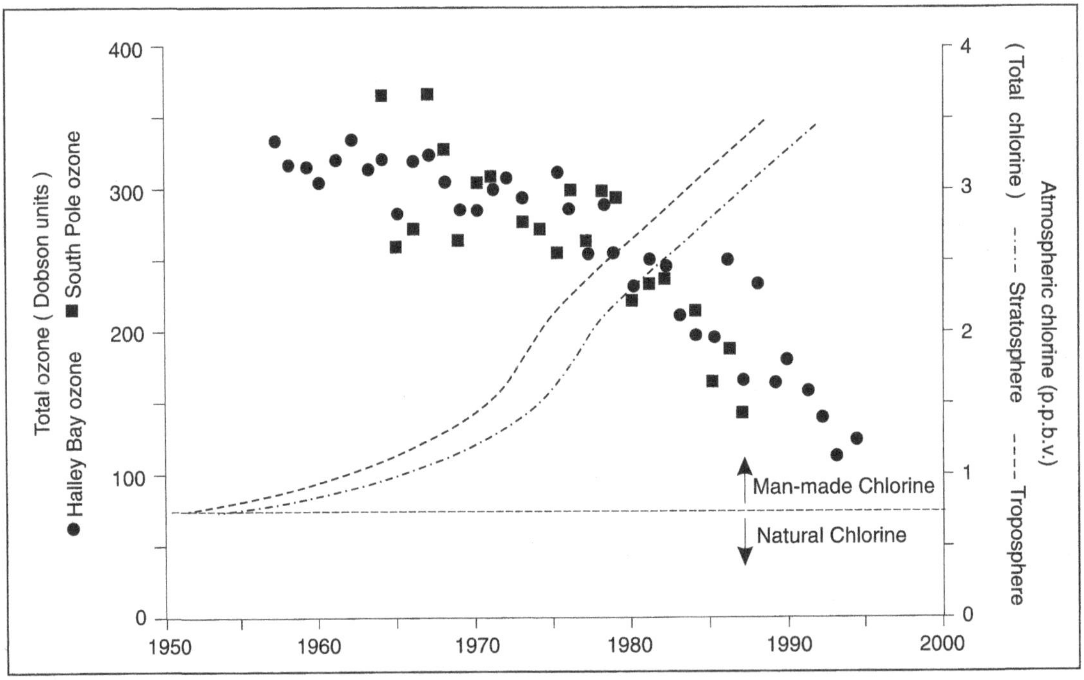

Fig 2.19 The decline of total stratospheric ozone above Antarctica 1956–1994, together with total chlorine abundances (source: redrawn after Jones and Shanklin, 1995).

variable pressure gradients off the coast between 60° S and 70° S; complex gradients over the steep coastline and low gradients over the continental interior (Phillpot, 1985). This results in a dominantly westerly airflow over the ocean, which drives persistent cyclonic storm systems along curved circumpolar paths, some of which skirt the coast of the continent (Rubin and Weyant, 1965). The wind speeds on some of the sub-Antarctic islands demonstrate the nature of the oceanic pressure gradient, the strong winds showing very little daily or seasonal variation (Phillpot, 1985). Closer to the continent there is a tendency for some cyclones to move polewards and impinge on the coastline of East Antarctica, but as a rule penetration of the polar plateau is limited, particularly in winter. The only significant topographic barriers to this circumpolar airflow are the peninsulas and mountains of West Antarctica, which create disruptions (see Figure 2.18B) and lead to locally increased windiness and precipitation as the cyclones rise and cool over the obstacles in their path. For this reason, summer fieldwork in the sub-Antarctic islands of the Scotia Sea is a windy and wet affair!

In addition to winds associated with atmospheric circulation, Antarctica is affected by strong and persistent local surface winds. The combination of a well-defined temperature inversion over a continental land mass that slopes in all directions towards the coast forces the cold and dense air of the inversion layer to flow under gravity in a shallow but high-velocity drainage wind. These katabatic winds affect only the lowest few hundred metres close to the ground, and their minimum velocity coincides with the top of the inversion layer (Tauber, 1960). As the katabatic winds drain off from the interior, the Coriolis force deflects them by about 45° to the left of the surface gradient, and this produces a narrow band of easterly moving winds that hug the coastline of the continent (Mather and Miller, 1967) (Figure 2.20). Local topographic conditions have a considerable effect on the strength of the katabatic winds, the highest velocities being produced where the slopes are steepest or where smooth icy surfaces keep turbulence to a minimum. Favourably oriented valley systems serve to channel the winds into a persistent pattern that makes some locations exceptionally windy, such as at Cape Denison in Terre Adélie, where the Australian explorer Mawson located his winter quarters in 1912. Another such site is a small island at the Drygalski Ice Tongue in the Ross Sea, where six men of Scott's northern party were stranded from February to September of 1912. Documented by Priestley (1974), an experienced meteorological observer, the wind experienced by the party was a cold, snow-laden katabatic wind blowing from the interior via the Reeves Glacier. It was strong and persistent but was essentially a surface phenomenon as stars could be seen through the snow drift at times (Bromwich and Kurtz, 1982).

> On 19 March (1912) the bitter plateau wind intensified, destroying the tent and pinning them beneath it for the entire day. At sunset the wind showed no signs of abating and, unable to walk against the wind, the men were forced to crawl and scramble across 2 km of windswept ice to the snow cave. From that time onwards the six men lived in a space roughly 4 m by 3 m, and only 1.7 m high.'

On 15 August Priestley (1974) wrote:

> . . . the wind remained as exasperating as ever, and none of us had met anything approaching it before. It had blown for 180 days without ever lulling for more than a few hours at a time.

Understandably, they decided to name their windy site Inexpressible Island.

Such persistent offshore winds also serve to clear the coast of sea ice, even in winter, and in favourable locations can produce coastal polynyas or ice free areas of ocean. A feature of the katabatic surface winds is the suddenness with which speeds can vary from near calm up to 50 m s^{-1} within a few minutes and bring associated rapid variations in surface pressure, temperature and humidity as well as high eddying walls of drift snow (Loewe, 1974; Phillpot, 1985). Ship-borne observations show that the effects of the katabatic winds are local and that wind speed drops markedly only a few kilometres offshore. Nevertheless, the passage of depressions in such areas leads to very strong winds with speeds of 50 m s^{-1} or more. In very marked contrast, in the stable and generally calmer conditions of the continental interior, the maximum gust reported at Amundsen–Scott station at the South Pole is 24 m s^{-1} (*ibid.*).

Precipitation

Three factors of geography interact to ensure that Antarctica is the driest of the world's continents: polar position, altitude and continentality. Unique among continents, Antarctica sits astride the pole and so lies at the high-pressure hub of a vortex pattern that generally involves air circulating around, rather than moving across, its bulk. The result is that few of the cyclones that track westwards around the coast penetrate far inland. Cold polar air also holds less moisture than warmer air, so low

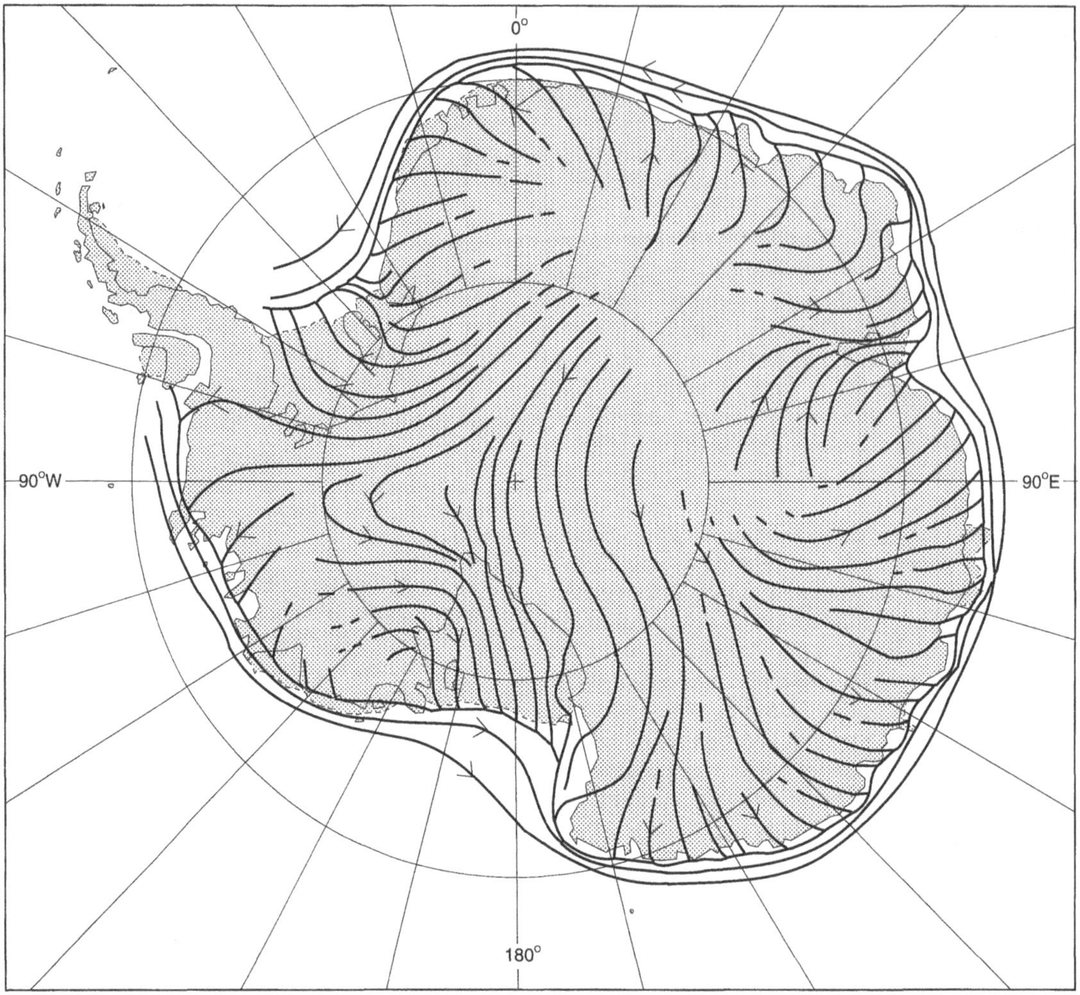

Fig 2.20 Katabatic drainage winds from the polar plateau deflect to the left to produce an easterly airflow at the coast (source: after Mather and Miller, 1967).

moisture levels characterise the Antarctic atmosphere. Any cyclones that penetrate the coast are also forced to rise to the altitude of the polar plateau, causing not only orographic precipitation on the coastal mountains but also precipitation-shadow areas downwind. Crucially, the bulk of the continent lies distant from the Southern Ocean. Since this is the main source of moisture, there is increasing aridity away from the narrow coastal fringe. This is also reflected in variations in cloudiness, and the coast is much more cloudy than inland due to the influence of the ocean and the frequent passage of coastal cyclones: the South Orkney Islands have an average cloud cover of 86%; the Weddell Sea coast 66%; and the South Pole 41%.

With the exception of summer rainfall at the coast, virtually all precipitation consists of snow. Strong winds and blowing snow make direct measurement difficult, so the totals are measured as water equivalents of annual snow accumulation at various sites on the ice sheet. Falling mostly in winter, the annual precipitation totals in East Antarctica are at a maximum in a narrow belt close to the coast, where 200–600 mm of water-equivalent precipitation occurs (Figure 2.21). These totals decline into the interior, where the vast bulk of the polar plateau receives less than 100 mm, and some areas less than 20 mm, per year (Doake, 1987). Since desert climates have an annual precipitation of less than 250 mm (Briggs and Smithson, 1985), Antarctica is easily the world's largest cold

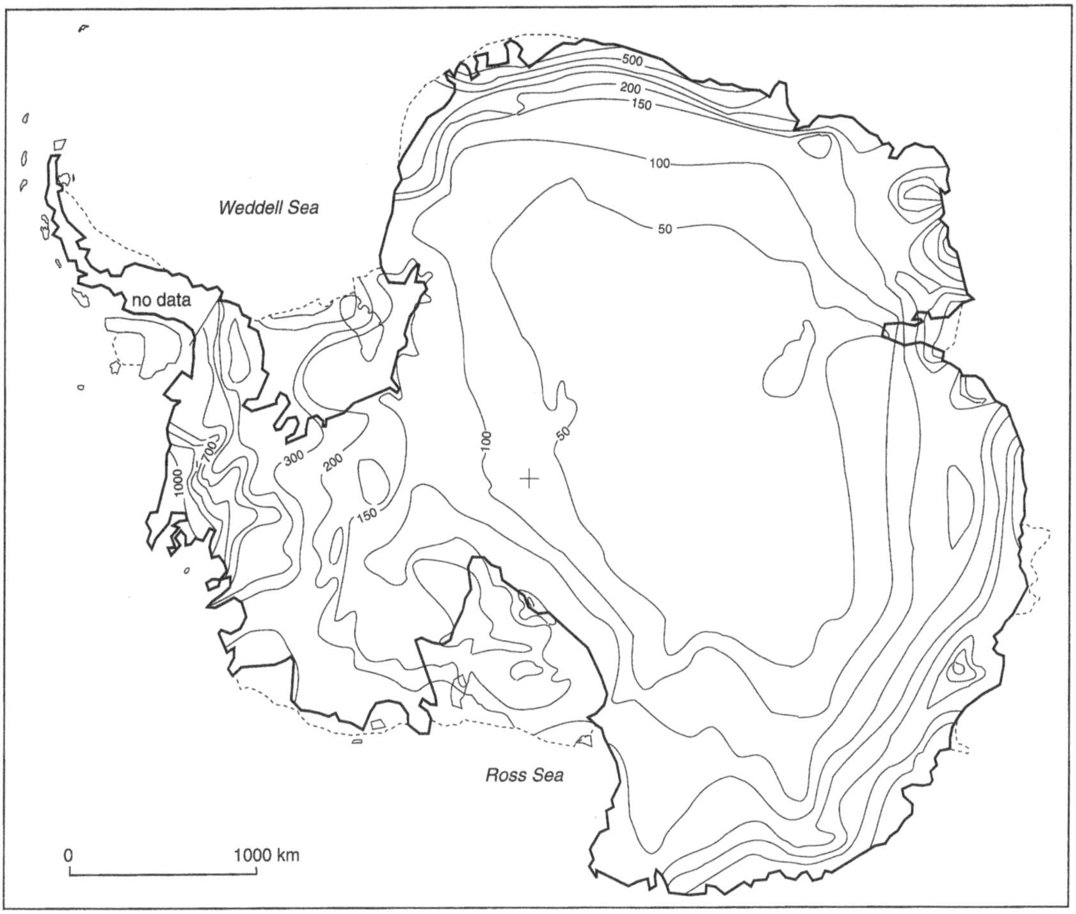

Fig 2.21 Mean annual accumulation over the Antarctic ice sheet in mm a^{-1} (ice) (source: modified from Drewry, 1992).

desert, its drier parts comparable with the Sahara in terms of precipitation amounts. Higher values of 100–300 mm per year occur in the interior of West Antarctica, rising to 1000 mm water-equivalent precipitation towards the Antarctic Peninsula (Giovinetto and Bentley, 1985; Giovinetto *et al.*, 1989; Drewry, 1991), where snowfall is augmented by rainfall. Hoar frost deposition on the cold mountain surfaces, which interrupt the westerly flow of moist oceanic air, is an important precipitation source in the Antarctic Peninsula. Estimates of total snowfall over the continent indicate that the surface accumulation is about 1500–2400 gigatonnes per year (Gt a^{-1}) (Drewry, 1991), 25% of which falls in the Antarctic Peninsula alone (see Figure 2.21) (Drewry and Morris, 1992). Representing only 6.8% of the continental area, the Antarctic Peninsula is clearly an important regional component of the accumulation of the ice sheet as a whole.

To a more limited extent, hoar frost deposition contributes to accumulation on the surfaces of the interior ice as a result of the downward movement of moist upper air replacing cold lower air lost via katabatic drainage (Rusin, 1964). Ice crystals are frequently observed falling in the inversion layer, and the resultant 'no-cloud' precipitation may give at least 2.8 mm a^{-1} of deposition over wide areas of the ice sheet (Loewe, 1962). High on the polar plateau, the number of days with ice crystal occurrence far exceeds the occurrences of snow blowing in from elsewhere, and both surpass the occurrences of *in situ* snowfall. It seems likely that hoar frost deposition and windblown snow are important sources of precipitation on much of the polar plateau. Using the data provided by precipitation maps, a review of five estimates of the net mass added to the ice sheet places the best estimate at about 2144 Gt a^{-1} (Jacobs, 1992). Coupled with estimates of ice loss from the margins, such estimates of ice gain form the basis for assessing whether the Antarctic ice sheet is increasing or decreasing in mass at present.

Precipitation trends

It should be noted that estimation of precipitation from snow gauges is very unreliable and the majority of the data gathered are from coastal stations, which are not representative of the ice sheet as a whole. Nevertheless, there are some signs that precipitation may be increasing at coastal stations. Ice cores from the east coast of the Antarctic Peninsula at Dolleman Island and James Ross Island demonstrate increases in the snow accumulation rate of 0.6–1.8% a^{-1}, representing a 20% increase since 1955 compared with the long-term average (1805–1940) (Drewry, 1991; Peel, 1992). Pourchet *et al.* (1983) identified an average precipitation increase of 30% for six stations, mainly in coastal locations, in Antarctica for the period 1965–1978 compared with the previous decade. However, 10% and 33% increases from the low levels of precipitation on the polar plateau were also noted at South Pole and Dome C, respectively, over this period. The trend of coastal precipitation increases is also supported by oxygen isotope data from ice cores along a 700 km stretch of the East Antarctic coast in Wilkes Land. Snow accumulation up to 1985 was 20% above the 1806–1985 long-term mean, most of this occurring since 1960. The increases in precipitation seem to be related to an intensification of coastal cyclonic activity (Morgan, *et al.*, 1991).

On the other hand, higher temperatures in the sub-Antarctic have been accompanied by reductions in precipitation: between 1964 and 1982 the 2.7 °C rise in temperature in Îles Kerguelen occurred simultaneously with a 500 mm drop in precipitation (Frenot *et al.*, 1995). The 10% decline in precipitation that has accompanied a 1 °C rise in temperature at Marion Island over the last 40 years (Chown and Smith, 1993) is, suggest Smith and Steelenkamp (1990), related to changes in the frequency and track of oceanic cyclones affecting the island. It appears that in the sub-Antarctic at least, declining precipitation has accompanied warming but that the degree of change may be related to the frequency and track of cyclones in the Southern Ocean. Further south the emerging picture is different. As coastal temperatures rise, the resultant pattern of enhanced coastal precipitation seems to have two major implications. Most importantly, it indicates that the initial response of the Antarctic ice sheets to global warming may be enhanced precipitation rather than greater melting in the coastal areas and possibly in parts of the interior, and that this will lead to a fall in sea level. Enhanced precipitation will increase accumulation on the ice sheets and eventually lead to ice expansion. These possible changes are discussed more fully in Chapter 5.

2.4.5 Summary

Occupation of a polar position bestows upon Antarctica some of the highest and lowest solar radiation values on Earth, but its high albedo and relative lack of atmospheric moisture results in much of this radiation being lost. Coupled with its altitude, this serves to make Antarctica the coldest and windiest of the continents, since an intensely symmetrical polar vortex results from the subsidence of cold dense air over the polar plateau and its rotation around the continent. Antarctic geography also plays a role in the climatic patterns outlined above because the symmetry of a uniformly high continental land mass surrounded by ocean allows an essentially symmetrical pressure system, which drives circumpolar cyclonic storm systems and precipitation patterns, to develop. On landfall, these cyclones are subject to orographic rise, which induces higher precipitation close to the coastline. Inland the totals fall rapidly since few of the storms manage to penetrate the polar plateau and much of the continent occupies a precipitation shadow and is remote from the source of moisture-laden winds. The symmetrical circulation pattern responsible for the dominance of the Antarctic westerlies effectively limits any strong poleward penetration of these moist cyclones. Strong and persistent katabatic drainage winds are a common feature of the continent. Deflection to the left as a result of the Coriolis force produces an easterly airflow close to the coastline. In recent decades, precipitation increases of up to 30% have been recorded from coastal stations in Antarctica and may be accompanied by a long-term warming.

2.5 Conclusion

The geology of Antarctica is dominated by the amalgamation (500 Ma ago) and break-up (180 Ma ago) of the Gondwanaland supercontinent and of the evolution of the fragments that eventually came to comprise the Antarctic continent. In essence the continent is composed of the ancient shield of East Antarctica, formed before 500 Ma, the deformed and metamorphosed rocks of the Pacific margin of Gondwanaland, which formed before 180 Ma and were subsequently uplifted into the Transantarctic Mountains, and the post-Gondwanaland development of West Antarctica. The break-up of Gondwanaland led to the opening of wide continental shelves between Antarctica and its former neighbours and the development of the Weddell Sea and Scotia arc as a complication of the rifting of the South American plate from the Antarctic plate.

The morphological inheritance of the Gondwanaland break-up and assembly of the Antarctic continent lies in the establishment of a high continental land mass in a polar location, traversed by a single and virtually unbroken chain of high mountains. Such a location and morphology has important implications for regional and global climate and for the development and maintenance of an ice cover. The possible mineral inheritance of Gondwanaland and its predecessor supercontinent reflects the spatial pattern of metallogenic rocks found in the supercontinental fragments elsewhere and therefore assumed to occur in Antarctica. The existence of wide continental shelves with substantial sedimentary cover is widely thought to indicate that hydrocarbons are likely to exist in Antarctica. However, for both minerals and hydrocarbons, the lack of geological information and the severity of the environment means that nothing is known of their economic potential.

As the pieces of the supercontinent dispersed and ocean basins opened on all sides of Antarctica, circumpolar oceanic currents were initiated in the expanding proto-Southern Ocean. Together with atmospheric circulation, these ocean currents serve to redistribute the polar cold to lower latitudes and to replace it with warmth. However, subsidence and Coriolis deflection of cold polar air also produces an intense polar vortex, which is largely responsible for the strong westerly circumpolar pattern of cyclones and winds in Antarctica. Loss of radiation in summer on account of the high albedo of ice and polar darkness in winter result in extremely low temperatures. Almost all the precipitation falls as snow, except in the coastal fringes and sub-Antarctic islands in summer. The polar plateau represents a cold desert on account of the low temperatures and limited penetration of moisture. Any precipitation that does occur very rarely melts in the very low temperatures. As a result, Antarctica is host to the largest accumulation of glacier ice on Earth. The nature of this ice mass and the terrestrial environment along its edges is examined in the next chapter.

Further reading

Dudeney, J.J. (1987) The Antarctic Atmosphere In Walton, D.W.H. (ed.) *Antarctic Science*. Cambridge University Press, Cambridge.

Schwerdtfeger, W. (1984) *Weather and Climate of the Antarctic*. Developments in Atmospheric Science No. 15, Elsevier, Amsterdam.

Splettstoeser, J.F. and Dreschoff, G.A.M. (eds) (1990) *Mineral Resources of Antarctica*. American Geophysical Union, Antarctic Research Series 51, Washington, DC.

Thomson. M.R.A. and Crame, C.A. (eds) (1991) *Geological Evolution of Antarctica*. Cambridge University Press, Cambridge.

Tingey, R.J. (ed.) (1991) *The Geology of Antarctica*. Clarendon Press, Oxford.

CHAPTER 3 The terrestrial environment

Objectives

Key themes and questions covered in this chapter:

- characteristics of the glacier systems of Antarctica
- current mass balance of the Antarctic ice sheet
- development of the Antarctic ice sheet: stability or instability?
- characteristics of the periglacial systems of Antarctica
- weathering and soil development
- nature of the terrestrial ecosystem
- Antarctic and sub-Antarctic vegetation cover and productivity
- terrestrial animals and the aquatic ecosystem

3.1 Introduction

This chapter aims to describe and explain the glacial and periglacial systems that characterise the Antarctic; the functioning of the Antarctic terrestrial ecosystem that allows life to cling to the peripheries of the continent; and the marked geographical variations that exist in both of the above. The Antarctic terrestrial environment comprises glacier ice and floating ice shelves, together with areas free of surface ice but often with a permanently frozen subsurface. Even where the ground is subject to seasonal melt, the environment is periglacial and subject to a cold climate. Such frost-riven, ice-free ground, together with limited areas of fresh water, provide an unlikely habitat for many species of Antarctic plants and animals. From any perspective, the vast amount of ice on both land and ocean is the single most impressive characteristic of the Antarctic environment. The statistics are staggering: the total area of Antarctica is almost 14 million km^2, of which glacier ice and ice shelf comprise 13.5 million km^2 (Drewry, 1991) (see Figure 1.3). The ice sheet has an average thickness of 2.3 km and a volume of 30 million km^3: sufficient to raise world sea level by up to 80 m should it all melt (*ibid.*). However, glaciers are intimately linked to climate, and their development and survival depend on a sensitive balance between the amounts of ice added and removed. It is important to summarise these processes before discussing the various elements of the Antarctic ice cover, its adjacent ice-free land and the ecosystems that occur there.

3.2 Glacier systems

3.2.1 Ice and glacier flow

The fundamental building block of glaciers is the snowflake, and the conditions that allow glaciers to develop are those that favour snow survival from one winter to the next with limited summer melting. Glaciers develop when summer temperatures are insufficient to melt all the winter snow, so a high ratio of annual precipitation falling as snow and low summer temperatures are fundamental requirements for glacier existence. Glaciers are found where mean annual air temperatures are below zero and where low amounts of solar radiation ensure limited summer melting (Sugden and John, 1976). On this basis, the distribution of glacier ice is largely restricted on a global scale to high latitudes, although at a local scale the pattern of glacierisation is affected by altitude, relief, distance from moisture sources and aspect. For example, the south coast of the sub-Antarctic island of South Georgia is more heavily glacierised than the north coast both as a result of its south-facing aspect and its orographically enhanced snowfall (Smith, 1960) (Figure 3.1).

Fig 3.1 The north coast of the sub-Antarctic island of South Georgia. Solar insolation is higher on north-facing slopes than on south-facing ones in the Antarctic and, as a result, not only are glaciers more restricted, but more heat and moisture are available for plant growth (source: authors).

The surviving snow becomes buried by ensuing snowfalls and undergoes transformation into a mass of loosely packed ice crystals with interconnecting air passages (Paterson, 1994). In the cool, humid conditions of the Antarctic Peninsula, the deposition of hoar frost or rime ice directly on ice or rock surfaces is locally important (Koerner, 1961). However, even on the polar plateau, where precipitation is low, Schwerdtfeger (1970) described hoar frost deposition due to moist air subsiding from altitude (see Chapter 2, section 4.4). Glacier ice begins to form at densities of 0.8 mg m^{-3}, when consolidation of the snow has advanced enough to isolate the entrapped air into individual bubbles. This process works fastest close to the melting point of the ice, and in warmer sub-Antarctic areas such as Îles Kerguelen probably takes place within a few years and at shallow depths. At cold continental sites, such as at Vostok at 3500 m altitude on the polar plateau, the annual mean temperature of –56 °C results in transformation to ice only after 2500 years and at depths of 95 m (Barnola *et al.*, 1987). Low temperatures and high snowfall are ideal for glacier development, so the Antarctic Peninsula is extensively glacierised (Figure 3.2), whereas South Georgia, though having a high snowfall, is also relatively warm and is thus only partially glacierised. At the other extreme, the cold desert conditions of the polar plateau are distant from moisture-laden winds and are much less favourable for snowfall. However, with mean summer temperatures that just climb above –30 °C for only two months (see Figure 2.15) (Schwerdtfeger, 1970), 100% survival of snowfall, combined with low flow rates, produces enormous glaciers. Where there are exceptionally dry conditions, such as in the dry valleys or 'oases' of East Antarctica, snowfall is insufficient to support glaciers even in relatively low-lying terrain (Figure 3.3).

A fundamental feature of glacier ice is that it is subject to gravity-driven flow and internal deformation at depth caused by the weight of overlying ice. Where slopes occur, the deformation operates in a downslope direction, but on a horizontal surface, ice build-up creates a surface slope steep enough to support internal deformation at depth (Sugden, 1982). The rate of deformation is related to ice temperature, so sub-Antarctic glaciers tend to flow faster than those in colder areas further south. The remaining mechanisms of glacier flow, pressure melting, slippage over a water layer and deformation of basal sediment, all require the basal ice to be close to melting. In general terms, 'warm' ice is at its pressure melting point and is found either where relatively high surface temperatures penetrate to the

Fig 3.2 The spectacular alpine landscape of the Antarctic Peninsula displays a variety of glacial landforms, tidewater glaciers and frozen fjords (source: aerial photograph courtesy of A. Alsop).

Fig 3.3 The Dry Valleys of south Victoria Land. A panorama of Balham Valley and Balham Lake. Such oases exist where local aridity reduces precipitation to very low levels and where the altitude of the surrounding ice sheet is not sufficient to allow ice inundation. The ground is not snow-covered, because the dry air evaporates more snow than falls. Annually at Lake Vanda, in Victoria Valley, there is 10 mm of water equivalent precipitation, a mean loss of water to the air from the surface of the lake of 300 mm and a resulting deficit of at least 290 mm (Chinn, 1990) (source: photograph courtesy of Trevor Chinn).

bed or where the weight of thick or sloping ice produces basal pressures high enough to create melting. On the other hand, 'cold' ice is found either where relatively low surface temperatures penetrate to the bed or where thin ice occurs, so such ice remains frozen to the bed (Sugden and John, 1976). In spite of low surface temperatures, the thickness of the Antarctic ice sheet results in much of the ice sheet being warm-based. The presence of warm basal ice allows glaciers to move past bedrock bumps by pressure melting on the upstream side, with the meltwater produced moving to the downstream side to refreeze. Sliding over a layer of basal meltwater sometimes accounts for up to 90% of total movement (*ibid.*). Deformation of the bed can occur where sediments saturated with water are placed under pressure by overriding ice. For example, Ice Stream B in West Antarctica has an exceptional flow rate of several hundred metres per year (Alley *et al.*, 1989) and is underlain by a layer of saturated till that may be subject to either rapid deformation under stress or continuous decoupling of the ice from the soft sediments (Clark, 1995).

3.2.2 Glacier mass balance

Glaciers can usefully be viewed as systems (Andrews, 1975) where the link between inputs (in the form of snow, rime ice, avalanches, etc.) and outputs (in the form of melting, icebergs, etc.) controls how rapidly and how far ice must be transported to allow the glacier to remain in balance. Using an idealised glacier (Figure 3.4), the amount of snow and ice accumulated each year in the upper part is greater than the amount melted away. In the lower part of the glacier, the amount of snow and ice lost each year is greater than the amount added. The equilibrium line separates the upper accumulation area from the lower ablation area (Sugden and John, 1976) and on an annual basis is the only point where input equals output. Ice transfer from the accumulation area to the ablation area results in glacier flow, the magnitude depending on the dimensions of the wedges added and removed annually. For example, in the sub-Antarctic South Shetland Islands, high winter snowfall must be counterbalanced by high rates of glacier flow and ablation if the glaciers are to remain in equilibrium. Where the melting offered by the ablation area is insufficient to cope with accumulation, the glacier enlarges its ablation area by extending into lower and warmer altitudes (see Figure 3.1). On the other hand, glaciers in Victoria Land are so distant from moisture-laden winds that snowfall is slight and

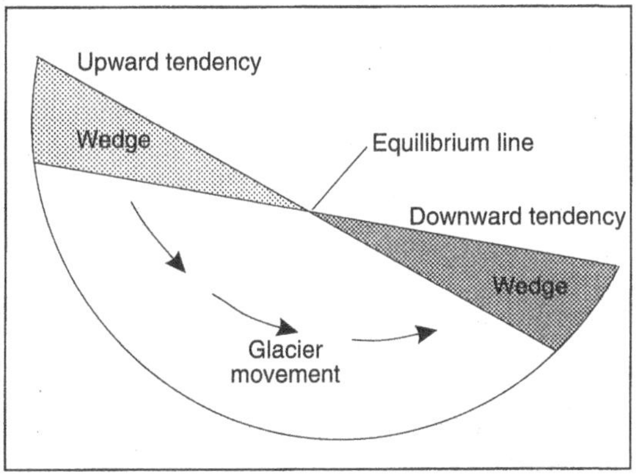

Fig 3.4 The glacier system. Addition of a net accumulation wedge above the equilibrium line and removal of a net ablation wedge below it drives the flow of the glacier (source: redrawn from Sugden and John, 1976)

both glacier velocity and ablation are low (Holdsworth and Bull, 1970). Where the equilibrium line occurs at a low altitude or at sea level, the available ablation area may be insufficient to allow melting on land. The equilibrium line around most of Antarctica is at or 'below' sea level (Drewry, 1991), and perhaps 75% of ablation is achieved by iceberg calving into the ocean and only 25% by basal melting where the snouts are floating (Jacobs, 1992) (see Figure 2.2).

Catchment area, surface topography and altitude, basal thermal regime, and bedrock topography, as well as the nature of the outflow areas, all constrain the behaviour of the ice in response to changes in mass balance. Any increase in accumulation will feed through to a response in the ablation area on a variety of time scales, dependent on catchment characteristics. Research suggests that, in view of the time lag involved, the West Antarctic ice sheet is currently responding to changes set in place at the end of the last glacial period and so significantly pre-date any human modification of climate (Alley and Whillans, 1991). In general terms, smaller glaciers in maritime areas respond to increased snowfall by thickening and advancing on a shorter time scale than large continental glaciers. This lag effect makes it difficult to relate present-day changes in accumulation rates to the corresponding effects on advance or retreat of the glacier terminus: for the Antarctic ice sheets, the response time is measured in several thousands of years (Sugden and John, 1976), whereas for glaciers in South Georgia it is likely to be of the order of a few decades (Gordon and Timmis, 1992).

3.3 The Antarctic ice cover

Other than the numerous small glaciers in the mountains and along the coastline, there is arguably one large glacier that caps the Antarctic continent and whose various components mostly coalesce. In spite of this, it is useful to identify the individual elements of the ice cover so that their relative importance to the glacial environment of Antarctica can be assessed and understood. The ice cover can be classified using a morphological classification suggested by Sugden and John (1976), who discriminate between ice that is unconstrained by topography (ice sheets, ice caps and ice shelves) and ice that is constrained by topography (valley and cirque glaciers and ice fields). The difference between ice sheets and ice caps is one of scale, the latter being less than 50 000 km^2 (Armstrong *et al.*, 1973).

3.3.1 The ice sheet

The Antarctic ice sheet is largely composed of an ice dome of massive proportions, drained at its margins by numerous outlet glaciers and ice streams. In ice sheets the ice builds up until the shear stresses produced by ice thickness and surface slope lead to internal deformation: thinner ice requires a steeper slope in order to maintain flow; thicker ice needs less of a slope (Paterson, 1994). Parts of the Antarctic ice dome reach altitudes of 4200 m, and in areas where the subglacial topography is below sea level are close to 3 km thick (Drewry, 1991); in other areas the ice conceals subglacial topography that in the Gamburtsev Mountains rises to 3000 m (Figures 3.5A and B). The surface of the East Antarctic dome approximates a parabola in places and is generally flat with gradients increasing towards the periphery, reflecting the outward flow of ice (see Figure 3.5B). The West Antarctic ice sheet is both lower and more variable in surface profile, partly because it is mainly grounded below sea level (Sugden and John, 1976; Van der Veen and Oerlemans, 1986).

The general pattern of flow within both the West and East Antarctic ice sheets is for velocities to increase towards the coast. An extremely useful means of conceptualising the way in which ice sheets flow is shown in Figure 3.6 (Sugden and John, 1976). In the centre of the Antarctic ice sheet, accumulation is limited and so flow velocities are restricted, whereas towards the periphery little melting takes place in the low temperatures and both accumulation and flow rates increase. As a result, a high proportion of ablation from the Antarctic ice sheet occurs by calving into the sea. In contrast, much of the margin of the Greenland ice sheet is land-based and so velocities decrease towards the periphery as melting reduces ice volumes. The simple concentric pattern of ice flow in Antarctica is confused by blocking and channelling of the ice flow by the coastal mountains and the Transantarctic Mountains. Where the outflow of ice is blocked by mountains, fresh snow is blown away at the surface to expose bare ice, which sublimates (Paterson, 1994). These so-called 'blue ice' areas make ideal runways for wheeled aircraft (Swithinbank, 1990, 1993b). Whillans and Cassidy (1983) and Cassidy (1991) also note the relationship between blue ice areas and concentrations of meteorites, which are carried hundreds of kilometres from crash sites on the ice sheet, only to accumulate at these blockages and subsequently be exposed by ablation.

Elsewhere, the constraining rock walls of the mountains are cut by numerous outlet glaciers, which drain the ice dome at velocities far in excess of those found elsewhere. Because of this, they extend well beyond the margins of the ice dome, generally have lower gradients, and produce large depressions in the surface of the feeder areas. The lower gradient of the 200 km long and 23 km wide Beardmore Glacier was favoured by the explorers Shackleton and Scott as a route to the polar plateau in spite of crevassing produced by a surface velocity of 372 m a^{-1} where the glacier emerges from the mountains and flows into the sea (Swithinbank, 1988a). The Lambert Glacier flows into Prydz Bay in East Antarctica and is one of the world's largest glaciers, reaching a staggering 700 km long and 50 km wide with a centreline velocity of 1200 m a^{-1} near the ice front (*ibid.*).

Rapidly moving ice streams 30–80 km wide and 300–500 km long occur within the generally slowly moving mass of the ice sheet. The best known occur in the West Antarctic ice sheet, where the spectacular, if a shade unimaginatively named, Ice Streams A, B, C, D and E flow into the Ross Ice Shelf (Figure 3.7A). Flow velocities in the heavily crevassed Ice Stream B are 0.3–2.3 m d^{-1}, compared with about 0.02 m d^{-1} on the ice sheet side of a 5 km wide boundary zone of chaotic crevassing (Englehardt *et al.*, 1990). Why such rapid flow should occur immediately adjacent to sluggish flow is not fully understood, but boreholes drilled into Ice Stream B suggest that the high flow rates are controlled by high basal water pressures and deformation of saturated basal sediments and/or continual decoupling from the soft sediments of the bed (*ibid*; Clark, 1995). However, since the beds of many ice streams extend beneath sea level and begin to float, subglacial water may also be controlled by hydraulic connection to the sea (Doake, 1987).

It has been known for over 20 years that water can collect in locations under the Antarctic ice sheet, and 77 subglacial lakes are known (Seigert *et al.*, 1996), the largest of which is near Vostok station in central East Antarctica. The sheer size of Lake Vostok has only recently come to light (Kapitsa *et al.*, 1996). Geophysical and satellite altimetry measurements indicate that 4 km under the ice of central Antarctica, Lake Vostok is 200 km long, in places 500 m deep, and covers an area of 10 000 km^2: comparable in size to Lake Ontario in North America. Geothermal heat flux and the heat produced by basal sliding raises the basal ice temperature to pressure melting point, and the thick blanket of ice above

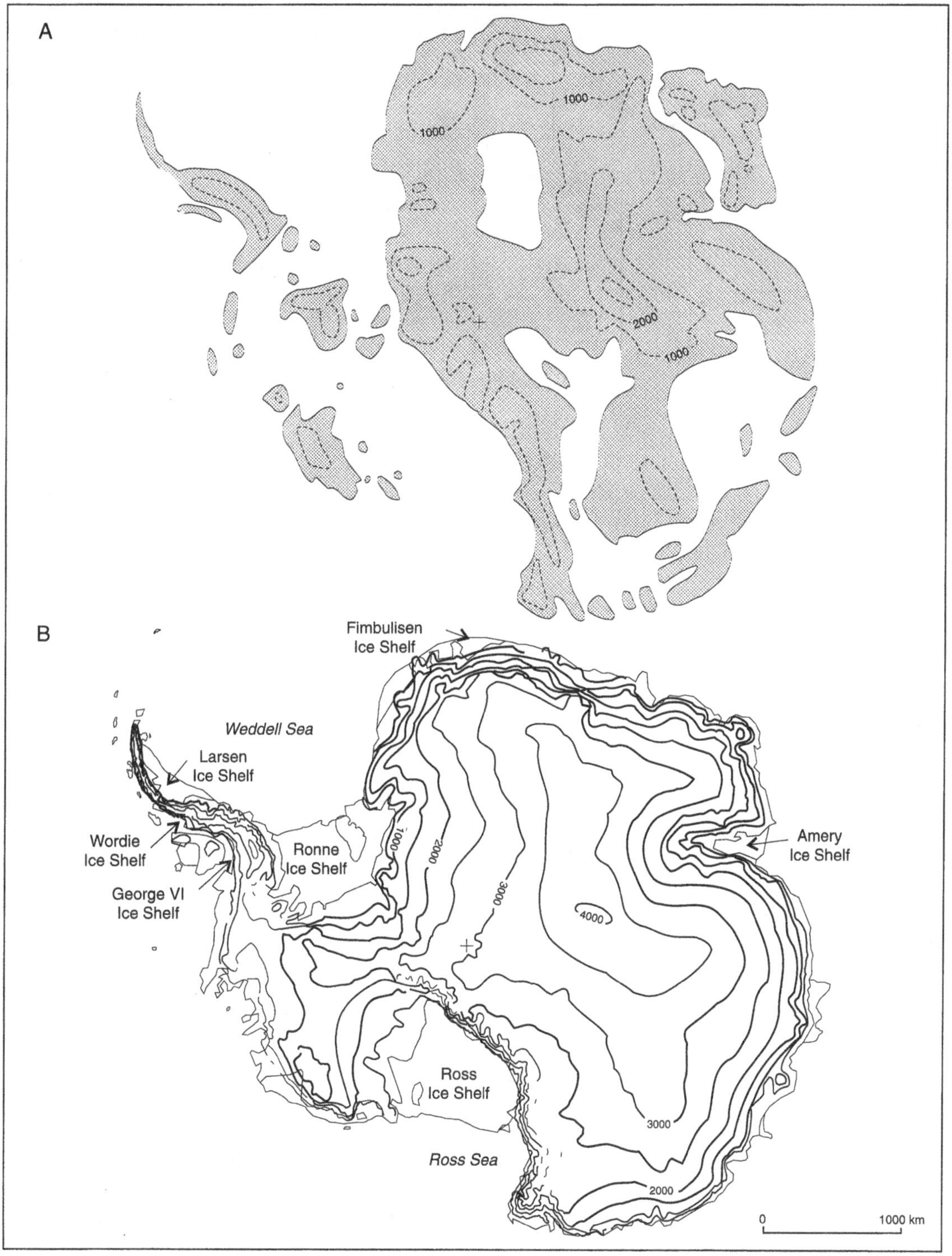

Fig 3.5 (A) Subglacial topography and sea level in Antarctica. The white areas are below sea level (source: data from Sugden, 1982). (B) Surface elevations of the Antarctic ice sheet in metres (source: data from Drewry, 1991).

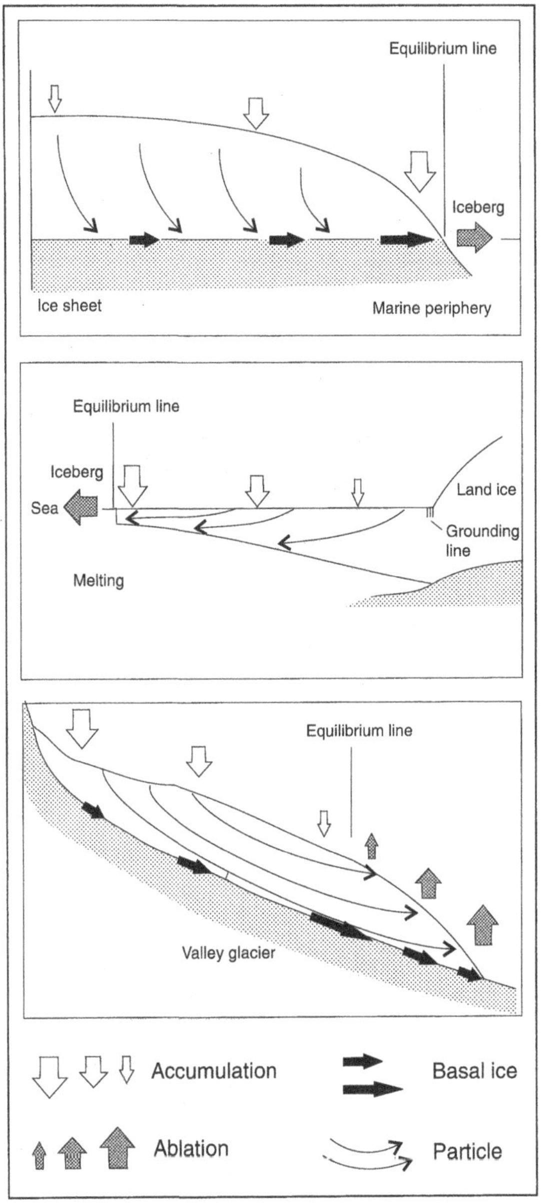

Fig 3.6 Particle paths and relative rates of accummulation, ablation and basal movement in ice sheets, ice shelves and valley glaciers (source: modified after Sugden and John, 1976).

provides thermal insulation from the extreme cold of the surface (Bentley, 1996). Although at both ends of the lake the bed lies about 700 m below sea level, radio echo-sounding indicates that it is largely composed of fresh water (Kapitsa *et al.*, 1996). In 1996, the international drilling programme above the site was stopped to allow assessment of the risks of contamination of such a pristine subglacial system. Around 70 other subglacial water bodies are known for the central East Antarctic ice sheet, so Lake Vostok may be part of a vast subglacial hydrological system (Ellis-Evans and Wynn-Williams, 1996).

3.3.2 Ice shelves

Ice shelves are floating ice sheets nourished by the seaward extensions of land-based glaciers or ice streams and from the accumulation of snow on their upper surfaces (Swithinbank, 1988a) (see Figure 3.6). Their development is favoured by rapid flow of cold ice into protected shallow embayments with localised shoals (Alley *et al.*, 1989). The two largest ice shelves owe their location and size to the geography and geometry of the two largest Antarctic embayments: the Ross Ice Shelf flows into the Ross Sea and the Filchner–Ronne ice shelves flow into and coalesce in the Weddell Sea. Ice shelves comprise around 47% of the Antarctic coastline (*ibid.*), but the extent of the others is limited by size of embayment so that they usually exist in a narrow girdle along the coast. Some of the smallest ice shelves exist in embayments at the tip of the Antarctic Peninsula close to the 0 °C summer isotherm, a line commonly taken to be the climatic limit for ice shelves (Mercer, 1978). More recently, the limit of viability of ice shelves has been placed at the –5 °C mean annual isotherm (Vaughan and Doake, 1996).

Ice shelf thicknesses vary, and the seaward edge may be an ice cliff of up to 50 m above sea level with 100–600 m below water. Most thicken inland because spreading is restricted: the Ross Ice Shelf is 1000 m thick at the grounding line where land ice meets floating ice subject to tidal rise and fall (Swithinbank, 1988a). Glacier accumulation upstream, plus local snowfall, controls the seaward extent of ice shelves. However, where the seabed shallows and touches the bottom of the ice shelf, the pinning effect produced allows a longer ice shelf to exist than is possible in deep water (Doake, 1987). Ablation of the Ross Ice Shelf indicates a roughly equal split between iceberg calving and melting from the lower surfaces (see Figure 3.6). Basal melting reaches 3 m a^{-1} near the ice front and plays an important role in the production of cold Antarctic Bottom Water (Jacobs *et al.*, 1986). As a result of melting and deformation under their own weight, ice shelves thin towards the front and, freed from the constraints of basal friction, achieve ice velocities of almost 3 km a^{-1}

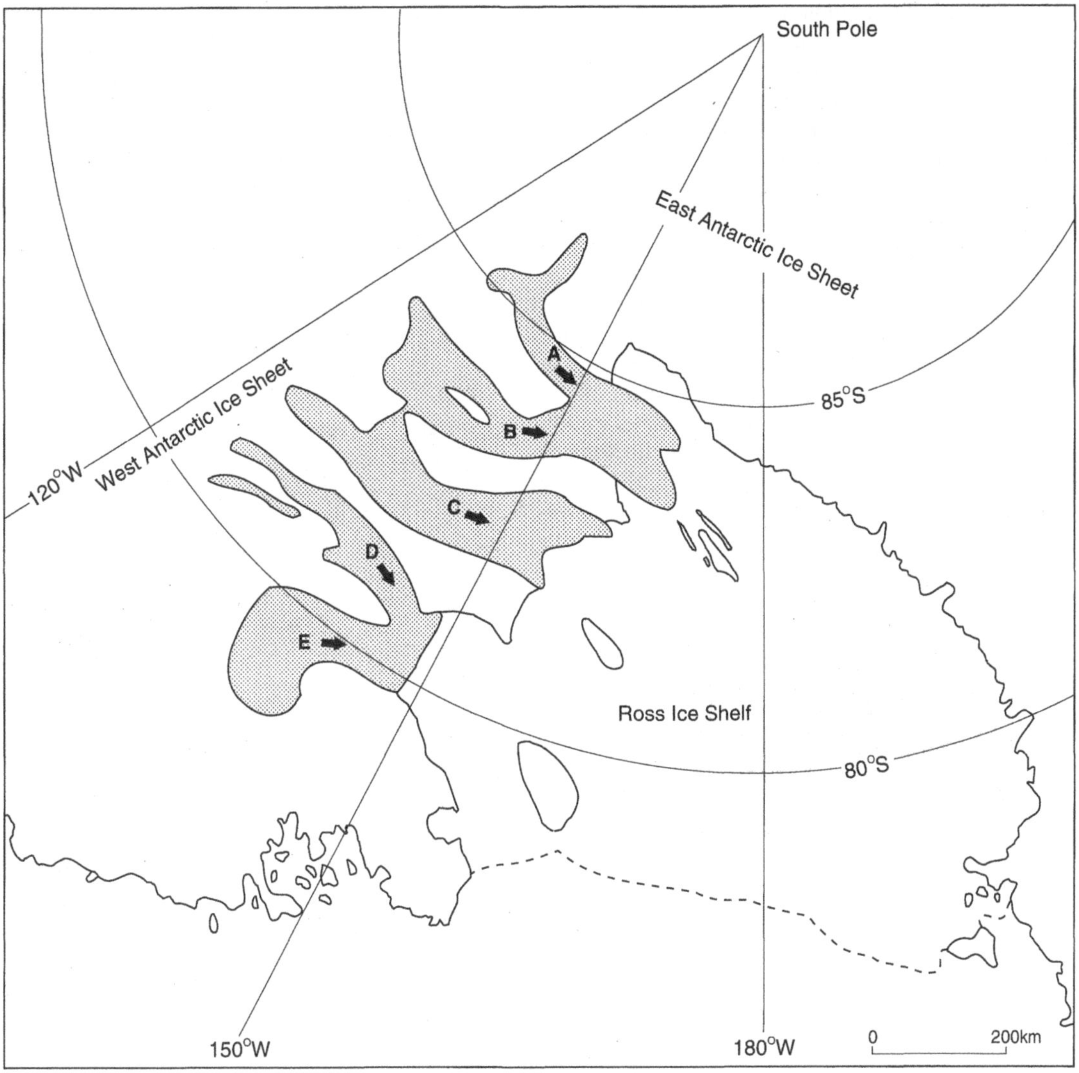

Fig 3.7 Ice streams A–E flow from the West Antarctic ice sheet into the Ross Ice Shelf at much greater velocities than adjacent ice (source: after Engelhardt *et al.*, 1990).

(Swithinbank and Zumberge, 1965). Since the ice streams that feed the Ross Ice Shelf are likely to be moving on a deformable bed of saturated glacial till, substantial glaciomarine sedimentation probably occurs at the grounding line, extending seawards for some distance (Alley *et al.*, 1989).

Ablation of ice shelves is also achieved by the calving of large tabular icebergs. One of the largest of recent years measured 100 km by 50 km and broke away in 1967 from Fimbulisen on the Queen Maud Coast as a result of collision with another tabular iceberg from further east. In September 1986, three giant icebergs covering an area equal to Northern Ireland broke off from the Filchner Ice Shelf in the Weddell Sea. They drifted north carrying with them three Antarctic bases: Ellsworth (USA); Belgrano 1 (Argentina) and Druzhnaya (Soviet Union)! However, calving is episodic by nature and the edge of an ice shelf may be advancing in places while other parts are retreating. For example, Jacobs *et al.* (1986) reported that the west of the Ross Ice Shelf was at its farthest north for 145 years, while the central section was at its furthest south. In addition, because they react to the regional temperature conditions of ice, sea and air, the retreat or break-up of one ice shelf is not evidence of the imminent break-up of all ice shelves (Zwally, 1991).

The behaviour of ice shelves is of concern on two counts. Ice shelves are often fed by ice streams flowing into constricted bays, so it is argued that they may perform a buttressing function for the ice sheet by holding back ice flow (Alley and Whillans, 1991). Recent research on ice streams complicates this picture, since it seems that the main part of Ice Stream B is thinning and draining ice 40% faster than replenishment, while its front is slowing and thickening (*ibid.*). The drainage is matched by a slow thinning of the interior of the West Antarctic ice sheet, where outflow currently exceeds ice accumulation by 20%. This effect is not expected from the buttressing model, where changes in the ice shelf should propagate up-glacier. The behaviour of ice streams seems to be mainly associated with mechanisms internal to the ice sheet, and since these are largely unknown at present, it is difficult to predict the contribution of both ice shelves and ice streams to the future stability of the ice sheet.

There has also been concern over disintegration of some of the northernmost of the Antarctic ice shelves. Time-series observations of the areal extent of nine ice shelves on the Antarctic Peninsula show that the five northerly ones have retreated dramatically in the past 50 years, while those further south show no clear trend (Vaughan and Doake, 1996). Doake and Vaughan (1991) report that the Wordie Ice Shelf on the west coast of the Antarctic Peninsula diminished in area from 2000 km^2 to 700 km^2 during a period of warming in the Antarctic Peninsula from 1966 to 1989 (Figure 3.8). Enhanced fracturing appears to be caused by increased amounts of internal meltwater as a result of rising temperatures, and the George VI, Larsen and Wilkins ice shelves, all of which have a history of recession or brine infiltration, may also be at risk (*ibid.*). The most recent event occurred in 1995, when a 200 m thick iceberg measuring 78 km by 37 km was lost from the Larsen Ice Shelf, rendering James Ross Island circumnavigable for the first time in recorded history. Such recession may well be in response to regional warming in the Antarctic Peninsula (see Chapter 2), but extrapolation to other parts of Antarctica is as yet unwarranted, since the climate of the more southerly ice shelves is more than 10 °C lower and as yet there are no significant trends in temperature or of sea ice extents (Zwally, 1991) (see Chapter 4).

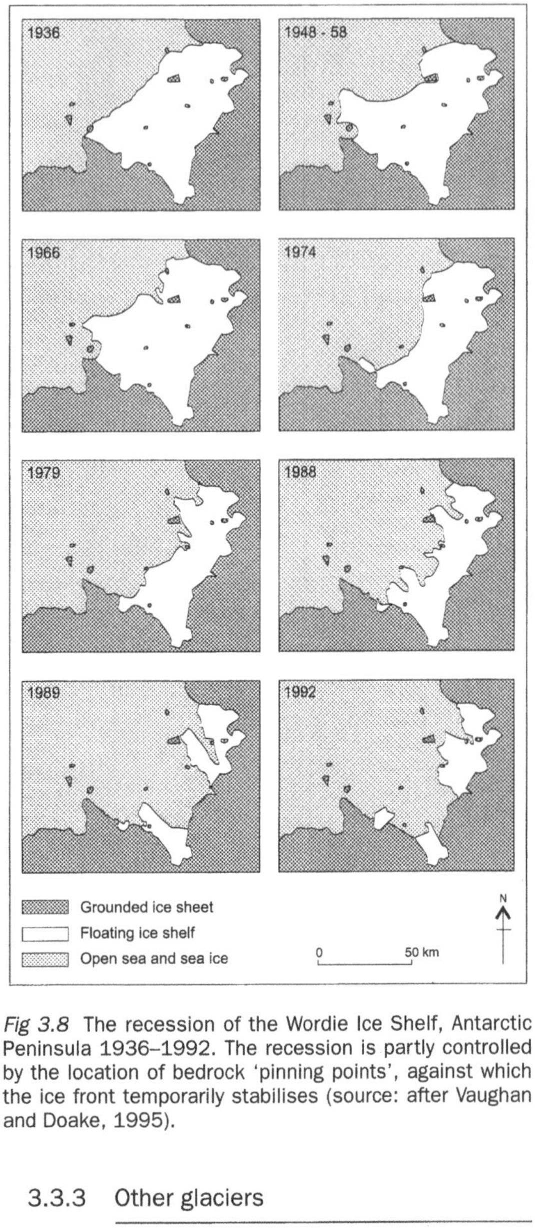

Fig 3.8 The recession of the Wordie Ice Shelf, Antarctic Peninsula 1936–1992. The recession is partly controlled by the location of bedrock 'pinning points', against which the ice front temporarily stabilises (source: after Vaughan and Doake, 1995).

3.3.3 Other glaciers

At a smaller scale, cirque and valley glaciers are constrained by topography that closely controls their shape and flow conditions, yet they still behave according to the mass balance constraints outlined above (see Figure 3.6). Cirque glaciers are smaller ice masses that occupy armchair-shaped hollows, more often than not bounded on three sides by sheer rock walls. Where conditions are suitable a cirque glacier may extend out from its basin to merge into, or join with, a valley glacier. Valley glaciers flow within rock walls and are generally steeply sloping and of relatively restricted length. In the Antarctic, equilibrium line altitudes are low and the ablation area of many glaciers extends down to the sea level. On a continental scale, cirque and valley glaciers have a relatively limited extent, being confined to areas where nunataks and rock valleys are common in the coastal and Transantarctic mountains and on the sub-Antarctic islands. For example, the 65 km long Axel Heiberg Glacier in the Transantarctic Mountains carries no ice from the polar plateau and is fed solely by snow falling in the mountains (Swithinbank, 1988a). Its steep and uninviting surface was the route selected by the Norwegian explorer Roald Amundsen to gain direct access to the polar plateau in 1911.

Numerous small valley and cirque glaciers characterise the 'Dry Valleys' area of Victoria Land (see Figure 3.3). The glaciers here are cold-based, have extremely low accumulation and ablation rates and thus move slowly (Chinn, 1988). By far the largest concentration of cirque and valley glaciers occurs in the Antarctic Peninsula (Haynes, 1995) and its offshore islands, where high snowfall, rime ice growth and low amounts of insolation are found in association with rugged mountains that extend relatively far north and intercept the westerly cyclones. Valley and cirque glaciers form the bulk of the ice cover of the sub-Antarctic islands of South Georgia, Kerguelen, Heard and Bouvet Islands (Smith, 1960; Clapperton *et al.*, 1978; Allison and Keage, 1986). Most of these glaciers are warm-based, judging from the amount of meltwater produced, the rapid flow rates and the amount of glacial erosion accomplished.

3.3.4 Mass balance of the Antarctic ice sheet

Assessment of whether the Antarctic ice sheet is in equilibrium or not depends upon quantifying the inputs and outputs to produce a mass balance, although Drewry (1991) warns that there may be as much as a 50% error in such a calculation. Precipitation maps (see Figure 2.21) show the mean annual surface accumulation over at least half of the ice sheet to be less than 100 mm water equivalent, but this increases to 1000 mm towards the coast of West Antarctica and the Antarctic Peninsula (Giovinetto and Bentley, 1985; Giovinetto *et al.*, 1989; Drewry, 1991). Using these data, the net mass added to the ice sheet is about 2144 Gt a^{-1} of accumulation (Jacobs, 1992). Until recently, estimates of the ablation of the ice sheet have been almost wholly related to iceberg calving calculations. Estimates of 2000–3000 Gt a^{-1} (Orheim, 1985) and 2200 Gt a^{-1} (Oerlemans, 1989) have been derived mainly from satellite and iceberg census data. Using the alternative, possibly more reliable, method of measuring the ice flux across the grounding line of ice streams, shelves and cliffs, Bentley *et al.* (1990) derived a figure of 1250–1550 Gt a^{-1}. Basal melting from ice shelves, long considered to be a minor loss, is now known be an important factor, exceeding 500 Gt a^{-1} (Jacobs, 1992).

Weighing up the accumulation and ablation estimates, Bentley (1990) calculated that the mass balance of the *grounded* ice sheet was in net surplus of 40–400 Gt a^{-1}, equivalent to a sea level lowering of 0.3–1.1 mm a^{-1}. Oerlemans (1989) suggested that the Antarctic ice sheet as a whole is in approximate mass balance. Jacobs (1992) used estimates for accumulation of 2144 Gt a^{-1}, for calving of 2114 Gt a^{-1} and for basal melting of 500 Gt a^{-1} to place the mass balance of the *entire* ice sheet in deficit of 470 Gt a^{-1}. Although there is uncertainty as to whether the Antarctic ice sheet is in net balance or slight deficit, individual drainage basins and parts of the ice sheet show different behaviour and different time scales of response as a result of their own specific characteristics. If the grounded part of the ice sheet is growing at 40–400 Gt a^{-1} and the entire ice sheet is shrinking at 470 Gt a^{-1}, then the floating ice shelves must be losing mass at 500–900 Gt a^{-1} (*ibid.*). Large calving events and disintegration over recent decades have characterised some ice shelves, but others have shown no appreciable change or are advancing, without major calving.

More recent attempts to model the mass balance of the ice sheet have emphasised the geographical importance of climatic zonation and the need to model transient shifts in temperature affecting sensitive areas such as the Antarctic Peninsula (Drewry and Morris, 1992). However, mass balance estimation is fraught with difficulties on account of the fragmented nature of both glaciers and topography. The general uncertainty surrounding the current mass balance of the Antarctic ice sheet awaits more detailed estimates and a clearer understanding of the dynamics of the ice sheet/ice shelf relationships. It seems possible that much will depend on the potential contribution of satellite altimetry to provide sequential topographies of the ice sheet interior at a scale and accuracy that has so far been elusive.

3.3.5 The development and fluctuations of the Antarctic ice sheet

Development of the first ice sheet

The development of the Antarctic ice sheet was one of the most important events of recent geological time because it fundamentally affected atmospheric and oceanic circulation patterns well beyond its immediate area in the Southern Hemisphere. Declining global temperatures during the course of the Tertiary were accompanied in the Southern Hemisphere by the tectonic isolation of a polar continent and the development of circumpolar oceanic and atmospheric circulation systems, which reduced heat exchanges with lower latitudes. Prior to about 55 Ma ago, parts of the Antarctic continent were vegetated and still had Gondwanaland connections (see Figure 2.7), although glaciers may have existed in the high mountains. At this time there were forests of *Araucaria* and *Nothofagus*, the southern beech, in the Antarctic Peninsula (Francis, 1991). This situation altered fundamentally during the Tertiary, although it seems likely that the decrease in temperatures was neither continuous nor steady (Sugden, 1992). The evidence for this change comes from various tectonic and palaeo-oceanographic events in the Southern Ocean as well as from the continent itself (Kennett, 1977). The phases are summarised in Figure 3.9.

Evidence of tectonic events provides important clues to the triggering of ice sheet growth. From the initial break-up 180 Ma ago, Antarctica was subjected to a gradual tectonic drift away from the fragmenting Gondwanaland, and by 55 Ma ago a widening ocean had opened around most of the continent. However, the development of full circum-Antarctic oceanic circulation remained blocked in two places. In the east, the ridge of the South Tasman Rise was removed 38 Ma ago by the ongoing northward drift of Australia (Kennett, 1977) (see Figure 2.6), and this probably reinforced the ocean cooling and the further build-up of ice in the mountains of the Antarctic continent, which had commenced about 40 Ma ago (Robin, 1988). In the west, a cuspate connection between South America and Antarctica still straddled the site of the Drake Passage, but as this extended, the Scotia Sea opened (Barrett *et al.*, 1991). The Drake Passage fully opened and deepened between 30 and 22 Ma ago and a westerly circum-Antarctic atmospheric and oceanic circulation pattern developed. Lower temperatures followed this change from a latitudinal to a meridional circulation, and proximity to the ocean and its storm tracks increased snowfall. The establishment of circum-Antarctic circulation is seen as a 'climate-ice switch' by Robin (1988), since the development of an oceanic polar front blocked the unrestricted flow of warm ocean waters to high latitudes and isolated Antarctica from moderating influences. The build-up of ice increased the albedo and height of the continent, reinforcing the glacial climate and expansion of ice (Sugden, 1996). This led to a substantial lowering of glacier equilibrium lines, an expansion of ice cover to sea level, and further reinforcement of the build-up of the East Antarctic ice sheet, a process that was complete by 14 Ma ago (Savin *et al.*, 1975; Kennett, 1977; Sugden, 1996).

The palaeo-oceanographic evidence centres on cores taken from the ocean floor, where both terrestrial sediments and the shells of microfauna that once lived in the oceans have accumulated. The terrestrial sediments show changes in sedimentation rates related to the strength of cold bottom currents, which are controlled by the extent and thickness of ice shelves. Such currents produce erosional unconformities, which, together with a dramatic increase of ice-rafted debris, suggest that although the ice sheet may have reached its present size about 13 Ma ago (Kennett, 1977; Doake, 1987; Robin, 1988), the first tidewater glaciers in Antarctica were present 30 Ma ago in Queen Maud Land and McMurdo Sound (Barker *et al.*, 1988; Barrett, 1988), between 35 and 42.5 Ma ago in Prydz Bay, East Antarctica (Barron *et al.*, 1988) and as early as 49.4 Ma ago on King George Island (Birkenmajer, 1988). A sizeable amount of ice may have existed before the full-scale development of the East Antarctic ice sheet. The full development of the West Antarctic ice sheet came after 10 Ma ago, with the ice sheet reaching its maximum size before 5 Ma ago (Robin, 1988). Independent evidence from the potassium-argon (K-Ar) dating of volcanic nunataks shows that subglacial eruptions occurred in the Transantarctic Mountains 15–18 Ma ago and in West Antarctica 14–27 Ma ago, and this points to ice sheet existence in these areas earlier than is suggested by the marine cores (Sugden, 1992).

The shells of microfauna that accumulate on the seabed and that contain varying oxygen isotope ratios ($^{18}O/^{16}O$) also provide evidence of climate change. Cold periods produce more evaporation of the lighter ^{16}O isotope from the oceans to be incorporated via precipitation into the ice. This leaves greater amounts of the heavier ^{18}O in the oceans. The oxygen isotope ratio thus reflects both the temperature of the ocean water and the quantity of ice on land. The

Ma	AGE				EVENT
	Cenozoic	Quaternary	Holocene		Warming trend, recession of sub-Antarctic glaciers, some expansion of Continental ice sheets
			Pleistocene		Minor fluctuation of ice sheets
		Tertiary	Pliocene	Late	Relative stability of East and West Antarctic ice sheets
				Early	**West Antarctic ice sheet fully formed**
			Miocene	Late	
10				Middle	West Antarctic ice sheet begins to form Arid interior environments, dry valleys **East Antarctic ice sheet fully formed**
20				Early	**Drake Passage opens, Circumpolar Current and Antarctic Convergence develop**
			Oligocene	Late	
30				Early	Last evidence of continuous *Nothofagus* Ice in East Antarctica. Change from chemical to physical weathering
			Eocene	Late	**Antarctic Tasman Rise open, ocean circulation commences**
40				Middle	**East Antarctic ice sheet begins to form** Tidewater glaciers in East Antarctica Glaciers in South Shetland Islands? General cooling trend
50				Early	
60			Palaeocene	Late	Warm period *Nothofagus* and *Araucaria* in Antarctic Peninsula **Final break - up of Gondwana, Antarctic separates from Australia**
				Early	

Fig 3.9 Cenozoic timescale of major events in the glaciation of Antarctica.

trend of oxygen isotope ratios against time reveals a gradual build-up of ^{18}O in the oceans over much of the Tertiary, and this suggests progressive cooling associated with the build-up of ice on land (Figure 3.10). Rapid changes in the temperature curve at 38, 14 and 5–6 Ma ago also appear to coincide with phases in the build-up of Antarctic ice. The coincidence of geophysical and palaeo-oceanographic evidence suggests that the East Antarctic ice sheet began to develop around 40 Ma ago and was of variable size until after 20 Ma ago, with the West Antarctic ice sheet forming after 10 Ma ago and reaching its maximum size before 5 Ma ago (Robin, 1988).

The East Antarctic ice sheet: stability or instability?

Taken together, the deep-sea isotope record suggests that the East Antarctic ice sheet reached its present size around 14 Ma ago and has since remained stable (Shackleton and Kennett, 1975; Robin, 1988), with the full development of the West Antarctic ice sheet occurring later, at about 5 Ma (see Figure 3.9). The isotope record shows no sign of the enhanced offshore deposit that any intervening warmth or melting of the ice sheet would have produced. This is supported by the virtually continuous record of ice-rafted debris deposited in the early to late Pliocene, which indicates an essentially intact East Antarctic ice sheet (Warnke *et al.*, 1996).

Recently, an alternative view has emerged suggesting that the East Antarctic ice sheet has been unstable over the Pliocene. This view centres on the interpretation of high-altitude semi-lithified glacial tills of the Sirius Group, which occur throughout the Transantarctic Mountains. The tills contain fragments of Pliocene (2.5–4.8 Ma BP) marine diatoms, together with fragments of *Nothofagus* and pollen characteristic of temperate mid-latitudes (Webb *et al.*, 1984), and are thought to be associated with volcanic ash dated at 3.0 ±0.5 Ma (Barrett, 1991). Such evidence has been used to suggest that during a warm period in the mid-Pliocene, a much shrunken East Antarctic ice sheet allowed growth of diatoms within a series of marine basins surrounding the Transantarctic Mountains. These basins were then overriden by ice about 3 Ma ago and the sediments elevated to their present positions by both ice transport and tectonic uplift. Beech forest cannot reestablish itself across oceans, so in order to have survived until 3 Ma ago, *Nothofagus* must have endured in coastal enclaves along the Ross Sea coast while the rest of the continent supported a waxing and waning East Antarctic ice sheet (*ibid.*). The evidence also suggests that the East Antarctic ice sheet did not form until the latest Pliocene, after 3 Ma ago, about the same time as the first Arctic ice sheets appeared (Barrett *et al.*, 1992).

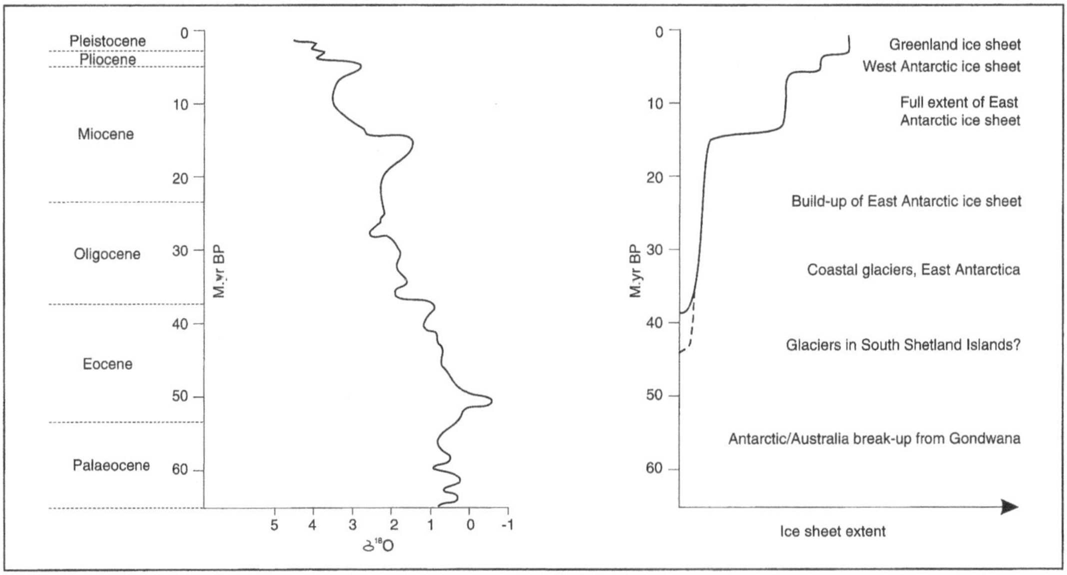

Fig 3.10 The inferred stepped build-up of glaciers in Antarctica over Cenozoic time compared with the deep-sea ^{18}O record (source: after Sugden, 1987).

However, there is extensive geomorphological evidence that supports the view of a stable East Antarctic ice sheet. Evidence from the Transantarctic Mountains and dry valleys suggests that the landscape is mainly a relict Miocene landscape with dissected and drowned preglacial fluvial valleys filled with Miocene and Pliocene sediments (Sugden, 1996). The Sirius deposits are confined to high-altitude plateau remnants that pre-date the dissection, so if the dissection was complete by the Miocene then the Sirius deposits are older still (*ibid.*). In addition, the glacier thickening that occurred in the Pliocene is insufficient to cover all the Sirius deposits (Marchant *et al.*, 1994). In several places, glacial landforms that have developed in the last 10 Ma are exclusively related to cold polar glaciers that were devoid of the meltwater that would have accompanied intervening warmth. Elsewhere, polar desert and periglacial features are capped by 13 Ma-old ashfalls, which are unlikely to have survived wetter conditions (Sugden, 1992). In a similar vein, Marchant *et al.* (1993, 1996) argue that ashfalls overlying desert pavements in the dry valleys of southern Victoria Land have been subjected to cold dry conditions for at least 4.3 Ma. Such an interpretation is supported by the cosmogenic (^{10}Be and ^{26}Al) dating of plateau surfaces and Sirius rocks in the dry valleys, which indicate minimum ages of 2.9 and 4.8 Ma for the Sirius Group at Table Mountain and Mount Fleming, respectively (Ivy-Ochs *et al.*, 1995).

The view of an unstable and fluctuating East Antarctic ice sheet is at odds with the offshore oxygen isotopic record, which points to the longevity of cold polar conditions and conflicts with the view that the modest warming of the Pliocene may have led to greater precipitation and ice expansion rather than increased melting and ice sheet collapse (Sugden, 1992; Marchant *et al.*, 1996). This latter interpretation gains support from evidence of the expansion of outlet glaciers of the East Antarctic ice sheet during the mid-Holocene warm period (Domack *et al.*, 1991), and from the current temperature-driven increases in precipitation now experienced around the coasts of Antarctica (see Chapter 2). However, recent re-examination of the Sirius diatoms now suggests that the mixture of marine and freshwater species has probably been borne aloft from sediments dried elsewhere and transported to Antarctica by atmospheric processes (Kellogg and Kellogg, 1996; Sugden, 1996). Since they range in age from 40 Ma to the present, it appears that the case for an unstable ice sheet is now much less secure and it is more likely that the East Antarctic ice sheet has been stable over much of its life (*ibid.*; Marchant and Denton, 1996).

A fluctuating West Antarctic ice sheet?

A similar controversy has now extended to the smaller West Antarctic ice sheet. Its later development has been thrown into question by the discovery of glacial lake sediments at 900–1300 m above the present ice sheet level near the top of Mount Murphy, a shield volcano on the Marie Byrd Land coast, where Miocene rocks show multiple transitions from subglacial to subaerial eruption activity (LeMasurier *et al.*, 1994). The subglacial rocks argue for the presence of a fluctuating Miocene ice cover in West Antarctica, while the younger high-altitude sediments are lithologically similar to the Sirius Group and contain recycled marine microfossils possibly derived from marine basins in the interior of West Antarctica following one or more deglaciations. LeMasurier *et al.* (*ibid.*) argue that Mount Murphy shows evidence of multiple intervals of deglaciation in West Antarctica between the lower lavas dated about 24 Ma BP and sediments dated about 3.5 Ma BP. However, as with the Sirius diatoms, if the Mount Murphy diatoms have also been emplaced atmospherically, then fluctuations in ice cover may have been limited to the Miocene. The case for a continent-wide synchroneity of ice sheet development is thus unproven and at odds with the marine isotopic record of a much younger West Antarctic ice sheet and long-lived stability of the East Antarctic ice sheet.

The significance of the Mount Murphy and Sirius Group sediments lies in how their interpretation affects our understanding of how the ice sheets have behaved in the past and thus how they might behave in the future. In particular, the East Antarctic ice plays an important role in moderating the global environment, and any suspicion that it might melt in the future has profound implications (Sugden, 1996). If the East Antarctic ice sheet has been stable over the Pliocene, then any future climate warming may initially result in increased precipitation and ice sheet expansion. If it has been subject to large temperature-driven fluctuations over the same period then future climate warming may result in ice sheet melting and sea-level rise.

Glacial periods in the last 500 ka

Fluctuations in the extent of the Antarctic ice sheets have continued into the Quaternary Period, which began about 2 Ma ago. Elsewhere, global temperature decline continued across the Pliocene–Pleistocene boundary, giving rise to the initiation of glaciation in non-polar locations. In the Antarctic, the Quaternary saw periodic fluctuations in the extent of the ice sheets, with a slow build-up over periods of 100 ka followed by rapid deglaciation and 10 ka of warmer, interglacial conditions. Developed by Milankovitch in the 1920s from earlier work, these cycles are produced by changes in solar radiation over periods of about 100 ka due to the eccentricity of the Earth's orbit; about 41 ka due to the changing axial tilt of the Earth; and about 23 ka and 19 ka due to the changing distance of the Earth from the Sun in any one season (Denton and Hughes, 1983). When the isotope data from deep-sea cores over the last 400 ka is matched against the trend of deuterium (a proxy for temperature) over the recently extended 3350 m Vostok ice core, convincing 100 ka cycles can be seen (Figure 3.11A and B). In addition, over this time scale tectonic activity in the Transantarctic Mountains was probably minor, so the extensive trimlines and moraines found there can be used directly for ice sheet reconstruction, especially in the later part of the Quaternary.

Over the last 500 ka of the Quaternary, the marine isotope record indicates that five major periods of cold affected the Antarctic, at 480–453, 352–336, 276–247, 190–127 and 30–11 ka BP (Bradley, 1985; Denton *et al.*, 1991) (Figure 3.11A). Four of these glacial–interglacial cycles can be seen in the Vostok core (Petit *et al.*, 1997). Remarkably little is known of the extent of the early Quaternary ice fluctuations in the Antarctic, but the cores show temperature

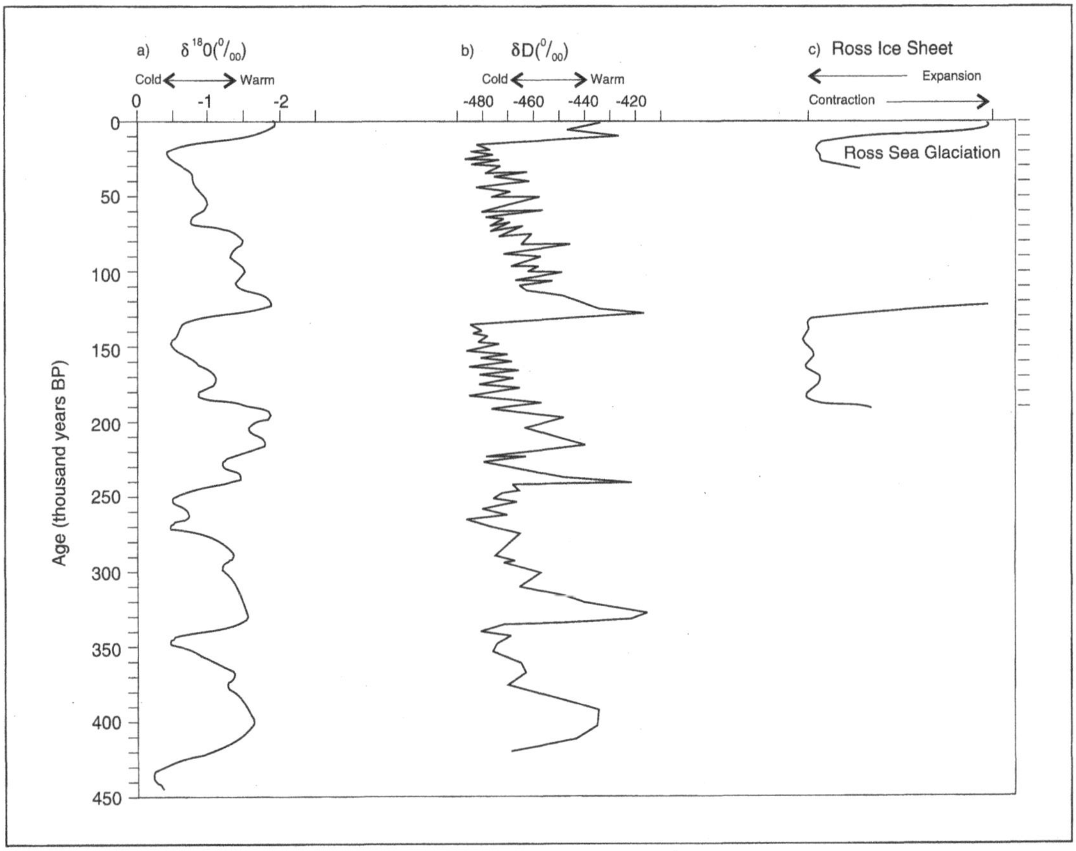

Fig 3.11 Comparison of (A) the global average deep-sea ^{18}O record; (B) the ^{18}O ice core record from Vostok; and (C) the glacial geomorphology record of the McMurdo Sound/dry valleys areas (source: compiled from Denton *et al.*, 1991; Petit *et al.*, 1997).

fluctuations that are broadly similar in magnitude over this period, so it is likely that the general pattern and extent of ice sheet expansion and contraction may have been broadly similar. More evidence is available for later glaciation, and the marine isotope record of cold conditions 190–127 ka ago is supported by the presence of widespread glacial tills in the McMurdo Sound area dated at 185–130 ka BP (Denton *et al.*, 1991) (Figure 3.11c). Detailed oxygen isotope analyses are available only for the upper 150 ka of the Vostok core (Jouzel *et al.*, 1987; Petit *et al.*, 1997), but there is clear evidence of a substantial glaciation spanning this period (Jouzel *et al.*, 1987) (see Figures 3.11, 3.12, 3.13).

Both marine and ice cores suggest that the 190–127 ka glaciation was probably slightly colder than the later glaciation and, due to the decreased precipitation associated with colder periods, led to a reduction in the surface elevations in central Antarctica of 200–300 m below those that occurred later (Robin, 1983). Past precipitation records based on the concentration of ^{10}Be from the upper part of the Vostok core support this view (Figure 3.12), showing very marked increases on the polar plateau during the two most recent interglacial periods centred on 120 and 6 ka BP. Temperatures fell markedly between 100 and 20 ka ago, when glaciation affected many other areas of the world, but this was characterised by a return to lower precipitation in the Vostok core (see Figure 3.12). Analyses of the concentrations of the atmospheric gases CO_2 and CH_4 in the upper Vostok core also mirror the trends in temperature over the last 160 ka (Figure 3.13).

Attempts to reconstruct the surface profile and extent of the last Antarctic ice sheet at 20 ka BP have employed the numerical models of the CLIMAP programme, adjusted to fit geological and geomorphological data. This allowed Hughes *et al.* (1985) and Denton *et al.* (1991) to show considerable thickening of ice of up to 400 m higher than present 250 km inland from the coast, but little or no change in the interior of the East Antarctic ice sheet. Together with the altitudes of glacial trimlines, erratics, till, striae and dated raised beach deposits, it appears that ice surface elevations reached between 1150 m and 600 m above present altitudes (*ibid.*). Such enhanced elevations also led to grounded ice in the area now occupied by ice shelves, and in the Ross Sea grounded ice produced prominent moraines tens of metres high and several kilometres across. Other

Fig 3.12 Past precipitation rates at Vostok relative to the mean Holocene value, derived from (A) ^{10}Be concentrations and (B) the isotopic temperature record (source: after Jouzel *et al.*, 1989).

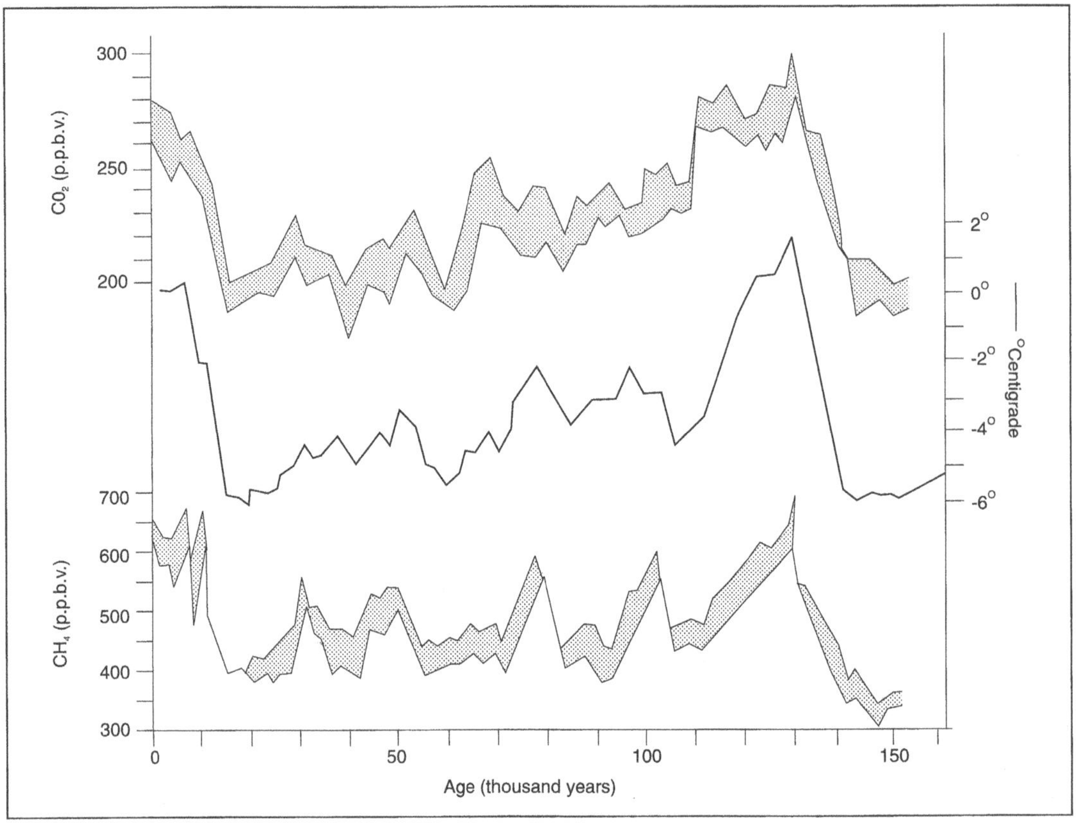

Fig 3.13 Atmospheric temperature change relative to present (°C) and atmospheric concentration of CO_2 (ppm) and CH_4 (ppb)) measured along the top 2083 m of the Vostok ice core. Shading indicates uncertainty in the data (source: modified from Jouzel *et al.*, 1993).

prominent moraines merge inland into those of the present ice surface, again suggesting that increasing aridity inland resulted in reduced elevations. Radiocarbon dates from the Dry Valleys also suggest that glaciers at this time had receded to behind their present positions (Stuiver *et al.*, 1981).

In the sub-Antarctic islands, the fluctuations of the mainland ice were paralleled by the expansion of local ice caps onto the adjacent continental shelves, especially at South Georgia, the South Orkney Islands and South Shetland Islands (Sugden and Clapperton, 1977). At the peak of the glaciation, these locally expanded ice caps penetrated into an ocean already subjected to a greatly expanded sea ice cover (Hays, 1978).

Interglacial periods in the past 500 ka

Interglacial conditions in Antarctica are shown by increases in temperature in the deep-sea core record at 453–352, 336–276, 247–190 ka BP, and centred on 120 and 6 ka BP (Denton *et al.*, 1991) (see Figure 3.11A). These are closely matched by fluctuations in the record from the Vostok ice core (Petit *et al.*, 1997). The interglacial conditions that characterised 247–190 ka and 120 ka BP are commonly believed to have been warmer than the present Holocene interglacial centred on 6 ka and so may give clues to the behaviour of the West Antarctic ice sheet during any coming CO_2-induced 'super interglacial'. At numerous locations around the world evidence exists to indicate global eustatic sea levels of 6 m above present at 120 ka BP and is strongly suggestive of a restricted ice cover in the Antarctic (Bloom, 1983). A modification to this view that has gained widespread currency is that put forward by Mercer (1978), who postulated that the 7–10 °C summer warming during an interglacial would result in collapse of the West Antarctic ice sheet due to elevated sea levels. Much of this ice sheet is based below sea level and buttressed by floating ice shelves that are in places pinned to bedrock. If these pinning points became flooded then the ice sheet would be less constrained and surge forward into the ocean, leading to the loss of buttressing support for the main body of grounded ice (Alley and Whillans, 1991). There is conflicting evidence for this idea, since, while glaciological modelling of glacier grounding lines supports the Mercer hypothesis (Thomas *et al.*, 1979; Lingle, 1985), Alley and Whillans (1991) show that the part of the West Antarctic ice sheet that flows into the Ross Ice Shelf is thickening at its front and thinning upstream: behaviour not accounted for by the classic buttressing model. Geomorphological evidence also suggests that the West Antarctic ice sheet may be able to survive in a warmer world without the presence of some ice shelves. Following a period of warmth, when it disappeared, the George VI ice shelf reformed in the last 8 ka with apparently little effect on the ice sheet (Sugden and Clapperton, 1980). However, this ice shelf may not be an important buttressing shelf.

The Holocene interglacial centred on 6 ka saw recession of the ice from 10 ka BP to present positions. Several dates show the recession to be complete from the Ross Sea by 6–5 ka BP (Denton *et al.*, 1991), from the Windmill Islands by 6 ka BP (Cameron, 1964), from the Vestfold Hills by 6.6 ka BP (Qingsong and Peterson, 1984), and from George VI Sound at the base of the Antarctic Peninsula by 7.2 ka BP (Clapperton and Sugden, 1982). In East Antarctica, synchronous ice sheet lowering occurred prior to 8 ka BP at Law Dome in Wilkes Land (Robin, 1983) and in Terre Adélie, a lowering of 200–400 m to present elevations had occurred by 7 ka BP (Young *et al.*, 1984). Raised beaches emplaced in the Bunger Oasis between 7.7 and 5.6 ka BP show that the East Antarctic ice sheet had receded to its present position by 5.6 ka BP (Colhoun, 1991). In the sub-Antarctic islands glacier recession was most marked after 9.7 ka BP, in South Georgia retreating to within the present limits during the mid-Holocene warmth at 6 ka BP (Clapperton *et al.*, 1989). In the South Shetland Islands glaciers had withdrawn to within present limits by 9.5 ka BP (Clapperton and Sugden, 1982). However, Holocene warmth brought varying responses. In areas constrained by precipitation, like the Dry Valleys of Victoria Land, enhanced atmospheric moisture due to both higher temperatures and grounded ice clearing from the Ross Sea led to glacier expansion (Denton *et al.*, 1991). Data from several East Antarctic outlet glaciers suggest expansion related to mid-Holocene warmth and enhanced precipitation (Domack *et al.*, 1991). At the base of the Antarctic peninsula, an ice-free George VI Sound seems to have favoured rapid valley glacier response and the reforming of the ice shelf after 6.5 ka BP (Clapperton and Sugden, 1982). Other than in the Dry Valleys close to the Ross Sea and along the East Antarctic coast, the overriding impression in the Holocene is of a broad synchroneity in glacier advance and retreat in several areas of Antarctica. The temperature curve from Vostok shows a marked warming of 0.5–1.0 °C from 10 ka BP, interrupted by cold phases at 3.5, 2.5 and 1.4 ka BP. Post 6 ka BP, the coincidence of Vostok data with glacier advance and retreat in South Georgia is striking, with the most extensive of Holocene advances occurring just before 2.2 ka BP (Gordon, 1987, Clapperton *et al.*, 1989).

3.3.6 Summary

The Antarctic ice cover represents the largest mass of glacier ice on Earth. Every year snow and ice accumulation adds about 2144 Gt of ice, and its outlet glaciers and ice streams send 2114 Gt of icebergs into the Southern Ocean. Since annual melting of floating ice accounts for 500 Gt, the mass balance of the entire ice sheet is thought to be currently in deficit. The East Antarctic ice sheet first developed about 40 Ma ago, reached its present size 14 Ma ago and has remained more or less in place since then. The West Antarctic ice sheet is younger and first developed 10 Ma ago, reached its present size before 5 Ma ago and has been subject to minor fluctuations since then. An alternative view that the East Antarctic ice sheet has waxed and waned over the Pliocene and that the present ice sheet formed only 3 Ma after an earlier collapse is not yet proven. In the latter part of the Pleistocene, deep-sea and ice cores show that glacial periods in Antarctica have largely matched the 100 ka cycles identified elsewhere. However, on the polar plateau, lower temperatures and precipitation during the last two cold periods resulted in ice sheet contraction. Elsewhere, glaciers in the maritime Antarctic expanded to their greatest extents due to higher precipitation and reduced melting. Interglacial conditions centred on 120 and 6 ka BP resulted in widespread glacier retreat. In dry areas, such as the dry valleys, enhanced precipitation led to expansion.

3.4 The ice-free areas

3.4.1 Introduction

There is a clear spatial pattern to the 2% ice-free land in Antarctica: it virtually all occurs in the Antarctic Peninsula, the Transantarctic Mountains or in scattered coastal oases. The primary coastal distribution is a result of latitude, the ameliorating influence of the ocean, low altitudes and the occurrence of peninsulas and islands that are remote or jut out from the uniform edges of ice sheets. The secondary distribution in mountain nunataks and oases is the result of altitude above the ice sheet profile and the occurrence of steep slopes or low precipitation, which prevent glacier development. Proportionately more ice-free ground is found in the sub-Antarctic islands. Some, such as Îles Kerguelen, support only a very restricted ice cover, whereas South Georgia is about 58% glacierised. On the continent, in spite of their extremely restricted area, the ice-free zones are of immense importance not only as habitats and breeding grounds for the flora and fauna of the Antarctic but also in providing evidence for Antarctic geological and glacial history. There is a strong coastal bias in the biogeographical distribution of terrestrial plants and animals, since only in a few relatively favoured areas can vegetation become established and animals successfully exploit the habitat. Similarly, not only the biogeography but also the breeding ecology of marine and bird life is conditioned by access to the coastal distribution of ice-free ground as the seasonal sea ice disperses (see Chapter 4). On account of the cold climate and proximity to ice, the ice-free areas of Antarctica are best described as periglacial or tundra environments. As a result of the size of the continental ice sheets and the proximity of the ocean, the Antarctic periglacial zone is much more restricted in extent than in the Arctic, where vast areas of tundra occur across the northern parts of North America, Europe and Asia.

3.4.2 Landscapes of glacial erosion in Antarctica

Glacial erosion is one of the most dominant landforming influences in the Antarctic. Spectacular examples of glacial erosion can be seen in the Transantarctic Mountains, not to mention the similarly impressive subglacial troughs and valleys that radio echo-sounding has shown to exist in the Gamburtsev Mountains beneath the East Antarctic ice sheet (Sugden, 1982). Using the classification of glacial erosional landscapes of Sugden and John (1976), it is possible to summarise the Antarctic landscape in terms of the type and severity of glacial erosion (Figure 3.14).

The classic glacial landscape is the *alpine* landscape of jagged pyramidal peaks and serrated arêtes that tower above and are separated by deeply incised valleys, some of which may still be occupied by valley glaciers. In Antarctica, the conditions for the development of alpine landscapes are best met where nunatak mountains rise above the ice sheet surface in the Transantarctic Mountains and the Sør Rondane Mountains of Queen Maud Land, the Ellsworth Mountains in West Antarctica and also in the offshore islands, where the ice is thinner and may be confined to the valleys. Almost the entire spine of the Antarctic Peninsula is characterised by the most spectacular alpine scenery on Earth, understandably making it a favourite destination for Antarctic tourists (see Figure 3.2). South Georgia is also an overwhelmingly alpine landscape, its mix of soaring peaks and deep valleys fronted by glacio-fluvial plains and lush sea-level vegetation making it ideal for Antarctic wildlife and tourists alike (see Figure 3.1).

Landscapes that everywhere bear the signs of extensive glacial erosion are known as *areally scoured* (*ibid.*) and comprise a surface of rock knolls with abraded surfaces that at the small scale is incredibly knobbly but at the landscape scale is a remarkably uniform and streamlined surface of erosion. Areal scouring requires the basal ice to be close to pressure melting point, allowing the ice to slip over its bed, so it is more likely to be found in a maritime environment or where the ice is thick. These conditions were achieved during earlier glaciations along parts of the Antarctic coast and in the sub-Antarctic islands (see Figure 3.14). Areal scouring in the form of heavily striated surfaces and roches moutonnées is reported from large areas of the Vestfold Hills (Qingsong and Peterson, 1984). Elsewhere, however, much of the coastline is still ice-covered and the eroded areas of the continental shelf have been flooded by sea-level rise during the Holocene. Nevertheless, radio echo-sounding of the offshore shelves of South Georgia, the South Orkney Islands and the South Shetland Islands has revealed extensive surfaces, with the irregular topography typical of areal scouring existing between deeply eroded troughs (Sugden and Clapperton, 1977).

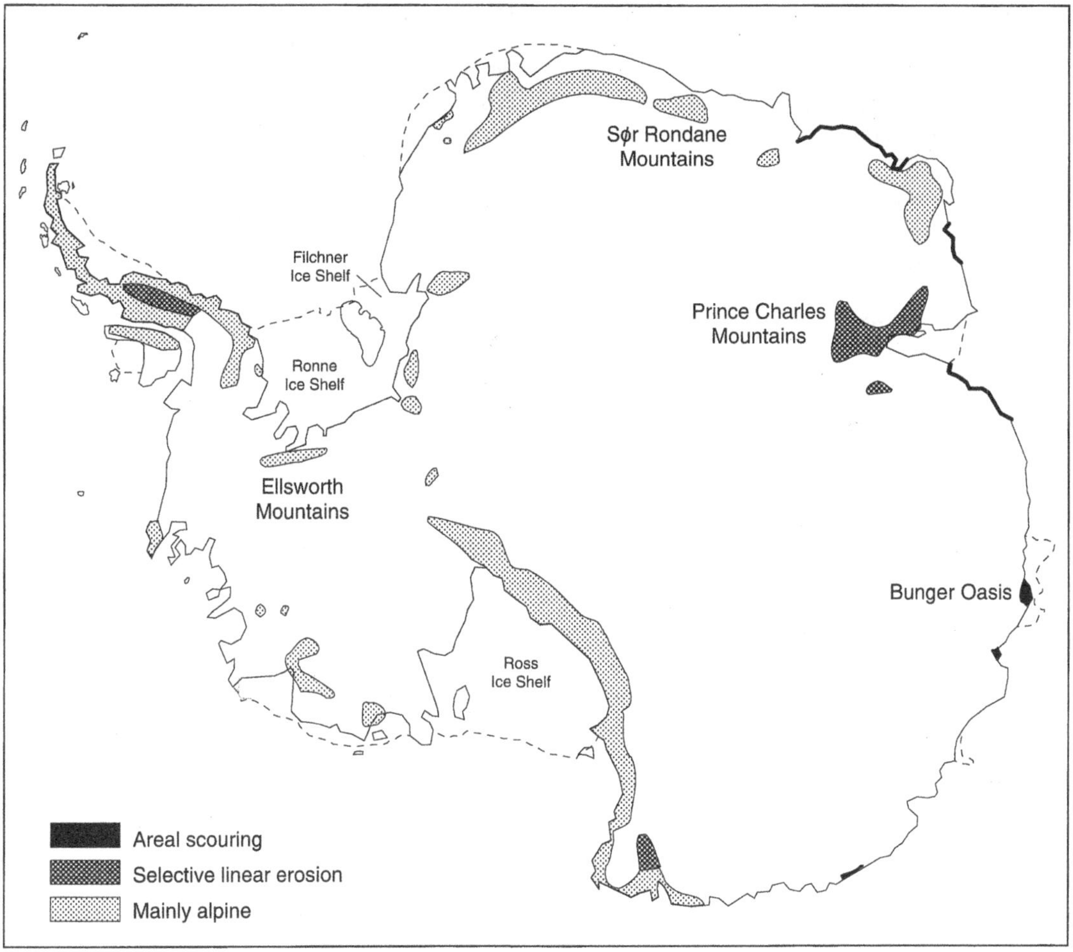

Fig 3.14 Estimated extent of landscapes of glacial erosion in Antarctica (source: redrawn from Sugden and John, 1976).

Landscapes characterised by deeply eroded troughs separated by virtually unmodified mountain plateaux are referred to as landscapes of *selective linear erosion* (Sugden and John, 1976) (see Figure 3.14). These occur where ice is at the pressure melting point on account of its greater thickness, often in the pre-existing valleys but remains frozen to the bed on the intervening uplands. In Antarctica, the most likely areas for selective linear erosion exist where thin ice covers a valley landscape, such as in the mountains of the central Antarctic Peninsula and the Prince Charles Mountains of MacRobertson Land. The submarine shelves of South Shetland Islands, the South Orkney Islands and South Georgia each has impressive linear troughs (Sugden and Clapperton, 1977), although here the intervening areas between troughs are scoured so that features that pre-date glaciation have been removed.

Where the ice is exceptionally thin or cold, or both, landscapes of *no erosion* reflect the protective function of cold ice. Some parts of the Dry Valleys of southern Victoria Land fall into this category and support glacial features which pre-date the last glaciations and are preserved by cold desert conditions (Marchant *et al.*, 1993) (see Figure 3.3). In the same way that patterns of glacial erosion change in space due to changes in ice conditions, they may also change in time and inevitably the above distinctions become obscured. Locations that were subjected to selective linear erosion at one stage in a glaciation or during a former glaciation may subsequently be subjected to increasing amounts of areal scouring, for example during a thickening of the ice cover or during climatic warming.

3.4.3 Glacial, glacio-marine and glacio-fluvial deposition

Glacial deposition in the Antarctic imparts only a muted signature to the landscape. Whereas a scatter of superficial erratic boulders is a common sight on almost all ice-free surfaces, extensive suites of glacial till and morainic belts are regionally rare and reflect the lack of both debris supply from periglacial activity on mountain slopes and the availability of flat ground at the ice margin for deposition. Even in favoured areas, such as the floors of deglaciated valleys or along ice-free peninsulas, groups of glacial moraines are small in size and of relatively limited extent in comparison with those of deglaciated mid-latitudes. Nevertheless, moraines are found high on the flanks of the Transantarctic Mountains and above the outlet glaciers that flow into the Ross Sea (Denton *et al.*, 1991). In the oases and dry valleys morainic evidence has been used to date ice sheet expansion and stability over very long time periods. Moraines are more prominent in the valleys of South Georgia (Figure 3.15), whose large percentage of ice-free ground and maritime climate supports active subaerial debris production and substantial morainic deposits, including the development of rock glaciers (Gordon and Birnie, 1986). In spite of greater till cover in the sub-Antarctic islands, it remains that much of the glacial deposition in Antarctica lies not on land but beneath the sea.

In marked contrast to the terrestrial record, substantial amounts of glacio-marine deposits are known from the growing number of stratigraphic cores in the shelves around Antarctica (Anderson *et al.*, 1984; Barrett, 1986; Barrett *et al.*, 1991; J.B. Anderson, 1991). Such cores have been concentrated in two locations that together provide the most complete stratigraphic coverage of the Cenozoic of Antarctica: the Palaeocene–Eocene sediments of the inner Weddell Sea basin and the Oligocene–Recent sediments of the Ross Sea basin (Webb and Harwood, 1991). Cores 700 m thick from 12 km offshore in the Ross Sea basin show four periods of glaciation associated with low relative sea level, which are considered by Barrett *et al.* (1991) to reflect episodes of major ice build-up on the continent. The sediments also include scattered subangular and subrounded faceted and striated dropstones, indicating glaciation on land and ice calving in the Ross Sea early in the Oligocene.

Fig 3.15 Well-defined and prominent moraines in Moraine Fjord, South Georgia. Except in favourable locations such as on the Ross Sea coast and in the sub-Antarctic islands, Antarctic moraines are generally small and scarce on account of limited supplies of debris (source: authors).

Glacio-marine sediments deposited into the sea in front of and below the floating snouts of tidewater glaciers show distinct zones related to the depositional environment: massive diamictons landward of the grounding line; weakly stratified diamictons in front of the grounding line fed by melt-out of englacial sediments and swept by cold and turbid undercurrents; muddier facies interrupted by iceberg-rafted dropstones further offshore; and eventually a deep-water zone of marine biogenic sedimentation as terrestrial sediment inputs decline. The occurrence of such patterns in the sediments offshore from present-day glaciers allows the reconstruction of the past seaward limits of more expanded ancestral glaciers such as the Ferrar Glacier in McMurdo Sound (Figure 3.16).

In many parts of the world, glacial tills and moraine are found in association with extensive deposits of glacio-fluvial sands and gravels. In Antarctica, the role of glacial meltwater is currently negligible since, in the dry interior, ablation mostly takes the form of sublimation, producing steep glacier fronts. At the coast, glaciers mainly ablate by calving and basal melting. The minor melt streams that do emanate from the surfaces of some glaciers and ice-cored moraines in summer are small and rarely produce substantial depositional features, even in the oases. The largest documented streams in Antarctica are the Talg and Tierney rivers in the Vestfold Hills of Princess Elizabeth Land (Qingsong and Peterson, 1984), and the Alph and Onyx rivers in Wright Valley, Victoria Land, all of which can have flow rates in excess of $2 \times 10^6\ m^3\ a^{-1}$. As no rain has ever been reported for Wright Valley, the streams are fed in the eight weeks of summer solely by glacier melt and can reach discharges that make them difficult to cross (Nichols, 1966; Howard-Williams and Vincent, 1986). Rather than contribute to increased discharge, snowfalls have the opposite effect because summer snowfall enhances the albedo, resulting in the loss of solar energy and reduced temperatures (Chinn, 1990). During the ten-month long winter the streams drain, evaporate, or freeze and sublimate. However, even ephemeral streams such as these are capable of producing small deltas, fans, terraces and floodplains from the unvegetated and loosely consolidated sediments of the Dry Valleys (see Figure 3.3).

Evidence of past meltwater erosion also exists, such as deeply eroded subglacial meltwater channels in the Battarbee Mountains of the Antarctic Peninsula (Clapperton and Sugden, 1982), and the dense network of meltwater channels that dissect the ice-free plateaux of the South Shetland Islands (John, 1972), but they are relatively uncommon on a continental scale. Much of the current meltwater activity appears to be related to snow melt rather than glacier melt, especially in areas like the South Shetland Islands, where networks of erosional channels and depositional fans have developed (Birnie and Gordon, 1980). In the warmer climate of South Georgia, the greater influence of glacier meltwater has constructed large glacio-fluvial plains and provide the sediment to construct some of the largest sand and gravel beaches found south of the Polar Frontal Zone (Hansom and Kirk, 1989) (Figure 3.17).

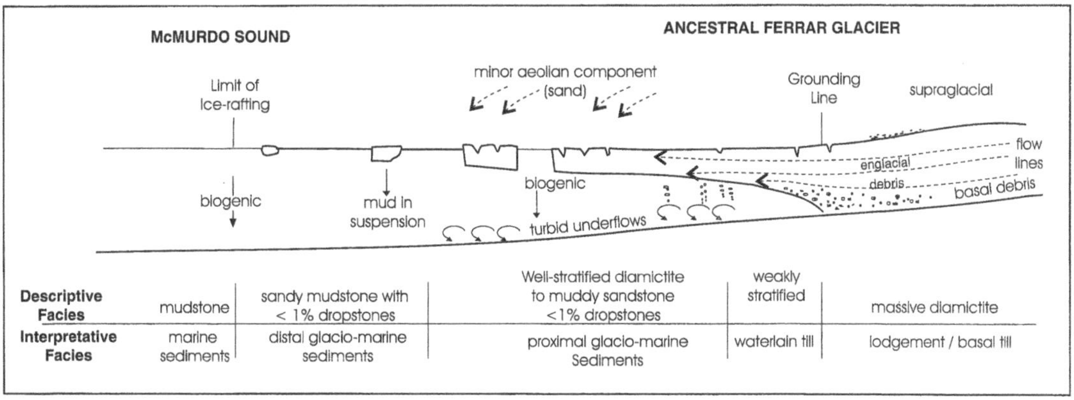

Fig 3.16 The range of sedimentary environments under floating glacier fronts varies from subglacial deposition close to the grounding line to the sedimentation of ice-rafted debris further offshore. The environments are based on a sedimentological model used to interpret cores taken from bed of the Ross Sea near the outlet of the ancestral Ferrar Glacier (source: modified from Barrett and Hambrey, 1992).

Fig 3.17 Glacio-fluvial outwash in St Andrews Bay, South Georgia, delivers large quantities of sand and gravels to some of the largest beaches in the Antarctic (source: authors).

3.4.4 Antarctic lakes

Because of the lack of surface streams in Antarctica, the few lakes that do exist are found mainly in coastal locations or on the offshore islands, where glacier recession has allowed basins to flood, producing initially nutrient-poor lakes (Gore, 1990) (Figure 3.18). Freshwater lakes are common in the sub-Antarctic islands, where glacier melt is augmented by snow melt and rainfall and where glacier ice and moraines act as dams (see Figure 3.15). More unusual are the Antarctic saline lakes. These form where sea water has become trapped by isostatic rebound following ice retreat; where evaporation has exceeded precipitation; or where fresh water is trapped between the land and an ice shelf.

Several lakes in the Vestfold Hills owe their genesis to isostatic uplift but have since become hypersaline due to evaporation, whereas others, such as those in the Dry Valleys, are related wholly to evaporation. Since fresh water will not mix with dense saline water, some lakes have become stratified (meromictic), with a nutrient-poor freshwater upper layer over a saline lower layer. The lower layers of Lake Vanda in Wright Valley, South Victoria Land are warmed to 25 °C, probably by geothermal heat and/or solar warming by heat transmitted via the vertically arranged crystals of the surface ice layer (Priddle, 1985; Fothergill, 1993). A different form of meromictic lake exists between the land and the floating ice shelves. A freshwater upper layer sits on top of sea water, which is linked to the sea beneath the ice shelf. Marine fish and invertebrates live in the tidal lower layer, a long distance from the open sea (Priddle and Heywood, 1980).

3.4.5 Periglacial landscapes

Weathering in the Antarctic

Other than in the maritime west, scarcity of water is a major constraint on weathering and periglacial landform development, and thus there are two fundamental types of rock weathering environment in Antarctica: the cold and dry ice-free valleys, mountains and nunataks of the continent; and the warmer, wetter and biologically more active maritime Antarctic Peninsula and offshore islands (Hall *et al.*, 1989). In the former, the Dry Valleys and oases of Victoria Land and the Vestfold Hills are examples of continental deserts that display features such as wind-polished ventifacts, desert pavements and sand dunes (Campbell and Claridge, 1988; Bockheim, 1990). The absence of leaching leads to surface concentrations of nitrate salts derived mainly from the atmosphere, the pressures exerted by salt crystallisation within pores and cracks being important to the decay of Antarctic rocks, whose surfaces may become heavily pockmarked. Coarser-grained rocks seem to be prone to disintegration, whereas fine grained rocks remain unaltered, often *in situ*, since the low temperatures prevent frost churning of the substrate (Claridge and Campbell, 1985). Removal of fines by wind produces coarse armoured lags called desert pavements, while the blown sand particles cause faceting and polishing of ventifacts (Campbell and Claridge, 1988). In the Vestfold Hills, strong katabatic winds armed with ice particles, snow and sand produce cavernous honeycombing of exposed outcrops (Qingsong and Peterson, 1984), but the occurrence of pits in sites protected from the wind suggests that salt crystal growth also plays an important role.

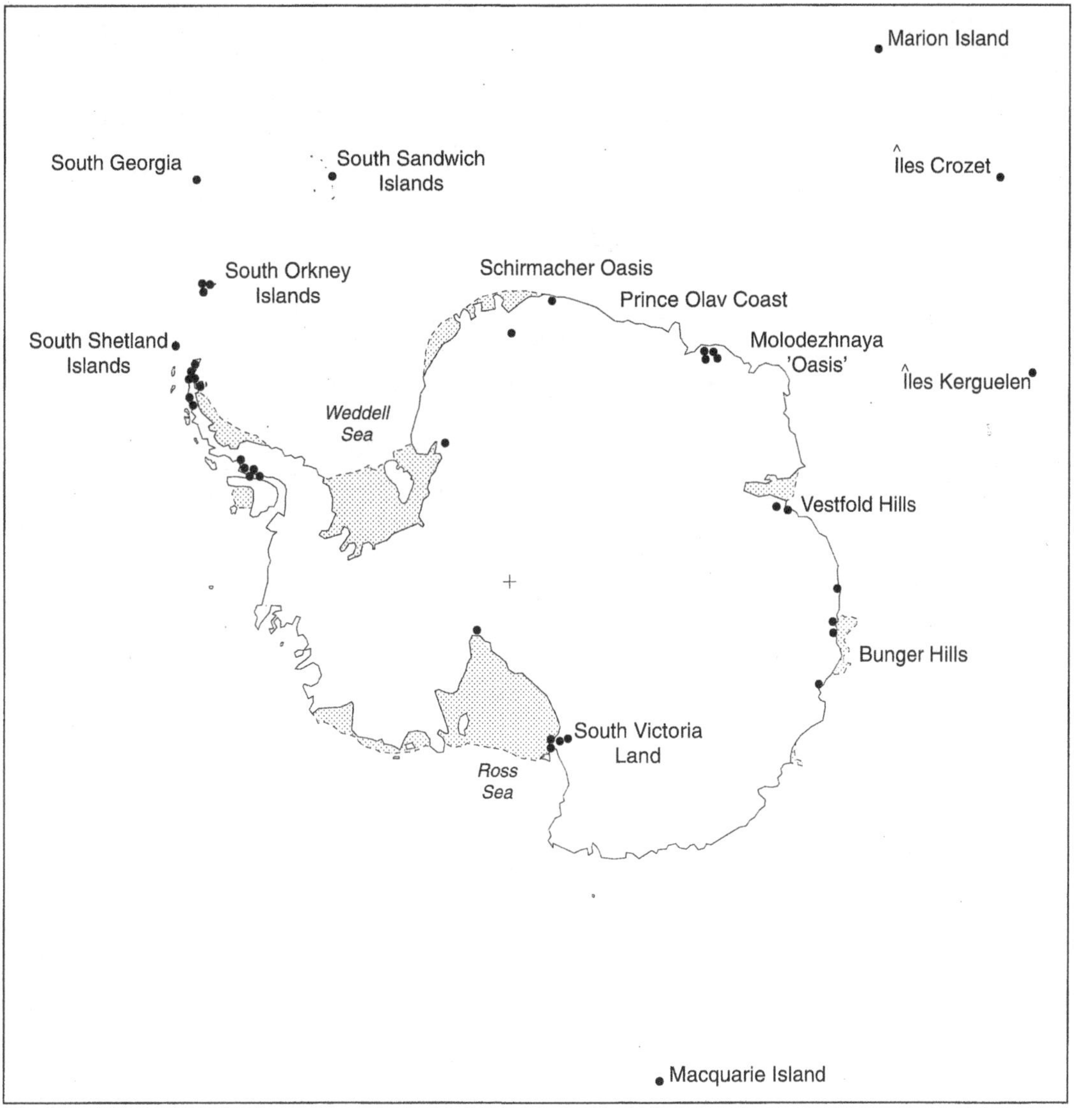

Fig 3.18 The distribution of subaerial Antarctic lakes (represented by dots on the map) is predominantly coastal and sub-Antarctic, although subglacial lakes also exist concealed beneath the ice sheet in locations where the ice is at pressure melting point (see Figure 3.20) (source: data from Priddle, 1985).

In the wetter and warmer conditions of the maritime Antarctic Peninsula and offshore islands, disruption by ice crystallisation is important. Higher temperature and precipitation conditions in South Georgia have led to surface oxidation being found everywhere (Gordon and Birnie, 1986). Here, the intensity of frost shattering, juxtaposed with the occurrence of glacially oversteepened slopes, has led to a highly active rockfall regime that encourages widespread supraglacial debris cover on the glaciers, extensive scree development, and the burial of ice to produce rock glaciers (Birnie and Thom, 1982). Occasionally, rockfalls of major proportions occur: the 1975 rockfall onto the Lyell Glacier in South Georgia represented about 100 years of 'normal' erosion (Gordon *et al.*, 1978) (Figure 3.19).

Fig 3.19 Debris from a major rock slide in 1976 covers the surface of the Lyell Glacier in South Georgia. Dust from the event blankets the snow-covered foreground, except where small avalanches have subsequently removed the dust cover (source: photograph by R.J. Timmis).

Permafrost and periglacial activity

In spite of the lack of moisture, much of the ice-free ground of mainland Antarctica is continuously below 0 °C and is thus underlain by permanently frozen ground (permafrost). Bockheim (1995) places the northern limit of permafrost at the –1 °C isotherm of mean annual air temperature. Earlier overestimates of the amount of permafrost in Antarctica included areas under the ice sheet that are now known to be subject to basal melting, so the probable extent of Antarctic permafrost is more limited than might be expected (Figure 3.20). However, most of the coastal fringe is underlain by permafrost, as are areas that have been exposed by retreating ice in the Holocene (*ibid.*). In Victoria Land, permafrost limits the depths to which beaches are scoured in summer (Kirk, 1972), although little is known of its northward or offshore extent or depth. All permafrost is capped by an active surface layer that melts in summer to a depth dependent mainly on temperature and surface materials. Other things being equal, the active layer should thicken northwards as temperature increases. The maximum depth to permafrost is 0.67 m on the Ross Sea coast (Nichols, 1966); 0.7 m in the South Shetland Islands (Thom, 1978); 2.0 m on Signy Island in the South Orkney Islands (Chambers, 1966); and on South Georgia no extensive permafrost has been noted (Gordon and Birnie, 1986). In comparison with the Arctic, permafrost in the Antarctic appears to be restricted not only in area but also in significance. Because of a regional lack of thick piles of unconsolidated sediments, thermal disruption of Antarctic permafrost is unlikely to result in the widespread thermal degradation and collapse that affects the surface of some Arctic permafrost when it is broken by either natural or human disruption.

In the active layer of freezing and melting above the permafrost surface, gelifluction, patterned ground phenomena, blockfields and scree are widely found. The action of frost shattering and churning produces a bouldery and stony periglacial landscape that develops a distinctive vegetational community where conditions permit. Cailleaux (1968) identified

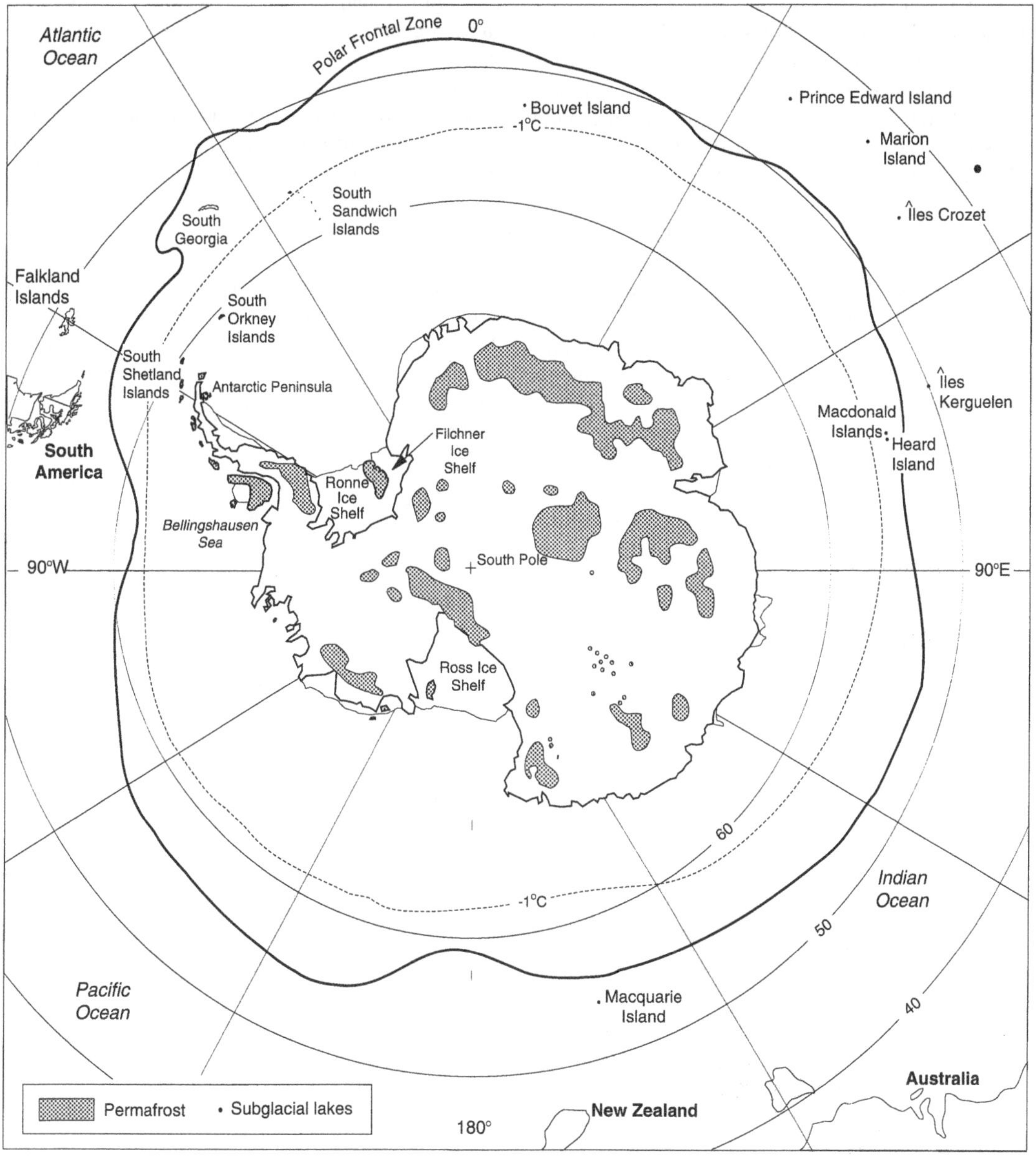

Fig 3.20 The geographical distribution of permafrost in Antarctica. The northern limit roughly corresponds to the –1 °C isotherm. All the ice-free areas are affected by permafrost. Subglacial areas affected by pressure melting and subglacial lakes lack permafrost (source: after Bockheim, 1995).

two periglacial areas, the first occurring in the colder and drier areas such as the Transantarctic Mountains and Dry Valleys, where large contraction wedge polygons occur. Repeated frost cracking of the ground in these arid areas leads to infilling of cracks with windblown sand. In uniform sediments, the orientation of the cracks is likely to be random and to propagate laterally until intersection with adjacent cracks produces polygonal shapes, which are widespread in the oases (Harrington and Speden, 1960; Qingsong and Peterson, 1984). Fossil periglacial forms may help to unravel local surface chronologies of deglaciation:

patterned ground and ice wedges have been used to record glacial fluctuations in Victoria Land because the ice wedge width gives an indication of the time since formation (Black and Berg, 1964). Growth rates of up to 5 mm a^{-1} show that all the patterned ground in this area has formed in the last 10 ka.

The second periglacial area is West Antarctica and the sub-Antarctic islands, where frost heaving, sorting and intense solifluction occurs (Dutkiewicz, 1982; Barsch and Stabilein, 1984). Gelifluction lobes are common on the Ablation Point massif in the Antarctic Peninsula, together with stone circles and tundra polygons (Clapperton and Sugden, 1983). Extensive patterned ground exists wherever superficial sediments occur in the South Shetland Islands, South Orkney Islands and South Georgia (Chambers, 1967; Clapperton, 1971) (Figure 3.21), and both Îles Crozet and Îles Kerguelen have well-developed stone stripes.

The soils of Antarctica

Antarctic soils form under the influence of similar chemical and physical processes to the soils of warmer climes, so their development is similar (Claridge and Campbell, 1985; Campbell and Claridge, 1987). What differs is that low temperatures and limited availability of free water result in slow rates of chemical and biological activity, so biological input is limited. Cryoturbation (frost churning) is characteristic of all Antarctic soils and, since the occurrence of fine materials such as silts and sands is limited on a continental scale, many soils are formed on glacial moraines that have been exposed by glacial down-wasting or on raised beaches abandoned by the sea. Spatial variations in the controlling factors of temperature and moisture availability allow three zones to be identified in the geography of Antarctic soils: the dry valleys and bare ground of the Transantarctic Mountains, the oases of coastal East Antarctica, and the maritime Antarctic Peninsula (Claridge and Campbell, 1985). The soils of the off-shore islands can be classified alongside those of the maritime Antarctic.

In the Transantarctic Mountains, permafrost remains close to the surface at all times, there is little available moisture, and weathering does not penetrate to any depth. As a result, some of the soils have developed on surfaces that may be as old as 400 ka (Armstrong, 1978) and are characteristically veneered by stone pavements over a thin layer of fine oxidised material (Claridge and Campbell, 1985). Bockheim (1990) identified good correlations

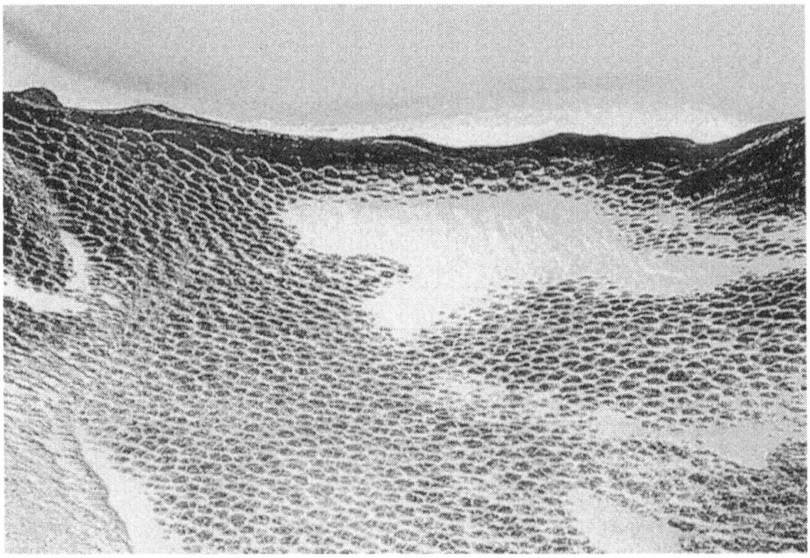

Fig 3.21 Periglacial polygons in the Antarctic Peninsula. Many ice-free surfaces in the Antarctic and sub-Antarctic are affected by periglacial activity. Not only does this produce patterned ground features such as circles and polygons, but the associated cryoturbation also serves to inhibit plant colonisation of such unstable substrates (source: aerial photograph courtesy of A. Alsop).

between the age of till surfaces and their soil characteristics at three sites in the Transantarctic Mountains, showing that the oldest soils were older than 250 ka. With temperatures almost always below freezing and virtually no precipitation, leaching is restricted and salts are not removed. Biological inputs are restricted to occasional mosses, lichens and algae and have little appreciable impact on soil development. The soils are thus essentially ahumic and alkaline with a high salt and carbonate content (Campbell and Claridge, 1985).

In the warmer conditions of the oases of coastal East Antarctica, summer air temperatures may reach 10 °C and moisture availability is higher, weathering penetration is greater and surface sodium salt accumulation is common from the oceanic winds. Mosses and lichens are more abundant and play a detectable influence on soil development, often producing 2–3 cm of organic sandy loam beneath the vegetation (Claridge and Campbell, 1985). Higher temperatures and increased moisture results in vigorous cryoturbation, which prevents soil horizons from developing. Many of the soils are relatively youthful, having formed on surfaces that have only recently become ice-free during the Holocene.

In the maritime Antarctic Peninsula and associated islands, the climate is moist and much less extreme than on the continent and this allows soil development to proceed, albeit with a high degree of cryoturbation (Hall and Walton, 1992). The soils are generally youthful in appearance, chemical weathering is much more active than elsewhere in Antarctica and clay-sized material is more abundant (Claridge and Campbell, 1985). The soils generally have lower amounts of salts and carbonates than the continental soils due to increased leaching in the wetter climate. Nitrogen input from bird guano and seal faeces is locally very important, and many soils contain considerable amounts of organic carbon and exchangeable cations that favour plant growth (Figure 3.22). Mosses, lichens and algae are common, and because decomposition is limited, peat accumulation up to 3 m thick can occur (Smith, 1993a). In the sub-Antarctic islands, such as South Georgia and Heard Island, higher temperatures and a plentiful supply of moisture result in well-developed podzols and brown tundra soils similar to those found in the Arctic. Organic material is more available, especially close to sea level, where organic contents of 90% can be achieved (Smith and Walton, 1975). On South Georgia, wet sites such as valley bottoms support peat, which has been developing for 9 ka, and well-developed brown tundra soils occur on well-drained slopes. Coastal soils near seal wallows and penguin colonies are highly enriched with nitrogen, phosphorus and other nutrients (Headland, 1984), and consequently the vegetation of these oceanic islands is the most luxuriant in the Antarctic.

Fig. 3.22 Guano from penguin and seal faeces produces a locally enriched habitat suitable for plant colonisation (source: authors).

The coastline of Antarctica

Biologically, the most important of the Antarctic ice-free areas is the coastline, 95% of which is composed of glacier ice, but this proportion falls in places such as Victoria Land, where 28% of the coastline is ice-free, and in South Georgia, where 85% is clear of ice (Hansom and Kirk, 1989). Of the permanent ice coasts it is possible to recognise two main types: ice shelves occupy 47% (Alley *et al.*, 1989), the remainder being grounded tidewater glaciers. Where ice shelves occur, ice cliffs comprise the coastline, with the bulk of sedimentation occurring adjacent to the grounding line. Where tidewater glaciers occur, the limited amounts of surface rock produce restricted amounts of glacially transported sediment, so the rate of sediment delivery to the Antarctic coast is low in comparison with glacierised coasts elsewhere. Since there is little meltwater activity, the Antarctic coast lacks the sand, gravel and enormous plumes of suspended sediment that are characteristic of Arctic coasts.

Where the climate is more benign, such as in South Georgia and Îles Kerguelen, large amounts of glacigenic sediment allow substantial beaches to develop (Clapperton 1971; Gordon and Hansom, 1986) (see Figure 3.17). Changes in coastal sediment types along the northern coastal bays of the Antarctic Peninsula are related to climatic variations (Griffith and Anderson, 1989) and it is likely that other environmental gradients spanning several degrees of latitude explain coastal development processes in Antarctica (Hansom and Kirk, 1989). The 5% of Antarctic shores not subject to permanent glacier ice can be categorised into hard rock coasts and soft depositional coasts, but it is important to recognise that ice remains an important element of the coastal regime. During the winter months the action of waves is seasonally excluded from the shore (see Figure 4.5) and, although the sea ice breaks up in summer, large areas of land-fast ice can persist in coastal bays for several years. In areas where the sea ice disperses, the seasonal development of an ice-foot, a multi-layered mass of ice that freezes onto the beach surface at the start of winter, ensures that the influence of ice in shaping coastal landforms remains strong (Figure 3.23A).

Hard rock cliffs are the most common of Antarctic shores, even in the sub-Antarctic island of South Georgia. On the mountainous coast of Victoria Land they acount for 22% of all coasts and reach 2000 m high (Gregory *et al.*, 1984a). Although data are sparse, there is evidence that frost action on cliffs may be important, at least in the maritime west, and cliff retreat rates of 1 cm a^{-1} have been calculated for the South Shetland Islands (Hansom, 1983a). Elsewhere, fluvial and slope processes are seasonally important, and narrow scree-fed, ice-cored beaches are discontinuously present at the foot of cliffs in Victoria Land (Gregory *et al.*, 1984a) and along the Antarctic Peninsula (Figure 3.23B). Low rocky shores are extensively found, especially in the Antarctic Peninsula and offshore islands, where wide sub-horizontal shore platforms have been eroded by the combined action of intertidal frost action and the persistent grounding of floating ice blocks (Hansom, 1983a) (Figure 3.23C).

Beaches are by far the most unusual shoreline features in Antarctica. In Victoria Land, beaches account for a mere 3% of the shores of an area that, by the standards of continental Antarctica, has a spectacular 28% non-ice shoreline (Gregory *et al.*, 1984a). This underscores the importance of such regionally rare beaches as principal wildlife habitats (penguin rookeries and seal haul-outs) and as preferred sites for human activities (Hansom and Kirk, 1989). Beaches in Antarctica are seasonally protected from wave interference by the presence of sea ice and ice-foot and only become mobile on break-up. Polar beaches are characterised by ice-push ridges, which develop from the bulldozing action of pack-ice driven onshore by winds and currents, by ice melt features such as summer pitting of ice-cored beaches and a range of ice-contact and solifluction features related to melting of the ice-foot and adjacent glaciers (Nichols, 1968; Hansom and Kirk, 1989). Where coarser sediments are found, the intertidal zone is often covered by a smooth, highly polished and striated mosaic of boulders (Figure 3.23D). Now widely recognised from several polar environments (Forbes and Taylor, 1994), the degree of development of Antarctic boulder pavements is related to the balance between the frequency of the floating ice blocks that compact them and the wave processes that tend to destroy them (Hansom, 1983b). Extensive sandy beaches form only in the sub-Antarctic islands of South Georgia, Heard Island and Îles Kerguelen, where the subaerial and glacial environment produces large volumes of sand and gravel.

3.4.6 Summary

Other than the nunataks of the Transantarctic Mountains and oases, the ice-free areas of Antarctica have a distinct coastal distribution, with the greatest

Fig 3.23 Different types of Antarctic coastal feature: (A) the seasonal ice foot protects all but the summer shoreline from wave erosion (Doumer Island, Antarctic Peninsula); (B) extensive cliffs, such as these in Victoria Land, are common along much of the ice-free coast; (C) wide subhorizontal shore platforms, such as these in the South Shetland Islands, are found where intertidal frost action and grounding ice blocks are common; and (D) where a local source is available, grounding ice blocks rearrange boulders into smooth boulder pavements (South Shetland Islands) (source: photographs (A) and (D) authors; (B) courtesy of Bob Kirk; and (C) courtesy of David Sugden).

occurrence in the Antarctic Peninsula and offshore islands. All of these areas can be described as periglacial and have been mostly uncovered by recession of the ice to reveal landscapes of glacial erosion. Much of the glacial deposition common to deglaciated areas of the Northern Hemisphere is lacking in the Antarctic, since debris production is limited and there are few sites for deposition to occur. Consequently, other than in the sub-Antarctic islands, most glacial deposition occurs in the sea beneath ice shelves or at the snouts of tidewater glaciers. Low temperatures ensure that surface streams and freshwater lakes are rare, except in the wetter maritime areas. Saline lakes occur in areas of high evaporation or isostatic uplift. Periglacial landscapes, weathering and soils in Antarctica reflect the two basic patterns of weathering environment. In the cold and dry regions, low amounts of moisture limit frost churning and the removal of salts by leaching. Low rates of chemical and biological activity are not conducive to soil development in spite of sometimes long periods of development. In the milder maritime Antarctic Peninsula and sub-Antarctic islands, high moisture content allows a high degree of frost churning but also more chemical and biological activity; locally high nitrogen inputs occur close to bird or seal colonies. Many of these animal colonies are to be found on the beaches and low rocky coastline of the continent and offshore islands, yet beaches are regionally rare.

3.5 Antarctic terrestrial and aquatic ecosystems

3.5.1 Introduction

The Antarctic biosphere comprises two ecosystems, terrestrial and marine, which, in spite of sharing some similarities related to isolation, cold and ice, also have fundamental differences. Looking seawards from the shore, as any summer visitor to the Antarctic will confirm, the marine ecosystem seems full of life and there is a rich species assemblage of plants and animals. These ecosystems are dealt with in Chapter 4. Looking landwards from the shore, the terrestrial and freshwater ecosystems of this frozen continent could not provide a more striking contrast. The terrestrial habitats and ecosystems are limited in extent and relatively simple in structure with few species. There are only a few exposures of barren rock, most water is locked up as ice and the extreme temperatures limit both absolute and relative productivity. A greater number of species is introduced if the sub-Antarctic islands are included but, generally, on account of the extreme environment and geographical isolation, the terrestrial ecosystems of the Antarctica are biologically impoverished and, on a continental scale, remain firmly at the bottom of the biodiversity league table.

3.5.2 The Antarctic terrestrial ecosystem

The number and distribution of species suggest that there are essentially two biogeographical zones in the Antarctic and one in the sub-Antarctic. The coastal fringe of the continent and the east coast of the Antarctic Peninsula comprise an extremely cold and dry continental zone, whereas the west coast of the Antarctic Peninsula, Bouvet and the offshore island groups in the Scotia Sea comprise a more northerly, milder and wetter cold maritime zone (Longton, 1988) (Figure 3.24). The much warmer and wetter sub-Antarctic islands are recognised as a cool biogeographic zone. The continental zone reveals an almost entirely coastal distribution on account of the severe environmental conditions encountered inland, which affect the main controls on life: temperature, light, water and nutrients. Nevertheless, on the 2% ice-free ground, life does exist and in places soil, a sparse vegetation and an attendant fauna occurs. South of the Antarctic Circle, day length varies from 24 hours of winter darkness to 24 hours of summer light (see Figure 2.12), and the extreme seasonality produced in the growth cycle affects most organisms.

Extreme temperatures, day length and the amount of solar radiation received directly affect the rate at which reactions occur, these slowing towards freezing point. This introduces another control, for not only do low temperatures slow down growth but at sub-zero temperatures the formation of ice within organisms may produce lethal effects (Everson, 1984). However, despite the view that the severity of the Antarctic macroclimate is the most important limitation to the abundance of terrestrial life, shelter and moisture availability also play important roles (Everson, 1987). Most Antarctic organisms live in the boundary layer close to the ground, where the microclimatic conditions are considerably more benign than would at first appear from the macroclimate statistics outlined in Chapter 2 (Kennedy, 1993) (Figure 3.25). Summer insolation together with thermal blanketing by snow during winter raises temperatures to the point where other factors play a primary limiting role (*ibid.*; Block, 1994). Moisture availability emerges as probably the ultimate control (*ibid.*) and a clear coastal spatial pattern emerges.

Almost all the water in both of the continental biogeographical zones exists for much of the time as ice and in this form is biologically unavailable. Most of the remainder is bound water held in the soil and, as a result of the relatively high salinity of Antarctic soils, large amounts of energy are required to release it (Vincent, 1988) (Figure 3.26). Other sources of water, such as the streams in the dry valleys, have neither spatial nor temporal permanence, so there exists a poor synchronicity between moisture availability and organism requirements (Kennedy, 1993). Discriminating between the influence of cold and desiccation in Antarctica is problematic, since there exist parallel north–south and coast–inland variations in both moisture and temperature gradients (see Chapter 2), but it is clear that on many slopes where water is locally available, vegetation colonises the damp sites leaving adjacent dry sites bare (Light and Heywood, 1975). Similarly, Kennedy (1993) shows that the peak populations of bacteria, algae and protozoa in soils coincide not with peak temperatures but with the release of water from thawing, arguing that moisture largely controls the spatial and temporal abundance of many organisms. It seems that on

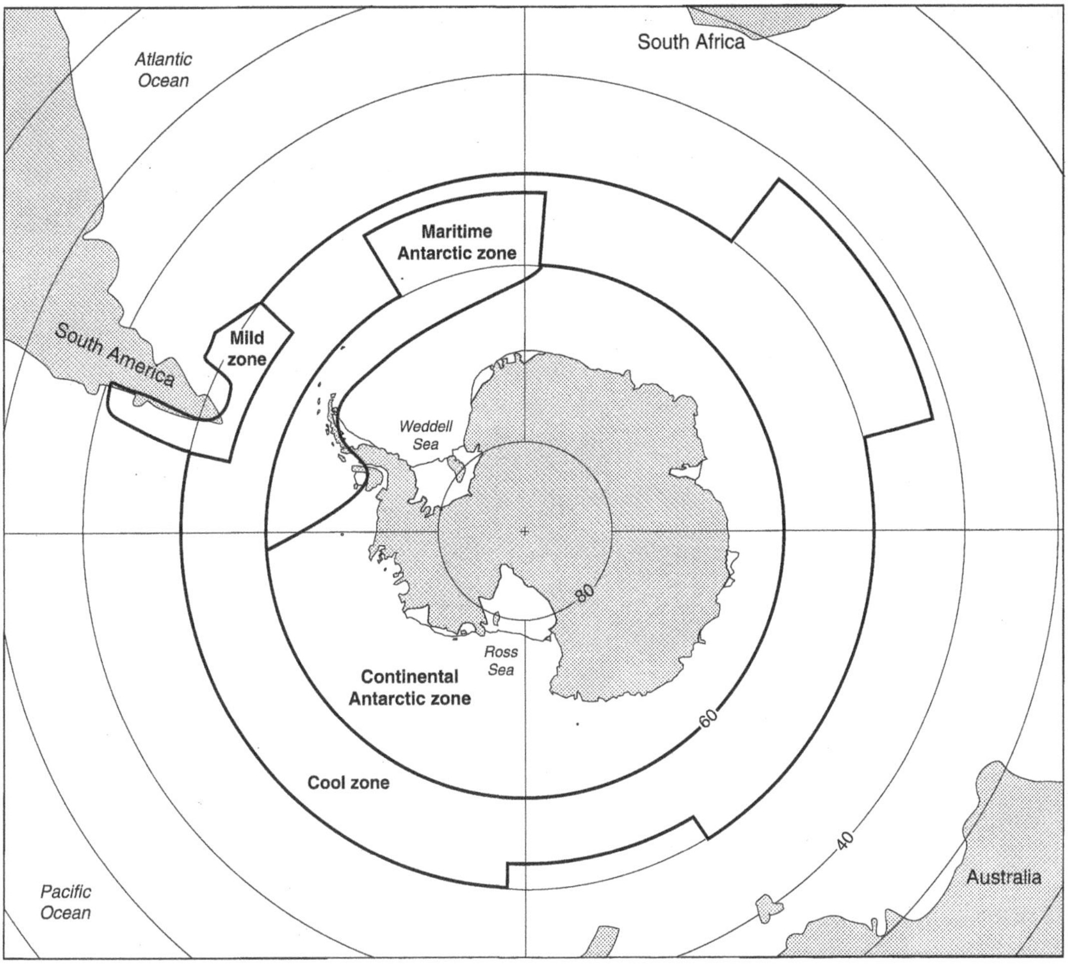

Fig 3.24 Terrestrial biogeographic zones in the Antarctic (source: modified from Longton, 1988).

the continent itself many of the limiting effects previously attributed to low temperature may actually operate through the water balance of organisms (*ibid.*), so dehydration resistance may be more important to survival than cold resistance .

One of the more successful attempts at correlating the biota with physical parameters has been that of Janetschek (1970) for south Victoria Land. His model shows that only where the depth of summer melting penetrates beyond the upper limit of the ice-cemented layer is free water available for biological processes (Figure 3.27). This condition is met close to the moist coast and well inland, where the ice-cemented layer is shallow enough to be reached by solar heating and melting. Between these two zones is a zone of complete aridity with no present-day precipitation, any moisture formerly held within the soil having been removed by thousands of years of evaporation. Rather than temperature *per se*, the vegetation distribution reflects moisture availability. On the other hand, water availability is greater in the milder sub-Antarctic, far fewer locations are likely to be moisture-limited, and so temperature may assume a greater relative importance. Specific adaptations of organisms to the Antarctic environment and the sensitivity of terrestrial systems to disruption and change are discussed in detail in Chapter 5.

Antarctic vegetation cover

The fossil record vividly indicates that throughout the Gondwanaland period and the separation of

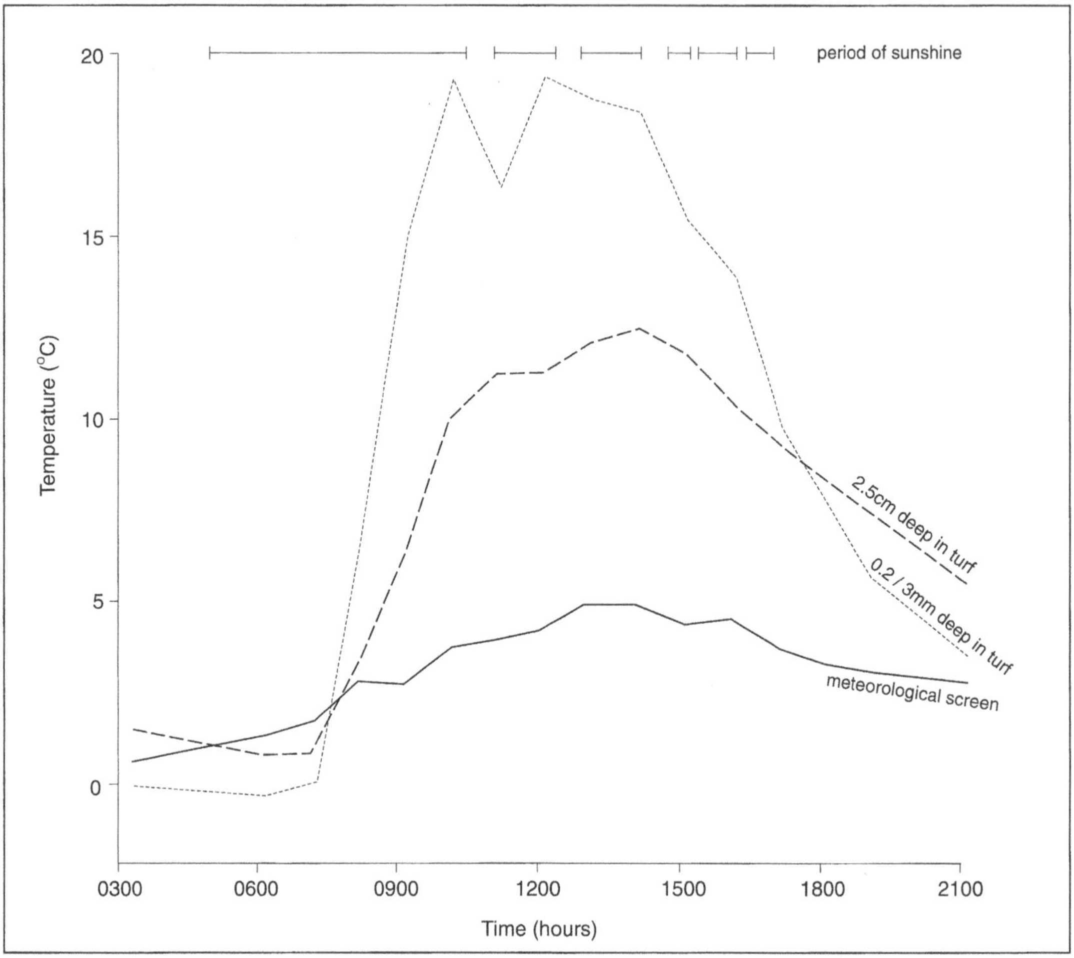

Fig 3.25 The effect of microclimate as shown by temperature fluctuations within a turf of *Polytrichum alpestre* on Signy Island, South Orkney Islands. The data were collected on 31.1.1965 at 20 m above sea level and show markedly higher temperatures than those recorded from a meteorological screen sited at sea level 100 m distant (source: redrawn from Longton and Holdgate, 1967).

Antarctica from its neighbours, a rich and varied vegetation cover was supported that only disappeared at the end of the Tertiary as glaciation began to develop (Elliot, 1985; Truswell, 1991) (see Figure 3.9). Other than a few, cold and icy nunataks, all of the continent and its oceanic islands were subsequently covered by Pleistocene ice. This means that later recolonisation of deglaciated terrain took place either by expansion from a few 'refugia' centred on ice-free nunataks (Dodge, 1973) or, more likely, by immigration from the adjacent continents (Lindsay, 1977) via transport by birds, winds and ocean currents. Only two species of flowering plant occur, and the present native flora of Antarctica is dominated by cryptogams: at least 200 species of lichens (predominantly in the drier, more exposed locations), 85 species of mosses, 28 macrofungi, together with many species of algae (Longton, 1985). The cryptogams are diverse and are subjected to the most severe environmental stresses of any Antarctic organism (Walton and Bonner, 1985a). Most species are restricted to either the continental zone or maritime zone, but a few species are found in both.

The two species of Antarctic flowering plants – the grass *Deschampsia antarctica* and the pearlwort *Colobanthus quintensis* (Stonehouse, 1972; Longton,

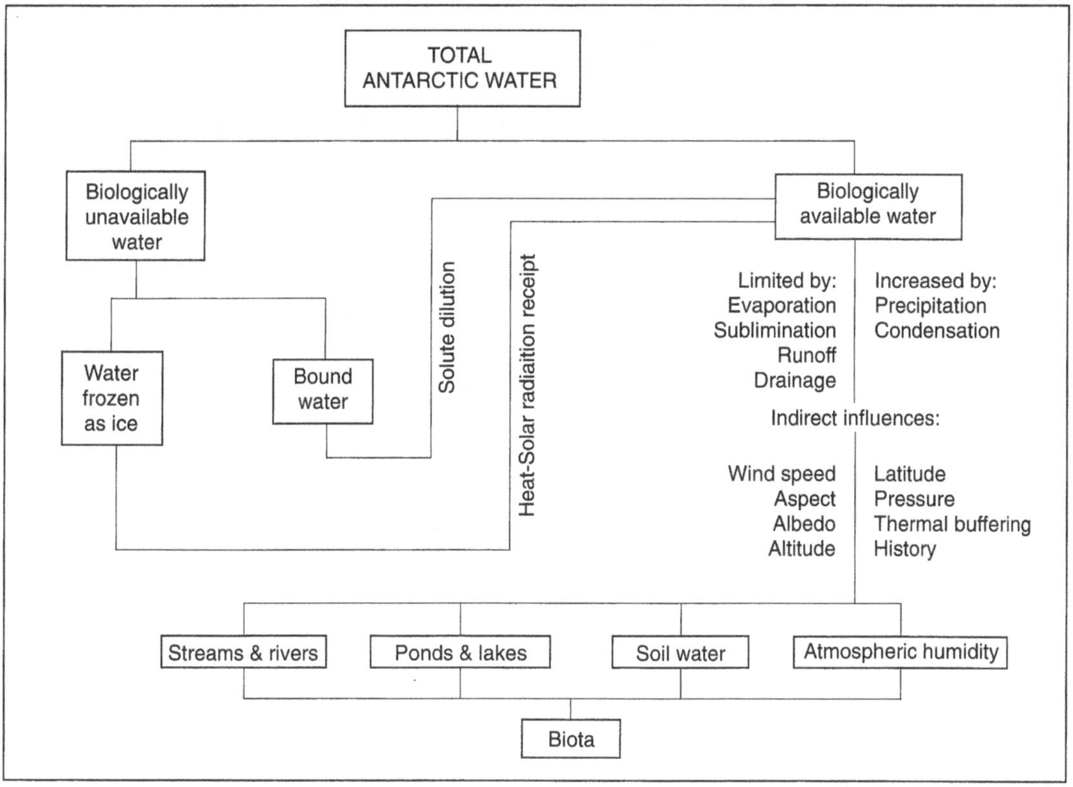

Fig 3.26 Biologically available and biologically unavailable water in the Antarctic (source: modified after Kennedy, 1993).

1985) are confined to the maritime zone, where the greatest diversity of species and the most extensive stands of vegetation are found. This zone has strong floristic affinities with southern South America and sub-Antarctic South Georgia (Longton, 1985). However, even in favourable and well-watered locations that have long since been deglaciated, continuous closed stands of vegetation are nowhere extensive. This is because the communities tend to be highly discriminating of the prevailing habitat conditions and the result is a mosaic of very small-scale communities. The essential climatic factors for good plant growth, higher temperatures and water availability, are both more prevalent in the western maritime zone than in the cold desert of the continental zone, where temperatures remain below freezing and snow comprises the bulk of the restricted precipitation. The lack of moisture and heat also means that the soils of the continental zone are immature, have a tendency to be saline and are impoverished of organic input.

The vegetation of continental Antarctica is sparse and is concentrated at a few favoured spots within the ice-free sites of coastal areas, the dry valleys and inland nunataks. Nowhere in this landscape of ice and rock is vegetation prominent, but where it survives, lichens, mosses and microphytic vegetation are the most important groups. Lichens are the most widespread on inland rock surfaces, with the prominent genera *Buellia*, *Caloplaca*, *Rhizocarpon* and *Xanthoria* being able to colonise most types of rock and boulder substrate. *Usnea*, *Alectoria* and *Umbilicaria* are locally abundant (Longton, 1985). Lichens dominate nunataks such as those of MacRobertson Land, where Filson (1966) recorded 26 lichen species but only two mosses. Mosses prefer sand and gravel substrates and, although cover is variable and usually is less than 5% on dry stony ground, short moss turf cushions of *Bryum antarcticum* can give up to 85% cover in moist hollows on Ross Island in the Ross Sea (Longton, 1985). Often the mosses occur as short turfs with shoot

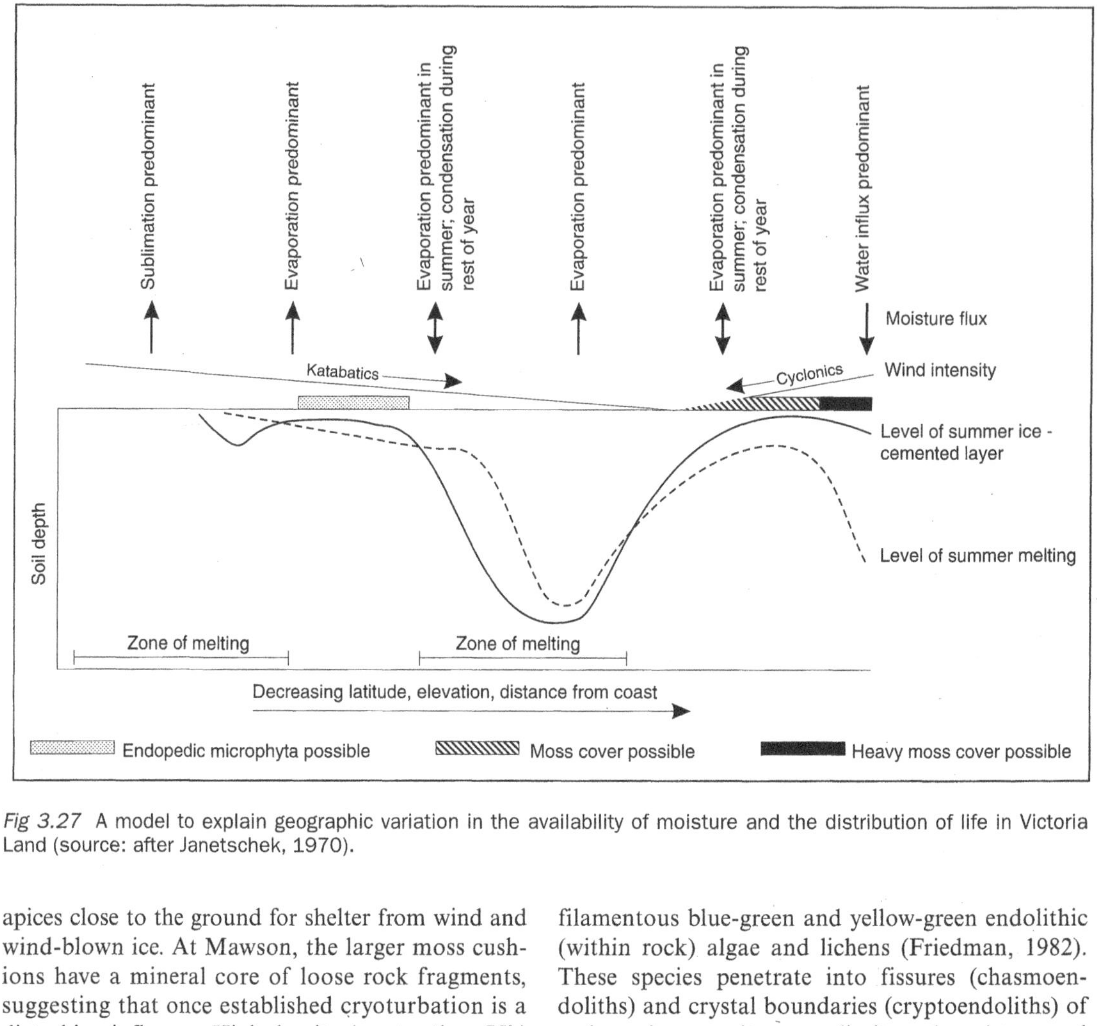

Fig 3.27 A model to explain geographic variation in the availability of moisture and the distribution of life in Victoria Land (source: after Janetschek, 1970).

apices close to the ground for shelter from wind and wind-blown ice. At Mawson, the larger moss cushions have a mineral core of loose rock fragments, suggesting that once established cryoturbation is a disturbing influence. High-density (greater than 75% cover) epilithic (surface) lichen communities seldom exceed 500 m^2 (Smith, 1993a). However, less than 5% cover is more typical even of favoured, but areally restricted, locations (Longton, 1985).

With increasing altitude and distance inland, the proportion of ice-free ground supporting vegetation that is visible to the eye declines. However, vegetation exists in many areas that are otherwise apparently lifeless such as gravel soils, bare rock and snow surfaces. The richest of the soil-based communities occur in coastal regions but bacteria and yeasts occur to at least 87° 21' S, although the limits for growth occur where summer heat fails to release ice-bound water (Janetschek, 1970) (see Figure 3.27). Recent studies suggest that the most widespread biotic communities of the continental Antarctic are the unicellular, filamentous blue-green and yellow-green endolithic (within rock) algae and lichens (Friedman, 1982). These species penetrate into fissures (chasmoendoliths) and crystal boundaries (cryptoendoliths) of rocks such as granite, granodiorite and sandstone and show zonation depending on light and moisture penetration from the surface (Figure 3.28).

Chasmoendolithic algae are widespread in coastal regions of Antarctica, underlying up to 20% of the rock surface in places by growing in fissures that are both perpendicular and parallel to the surface (Longton, 1985). Species zonation occurs, with blue-green algae at depth and green algae, fungi and lichens nearer the surface. In the Dry Valleys of southern Victoria Land, cryptoendolithic lichens and associated fungi and algae appear to be the dominant form of life (Kappen and Friedmann, 1983). Within the 1–2 mm thick abiotic surface crust of the rocks are successive zones of black lichen (1 mm thick), white fungus (2–4 mm thick) and the green algae *Trebouxia* (2 mm thick) (see Figure 3.28) (Longton, 1985). Organic

Fig 3.28 Endolithic lichens and algae inhabit sheltered layers of sandstone rock in Victoria Land, where plants need protection to survive. Beneath the abiotic surface layer is a black layer of lichen, a white powdery layer of fungus and a green band of algae at depth (source: photograph by C.J. Gilbert, courtesy of BAS).

secretions, including acids, from these cryptoendolithic lichens in the Dry Valleys appear to contribute to the biogenic exfoliation of the surface crusts of host rock (Friedman, 1982). Some species of cryptoendolith can recolonise the rock surface by this method, and the genera *Acrospora*, *Lecidea* and *Buellia* have been recognised in this way. Since these genera are well-represented in the common forms of Antarctic surface lichens, their cryptoendolithic status has been interpreted as an evolutionary adaptation to deteriorating environmental conditions since the Tertiary (*ibid.*). During the period of spring and summer melting, snow banks in northern coastal areas and the sub-Antarctic islands become coloured by red, green or yellow concentrations of unicellular snow algae, including species of *Chlamydomonas*, *Raphidonema* and *Ochromonas*. Fogg (1977) considers the rapid appearance in spring and summer of these cryoplankton communities to be related to progressive concentration by melting rather than by high productivity.

Vegetation in the maritime Antarctic is characterised by the dominance of either a cryptogam tundra vegetation of lichens, mosses and algae developed on rock surfaces or a herb tundra vegetation developed mainly on soils. Of the cryptogamic communities on rock surfaces, crustose lichens such as *Verrucaria*, *Caloplaca* and *Xanthoria* respectively favour zones increasing in altitude above the littoral zone of coastal sites, where salt spray or bird guano is plentiful (Figure 3.29A). Brightly coloured crustose lichens are favoured by the nitrogen-enriched conditions adjacent to penguin colonies (Smith, 1993a). On Bouvet Island, species typical of the shore zones reach 200 m altitude where salt spray is driven up high cliffs (Engelskjon, 1981). With increasing distance from the sea there is a transition from the crustose lichens of the maritime communities to the fruticose lichens and moss cushions of the more montane communities; the fructose lichen *Usnea antarctica* often adding a hint of green to an otherwise monochrome inland landscape of boulders, cliffs and scree (Longton, 1985). *Usnea* also occurs in association with small cushion-forming mosses, which can become so abundant as to form a continuous carpet of vegetation in low-lying sites (Figure 3.30). The most distinctive of the Antarctic vegetation communities are the stands of tall mosses such as *Polytrichum alpinum*, *Chorisodontium aciphyllum* and *Polytrichum alpestre*. These stands can range in size from small discrete turfs to banks up to 50 m long and 2 m deep, especially on well-drained and gently sloping surfaces of scree or boulders (*ibid.*). Algae are also common, especially where nitrogen enrichment favours growth close to seal or penguin colonies.

The vegetation of the maritime zone is also characterised by the Antarctic herb tundra formation, although it is best-developed where soils are more established. The moister soils support both of the native flowering plants of Antarctica, although they are seldom abundant. *D. antarctica* is more widespread, its swards often coalesce to reach several metres wide in sheltered low-altitude sites in the South Orkney Islands and Antarctic Peninsula. *C. quintensis* forms compact cushions up to 25 cm wide. Both species of flowering plant are commonly found in association with mosses and lichens. However, even in the north of the maritime zone on Signy Island in the South Orkney Islands, Tilbrook (1970) estimated that herb tundra cover is areally limited and amounts to only 0.001%. However, both *D. antarctica* and, to a lesser extent, *C. quintensis* are also pioneer colonists of deglaciated terrain, and rapid 25-fold and fivefold population increases, respectively, have occurred recently on Argentine Island, Antarctic Peninsula, possibly in response to increasing summer air temperatures (Fowbert and Smith, 1994). Elsewhere, deglaciated surfaces are usually heavily disrupted by periglacial activity, and the continuity of the plant communities is broken by small-scale gradients in moisture, soil texture and micro-relief (Smith, 1993a). The resultant mosaic of mosses, lichens and occasional herbs colonising a varied surface of glacial till, glacio-fluvial gravels, boulder fields and scree is often referred to as

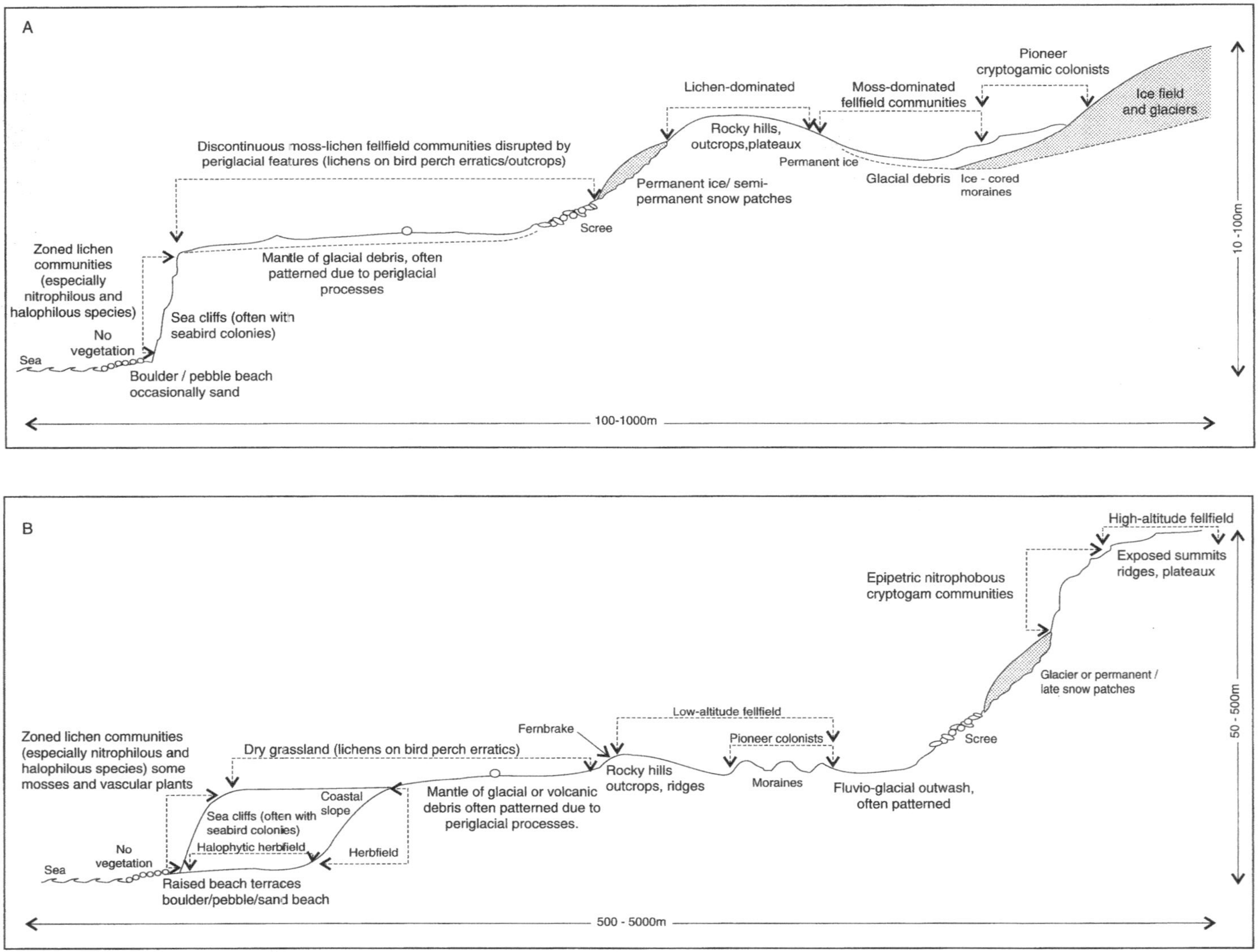

Fig 3.29 Schematic profiles across typical coastal areas illustrating the distribution of the main dry ecosystems in (A) coastal Antarctica; and (B) a sub-Antarctic island (source: redrawn after Smith, 1993).

Fig 3.30 *Usnea* colonisation of raised gravel beaches in the South Shetland Islands. Together with a field hut, a touch of colour is added to an otherwise monochrome scene (source: authors).

'fellfield' vegetation: a floristically diverse community comprising many of the epilithic species but often with certain mosses and occasional liverworts locally forming dense closed stands several square metres in area or more open stands covering up to several hectares (*ibid.*).

In the maritime Antarctic, the more acidic soils in hollows and areas protected by deep winter snow cover are typically dominated by mosses such as the several species of *Andreaea*, with increasing numbers of liverworts in the north on the islands of the Scotia Sea. In reality, there are few distinct vegetation boundaries in the maritime Antarctic, rather a northward gradient of enhanced floristic diversity that parallels the ameliorating climatic gradient; vegetation cover reaches its greatest extent in the South Orkney Islands. Areal data are limited, but the unglaciated area of Signy Island supports 12% and 6% cover of short and tall mosses, respectively, 50% is sparsely covered by lichens and 26% is bare rock (Tilbrook, 1970). This cover decreases south along the west coast of the Antarctic Peninsula, whereas the east coast has a sparser vegetation with a continental bias. However, even in favourable, well-watered locations that have been deglaciated for several decades, continuous closed stands of vegetation are nowhere extensive. Plants tend to be highly discriminating of the prevailing habitat conditions and the result is a discontinuous mosaic of small-scale communities.

Sub-Antarctic vegetation cover

Moving north from the maritime Antarctic zone, the vegetation of the sub-Antarctic islands of South Georgia, Marion Island, Heard Island, Îles Crozet and Îles Kerguelen has a tundra-like appearance, consisting mainly of herbs and cryptogams. Whereas continental Antarctica has two native species of flowering plant, the sub-Antarctic islands have 70 native vascular taxa, with South Georgia alone boasting 26 (Smith, 1993b). Several geographical factors distinguish the sub-Antarctic as a more favourable environment for vegetation growth and development than the land further south. Almost all of these factors stem from the direct or indirect benefits of a more northerly location and an essentially equable oceanic climate. The islands have high temperatures by Antarctic standards; this aids growth and has also resulted in the early deglaciation of surfaces between 18 and 11 ka BP. All of the islands have a locally more varied substrate than the continent itself and have relatively abundant ice-free areas of coastal and inland cliff, rock and boulder surfaces, glacial tills, raised sand and gravel beaches, volcanic ash and substantial glacio-fluvial plains. Mid-oceanic islands with such an abundance of available surfaces, often close to sea level, are ideal sites for colonies of marine mammals and birds, and this has resulted in locally high inputs of organic material into the soil. The combination of numerous available sites, a lengthy period of potential colonisation, a range of suitable substrates, milder conditions and availability of moisture and nutrients has produced a floristic diversity and vegetation cover unparalleled south of the Polar Frontal Zone. Vegetation cover can be almost continuous over extensive areas of as much as 25 km^2 (Smith, 1993b) (Figure 3.31). In the cool oceanic climate, where precipitation is high, many islands have developed 3–6 m thick peat bogs whose basal layers date to between 9.5 and 11 ka BP.

Although the sub-Antarctic islands are widely separated from each other, there is a remarkable degree of similarity in species composition and community structure. South Georgia has a very strong South American and Fuegan floristic affinity (64% of the taxa) and all of the sub-Antarctic islands have a common Fuegan–New Zealand element (Smith, 1984), probably on account of the prevailing westerly winds and the possibilities of colonisation from the west by birds, or by air and ocean transport. Common elements also include proximity to the coast, with several of the principal communities extending several hundred metres inland, and the highly varied substrate, which gives the landscape the appearance of a vegetation mosaic (Smith, 1993b) (see Figure 3.29B). In response to the high rainfall, there is widespread development of bog and mire (Walton, 1985) but there is also a range of vegetation communities, including at low altitudes herbfield, short and tall grasslands, and fernbrake (mainly on Marion Island) and at higher altitudes open fellfield communities (Smith, 1993b).

Herbfield communities exist on well-drained ground and are dominated by the deciduous dwarf shrub *Acaena magellanica*, which forms extensive and dense swards often covering several hectares. Close to the shore the herbfield is characterised by *Acaena* and tussocks of *Parodiochloa flabellata* (until recently known as *Poa flabellata*) (Figure 3.32). This waist-high grass occurs mainly on abandoned rock platforms and raised beaches, where, favoured by the nutrient inputs from penguin and seal colonies, it produces luxuriant and rapid growth. Decomposition is slow, however, and erosion of the peaty soil between the plants by seals and penguins often produces vegetation-capped pedestals taller than a person on South Georgia (Headland, 1984). Where the substrate is wet or peaty, the grassland communities are dominated by *Agrostis magellanica* or *D. antarctica*, but where it is well-drained, *Festuca contracta* grassland of up to 75% cover dominates, often with horizons of brown tundra soils beneath (Smith and Walton, 1975). There is commonly a continuum of *Acaena* on moist sites, *Festuca* on better-drained and sheltered north-facing slopes, and shorter more open grass with mosses ad lichens, such as *Cladonia,* on exposed and windy slopes (Smith, 1993b).

The fernbrake communities of Marion Island occur on well-drained ash slopes, where the shade-sensitive fern *Blechnum penna-marina* forms mats, with *Acaena* increasing in cover where shelter exists (Huntley, 1971) (see Figure 3.29B). Fellfield

Fig 3.31 Where conditions of temperature and moisture are favourable, enhanced vegetation cover can produce vegetation productivity levels that can support introduced mammals, such as these reindeer at St Andrews Bay, South Georgia (source: authors).

Fig 3.32 Stands of *P. flabellata* in coastal sub-Antarctic locations are heavily used, and well-manured, by breeding seals and by birds such as these light-mantled sooty albatrosses in South Georgia (source: authors).

communities are common throughout the sub-Antarctic and comprise discontinuous patchy communities of mosses, short grasses and lichens capable of surviving in the often very dry conditions found at altitude on windswept and stony soils. Such surfaces are often subjected to intense periglacial activity and develop stone circles, polygons and stripes, which inhibit the colonisation and establishment of plants (Heilbron and Walton, 1984) (see Figure 3.21). On level surfaces dense cover is possible, with *Acaena*, *D. antarctica* and *Festuca* being the most common, together with some mosses and lichens. On slopes where periglacial activity has sorted the substrate into different stone calibres, the faces of downslope-moving turf-banked solifluction lobes and terraces are often picked out by crescents of lichen-rich *Festuca* communities (Figure 3.33).

The productivity of Antarctic vegetation

The annual productivity of Antarctic plants is thought to be low, but there are few data outside Signy Island in the maritime Antarctic (Smith, 1993a and b). Even for a relatively mild and moist sub-Antarctic location such as South Georgia (latitude 54° S), most growth parameters have values that are appreciably lower than those for Disko Island, West Greenland (latitude 69° N) (Smith and Walton, 1975). The reasons for low plant metabolic rates and growth in the Antarctic are related primarily to low temperatures and lack of moisture, but the duration of snow cover is also important. In the northern maritime Antarctic the growing season may be only 100–135 days, in addition to several weeks photosynthesis and growth under thin snow, whereas in coastal continental Antarctica it is closer to 60 days (Collins and Callaghan, 1980).

Geographically, net annual dry weight production should be highest in the north and lowest in the south, but there is great variability: the 100–300 g m^{-2} produced by fellfield moss communities on Signy Island falls to 10–50 g m^{-2} for moss production in the coastal continental zone further south (Smith, 1993a). Higher rates of production have been reported for mixed *Usnea* stands: 1750 g m^{-2} on Signy Island and 1900 g m^{-2} on the South Shetland Islands, falling to 900 g m^{-2} in continental Antarctica (Kappen, 1985). However, there can be marked annual variation in the production of any one species. *P. alpestre* has shown differences of 430–600 g m^{-2} in consecutive years, and within one season 404g m^{-2} of production was recorded at Argentine Island on the Antarctic Peninsula, compared with 342 g m^{-2} at Signy Island further north. The higher value at the southern site probably reflects greater mean daily sunshine duration and higher temperatures in that year (Smith and Walton, 1975). The most productive dry coastal ecosystems on Signy

Fig 3.33 Turf-fronted solifuction lobes and steps characterise many slopes in the maritime Antarctic and sub-Antarctic (source: authors).

Island rarely exceed a standing crop of 5000 kg ha^{-1} and most are closer to 500 kg ha^{-1}. At most coastal continental sites the standing crop probably varies between 0 and 10 kg ha^{-1}, with net annual dry weight production of between 1% and 10% of the biomass (Smith, 1993a).

On account of the wet climate of the sub-Antarctic islands, net primary productivity can be relatively high by Antarctic standards: locally on South Georgia, 100% closed stands of *Acaena* at 10 m altitude can reach a dry weight production of 1605 g m^{-2}, about a third of which is below ground (Smith and Walton, 1975). This below-ground adaptation allows many sub-Antarctic plants to build up reserves for overwintering (see Chapter 5). Equivalent data for the 60% cover of a *P. flabellata* community at 3 m altitude in South Georgia reach a total net annual production of 6025 g m^{-2}, but only 10% of this production is below ground. Individual plants of *P. flabellata* can achieve above-ground standing crops of 15 kg m^{-2} of living material and 10 kg m^{-2} of dead material (Smith and Walton, 1975). At 35 m altitude in South Georgia, 100% cover of *Festuca* grassland achieves a much lower annual net production rate of 840 g m^{-2}. By comparison, on Macquarie Island, 100% grassland cover at 45 m altitude yields 5581 g m^{-2} (Smith and Walton, 1975). At 235 m altitude on Macquarie Island, 100% cover of herbfield yields 1014 g m^{-2} compared with 1605 g m^{-2} at 10 m altitude on South Georgia. Productivity is higher on South Georgia than on the continent itself, but productivity on Macquarie Island is even higher. The reasons are geographical: several species on Macquarie Islands (and also Marion Island) are known to continue to photosynthesise and grow throughout most of the year, while on South Georgia biological activity is seasonally constrained on account of a more severe climate.

Antarctic terrestrial animals

The biogeography of terrestrial animal life in Antarctica is fundamentally dominated by its polar climate and isolation. The extreme climate has led to a sparse and slow-growing vegetation, and this partly explains the absence of the large grazing herbivores that are so characteristic of the Arctic tundra. The isolated location has also precluded any potential colonisation, even of the more favoured coastal locations rich in prey items such as seabirds and seals. Expansion of Pleistocene ice swept any terrestrial animals from the continent and sub-Antarctic islands at a time when tectonic drift from the former Gondwanaland had created a wide ocean barrier. Thus, in spite of glacier recession early in the Holocene, allowing the development of habitats that in some areas can now support larger herbivores, such as in South Georgia and Îles Kerguelen, colonisation by natural means has always been unlikely. Without herbivores as a source of prey, there can be no large Antarctic terrestrial carnivores. The native terrestrial animal life of the Antarctic is restricted to the invertebrates, including nematodes, tardigrades (water bears), insects and mites a few millimetres in size, which are preyed upon by carnivorous insects of similar size (Stonehouse, 1972). Even amongst the invertebrates, several groups important to warmer regions are missing or poorly represented, such as earthworms, and molluscs (Stonehouse, 1989).

The number and diversity of species in the Antarctic is low compared with lower latitudes but increases northwards towards the sub-Antarctic (Sømme, 1985) (Table 3.1).

The distribution of the invertebrates has also been determined partly by past links to the former Gondwanaland continents and later immigration (*ibid.*). For example, there are strong affinities in the

Table 3.1 Numbers of insect species found in Antarctica and on South Georgia

	Continental zone	Maritime zone	South Georgia	Total
Collembola (springtails)	12	8	16	36
Mallophaga (biting lice)	18	32	35	85
Anoplura (sucking lice)	3	3	1	7
Thysanoptera (thrips)	–	–	1	1
Hemiptera (bugs)	–	–	2	2
Coleoptera (beetles)	–	–	7	7
Siphonaptera (fleas)	1	1	1	3
Diptera (flies and midges)	–	2	13	15
Hymenoptera (parasitic wasps)	–	–	1	1
Total	34	46	77	157

Somme, 1985

distribution of some mites with those found in their closest Gondwanaland neighbour. However, the effect of long isolation may be reflected in a high proportion of endemic species throughout the Antarctic and a higher proportion of immigrants at lower latitudes. Transport by birds seems the most likely form of immigration: for example, Bouvet is 1700 km from its nearest neighbour and its rocks are only 3 Ma old, yet it has three species of Collembola and six species of mite.

Four main species of protozoan unicellular organisms exist in freshwater or moist habitats on land. They mostly prefer moss carpets and ground covered by plants and so their numbers decline southwards. Sixty-five species have been established from the Antarctic Peninsula, Signy Island and South Georgia (Smith, 1978). Found almost everywhere where organic materials (living or dead) occur, there are about 70 species of nematodes in the Antarctic. Their numbers in the soils and vegetation on Signy Island may reach several millions per square metre (Sømme, 1985). The distribution of these species appears to be restricted to small geographical areas: 34 of the 40 species of nematodes known from the maritime Antarctic occur nowhere else (*ibid.*). Tardigrades are small animals less than 1 mm in length that live and graze mainly on mosses and plant material, often at densities of up to 14×10^6 m^{-2}, and 23 species have been reported from the Antarctic (Jennings, 1976). One of the reasons for their success may be an ability to withstand periods of severe desiccation. Rotifers are small animals less than 0.5 mm long that live in moist habitats like moss cushions at densities of up to 10^5 m^{-2} (Jennings, 1979).

Of the 67 continental and maritime insect species, 45 are parasites on the warm-blooded birds and seals: biting lice and fleas are mainly found in close contact with birds, either at the base of their feathers or in their nests, and sucking lice are found on some species of seal (Sømme, 1985). The only free-living insects on the continent are the primitive and wingless Collembola (springtails) up to 2 mm in length. On Signy Island, Tilbrook (1977) recorded populations of *Cryptopygus antarcticus* in a moss carpet of up to 9.5×10^4 m^{-2}. Two species of non-biting midge are known in the maritime zone and one of these has lost the ability to fly through wing reduction, a frequent adaptation in insects from cold and windy areas (Sømme, 1985). There are about 70 species of mite in the continental and maritime zones and a similar number on South Georgia (Gressitt, 1970). Some are parasitic on birds and seals but the bulk of the free-living species belong to the soft-bodied prostigmatid mites or the hard-bodied cryptostigmatid beetle mites. One of the latter might be the world's hardiest animal: *Nanorchestes antarcticus* is 0.3 mm long and inhabits continental nunataks as far south as 85° S (Sømme, 1985). The only spiders in the Antarctic live on South Georgia, where four native and one introduced species occur.

Most invertebrates in Antarctica graze on bacteria, algae and fungi (Block, 1994). In spite of its abundance, moss does not seem to be consumed in great quantities and experiments suggest that some species, such as *Cryptopygus antarcticus*, have a preference for micro-algae growing within the moss carpet (Sømme, 1985). The mite *Alaskozetes antarcticus* is both a grazer and a detrital feeder (Figure 3.34). With the number of arthropod species in any community being about ten, the terrestrial communities of Antarctica are amongst the simplest in the world (Usher and Edwards, 1986a), and a unique feature of maritime Antarctic communities is the low level of invertebrate predation (Block, 1994). At Signy Island, the community contains a single predatory species of mite, the fast-moving *Gamasellus racovitzai*, which mainly consumes the springtails, *Crypotopygus antarcticus* and *Parisotoma octooculata* (Usher *et al.*, 1989). In the moss communities of Signy Island, the food web is thus fairly simple with protozoan, rotifer, tardigrade, nematode, mite and Collembola primary consumers feeding on the microflora of bacteria, yeasts, algae and fungi (Davis, 1981). They themselves are preyed upon by *Gamasellus*, one species of nematode and one species of tardigrade. Such communities may lack species redundancy and so are sensitive to environmental change and disturbance. For example, the nematode community structure in the soils of the dry valleys region is limited to between one and three species, and the loss or decline or even a single species may result in a high degree of disturbance to the soil ecosystem (Freckman and Virginia, 1997).

All of these species have developed strategies for survival in harsh conditions. Choice of habitat is greatly influenced by the availability of vegetation for both food and shelter and is consequently controlled by the occurrence of water. Where moisture is limited, such as on bare rock surfaces or on the loose mountain screes of the Antarctic Peninsula, no invertebrate fauna is found. However, where moisture is available from melting ice in summer, even inland nunataks may support bacteria, algae and fungi, which in turn may support a few species of Collembola and mites (Janetschek, 1970). Only a few insects can withstand the formation of ice in their tissues (freezing-tolerant);

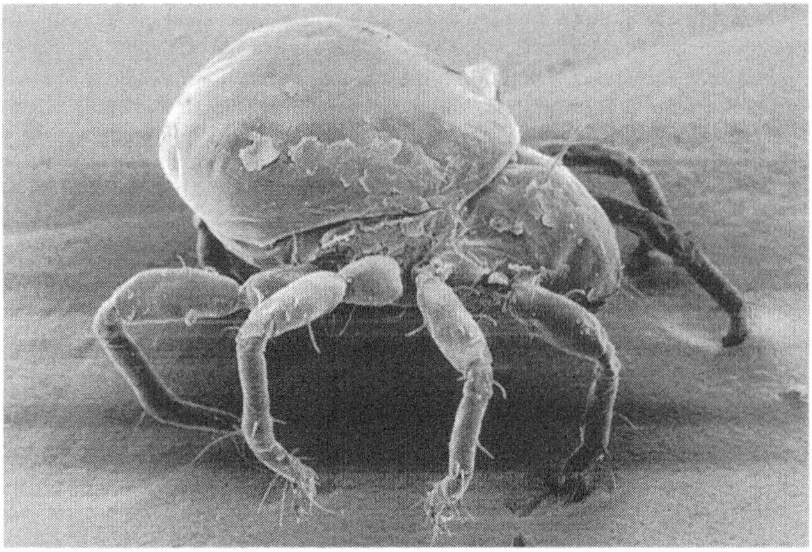

Fig 3.34 Alaskozetes antarcticus: at 1 mm long, is one of the largest of Antarctic terrestrial animals (source: photograph by C.J. Gilbert, courtesy of BAS).

most insects and terrestrial arthropods are killed by ice (freezing-susceptible). Many have also adapted to withstand freezing temperatures by depending on supercooling their body fluids and developing cryoprotectant 'antifreezes', the details of which are examined further in Chapter 5.

In spite of the occurrence of a vegetation cover that is extensive enough in some areas to support a tundra ecosystem complete with invertebrates, the lengthy period of isolation of Antarctica (a minimum of 20 Ma), the severity of its glacial climate and distance from its nearest neighbours have resulted in a total absence of native land vertebrates. However, the suitability of the climate and vegetation of the sub-Antarctic islands for the survival of land vertebrates is shown by the success of both deliberately and accidentally introduced exotic species. At present, eight species of mammal, including reindeer, cats, rats and mice, occur in established populations on six sub-Antarctic island groups, and only Heard Island, Bouvet Island, MacDonald Island and the South Sandwich Islands are free of introduced mammals (Leader-Williams, 1985). The impact of exotic introductions on the habitats of the sub-Antarctic islands is examined in Chapter 8.

Terrestrial aquatic ecosystems

Both subglacial and subaerial lakes exist in Antarctica. Although subaerial lakes are geographically restricted to coastal and peripheral areas of the Antarctic (see Figure 3.18), their ecosystems have attracted scientific interest on account of the effects of low temperature on their biological systems. Freezing of the upper layers during all or part of the year produces a closed system in the unfrozen part, which is subject to strong thermal stratification and low amounts of solar radiation and photosynthesis. The relatively barren catchments also mean that the supply of biologically useful nutrients is limited (Priddle, 1985). Both freshwater and saline lakes occur, but freshwater lakes are the more common. Most of these occur along the ice edge and are of a temporary nature as the ice dam that supports one side fluctuates. As the ice retreats, deeper lakes form in rock basins or where drainage is blocked by moraines. These tend to follow an evolutionary sequence of nutrient enrichment as their catchment areas become deglaciated, vegetation cover becomes established and runoff containing dissolved ions increases (Priddle and Heywood, 1980). Signy Island, in the South Orkney Islands, has several small lakes sited in fairly well-vegetated, mossy catchments that are home to blue-green algae, microbes, bacteria, crustaceans, nematodes and worms (Priddle, 1985). Most of these lakes also have varying amounts of marine inputs, directly by way of salt spray or by enrichment by seal faeces. By virtue of their isolation and relatively recent deglaciation, neither the Antarctic nor sub-Antarctic lakes have mollusc or fish populations (Weller, 1975).

In the water column of freshwater proglacial lakes on the continent, there are few planktonic organisms and most of the biological activity is found in the benthic community, where mosses overgrown with algae are found. In contrast, below the depth scoured by seasonal ice, even the nutrient-poor Moss Lake in the South Orkney Islands displays extensive luxuriant stands of aquatic mosses (Priddle, 1985). Here, the benthic plant communities are perennial and can grow for most of the winter under much reduced light conditions, although annual production remains low. Planktonic algae are scarce, bacteria use most of the phytoplankton production and the fauna is poorly developed and restricted to bottom dwellers. Such lakes mostly have low productivity as a result of strongly seasonal solar radiation coupled with low nutrient availability. However, where the lakes are organically enriched by seal and bird manure larger phytoplankton communities develop. At Heywood Lake, in the South Orkney Islands, a sharp rise in seal numbers in the 1960s seems to have resulted in enhanced summer total cell volumes and phytoplankton abundance, although overall phytoplankton levels are well below those of the marine habitat (Light and Heywood, 1973) (Figure 3.35). Similarly, many lakes on South Georgia and Marion Island become biologically enriched by inflow from penguin colonies, creating good conditions for phytoplankton growth, with carbon fixation reaching 0.8 g C m^{-2} d^{-1} in summer, producing high turbidity (Grobelaar, 1974). In such lakes, the phytoplankton crop supports an enhanced zooplankton biomass that is highly seasonal and produced from only a few species.

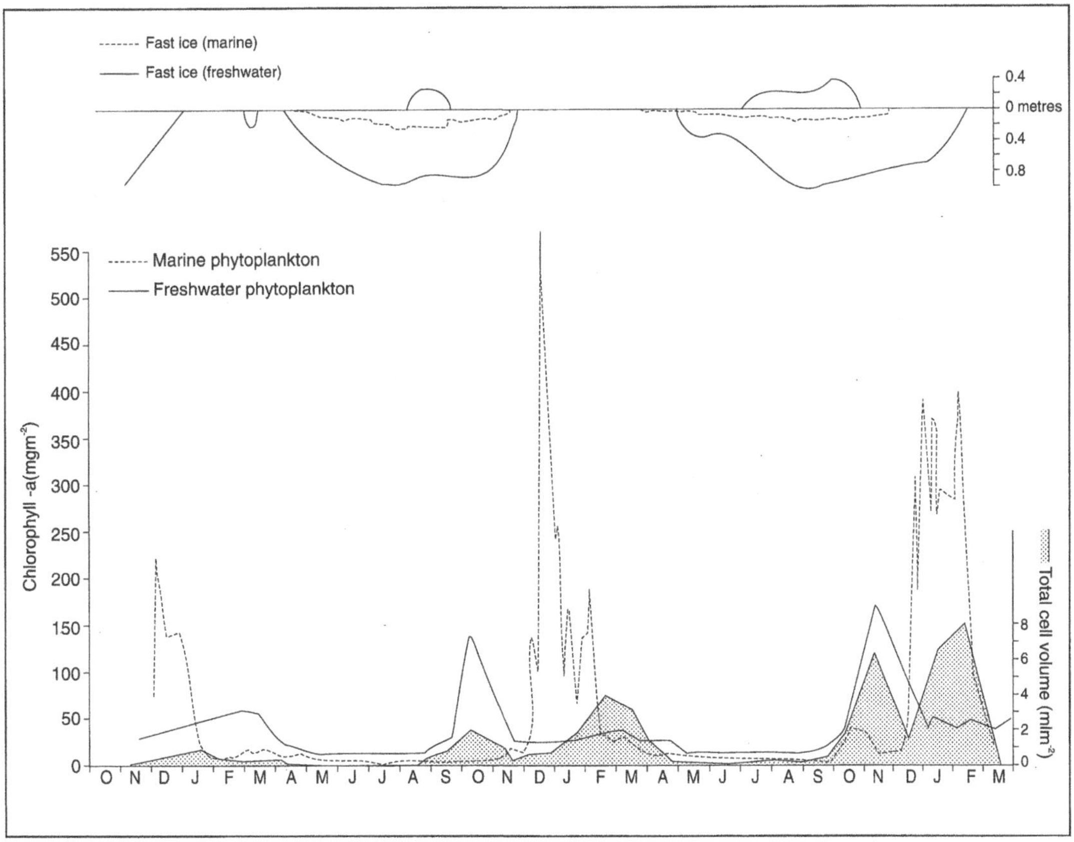

Fig 3.35 The strongly seasonal pattern of chlorophyll-*a* production in two Antarctic aquatic habitats. At Signy Island, the timing of peaks in Borge Bay (marine) is related to the disappearance of sea ice, whereas the early disappearance of surface snow and the later melting of lake ice at Heywood Lake (freshwater), produces twin peaks (source: modified after Clarke, 1985).

Two types of saline lake in Antarctica provide an aquatic terrestrial habitat: those that have been abandoned by the sea; and those that are the result of evaporation (Burton, 1981). As a result of isostatic uplift of marine bays, several of the first type are found in the Vestfold Hills, where salinity levels of up to 5.5 times that of sea water prevent freezing even at –28 °C. However, from a biogeographical viewpoint, the virtually sterile environment of high salinity and low temperature supports little life other than sparse algae and bacteria. Of the second type, the most interesting is Lake Vanda in the Dry Valleys of Victoria Land. The Onyx River flows into Lake Vanda, contributing salts that concentrate at the bottom of the lake, which is capped by nutrient-poor fresh water that collects at the top beneath 4 m of permanent ice (Figure 3.36). Solar heating raises the temperature of the lower layers to 25 °C (Fothergill, 1993) but in spite of this, the dense and saline bottom water does not mix with the less dense surface water (Priddle, 1985). The result is an upper layer that resembles other Antarctic freshwater lakes, with phytoplankton growth limited by ice cover and a surprisingly diverse benthic community of algae, protozoa and invertebrates (*ibid.*). Below this, various different bacteria and algae inhabit each successive layer. For example, conditions at the boundary zone are ideal for nitrifying bacteria, which produce high levels of nitrous oxide; below this is a peak in algal growth; and below this still, conditions are ideal for denitrifying bacteria to break down the nitrous oxide produced at shallower depths (*ibid.*).

Very little is known about the life contained within the subglacial lakes beneath the East Antarctic ice sheet. However, there is a very strong possibility that Lake Vostok may be a unique habitat for ancient bacterial life. Microbiological studies of the Vostok ice core have revealed a variety of microbes including yeasts, which remain viable in ice for up to 3 ka, and viable fungi up to 38 ka old that have been transported via the atmosphere from elsewhere (Kellogg and Kellogg, 1996). The oldest viable forms are spore-forming bacteria that are 200 ka old, and algae and diatom shells have been observed at 1525 m and 2395 m down the core and aged 110 ka and 180 ka, respectively (Ellis-Evans and Wynn-Williams, 1996). It seems clear that there is a ready source of microbes potentially to seed the subglacial lake. The main bio-

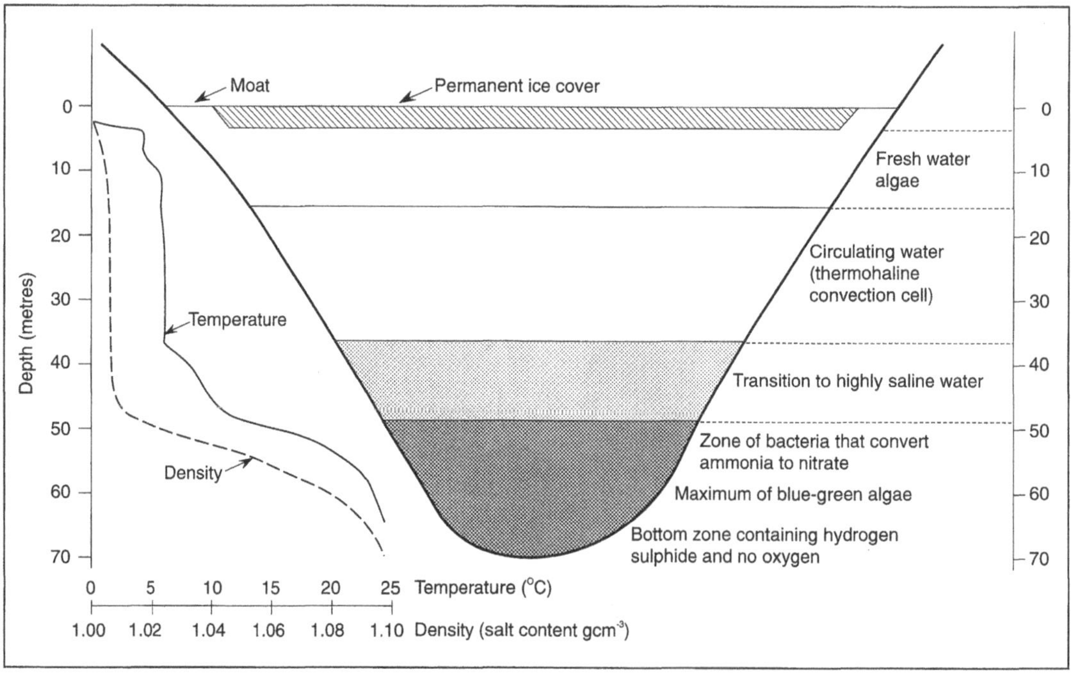

Fig 3.36 Stratification and life in Lake Vanda in the dry valleys of Victoria Land. Beneath the frozen ice cover a cold and biologically poor freshwater layer overlies a warm and biologically rich saline layer. See also Figure 3.3 (source: redrawn from Chinn, 1990).

logical significance of life in Lake Vostok lies in the gene pool of its microbes, because it should contain characteristics developed well before 200 ka ago that have since been isolated from external influences.

3.5.3 Summary

It is clear that in spite of a seemingly huge and uniform ice cover over the Antarctic continent, great spatial variations in the nature of the environment profoundly influence Antarctic biogeography. The marked gradients in temperature and water availability that exist both latitudinally and inland from the coast present severe limitations to life, and in general there is a decrease in primary productivity towards the pole. This is reflected in the changes in dominant plant types from the dense swards of grasses found on the lowlands of the sub-Antarctic to the mosses and lichens of the Antarctic Peninsula and finally to the sparse lichen cover of the few ice-free areas inland. In comparison with aquatic systems, grazing by herbivores of the dry terrestrial ecosystem is negligible and most of the organic matter is utilised by an invertebrate decomposer community that is severely restricted by seasonality, low temperatures and lack of moisture. Because of this, most of the Antarctic terrestrial ecosystem is of fragile construction, with plants and animals either living at the limits of their range or adapted to cope with the strictures of extreme conditions. However, this also renders them susceptible to changes in the physical environment: limited ground cover, slow growth and colonisation rates are a disadvantage where exposed surfaces of rock or frost-churned ground make establishment difficult. Conditions in the aquatic environment are also limited by light, temperature and nutrient availability, with the most productive systems occurring as a result of enrichment by seals or birds. In spite of this input, the regime of aquatic ecosystems remains highly seasonal. The sensitivity to change and fragility of terrestrial ecosystems are examined further in Chapter 5.

3.6 Conclusion

The terrestrial environment of the Antarctic is dominated by two great ice sheets that have coalesced to form the largest body of ice on Earth. The larger and more stable East Antarctic ice sheet seems to have reached its present size about 14 Ma ago, with the West Antarctic ice sheet reaching its present size about 5 Ma ago. Understanding the development of the ice sheets is important, since it affects our interpretation of how the ice sheet might behave during periods of warmth such as the present. It appears that there is a strong geographical component to the effects of warming: during the last two cold periods lower temperatures and precipitation resulted in ice sheet contraction on the polar plateau; and at the same time expansion of glaciers occurred in the maritime Antarctic due to higher precipation and reduced melting. Convincing evidence also exists of the reverse trend during periods of past warming: enhanced precipitation ultimately leading to expansion of continental ice contemporaneously with enhanced melting and contraction of maritime glaciers. The way in which glaciers will respond to future warming depends on their location, since this controls whether enhanced melting or accumulation predominates.

Nunataks apart, the ice-free but periglacial areas of Antarctica have a distinct coastal distribution, with the greatest extent in the Antarctic Peninsula and offshore islands. Landscapes of glacial erosion have an extensive distribution, but most glacial deposition occurs in the sea beneath ice shelves or at the snouts of tidewater glaciers. Low temperatures ensure that surface streams are rare except in the wetter maritime areas, and this pattern is also largely true of freshwater lakes. The geographical distribution of periglacial landscapes, weathering and soils in Antarctica reflects the two basic patterns of weathering environment. Low amounts of moisture and chemical and biological activity are not conducive to soil development in the cold and dry regions, while in the milder maritime Antarctic Peninsula and sub-Antarctic islands high moisture contents allow more chemical and biological activity and the development of soils.

Geographically controlled climatic gradients also affect biological activity and again two fundamental regions emerge: the dry and cold interior is home to sparsely distributed lichens and mosses but few animals; the wetter and warmer maritime Antarctic and the sub-Antarctic islands provide more varied and favoured habitats where vegetation cover can be extensive and, in places, as luxuriant as any vegetation in the Arctic tundra. Nevertheless, it remains that in continental Antarctica, the vigour and diversity of the terrestrial ecosystem is seriously limited by isolation, extremely low temperatures and lack of available water. It is thus highly sensitive to small changes in any of the variables that affect survival. This view of a barren and largely sterile terrestrial environment supporting the most fragile and sparse of ecosystems stands in stark contrast to the view presented in subsequent chapters of the environment and life of the Southern Ocean.

Further reading

Bonner, W.N. and Walton, D.W.H. (eds) (1987) *Key Environments: Antarctica*. Pergamnon Press, Oxford.

Denton, G.H., Prentice, M.L. and Burckle, L.H. (1991) Cainozoic history of the Antarctic ice sheet. In Tingey, R.J. (ed.) *The Geology of Antarctica*. Clarendon Press, Oxford, 365–433.

Doake, C.S.M. (1987) Antarctic Ice and Rocks. In Walton, D.W.H. (ed.) *Antarctic Science*. Cambridge University Press, Cambridge, 140–190.

Everson, I. (1987) Life in a Cold Environment. In Walton, D.W.H. (ed.) *Antarctic Science*. Cambridge University Press, Cambridge, 71–138.

Hatherton, T. (ed.) (1990) *Antarctica, the Ross Sea Region*. DSIR Publishing, Department of Scientific and Industrial Research, Wellington, New Zealand.

Stonehouse, B. (1989) *Polar Ecology*. Blackie, Glasgow and London.

Sugden, D.E., Denton, G.H. and Marchant, D.R. (1995) Landscape evolution of Dry Valleys, Transantarctic Mountains: tectonic implications, *Journal of Geophysical Research*, 100, 9949–67.

Sugden, D.E., Denton, G.H. and Marchant, D.R. (1993) The case for a stable East Antarctic Ice Sheet. *Geografiska Annaler*, 75A (4), 151–351.

CHAPTER 4 The Southern Ocean

Objectives

Key themes and questions covered in this chapter:

- the structure and circulation patterns of the Southern Ocean
- the nature, development and spatial variability of the sea ice cover
- the role of nutrients and sea ice in the growth of phytoplankton
- the 'Antarctic Paradox' and biogeochemical cycling in the Southern Ocean
- the biogeography of zooplankton
- the distribution of Antarctic fish, birds and mammals
- ecosystem response to human predation

4.1 Introduction

The aim of this chapter is to examine the geography, characteristics and functioning of the main physical and biological systems of the Southern Ocean. In common with the Arctic Ocean, the principal result of the polar location of the Southern Ocean is that marked fluctuations in air temperature and insolation drive a strongly seasonal regime in both physical and biological processes. For example, the sea surface is covered for varying parts of the year by a thin layer of floating ice and this affects the timing and duration of the growth period of marine organisms. However, in contrast to the Arctic, the Southern Ocean surrounds a polar continent and is unique amongst the world's oceans in that it encircles the globe in a virtually unbroken swathe of water 2500 km wide. Even at its narrowest point it is still 1100 km wide in the Drake Passage between the Antarctic Peninsula and South America. The Southern Ocean is generally taken to mean the ocean area south of 50° S, so it also includes sub-Antarctic areas. It is traversed between latitudes 50° and 60° S by the Polar Frontal Zone (until recently this was thought to be one frontal zone called the Antarctic Convergence), a major oceanographic and biological boundary that separates the cold Antarctic waters from waters that are up to 4 °C warmer to the north (Macintosh, 1946; Knox, 1994) (Figure 4.1). From the Polar Frontal Zone south, the sea seems to teem with life and some areas of the Southern Ocean, especially the coastal and inshore waters and the Scotia Sea, are biologically very productive (Fogg, 1977).

The geography of the Southern Ocean is influenced by the symmetry of the continent, its ice masses and the atmospheric circulation pattern (see Chapter 2). Because there are no land barriers to interrupt the response of currents to wind forcing, the Southern Ocean circulation displays a consistent circumpolar pattern that is also detectable in its biogeography and, in contrast to the Arctic, most Antarctic marine species have a circumpolar distribution (Knox, 1994). The interaction of cold circumpolar currents and the upwelling of warmer currents from lower latitudes provides an ideal habitat for the growth of phytoplankton (free-floating microscopic plants) and of the zooplankton animal species that prey on the phytoplankton. The phytoplankton and zooplankton form the basis of a food web that is short but also relatively complex at all levels (Bonner and Walton, 1985), and the range of predators includes vast numbers of birds, seals and whales. Attracted by such bounty, sealers and whalers came to the Southern Ocean and were responsible for the near extinction of several species.

4.2 Sea-floor geomorphology of the Southern Ocean

Antarctica is surrounded by a continental shelf of 200 km mean width and 500 m mean depth (see Chapter 2). It is deeper than most continental shelves, and close to the continent this is mostly due to isostatic depression by the ice sheets. From the outer part of the continental shelf, the continental slope extends out to depths of 3000–6000 m, where it gives way to virtually flat abyssal plains. One notable exception to this is the deep and active subduction trench of the South Sandwich Islands which reaches 9000 m. However, several oceanic plateaux and ridges, such as the Falkland Plateau and the Kerguelen–Gaussberg Plateau, rise above the abyssal plain. The plateaux are important because the relatively shallow depth enhances their fisheries potential and the thick covering of sediments makes them potential reservoirs for hydrocarbons. The final significant landform of the Southern Ocean floor lies midway between the continental shelves of Antarctica and the continents to the north. The mid-ocean ridge systems discussed in Chapter 2 all but surround the continent and represent present-day sea-floor spreading sites. Their activity through the last 180 Ma has controlled the evolving geometry of the Southern Ocean and contributed to the development of oceanic circulation: the full opening of a deep ocean seaway south of the South Tasman Rise by 38 Ma ago allowed circulation between the Indian and the Pacific Oceans (Kennett, 1978); and the opening of the Drake Passage at 30–22 Ma BP allowed circumpolar circulation to become fully established in the proto-Southern Ocean.

4.3 Water masses, circulation and floating ice in the Southern Ocean

4.3.1 Water masses and circulation in the Southern Ocean

The circulation of the Southern Ocean reflects the essentially symmetrical shape of the continent and the relatively unobstructed nature of the ocean floor. The only element of equatorward flow is due to the Antarctic Peninsula, which locally deflects flow northwards. Kort (1962) identified the interaction of four main water masses within the circulation pattern (Figure 4.1). Water cooled to 0 °C by both the coastal ice masses of the continent and the cooling effect of sea ice development sinks and flows down the Antarctic continental shelf and then north as Antarctic Bottom Water (see Figures 1.2 and 4.1). An equal volume of warmer (0.5–2.0 °C) and nutrient-rich Circumpolar Deep Water flows southwards from the world's warmer oceans to replace it (Gordon, 1988) and upwells at the Antarctic Divergence. On the surface, low-salinity Antarctic Surface Water chilled to − 1 °C by contact with the cold air and ice of the continent flows north. At its northern extent, the Antarctic Surface Water meets and sinks below the warmer southward-moving Sub-Antarctic Water, giving rise to abrupt falls in sea temperature of up to 4 °C. If the weather is calm and the sea smooth, this zone is often marked by a fog bank or a faint zone of turbulence in the water (Stonehouse, 1972). The Antarctic Surface Water continues north at intermediate depths as Antarctic Intermediate Water. It is detectable north of the equator in the Atlantic and it also cools the Indian and Pacific Oceans, emerging as coastal water in Australia and New Zealand. From a global perspective, the redistribution of heat that occurs as warm water from the world's warmer oceans is exchanged with cold water produced in the Southern Ocean is crucial to the maintenance of the global heat balance. Of equal importance to the productivity of the world's oceans are the exchanges of nutrients that occur as the nutrient-rich warm water upwells and mixes in the Southern Ocean, and as the mineral-rich cold Antarctic waters extend north into the world's oceans. This interchange of heat and nutrients lies at the heart of the oceanic ecosystem.

The circulation pattern of the various water masses is driven not only by temperature and salinity differences but also by the atmospheric circulation around the continent. Winds from the west produce an important eastward flow of water called the Antarctic Circumpolar Current (ACC) (Figure 4.2). The ACC is the largest ocean current system on Earth, transporting about 1.3×10^{8}t s^{-1} of water (several times larger than the Gulf Stream). The ACC is bounded by the Sub-Antarctic Front (SAF) on the northern side and by the Antarctic Polar Front (APF) on the southern side. A third front, the Southern Polar Front, exists in some regions of the ACC. The fronts contain fast-flowing oceanic jets of water that extend to the sea floor and, although

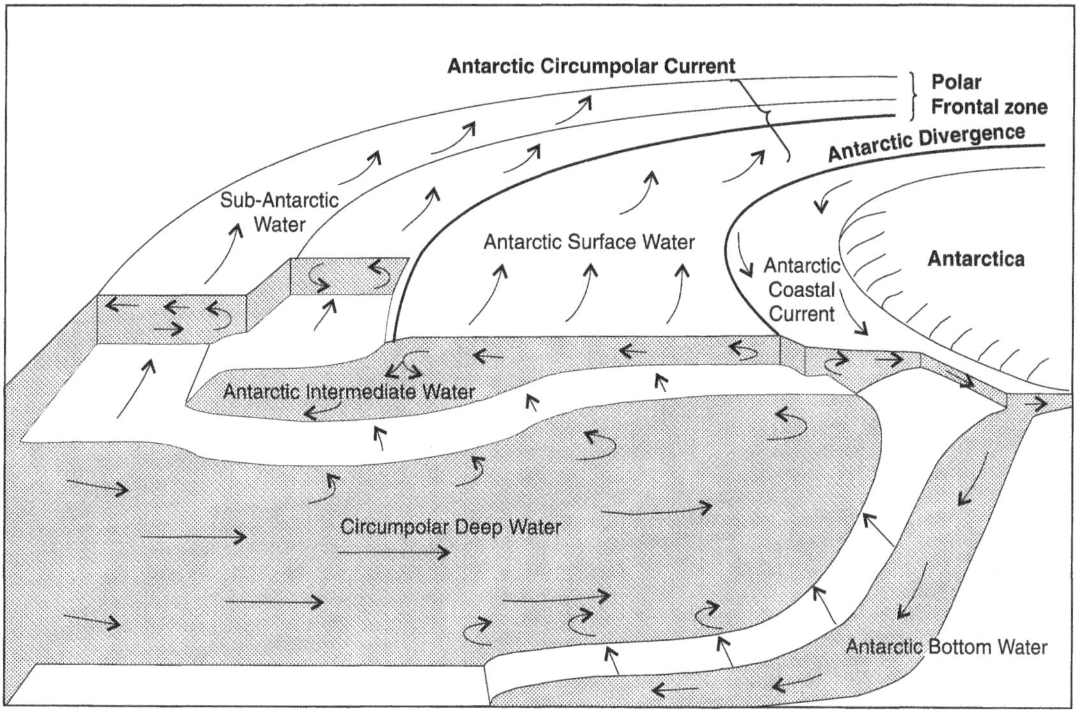

Fig 4.1 Schematic diagram of Antarctic water masses and the meridional and zonal flow of water (source: modified after Kort, 1962).

accounting for only a small part of the cross-sectional area of the ACC, they are responsible for the majority of its transport (Klinck and Hoffman, 1986). The northern flank of the Antarctic Circumpolar Current (see Figure 4.1) in particular is characterised by fast-flowing 30–100 km wide eddies, meanders and rings, which may also migrate up to 100 km in 10 days as suggested by various forms of satellite imagery (Nowlin *et al.*, 1977) and mathematical models (Webb *et al.*, 1991) (Figure 4.3). It is important to recognise that the SAF, the APF and the intervening water of the ACC are now collectively known as the Polar Frontal Zone.

The combined effect of the Earth's rotation and friction on such oceanic currents is to produce a deflection of water, known as Ekman transport, to the right of the wind direction in the Northern Hemisphere and to left of the wind direction in the Southern Hemisphere (Gross, 1977). Ekman transport in the ACC produces a northeastward tendency to the surface circulation away from the continent (see Figure 4.1). A spectacular confirmation of this outward and eastward flow was provided in March 1962, when a submarine eruption close to Zavodovski Island, in the South Sandwich Islands, released millions of tons of pumice into the sea. The pumice began to reach Macquarie Island, Tasmania and New Zealand in late 1963, having travelled east at 11–17 cm s^{-1} (Deacon, 1984). Closer to the continental shores of Antarctica, katabatic drainage winds from the ice sheet are leftwards deflected by the Earth's rotation (Chapter 2) and this forces a westward-flowing Antarctic Coastal Current, or east wind drift, with a southwesterly onshore component (see Figures 4.1 and 4.2). This divergence in surface winds approximately coincides with a divergence in currents at the Antarctic Divergence and results in the replacement of the water moving away by the upwelling of warm and nutrient-rich subsurface Circumpolar Deep Water (*ibid.*) (see Figure 4.1)

Although the circulation of the currents in the Southern Ocean is fairly symmetrical around the continent, ocean-floor geomorphology and coastal outline play a role not only in controlling the position of the boundaries of currents but also by introducing circulatory currents (gyres), which have

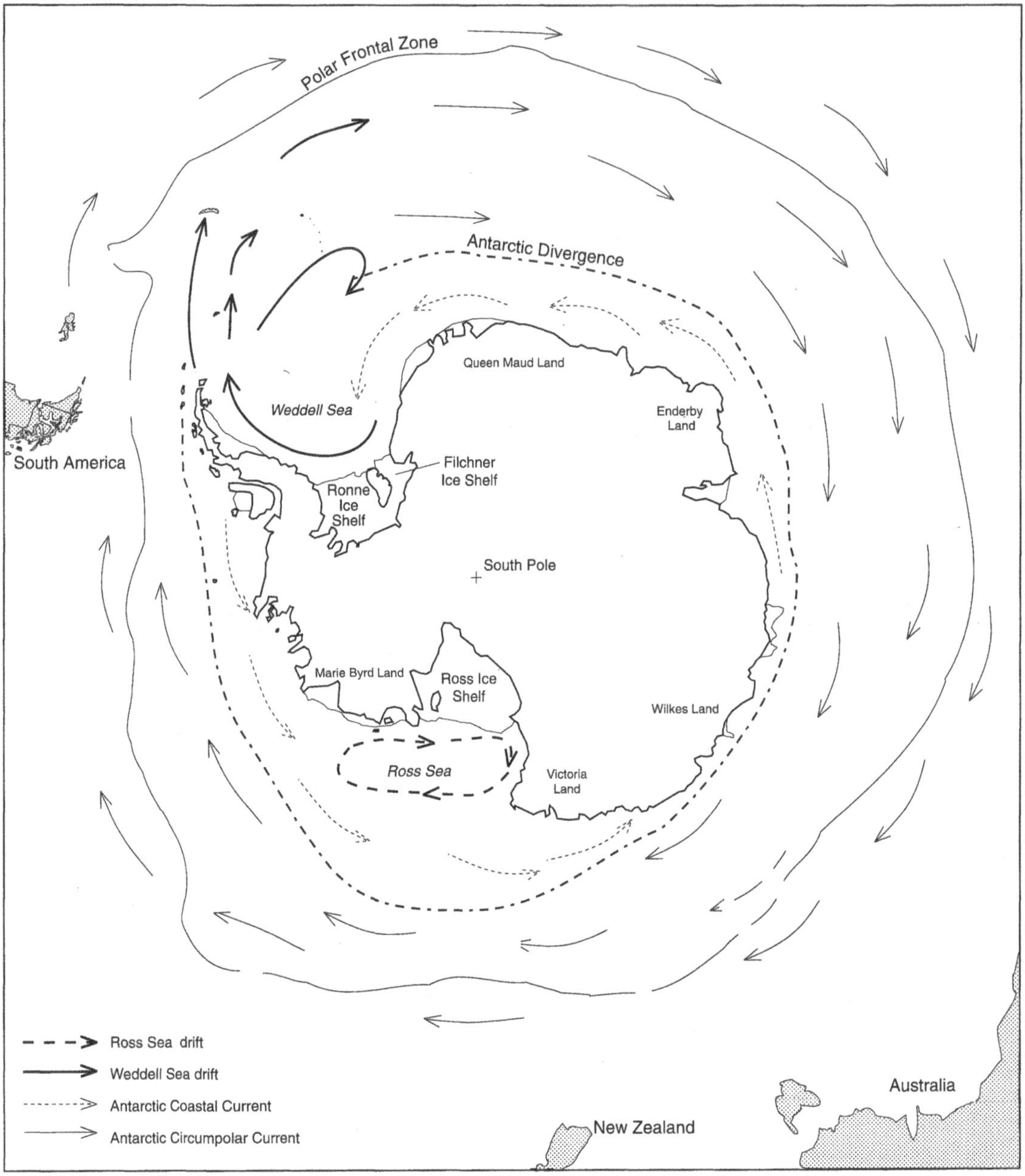

Fig 4.2 The general pattern of water circulation in the Southern Ocean and the location of the Polar Frontal Zone and Antarctic Divergence (drawn from various sources).

north–south components (Gordon, 1988). For example, ocean-floor bathymetry affects both the limits and patterns of the ACC, as demonstrated by the FRAM model (see Figure 4.3). In most areas, the position of the Polar Frontal Zone is approximately stable and appears to be related to the location of submarine ridges or escarpments. South of Australia for example, the position follows the line of the mid-ocean ridge (Deacon, 1984). In the Drake Passage, the ACC is highly constricted within the fracture

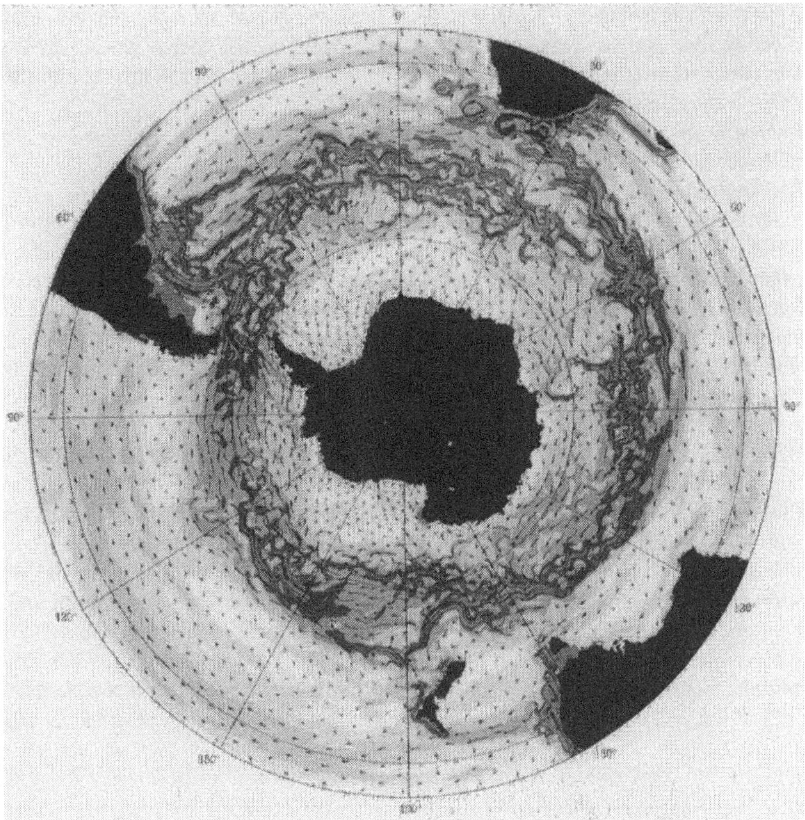

Fig 4.3 A spectacular image of the modelled velocity fields in the surface layers of the Southern Ocean, as modelled by the Fine Resolution Antarctic Model (FRAM). The large meanders and eddies in the region of the Polar Frontal Zone are particularly clear. These meanders extend to the sea floor (source: reproduced with permission from the FRAM Atlas of the Southern Ocean, NERC).

zones of the southeast Pacific Basin (Gordon, 1988). To the east of the Drake Passage, temperature measurements suggest that the position of the Polar Frontal Zone has not altered since it was surveyed by James Clark Ross 150 years ago and is probably controlled by the Falkland Plateau and Burdwood Bank (Deacon, 1984).

Bathymetry and coastal topography also control the location of several large gyres, notably in the Weddell and Ross Seas (see Figure 4.2). The largest of these was revealed in January 1915, when Shackleton's *Endurance* became trapped in Weddell Sea ice close to the eastern end of the Filchner Ice Shelf. Throughout the winter the ship drifted west and then north with the frozen sea ice until crushed in October close to the Antarctic Peninsula. The crew drifted north for a further six months on an ice floe to within 80 km of the South Shetland Islands, where they made a successful landing (Shackleton, 1919). The 8 cm s^{-1} current that had transported the expedition to safety was the Weddell Sea Drift, a part of the Antarctic Coastal Current that is deflected north by the Antarctic Peninsula to join the ACC in the vicinity of the South Shetland Islands (Deacon, 1984). The Ross Gyre is a smaller clockwise flow caused by the deflection of the Antarctic Coastal Current as it meets the western side of the Ross Sea embayment (*ibid.;* Wadhams, 1991) (see Figure 4.2). A more poorly formed gyre exists east of the Kerguelen Plateau (Knox, 1994).

In addition to redistribution of heat and cold, the biological importance of the water masses, oceanic circulation pattern and its various gyres, upwellings and eddies lies in the way in which nutrients are exchanged and distributed between the Southern Ocean and its neighbours. Reflecting their source areas, the concentration of nutrients, such as nitrate and phosphate, are lowest in the surface waters flowing from the Antarctic continent and highest in the Circumpolar Deep Water flowing from the lower latitudes, while the concentration of silicates is highest in the bottom waters that flow down the continental slope (*ibid.*). The exchange benefits the nutrient status and thus biological productivity of all the Southern Hemisphere oceans. However, in general the concentrations of nutrients in the surface waters south of the Polar Frontal Zone are much higher than those found in other oceanic waters (El-Sayed, 1978; Knox, 1994). Such nutrients are upwelled at the Antarctic Divergence and then flow north along the surface to downwell at the Polar Frontal Zone. In other regions of upwelling these nutrients are utilised by marine plants or phytoplankton for growth and are stripped out of the top 50–100 m of water. In the Southern Ocean, concentrations of nitrates and phosphates do not appear to be depleted by the time the waters reach the Polar Frontal Zone, and only fall to low values to the north of it (*ibid.*). Silicate concentrations, however, drop to low values in the vicinity of the Polar Frontal Zone and so may limit the growth of some phytoplankton species. The biological significance of the apparent surfeit of nutrients is discussed later in this chapter (Section 4.4.3).

4.3.2 Sea ice

Climatically, one of the most important features of the Southern Hemisphere is the dramatic seasonal variation in the extent of the floating sea ice that forms a lid on the surface of the ocean. During the summer, the sea ice covers 3.5×10^6 km^2 in February and is largely confined to the coast but, influenced by falling winter temperatures and the development of cold surface winds blowing out from the polar vortex (see Chapter 2), the sea begins to freeze in a girdle that progressively encircles the continent, eventually to cover 19×10^6 km^2 in September (Zwally *et al.*, 1983; Gloerson *et al.*, 1992) (Figure 4.4). This seasonal sea ice zone extends up to 22 00 km from the coast and is the entire area bounded by the winter maximum and summer minimum extents. Within this, the marginal ice zone (MIZ) is the 100–200 km wide complex interface between the sea ice and the open ocean, a zone that moves with the regional location of the seasonal ice edge. The reduced

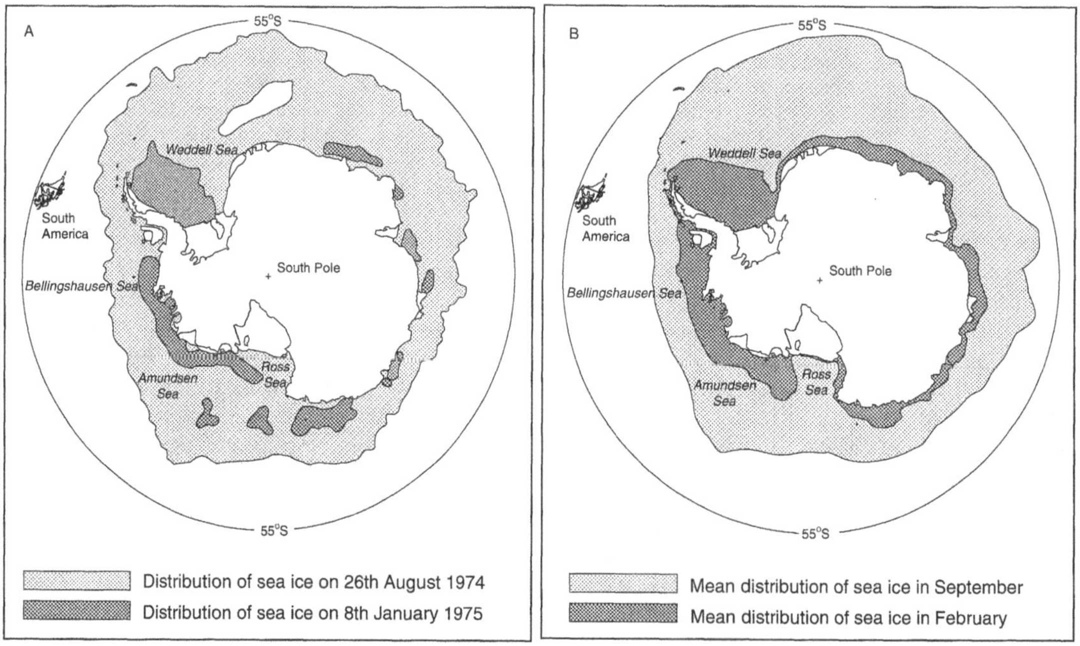

Fig 4.4 Seasonal variation in sea ice distribution and extent as shown from satellite passive microwave observations. (A) 1974–1975: note the extent of the Weddell Sea polynya, which occured in the mid-1970s; and (B) averaged over 1978–1987 (source: after Zwally and Gloerson, 1977; Gloerson *et al.*, 1992).

light penetration and increased albedo of the frozen ocean surface has the effect of further depressing temperatures and imparts a stronger seasonality to ocean life than would otherwise occur. Furthermore, an ocean seasonally covered with ice also has much reduced heat and moisture exchanges with the atmosphere, sunlight is reduced and photosynthesis is curtailed, affecting biological production at all levels (Clarke, 1985). The thermal regime of the upper layers of the Southern Ocean is dominated by the cooling effect of melting sea ice, and this has a profound effect on the climate of the Southern Hemisphere.

Freeze-up and break-up of sea ice

The onset of the Antarctic winter is heralded by the sea beginning to freeze first in small calm bays and inlets, then eventually spreading to more exposed waters (Figure 4.5). No amount of surface agitation by wind or wave can prevent freezing, and each winter the sea develops a thin, *ca* 1 m, lid of floating ice. Freezing point is dependent on salinity, and while fresh water freezes at 0 °C, water of a salinity of around 34 parts per thousand (‰) freezes at –1.91 °C. At the surface salinities normally found in the Southern Ocean (33.8–34.4‰ (Webb *et al.*, 1991)), the density inversion characteristic of the freezing of fresh water does not take place and the water reaches freezing point over a depth of about 2 m more or less simultaneously. The freezing process also releases salt and new sea ice has a salinity of only 4–6%. As the ice thickens, brine trapped between the crystals drains by gravity, meltwater flushing and thermal expulsion so that first-year ice and multi-year ice, respectively, have lower brine contents than new ice (Gow *et al.*, 1982). One of the few benefits to mariners trapped within multi-year ice is that it is almost salt-free and perfectly drinkable.

The normally accepted pattern of freezing is for sea ice growth to commence as congelation ice (Table 4.1) and pass through a well-known series of stages. At first, a suspension of small ice crystals or platelets called frazil ice forms in the water column and floats to the surface to form a thin layer of randomly oriented crystals only a few centimetres thick (Wadhams, 1991). These soon freeze together into a skin called ice rind or nilas, and long columnar crystals of ice begin to form beneath. The constant movement of water and wind break the thin layer into pancakes a few centimetres in diameter with upturned edges due to collisions. As more crystals form, the pancakes begin to coalesce to 3–5 m in diameter and 50 cm thickness, and this dampens further movement (Figure 4.6).

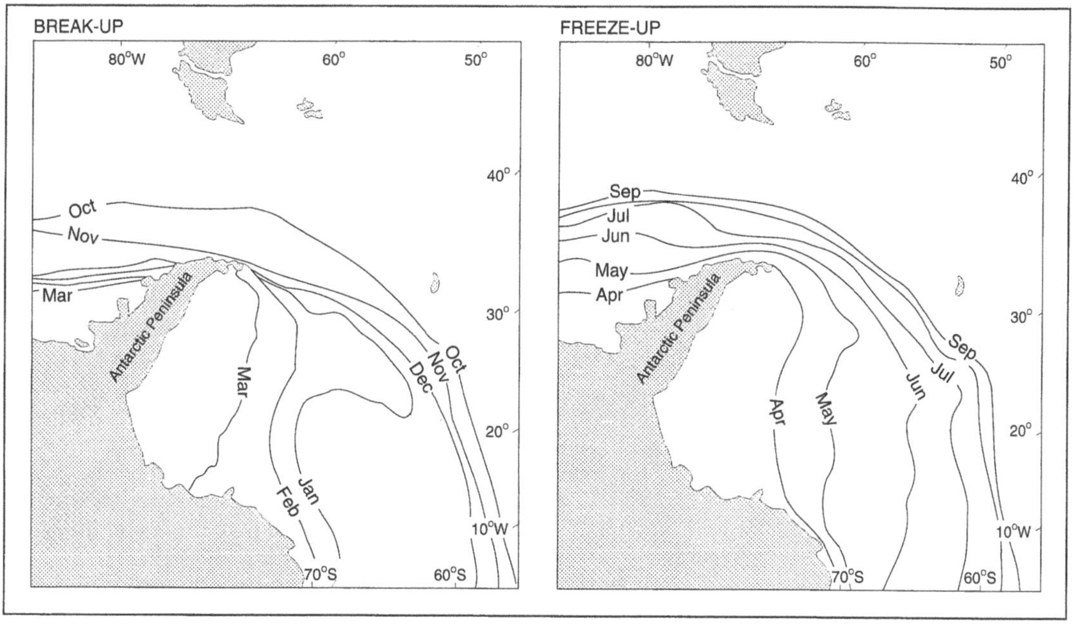

Fig 4.5 The progress of freeze-up and break-up in the Antarctic Peninsula area, shown by the average extent of sea ice mapped from the Antarctic Ice Charts 1973–1982, Naval Polar Oceanography Center, Washington DC (source: Hansom and Kirk, 1989).

Table 4.1 Sea ice terminology

Sea ice	ice formed at sea by the freezing of sea water
Congelation ice	a coherent layer of frozen sea water
New ice	recently formed ice
First-year ice	sea ice of not more than one winter's growth
Old ice	sea ice that has survived at least one summer's melt
Frazil ice	platelets of ice suspended in sea water
Ice rind or nilas	a thin elastic crust of sea ice a few cm thick
Pancake ice	predominantly circular pieces of ice with raised rims due to pieces striking against each other
Anchor ice	submerged ice attached to the bottom
Infiltration ice	ice formed by sea water seeping into an existing layer of sea ice, often as a result of depression of this ice by the weight of surface snow
Fast ice	sea ice that forms and remains frozen to the shore
Pack ice	used in a wide sense to include any sea ice other than fast ice
Ice floe	piece of floating ice, 10 m–10 km across
Pressure ridge	ridge of broken ice caused by collision of floes, usually of old ice and with ice keel below water
Polynya	area of open water in sea ice
Lead	narrow passage through sea ice

Source: Armstrong *et al.*, 1973; SCAR, 1994

Fig 4.6 First-year sea ice in the Weddell Sea, 1904. Photographed from the mast of the *Scotia* during the Scotia Expedition, led by W.M. Bruce, 1902–1904 (source: courtesy of Glasgow University Archives and Business Record Centre).

However, recent studies in the Weddell Sea show that the above sequence may not be the principal way that ice is generated in the Antarctic, since most of the ice consists not of columnar ice but of randomly oriented frazil crystals (Gow *et al.*, 1982). The high energy and turbulence of the wave field in the Southern Ocean forms a dense suspension of frazil ice, which wave compression then congeals into pancake ice (Wadhams, 1991). In this process, brine is trapped throughout the pancakes, which are then frozen into groups, but, in the Weddell Sea, overall freezing is not achieved until the wave field is damped sufficiently. The top surface of such pancake ice is rough and jagged, unlike the smoother surfaces of ice formed in calmer waters (Wadhams *et al.*, 1987). Pancake ice in the Weddell Sea is only about 60 cm thick, but in the Ross Sea heavy winter snow adds 2–4 m to its thickness, leading to snow loading of the thin ice layer and resultant partial sinking. Seepage of sea water and the formation of infiltration ice is the result.

The springtime break-up and melting of the Antarctic sea ice happens first in the north (Heap, 1965) but progresses south as temperatures rise (see Figure 4.5). Small pockets of trapped brine melt out first and help to both raise the temperature and lower the albedo of the adjacent ice. Early in the thaw, snow cover ensures that a large amount of the incoming solar radiation is reflected, but as the spring advances, and the hours of sunshine increase, surface melt pools reduce the surface albedo from 90 to 40% (Sugden, 1982). The ice temperature gradually rises to melting point, and since this process weakens the coherence of the sea ice, waves and winds along the ice edge become progressively more successful in penetrating and fragmenting the ice. The process is aided by the northeasterly component of the ACC, which constantly spreads the sea ice into lower latitudes, where it melts. In this way, the vast majority (about 87% (Goody, 1980)) of the Antarctic sea ice breaks up and melts each spring and summer and so by definition it is mainly first-year ice. Only a small proportion survives to become multi-year ice.

In the areas least affected by the ACC, or where a semi-enclosed gyre occurs, quantities of ice survive the melt and break-up to become multi-year ice. This occurs in the Weddell Sea, the Ross Sea and to a lesser extent the Bellingshausen Sea. Multi-year ice differs from its younger counterpart in that it is characterised by a confusion of pressure ridges, formed when adjacent floes collide. These ridges and corresponding keels may reach 6 m draught in the Antarctic (Wadhams, 1991) (Figure 4.7), but they rarely reach the 10 m high ridges and 30 m deep keels encountered in the Arctic, where multi-year ice is common as a result of the enclosed nature of the ocean (Weeks *et al.*, 1989). Weddell Sea multi-year ice tends to be found only in the

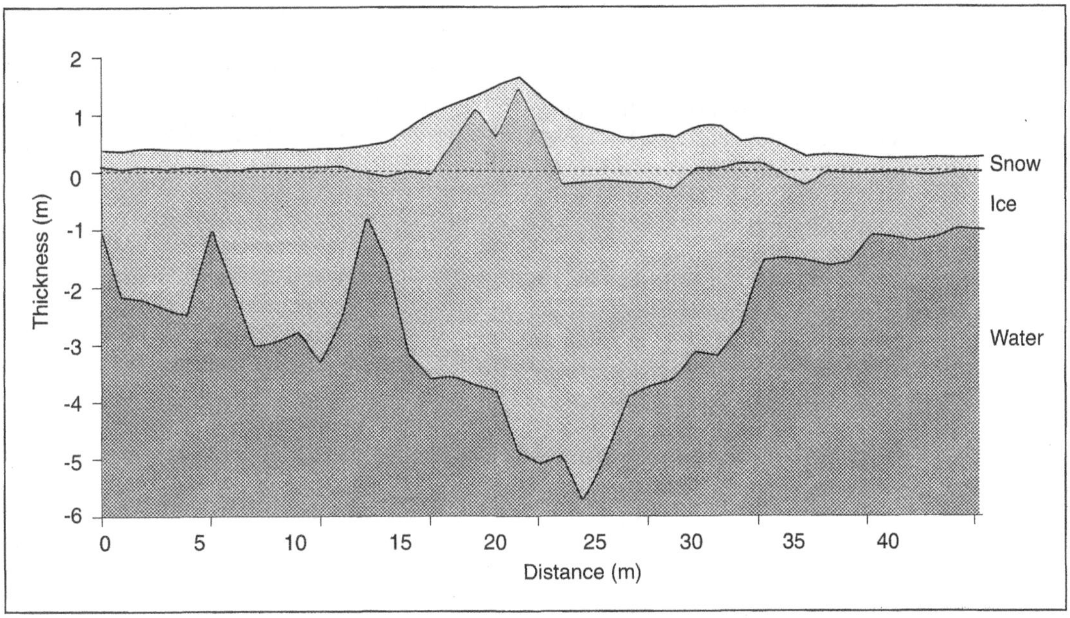

Fig 4.7 Cross-section through a typical Antarctic pressure ridge in the western Weddell Sea (source: redrawn after Wadhams, 1994).

northward drift west of about 40° W as the ice is forced by the Weddell Sea Gyre against the east coast of the Antarctic Peninsula. Even where it remains undeformed, its average 1.17 m thickness may be twice the first-year ice values found in the eastern Weddell Sea (Wadhams and Crane, 1991). It may be no coincidence that Shackleton's *Endurance,* having successfully weathered the winter held within a frozen lead in first-year ice, was finally crushed by a developing pressure ridge in October as the ice that carried the vessel was forced up against the Antarctic Peninsula. Other vessels beset in Antarctic sea ice have avoided this fate because they were not subject to pressure ridges. Filchner's *Deutschland* was beset in Weddell Sea ice in 1912, but with a track well to the north and east of *Endurance*, multi-year ice was avoided and the ship survived. The absence of a major barrier to ice dispersal similarly aided de Gerlache's *Antarctic* in 1898–1899. Frozen within the ice of the Bellinghausen Sea, the vessel was finally swept west in the summer of 1899–1900 to be safely released within the outward-flowing pattern of the ACC well to the west of Peter I Øy (Heap, 1965). Bruce's *Scotia* became beset in Weddell Sea ice in the late summers of 1903 and 1904, although it was successful in breaking free before winter (Figure 4.8).

Fig 4.8 Scotia beset in sea ice at 74° S in the Weddell Sea, April 1904. In October 1915, Shackleton's *Endurance* was lost in pressure ridges in the western Weddell Sea (source: courtesy of Glasgow University Archives and Business Record Centre).

Extent, variability and distribution of sea ice

Complete spatial and temporal coverage of Antarctic sea ice extents has been available since early 1973, when satellites began to transmit Scanning Multichannel Microwave Radiometer (SMMR) passive microwave data and, since 1987, Special Sensor Microwave/Imager (SSM/I) data (Hanna, 1996). Growth of the sea ice is rapid between March and July, with the full winter girdle of ice eventually doubling the ice-covered area of the Antarctic region (see Figures 4.4 and 4.5). Yet even when the sea ice is at its maximum extent there exists 18–21% by area of ice-free ocean within leads and polynyas (Gloerson *et al.*, 1992). Two types of polynya occur: open-ocean and coastal. A very large open-ocean polynya occurred in the Weddell Sea during three consecutive winters of 1974–1976 (Zwally *et al.*, 1983) (see Figure 4.4). Over this period the polynya migrated, so that its 1976 position barely overlapped with its 1974 position. Measuring 1000 × 350 km at its largest, the Weddell polynya's original appearance is thought to be related to the upwelling of warm subsurface water over a seabed topographic high known as the Maud Rise. This prolonged deep-ocean convection kept the area ice-free. Its termination in 1978 may have been achieved by enhanced convergence of the sea ice, which on melting led to a weakening of the convection (Comiso and Gordon, 1987).

Coastal polynyas are smaller and more frequent and regular in occurrence since they exist at several well-defined coastal locations where topography favours katabatic winds. These winds blow sea ice away from the coast as fast as it can form, re-exposing the surface to allow more freezing and so removing large amounts of latent heat from the surface waters (Figure 4.9). Coastal polynyas have been likened to sea ice factories that manufacture ice on a grand scale: a small (50 × 50 km) recurring polynya in Terra Nova Bay in the Ross Sea (Kurtz and Bromwich, 1985) is thought to produce 10% of the sea ice generated over the entire Ross Sea area. While it is clear that coastal polynyas are sites of enhanced ice production, they also play an important role in the production of the saline, cold and dense water that comprises Antarctic Bottom Water (Zwally *et al.*; 1985; Gordon, 1988; Wadhams, 1991) and are of great biological importance (see Section 4.4) (Massom, 1988).

The break-up of the sea ice cover is at its most rapid from November to January, and by February the only major areas of ice remaining are in the western Weddell Sea, the southern Bellingshausen and Amundsen Seas and the southeastern Ross Sea, although a narrow fringe of ice also exists around most of the rest of the continent (Parkinson, 1992) (see Figure 4.4). In contrast to the cycle of Arctic sea ice, there is a marked asymmetry in the development cycle of Antarctic sea ice, with growth being much slower than decay. The Ross Sea sector shows more rapid growth than other regions in Antarctica (Doake, 1987), perhaps related to rapid ice production at coastal polynyas. It also shows a tendency early in the summer season to be swept clear of sea ice by strong and persistent katabatic winds draining from the interior. Over the years, this characteristic has provided

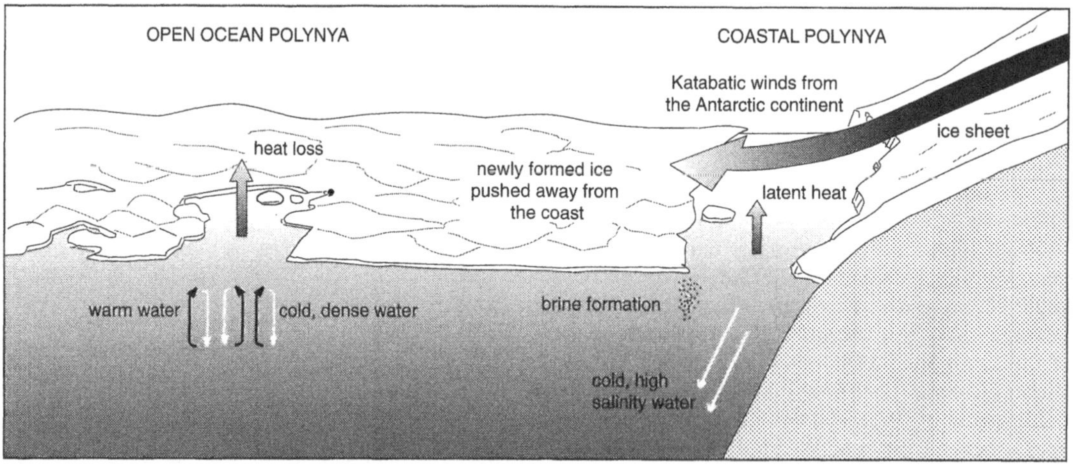

Fig 4.9 Different types of polynya perform several roles in the production of both sea ice and Antarctic Bottom Water (source: redrawn after Open University, 1989).

the ice-free water access to the continent so vital to early exploration and the support of science.

Interannual variability in sea ice extent is an important, but as yet poorly understood, consideration in the context of global climate change. Gloerson *et al.* (1992) found a 5% natural interannual variability in winter sea ice extent in the Antarctic that was enhanced near the ice edge. Using SMMR data, Gloerson and Campbell (1988) found no significant change in the amount of Antarctic sea ice cover but a reduction in Arctic sea ice cover between 1978 and 1987. In spite of the continually expanding length of the satellite database, statistical analyses of both SMMR and SSM/I data have so far failed to detect any overall trend in Antarctic sea ice extent in spite of some small local trends (Gloerson and Campbell, 1991; Johannessen *et al.*, 1995). However, using whaling records, de la Mare (1997) identified an abrupt 25% decline in sea ice extent between the 1950s and 1970s, apparently undetected by satellite analysis. It is also known that in some areas, such as in the Bellingshausen Sea, the large interannual variability in sea ice extent has a profound effect on the climate of the western Antarctic Peninsula and is likely to be linked to atmospheric circulation (Wadhams and Davis, 1997). The detailed analysis of sea ice extent and interannual variability is important since it provides baseline data from which to assess future changes in extent consequent upon climatic change. If future sea ice edges remain within their past limits then claims of climatic change affecting sea ice might be difficult to sustain, whereas changes in areas of low variability of sea ice cover might be of more significance (Parkinson, 1992). The possible effect of changing extents and variability of sea ice on climate is discussed further in Chapter 5.

4.3.3 Icebergs

Amidst the Antarctic sea ice and in the surrounding ocean it is not uncommon to sight larger blocks of floating ice that have broken off from the coast of the continent. These are distinguishable from sea ice not only on account of their sheer size and height, although small pieces of icebergs lie close to the water, but also on account of their colour and density. Icebergs are pieces of ice calved from the snouts of tidewater glaciers or ice shelves, so they are composed of clear or translucent freshwater ice together with any of the associated debris that the parent glacier may have been transporting. Most Antarctic icebergs begin as great tabular pieces calved from the fronts of ice shelves. As they progress away from the continent under the action of wind and currents, melting occurs below the water line, causing them to tilt or keel over completely (Deacon, 1984). Many break along lines of weakness or crevasses inherited from the parent glacier, and eventually large tabular icebergs are reduced to water. The longevity of icebergs depends on the nature of the parent ice shelf and its proximity to warm ocean currents. Short-lived ones tend to be heavily crevassed or located where access to the ACC is unrestricted. The longest-lived and largest icebergs are usually tabular in shape and associated with the Weddell Sea Gyre, but other important source areas are the Ross Sea and the Amery Ice Shelf (Doake, 1985, 1987). Icebergs in the Weddell Sea follow the westward-moving Antarctic Coastal Current before turning north, eventually to join the eastward-moving ACC, where they break up and melt. The total mass of icebergs calved into the Southern Ocean may be as high as 2000–3000 Gt a^{-1} (Orheim, 1985), so it is not surprising that they have been suggested as good sources of fresh water for those countries where it is in short supply; this aspect is covered further in Chapter 7.

4.3.4 Summary

It is clear that both the distribution and spatial interaction of Antarctic atmospheric and topographic factors fundamentally affect the distribution and extent of ocean currents, sea ice and icebergs within the Southern Ocean. On the one hand, there is an outflow of cold water at depth in the Antarctic Bottom Water and on the surface in the ACC. The surface outflow annually carries with it 87% of the sea ice produced over the winter, together with large numbers of icebergs. The remaining sea ice and water circulates within the katabatically driven Antarctic Coastal Current and may be swept into topographically controlled cyclonic gyres in the Weddell and Ross Seas, eventually to be returned to the outward flow. On the other hand, the cooling influence on the world's oceans and atmosphere is counterbalanced both aloft and at depth by the inflow of warm air and water from lower latitudes. The interchange of heat and water is fundamental to the world's energy balance, but with it comes an interchange of minerals and nutrients held within the water column that fuels the Antarctic marine ecosystem.

Since most of the sea ice is lost annually, the incidence of the thicker and deformed multi-year ice is much more restricted than in the Arctic. An important feature of the Southern Ocean is the pattern and variability of the seasonal sea ice zone. The presence or absence of sea ice is important to the climate and the timing of biological activity in the Southern Ocean. On a regional scale, the absence of sea ice in the coastal polynyas allows the sea surface to be exposed to cold winds, allowing the removal of heat and the production of large amounts of sea ice. Such areas of open water are of great biological importance.

4.4 The Antarctic marine ecosystem

4.4.1 Introduction

At the core of the Antarctic marine ecosystem is the production of phytoplankton, freely floating microscopic plants that form the first level of the marine food web and the basis upon which all other species directly or indirectly depend for their food (Figure 4.10). Phytoplankton that remains uneaten by grazing zooplankton falls to the seabed to be consumed and utilised by benthic grazers. At the second level and grazing on the plant growth of the first level are the zooplankton, a group dominated by the copepod crustaceans. Copepods account for 73% of Antarctic mesoplankton biomass (El-Sayed, 1985) and are followed in abundance by the chaetognaths (arrow worms) and the euphausiids (krill). Although other species of zooplankton are locally more abundant, especially in the northern parts of the ACC, the shrimp-like krill demands attention since it forms the most important group biologically and economically in the Antarctic marine ecosystem. Vast numbers of krill occupy an important position in the food web and contribute to the nutrition of much of the predator biomass. The productivity and distribution patterns of phytoplankton, and of the Antarctic krill, are thus crucial to the stability of the food web that they support. Much of the recent research focused on these and other species in the Southern Ocean has come from the international BIOMASS (Biological Investigations of Marine Antarctic Systems) programme, which was started in the mid-1970s to elucidate the biology and productivity of the key species of the marine ecosystem.

There appear to be three main pelagic zones in the Southern Ocean: the ice-free zone of open water rich in nutrients but relatively poor in primary production; the seasonal sea ice zone rich in phytoplankton and zooplankton, and high in primary production; and the permanent sea ice zone rich in benthic feeders and fish. Specific adaptations of organisms to the Antarctic marine environment and the sensitivity of these environments to disruption and change are discussed in more detail in Chapter 5.

4.4.2 Phytoplankton

Phytoplankton use light energy to fix inorganic carbon and nutrients in the surface waters by photosynthesis and so provide an important link in the movement of CO_2 between atmosphere and ocean. The cold Antarctic surface waters have a high capacity to absorb gases from the atmosphere and become a sink for CO_2. Ocean current circulation patterns also enrich these waters with nutrients, particularly at zones of mixing and upwelling (Treguer, 1994). However, the photosynthesis performed by Antarctic phytoplankton also accounts for a significant, but unquantified, part of atmosphere–ocean CO_2 exchange. Until recently, this was thought to be dominated by the larger, more conspicuous and robust diatoms, with only smaller numbers of small-celled algae, but there is evidence that these other groups may be more important to the total productivity of the Southern Ocean than previously thought (El-Sayed, 1985; Knox, 1994).

All of these unicellular algae appear to be circumpolar in distribution, but dinoflagellates are dominant north of the Polar Frontal Zone and, in general, diatoms are dominant to the south (El-Sayed, 1985). The distribution of phytoplankton is related to the ocean currents and the location of the sea ice margin. Early work by Hart (1942) demonstrated that marked spatial and seasonal variation occurs in phytoplankton production, although pigment values give only a sketchy index of biomass and production values (Figure 4.11). Early work also seems to have underestimated the importance of the timing and mode of retreat of sea ice. The onset of maximum production changes from early spring to late summer with increasing latitude, and the period of maximum production decreases with latitude, although variations in oceanographic and, in particular, sea ice conditions affect the timing and magnitude of the peaks (Knox, 1994). At the Antarctic Divergence, upwelling of warmer, nutrient-rich Circumpolar Deep Water contributes to luxuriant phytoplankton growth, which is also related to increasing light availability in

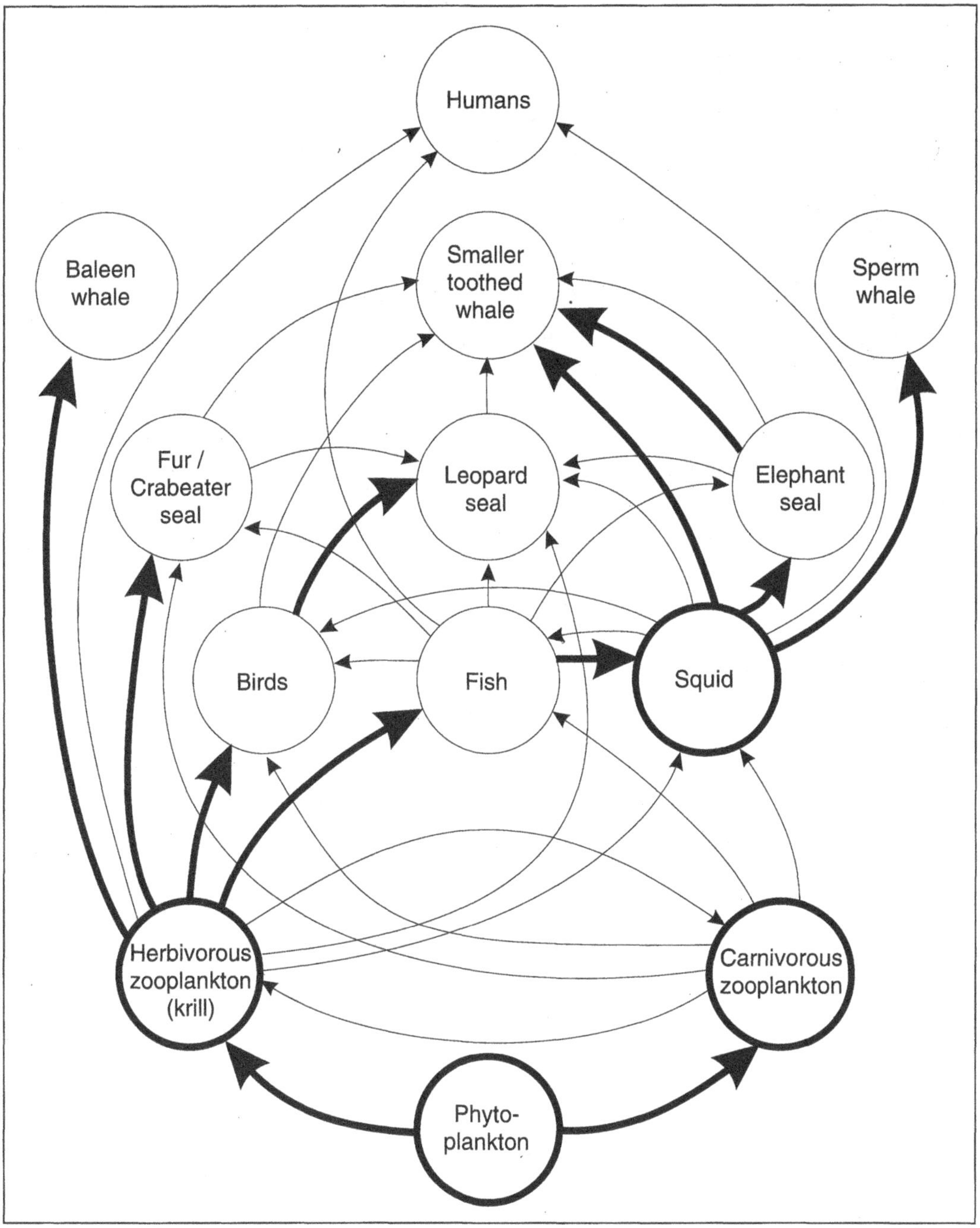

Fig 4.10 Main elements of the Antarctic marine food web. The solid lines represent the more important links in the food web (source: after Mackintosh, 1965).

the early summer and the break-up of the sea ice (El-Sayed, 1985; Hempel, 1985). As in the open ocean, production within coastal bays is closely related to the break-up of sea ice and the penetration of sunlight (see Figure 3.35). Using data gathered weekly at Signy Island between 1988 and 1994, Murphy *et al.*

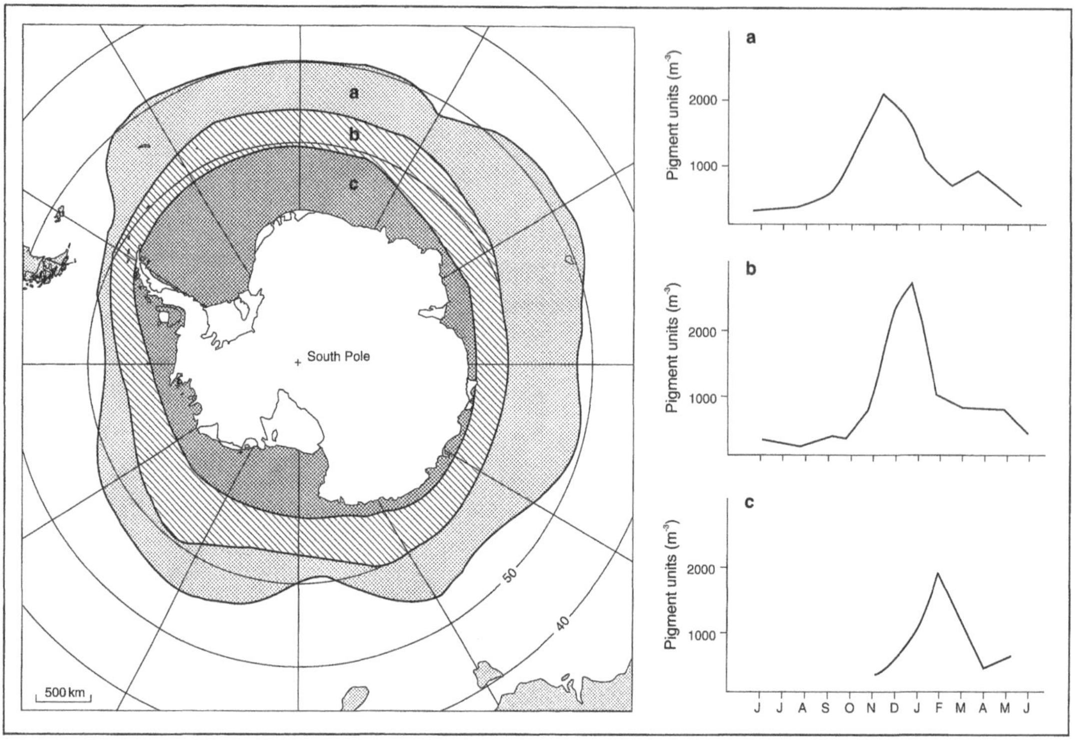

Fig 4.11 The geographical zonation of phytoplankton in the Southern Ocean: (a) the northern zone from 50°–55° 50' S; (b) the intermediate zone from 55° 50'–66° S; and (c) the southern zone. Seasonal changes in plant pigments show the delay in phytoplankton biomass peaks further south (source: modified from Hart, 1942; Knox, 1994).

(1995) and Clarke and Leakey (1996) clearly demonstrate not only the occurrence of marked seasonality in phytoplankton production but also the occurrence of significant year-to-year variability. Interannual variability occurs at Signy Island in seawater temperature, winter sea ice, phytoplankton production and nutrient utilisation, which may be driven by subdecadal cyclicity in atmospheric and oceanographic processes (*ibid.*).

The potential causes of phytoplankton blooms in the MIZ appear to be fundamentally related to the production of a stable surface layer of low salinity and less dense meltwater within which light levels are favourable to growth. Three zones of phytoplankton growth have been identified by Sullivan *et al.* (1988): under the sea ice, growth is low as a result of low light levels and the presence of a deep mixed layer of water; in the vertically stable area just seaward of the ice edge, growth reaches the maximum possible for the temperature; and with increasing distance from the ice edge, vertical stability is reduced by both increased vertical mixing and lateral inflow, so that growth again becomes limited by light in the deep mixed layer (Figure 4.12A). These ice edge 'blooms' are controlled by two principal factors. The first relates to the growth and decline of the ice edge phytoplankton, which occur as a result of factors that affect not only growth, such as light and vertical stability, but also losses such as grazing, vertical mixing, advection and sinking. The second relates to the movement of the bloom over wide geographic areas over the few months that the ice edge is forced to recede as a result of atmospheric influences (Sullivan *et al.*, 1988).

Of vital importance to primary productivity is the microalgae community that grows on the lower surface and within the ice layers of the underside of the sea ice. This 'sympagic' or sea ice microbial community consists of ice algae and zooplankton that have become frozen into the developing platelets and frazil ice at the onset of winter and are released in summer as the thaw and break-up commences (Spindler and Dieckmann, 1994) (Figure 4.12B). The community is

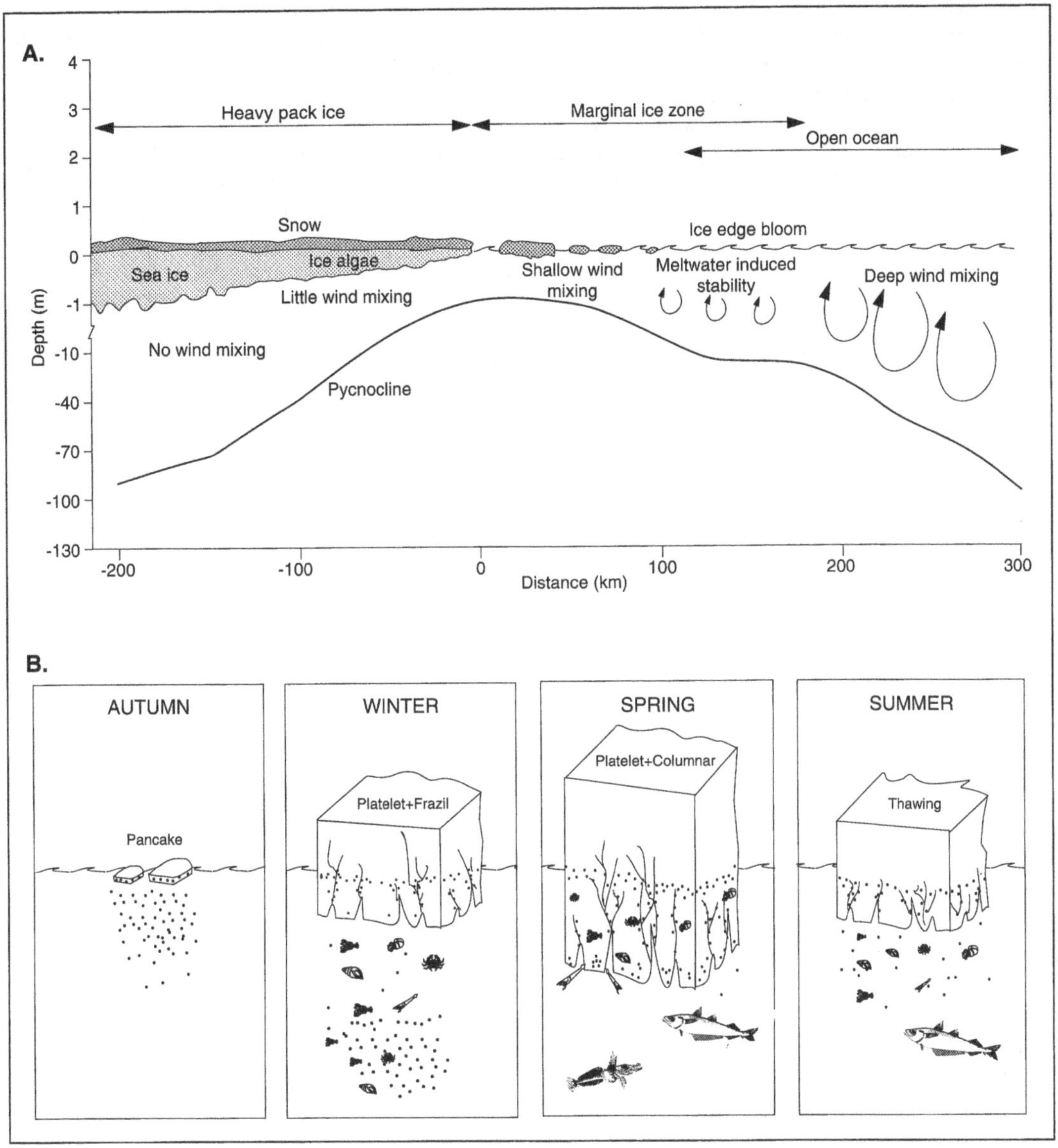

Fig 4.12 (A) Schematic model of the conditions necessary for the development of an ice-edge bloom of phytoplankton (source: redrawn after Sullivan *et al.*, 1988). (B) Seasonal development of the sea ice, or sympagic, community. The dots represent microalgae and small zooplankton (source: modified after Spindler and Dieckmann, 1994).

dominated by microalgae that are adapted to allow some degree of photosynthesis to occur even at the very low light levels encountered beneath the sea ice. Photosynthesis can occur at light levels of approximately 0.1% of those encountered at the surface, in spite of the very low surface incident radiation as a result of low sun angles for the winter and early spring periods. It is also known that sea ice crystals act as light guides, thereby enhancing the photon flux to the algae, which are held suspended at low, but sufficient, light levels to remain productive in winter (Knox, 1994). These factors contribute to the series of extensive algal blooms that follow the break-up and retreat of the ice edge from north to south, and together with the ice-edge 'blooms' form a single peak of phytoplankton production. In the Weddell Sea, the pancake ice zone is 270 km wide and provides an important winter habitat, but perhaps the key feature of the MIZ is that it varies in width and timing between areas.

The sea ice biota and the phytoplankton in the water column of the MIZ are together a rich source of biomass, and although the former is difficult to estimate, the latter was estimated by Smith *et al.* (1988) to account for 20–40% of the total primary production of the Southern Ocean. Recent modelling of primary production in the sea ice suggests that it contributes up to 4% of the total biogenic carbon production of the Southern Ocean and up to 25% of the production of the ice-covered Southern Ocean (Arrigo *et al.*, 1997). Of this production, 75% is associated with first-year ice (nearly 50% is produced in the Weddell Sea) on account of the thick snow cover resulting in enhanced infiltration and nutrient exchange (*ibid*). However, estimating how much of the MIZ phytoplankton blooms are derived from algae that have been living in the ice over winter is not straightforward. Not only does much of the production sediment to the seabed rather than being grazed (Clarke and Leakey, 1996) and so plays little part in seeding the MIZ spring blooms, but also there are species in the blooms that do not occur within the ice. Nevertheless, the consensus view is that in spite of its light-shading effects the sea ice cover enhances the productivity of this area of the Southern Ocean, particularly adjacent to ice edges, leads and polynyas (Niebauer and Alexander, 1989). It is also likely that light absorption by the algae within the sea ice speeds up melting and break-up: a good example of biological feedback.

Primary production in the Southern Ocean is highly variable. Early research, which suggested that Antarctic seas were highly productive, focused on the rich marine life, particularly of abundant birds and seals. However, more recent work suggests that low phytoplankton growth rates occur, except in a few special areas or 'hot spots'. Low values for primary production rates are found in the Drake Passage and Bellingshausen Sea as well as in the open oceanic regions of the Southern Ocean, where El-Sayed (1985) reported values of just over 0.1 g C m^{-2} a^{-1}. On the other hand, high values have been found in the Scotia Sea near Deception Island (3.61 g C m^{-2} a^{-1} (Mandelli and Burkholder, 1966)), in the Gerlache Strait (3.2 g C m^{-2} a^{-1} (El-Sayed, 1965), and in the waters close to the Antarctic Peninsula and the South Orkney Islands. Such high values are comparable with those reported from the areas of rich upwelling off Peru and southeast Africa amongst others and contributed to the belief that Antarctic waters were highly productive. The reasons for large spatial variations in phytoplankton productivity are unclear but may be related to the distribution of nutrient-poor waters and the lack of local upwelling of nutrient-rich waters. Primary production is also thought to be highly variable in the MIZ, and reported high values may reflect the sampling of spring phytoplankton blooms. Averaged over the entire Southern Ocean, the estimates of primary productivity suggest that, in spite of locally highly productive spots, in the MIZ and in coastal regions, the Southern Ocean is only moderately productive (Clarke and Leakey, 1996; Priddle *et al.*, 1997). However, the study of the phytoplankton of the Southern Ocean extends beyond a concern for plant growth *per se* and increasingly appears to be important for understanding biogeochemical cycling in the Southern Ocean and possibly, via the CO_2 cycle, the role of the Southern Ocean in climatic change (Treguer, 1994).

4.4.3 Biogeochemical cycles in the Southern Ocean

For most of the world's oceans, primary production is usually limited by the availability of dissolved nutrient salts, which supply nitrogen, phosphorus and silicon, whereas on land the photosynthesis of plants is usually limited by the concentration of CO_2 in the air. However, some ocean regions are rich in nutrients but paradoxically poor in phytoplankton growth (Hempel, 1985), the Southern Ocean being one of the largest of these high nutrient–low chlorophyll (HNCL) environments. This vast underutilisation of available nutrients in the Southern Ocean confounds most analyses and has been dubbed the 'Antarctic Paradox'. Why should 80% of the nutrients entering the system remain unused by phytoplankton (Treguer and Jacques, 1992)? Further, since the availability of nutrients is not a limiting factor in the Southern Ocean, the reasons for the abnormally low rate of nutrient utilisation and photosynthetic uptake of CO_2 displayed by Antarctic phytoplankton remain problematic. Present understanding of the major processes that control the exchange of CO_2 and nutrient cycling in the Southern Ocean highlight two surface zones of importance, at the Antarctic Divergence and the Polar Frontal Zone (Treguer, 1994). Figure 4.13 attempts to summarise these processes into the upwelling of Circumpolar Deep Water as a source of CO_2 and nutrients; the

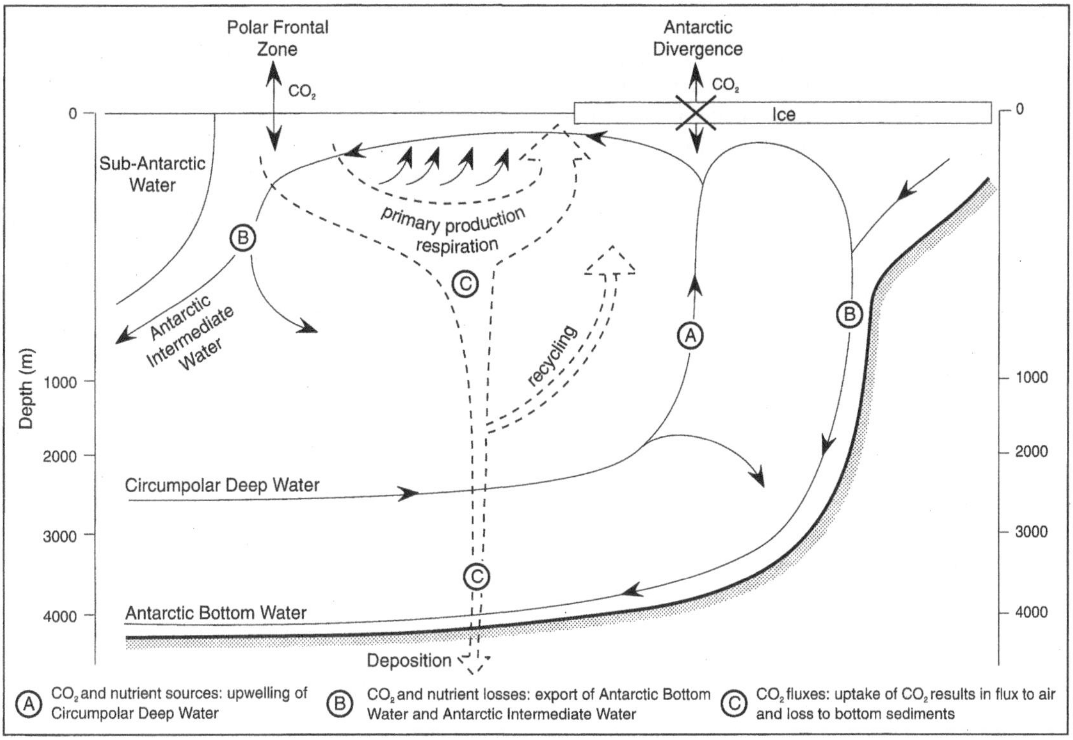

Fig 4.13 Atmosphere–ocean CO_2 exchanges in the Southern Ocean and the transport of nutrients: sources of CO_2 and nutrients in the upwelling of Circumpolar Deep Water at the Antarctic Divergence (A); export of Antarctic Bottom Water and Antarctic Intermediate Water also export CO_2 and nutrients (B); uptake via primary production, and respiration of dissolved CO_2 results in net loss to bottom sediments (C). Biological fluxes are shown in dashed lines (source: modified after Treguer, 1994).

export of Antarctic Bottom Water and of Antarctic Intermediate Water as losses of cold waters that are potential sinks for CO_2 and involve loss of nutrients; and biological fluxes via primary production and respiration, which return some CO_2 to the atmosphere and lose some organic carbon to deep-sea sediments. The winter extension of the sea ice zone dramatically reduces the exchange of CO_2 between atmosphere and ocean (*ibid.*). If the main elements of this cycle are broadly accurate, there appear to be at least four main hypotheses that might explain the low nutrient and CO_2 uptake of the Southern Ocean (*ibid.*). It is important that the following hypotheses are not seen as mutually exclusive: the very rapid changes that take place in environmental conditions in the open ocean mean that the system is likely to be driven more by interactions between the various factors rather than by any one factor alone.

The irradiance-mixing regime of the Southern Ocean

Phytoplankton favour the surface mixed layer or SML (about 40–100 m deep) if light levels will support net growth. If the critical depth (z_{crit}) at which light is sufficient for growth is deeper than the SML then growth takes place down to z_{crit}. However, since z_{crit} is defined by light attenuation with depth, and since light attenuation is in part attributed to absorption by phytoplankton, as phytoplankton growth proceeds, z_{crit} becomes shallower. Where an ocean is well mixed by strong winds, as is the case with the Southern Ocean, the SML may be deeper than z_{crit} and some of the phytoplankton may be below the light level needed for growth. The overall production of the SML declines. This process is associated with the passage of storm systems.

Macronutrient limitation

Silicate, nitrate and phosphate are important macronutrient requirements for cell growth. The availability of nitrates and phosphates in the surface waters shows no real gradients south of the Polar Frontal Zone, unlike that of silicates, which increases. However, in the vicinity of the Polar Frontal Zone and in coastal zones such as the Ross Sea there is evidence of silica depletion, and this has fuelled speculation that diatom growth is constrained by silica availability and so primary production is limited in those areas where otherwise it should be high (Nelson and Treguer, 1992). Nitrates comprise 80–95% of the inorganic nitrogen supply in the Southern Ocean but nitrogen can only be taken up and used in plant cells by enzymatic processes, which require energy. Not only must energy be expended in converting nitrate into ammonium before it is of use to the plant, but also the enzyme responsible for the process uses iron, and since this is in short supply in the Southern Ocean (Martin, 1992), it is one of the main reasons why phytoplankton cannot utilise all of the available nitrate (Priddle *et al.*, 1997). Ammonium is also available in the water and costs less energy to utilise but, at lower concentrations, it is relatively scarce. In spite of this, ammonium appears to be the dominant nitrogen supply for phytoplankton in the iron-poor Southern Ocean (*ibid.*).

Micronutrient limitation

As noted above, the role of iron in nitrogen nutrition is important for cell growth and, unlike phosphate, nitrate and silicate, the concentration of iron in the Southern Ocean is low and likely to be limiting to phytoplankton production. Although there is debate about the relative importance of iron, recent results from the Southern Ocean strongly support the iron-limitation hypothesis (de Baar *et al.*, 1995). Upwelling of deep water within the ACC carries with it enough iron to sustain moderate primary production but not blooms, yet at the Polar Frontal Zone the fast-flowing jet of iron-rich water allows large blooms. Such a sharp delineation between plankton-rich Polar Frontal Zone water and adjacent iron-poor water indicates that iron availability is critical in allowing blooms to occur (*ibid.*). In addition, iron depletion affects basic metabolism and photosynthesis. There is great interest in the possibility of seeding the Southern Ocean with iron to stimulate productivity and thus create an enhanced sink for atmospheric CO_2.

Herbivore grazing

Grazing is different from the controls listed above in that it acts not at the cell-specific level of primary production but as a mechanism whereby biomass is removed. Smetacek *et al.* (1990) show that krill swarms can deplete phytoplankton amounts in both the open ocean and the MIZ. Experiments with artificially fertilised patches of phytoplankton also show zooplankton to be a highly efficient grazing control on phytoplankton amounts (van Scoy and Coale, 1994). However, positive feedback occurs, since grazers that are present in sufficient biomass to limit phytoplankton production are themselves important sources of ammonium and may thus act to stimulate phytoplankton growth. On South Georgia, this feedback is suggested by the failure in some years of the main grazer, the Antarctic krill, which is transported to the island in the ACC. In a poor krill year, when abundance of krill is only 15% of normal, dense phytoplankton blooms and nitrate depletion occur. In contrast, in normal krill years, phytoplankton biomass remains relatively low and nitrate levels high (Priddle *et al.*, 1997). There is also a diurnal cycle in ammonium concentration in years when krill are abundant around South Georgia. Low daytime values resulting from phytoplankton uptake are replaced at night by high values due to zooplankton excretion, suggesting that the resupply of ammonium may enable greater phytoplankton growth (*ibid.*). In spite of the fact that they limit primary production, grazers may also play a role in carbon export because they convert phytoplankton carbon into rapidly sedimenting faecal pellets, which collect on the ocean floor.

Whether iron or ammonium are limiting factors or not, it remains that the Southern Ocean is vitally important to the global import and export of nutrients and that interference with the cycle may have widespread impacts well beyond the Antarctic. Thus, any suggestions that the primary productivity of the Southern Ocean might be enhanced by the artificial seeding of iron must be carefully assessed. Further, because photosynthesis affects the amount of CO_2 uptake (see Figure 4.13), and CO_2 is related to climatic change, then phytoplankton may hold one of the keys to understanding climatic change both in the past and in the future. This aspect is considered further in Chapter 5.

4.4.4 Zooplankton

Zooplankton species occupy the next level of the food web and since they mainly graze on phytoplankton, their distributions and life cycles are related to phytoplankton availability (see Figure 4.10). The herbivorous zooplankton consists largely of protozoans, the larval and juvenile stages of pelagic crustaceans such as copepods and euphausiids, and appenducularians (doliolids and salps). The carnivorous zooplankton is dominated by copepods, chaetognaths (arrow worms) and pelagic polychaetes. Larval fishes also comprise a significant component of the zooplankton (Knox, 1994). In terms of biomass, the most important are the copepods and euphausiids: the pelagic crustaceans. In some areas three species of copepod, *Rhincalanus gigas*, *Calanus propinquus* and *Calanoides acutas,* contribute up to 73% of the biomass (Voronina, 1966). Elsewhere, copepods and salps account for biomass peaks in the 1–5 mm and 7–8.5 mm ranges, respectively (Knox, 1994). With a high production:biomass ratio, copepods contribute most to total zooplankton production in the Southern Ocean, but they do not constitute a major food resource for the large predators; rather they are preyed upon in some areas by the mesopelagic fishes (Rowedder, 1979) and in other areas by the Antarctic herring, *Pleuragramma antarcticum*.

Among the eleven species of euphausiids, one has recieved most attention. The 6 cm long shrimp-like Antarctic krill, *Euphausia superba* (Figure 4.14), has a widespread abundance and forms an important part of the Southern Ocean ecosystem. Many higher species feed on it exclusively during their breeding period (Priddle *et al.*, 1988) (see Figure 4.10). However, other krill species such as *E. crystallorophias* are also important, particularly within the seasonal sea ice zone and in the waters close to the continent. While the term 'krill' is generally taken to mean *E. superba*, it is also used to include all of the euphausiids in the Southern Ocean. There are problems in determining the age of krill but, in general, growth to 50 mm is achieved in two years and the adults reach four–seven years old and are well adapted to life both at sea and within the seasonal sea ice zone (Knox, 1994). Under the sea ice, krill graze as individuals on a wide variety of phytoplankton and even other krill, but they swarm at the ice edge to take advantage of spring phytoplankton blooms, which then sustain heavy grazing (Hempel, 1991). Swarming appears in part to be a grazing strategy in the open ocean too, and krill occur in swarms distributed over the top 150 m of the water column, although very occasionally they can be large and dense enough to discolour the sea surface pink over several hundred metres (El-Sayed, 1985). However, they can also be almost completely absent (Murphy *et al.*, 1988). Individually, they are good swimmers and may change direction together in a coordinated fashion much as birds do in flocks. There is a tendency for euphausiids in general to migrate to the surface at night and remain in deeper water during the day. This daily movement and swarming behaviour may make the species an easy target for predators such as baleen whales, seals or fishing vessels.

Fig 4.14 The Antarctic krill, *Euphausia superba* (source: photograph by C.J. Gilbert, courtesy of BAS).

The spatial distribution of krill on a Southern Ocean scale is very patchy and while low numbers occur throughout the Southern Ocean, concentrations are much higher in coastal waters (Amos, 1984). The two main circumpolar currents provide a framework for identifying areas of high krill biomass (El-Sayed, 1985; Priddle *et al.*, 1988) (Figure 4.15). There is general agreement that krill are most abundant in 'hot spots' in the Antarctic Coastal Current in the Bellingshausen Sea, and off the coasts of Wilkes Land, Enderby Land and Queen Maud Land in East Antarctica (*ibid.*), adjacent to the upwelling zone of the Antarctic Divergence. The second group of 'hot spots' is related to where the Antarctic Coastal Current is forced to flow into the gyres of the Weddell and Ross Seas and interact with the waters of the ACC. The zone of mixing of the largest of these gyres within the Scotia Sea coincides with the largest area of high krill concentration in the Southern Ocean (see Figure 4.15). The presence of these concentrations might suggest regional stocks or independent populations (Siegel, 1988), but so far there is no reason to alter the concept of a circumcontinental population of krill linked to oceanic circulation (Priddle *et al.*, 1988). Modern molecular

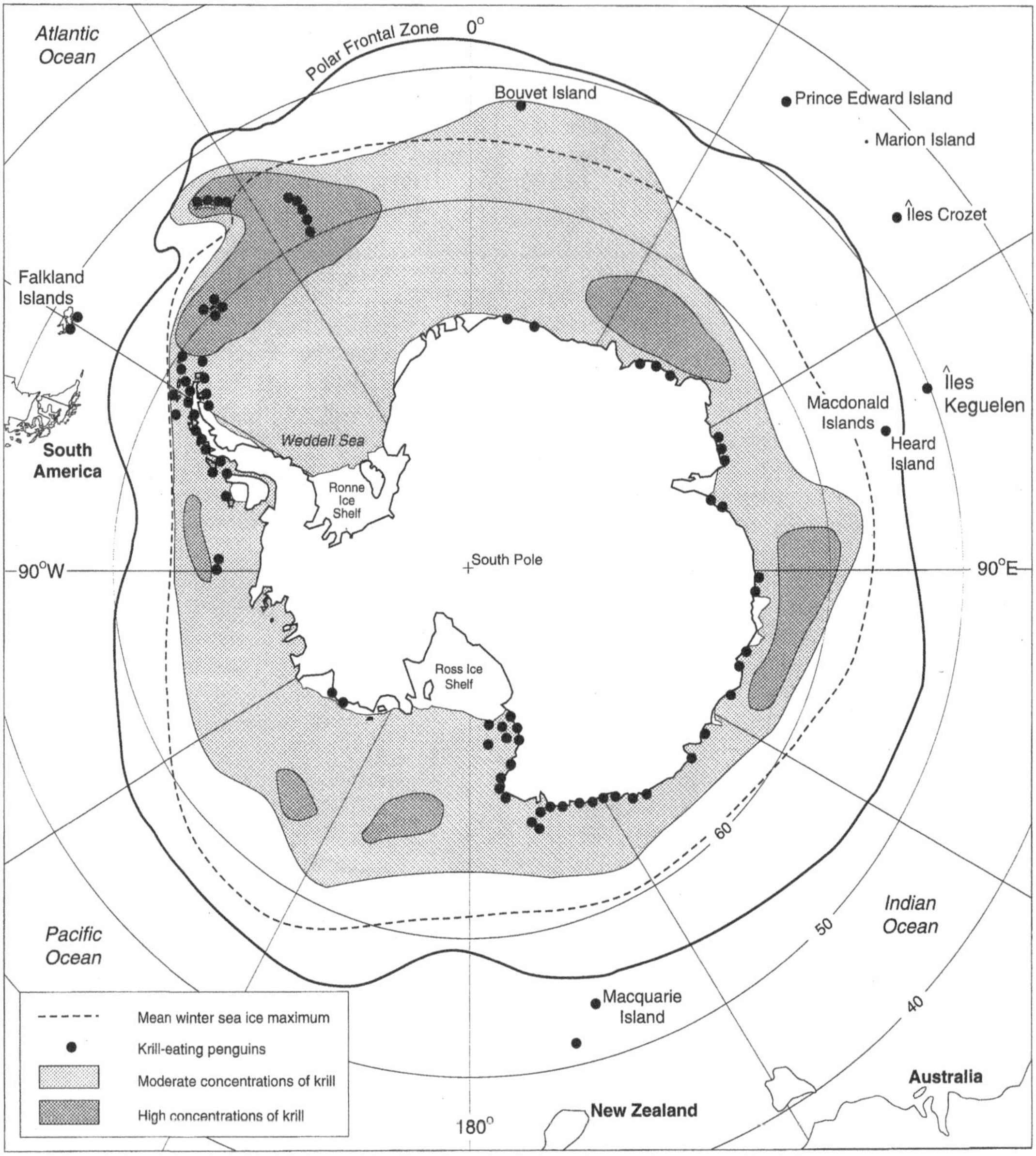

Fig 4.15 The distribution of krill in the Southern Ocean and the locations of colonies of one of its main predators: penguins (source: compiled from Laws, 1985; Nicol and de la Mare, 1993).

genetic techniques now offer a better chance of resolving any potential differences.

For a species so central to the functioning of the Antarctic ecosystem, the life cycle of the krill is not particularly well known. The consensus view appears to be that the effects of water circulation on the eggs and larvae are important (Mackintosh, 1972; Knox, 1994). Following spawning at 80–200 m depth, the eggs sink and hatch, and the krill then rise to the surface through a series of larval and post-larval stages. Although adult krill are strong swimmers, eggs and larvae are passively moved by currents and are subject to the same interchanges as the water. For example, larvae within the Antarctic Coastal Current may be carried west then north, where they join the ACC and are carried east (Miller and Hampton, 1989) as juveniles and adults (see Figure 4.1). Any spawning en route results in the eggs sinking and being returned to the ACC via deep-water upwelling at the Antarctic Divergence. However, there are two recent and important modifications to this view.

First, there is increasing emphasis on the importance of the gyres and eddies along the boundary zones between the ocean currents (see Figure 4.3). The larger-scale topographically induced gyres also coincide with the densest concentrations of krill, especially in the Weddell Gyre and Scotia Sea, and it may be that meanders in the gyres also serve to concentrate krill into dense patches in quiescent zones between faster flows (Knox, 1994). Second, increasing recognition of the importance of the sea ice to the distribution of krill can be extended to suggest that krill may migrate annually. For example, in the Weddell Sea the bulk of the winter population of krill is found beneath the sea ice (Hempel, 1988), with individuals of up to 6 cm regularly found. Since krill shrink when starved, this indicates that food is available through the winter and probably consists of copepods, cryopelagic invertebrates and sea ice and benthic microalgae (Knox, 1994). In the MIZ in the Weddell Sea, Daly (1990) found that krill larvae were abundant at the ice edge and on the underside of the sea ice. Gut studies show that almost all of the krill larvae feed both night and day on sea ice microalgae (Knox, 1994). Support for this view comes from direct observations of large concentrations of krill on the rugged underside of the sea ice (Marschall, 1988) grazing not only on the meadows of ice algae held within frazil ice but also on other grazers (see Figure 4.12). Under thicker ice further from the ice edge zone, krill are as rare as ice algae (Hempel, 1991). This view suggests that some krill may undertake seasonal migrations to the north in summer and to the south in winter where the presence of ice allows the krill to remain in place. On the other hand, Sun *et al.* (1995) suggest that since krill occur over a large area, the wintering behaviour of some populations are bound to be different from others. For example, around South Georgia krill live in an ice-free environment all year, whereas krill in the Weddell Sea rarely experience ice-free conditions. It is as yet unclear whether the under-ice krill population represents a significant proportion of the total population of krill in the Southern Ocean (Everson, 1992) or whether it forms only part of the more general ocean current-driven life cycle of krill.

In spite of over a decade of intensive study under the international BIOMASS programme directed at understanding the biology, distribution and biomass of krill (together with other species), there is still uncertainty about how many krill there are. Assuming that krill make up 50% of the zooplankton, Everson (1992) showed a total standing stock of Southern Ocean krill of 750×10^6 t, and other estimates range from 125 to 5000×10^6 t (El-Sayed, 1985). Although such figures must be viewed with caution, they do indicate that krill stocks are very large indeed. Estimates of krill production based on consumption by predators amount to 470×10^6 t from a standing crop of $1175–1985 \times 10^6$ t a^{-1} (Everson, 1992). However, because krill are dependent on ocean currents and sea ice, it is not surprising that distributions are subject to large interannual mesoscale variations. For example, observed winter abundances of krill around South Georgia in 1983–1984 were found to be 3–12% of their summer values, the shortfall being attributed by Priddle *et al.* (1988) to southward deflection of the Polar Frontal Zone and its krill-rich eddies. Similarly, the seasonal sea ice zone is a highly dynamic area with its own cycles of variability and these are likely to influence krill distributions. Until the factors influencing krill biology and abundance are fully known, it is clear that a considerable threat to the Antarctic ecosystem could be posed by large-scale commercial fishing (see Chapter 7). This is particularly relevant in the context of the so-called 'krill surplus' hypothesis. In general terms, reduction in the stocks of whales as a result of whaling activities is popularly thought to have produced a 'krill surplus' that might be available as the basis of a commercial krill fishery. However, there are two main problems with this view. The first problem is that there has been no measure of krill biomass over the period of whale exploitation to allow meaningful comparisons. In addition, the krill surplus is probably illusory since any surplus is likely to have been largely taken up by other competitor species, and the various elements of the Antarctic ecosystem appear to have already adjusted to maintain a constant level of krill consumption (Berkman, 1992). This aspect is reviewed in more detail in Section 4.4.5.

4.4.5 Zooplankton feeders

There is no sharp dividing line between the pelagic animals that constitute the plankton and those that constitute part of the next or third trophic level, because many of the nekton species of cephalopods and fishes spend the early stages of their life cycle as plankton. However, as adults they swim well and generally predate the zooplankton, which is consumed by a host of carnivorous species both large and small (see Figure 4.10). Zooplankton are eaten in vast quantities by the cephalopods, fish, birds and mammals of the Southern Ocean. There are also important benthic communities that characterise the seabed and the coastal zones of both ice-affected and ice-free shores. Many of these communities feed on the detritus and body parts of zooplankton such as krill that fall to the bottom. However, much less is known about the fauna and flora of these communities, so the focus here is on the ecologically important groups of cephalopods, fish, birds and mammals.

Cephalopods

In the Southern Ocean, by far the most ecologically important of the cephalopods are the oceanic squids (Clarke, 1985). Surprisingly little is known of their biology and most of what is known comes from their ecological relationships with predatory species. Squid grow rapidly to a metre in length and some species reach over 5 m, so they must themselves have a large predatory effect on both the fish and krill populations that sustain them (*ibid.*; Knox, 1994). Very little is known of the reproductive and life cycles of cephalopods, but it appears likely that some complete their entire life cycle in 1–2 years. Egg masses containing large numbers of eggs are laid on the continental slopes deeper than 1000 m, and this is followed by rapid growth, a single spawning, deterioration and death (Voss, 1973). The ecological relationships are better known and depend on squid feeding habits, depth distribution and abundance. There are several species of squid, each of which differs in its habitat. The giant squid, *Mesonychoteuthis hamiltoni*, spends its life at depth in the deep oceans, reaches over 5 m in length, weighs 184 kg and forms 75% of the diet of sperm whales; others, such as *Galiteuthis glacialis* and *Alluroteuthis antarcticus*, spend at least part of their lives near the surface and are preyed upon by seals and albatrosses (Rodhouse, 1990). Large quantities of squid beaks are found in the stomachs of sperm whales (Nemoto *et al.*, 1988) and as many as 18,000 have been found in a single stomach, many species of which have rarely or never been caught in nets (Clarke, 1985). Close to South Georgia, the abundant *Martialia hyadesi* is an important part of the diet of wandering albatrosses (Priddle *et al.*, 1988), and Prince (1985) found that 50% of the diet of the grey-headed albatross was of two species, *Todarodes sagittatus* (91% by weight) and *Mesonychoteuthis* sp. (4%). Of the estimated 3.7×10^6 t a^{-1} of squid taken by seabirds and seals in the Scotia Sea, 76% is consumed by elephant seals, mainly around South Georgia (Croxall *et al.*, 1985).

Laws (1985) relates an increase in squid numbers this century to the decline in numbers of its main predator, the sperm whale. Seabirds, seals and the great whales together are estimated to consume of the order of 34×10^6 t a^{-1} of squid: the stocks necessary to support this level of predation reach well over 100×10^6 t a^{-1}, a large biomass by oceanic standards (Clarke, 1985). Squid are already fished commercially in the northern part of the Southern Ocean offshore of Patagonia and the Falkland Islands and to the south of New Zealand, and it may only be a matter of time before the stocks in the southern part of the Southern Ocean become a quarry. Following the catch of commercial quantities of *Martialia hyadesi* from two squid-jigging boats near South Georgia in 1989, Rodhouse (1990) considers this species to have the most potential for commercial fishery. However, since squid are generally fast-growing and short-lived, they are prone to extreme fluctuations in population size and this makes them particularly susceptible to overfishing because recruitment is dependent on the breeding success of one generation (Knox, 1994). Any overexploitation of squid would pose problems for a number of predator species that depend on squid for a major part of their breeding season diet.

Antarctic fish

For its size, the ocean surrounding Antarctica does not support a great variety of fish species; of the 20,000 or so species known globally, only about 120 inhabit the Southern Ocean south of the Polar Frontal Zone (Knox, 1994). The Polar Frontal Zone, together with the great depth of the Antarctic continental shelf, seems to represent a boundary for the evolution and composition of shallow and coastal water species, whereas it forms no distinct barrier for deep-water pelagic fish distributions (Kock, 1985, 1992). Thus more than 85% of coastal species of Antarctica are endemic, as opposed to 25% of the deep-sea fish. The dominant group of Antarctic fishes is the suborder Notothenioidei, whose four families make up 60% of species and over 90% of individuals. These are the Harpagiferidae (plunder fish), the Bathydraconidae (dragon fish), the

Nototheniidae (comprising 30 species of Antarctic cod) and the Channichthyidae (comprising 16 species of ice or white fish). With relatively few non-notothenioids in the Southern Ocean, the notothenioids fill the ecological roles normally filled by a variety of fish. The most important numerically, and therefore of greatest ecological and potential economic importance, are the Nototheniidae and the Channichthyidae. In contrast to the other oceans, the Southern Ocean does not appear to contain dense stocks of fish swimming close to the surface and, in general, deep-sea species appear to be less important than coastal species. The number of species and abundance of fish is greatest in the Atlantic sector of the Southern Ocean around South Georgia, the shelf areas of the Scotia Sea and on the Kerguelen Plateau. South Georgia appears to be an important transition zone for fish species between the Patagonian and Antarctic regions, and many species are common to these areas. There is also some species exchange via the ACC between South Georgia and Kerguelen (Kock,1985).

The Nototheniidae and the Channichthyidae are mainly bottom or coastal dwellers and feed on the shelves around the continent and the sub-Antarctic islands and so are often associated with habitats at depths of 400 m or more (Kock, 1992). Large numbers of these two families are secondarily pelagic since they spend their first year or years in mid-water associated with drifting or fast ice or feeding at the edges of krill swarms (Targett, 1981). In this way, the notothenioid Antarctic herring, *Pleuragramma antarcticum*, has successfully adapted to temporary or constant life in the pelagic zone to become the most abundant pelagic fish in the coastal waters close to the continent (Hubold, 1985). Many others alter their ecological niche at least once during their life cycle. The marbled *Notothenia rossii* spends its first year far from the coasts, grazing on krill, but then migrates inshore and spends five–six years maturing on benthic fauna before moving back into the pelagic zone to feed on krill again (Kock, 1985). One of the most frequently encountered mesopelagic or mid-water fish are the Myctophidae (lantern fish). Efremenko (1983) found eight species of these to be abundant in the Scotia Sea, spawning at depths of below 200 m with larvae found from the surface down to 1000 m, depending on season. The Myctophidae form the diet of many warm-blooded predators and have been the subject of a major fishery.

Many Antarctic fish are relatively small: most reach only 25 cm in length; only twelve species grow to more than 50 cm in length, and a few, such as *Dissostichus elegeniodes* (Patagonian toothfish), reach a length of 185 cm, a weight of 52 kg and an age of 22 years (Knox, 1994). With the low temperatures encountered in the Antarctic, most fish have evolved to be most efficient in cold water with a variety of adaptations such as freezing resistance, buoyancy, and cold and metabolic adaptations. Some of these adaptations to life in cold environments are shared by other species and are examined more fully in Chapter 5. Most Antarctic fish have slow seasonal growth patterns, low metabolic rates and relatively long life spans. As a consequence they do not reach sexual maturity until three to eight years (Everson, 1987; Knox, 1994). Most Antarctic fish produce relatively low numbers of large yolky eggs, and the larvae hatch at an advanced stage. In notothenioids there is also evidence of an increase in egg size and a decrease in fecundity towards higher latitudes (Kellerman, 1990). In some species, eggs take up to two years to be produced, and such fish depend on a large spawning population to maintain numbers (Everson, 1987). Slow growth and low fecundity make most Antarctic fish stocks very vulnerable to overfishing (Fifield, 1987).

Reliable estimates of biomass and productivity of fish stocks are scarce, since the numbers of many species show a high degree of interannual variability independent of fisheries exploitation. This appears to be in spite of the opportunistic nature of predation shown by most Antarctic fish species. For example, *C. gunnari* feeds exclusively on krill when they are available and, although it will substitute other prey when krill are in short supply, large variations in its numbers are considered to be directly related to krill abundance (Everson, 1992). Where krill are continuously available, such as in the southern Scotia Sea, they make up 90–100% of the diet of *N. rossii* and *C. gunnari* (Kock, 1992). Where krill are not so continuously available, such as off South Georgia, the proportion consumed by these two species varies considerably between 15–90% and 18–94%, respectively. Commercial fishing in recent years has reduced the significance of these species as krill predators.

As well as their importance as krill consumers, 1.5×10^7 t a^{-1} of fish are themselves consumed by birds and seals. Conservative, order-of-magnitude estimates of the standing stock biomass of demersal fish in the 1970s were 130,000 t for Kerguelen (Hureau, 1979) and 500,000 t for South Georgia (Everson, 1987), this representing maximum sustainable yields of 20,000 t and 50,000 t, respectively.

However, reported fish catches, even of single species, have often exceeded these estimates. The 1969–1970 catch of *N. rossii* around South Georgia yielded 400,000 t (*ibid.*) before declining catches deflected effort towards *C. gunnari*. A similar trend of depletion of the *N. rossii* stock followed the move of the Soviet fishing fleet to Îles Kerguelen in 1970–1972 (*ibid.*). Biomass estimates indicate that populations of *N. rossii* are currently about 3% of the unexploited stock level. Overall, it is clear that Antarctic fish, with their relatively high age of sexual maturity, low growth rates and low fecundity, cannot sustain the levels of commercial fishing experienced during the 1970s and 1980s (Knox, 1994). This is considered further in Chapters 7 and 9, where the fishing industry and its regulation are examined.

Antarctic birds

The birdlife of Antarctica is fundamentally influenced by the high, windy and ice-covered nature of the continent, where there is little ice-free ground available for nesting. Set amidst a circumpolar ocean, the continent is also remote from other large land masses, and this, together with the pronounced circumpolar patterns of wind and water circulation, serves to ensure that the birds are almost exclusively marine species. Of the 43 species of Antarctic birds that breed south of the Polar Frontal Zone, only five are land-based and only three of these are not dependent in one way or another on the sea for their food (Stonehouse, 1972). Only twelve species actually breed on the continent, so the distribution of bird species shows a strong bias towards the sub-Antarctic islands. The relative uniformity of the Antarctic oceanic environment has also favoured a general uniformity of avifauna over large geographical areas, with only four examples of differentiation into subspecies (Knox, 1994).

Few summer visitors can avoid the impression that birds are the most numerous animals in the Antarctic. Seabirds are in abundance everywhere and with large colonies of penguins found around the margins of the continent and offshore islands, the popular view is that the Southern Ocean supports a great variety of seabirds. However, this abundance is largely a summer phenomenon, as many of the species congregate on land only to breed and subsequently disperse during the winter, when few birds are seen on either the coast or open ocean. Most of the birds that breed in the Antarctic prey largely upon marine animals such as krill and larval fishes, which themselves are dependent on a strong summer pulse of phytoplankton to sustain them. The others are dependent on similar summer abundance of prey, often of other birds or of the carrion associated with the breeding colonies of birds and seals. Consequently, most Antarctic birds must feed and breed when large stocks of food are available, since there will be little to be had in winter when the temperature falls and the ocean freezes. After a short breeding season, most disperse northwards (Knox, 1994).

The marine birds include seven species of penguin, 24 species of petrel, five species of gull, skua and tern, and two species of cormorant. The land-based birds comprise two species of pintail duck, one on South Georgia and the other on Îles Kerguelen, two species of sheathbill, scavengers of penguin rookeries and seal colonies, and the South Georgia pipit, which is the only Antarctic songbird (Stonehouse, 1989). About 90% of the total mass of this birdlife, and 90% of the food that they eat, is accounted for by penguins (Fifield, 1987; Knox, 1994), but in terms of sheer numbers, the petrels are the most abundant (Stonehouse, 1985). Most of these birds are secondary consumers occupying the third trophic level in the Antarctic food chain, as they are dependent on zooplankton, mainly krill, for their food (see Figure 4.10). However, many birds consume tertiary feeders such as fish and squid as a part or whole of their diet (Siegfried, 1985). In addition, several species, such as the skuas and giant petrels, are predatory on the smaller seabirds and penguin chicks as well as feeding on any carrion that becomes available such as the carcasses of penguins, seals and whales.

Of the secondary consumers, the most important are the Sphenisciformes or penguins. These flightless, stocky birds comprise seven species that breed south of the Polar Frontal Zone. There are essentially two circumpolar patterns to the distribution of penguins: a coastal continental pattern comprising the emperor, Adélie, chinstrap and gentoo (although the latter two are found only at the northernmost parts of the Antarctic Peninsula); and a sub-Antarctic pattern based on the oceanic islands and comprising the king, macaroni, rockhopper and gentoo (see Figure 4.15) (Knox, 1994). All polar penguins nest in open, densely packed colonies or rookeries, which often number tens of thousands of breeding birds (Figure 4.16). The colonies may extend over several hectares and are usually found along the coast, although some of the larger ones extend inland and on to higher ground (Stonehouse, 1985). Once the penguin droppings and feathers are cleared, the routes over cliff

Fig 4.16 Part of a macaroni penguin colony, Cooper Bay, South Georgia (source: authors).

and scree to rockhopper and macaroni rookeries can be seen to be polished and smoothed into pathways by the attentions of generations of penguin feet! For the bulk of penguins, the location of rookeries is determined by the availabilty of suitable sites and access to good feeding grounds. Penguins are important consumers of krill, and in the Scotia Sea they take 76% of the 1.1×10^7 t a^{-1} of krill consumption ascribed to birds (Croxall *et al.*, 1985). Two species, macaroni and chinstrap penguins, take 7.7×10^6 t a^{-1} of this. However, a dependence on krill also exposes such species to reduced breeding performance when krill-poor years occur. On South Georgia, macaroni and gentoo penguins, black-browed albatrosses and Antarctic fur seals were affected in this way in 1984 and 1987 (Croxall, 1992). In contrast, fish- and squid-eating species such as grey-headed and wandering albatrosses were unaffected. Croxall *et al.* (1988) highlighted the differences between South Georgia and the South Orkney Islands, where winter ice conditions appeared to be more important than food supply in determining breeding success.

Penguins are themselves subject to predation at the nest by skuas, sheathbills, dominican gulls and giant petrels, which remove undefended eggs and weak chicks, and at sea by killer whales and leopard seals. However, such predation is generally considered insufficient to influence stock size significantly (Stonehouse, 1985), so their numbers can be related almost directly to environmental changes or variations in food supply. In general terms, the enhanced availability of krill in the Southern Ocean following the removal of the baleen whales has led to an increase in the populations of most of the species that eat krill, including penguins. However, penguin numbers are subject to fluctuation. Breeding Adélies at Cape Royds suffered massive declines in the 1950s and early 1960s, which were probably related to human disturbance from the bases at McMurdo Sound (Thomson, 1977) (Chapter 8). A similar trend in Adélie and emperor numbers was noted close to the base at Point Geologie, Terre Adélie, while Adélie colonies in adjacent, more remote areas showed a doubling of numbers (Jouventin and Weimerskirch, 1990). In seasons when sea ice persists locally, the lack of open water, leads or polynyas makes the gathering of food very difficult for emperor, Adélie and chinstrap penguins, whose colonies may lie kilometres from the ice edge in early spring. At Point Geologie, this is the time when emperor chicks are growing most rapidly and mortality rates as high as 90% have occurred due to persistent ice in late spring. Such environmental pressure, together with human disturbance, is likely to have contributed to the crash in pairs of breeding emperors at Point Geologie of 5500 to 2300 between 1962 and 1989 (*ibid.*). The effect of environmental change and human upset on penguin numbers is discussed further in Chapters 5, 7 and 8.

The proportions of plankton, squid and fish eaters changes in a general way southwards (Abrams, 1985), with squid predominant in the diets of birds north of the Polar Frontal Zone, zooplankton reaching a maximum in the ACC, and an equal share of zooplankton, squid and fish in the diet of birds close to the continental shores in the Antarctic Coastal Current (Figure 4.17). The relative proportions will also vary with species. In all, 55% of pelagic seabirds (excluding penguins) feed on krill. Most of these are petrels, which feed on the ocean surface and are most abundant south of the Polar Frontal Zone, but some albatrosses, especially the black-browed albatross, also eat krill (Siegfried, 1985). Squid are the dominant food north of the Polar Frontal Zone and form the diet of 20% of the pelagic seabirds, mainly of the larger birds such as albatrosses and the larger petrels. At South Georgia, squid constitute 49% and 47% by weight of the diet of grey-headed and light-mantled sooty albatrosses, respectively, with fish making up 35% and 37%, respectively (Prince, 1985). A further 20% of pelagic seabirds are classified as mixed-diet, scavenging and surface-seizing whatever is available near the Polar Frontal Zone and close to the continent. Fish eaters account for a very low proportion, this apparently reflecting the lack of pelagic fish at the surface (*ibid.*).

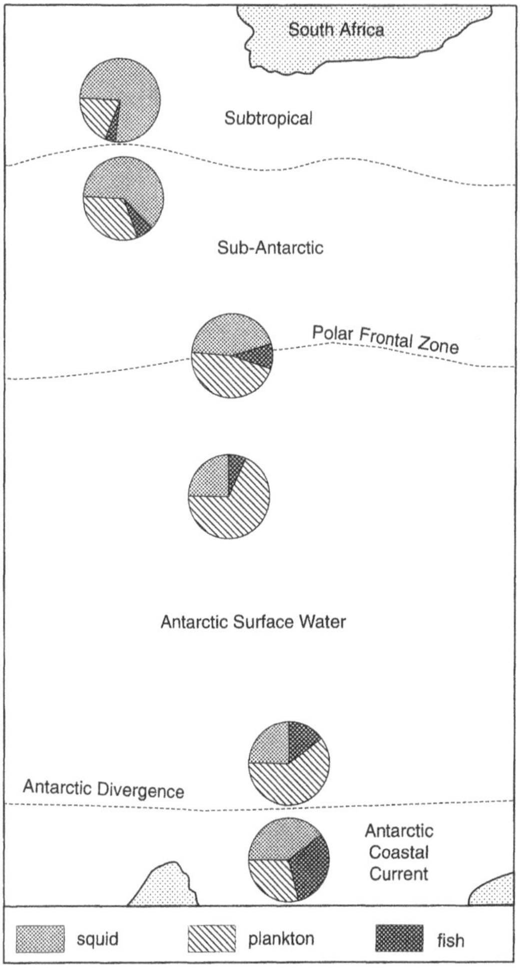

Fig 4.17 The relative proportions of seabird food types for different zones of the Southern Ocean (source: modified after Abrams, 1985).

Fewer species still obtain their food by kleptoparasitism or piracy. Both species of skua fall into this category, but pirating food from other birds after an aerial chase, as occurs in the Arctic, is relatively rare in Antarctica, where the skuas tend to be opportunistic scavengers. They also nest close to penguin colonies and will take eggs, young and injured birds as well as any bird or seal carrion, all traits that they share with the sheathbills. Other opportunistic predators are the giant petrels, which gorge on krill at sea but also use their powerful beaks to tear apart any seal or penguin carcass that they find. These birds, together with the pintado petrels, showed a great increase in numbers during the whaling era on account of the vast quantities of offal produced (Fifield, 1987). In general terms, due to the relative availability of krill, those species that are krill eaters such as penguins appear to have undergone increases in population over recent years (Croxall *et al.*, 1988) (Figure 4.18).

Within these broad trends there also exists a geographical zonation in the species encountered as a function of their foraging range around breeding sites. For example, there is a very distinct zonation around sub-Antarctic South Georgia related to the distances that foraging birds cover to secure enough of the appropriate type of prey (Figure 4.19A). For the gentoo, macaroni and chinstrap penguins, whose diet is mainly krill, the range is much less than for the king penguin, whose diet is squid and fish found in deeper water. Petrels and albatrosses range widely in search of food. As a result, there is also a geography of predator impact around such breeding sites. For the birds of South Georgia, this impact falls off rapidly at about 150 km from the island (Figure 4.19B). For some species, such as the wandering albatross, recent satellite tracking has shown that the foraging range is much larger than previously thought, the birds travelling great distances to locate their food. Unfortunately, this has brought them into direct contact with commercial squid fisheries, and many have been fouled by hooks (see Chapter 7). Wandering albatross numbers on Îles Crozet and Îles

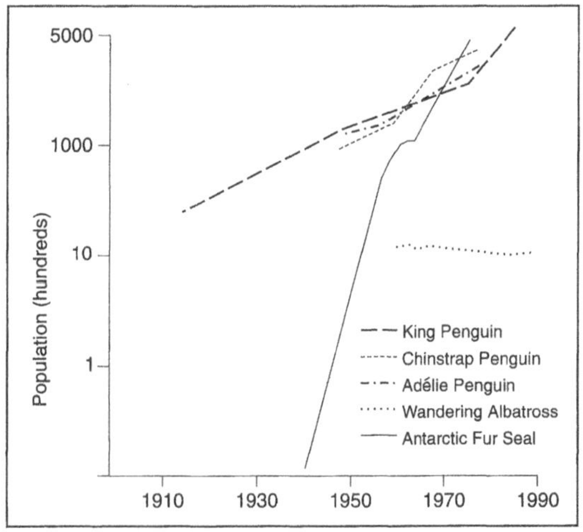

Fig 4.18 Changes in the population sizes of: Adélie and chinstrap penguins, Signy Island, South Orkney Islands; and of king penguins, wandering albatrosses and fur seals, South Georgia. Note the logarithmic scale on the y-axis (source: data from Croxall *et al.*, 1988).

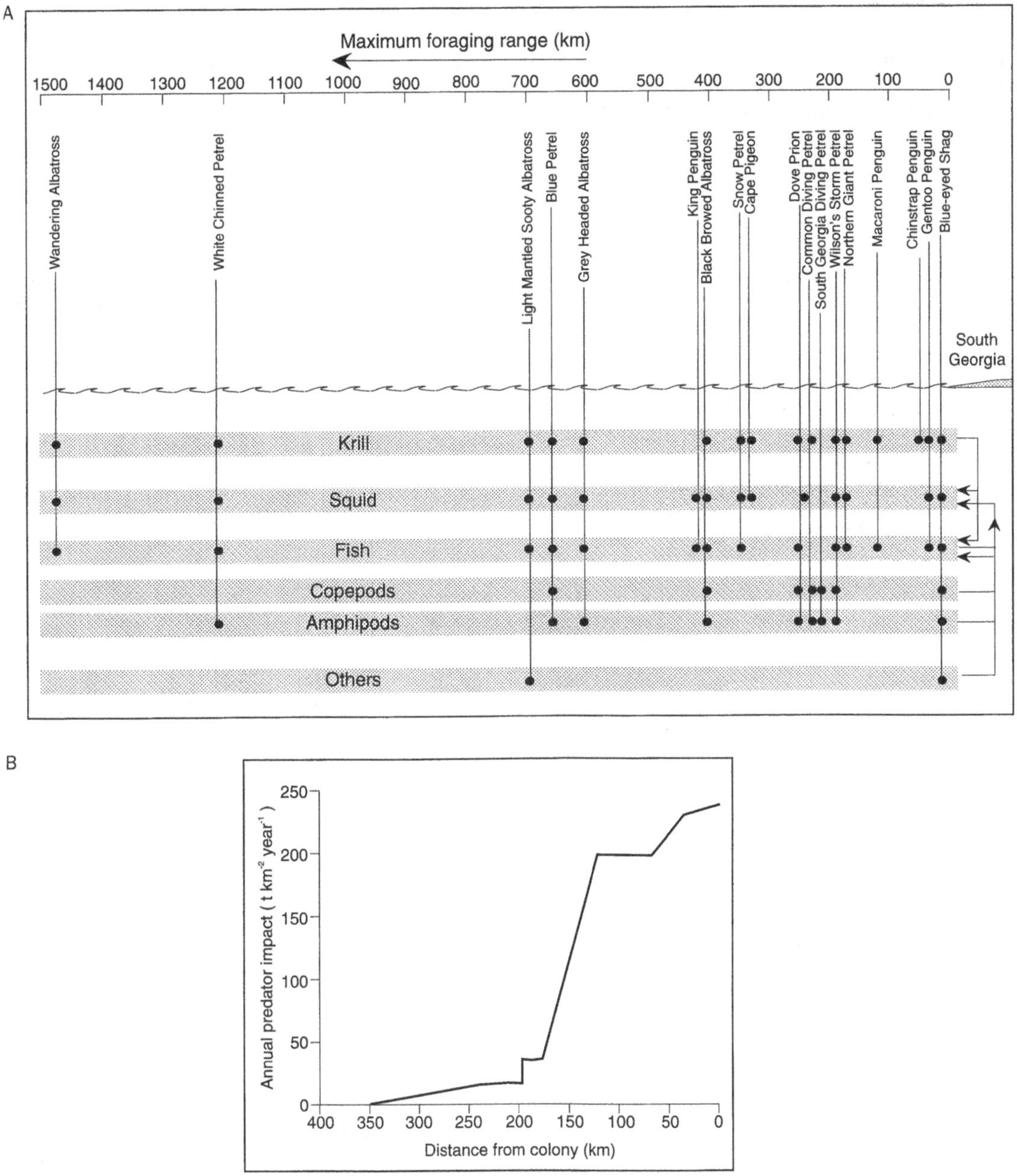

Fig 4.19 (A) Spatial operation of part of the South Georgia marine ecosystem. The maximum foraging areas and prey items of seabirds are shown on the x-axis. The y-axis shows predation among the prey items themselves, e.g. krill are eaten by fish and squid as well as by birds (source: redrawn from Murphy, 1995). (B) The annual impact on the local marine ecosystem as a result of seabird predation from South Georgia (source: redrawn after Murphy, 1995).

Kerguelen have been declining over the last two decades, and Jouventin and Weimerskirch (1990) suggest that this is a result of fishing activities. Similar trends have occurred on South Georgia.

Antarctic marine mammals

A long tectonic isolation from Gondwanaland and the occurrence of several intervening continental glaciations have ensured that no terrestrial mammals now occur in Antarctica. However, the oceans present no barriers for marine mammals, which have been highly successful in colonising the area south of the Polar Frontal Zone. All of the Antarctic marine mammals feed exclusively at sea, but seals must return to land or ice to breed, whereas whales mostly breed in coastal waters in lower latitudes and move into Antarctic waters in summer to feed. Mammals represent some of the most important secondary consumers in the Antarctic food web, consuming 1.27×10^8 t a^{-1} of krill (Croxall, 1992; Berkman, 1992) (see Figure 4.10). Most species of seals and whales feed on krill, but several also feed on krill feeders such as squid, fish, penguins, and even other seals or whales. Excluding humans, many such species have no natural enemy, and species such as the leopard seal and the killer whale are the top predators of the Antarctic food web. Adaptation of these mammals to life in cold environments is discussed in Chapter 5.

The two broad groups of Antarctic seals have different spatial distributions: the Phocidae or true seals have, with one exception, colonised the shores of the Antarctic itself; and the Otariidae or eared seals have successfully colonised the shores of the oceanic islands (Bonner, 1985a), although the Antarctic fur seal, *Arctocephalus gazella*, has recently reached parts of the Antarctic Peninsula. Most of the first group are ice-breeding species and include the crabeater seal, *Lobodon carcinophagus* (Figure 4.20A); the leopard seal, *Hydrurga leptonyx* (Figure 4.20B); the Ross seal, *Ommatophoca rossii*, and the Weddell seal, *Leptonychotes weddellii* (Figure 4.20D).With the exception of the Weddell seal, which forms small pupping colonies, all of these are fairly solitary animals that pair up on the sea ice only during the breeding season (Knox, 1994). Based on surveys carried out in the late 1960s, the crabeater seal numbered a spectacular 1.5×10^7 individuals, making it one of the most abundant mammals in the world (Bonner, 1985a). By

Fig 4.20 Some common types of Antarctic seal: (A) the crabeater is the most numerous of Antarctic seals and favours a sea ice habitat; (B) leopard seals are predatory on most other seal species, penguins and fish, although 50% of their diet is krill; (C) elephant seals, the largest of the seals, favour the sub-Antarctic islands but are known to travel great distances in search of food; (D) the Weddell seal breeds and hauls out mainly on fast ice close to the shores of the continent (source: authors).

1982, Laws (1984) estimated the number at 3×10^7, but recent estimates place numbers at a more modest $1.1–1.2 \times 10^7$ (Erickson and Hanson, 1990). Rarely seen on the shore, the 2 m long crabeater weighs around 220 kg and prefers drifting pack ice, where it feeds, in spite of its name, almost exclusively on krill (see Figure 4.20A, 4.21) strained between interlocking cuspate teeth. Laws (1977a) estimated that crabeaters consume at least 6.3×10^7 t a^{-1} of krill, a figure that exceeded the total removed by the remaining baleen whales. It is preyed upon in its first year by the 3 m long, 350 kg leopard seal, and many crabeaters that survive its massive canines carry the tell-tale parallel scars on their flanks.

The 315,000-strong population of leopard seals (Erickson and Hanson, 1990) is found extensively amongst the pack ice as well as on the shores of the oceanic islands (see Figure 4.20B). The feeding habits of this ferocious predator also extend to penguins, and most rookery landing places have a leopard seal in attendance. However, leopard seals also take up to 50% krill, 24% birds (mainly penguins), other seals and fish and cephalopods in equal quantity in their diet, the proportion varying by season depending on the availability of prey (Bonner, 1987; Knox, 1994). The Ross seal numbers 152,000 (Erickson and Hanson, 1990) and inhabits the heavy pack ice close to the continent. It reaches about 2 m long and 200 kg in weight and feeds mainly on squid (57%) and fish (34%) (Bonner, 1987; Knox, 1994). There are close to 10^6 Weddell seals (Erickson and Hanson, 1990). The Weddell seal reaches about 3 m and up to 400 kg and inhabits the fast ice close inshore, where it feeds mainly on fish (58%), together with crustaceans and squid (Figure 4.21).

The land-breeding species include one of the true seals, the elephant seal, *Mirounga leonina*, which prefers the oceanic islands south of the Polar Frontal Zone, where it is found in great numbers (see Figure 4.20C). This distribution it shares with the most common of the southern eared seals, the Antarctic fur seal. The elephant seal is the largest of seals, and full-grown bulls reach 4.5 m in length and 4000 kg in weight, but the cows are much smaller at 900 kg and 2.8 m long (Bonner, 1987). The 600,000-strong elephant seal population (Laws, 1977a) is circumpolar in distribution and is seasonally migrant. The elephant seal favours the oceanic islands of the Scotia Sea, Îles Kerguelen, Îles Crozet and Heard Island and tends to avoid breeding sites heavily affected by sea ice. About 75% of its diet is squid, with fish comprising the rest, and they are estimated to take 75% of all squid taken by predators in the Scotia Sea (Croxall *et al.*, 1985). The Antarctic fur seal (*A. gazella*) inhabits the waters of the oceanic islands close to the Polar Frontal Zone, and 95% of the world population is found on South Georgia (Croxall, 1992), although historically the islands of

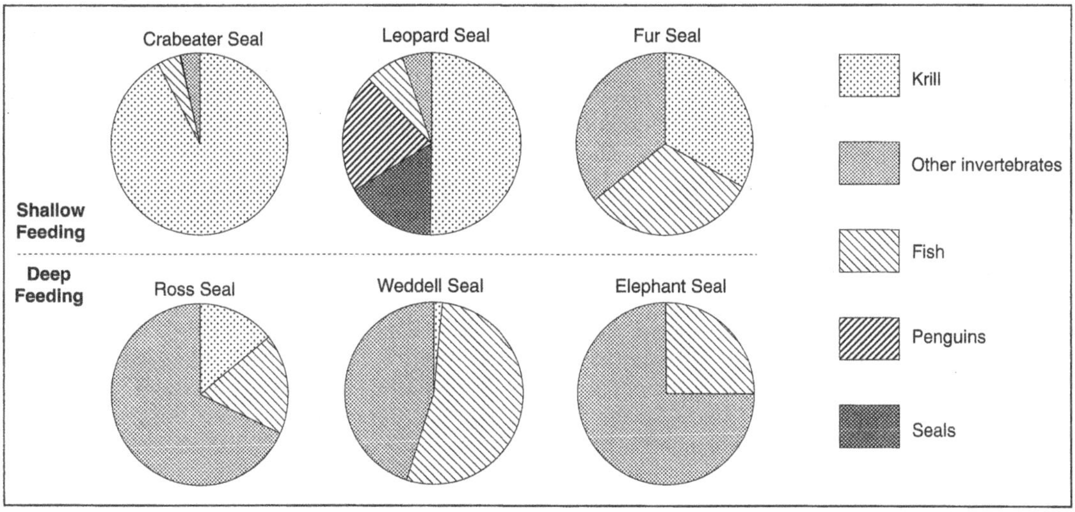

Fig 4.21 Percentage composition of stomach samples of Antarctic seals showing the similarities in diet amongst the shallow-feeding seals and deep-feeding seals as well as the differences between these two groups (source: modified after Laws, 1984).

the South Shetlands, South Orkneys and South Sandwich group also supported large populations. Fur seals are not large, the 200 cm long bulls weigh 125–200 kg and the 140 cm long cows weigh only 50 kg (Bonner, 1985a). The diet of the $0.7–1.0 \times 10^6$ Antarctic fur seals can reach 80% krill and about 10% each of fish and squid in summer (Croxall *et al.*, 1985), but Laws (1984) places the overall diet as equally split between fish, krill and squid (see Figure 4.21). However, the fur seal does not compete for krill with the crabeater seal, since their ranges do not overlap. Similarly the ranges of the squid-eating elephant and Ross seals do not overlap.

All the seal populations appear to have been subject to historical fluctuations. Seals such as the fur, elephant and, to a very minor extent, crabeater have been subject to exploitation by humans in the past, and this is reviewed in Chapter 7. Of these, *A. gazella* came close to extinction, with no seals seen at South Georgia between 1907 and 1919 (Bonner, 1985a). Since then there has been an increase on South Georgia, so that the population now exceeds pre-exploitation levels (Croxall, 1992) (see Figure 4.18) and is still growing at 10% a^{-1}. There is also evidence of breeding and a 2–6% rate of population increase, mainly derived from the South Georgia stock, in the South Orkney Islands and the South Sandwich Islands (Knox, 1994). In the South Shetland Islands, numbers were reported to be increasing in the 1970s at 34% a^{-1}, although reinforcement by immigration is likely (Aguaya, 1978; Bonner, 1985a). The fur seal has now reached Adelaide Island and adjacent parts of the Antarctic Peninsula. In some populations, the numbers of elephant seals appears to have either held constant or increased since the end of sealing in the 1950s (Croxall, 1992), but other populations are declining. Variations may be due to fluctuations in environmental factors or in the availability of prey species between regions, possibly in part as a result of the squid fishery. The steady rise in fur seal numbers this century appears to be related to a krill surplus caused by the demise of the great whales. Similar increases were noted in crabeater numbers in the late 1960s and 1970s, but the estimated numbers now seem to have declined by up to 53% (Erickson and Hanson, 1990). This apparent fall in crabeater numbers probably reflects earlier overestimation of seal stocks (*ibid.*), but a minor component may also be the result of increasing competition for food due to the revival of whale consumption of krill. Long-term population trends for Weddell seals suggests stability (Croxall, 1992) and thus, possibly, little change in competition for its prey of relatively deep-water fish close to the continent.

In addition to these trends there is also evidence of cyclicity in those seal populations that breed on sea ice. For example, there are strong age cohorts on a four–five-year cycle in the crabeater seals of the Antarctic Peninsula, four–five-year fluctuations in numbers of leopard seals on Macquarie Island and four–six-year fluctuations in reproductive success in Weddell seals in McMurdo Sound (Croxall, 1992). Testa *et al.* (1991) related these cycles to the Southern Oscillation Index, a measure of the strength of El Niño/Southern Oscillation or ENSO events. El Niño is a climatic fluctuation in the Pacific that affects the pattern of ocean–atmosphere interactions every two–ten years: the strength of the easterly trade winds weakens and they may become westerly in the western Pacific (Open University, 1989). This affects the sea temperature distribution within the tropics and it may also influence the Southern Ocean. The mechanism by which ENSO affects the Southern Ocean is unclear but it is likely to operate by altering sea ice variability and ocean currents. Guinet *et al.* (1994) also identify a possible link between ENSO events and the breeding success of fur seals on sub-Antarctic islands. None of the above linkages is yet proven but it appears possible that in some species of seal there may be indirect linkage between population trends and the strength and nature of ocean–atmosphere interaction and sea ice variability.

By far the largest of the tertiary consumers commonly found in the Southern Ocean are the seven species of krill-eating baleen or whalebone whales and the two species of toothed whales. The blue whale, *Balaenoptera musculus*, the world's largest animal, reaches 30 m and 150 t; the fin whale, *B. physalus*, grows to 25 m and 90 t; the sei whale, *B. borealis,* reaches 18 m and 30 t; the minke whale, *B. acutirostrata*, seldom exceeds 11 m and 19 t; the humpback whale, *Megaptera novaeangeliae,* reaches 16 m and 60 t; and the right whale, *B. glacialis*, reaches 18 m and 90 t (Gambell, 1985). The pygmy blue whale, *B. musculus brevicauda*, is found around Marion Island, Îles Crozet and Îles Kerguelen. The largest of the toothed whales is the sperm whale, *Physeter macrocephalus*. Bulls grow to 18 m and 70 t and cows to 11 m and 17 t, but only the males are found in Antarctic waters. The second most important toothed whale is the killer whale, *Orcinus orca,* which reaches 9 m and 8 t. Several small and medium-sized toothed whales are also found, the largest of which is the southern bottlenose whale, *Hyperoodon planifrons*, and Arnoux's beaked whale, *Berardius arnuxii.* They form an important but seasonal element in the food web. Most baleen whales move into Antarctic waters in early summer to take advantage of the rich krill harvest but, after about three or four months of intensive feeding, they return to warmer waters to breed during the long Antarctic winter. All of the baleen whales take krill sieved through the massive plates of horny baleen that grow from the upper jaws. The system is highly efficient and allows the southern whales to consume an estimated 62×10^6 t a^{-1} of krill (Berkman, 1992; Croxall, 1992) (Table 4.2). A mature blue whale can consume 4 t of krill in a day. Not surprisingly, the greatest densities of whales coincide with those of krill in the Scotia and Bellingshausen Seas, the Ross Sea and offshore of Wilkes Land and Queen Maud Land (Figure 4.22).

Whales tend to congregate at first along the receding edge of the pack ice, where krill graze on planktonic blooms. Two techniques of feeding are employed, swallowing and skimming. Blue, fin, sei, minke and humpback all force mouthfuls of sea water out through the baleen by raising the tongue in the mouth, thus entrapping the krill. Humpbacks, often in pairs, have also been observed to encircle a krill swarm with a ring of bubbles before surfacing from depth in the middle of the swarm with mouth agape (Gambell, 1985). Sei whales also employ skimming, where the whale swims through krill swarms with head above the surface and mouth open, a technique habitually used by the right whale (*ibid.*). In addition, the sei whale eats copepods in large numbers (Knox, 1994). Sperm whales are also primarily tertiary consumers and feed almost exclusively on medium-sized (1 m) squid in the Antarctic, although some fish are taken. Of the Antarctic whales perhaps the whale with the most fearsome reputation is the killer whale. Hunting in packs often numbering 30 or 40, killer whales feed on squid, fish, penguins, seals and other whales, especially minke. Like the leopard seal, feeding for this top predator is opportunistic since any suitable prey, live or dead, is likely to be taken (Gambell, 1985). This is supported by whalers, who report packs of killer whales following whaling ships and tearing chunks of flesh from recently killed baleen whales (Stonehouse, 1972). As a deterrent, individual killer whales were sometimes shot by the whalers, only to be promptly devoured by the rest of the pack (McLaughlin, 1962).

The rise and fall of the pelagic whaling fishery is reviewed in Chapter 7 but it is necessary here to assess the significance of the great whales to the Antarctic ecosystem and the implications of their demise. There can be no doubt that the removal of the great whales represented the single largest perturbation that the ecosystem of the Southern Ocean has

Table 4.2 (A) Estimates of initial Antarctic whale stocks and food consumption; (B) estimates of present Antarctic whale stocks and food consumption. Estimates such as these are notoriously unreliable because of different methodologies for data collection. For example, Best (1993) estimates present Antarctic and Blue numbers at 6.8×10^3 and 0.71×10^3 respectively and Gambell (1998, personal communications) places Minke numbers at 761×10^3 and Blue at 0.46×10^3

A.			Food consumed (10^6 t a^{-1})		
Species	Population (10^3)	Biomass (10^3 t)	Krill	Squid	Fish
Blue	200	17.6	72	0.74	1.5
Fin	400	20.0	81	0.84	1.7
Humpback	100	2.7	11	1.11	0.2
Sei	75	1.4	6	0.06	0.1
Minke	200	1.4	20	0.20	0.4
Sperm	85	2.6		10.2	0.5
Total	1060	45.7	190	12.2	4.4
B.					
Blue	10	0.83	3.4	0.04	0.07
Fin	84	4.03	16.4	0.17	0.34
Humpback	3	0.08	0.3	0.00	0.00
Sei	41	0.71	2.9	0.03	0.06
Minke	400	2.80	39.6	0.40	0.82
Sperm	43	1.16		4.63	0.24
Total	581	9.61	62.6	5.27	1.53

Source: Berkman, 1992; Croxall, 1992.

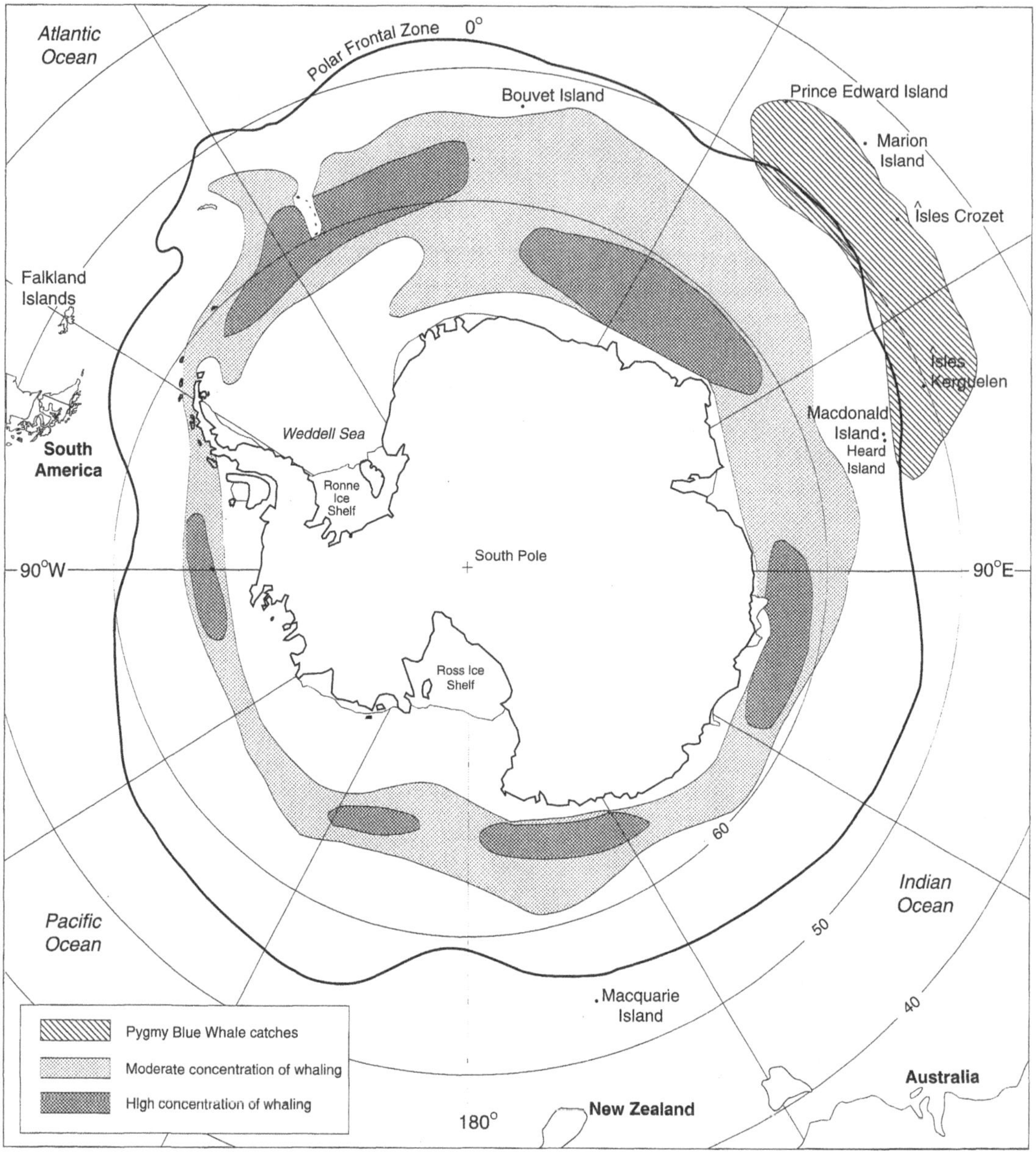

Fig 4.22 The distribution of Antarctic whales as reflected in the areas used most intensively during the period of pelagic whaling (source: after MacKintosh, 1965).

experienced in modern times. Initial Antarctic whale stocks are estimated to have been about 1.05×10^6 individuals, which consumed 1.9×10^8 t a^{-1} of krill as well as sizeable quantities of fish and squid (Berkman, 1992; Croxall, 1992) (see Table 4.2). As a result of whaling, humpback and blue whales were reduced to 3% and 5%, respectively, of their pre-exploitation levels and sperm whales to 50% (*ibid.*) (Figure 4.23). Minke whales, exploited only at the very end of commercial whaling, have largely avoided the catastrophic losses of the larger whales and are now the most common whale species (Croxall, 1992).

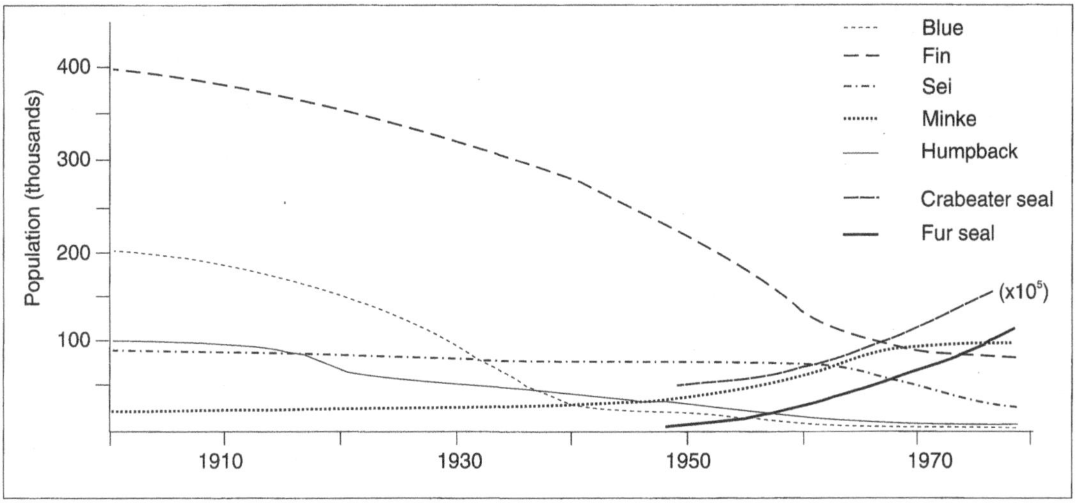

Fig 4.23 Changes in some whale and seal populations in the Southern Ocean, 1900–1980. Note the phased timing of the decline of humpback, blue, fin and sei whales and the increase in numbers of minke whales and crabeater and fur seals (source: data from Knox, 1994).

However, as a consequence of the removal of the large whales, the annual predation on krill, squid and fish decreased from over 20 to less than 5×10^7 t a^{-1} (see Table 4.2), and the effects of this release of food radiated through the Antarctic marine ecosystem to the benefit of competitor species.

The above reduction in the stocks of whales is popularly thought to have produced a 'krill surplus' that might sustain a commercial krill fishery. However, this is an unlikely scenario, since any surplus is likely to have been largely taken up by other competitor species (Laws, 1985). For example, there have been large increases in crabeater seals, which now consume 1.5 times as much krill as did the whales; the rebound of the fur seal population following overexploitation has been faster than expected due to availability of krill; and 2–10% a^{-1} increases in numbers of krill-eating penguins have been related to the decline of the great whales (Croxall and Prince, 1979; Croxall, 1992) (see Figures 4.18 and 4.23). The various elements of the Antarctic ecosystem appear to have already responded and maintained a constant total level of krill consumption (Berkman, 1992). Over this period, changes have also occurred in the biological characteristics of both whales and some seals as a response to decreased whale abundance. The pregnancy rates of Antarctic blue, fin, sei and minke whales have increased, while the mean age at sexual maturity has decreased over 40 years (Gambell, 1985; Knox, 1994) (Figure 4.24). Some of these trends can be recognised in advance of any direct exploitation of the species themselves and show interspecific responses to the earlier removal of competitor whales and the release of food (Gambell, 1985). Crabeater age at sexual maturity has fallen from five to three years over the 30 years 1941–1971 (Knox, 1994). The response of individual species to removal of either food or competitors is thus neither simple nor, given the present imperfect level of knowledge, easily predictable. However, it remains clear that a 'krill surplus' may not now exist and that any commercial catch of krill will increase competition between the krill eaters.

4.4.6 Summary: the geographical zonation of the marine ecosystem

The Antarctic marine system is fundamentally controlled by the unrestricted flow of ocean currents circulating around the continent, together with the effects of the seasonal pulse of sea ice freeze-up and break-up. These currents bring together nutrients from both the warmer oceans and the Antarctic continent, the resultant mix providing much of what is necessary for growth as the spring sunlight warms the surface layers of the ocean, particularly in the MIZ. The resultant phytoplankton blooms lie at the heart of the Antarctic marine food web and sustain large but spatially variable populations of grazing

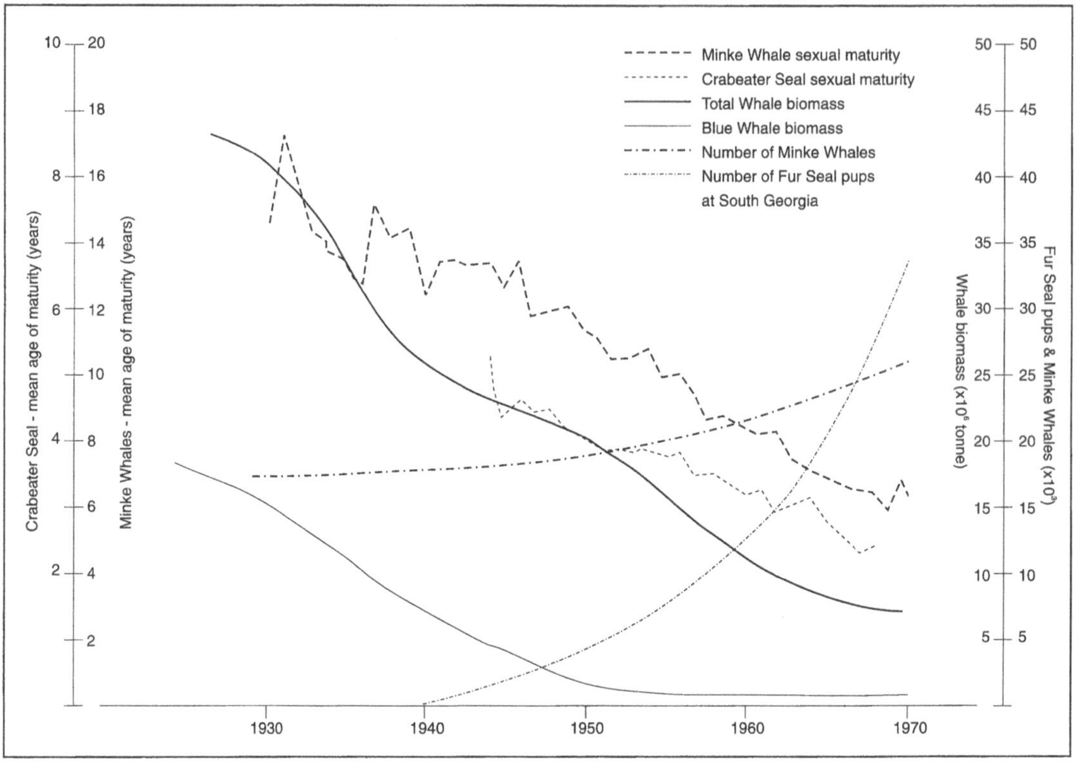

Fig 4.24 Interspecific relationships between krill-feeding animals and baleen whale biomass in the Antarctic marine system (source: redrawn after Knox, 1994).

zooplankton, the most important of which is krill. There are distinct areas and times where phytoplankton are plentiful, these being broadly coincident with the densest congregations of krill.

In spite of within-zone and longitudinal variations, three fundamental circumpolar zones in species composition can be recognised, which broadly follow the zones of phytoplankton abundance shown in Figure 4.11: the ice-free zone of open Antarctic waters; the seasonal sea ice zone; and the permanent sea ice zone (Knox, 1994). Although overlap exists, particularly for opportunistic feeders like the leopard seal and the petrels, the distribution of Antarctic species shows a fair degree of zonation in response to availability of prey items (Figure 4. 25). In general, the ice-free zone of the ACC is nutrient-rich but, apart from a few areas, poor in primary production. Ecologically, the most important of the zooplankton are the copepods and euphausiids, while whales and petrels dominate the higher fauna. The seasonal sea ice zone, the mean northern boundary of which is about 60° S, occupies most of the area of the Antarctic Coastal Current and is particularly rich in krill and krill feeders such as penguins and seals. Krill are thus available to the penguin and seal populations that breed in this zone and to the baleen whales that penetrate it (*ibid.*). In the permanent sea ice zone, including the polynyas, phytoplankton production is limited to a short summer period, and *E. superba* is largely replaced by the smaller *E. crystallorophias*. Fish are abundant, as are fish-eating seals, but the krill-eating mammals are less so and some of the primary production remains unconsumed, falls to the bottom and is incorporated into sediments or consumed by benthic feeders.

4.5 Conclusion

In response to a circumpolar atmospheric circulation pattern and the absence of land barriers, the currents of the Southern Ocean follow a consistent and largely circumpolar circulation pattern. The radial outflow of cold air and water from the continent is

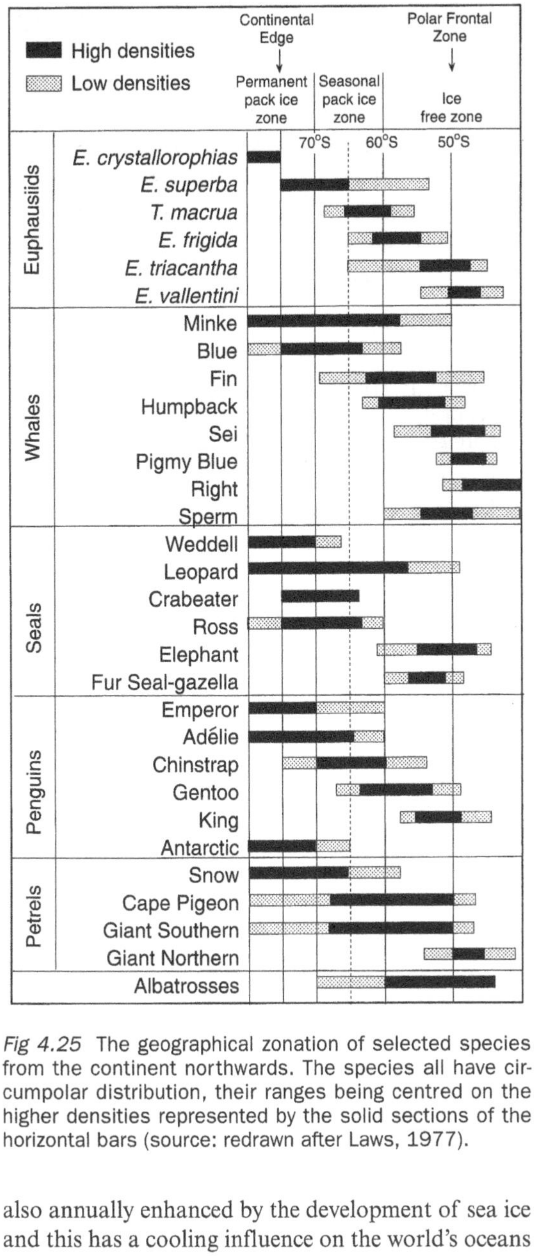

Fig 4.25 The geographical zonation of selected species from the continent northwards. The species all have circumpolar distribution, their ranges being centred on the higher densities represented by the solid sections of the horizontal bars (source: redrawn after Laws, 1977).

also annually enhanced by the development of sea ice and this has a cooling influence on the world's oceans and atmosphere, which is counterbalanced by the inflow of warm air and water from the lower latitudes. This interchange is fundamental to the world's energy balance but it also introduces several environmental factors important to life in the Southern Ocean. The winter expansion of the sea ice zone adds to the strong seasonality of marine biological production, and with the interchange of ocean currents comes a mixing of minerals and nutrients from both north and south that contributes to both Antarctic and mid-latitude marine ecosystems. The mixing is greatest at the zones of contact of the ocean currents, especially where they are deflected northwards by land, such as the Antarctic Peninsula. Such nutrient-rich waters provide an ideal habitat for phytoplankton growth and the populations of zooplankton that graze on it. These form the basis of a food web that is short but also relatively complex at all levels, with a range of upper predators including large numbers of birds, seals and whales. The areas of most intense oceanic mixing are the most biologically productive, but elsewhere the Southern Ocean is much less productive and in general appears to be no more productive than other oceans. Krill occur in such quantities that a large proportion of the birds and marine mammals found in the Southern Ocean prey upon them to some extent. Mapping the distributions of zooplankton species and the species that prey on them reveal distinct biological zonations in the Southern Ocean ecosystem.

The Southern Ocean ecosystem has also been subject to dramatic perturbations in the recent past, and the impacts, first of sealing and later of whaling, had a widespread effect on the species exploited and on their prey and competitor species. Several species have been reduced to extremely low numbers by exploitation. However, some species show a remarkable ability to recover their populations. This is especially the case with fur seals, whose life cycle is relatively short and so can be expected to show a rapid response: populations on South Georgia now exceed the estimated pre-exploitation levels. Amongst the whales the reduction of mean age at sexual maturity and the increasing pregnancy rates of some species show an interspecific response to the earlier removal of competitors for food. However, the so-called 'krill surplus', often suggested as the basis of a krill fishery, has probably already been utilised in the expansion in the populations of other krill eaters.

All this serves to indicate that, contrary to popular belief, the Southern Ocean is far from being a pristine marine environment and is an ecosystem highly modified by humans. However, judging from its response to past human predation of marine species, it also seems to be a fairly robust system. This is perhaps not surprising, since the effect of harvesting species that occupy the higher levels of the food web and that are not themselves preyed upon is to increase the availability of food and enhance the productivity of competitor species (Knox, 1994). Unfortunately, this may not be the case for the harvesting of species at lower levels in the food web such as krill or finfish. The low growth and fecundity of Antarctic finfish render them susceptible to overfishing, and the resultant effects on their predators is not well-known. Similarly, the interspecific responses to the harvesting of an ecologically important species such as krill would be widespread and impact unpredictably on the krill-competitor species, as well as on the krill eaters and their dependants (Chapter 5), particularly since it is still unknown even whether krill populations are discrete. When stressed by the harvesting of a species so important to the food web, the response of the Southern Ocean ecosystem is likely to prove much more sensitive and less robust than has been the case so far. Much may depend on the ability of Antarctic species to react to changes in conditions, not only in the short term as a result of human predation, but also in the longer term in response to climate change. The nature of these adaptations and interactions between the natural and physical systems of the Antarctic are examined in the following chapter.

Further reading

Deacon, G. (1984) *The Antarctic Circumpolar Ocean*. Cambridge University Press, Cambridge.

Knox, G.A. (1994) *The Biology of the Southern Ocean*. Cambridge University Press, Cambridge.

Sahrhage, D. (ed.) (1988) *Antarctic Ocean and Resources Variability*. Springer-Verlag, Berlin.

Stonehouse, B. (1989) *Polar Ecology*. Blackie, Glasgow and London.

Sugden, D.E. (1982) *Arctic and Antarctic*. Blackwell, Oxford.

Wadhams, P. (1994) The Antarctic Sea Ice Cover, in Hempel, G. (ed.) *Antarctic Science: Global Concerns*. Springer-Verlag, Berlin, 45–59.

Walton, D.W.H. (ed) (1987) *Antarctic Science*. Cambridge University Press, Cambridge.

CHAPTER 5

Interactions and changes within the physical and biological systems of the Antarctic

Objectives

Key themes and questions covered in this chapter:

- the physiological characteristics and ecological and behavioural strategies of organisms in the Antarctic
- potential impacts of climate change on the physical systems of the Antarctic
- potential biological impacts of climate change
- the sensitivity of Antarctic ecosystems to past and future change

5.1 Introduction

The nature, functioning and spatial patterns of the major Antarctic systems have been discussed in the preceding chapters and it is one of the aims of the present chapter to examine the ways in which these systems interact and influence Antarctic flora and fauna. A second aim is to investigate how the relationships between the physical and natural systems of the Antarctic might change in response to changes in the climatic factors that so fundamentally affect the environment. On the apparently inhospitable land and in the surrounding ocean the fact that life not only survives but in many cases thrives raises questions about the nature of the variables that control the survival and success of organisms in the Antarctic environment. Both the structure and functioning of the terrestrial ecosystems are bound up with the seasonal rhythm of sunlight, snow and ice cover, water availability and temperature, and whereas the marine ecosystem is less affected by temperature fluctuations, it is still subject to marked seasonality in ice cover and light.

The biogeography of the Antarctic is influenced by spatial variation in environmental controls, and this conditions the way in which flora and fauna adjust to cope. However, very similar animals may adopt quite different strategies to cope with the extremes of climate. For example, the emperor penguin has evolved the ability to incubate on sea ice during the winter and this allows it to breed every year, its chicks using all of the short summer for growth (Croxall, 1997). Its close relative, the king penguin, incubates in summer on land and then feeds its young over the subsequent winter. Birds that breed successfully in one summer can then breed late in the summer of the following year, but many eggs and late chicks are abandoned to die as food becomes limiting. On average, breeding of king penguins is successful once in every two years (Stonehouse, 1985). The latter strategy requires year-round access to an ice-free sea for food, whereas the former requires the bird to fast for part of the winter at sites remote from the open ocean. These very similar birds have developed successful but very different breeding strategies in response to geographical variation in environmental conditions. How might these, and other organisms, cope with any future climate change? In order to assess the potential biological impacts of such changes, we need to know how biota have altered their life-style to cope with the Antarctic environment as well as how climate change might affect the physical systems that underpin the habitat.

5.2 Adaptations of organisms to the Antarctic physical environment

Ecologists recognise two broad ways in which organisms cope with the strictures of environmental stress:

the development of physiological processes that allow organisms to withstand extreme temperatures and regulate body temperature, and ecological and behavioural adaptations related to the ways in which organisms take advantage of favourable conditions and protect themselves from adverse conditions. However, possession of particular physiological characteristics may also provide organisms with ecological advantage.

5.2.1 Physiological processes

Since the temperature of most Antarctic environments is well below freezing for long periods of the year, a major physiological requirement for most species is both to avoid freezing and to cope with the unavailability of water. On land, the latter is probably the ultimate control on organism activity (Kennedy, 1993) but is obscured in Antarctica since the latitudinal decrease in temperature coincides with a latitudinal decrease in precipitation. Clearly both plants and animals need to adopt strategies to survive under such constraints. Organism survival in the Antarctic has been studied from two main perspectives: photosynthesis and respiration in plants and resistance to cold, principally in animals (Block, 1994).

Photosynthesis and desiccation

Photosynthesis in continental Antarctic lichens occurs down to –18 °C, the frozen plants reaching their pre-freezing metabolic rates immediately after thaw and rehydration (Kappen, 1993). Experiments with liquid nitrogen show that crustose lichens can even withstand freeze-drying down to –196 °C. The two flowering plants of the Antarctic, *Deschampsia antarctica* and *Colobanthus quintensis,* can photosynthesise to –5 °C (Smith, 1984). Mosses also appear to be able to photosynthesise at subzero temperatures (down to –10 °C for the maritime Antarctic moss *Polytrichum alpestre*) and, even under the much reduced light levels beneath snow cover, up to 30% of annual growth can occur (Everson, 1987). However, mosses generally appear to be less well able to survive extremes of temperature and moisture than lichens and take more time to recover metabolic activity on thawing and rehydration (*ibid.*). The physiological ability of these plants to survive low temperatures, freezing and desiccation, only to resume photosynthesis rapidly when conditions present themselves, represents an important colonising advantage in a harsh environment. In the dry valleys of South Victoria Land the severity of the surface environment has resulted in some lichens, free-living algae and fungi filling an ecological niche within the mineral grains of sandstone rock at up to 9 mm below the surface (see Figure 3.28). They survive in this microhabitat by utilising light penetration through the quartz crystals in the rock, moisture retention from condensation in the interstitial spaces, and the warming effect of relatively high, if brief, surface temperatures. However, when conditions result in the lichens being exposed on the surface, for example by weathering, they can produce the small thalli typical of surface lichens. It therefore appears likely that these cryptoendolithic lichens represent highly adaptive forms that have evolved in response to deteriorating Antarctic environmental conditions since the Tertiary (Friedmann, 1982).

Avoidance of freezing

There are two principal ways in which Antarctic fauna avoid freezing. The poikilotherms or cold-blooded animals have mainly adopted strategies that involve closely following the temperature of their environment and developing ways of preventing body fluids from freezing. The homeotherms or warm-blooded animals generally maintain higher body temperatures by minimising heat loss (Everson, 1987). Of the former group of animals, at least four mechanisms appear to be used to avoid the harmful effects of freezing: tolerance of freezing itself; 'supercooling' without the use of special substances; internal production of polyhydric alcohols (polyols) or sugars to aid supercooling; and internal production of 'antifreezes' (*ibid.*). Freezing tolerance is the exception rather than the rule in the Antarctic and has only been demonstrated in the midge *Belgica antarctica*, whose larval stage can withstand the freezing that is lethal for the adult (Block, 1994). Supercooling without the use of special substances is used by a small number of species that rely on body fluids remaining liquid at low temperatures. It is found in some freshwater species, where the ionic concentrations of animal body fluids is higher than that of the surrounding water. If the water does not freeze to the bed then the animal will survive in the unfrozen water. In the ocean many species of polar fish survive in water that is up to 1 °C colder than the equilibrium freezing point of their body fluids, so they must avoid any possibility of actual contact with ice. Such mechanisms demand that the biogeography of these fish must be a deep-water one well away from the ice-infested shallow water zone

(Kock, 1992): a behavioural response to a physiological requirement.

Alternative mechanisms are used by other organisms. Freeze-avoidance by the production of polyols or sugars to aid supercooling is known to occur in some Antarctic plants and many species of terrestrial arthropods. The 1 mm long mite *Alaskozetes antarcticus* uses polyols to supercool to –25 °C in amounts that appear to be related to feeding activity and temperature (Sømme, 1985). These sugars are hydrophilic, serving to reduce the amount of free water and thus the chances of ice nucleation in the animal. However, this may not be specific to the polar environment, since the use of polyols is also characteristic of many temperate relatives of *Alaskozetes* (Block and Convey, 1995). For some Antarctic fish, the seasonal sea ice zone offers good feeding opportunities and in order for such fish to survive in this zone they have developed ways of producing enhanced levels of sodium and chlorine salts in the body fluids in order to prevent freezing. These salts account for 40–50% of the freezing point depression, with the remainder balanced by production of antifreezes such as glycoproteins or peptides (Kock, 1992) (Figure 5.1). The mechanism is as yet unclear, but glycoproteins appear to work either by interacting with the surface of the ice crystal lattice and preventing ice growth penetrating the surrounding free water or by dividing the surface of the ice crystal into many small microcrystals, each of which has a lower melting point than the larger crystal (Eastman and de Vries, 1986). Either way the advantage of glycoproteins would be lost to fish if they were excreted in the normal way via the kidneys, so in the notothenioids, for example, urine can be secreted rather than filtered. Such renal retention leaves the antifreeze in circulation and thus saves the energy that would have been lost in its replacement (Kock, 1992).

The parasitic insects of the continent ingeniously manage to combine protection from the Antarctic climate with proximity to their source of food. Biting lice are the most numerous of these, living year-round among the bases of feathers. Sucking lice are found on some seal species, and the Antarctic flea (*Glaciopsyllus antarcticus*) is found in the nests of fulmars and petrels. The close contact with their warm-blooded hosts protects them from the extremes of the Antarctic climate (Sømme, 1985).

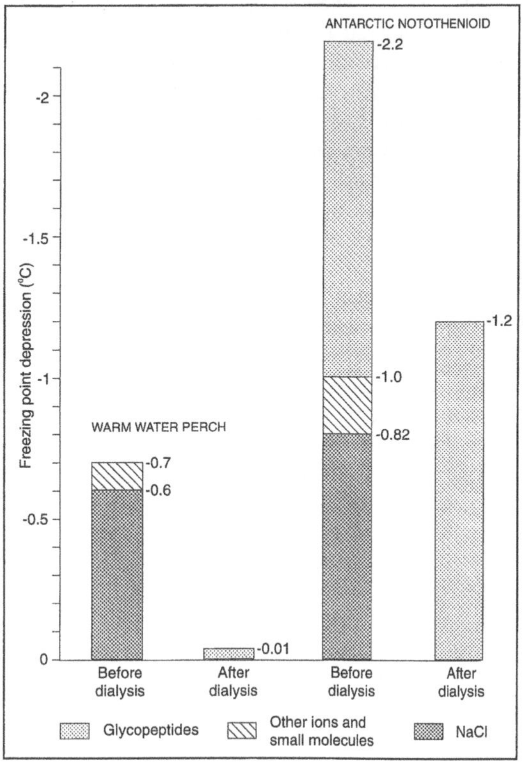

Fig 5.1 The blood plasma collected from both warm-water perch and an Antarctic notothenioid before and after dialysis to remove blood solutes of low mass. After dialysis, the freezing point of the perch plasma rose slightly, indicating limited freezing-point depression as a result of sodium chloride. After dialysis, the freezing point of notothenioid plasma rose but still remained low, the continued freezing-point depression being the result of the presence of larger glycopeptide molecules (source: redrawn after Eastman and de Vries, 1986).

Metabolic rates

Other mechanisms are employed by Antarctic animals to mitigate the effects of cold on their body functions. Some invertebrates have the ability to undergo almost instantaneous metabolic compensation for temperature changes in the Antarctic and so allow them to exploit the short summer fully by rapid increases in activity and growth (*ibid.*). The metabolic compensation of some marine invertebrates allows them cover the full temperature range of their environment and so to maintain the same level of activity under all conditions. In general, because chemical reactions and metabolic activity are temperature-related, the cold tends to slow animals down and thus their energy demands are less. The respiration rate of Antarctic fish is less than that of fish from warmer waters. The icefish (Figure 5.2) has only half the metabolic rate of other Antarctic fish. The nature of the Southern Ocean has also resulted in physiological processes in fish that save vital energy. Most Antarctic fish have only half the haemaglobin levels of fish in warmer seas, and the icefish has no haemaglobin at all (Kock, 1992). Blood haemaglobin is normally used to extract oxygen from the gills and carry it to the parts of the body where it is needed. However, cold water holds more oxygen than warm and together with its stormy and well-mixed nature, the Southern Ocean is rich in oxygen (Kunzmann, 1991). With low energy requirements and an oxygen-rich oceanic environment, the icefish has no need to expend energy producing haemaglobin and has developed other ways of extracting it. Matched against comparable red-blooded notothenioids, the oxygen-carrying capacity of icefish blood is a mere 10%, but this is compensated for by larger gills to allow efficient oxygen uptake. It also has two–four times the blood volume, a faster circulation and a heart that is two–three times larger and beats six–fifteen times faster than that of comparable fishes (Kock, 1992). In the cold Antarctic waters, lack of haemaglobin appears to present no real disadvantage to the surprisingly active icefish.

Growth rates and reproductive effort in Antarctic fish are generally low. Since metabolic activity is broadly temperature-related, this originally led to claims that Antarctic fish had higher metabolic rates than would be predicted by temperature alone. This was thought to be a form of cold adaptation that required expending more energy to maintain a basal metabolic rate, and less energy on the other functions of growth and reproduction. However, recent experiments suggest that many Antarctic fish show little evidence of this form of cold adaptation in their metabolic rates (*ibid.*). Although controversy remains over the issue, the consensus appears to be that most Antarctic fishes can compensate for the effects of low temperatures on metabolic rates and that living in cold water may be an advantage compared with the metabolic costs of living in warmer waters: for any given energy intake less energy is required for the maintenance of basal metabolism in cold waters and so more can be directed towards growth and reproduction (*ibid.*). Irrespective of this, growth rates are slow in Antarctic fish and this raises serious implications for the sustainability of Antarctic fisheries (Chapter 7).

Fig 5.2 The icefish, *Champsocephalus aceratus*, on the sea bed. The species is well-adapted to life in cold water, with a large heart and rapid circulation to offset low levels of blood haemoglobin (source: photograph by D.G. Allan, courtesy of BAS).

Thermoregulation

The second major physiological mechanism used by Antarctic animals to avoid the cold relates to the maintenance of higher body temperatures and a metabolism unhindered by outside temperature. The thermoregulation that this requires is used by most Antarctic homeotherms. Marine mammals and birds pay for the cost of enhanced basal metabolism by higher levels of energy consumption in conjunction with enhanced body insulation designed to minimise heat loss. Some of these physiological characteristics are also associated with adoption of behavioural and ecological strategies that avoid exposure to extreme conditions (Everson, 1987). For example, all of the whales and phocid seals have developed a thick layer of subcutaneous fat or blubber, which insulate them effectively from the cold. The thickness of the layer depends on the species: for blue whales it reaches 20 cm thick; and for a bull elephant seal, about 12 cm thick. Since there are limits to the optimum thickness of the insulating layer, larger species such as the blue whale should in theory carry lower proportions of body weight as blubber. However, this is rarely the case, since blue whale blubber built up during the summer Antarctic feeding period also serves as a food reserve for the winter months spent outwith the Antarctic (Gambell, 1985). In contrast, the much smaller minke whale spends the whole year in the Antarctic yet has a perfectly adequate, but relatively much thinner, layer of blubber (Everson, 1987). In any case, there may also be a tendency to overemphasise thermoregulation in marine mammals, since water temperatures in the Southern Ocean are never very far below 0 °C at any time.

Another tendency seen in many polar animals is to favour shapes and sizes that minimise heat loss from the body surface. From the perspective of heat loss, a sphere provides the minimum surface area per unit volume and for most marine mammals rounded shapes with minimal appendages serve to reduce heat loss. However, a compromise must also be achieved between a rounded shape and hydrodynamic efficiency: fat spherical seals are more easily caught and eaten and, as a result, spherical seals have become extinct! In response, seals and whales have developed rounded but streamlined shapes with relatively small and hydrodynamically efficient flippers. The large size of the marine mammals such as the whales and large seals is efficient from a heat loss perspective since the larger the body size, the lower the proportion of surface area through which the heat can escape. However, as Everson (*ibid.*) points out, this poses a problem since all such animals must start off small before they grow large, so the rapid early growth rate displayed by all of the Antarctic mammals is advantageous in minimising heat losses when young (Figure 5.3). It is possible that the predominance of large seabirds in the Antarctic is also related to heat efficiency advantages (Figure 5.4).

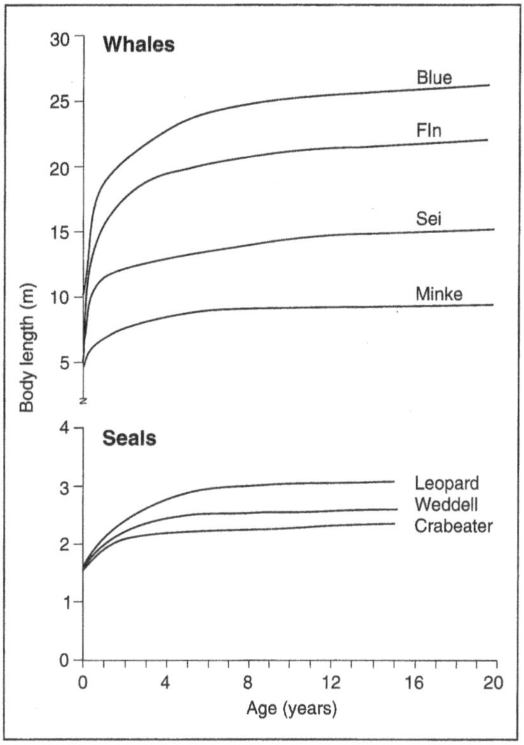

Fig 5.3 The growth curves for four species of Antarctic whale and three species of ice-breeding seal all show very rapid initial growth rates to minimise cold stress and predation (source: data from Gambell, 1985; Everson, 1987).

In addition to a layer of subcutaneous fat, another mechanism to reduce heat loss is used by the eared seals and birds. The fur seals have developed a thermally efficient layer of very fine fur beneath the longer and coarser guard hairs (Figure 5.5A). The fine fur traps air close to the skin and provides the diving seal with an insulating layer similar in function and effectiveness to the air contained within the neoprene of a scuba diver's wetsuit. In earlier centuries, this dense and fine fur was highly prized and almost brought about the extinction of the species from sealing activities. Birds, such as penguins, have also developed a layer of closely overlapping feathers to prevent water penetration (Figure 5.5B). Tufts of down that grow from the base of each feather create a fine undershirt close to the body, which traps a thermally efficient layer of air (Watson *et al.*, 1975). The system is so effective that when air temperatures are below zero, snow falling on the backs of incubating penguins does not melt.

As any human who has overwintered in the Antarctic will testify, breathing cold air that is then warmed inside the lungs leads to heat loss. However, the nasal passages of seals and penguins allow the recovery of heat added to the air whilst inside the body. Similar heat exchange adaptations are used by seals, whales and penguins as part of their blood circulation systems. Most of these systems operate by close positioning of the arteries and veins to allow the heat of the arterial blood pumped from the heart to be recovered by raising the temperature of the returning venous blood (Figure 5.6). In this way, mammal body temperature remains high and heat loss at the skin is minimised. A by-product of this thermal efficiency is that all of these animals tend to overheat if exerted or if outside temperatures rise, so the ability to control the amount of heat loss becomes important. Whales and phocid seals achieve this in the water by enhancing the supply of blood to the blubber layer and extremities when heat needs to be lost and restricting the flow of blood when cold conditions demand conservation (Everson, 1987). Penguins tend to overheat readily and gape and ruffle feathers when the sun shines. The feet and wings of penguins and the flippers of fur seals are well supplied with blood vessels so that when the animal is hot they can be turned on like radiators and flushed with blood to help to increase heat loss (Stonehouse, 1972).

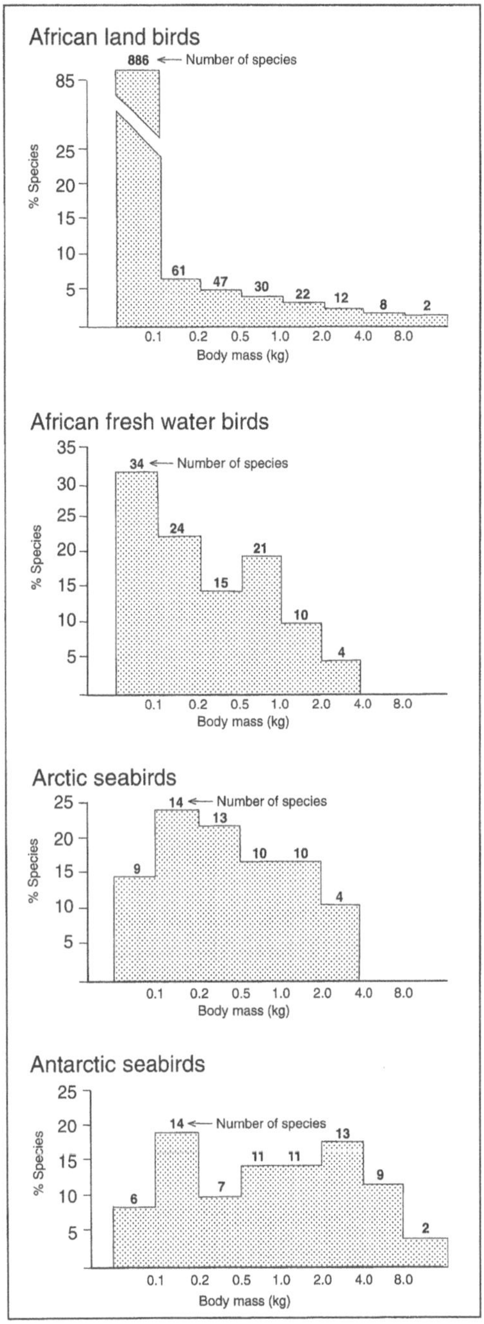

Fig 5.4 The distribution of large body sizes of Antarctic birds can be clearly seen in this comparison with other regions. The number of species in each weight class also shows that the Antarctic has greater numbers of seabird species than the Arctic, perhaps reflecting the lack both of competition from terrestrial birds for breeding sites and of predation by terrestrial birds and animals (source: redrawn from Seigfried, 1985).

A

B.

Fig 5.5 (A) The coarse guard hair of the Antarctic fur seal overlies a woolly coat of thermally efficient fine fur. (B) The stiff, oily and strong outer feathers of Antarctic penguins conceal a fine undershirt of down (source: authors).

5.2.2 Ecological and behavioural strategies

The seasonal rhythm of sunlight and temperature change and freezing and melting exert a control over the Antarctic environment with which all organisms have to cope in order to survive. For example, the geographical patterns of plant species and cover mentioned in Chapter 3 reflect these environmental factors. To an extent, the physiological characteristics of some species allow them to operate in conditions

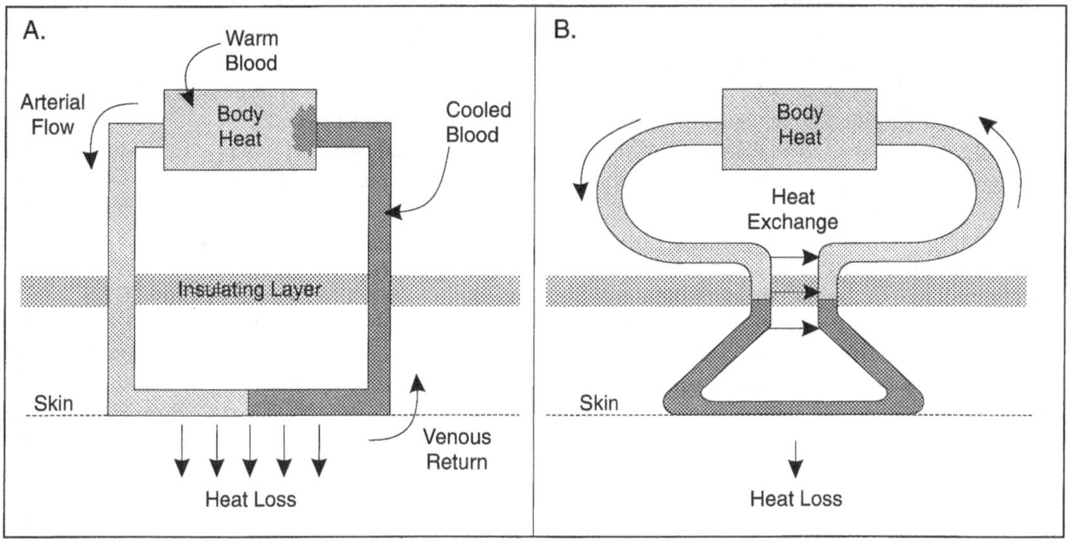

Fig 5.6 A comparison of the heat exchange mechanism of (A) a mid-latitude mammal and (B) Antarctic whales and seals. In the latter, warmed by the body heat of the animal, the outgoing arterial flow from the heart passes in close proximity to the returning cold venous flow, and heat is exchanged in the insulating blubber layer. The net effect is to maximise this exchange and minimise heat loss (source: modified from Everson, 1987).

that would otherwise prevent or seriously limit life, but most species adopt strategies in life-style that allow them to enhance their chances of survival in particular situations. These can be broadly grouped under the banner of ecological and behavioural strategies, since they relate to ways in which organisms, either as individuals or as groups, exploit environmental gradients and ecological niches to suit their life-styles and maximise their chances of survival.

Shelter

Given the extreme nature of the Antarctic terrestrial environment, there is a clear bias for the colonising success of plants to be related to the availability of sheltered locations or other favourable sites. In many ways, shelter is one of the most important factors controlling the success of individual organisms and, from this viewpoint, animals have a major advantage over plants in that they can actively seek out the most favourable microclimate for the prevailing conditions (Walton, 1985). Lack of shelter means that Antarctic vegetation is dominated by cryptogams (bryophytes, lichens, algae, etc.), which by their low-growing nature are well-suited to severe environments (Stonehouse, 1979). Even the two flowering plants are of low stature and the fellfield species are typically dwarf, usually with high-biomass root systems. Strong and persistent winds blow all else away, tents included! The critical criterion is for the plant to remain within the boundary layer, the layer of air close to the ground surface where the wind velocities and blown ice effects are reduced by the topography of the site. This can be only a few millimetres thick on very exposed rock surfaces to several centimetres thick in sheltered sites or where the surface is rough. All lichens growing in extreme habitats are drought-resistant, but the crustose forms are able to tolerate greater extremes of drought, low temperature and wind/ice-crystal abrasion than the macrolichens (fruticose and foliose species). Thus crustose lichens are more able to cope with the environmental conditions encountered in the arid interior. Only in relatively benign spots close to sea level or in the sub-Antarctic islands are the larger vegetation types found. For example, the waist-high tussock grassland of the sub-Antarctic islands is exclusively found close to sea level and often in the shelter of raised beaches or cliffs, its growth usually aided by heavy manuring by seals, penguins and burrowing petrels (see Figure 3.32).

Available nitrogen, notably ammonium, is limiting in many continental habitats, so where it is locally plentiful near nesting colonies of birds, such as snow petrels on inland nunataks, lichen abundance and diversity increases (Longton, 1985). Areas adjacent to penguin colonies are rich in guano-derived nitrogen and usually support a relatively lush version of the adjacent vegetation types. Many of these nitrophilous lichens are brightly coloured, especially with the various hues of orange (carotenoids), and these biologically influenced or dependent communities can be seen from a great distance as colourful splashes against adjacent surfaces. In comparison with areas of bare rock, greater numbers of lichens have been found growing on rocks in areas covered with thin snow for most of the summer. Water and temperature levels are higher than in the open, and light is adequate for photosynthesis (Ahmadjian, 1970). In areas where strong westerly or katabatic winds prevail, upwind-facing aspects are subjected to more wind stress than leeward faces and as a result upwind aspects are generally less favourable sites for survival: where katabatic winds dominate, lichens grow on the leeward side of rocks. In high latitudes, aspect takes on much greater importance since, in the Antarctic, south-facing slopes receive less solar radiation than other aspects. In East Antarctica, Schofield and Rudolph (1969) found that the growth of lichens, mosses and algae was favoured by east- and north-facing aspects. In sub-Antarctic South Georgia, the densest and most complete vegetation cover occurs on the northeast side of the island, partly as a result of aspect but also because this side has been extensively ice-free for the last 9 ka (Clapperton *et al.*, 1978) and thus has been colonised for much longer than the southwest side.

Responses to light and food availability

Geographical location determines whether light levels are sufficient to reduce primary production and photosynthesis to very low levels in winter or whether a higher level of photosynthesis is possible for most of the year, such as on the sub-Antarctic islands. The increase in primary production is largely controlled by the onset of springtime light and warmth. In the marine ecosystem, as soon as the seasonal sea ice begins to break up, the waters beneath are exposed to light and this signals a rapid and dramatic bloom of phytoplankton in the surface waters (Hempel, 1991) (see Figures 4.11 and 4.12). In three consecutive years, this sharp increase in phytoplankton occurred within a period of only ten days at Signy Island (Clarke, 1985). However, many of the microalgae species that inhabit the underside of the sea ice can photosynthesise at very low light levels and can remain productive, albeit at much reduced levels, over most of the winter (Lizotte and Sullivan, 1991). As a result, their contribution to the primary production of the Southern Ocean, as a source of food for overwintering krill and as a potential seed population for the spring phytoplankton blooms, is probably underestimated.

The annual cycles of summer food abundance and winter shortage require herbivores to be able to respond rapidly to food availability by intensive grazing, rapid reproduction and growth. Some species, such as the copepod *Rhincalanus gigas,* feed intensively in summer to build up a store of lipids to tide them over the winter famine (Everson, 1987). Krill have no such lipid store and how they survive the winter is unclear. What is certain is that their diet must be different in the winter, when phytoplankton in the water is absent. Smetacek *et al.* (1990) support the view that krill overwinter by grazing on the crop of microalgae held in the underside of the sea ice (Figure 4.12B), but cannibalism also occurs. An alternative is that krill may migrate to the lowest levels of the water column to feed on detritus and other zooplankton (Nicol, 1994): the sheer volume of the springtime phytoplankton bloom results in a substantial uneaten proportion, which sinks to the bottom. Occurring in both lakes and the sea, this rain of phytoplankton provides a food source for both freshwater and marine copepods and benthic feeders as well as possibly sustaining oceanic krill in winter (Everson, 1987).

Postponed development

All of the Antarctic and sub-Antarctic vascular plants are perennials, as are almost all the mosses and lichens (only a few mosses that colonise unstable soil are annuals) (Smith, 1984). All of the flowering plants produce seeds, and many Antarctic lichens produce spores, but as climatic severity and exposure increase, mosses and lichens fail to reproduce sexually and vegetative reproduction becomes the main reproductive strategy (*ibid.*). Thus, although some plants overwinter as spores or seeds, most can survive in vegetative form using mechanisms such as vegetative reproduction, the ability to photosynthesise at low temperatures, well-developed food storage organs and, in the sub-Antarctic, preformation of flowers in the season before they are needed (Walton, 1984). In order to survive the reduction of photosynthesis in winter in vegetative form, plants must rely on stored food reserves. This strategy results in spreading the products of summer photosynthesis over the winter period and results in slow annual growth rates and a low seed output. Many sub-Antarctic plants immediately store the products of summer photosynthesis in shoot bases, roots or rhizomes rather than in production of new green tissue. In South Georgia, only 25% of the total standing crop of the dwarf shrub *Acaena* is above ground and these reserves allow the plant to overwinter successfully and either capitalise on early springtime radiation by rapidly producing new photosynthetic tissue (Clarke, 1985) or, if the weather is poor, simply to survive. For example, the total size of the fructose store of the tussock grass on South Georgia is sufficient to enable the plant to survive through two consecutive poor growing seasons. In the sub-Antarctic, wind pollination and self-compatibility are major reproductive specialisations, and the seeds of some South Georgian flora, such as *Acaena* and *Uncinia*, are especially suited to long-distance dispersal on bird plumage and by wind (Walton, 1982a).

The nature of the Antarctic substrate is a major disadvantage for sexual reproduction in plants. In most Antarctic soils, the predominance of freeze–thaw cycling and cryoturbation produces a mobile rooting substrate and reduces the chances of seedling survival. In sheltered and enriched lowland sites, sexual reproduction is possible for some and seeds are produced but, overall, sexual reproduction is low in comparison with Arctic and alpine regions (Callaghan and Lewis, 1971). However, insights into the importance of soil propagule banks for long-term colonisation have resulted from the use of cloches placed over initially barren fellfield soils (Smith, 1993c; Kennedy, 1995). Beneath these sealed greenhouses, soil invertebrate populations and plant development and cover are greatly enhanced and indicate the colonisation potential of the soil propagule bank. It appears that a substantial reservoir of spores lies dormant in Antarctic soils ready to germinate when conditions are right, so it seems reasonable to assume that increases in temperature as a result of global climate change may bring about a similar community response (see Section 5 4.2, *Impacts of changes in temperature*, p. 156).

In common with many Antarctic plants, some of the terrestrial invertebrates have the ability to postpone part of their development to another season (Sømme, 1985). Extending the life cycle in this way is a useful strategy for species living under extreme conditions, since it helps to reduce dependence on success in any one year, and so many insect species overwinter in more than one stage. The collembolan *Cryptopygus antarcticus* has generations that span more than one year, with opportunistic growth when conditions are favourable. Freezing tolerance has been demonstrated in the larval stages of the midge *Belgica antarctica*, which survives for several years before emerging as an adult with a lifespan of only a few days. *Alaskozetes* (see Figure 3.34) may have one life history stage per annum, so most individuals must overwinter as an adult once in their lifespan (Block, 1980). This is problematic since, in spite of their ability to supercool by producing cryoprotectants, the mites perish if freezing occurs, because food particles in the gut invariably act as ice nucleators. In response to temperatures falling regularly below zero, *Alaskozetes* has adapted behaviourally by simply ceasing to feed, not because of lack of food but in order to avoid freezing (*ibid.*) (Figure 5.7). Not only can the mite live without food but it also has the physiology to survive for up to a month without oxygen in complete anaerobiosis (Sømme, 1985): in winter, the small crevices in which *Alaskozetes* shelters can become entirely enclosed by ice and, due to its own respiration, entirely devoid of oxygen – yet the mite survives.

Energy efficiency

The seasonal rhythm of food availability permeates the functioning of the entire Antarctic ecosystem in terms of the timing of biological productivity and the strategies adopted by individual species in their drive to grow and reproduce. Species that are energy-efficient can benefit from better utilisation of scarce resources. Broadly speaking, energy consumed as food is devoted to reproduction, basal metabolism, growth and activity in decreasing order of precedence (Clarke, 1980). Assuming that adequate energy is available to satisfy the basic functions of metabolism, the principal consumers of energy are growth and reproduction, so energy-efficient mechanisms in these areas are likely to bring important advantages. Like the Antarctic plants (see Figure 3.35), growth in Antarctic animals is often strongly consonant with the seasonal patterns of energy and nutrient flow but, averaged over the year, growth rates are slow (Clarke, 1985). For example, the ecologically important zooplankton species such as krill are essentially herbivores, so growth occurs mainly in summer, when food is plentiful, and slows in winter, when food becomes scarce.

Low temperatures, lack of food and the resultant low metabolic rate of many benthic animals not only mean that growth rates are low in absolute terms but also that animals live a very long time and can thus

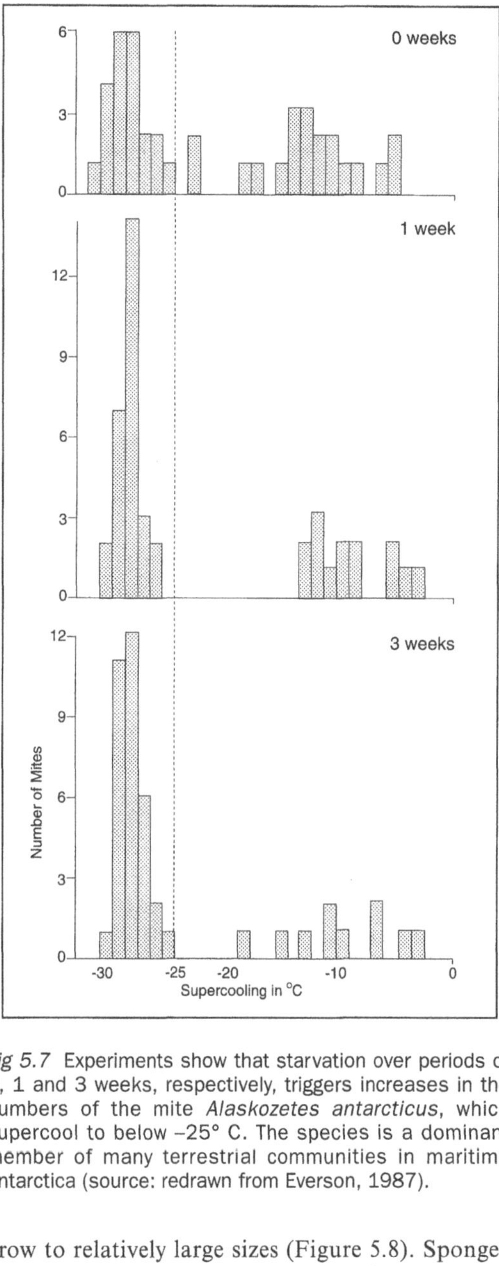

Fig 5.7 Experiments show that starvation over periods of 0, 1 and 3 weeks, respectively, triggers increases in the numbers of the mite *Alaskozetes antarcticus*, which supercool to below –25° C. The species is a dominant member of many terrestrial communities in maritime Antarctica (source: redrawn from Everson, 1987).

grow to relatively large sizes (Figure 5.8). Sponges may be several centuries old, and the limpet *Nacella* can survive at least 100 years (Shabica, 1971). Many of the seabed creatures, such as the isopod *Glyptonotus antarcticus,* can grow to 20 cm long, about three times the size of its relations elsewhere. Other giants include sea spiders, amphipods, ragworms and metre-high glass sponges. Antarctica is home to the highest proportion of giant amphipod

Fig 5.8 A large ten-legged sea spider, *Decolopodium antarcticum*, on the seabed (source: photograph by D.G. Allan, courtesy of BAS).

species and the lowest proportion of dwarf amphipod species on Earth (de Broyer, 1977). However, gigantism is not found in Antarctic animals that utilise a high level of calcium in their bodies. These have difficulties growing in cold water, so molluscs with shells are very small indeed (Picken, 1985). For example, the mollusc *Yoldia eightsi* takes 20 years to grow to 2 cm at Signy Island (Everson, 1987) and the largest known specimens of more than 4 cm in length must be many decades old (Davenport, 1989). The Antarctic fish species that feed on benthos also display low growth rates, and the time taken to reach sexual maturity appears to be seven or eight years. For the krill-feeding pelagic species, such as *N. rossii* and *C. gunnari* (icefish), food is more generally available and this results in faster, but more seasonal, growth rates, with time to maturity being six and four years, respectively (Kock, 1992).

The warm-blooded nature of birds, seals and whales ensures that their metabolism is largely independent of the direct effect of temperature, but these animals have also adopted strategies in behavioural ecology to maintain body temperature. In cold and windy conditions, emperor penguins are vulnerable to cold stress, which they minimise by huddling (Figure 5.9). Huddling behaviour in emperor males allows them to maintain a metabolic rate 25% below that measured in an individual bird (Ancel *et al.*, 1997). Unique amongst penguins, huddling in emperor penguins is possible because of the absence of territorial behaviour and terrestrial predators. For the emperor penguin adults, laying and incubating on sea ice also represents an energy saving since the sea beneath is much warmer than the adjacent land (Stonehouse, 1985). Another strategy to maintain

Fig 5.9 The huddling behaviour of emperor penguins allows the chicks to conserve vital body heat when temperatures drop (source: photograph courtesy of A. Alsop).

body temperature is the rapid gain of an optimal size by feeding strategies (Everson, 1987). All the whale and seal species have large young and very rapid initial growth rates (see Figure 5.3) in order to minimise the period spent as a small individual and therefore susceptibility to cold stress and predation (*ibid.*). Both seals and whales produce large offspring at birth. Blue, fin and sei whale calves achieve about 75% of their maximum length and 40% of maximum weight in their first year. Adaptations in the cow elephant seal allow the production of a highly nutritious milk to fuel rapid pup growth: by the second week of lactation, the cow is producing milk with 40% fat content (Knox, 1994). Growth slows as both seals and whales reach maturity, but since the time taken to reach sexual maturity in whales is dependent on body size, the times taken by different species vary greatly (Gambell, 1985).

Food availability is clearly a major control on growth opportunities and therefore on the time taken to reach sexual maturity. In common with crabeater seals (see Figure 4.24), the mean age at maturity of fin, sei and minke whales in the Antarctic seems to have been declining for some time (*ibid.*) (Figure 5.10). Fin whales born up to the mid-1930s were mature at ten years, declining to six years by 1950 and four years by 1960, while sei whales born up to 1935 were mature at eleven years, declining to seven years by the early 1970s (Masaki, 1978). Direct observations of fin whales caught in 1957, 1961, 1964 and 1968 gave mean ages at sexual maturity of 11.5, 10.6, 9.2 and 6.0 years, respectively (Ohsumi, 1972). Such dramatic decline in the ages of sexual maturity are related to increased feeding and are likely to be a response to the enhanced food supply resulting from the depletion of blue and fin stocks by whaling since the 1930s (Gambell, 1985). The pregnancy rates of blue, fin and sei whales have also increased in response to whaling. In the early 1930s, about 25% of the mature female blue and fin whales caught were pregnant, but by the 1970s the rates for blue, fin and sei were closer to 55%; this represents a halving of the time between successive births in all three species (*ibid.*) (see Figure 5.10). Whales and seals clearly have the ability to adapt to changing conditions by altering behavioural responses.

Reproductive strategies

Seasonality of food availability and temperature also affect the reproductive strategies of Antarctic animals. Two extremes exist at each end of a continuum of strategies adopted by animals to optimise their energy use in reproduction (Odum, 1983). *K* strategists rely on maximising competitive ability in conditions where food is limiting, whereas *r* strategists rely on maximising the rate of population increase. Theory suggests that in the Antarctic, where food availability is largely limited to the summer months, *K* strategies should be favoured. For animals whose

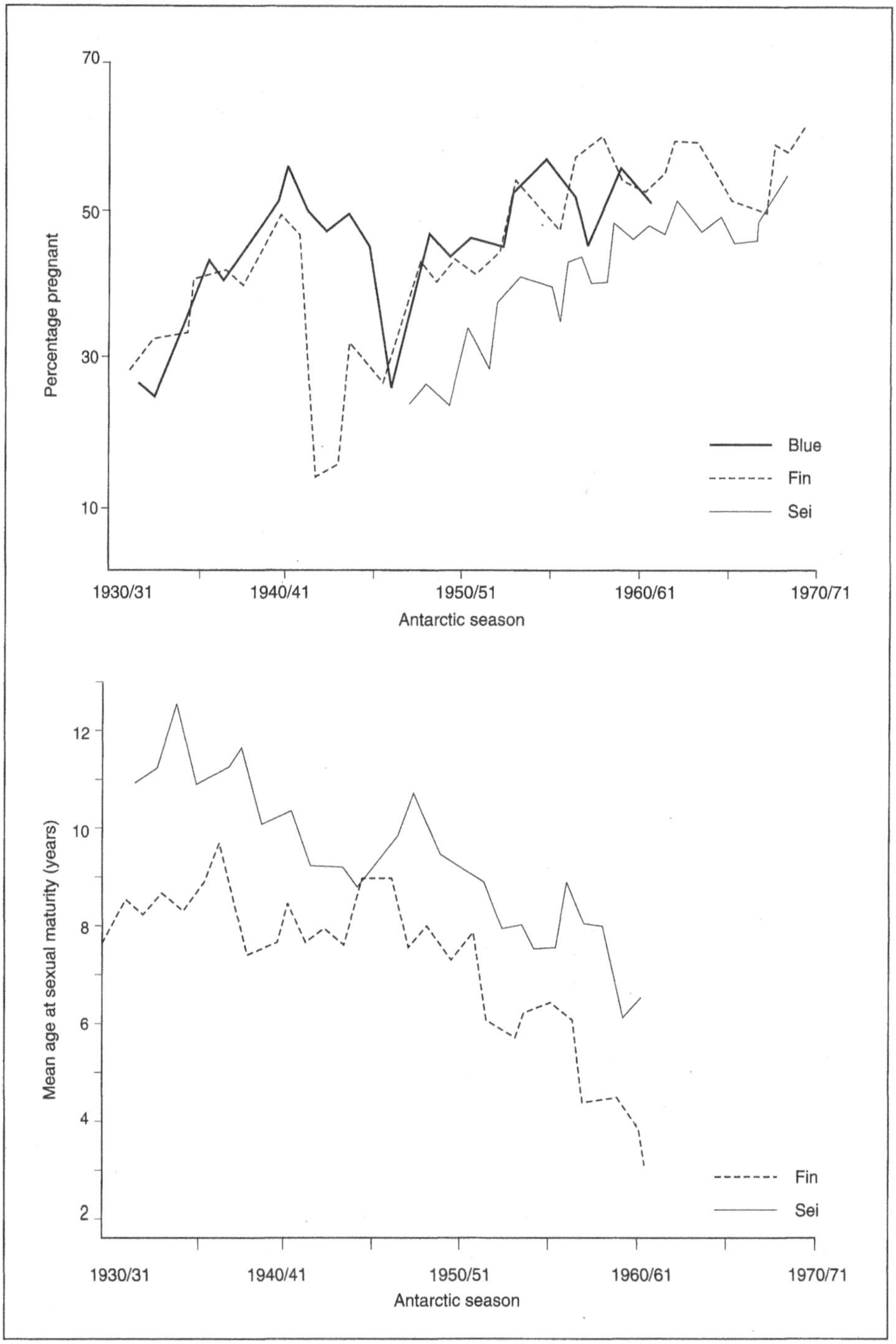

Fig 5.10 The pregnancy rates of some whale species have increased simultaneously with a halving of the time taken to reach sexual maturity, probably in response to whaling and the resultant enhanced food availability for the remaining whale population (source: data from Laws, 1977).

body temperature remains close to that of its habitat, such as the invertebrates or fish, a typical *K* strategy would be for slow growth, delayed sexual maturity, long lifespan, low fecundity and the production of few large and yolky eggs (Everson, 1987; Kock, 1985; 1992). This is the case for many Antarctic fish: notothenioid eggs commonly develop over two years and are up to 4.5 mm in diameter when spawned in autumn. A prolonged winter incubation ensures that the larvae are well-developed, with larger jaws and pectoral fins. Larger prey items are thus possible and the advanced swimming capability allows for more successful foraging and avoidance of predators (Kock, 1992). Winter spawning also ensures that the spring larvae can feed on the phytoplankton blooms. Only very few Antarctic fish species display *r* strategies and lay many small eggs, leading to large numbers of smaller, less well-developed larvae. The higher post-larval mortality incurred is compensated for by higher fecundity (Kock, 1985).

Fish as a group are not noted for parental care and generally invest little time in their young. However, in the Antarctic the level of reproductive energy investment is such that in some species, for example *Harpagifer bispinis* and *Trematomus bernacchi,* the eggs are guarded in well-defended nests, with protection continuing beyond hatching (Moreno, 1980). Post-natal care appears to be relatively common in animals living in extreme environments, and even molluscs and crustaceans tend their young as a mechanism for enhancing reproductive success. Most produce a few large yolky eggs, and brood protection, common in the Antarctic, is found in many species of lamellibranchs, echinoids, amphipods, isopods and polychaetes, and in more than 80% of Antarctic sponges (Picken, 1985). Energetically, the production of large eggs that develop slowly, followed by a brooding phase, places a significant burden on the adult. For example, the female scavenging shrimp, *Chorismus antarcticus*, must feed to produce large eggs in one summer, spawn in autumn and then brood over the winter to allow her larvae to be developed enough to utilise the spring phytoplankton blooms (Clarke, 1985). The isopod *Serolis cornuta* broods for 18 months, so its breeding cycle is biennial.

In such strategies, parallels can be drawn with many Antarctic birds, which have both larger body sizes and fledging periods that are very long in comparison with seabirds elsewhere (see Figure 5.4). Fledging time is a function of size, and larger birds generally produce larger eggs and take the longest to fledge (Croxall, 1984). Fledging in the wandering albatross takes eleven months, so these birds breed only every second year. On average, the king penguin breeds successfully on South Georgia once every two years. Since the chicks need 10–13 months to fledge and must be fed by the parents throughout the winter, their geographical distribution is conditioned by a need for year-round access to the ocean food supply and is thus restricted to the sub-Antarctic islands. Grey-headed and light-mantled sooty albatrosses will not breed in the year after successful rearing because of the time taken to reach breeding condition (Clarke, 1985). All of these large birds have an energy price to pay for the long fledging periods required to ensure the success of their offspring, but few surpass the difficulties faced by the largest Antarctic bird, the emperor penguin.

Breeding on sea ice at high latitudes, where winter food is distant and the summer feeding season is short, the emperor penguin copes with polar winter conditions and long fledging times by laying as late in the summer as possible and incubating over the coldest winter months (Stonehouse, 1985). During this time, the males incubate alone for the first 65 days by existing on their reserves of body fat until relieved by well-fed females returning from the sea. Emperor penguins are both morphologically and physiologically well-adapted for winter fasting, with a low surface-to-volume ratio, a blood vessel heat-conservation mechanism twice as extensive as that of the king penguin and flippers and bill proportionately 25% smaller than other penguins (Croxall, 1997). However, without the energy saving from huddling behaviour, the males would reach the metabolic status that would trigger re-feeding three weeks earlier and the egg would be abandoned before the return of the female (Ancel *et al.*, 1997). Chicks are born in late winter or early spring and fed by the adults in order that they be ready for the open sea in January, when the sea ice melts and food is plentiful. The energy demands are such that the adults lose 25% of their weight during courtship and the males lose a further 15% when incubating the eggs: a substantial energy cost to the individual, especially since only 19% of the chicks will survive the first year. However, the unique adaptation to lay and incubate on ice is a successful strategy since it allows a chick to be reared every year. On the other hand, the constant feeding strategy adopted by the king penguin allows on average a chick every second year but results in 80% of the chicks surviving their first year.

All the homeotherms have the ability to avoid the worst of Antarctic conditions by migration, selection of breeding sites and adjusting breeding cycles to benefit from changing conditions. The large baleen whales, which avoid the Antarctic winter by breeding in tropical and subtropical waters, are a good example of a strategy of migration to exploit the summer glut of food in the Southern Ocean (Gambell, 1985). In the autumn, a year after mating, the female gives birth to a single calf, which is suckled over the winter and during the spring migration until weaned at six months in the Antarctic feeding grounds (Payne, 1979). Since little feeding seems to take place outside Antarctic waters, the animals must provide for themselves and suckle their young by utilising reserves of blubber built up in the Antarctic summer. The reproductive cycle, which has all the attributes of a *K* strategy, spans two or three years in such whales and represents a substantial investment of energy into the successful production of a single calf. The rapid initial growth of the calf is based on the highly nutritious milk of the cow and is gained at a cost to her blubber reserves. The blubber of pregnant blue and fin whales is about 25% thicker than in non-pregnant females, and nearly all the extra fat is utilised in lactation (Lockyer, 1981). In most female baleen whales, this extra blubber requirement for reproduction requires them to remain feeding in the Southern Ocean for longer in order to build up fat reserves and is one of the reasons why females are larger than males.

Many Antarctic seals also migrate for their annual breeding period. Elephant seals spend a very small amount of their year onshore, their blubber making them well-adapted for life at sea. In common with the fur seal, thousands arrive on the sub-Antarctic breeding beaches in early spring, with the males arriving first in order to claim territory (Bonner, 1985a). In these two harem-forming species, the males that hold territories for longest have the best chance of mating with more females, so the males have adapted to be much larger than the females and only the largest and strongest are successful in breeding (Knox, 1994) (Figure 5.11). Adaptations other than size include selection for male aggressiveness, larger canines, protective layers of skin or fur and a capacity for prolonged fasting (*ibid.*). The energy cost of defending the harem is high and mating allows no time to feed. The result is that in spite of the impressive adaptations, large 'beachmaster' males last only a few seasons before dying or yielding to younger, fitter males (McCann, 1980).

In contrast, the ice-breeding species show a reversed sexual dimorphism, with larger females than males (Bonner, 1985a). In these seals, the unpredictability of the sea ice environment and ice floe size has selected against the development of colonial breeding, and in the absence of a need to defend territory, males do not need to be large. In fact, beneath the ice smaller more agile males can avoid predators better (Knox, 1994). Perhaps because ice is an unreliable breeding platform, all of these seals have very short suckling periods, rapid growth and early independence of the young in a short summer season (Bonner, 1985a). In common with the baleen whales, the female ice-breeding seals have adapted to be larger than the males on account of the extra blubber needed to produce milk. The crabeater seal is weaned in just four weeks of suckling on very nutritious milk. However, the energy demands on the females are severe, and a 50% loss in initial weight is sustained over the suckling period (*ibid.*).

Fig 5.11 In the harem-breeding species, selection favours large and aggressive bulls, which become beachmasters. Each of the bull elephant seals participating in this encounter on a beach in South Georgia weighs about 4 tonnes (source: authors).

Whereas elephant seal females also live on their blubber reserves and consequently suffer an energy cost while suckling the single young, the limited fat of the fur seal requires the female to feed throughout the suckling period. Females forage for periods of three–six days and then suckle for two–five days throughout an extended 3.5-month suckling period (Everson, 1987). The energy demands on the female are high since she has to provide for herself as well as keeping up her milk production, so her foraging must be efficient. Studies using radio transmitters of the feeding patterns of fur seals have shown that 75% of all feeding dives occur at night and to much shallower depths than daytime ones (*ibid.*). This adaptation neatly exploits the nightly migration of krill

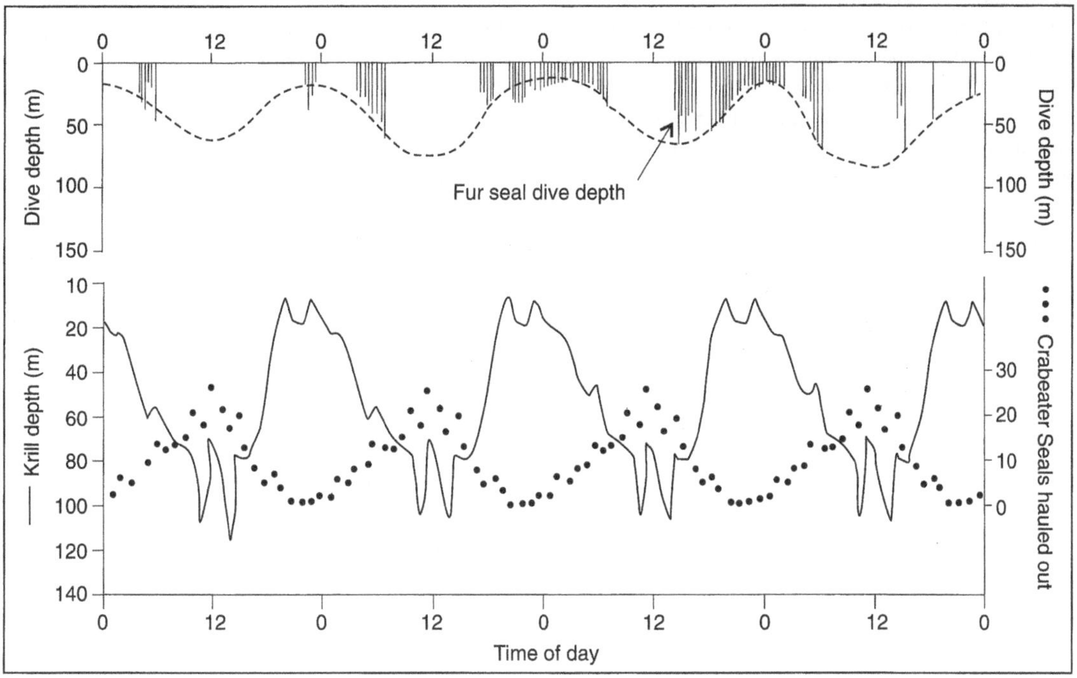

Fig 5.12 Radio transmitters reveal that the feeding patterns of Antarctic fur seals are closely related to the nightly migration of krill to the surface. Similarly, the numbers of crabeater seals hauled out on to sea ice reach a minimum when krill are near the surface (source: redrawn after Laws, 1984; Knox, 1994).

towards the surface and maximises the return in terms of feeding strategy (Figure 5.12). Crabeater seals have a similar relationship with krill and tend to spend their daylight hours hauled out on sea ice. For the fur seal, the foraging benefit can be passed on in the form of rapid growth rates of the young. In addition, since the thermal efficiency of the fur seal's coat declines very rapidly with depth, shallow dives are energy-efficient. During the suckling period the foraging range of the females is limited and necessitates local access to plentiful krill. As a result, the geographical distribution of the Antarctic fur seal is a function of its need for breeding beaches close to abundant krill. Although now extending in range into the northern part of the Antarctic Peninsula, the above conditions are best met on island locations in the sub-Antarctic: 95% of all Antarctic fur seals breed on South Georgia alone.

5.2.3 Summary

Antarctic organisms have adapted to the severity of their environment in two main ways: physiological adaptation to withstand freezing; and ecological and behavioural adaptation to take advantage of favourable conditions whilst protecting themselves from adverse conditions. Antarctic plants can withstand prolonged desiccation and very low temperatures to resume photosynthesis rapidly when conditions improve. They have developed strategies to reproduce vegetatively that make them successful even in extreme conditions. Many Antarctic animals are also physiologically adapted to cope with the effects of low temperatures by using mechanisms such as ice avoidance, supercooling and antifreeze production. It appears fairly clear that in spite of a few exceptions, most Antarctic animals have a tendency to display *K* strategies to cope with the seasonal limitations in food availability on their reproductive activity. This requires the animals that exist close to the temperature of their environment to adopt strategies such as slow growth, deferred maturity, long lifespans, low fecundity, large yolky eggs and post-natal care for the young.

On the other hand, physiological adaptations in the homeotherms centre on thermoregulation mechanisms to maintain their core temperatures and heat-exchange mechanisms to control body heat within certain limits. Because a stable thermal environment within their bodies is physiologically possible, the homeotherms are relatively independent of the constraints of environmental temperature, and this allows them to remain active throughout the year. They have adopted strategies to cope with the severe Antarctic winter such as migration and rapid juvenile growth rates. They are also less controlled by their immediate habitat and can take advantage of favourable conditions when and where they might occur in the pursuit of food or in the selection of breeding sites. In theory, such adaptive flexibility should provide them with an advantage in the face of any fundamental change in environmental factors such as climate, sea ice extent, oceanic productivity or human exploitation of prey. The current debate concerning the extent to which such climate change is currently underway is matched by a parallel debate over the precise magnitude and effects of change in the Antarctic. Any change will impact on the physical systems and natural habitats and these are reviewed and assessed next.

5.3 Potential impacts of climate change on physical systems

5.3.1 Introduction

Almost all the above adaptations of organisms to the seasonality and severity of the Antarctic environment have come about via selection that favours characteristics that in one way or another enhance success in extreme conditions. In this respect, several types of strategy seem to have allowed Antarctic plants and animals to cope well not only with seasonal climate changes but also with the medium-term climate changes over the Holocene (see Figure 3.9). Most terrestrial life in the Antarctic and sub-Antarctic dates from the time of deglaciation of the continental margin or islands since, other than a scattering of nunataks and ice-free oases, most of the continent has been ice-covered at least since the mid-Miocene. For much of continental Antarctica, this places the colonisation of many surfaces to within the last 10 ka, and in the sub-Antarctic to within the last 15 ka, with the most rapid release of land from ice occurring within the last 10 ka. Such contraction of land ice also signalled a contraction in the extent of sea ice (Sugden and Clapperton, 1977), so the late Quaternary climatic amelioration also affected the marine environment with the southward penetration of warmer water (Hays, 1978). Clearly, such environmental change in the past will have affected the biotic world in a variety of ways, and the effect of any present and future changes can be expected to force similar changes. In order to assess the magnitude and direction of present and potential impacts of climate change, it is first necessary to review the evidence for present and potential trends in Antarctic climate and related physical systems. Such trends as are identified in the physical systems are then used to assess the impacts of change on biological systems.

5.3.2 Present and potential climate change in the Antarctic

Chapter 2 reviewed the important factors affecting Antarctic climate and it is the purpose here to discuss the ways in which climate changes might be detected, modelled and assessed to identify their effects on the associated terrestrial and marine systems. There are various ways in which changes in climate can be detected, and these centre on the existing network of ground observational sites (scientific stations, ship cruises, etc.) and an expanding list of remote-sensing platforms (aircraft and satellite). Four direct indicators of climate change and four indirect climate indicators have been identified by SCAR as part of a wider plan for an international research programme to assess the role of Antarctica in global change (SCAR, 1993a). The direct climate indicators are changes in surface and atmospheric temperatures, and in the dynamics and composition of the atmosphere. The indirect climate indicators show the results of climate change on physical and natural systems such as ice sheet mass balance and sea level, sea ice conditions and ocean circulation, surface hydrology, and ecosystem sensitivity (*ibid.*). Rycroft (1994) summarises the ways in which these indicators might interact to produce change via their driving mechanisms (Figure 5.13).

5.3.3 Direct indicators of climate change

Actual and modelled surface temperature

Surface temperatures in Antarctica have been the subject of much scrutiny because of possible effects on the stability of the Antarctic ice sheet. Cloud-free

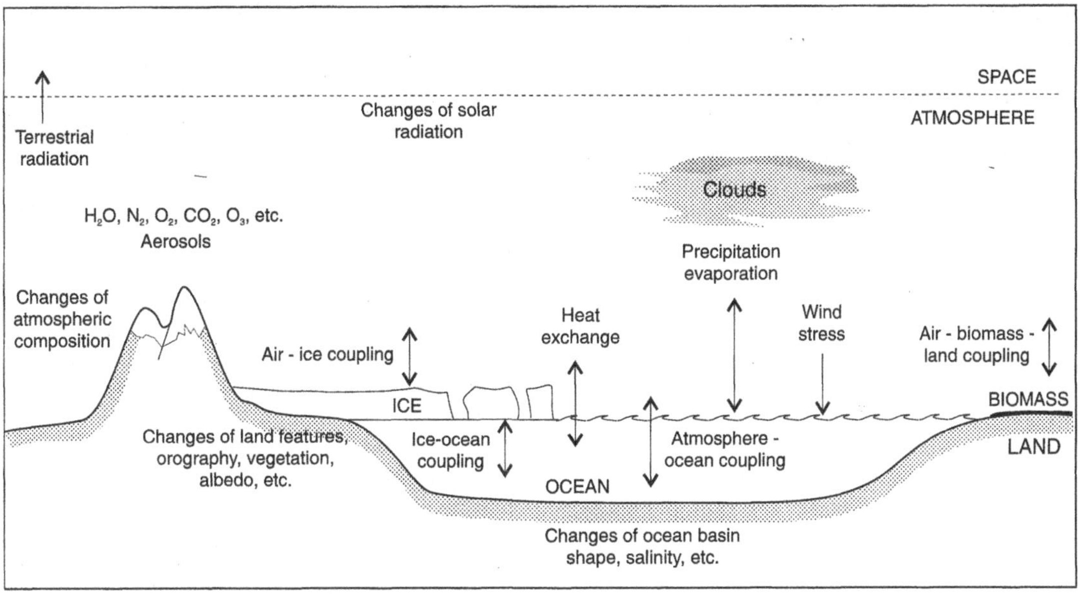

Fig 5.13 Processes affecting the climate at the Earth's surface and some of the changes that are likely to force climate change (source: after Rycroft, 1994).

surface temperature estimates are at present derived from satellite instruments such as the Advanced Very High Resolution Radiometer (AVHRR) and the Temperature–Humidity Infrared Radiometer (THIR) (Weller, 1992). However, few Antarctic climate stations have records extending back longer than 40 years, so it is difficult to identify any trends that might parallel the 0.5 °C increase in global temperature identified over the past century or so. One region where a warming trend of 2.5 °C over the past 50 years has been identified is the western coast of the Antarctic Peninsula (see Figure 2.17). An increase in decadal mid-winter temperatures of about 4 °C has been reported from Faraday Station on the Antarctic Peninsula, and warming trends have been recorded from two coastal stations in East Antarctica (Adamson and Adamson, 1992; Stark, 1994). While it appears that there is some evidence for an increase in temperatures at some Antarctic stations (Jacka and Budd, 1991), the distribution of surface stations is too sparse and the record too short (most data start about 1957) and variable to show a statistically significant trend (Allison, 1992). Such inconclusive data are at odds with the long-term data assembled by Jones (1990, 1995), who estimates Antarctica to be about 1 °C warmer now than in the early years of this century. The predictions of computer models also suggest that any greenhouse-induced warming will be greatest at high latitudes (Allison, 1992).

Computer models, or Atmospheric General Circulation Models (AGCMs), are based on equations that aim to emulate in time the motion of the atmosphere, its temperature, water content and pressure (Cattle, 1991; Cattle *et al.*, 1992). AGCMs are usually run in conjunction with interactive models of the land or sea surface, although because of the long timescales of response of both the Antarctic ice sheet and the deep circulation of the Southern Ocean, the capacity of AGCMs to portray accurately all the characteristics of likely changes is relatively modest (Stauffer *et al.*, 1989; Cattle, 1991). Nevertheless, a sufficient number of studies have now been carried out to assess how well the evolving models represent Antarctic climate (Schlesinger, 1986; Cattle and Roberts, 1988; Mitchell and Senior, 1989; Cattle, 1991). The most comprehensive comparison of the relative performance of nine AGCMs against observed climate showed that the simulated surface air temperatures over Antarctica were 10 °C higher than the actual temperatures in summer and 15 °C higher in winter (Schlesinger, 1986).

Much depends on how the effect of the greenhouse gas CO_2 is modelled. An instantaneous doubling of CO_2 concentrations (used by equilibrium response models) shows maximum warming over the polar regions in winter, and in the Antarctic context, annual mean surface temperature changes of up to 6 °C are predicted over the sea ice area (Wilson and Mitchell, 1987; Houghton *et al.*, 1990; Cattle *et al.*, 1992) (Figure 5.14). However, there are problems with inaccurate modelling of coastal katabatic winds and sea ice characteristics and concentrations, and most of the models examined by Cattle (1991) overestimate sea-surface temperatures and thus sea ice extent unless corrected by observed temperatures. More modest temperature rises in the Antarctic are predicted by transient response models, which run over timespans of about 100 years and slowly increase CO_2 concentrations annually (Stauffer *et al.*, 1989; Cattle *et al.*, 1992; Manabe *et al.*, 1992) (Figure 5.15). Much still depends on future trends in the amount of greenhouse gases and how any changes are modelled (see the following section).

Some model experiments concentrate on varying the extent and concentration of sea ice in the Antarctic in order to assess the sensitivity of atmospheric circulation to actual processes. Using anomalies in sea ice extent and concentration, Simmonds and Budd (1990) produced a July warming of up to 6 °C with an ice concentration of 50%. What these experiments show is that any decreases in the concentration of sea ice that may accompany a climate change have a warming effect in the Antarctic independent of the location of the ice edge (Simmonds, 1992). Further, this can be expected to elevate evaporation rates, resulting in the cyclonic transport of moisture southwards to enhance precipitation over Antarctic coastal areas. Assuming no change in the location of the ice edge, the net July accumulation over the continent could be increased by 13%, 24% and 67% when the open water fraction of the sea ice is increased by 5%, 20% and 50%, respectively (Simmonds, 1992) (Table 5.1). Experiments by Mitchell and Senior (1989) also show increases in winter of 0.3 mm d^{-1} at the coast to 0.05 mm d^{-1} in the interior, with much reduced gradients in summer.

From such simulations it appears that the direction of change in both temperature and net precipitation is likely to be positive but markedly more limited over the continent than in coastal areas. By 2050, such increases may contribute to a sea-level lowering of 18–91mm (Simmonds, 1992). For the coastal fringe, it appears that these modelled results may not be as fanciful as might at first appear. Using ice cores taken from within 300 km of the East Antarctic coast, Morgan *et al.* (1991) found an increase in accumulation of about 20% since 1960. Balanced against enhanced ablation, increasing precipitation in a warmer world may cause the ice sheet to grow more or, at least, to shrink less.

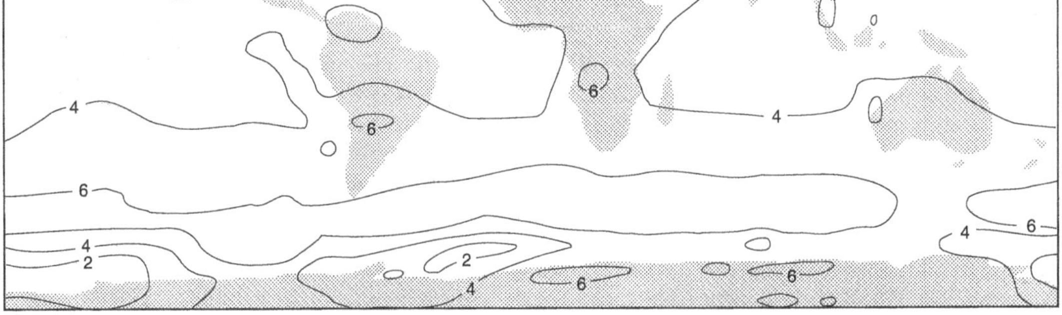

Fig 5.14 Geographical variation in the annual mean surface temperature increase in °C over the Earth that would result from an instantaneous doubling of atmospheric CO_2 concentration (redrawn after Wilson and Mitchell, 1987).

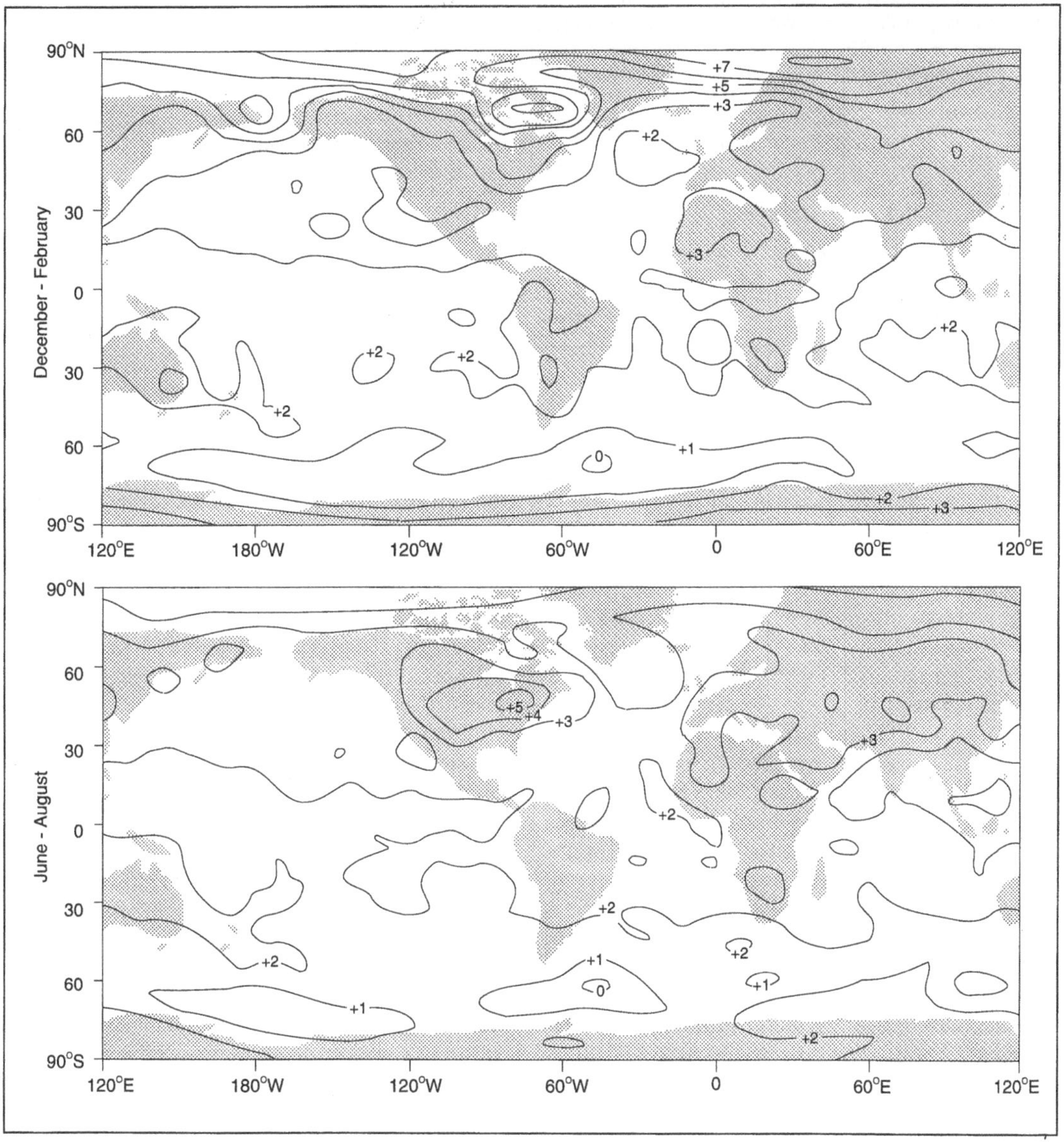

Fig 5.15 Temperature changes in °C after 60–80 years due to increases of 1% per year in atmospheric CO_2 (source: redrawn after Manabe *et al.*, 1992).

Changes in atmospheric composition: carbon dioxide, ozone and methane

The temperature, dynamics and composition of the atmosphere have also attracted attention from SCAR (1993a), who highlighted the lack of data on radiation fluxes at the top of the atmosphere in polar regions and especially on the role of radiative forcing by clouds. Satellite-derived data appear to hold the best promise of the synoptic view that is required in order to monitor change. The general circulation of the atmosphere, as well as cloud cover and precipitation, will be greatly affected by any changes in global atmospheric temperature (Weller, 1992). Furthermore, the distribution of storm tracks over the Southern Ocean has a great effect on energy fluxes and may explain much of the variability seen in temperature

Table 5.1 July averages of surface temperature and net accumulation over the sea ice, coastal areas and interior of Antarctica as a result of a 5%, 20% and 50% reduction in sea ice concentration

	Open water fraction of sea ice concentration 100%	95%	80%	50%
	Surface temperature (°C)		Temperature change (°C)	
Sea ice	– 19.2	2.0	6.3	12.5
'Coast'	– 39.75	1.3	2.5	5.4
Continent	– 48.75	0.9	1.8	4.2
	Precipitation minus evaporation (mm d^{-1})			
Sea ice	1.76	– 0.13	– 0.28	– 0.64
'Coast'	0.50	0.09	0.19	1.55
Continent	0.46	0.06	0.11	0.31

Source: modified from Simmonds, 1992.

and sea ice extents. However, the greatest amount of concern centres around changes in atmospheric composition, especially changes in the so-called greenhouse gases of carbon dioxide (CO_2), ozone (O_3) and methane (CH_4).

Modelling of the changes in CO_2 concentrations has produced a range of predictions concerning climate change, indicate that the equilibrium response models (e.g. Figure 5.14) represent the upper limit of expected temperature changes and that changes as a result of 1% a^{-1} increases in CO_2 concentrations in the Antarctic can be expected to be lower (e.g. Figure 5.15). The transient response models show a progressive warming in the northern high latitudes, but the action of deep circulatory currents in the Southern Ocean results in a much reduced Antarctic warming (Stauffer *et al.*, 1989; Cattle, 1991; Cattle *et al.*, 1992). CO_2-driven heat is taken down into the deeper ocean by downwelling in the Polar Frontal Zone and replacement cold water is upwelled at the Antarctic Divergence. The temperature inertia introduced by the Southern Ocean results in mean surface warming over the seasonal sea ice zone being limited to 1 °C for a 1% a^{-1} CO_2 increase over 100 years (Cattle, 1991) (Figure 5.16), and this may well lead to enhanced sea ice melting and adjacent coastal precipitation. However, the 2 °C increases predicted over the continental ice sheets are unlikely to have a major impact on melting in view of the very low mean annual temperatures in this area. As a result of better modelling of sea ice concentrations, thickness and extent, together with more accurate depiction of the deep circulation of the Southern Ocean, estimates of the amount of CO_2 warming in the Antarctic are now much reduced and so too are the indirect impacts.

As early as 1974, Molina and Rowland had predicted that anthropogenic emissions of chlorofluorocarbons (CFCs) would lead, over several decades, to a reduction in the amount of stratospheric O_3. CFCs, used extensively by industry as coolants in refrigerators, are chemically inert compounds in the lower atmosphere but are broken down by solar ultraviolet radiation in the upper atmosphere to release substantial amounts of chlorine gas, which catalyse the destruction of O_3 (a single chlorine molecule can eliminate 10^6 O_3 molecules). Although O_3 loss is at present restricted to certain altitudes and temperatures, depletion could occur over larger areas and for longer periods if (1) the temperature of the stratosphere decreases over a larger area; (2) temperatures remain low for a longer time; or (3) the polar vortex is delayed in its weakening; or (4) any combination of these (Voytek, 1989). Using AGCMs that include chemical processes, computer modelling of O_3 transport and chemistry is now possible. These show that the winter reduction in atmospheric O_3 over Antarctica may not be acting independently of global temperature change. Brasseur and Hitchman (1988) point out that CO_2 warming of the lower atmosphere by 2–4 °C will lead to a cycle of stratospheric cooling of 8–15 °C by the middle of next century and enhanced O_3 destruction. Models suggest that a decrease of 1% in total column O_3 gives rise to a 2% increase in UV radiation at the Earth's surface.

This effect is not limited to Antarctica, because low-O_3 air has broken away in the past to cross the Southern Ocean to Australia and New Zealand (Atkinson *et al.*, 1989; Ellis, 1991) and to southern South America (Frederick *et al.*, 1994). As a result of O_3 depletion, springtime levels of incident UV-B radiation are greatly increased during the period of

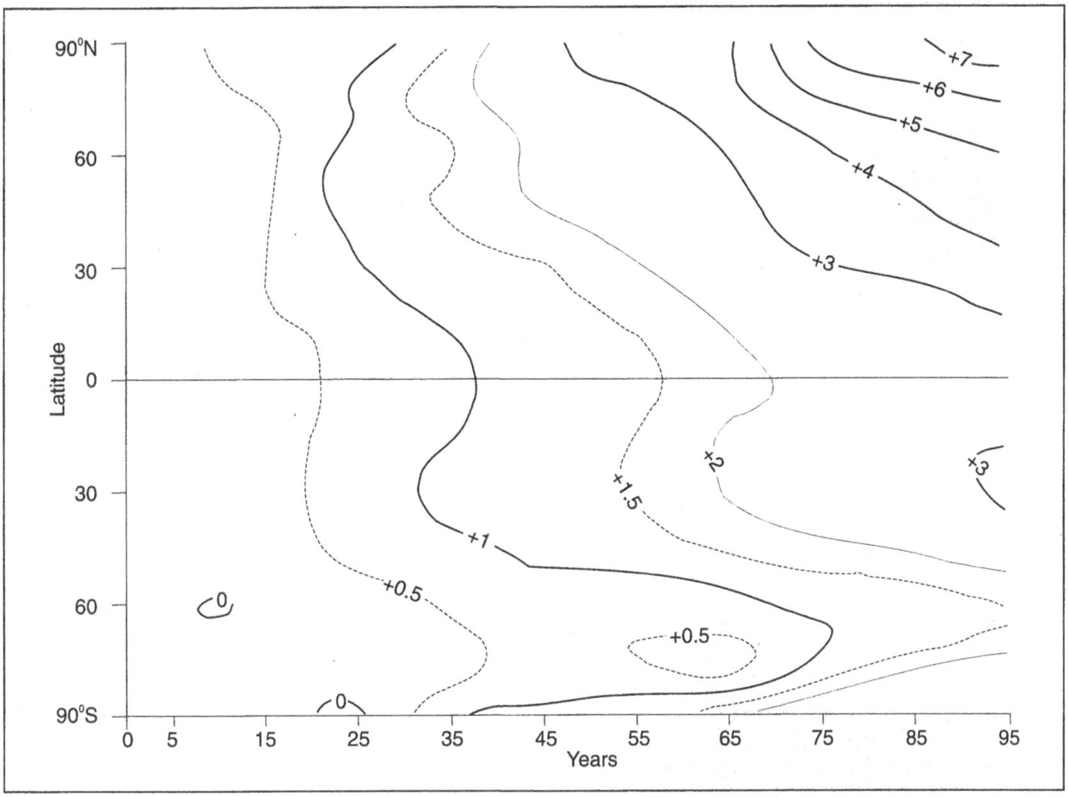

Fig 5.16 The spatial and temporal distribution of changes in surface air temperature in °C as a result of a 1% per year increase in atmospheric CO_2 concentrations (source: after Stouffer *et al.*, 1989).

the so-called 'ozone hole', and in 1990 summer equivalent levels of radiation occurred about two months early at McMurdo Station (Frederick and Snell, 1988; Lubin *et al.*, 1992; Stamnes *et al.*, 1992). There is also evidence that O_3 depletion is now extending through to late summer (Jones and Shanklin, 1995). Reductions in O_3 concentration have important implications for biological life on earth because harmful UV light is absorbed by O_3 (Ellis, 1991). The biological effects of O_3 depletion on terrestrial and marine organisms are covered later in this chapter.

There have also been associated changes in other atmospheric gases such as CH_4. Though much lower in concentration than CO_2, molecule for molecule CH_4 traps 25 times more of the Sun's heat in the atmosphere than does CO_2. CH_4 is produced by bacteria in the absence of oxygen, and concentrations have been rising by 1% a^{-1} since at least 1950 (Pearce, 1989) and have more than doubled since industrialisation (Staffelbach *et al.*, 1991). The world's wetlands are the main source, and these have been increasing as a result of increasing acreages of rice paddies, burning of forests and grassland and the spread of Northern Hemisphere peat bogs (Pearce, 1989). CH_4 levels in the Quaternary were about 0.3–0.4 parts per million (ppm), rising over the Holocene up to AD 1700 to 0.7ppm and since 1977 they have increased to concentrations of 1.7ppm (Ellis, 1991). Shallow ice core data from the Antarctic Peninsula also show increasing amounts of CH_4 since about AD 1700 (Stauffer *et al.*, 1986). As a greenhouse gas, CH_4 appears to be greatly underestimated yet, like O_3, it too may act in concert with CO_2 increases. CO_2-driven warming is thought to be responsible for a 2–4 °C warming of the Northern Hemisphere permafrost over the twentieth century, which has resulted in increased tundra wetness, expansion of peat bogs and production of more CH_4 (Pearce, 1989).

5.3.4 Indirect indicators of climate change

Several indirect indicators of climate-driven environmental change can be identified, centring mainly on ice sheet mass balance and sea level, sea ice conditions and ocean circulation, and ecosystem sensitivity. All of these areas of potential impact are inter-linked, changes in one forcing a direct response in another and numerous indirect responses elsewhere. The first two areas relate to responses in the physical systems, and these are examined below and their inter-linkages highlighted where appropriate. Biological effects are treated separately in Section 5.4.

Changes in ice sheet stability and sea level

The Antarctic ice sheet is now less extensive than it has been during much of the Quaternary Period, yet it still contains 90% of the world's ice and 70% of the world's fresh water (Ellis, 1991). It represents a vital link in the global water balance and if it were to melt catastrophically it would raise world sea level by more than 60 m (SCAR, 1993), although estimates range up to 80 m (Drewry, 1991). As noted in Chapter 3, there are two parts to the ice sheet. The West Antarctic ice sheet is grounded below sea level for the most part and is thus sensitive to sea level change and global warming. If it melted then a sea level rise of 5–7 m would result (Denton *et al.*, 1991; Cattle *et al.*, 1997). The East Antarctic ice sheet is more firmly based on bedrock at altitude, making it more stable and less susceptible to sea level forcing. If it melted however, it would raise sea level by about 60 m (Denton *et al.*, 1991). It is extremely unlikely that catastrophic collapse would occur, since this would require a massive elevation in the equilibrium line of the ice sheet (see Chapter 3). It is more likely that over the next 100 years fractional changes in ice volume induced by, for example, the modelled 1% a^{-1} increase in CO_2 concentration will have the potential to produce melting effects that will influence sea level (Drewry, 1991). Many glaciers in mid-latitudes are now showing loss of ice mass, so great importance is attached to estimates and projections of possible changes to the Antarctic ice mass and its stability, especially since the evidence from ice cores, deep-sea cores and relict landforms shows that the ice volume has changed in the past. The two main strands in the current debate over the sensitivity of the Antarctic ice sheet to climatic change focus on attempts to (1) improve the accuracy of estimates of ice volume changes and the response of the ice sheets over different time scales; and (2) enhance the understanding of the palaeoenvironmental history of the ice sheet.

Estimates of the mass balance of the Antarctic ice sheet depends on the accuracy of the estimates of precipitation input and output in the form of basal melting of ice shelves (estimated to be 25% of total) and iceberg calving (estimated to be 75% of total). This raises an important perspective on the effects of any climate warming, since temperature increase is unlikely to be directly matched by a corresponding increase in calving rate, although bottom melting of ice shelves will rise. However, the warmer Antarctic coasts can expect to see enhanced atmospheric moisture and precipitation, and although this is already happening (Drewry, 1991), it is not known whether this is a long-term change. Enhanced accumulation will eventually result in enhanced ice discharge over timescales of 500 years or more, and this may contribute 2 mm a^{-1} to sea level rise (*ibid.*). For Antarctica as a whole, Fortuin and Oerlemans (1992) predict that for a 1 °C uniform warming, higher accumulation and stronger evaporation in coastal areas will result in –0.27 mm a^{-1} of sea level change. The present estimates of the mass balance of the ice sheet suggest that the ice sheet is in slight deficit of 469 Gt a^{-1} (Jacobs, 1992), although others suggest approximate equilibrium or a modest surplus of 40–400 Gt a^{-1} (Bentley and Giovinetto, 1991). However, a serious problem is that iceberg calving may be very episodic, and if the mean time between major calving events exceeds 100 years, then historical data are of limited value (Warrick *et al.*, 1995).

Considerable uncertainty exists over the accuracy of mass balance estimates, particularly since most models do not yet fully incorporate sub-ice topography, low-altitude ablation and local effects, which are significant in mountainous areas such as the Antarctic Peninsula. One-quarter of the total annual accumulation in the Antarctic falls in the Antarctic Peninsula, so this poorly modelled region contributes a greater proportion to the mass balance of the continent than its 6.8% area would suggest (Drewry and Morris, 1992). Drewry and Morris developed a model to cope with the distinctive contribution of the Antarctic Peninsula and predicted that a 2 °C rise in mean annual temperature over 40 years would result in a 1 mm ablation-driven rise in sea level, offset by a 0.5 mm fall in sea level associated with elevated levels of precipitation. Warrick and Oerlemans (1990) imply a peninsula-contributed fall of 1.6 mm as part of a global sea level fall of 24 ± 24 mm over the next

40 years. For Antarctica as a whole, Huybrechts and Oerlemans (1990) estimated that with a warming of up to 5 °C, there would be a steady increase in accumulation, adding 640 Gt a^{-1} to the ice and lowering sea level by 1.8 mm a^{-1}. Huybrechts (1993) also modelled the response of the East Antarctic ice sheet to warming in order to establish the conditions for complete meltdown. With surface ablation set as an important control on melting, the model indicated that temperature increases of less than 5 °C would lead to increased precipitation and expansion, whereas complete meltdown required an unlikely rise of 15–17 °C (*ibid.*).

Most of this modelling assumes a linear ice sheet response to rising temperature, but there is also evidence to suggest that this may not occur. Non-linear response can occur due to the favourable location of a high-level plateau for ice sheet development (e.g. parts of East Antarctica) or the configuration of a bay to allow accelerated calving and ice sheet collapse (e.g. Hudson Bay and the collapse of the Laurentide ice sheet, 8 ka ago) (Sugden, 1991). The unique topography of the Antarctic may therefore introduce instabilities into the way in which the ice sheet responds to change, and this is of particular importance in the case of the West Antarctic ice sheet grounded below sea level. Payne *et al.* (1989) produced a stepped model of ice growth for the Antarctic Peninsula where growth was achieved when sea level fell enough to allow calving to be reduced, and collapse was achieved when sea level rise allowed calving rates to climb and penetrate along pre-existing troughs deep into the centre of the ice sheet (Figure 5.17). Such work highlights the potential importance of subglacial topography in the response

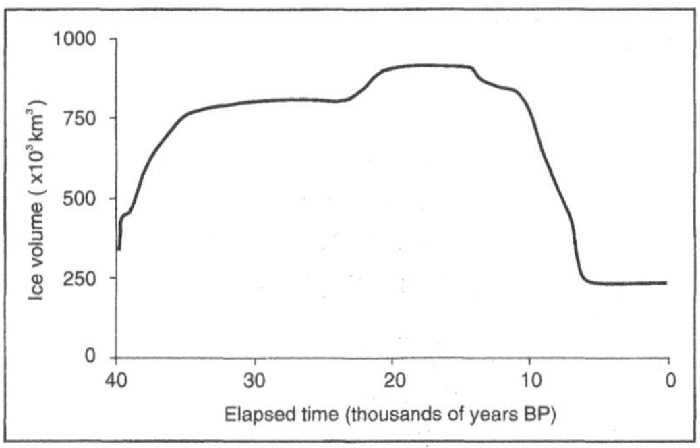

Fig 5.17 Over the past 40 ka, the Antarctic Peninsula ice sheet has shown stepped changes in its volume as a result of topographic controls (source: data from Sugden, 1991).

of ice sheets to climate change. If the ice sheet response is non-linear then the resultant sea level change will reflect this and be stepped rather than regular. There is also evidence from the past of stepped responses of the Antarctic ice sheet to climate change and this is addressed below.

The second important strand in understanding how ice sheets respond to climate change involves past fluctuations as a means to gauge future behaviour. There are essentially two main theories that claim to indicate the way in which the Antarctic ice sheet has responded to Quaternary and earlier climate changes. The consensus view is that East Antarctica is a high, cold and stable land-based ice sheet that is thought to have existed close to its current configuration for the last 14 Ma (Sugden *et al.*, 1993). This timescale includes the relatively warm Pliocene and if correct bears on the way in which ice sheets may respond to the current warming trend: not by contraction or meltdown but by stability and, possibly, expansion. There is also a 'dynamicist' view, which contends that for much of the past 40 Ma the Antarctic region was characterised by waxing and waning temperate ice sheets and that the present polar ice sheet may have developed only after 3 Ma BP following a period of Pliocene warmth (Wilson, 1995), a view echoed by work in West Antarctica that suggests multiple intervals of near-complete deglaciation between 24 and 3.5 Ma ago (LeMasurier *et al.*, 1994). Although put into question by the uncertain age of the Sirius sediments (see Chapter 3, Section 3.3), this alternative view indicates that in a warmer world, substantial melting is possible and the stability of the Antarctic ice sheets is in question.

Although the East Antarctic ice sheet is likely to have been stable for at least 14 Ma, the West Antarctic ice sheet has for long been suspected to be unstable and may respond to climate change in an unstable or stepped manner and may thus be susceptible to rapid collapse (Mercer, 1978: Hughes, 1987). What remains unclear is the extent to which outward flow might be enhanced by the weakening and thinning of major ice shelves as a result of enhanced basal melting and the reduction of any 'buttressing' effect on up-glacier ice by unpinning from grounding points. Budd *et al.* (1987) show that once ice shelf thinning is sustained there is a rapid retreat of the ice front, high sliding velocities in ice streams and ice volume decrease by drawdown. Certainly the West Antarctic ice streams flowing into the Ross Ice Shelf appear to be undergoing significant changes (Shabatie *et al.*, 1988), and an 8% increase in the velocity of Thwaites Glacier in

West Antarctica was measured between 1984 and 1990 (Ferrigno, 1993). However, the picture is unclear, since the main part of Ice Stream B is thinning, as is its catchment area in the West Antarctic ice sheet, whilst its front is slowing and thickening. The buttressing model should produce changes in the front at first, which are then propagated upstream (Alley and Whillans, 1991). The role played by sea ice in ice sheet response by influencing air and water temperatures may also be important, and recent work by Harwood *et al.* (1994) has indicated that a coherent girdle of sea ice surrounding the Antarctic is critical in maintaining the stability of both West and East Antarctic ice sheets. An indication of the possible nature of changes to the ice shelves in a warmer world is provided by the seasonality of water flows into the sub-ice cavity: floating ice shelves are vulnerable to climate change at their lower as well as their upper surfaces. In the Filchner–Ronne Ice Shelf, the inflow of dense but relatively warm sea water from the formation of sea ice in the north serves to melt the lower surface of the ice shelf. Any climate warming that led to lower rates of sea ice production would cause a reduction in the flux of this warmer water and so lead to a reduction in the total melting beneath the ice shelf (Nicholls, 1997). If this seasonality of inflow and basal melting is an accurate analogue for climate warming then the response of an ice shelf to a warming of the climate may be basal thickening, reinforcing rather than threatening its longevity (*ibid.*). At present, whereas the more northerly ice shelves are disintegrating, the larger and more southerly ones are more secure (Vaughan and Doake, 1996).

Overall, the evidence pointing to the direction of the connection between climate change and the response of the Antarctic ice sheets is inconclusive. Modest climate warming over the seasonal sea ice zone appears to suggest that precipitation increases are likely beyond those which have already begun. This will result in lowering of sea levels until the resultant enhanced ice discharge eventually leads to sea level rise, possibly 500 years or more later. Whether this leads to reduction of the role played by ice shelves in buttressing the stability of the West Antarctic ice sheet is less clear, because the behaviour of the shelves and the ice streams that feed them is not well-known. Neither is the possible response of the major ice shelves to future warming and sea level rise. Local warming of the Antarctic Peninsula appears to be responsible for the disintegration or retreat of five out of nine of the smaller and most northerly ice shelves (see Figure 3.8), but the evidence for more widespread warming is absent and there is little evidence of a reduction in the area of the larger ice shelves further south. Taken together, an immediate collapse of the West Antarctic ice sheet over the next century can probably be discounted, as can any major change in the stability of the East Antarctic ice sheet. More likely is that local warming may lead to further disintegration of the most marginal ice shelves. However, also likely is a continued increase in accumulation along the coastal fringes of the continent, leading to locally greater ice volumes and outward flow.

Impacts of climate change on Southern Ocean sea ice conditions

The spatial and temporal variation of sea ice affects many aspects of the Antarctic environment. Not only does sea ice increase the albedo of the ocean surface, so that surface warming is reduced, but it also represents an insulating layer sandwiched between air and sea whose influence in reducing heat flux is particularly strong in winter (Parkinson, 1992). Sea ice presence also reduces the amount of evaporation from the ocean surface and so may affect precipitation well away from the immediate area (Simmonds, 1992). Brine rejection on freezing contributes to the formation of Antarctic Bottom Water and density-driven convective overturning in the Southern Ocean. The winter sea ice substrate offers a habitat to phytoplankton and zooplankton that is now recognised to be of critical importance to the productivity of the springtime Southern Ocean. The way in which climate change may impact on the nature and variability of the sea ice and water mass modification and ocean circulation has important implications for the marine ecosystem.

Much of the present understanding of sea ice extent and variability comes from over two decades of satellite monitoring using microwave radiometer measurements, which started in 1973. With the earlier imagery there are problems associated with the way in which remotely sensed data are used to identify ice surface flooding and the various types of thin and new ice that occur near the advancing ice edge (Wadhams, 1991). However, more recent satellites such as ERS-1 (1991), ERS-2 (1995) and future probes ENVISAT (1999) and RADARSAT will variously detect ice extent, type, motion, concentration and possibly thickness (*ibid.*; SCAR, 1993a). A novel approach to identifying longer-term trends in sea ice extent has been to examine whaling records for circumpolar coverage of whale catches between 1931 and 1987, since the southern limit of whaling was constrained by sea ice (de la Mare, 1997).

Most research identifies the extent and concentration of sea ice as an important factor in determining temperature and precipitation shifts in a warmer world (Simmonds, 1992; Priddle *et al.*, 1992), yet models show large disagreements in their predictions for the Antarctic seasonal sea ice zone. Two problems cloud any attempt to assess the impact of climate change on sea ice. Deriving a synoptic view of the extent of the seasonal sea ice cycle is extremely difficult because of its great variability: so much natural noise exists in the sea ice data that long-term trends in the sea ice system are difficult to detect (Parkinson, 1992; Hanna, 1996). In addition, the precise nature of the air–sea–ice interaction is not well-known in the Antarctic, and this produces differences in the modelled predictions of the impact of climate change (Wadhams, 1994).

Attempts have been made to link changes in sea ice extent to climate change, but such is the regional and interannual variability that trends are difficult to identify (Parkinson, 1992; Zwally, 1994; Hanna, 1996; Murphy and King, 1997). Sea ice data from 1966 showed an increased ice cover to 1973, this spanning a period of temperature increase (Zwally *et al.*, 1983). Zwally (1994) found a 10% decrease in overall area from 1973 to 1976, but from 1979 to 1986 there was no overall change. Similarly, Johannessen *et al.* (1995) using the more recent Special Sensor Microwave/Imager (SSM/I) failed to detect any overall areal trend between 1987 and 1994. Regionally, attempts to show a correlation between temperature and reduced sea ice have also been ambiguous. Weatherly *et al.* (1991) found correlations between warmer temperatures and sea ice decreases, and Jacka and Budd (1991) identified similar relationships for 1973–1989. However, these estimates may be anomalous, since they include the Weddell Sea polynya (see Figure 4.4A). Neither Gloerson and Campbell (1991) nor Parkinson (1992) found any change in monthly average sea ice extent in the Antarctic between 1978 and 1987. However, indications of no real change in sea ice extent may be misleading if the satellite sensor fails to detect different types, concentrations and thicknesses of ice accurately. As well as better satellite data, there is a need for alternative ways of monitoring sea ice extent such as more winter shipborne surveys, drifting ice camps and the use of 'smart' buoys with a range of sensors to monitor ice (Wadhams, 1994). For example, from the whaling records, de la Mare (1997) identified abrupt mid-twentieth-century declines in sea ice extent, with the average summer sea ice extent moving southwards by 2.8° latitude between the mid-1950s and early 1970s, a decline in ice extent of 25%. With no change in whale catching technology over this period of shift, and no change in the relationship of catching and sea ice edge, it appears that the abrupt southward shift in ice edge is a real, statistically significant, trend.

How does sea ice interact with climate? It is clear from the results of coupled ocean–atmosphere climate models that the role of sea ice in climate change is not well-understood (Murphy and King, 1997). Some of the greatest discrepancies in the models occur in the seasonal sea ice zone: compare the mean annual 2–6 °C temperature change over the Antarctic seasonal sea ice zone on CO_2 doubling predicted by Wilson and Mitchell (1987) (see Figure 5.14) with the 0–3 °C change in 60–80 years of Manabe *et al.* (1992) (see Figure 5.15) and the 1 °C change after 60 years of Stauffer *et al.* (1989) (see Figure 5.16). At the heart of these discrepancies lies a lack of information about the way in which the Southern Ocean interacts with climate, especially in the seasonal sea ice zone. Some models use deep circulation lag effects, which allow the atmosphere to warm slower due to heat absorption in the ocean (Wadhams, 1994). Others use feedback mechanisms to allow indirect adjustments to come into play so that the concentration of leads and polynyas, upper ocean structure and density adjust to minimise the effects of the changes (Martinson, 1990). It may also be that first-year sea ice is more resistant to warming than previously thought via these feedbacks (Squire, 1991). The relative amounts of open water leads, polynyas and new ice can be expected to greatly affect not only the heat flux but also the precipitation/evaporation balance as modelled by Simmonds (1992). The amount of thermal inertia in the Southern Ocean also appears to have an important, but as yet unclear, influence. The abrupt change in sea ice extent between the 1950s and 1970s identified by de la Mare (1997) also poses problems for the modelling of a system that appears to be roughly stable but then switches abruptly to again appear roughly stable. It appears possible that rapid changes in sea ice, ocean and atmospheric processes may take place naturally and may be unconnected to human-induced changes, but we do not yet know (Murphy and King, 1997).

5.3.5 Summary

The direct indicators of climate change produce evidence of warming in the Antarctic Peninsula but inconclusive evidence for warming elsewhere on the continent, possibly on account of sparse and inadequate data coverage. Computer model predictions mostly indicate some warming in the future with associated increased precipitation in the seasonal sea ice zone and along the continental coasts. The best estimates indicate a warming of 1 °C over the seasonal sea ice zone for a 1% a^{-1} CO_2 increase over 100 years. These temperature increases may lead to some enhanced melting of the margins of the ice sheets, but the amount of increase is unlikely to affect the East Antarctic ice sheet substantially, and a more important effect may be enhanced snowfall in coastal areas, which may lead to local ice sheet thickening. One effect could be on the extent of sea ice, since the seasonal girdle of sea ice enhances winter cooling. The status of the floating ice shelves that flow from the continent into the sea is also the subject of concern, since these may perform a vital buttressing function without which it is possible that the ice sheet may collapse. Recent local warming in the Antarctic Peninsula is thought to be responsible for the disintegration of smaller ice shelves but, further south, there is as yet no sign of a cooling trend or of a more widespread reduction in ice shelf area. However, the increase in temperature seen in the Antarctic Peninsula is greater than that observed in the Southern Hemisphere and elsewhere in Antarctica and suggests that it is an area of high climatic sensitivity. It may thus provide a timely indication of more widespread Antarctic warming in the future (Stark, 1994). Although there has been no long-term variation in sea ice extent from satellite data, a marked decline in summer extent and an abrupt southward shift in ice-edge location in the period 1950s to 1970s has been recorded from whaling records, which poses problems for modelling the system.

5.4 Biological impacts of climate change: ecosystem sensitivity

5.4.1 Introduction

The relationship of both terrestrial and marine organisms and ecosystems to climate (see Chapters 3 and 4) and the effect of change on Antarctic physical systems have been summarised above. It remains to examine the possible implications of enhanced global warming on organisms and ecosystems, and this has been identified as a major research priority (SCAR, 1993a). It should be emphasised that much of this research is necessarily speculative: the data needed for the prediction of growth, productivity and community composition under different climates are often inadequate, so predictions are based on extrapolations of present biogeography, palaeobiogeographical reconstructions, ecosystem modelling and ecophysiology (the study of the effects of environmental variables on growth) (Adamson and Adamson, 1992). There are also important feedbacks to be recognised and in this respect the marine ecosystems are particularly sensitive: for example, the functioning of the marine ecosystem is especially important in global carbon fixation, providing important feedbacks in the modulation of atmospheric CO_2 and possibly control the development of ice ages. Several factors may drive the response of ecosystems to climate change. These include higher temperatures; changing precipitation; enhanced levels of CO_2 and other greenhouse gases; O_3 depletion and enhanced UV-B irradiation; salinity change enhanced sea ice variability; glacier retreat from habitats; and changes in food chains and animal foraging range. These factors can be resolved into impacts related to climate change and O_3 depletion on (1) the terrestrial ecosystems of Antarctica and (2) the ecosystems of the Southern Ocean.

5.4.2 Likely impacts of climate change and O_3 depletion on the terrestrial ecosystems of Antarctica

Impact of increases in temperature

The present distribution of Antarctic terrestrial communities is fundamentally restricted by the lack of ice-free ground and, in spite of the possibility of an initial increase in precipitation and eventual glacier expansion, the major long-term effect of a future temperature increase may be a reduction in ice cover, particularly in the northern parts of the maritime Antarctic (Smith, 1990; Hall and Walton, 1992), where the relatively low-density ice is sensitive to small changes in summer temperatures. Any reduction in ice cover will lead to exposure of the substrate and enhanced mechanical and chemical weathering, associated rock fracturing, and water penetration (Hall and Walton, 1992). Thaw of the permafrost upper layer and an increase in the soil active depth may result in an increase in the rate of soil formation and enhanced opportunities for colonisation (Kennedy, 1995).

Increases in temperature can also be expected to increase the rate of plant colonisation and growth. The present biogeographical distribution of species shows a southward reduction of diversity as a result of low temperature, often acting through secondary factors such as water availability and freeze–thaw stress (Kennedy, 1995). Phanerogams cease at 68° 42' S, mosses at 84° S, and lichens are not known beyond 87° 21' S. The biogeography of Antarctica may be expected to change with climate warming, and those species whose environmental tolerance limits them to warmer areas may extend their ranges into higher latitudes and altitudes. Two lines of evidence support this view. Where temperatures are higher in the sub-Antarctic islands and geothermally heated areas of the continent, the environment supports a considerably richer and more luxuriant flora and fauna than elsewhere in the Antarctic. In addition, experiments using cloches show low temperature to be limiting to several species (Hall and Walton, 1992). In the above situations the community response to elevated temperatures is convincing. Expansion of lichens and mosses on to previously uncolonised glacial moraine on Signy Island and the observed 25-fold and five-fold expansion of *D. antarctica* and *C. quintensis,* respectively, between 1964 and 1990 in the Argentine Islands, Antarctic Peninsula, is attributed to increasing summer air temperatures (Smith, 1990; Fowbert and Smith, 1994) (Figure 5.18). Similarly, on Heard Island and Îles Kerguelen, glacier recession over the

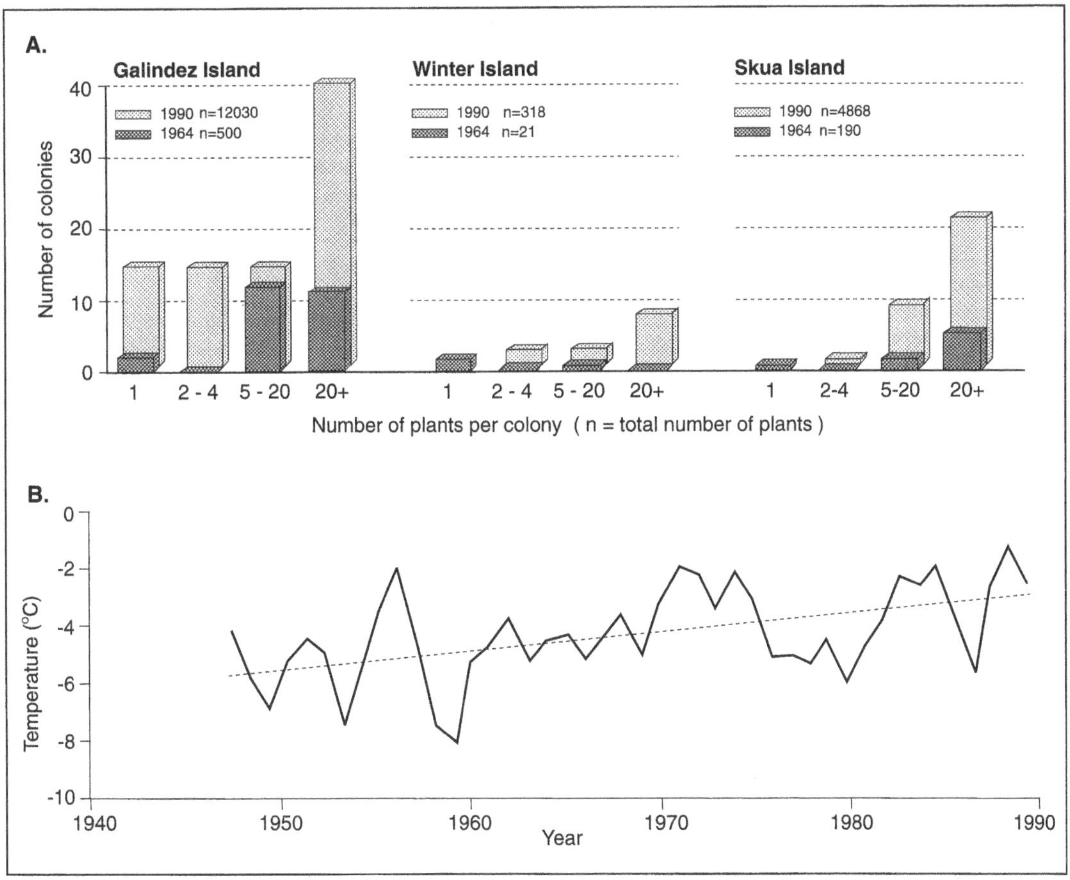

Fig 5.18 (A) The expansion of *D. antarctica* in the Argentine Islands, Antarctic Peninsula, from 1964–1990. (B) The increase of mean annual temperatures at Faraday Base, central Antarctic Peninsula, shows a statistically significant warming of 2.67 °C over 44 years (source: data from Fowbert and Smith, 1994; Stark 1994).

past 40 years has allowed expansion of the vegetated area, and changes in the distributions of some vascular plants may be related to climatic amelioration (Scott, 1990). The deployment of cloches over barren fellfield sites on Signy Island and the resultant elevation of temperatures has resulted in dramatic increases in vegetation cover and soil invertebrate populations, and in successful germination of spores lying dormant within the soil (Kennedy, 1995).

The effects of global warming are also felt by plants and animals through increases in the length of potential growing season. Many vegetated habitats in the Antarctic have a mean annual temperature of less than 0 °C and only a very short summer season, resulting in low rates of production, decomposition and recycling. For much of the year, water availability is restricted by freezing. Depending on location and latitude, climate warming will result in a reduction in the number of degree-days below zero and in an increase in the length of the growing season. Increases in temperature can be expected to increase the rate of plant photosynthesis, respiration, development and flowering frequency. Oquist (1983) points out that most plants have the ability for photosynthesis and growth in the range 10–35 °C. Although Antarctic plants have adopted a variety of mechanisms to allow photosynthesis and growth at temperatures well below this range, this has not involved sacrificing the ability to metabolise under warmer conditions (Adamson and Adamson, 1992; Kennedy, 1995). Since the optimum temperature for photosynthesis is above that usually found in Antarctic habitats, climate warming should increase the rate of photosynthesis of most Antarctic plants (Smith, 1990). However, there are constraints on such an increase in the productivity of Antarctic vegetation: Adamson and Adamson (1992) indicate that if increased temperature promotes respiration more than photosynthesis, then productivity will fall. Any temperature increase should also result in a higher frequency of sexual rather than vegetative reproduction in plants, since warmer conditions increase the success of production, germination of both spores and seeds as well as seedling survival (Smith, 1990; Kennedy, 1995). This should promote recruitment and expansion of these plants.

The response of invertebrates to temperature increase is not clear, since many depend on temperature cues for transformation into a different stage in their life cycles. Organisms that respond to unseasonably high temperature cues may find themselves subsequently exposed to severe conditions in a form not adapted to cope. For some, the production of cryoprotectant polyols is triggered by temperature, so changes that affect the cold-hardiness of species place them at risk when warmer spells are followed by cold snaps (Callaghan *et al.*, 1992). In general, species that are closely adapted to their environment are at greater risk from climate change through disruption of their life cycles. The need for prolonged life-cycle adaptations, characteristic of many Antarctic species, currently present a barrier to potential immigrant species that would be lessened with warming. On the other hand, many resident species have flexible life-styles and the potential to adapt to new climatic conditions: growth and development in *Alaskozetes* only takes place when environmental conditions allow (Block, 1980). Such residents may experience population increases due to enhanced temperatures and may expand into other areas previously unsuited to colonisation. Experiments on Signy Island show greatly increased arthropod populations as a result of summer temperatures within cloches that are 2.5 °C higher than outside (Kennedy, 1994). Potentially, this benefit may be offset by increasing competition for scarce food resources as a result of immigrant species that have been previously excluded by virtue of their inability to cope with severe conditions.

Impacts of changes in precipitation

One of the consequences of higher concentrations of atmospheric CO_2 and higher temperatures is a predicted 20–33% increase in the global concentration of water vapour (Trenbeth *et al.*, 1987). Reduction in the extent of sea ice and the enhanced penetration of moister air into the Antarctic region may result in increases in precipitation of 5–20% (Schlesinger and Mitchell, 1987). In much of continental Antarctica, plants are limited in productivity and distribution more by aridity than by any other factor, and any increase in water availability will lead to an increase in plant cover and in the complexity of communities (Kennedy, 1993; 1995). The response of invertebrates is also likely to be positive, since many species suffer from a lack of biologically available water during the winter period. In wet conditions following drought, mosses and lichens rapidly rehydrate and resume photosynthesis and so should be favoured by any increase in moisture. Also favoured should be the flowering plants *D. antarctica* and *C. quintensis*, since both are associated with moister soils under present conditions. Free water is also essential for successful sexual reproduction in cryptogams, an important factor in long-distance dispersal (Callaghan *et al.*, 1992). Given the geographic isolation of many potential colonisation sites in Antarctica, such increases in moisture will provide the cryptogams with an important colonising advantage over other species. Native species will also face competition from other areas: as moisture increases, the southward and inland spread of flowering plants now restricted to the maritime Antarctic might be possible.

Variations in the amount and duration of winter snowbanks are important both for availability of shelter and protection from cold, wind and wind-blown ice during the winter and for the spring release of moisture vital to the survival of many plants and animals. Impacts may vary spatially, depending on the amount of increased snowfall balanced against the increase in temperature. Enhanced snowfall in warmer winters may result in longer-lying snowbanks in some places on continental coasts but, on average, the snow-free period may be considerably extended by warmer summers. On the Antarctic continent, enhanced wetness in the presence of low temperatures will promote cryoturbation, and soil instability, already an important limiting factor in successful colonisation, may increase (Kennedy, 1995).

Higher temperatures in the maritime Antarctic may offset the effect of higher snowfalls and lead to earlier and more extensive snow melt, longer growing seasons, greater water availability and enhanced colonising opportunities due to glacier recession. In the sub-Antarctic, higher temperatures may also bring enhanced colonising opportunities and longer growing seasons. In such locations, where moisture is not a limiting factor for plant growth, increases in precipitation could lead to an increase in the area covered in peat bogs at the expense of drier communities (e.g. fellfield communities). Climate change may not necessarily lead to greater moisture availability at all sites in Antarctica. For example, retreat of glaciers and snowfields may produce locally dry conditions, and melting of permafrost may increase soil permeability and promote desertification (Callaghan *et al.*, 1992; Kennedy, 1995). However, overall it is likely that the effects of increased moisture availability will impact positively on the structure and function of terrestrial ecosystems in Antarctica.

Impact of increases in atmospheric CO_2

Levels of atmospheric CO_2 in the pre-industrial period of 280ppmv increased to about 380ppmv in 1994, but there is concern that the predicted increase to 470ppmv by 2050 (Houghton *et al.*, 1990) will bring varying fortunes. In the terrestrial ecosystem, since many invertebrates show a sufficient amount of flexibility in their life cycles, it is possible to speculate that they might respond to future climate changes just as successfully as they have done in the past (Callaghan *et al.*, 1992). Soil invertebrates already survive in conditions of enhanced CO_2, especially over winter, when they might be encased in ice-enclosed soil, so they can be expected to be little affected. Similarly, mosses, lichens and higher plants mostly react to higher levels of CO_2 by increasing the rate of photosynthesis (the CO_2 fertilisation effect); enhanced productivity; reduction of time to maturity; enhanced efficiency in use of water, nutrients and light; and changes in seed output, senescence and plant architecture (*ibid.*). However, increased productivity via the CO_2 fertilisation effect can take place only if water, nutrients and light are also non-limiting (Kennedy, 1995). In this respect, the CO_2 fertilisation effect should favour sub-Antarctic and coastal environments to a greater degree than the plant communities of the continent since at dry polar sites, growth is more often limited by water availability than sub-optimal CO_2 concentrations (Adamson and Adamson, 1992). The earlier melting of snow cover may also allow enhanced water and light, but in the nutrient-poor soils of Antarctica, increased growth in one year may lead to reduced nutrient availability in the following year unless microbial activity stimulates nutrient cycling (Kennedy, 1995). One important effect lies in the changed proportion of carbon (C) relative to nitrogen (N) in plants under enhanced CO_2 conditions. Grazing animals need 20–80% more plant biomass in order to extract a given amount of protein from plants grown in a CO_2-rich environment than from air-grown plants, so if CO_2 levels become higher in the future then overgrazing of some species may allow non-grazed species to become more productive and competitive (Callaghan *et al.*, 1992). Further, increased C:N ratios, which force grazers to consume more plant material, will alter the routing of primary production through grazing rather than the detrital pathway, and this may reduce invertebrate growth rates and recruitment (Kennedy, 1995). Overall, where water, light and nutrients are non-limiting, the effects of CO_2 increases should favour enhanced productivity in Antarctic plants.

Impacts resulting from O_3 depletion

Changes in the amount of atmospheric O_3 may herald a range of adjustments that act in concert with changes forced by climate warming *per se*. The O_3 layer over the Antarctic effectively acts as a shield against biologically harmful UV-B radiation, so Antarctic ecosystems have evolved under UV-B radiation levels that increase naturally during spring due to increasing elevation of the Sun and increasing day length. However, over the last decade up to 70% reduction in springtime stratospheric O_3 has occurred over Antarctica (see Figure 2.19), and continued depletion is predicted into the next century. Comparing the solar irradiance ratios of dates before and after the existence of a major O_3 hole, Frederick and Alberts (1991) showed that present-day irradiance is double previous levels (Figure 5.19). This will enhance the potential impact on the Antarctic biosphere of the springtime flux in UV-B radiation, particularly for terrestrial vegetation and for the shallow springtime phytoplankton blooms of the seasonal sea ice zone (Bidigare, 1989; Smith *et al.*, 1992). In effect, the duration of high levels of biologically effective irradiance normally associated with high solar angles in summer is increased (Lubin *et al.*, 1992). The timing and duration of the O_3 minimum and the year-to-year variations are also important, since the effects will be greater later in the season as solar angle and daylight duration increase; for example, at Palmer Station in 1990 UV enhancement persisted through the period of peak irradiance in December (Frederick and Alberts, 1991). Significantly there is

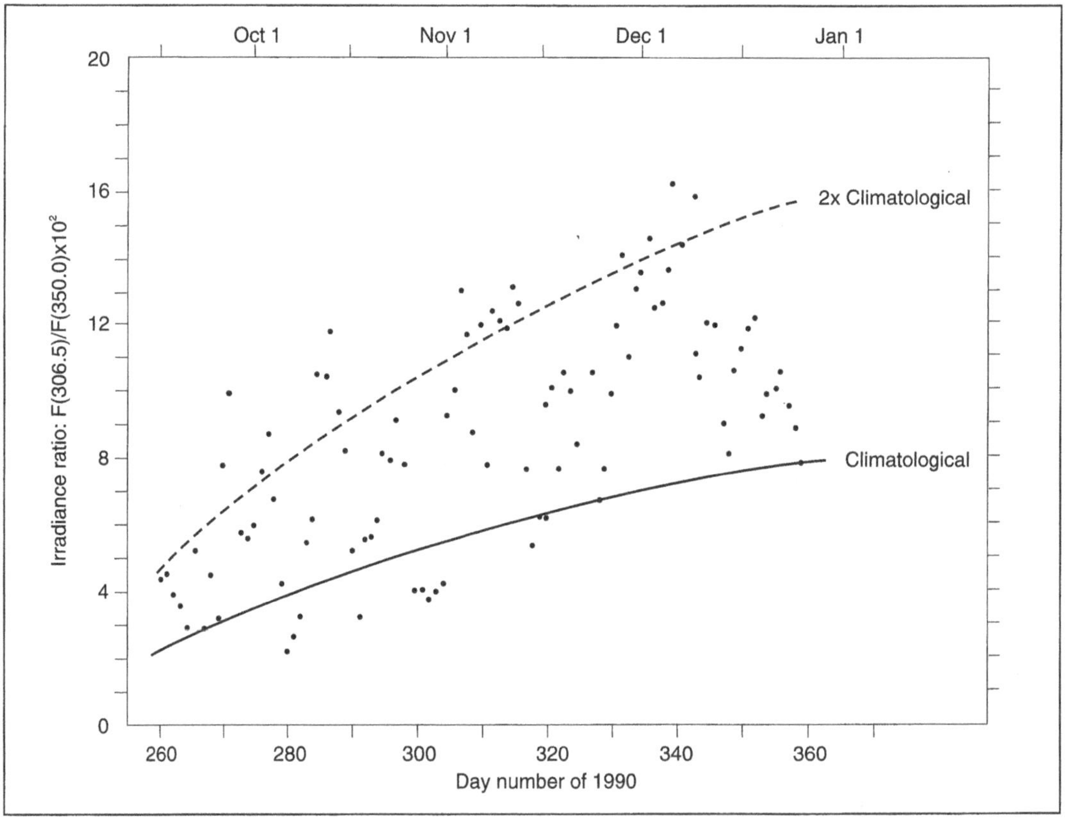

Fig 5.19 Ratio of noontime solar irradiance at 306.5 nm, to that at 350 nm (an index of biologically effective radiation) for the spring of 1990 at Palmer Station, Antarctic Peninsula. Typical O_3 irradiance ratios calculated for the years prior to the 'O_3 hole' are represented by the 'climatological' curve. The dashed line is double the climatological irradiance ratio: if there were no O_3 depletion, the points would scatter on either side of the climatological curve. Approximately 20% of the days have ratios more than double the predicted climatological value (source: modified from Fredrick and Alberts, 1991).

now evidence that O_3 depletion is extending through to late summer (Jones and Shanklin, 1995).

The biological effect of O_3 depletion depends on the action spectrum of the biological component involved (e.g. damage to DNA or inhibition of photosynthesis), the sensitivity of the species concerned, the ratio of UV-B to photosynthetically active radiation (which is used in photorepair), as well as a range of environmental and meteorological factors. Observations at Palmer Station (Lubin *et al.*, 1992) show that under conditions of modest O_3 depletion, as in 1988, levels of UV radiation potentially damaging to DNA could be reached two months early. In 1989 and subsequent years, severe depletion led to UV levels being reached that were inhibiting to photosynthesis in plankton. Excess UV-B radiation inhibits photosynthesis and interferes with normal metabolic functioning. Mosses, liverworts and lichens may all be affected, and Adamson *et al.* (1988) reported inhibition of photosynthesis and possible destruction of chlorophyll in moss exposed to bright sunlight.

In the long term, excess UV-B radiation may result in morphogenetic aberration, impaired growth and ultimately, because susceptibility to UV damage is species-specific, changes in community composition (Young *et al.*, 1993). Dominance patterns may change as UV-tolerant species are favoured and avoidance strategies such as migration and the synthesis of protective pigments are adopted by plants. Studies of plant response to increases in UV-B radiation suggest that many are already well-adapted by the production of photoprotective pigments and will increase pigment production in the most exposed sites. Adamson and Adamson (1992) suggest that different levels of UV-absorbing pigments in lichens and mosses would render them less susceptible to damage than algae. Post and Larkum (1993) found that although the alga *Prasiola* contained relatively high levels of UV-absorbing compounds it was nevertheless susceptible to reduced productivity under enhanced levels of UV-B. However, extensive snow cover during the critical spring period may also mitigate impacts (Karentz, 1991). The available experimental data suggest that the different species of cyanobacteria, a significant component of the terrestrial and freshwater ecosystems of bare rock, soil, lakes and streams, also show different sensitivities to enhanced UV radiation, which in part reflect a range of protection strategies and repair mechanisms. In spite of this, there has been only limited research on the impacts of increased UV-induced stress on Antarctic terrestrial or limnological systems. By analogy with observed effects in other regions (Caldwell *et al.*, 1995; Johanson *et al.*, 1995; Manning and Tiedemann, 1995), the principal impacts on plants will be likely to include loss of production and species diversity (Voytek, 1989; 1990).

The sensitivity of terrestrial invertebrates to enhanced UV-B radiation is relatively unknown but is thought to be outweighed by the negative effects on the plants that support them: where vegetation is damaged, grazing species will suffer. Any reduction in productivity as a result of enhanced UV-B radiation is likely to present consumers with reduced food resources and this will mitigate the stimulatory effects of global warming.

Interrelationships and indirect effects

It is clear that changes in temperature, precipitation, CO_2 and O_3 often act together, and because of the complexity of feedback mechanisms the precise magnitude and direction of response in ecosystems is as yet unclear. For example, since the late 1960s, temperatures have increased at most of the sub-Antarctic and maritime Antarctic islands (Smith, 1990; Smith and Steenkamp, 1990; Gordon and Timmis, 1992). This has impacted on the relatively species-poor and simple terrestrial ecosystems at both Signy Island (Smith, 1990) and Marion Island (Smith and Steenkamp, 1990). At Marion Island, mean annual air and sea-surface temperatures have increased, notably since 1968, whereas precipitation has been lower. Smith and Steenkamp consider that the principal impact on the ecosystem has been on primary and secondary production and nutrient cycling. Higher temperatures and increased levels of CO_2 are likely to increase vegetation productivity and nutrient demand, particularly in the higher plants (Callaghan *et al.*, 1992). Soil microbiological activity is more affected by waterlogging than by temperature controls, but in spite of this and of the low nutrient status of Antarctic soils, the increased nutrient demand of plants may be met by increased temperature-driven activity of the soil macro-invertebrates that break down plant detritus. However, there may also be increasingly negative effects on CO_2 and temperature-induced growth as a result of increased UV-B radiation damage to plant tissue early in the growing season (Fowbert and Smith, 1994). It seems clear from warming in the sub-Antarctic and maritime areas of the Antarctic that in the short term the number of suitable habitats is likely to increase with climate warming, although this may be curtailed due to the effect that enhanced snowfalls may have on increased glacier ice extent. Increased snow melt around the coast of the continent will provide more moisture for plants currently restricted by water availability, and higher temperatures and longer growing seasons will allow more annual growth (Adamson and Adamson, 1992).

Indirect impacts on other components of the terrestrial ecosystem are more uncertain, but changing atmospheric circulation patterns are likely to affect the feeding grounds and breeding patterns of the large seabird populations more than direct temperature changes alone. Any such change will impact on those vegetation and invertebrate communities that derive a significant input of nutrients from bird guano, such as onnithogenic soils and small lakes in the South Shetland Islands, South Orkney Islands and the sub-Antarctic islands. In the short term at least, it is likely that the numbers and size of such lakes and streams will increase with climate warming as ice and snowbanks are subjected to melting. For many maritime and coastal continental Antarctic environments even a small increase in summer temperatures will lead to a diminution of their currently marginal status for vegetation growth and production. Further, as the most likely scenario is that coastal continental areas may also benefit from an increase in precipitation as a result of warming, then two of the main limiting factors on successful vegetation colonisation and growth begin to weaken.

Importantly, all the predictions are that climate change will be relatively rapid and that animals and plants are more likely to move than adapt (Young, 1991). This means that changes to the communities within ecosystems will occur as a result of alterations in relative abundances of resident species together with the immigration of new species, and this will impact differently depending on species and location. Higher plants tend to migrate more slowly than the predicted rate of climate change, so many species may be trapped in supra-optimal environments where competition from immigrants may disadvantage them. The spores of mosses and lichens can migrate faster, and although these will enjoy an enhanced colonising advantage, they have fewer mechanisms to respond to changing environments once established (Callaghan *et al.*, 1992). Smith and Steenkamp (1990), Walton (1990a) and Kennedy (1995) also recognise that changes in climate could favour the introduction of new biota by seed dispersal by wind, birds and the growing volume of human air and sea transport. Immigration is likely to have a dramatic effect on the existing Antarctic ecosystems, whose very simplicity renders them vulnerable to competition from alien species: in the absence of predators and pathogens the success of aliens may see the local extinction of Antarctic species (Walton, 1990a; Kennedy, 1995).

On the other hand, it is possible that the potential impact on the Antarctic communities of immigrants from outside may be overstated. Palynological evidence from South Georgia and other sub-Antarctic islands indicates that in spite of periods of both warmer and colder climate over the past 10 ka (Clapperton and Sugden, 1988; Clapperton *et al.*, 1989), there has been no gain or loss of species (Barrow, 1978). On South Georgia, the 800-year period starting 5.6 ka ago was probably warmer than at present (Clapperton *et al.*, 1989), yet no new species appeared. Similarly, moribund vegetation dated to at least 2.5 ka BP now being exposed by receding ice in the South Orkneys and the Antarctic Peninsula is identical to what is now present there. It appears that in the mildest and most accessible parts of the Antarctic for plant colonisation, the varying climatic conditions experienced during the Holocene have not resulted in any detectable impact on the species present. In spite of this cautionary evidence, the present weight of scientific opinion is that the biogeography of the Antarctic is likely to change with a warmer climate but, in the short term at least, that this is unlikely to be a simple shift of vegetation zones southwards. Enhanced growing conditions in continental Antarctica are likely to be reflected first in enhanced moss and lichen cover and colonisation rather than the immediate invasion of higher plants. Such higher plants are likely to expand their cover initially in areas where they already have small colonising communities, such as is already occurring in the Argentine Islands (Fowbert and Smith, 1994; Smith, 1996).

5.4.3 Impacts on the Antarctic marine ecosystem

The Antarctic marine ecosystem has been referred to as the largest identifiable ecosystem (or ecosystems) on Earth (Young, 1991). As a result, the direct and indirect effects of any future climate change have great importance both for its individual elements and for the functioning of the ecosystem as a whole. Clearly, much revolves around the pivotal roles of phytoplankton and zooplankton in the functioning of the system, but great importance is also attached to the role of sea ice extent and concentration in influencing the locations of plankton blooms in spring and in the distribution, feeding and breeding of fish, birds, seals and whales. There are also a multitude of interannual, indirect and interspecific effects, which may be as important as any direct physical impact.

Direct effects of increased temperature

The temperature of Antarctic sea water varies from −2 to +2 °C close to the continent and from +3 to +6 °C close to the Polar Frontal Zone, with little seasonal variation. These temperatures are sub-optimal for the maximum growth of many planktonic species (Vincent, 1988), so any increase in sea water temperature will promote growth. Some species of cyanobacteria exist as part of the marine phytoplankton in the temperate and tropical oceans, where they are responsible for 25–90% of the phytoplankton biomass and 25–80% of the primary productivity (Marchant, 1992). However, their abundance is related to temperature and declines to only 18% of the phytoplankton biomass and 10% of primary productivity of the Southern Ocean (Hosaka and Nemoto, 1986), so under the present climate, they play only a minor role in the Antarctic marine ecosystem. An increase in water temperature would lead to an increase in the abundance of cyanobacteria and thus in its relative importance in the marine food web (Marchant, 1992). As a consequence of this changing composition, the species utilising the cyanobacteria and the zooplankton that feed on them may change in an unknown fashion. In fact other phytoplanktonic species, such as the coccolithophorids, are also temperature-dependent and decline in abundance to zero south of 60° S. Increasing temperatures will lead to an extension of their range far into the Southern Ocean, again with unknown consequences.

Most Antarctic invertebrates and fish have evolved to be efficient at low and stable temperatures, and the most likely effect of increasing temperatures is a change in their patterns of distribution. For example, *E. superba* live in waters of between −2 and +5 °C, and because they live close to this upper limit in the Scotia Sea, any increase in water temperature would be likely to lead to the northern limit of their range moving southwards (Everson, 1984). Since krill can also be found in great numbers close to the shores of the continent itself, the geographic range of krill will diminish in a warmer world. There may also be indirect changes in the behaviour and distribution of krill, since swarming is often related to specific conditions within the currents and eddies of the Polar Frontal Zone, the nature of which may change in the future. Increases in water temperature on the shelf areas around the continent and sub-Antarctic islands could also make it too warm for some commercially important species, such as the icefish, which spend part of their life cycle in such waters. The icefish, like many species of Antarctic fish, cannot survive in warmer waters or in water that fluctuates in temperature and has little ability to acclimatise to warmer conditions (Marchant, 1992). The reproductive strategies that are so successful in a cold stable sea, such as low fecundity, large and few eggs, and the brooding of young (see Section 5.2), do not favour efficient dispersal into new areas. Further, the very geography that favours the present strategies of many species of Antarctic fish and benthos also limits the opportunities for dispersal, because of the tracts of deep water between the Antarctic and neighbouring shallow continental shelves. On the other hand, the Polar Frontal Zone undoubtedly acts as a barrier to the immigration of species into the Antarctic (Clarke and Crame, 1992).

There are few physiological problems for birds and mammals associated with temperature increases since they are generally well-equipped to maintain relatively constant body temperatures independent of outside conditions. Many of them can cope with the slightly warmer conditions occasionally experienced at present. However, higher temperatures early in the spring may lead to the earlier nesting of birds and, if followed by colder conditions later in the season, may disadvantage incubation or chick development. Alteration in temperature would affect fledging times if more food becomes available at different times of the year. Such direct effects are important for the success of these animals in warmer conditions, but indirect effects affecting both temperature and other factors are also likely to be important. It has been suggested that an observed increase in Adélie penguin numbers in the Ross Sea area, particularly since 1981, and a southward extension of their breeding range in McMurdo Sound, is related to recent climate warming (Taylor and Wilson, 1990; Taylor *et al.*, 1990). Enhanced temperatures may have improved nesting success, reduced adult mortality and allowed increased availability of winter feeding. However, the changes might also be related to reduction of whale stocks and increased availability of krill (Laws, 1985), as has been proposed elsewhere for increased penguin and seal populations (Conroy, 1975; Croxall and Prince, 1979; Smith, 1988a). If this is the case, then a substantial time lag is involved.

The discovery by Hiller *et al.* (1988) and Steele and Hiller (1997) that petrels were able to breed on the Antarctic continent throughout the last glacial maximum throws into doubt simple extrapolations of variations in breeding range consonant with temperature change.

Impacts resulting from O_3 depletion

Of particular concern are the effects of O_3 depletion in the upper layers of the oceans, especially in the MIZ, where algae are exposed to UV-B radiation during the spring break-up and where spring and summer phytoplankton blooms at the surface may account for 20–40% of the total primary production of the Southern Ocean (Hempel, 1991). The high productivity of surface waters is favoured by high nutrient and sunlight levels, but this is also the zone potentially at greatest risk from enhanced UV-B radiation, which penetrates with marked effects to 30 m depth (Karentz and Lutze, 1990; Vosjan and Pauptit, 1992). Under normal conditions the flux of UV-B radiation is sufficient to inhibit primary production in the upper 20 m of the water column by 8–12% (Holm-Hansen *et al.*, 1993; Prezelin *et al.*, 1994) (Figure 5.20). Organisms that occur within or at the base of the sea ice (diatoms), under the sea ice (*Phaeocystis*) and in the well-mixed waters north of the MIZ are less likely to be affected because of the high attenuation of UV in such environments (Bidigare, 1989). Krill, which are thought to spawn at depths greater than 50 m and which migrate to the surface layers at night, are likely to be exposed to very low doses of UV radiation. More significant effects on the grazing or predatory species are likely to occur via the food chain, for example through reduced primary production and changes in species composition, and a particular concern is the way in which phytoplankton, the preferred food of krill, could be affected (El-Sayed *et al.*, 1990; Voytek, 1990).

Evidence and observations from both Antarctic waters and higher latitudes (El-Sayed *et al.*, 1990; Häder and Worrest, 1991; Karentz *et al.*, 1991 a and b; Helbling *et al.*, 1992; Karentz and Gast, 1993; Marchant, 1994) suggest that deleterious effects from enhanced radiation, including physical damage, decreased rates of photosynthesis, nutrient uptake and productivity, and possibly DNA damage are likely to occur in marine organisms. It is known that small but potentially damaging amounts of UV-B radiation can penetrate the exteriors of some marine vertebrates, but further work is needed to assess the

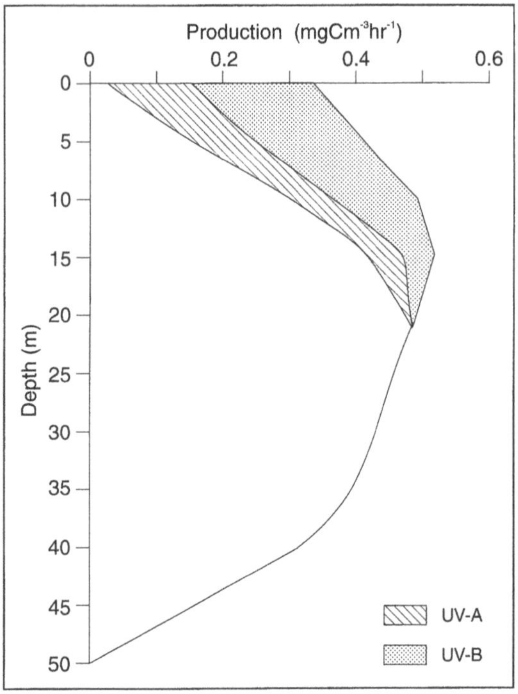

Fig 5.20 The depth of impact of UV-B and UV-A radiation under the Antarctic O_3 hole on marine primary production in the MIZ of the Bellingshausen Sea. At the time of sampling, the atmospheric O_3 concentration was 180 DU (source: after Prezelin *et al.*, 1994).

extent and longevity of any damage (Karentz and Gast, 1993). The effects of sea ice cover in reducing UV-B transmission may be critical in protecting key components of the marine ecosystem, such as the sea ice algae (Perovich, 1993; McMinn *et al.*, 1994; Ryan and Beaglehole, 1994), but the springtime depletion of O_3 coincides with the southward retreat of the sea ice. The transparency of sea ice to UV-B transmission is also likely to be greatest in the spring (Trodahl and Buckley, 1989). Additional uncertainties include variations in cloud cover, species tolerance levels, the degree of mixing in the water column and the time spent by phytoplankton in surface waters.

A further complicating factor is the existence of photorepair mechanisms, photoprotective strategies and differences in UV-absorbing compounds that vary among species (Karentz *et al.* 1991a and b; Davidson *et al.*, 1994; Karentz, 1994; Vernet *et al.*, 1994). Changes in species composition as a result of enhanced UV-B receipt are therefore considered to be a possibility (Karentz, 1991), although there is as yet no firm evidence of this. Experimental studies on diatoms (Davidson *et al.*, 1994) and phytoplankton (Holm-Hansen *et al.*, 1993) show high tolerance and minimal effects. In the fjords of the Vestfold Hills area, McMinn *et al.* (1994) found no significant compositional changes in the diatom community over the past 20 years, which they ascribed to the protective effects of the sea ice cover and the timing of the phytoplankton blooms in this area after the period of maximum O_3 depletion. Behrenfeld *et al.* (1994) suggest that UV-B inhibition of photosynthesis may not be detectable where there are competing limiting stresses, e.g. nutrient limitation. The effects of enhanced UV-B radiation may therefore become evident only in nutrient-rich areas of the ocean or during spring blooms. Moreover, the natural variability in species abundance, distribution and production may also mask any impacts until they become extreme.

Some studies have assessed the loss of primary production in Antarctic waters arising from enhanced UV-B radiation (Cullen *et al.*, 1992). In the Bellingshausen Sea, Smith *et al.* (1992) estimated a decrease in phytoplankton primary production by a minimum of 6–12% in the MIZ over a six-week period due to a 33% fall in column O_3 abundance. They also estimated an annual loss of productivity of 2–4% over the entire MIZ compared with an estimated natural annual variability of ±25%. Allowing for the extent and duration of reduced O_3 concentrations during the spring, and the seasonal and spatial variation in both ice cover and rates of primary production, others have estimated less than a 0.2% decrease in annual primary production over the entire Southern Ocean (Holm-Hansen *et al.* 1993; Helbling *et al.*, 1994). These results suggest a relatively small loss in primary production due to O_3 depletion. However, they do not include possible cumulative impacts and the effects of interspecies variations. Moreover, the extrapolation of longer-term consequences from short-term observations must be viewed with caution.

Possible impacts on the food web and its individual components have been considered by Voytek (1990). There is particular concern about how the effects of any reduction in primary production might be transferred through the food web to krill and other species at higher levels (El-Sayed *et al.*, 1990). However, on the present evidence, claims of an eco-disaster (Roberts, 1989) may be overstated, but there may well be changes in species composition as more tolerant species expand at the expense of less tolerant ones. The implications for krill feeding and the wider food web are difficult to predict (Karentz, 1991; Karentz *et al.*, 1991a). According to Voytek (1990), a further possible consequence of decreased productivity would be a reduction in global carbon fixation, since phytoplankton form an important component of the natural removal of CO_2 from the atmosphere. Such a change could enhance the greenhouse effect and global warming. However, Smith *et al.* (1992) cautioned that the estimated loss from the MIZ is likely to be insignificant in a global context, and modelling by Peng (1992) showed that the effect of O_3 depletion on the carbon cycle is likely to be insignificant.

Interrelationships and indirect effects

There are three major aspects of the Southern Ocean ecosystem that are important when considering the indirect effects of climate change. Each of these aspects is spatial in nature, the undoubted biological importance of all three arising from their very distinctive geographies. They are the enhanced productivity along the Polar Frontal Zone, where mixing between different water bodies occurs; the importance of the sympagic community of sea ice algae to oceanic productivity as the sea ice moves seasonally from north to south; and the importance of the distribution and nature of sea ice as a habitat for seals and penguins (Young, 1991).

One of the most likely reasons for an increase in water temperature would be a shift in the position of the Polar Frontal Zone, which, although deflected and partly controlled by submarine topography, is also influenced by the position, intensity and frequency of the westerly-tracking cyclones that revolve around Antarctica. Large-scale fluctuations in krill in the mid-1980s were noted as part of the BIOMASS studies around South Georgia, indicating that episodes of krill scarcity in the Scotia Sea may occur two or three times per decade. Priddle *et al.* (1988) considered the changes to be unrelated to krill biology but more likely to be related to a southward shift in the Polar Frontal Zone due to enhanced southward airflow (Figure 5.21). A southward shift in the Polar Frontal Zone tends to disrupt the pattern of the zone's numerous well-defined eddies and meanders (see Figure 4.3), leading to a release of krill to the east of the normal area in the Scotia Sea (Knox, 1994). Thus krill may be taken beyond the range of land-based breeding predators and only return within range when the Polar Frontal Zone returns to its more regular position.

If this is the case, then in spite of the total amount of krill remaining approximately constant, large

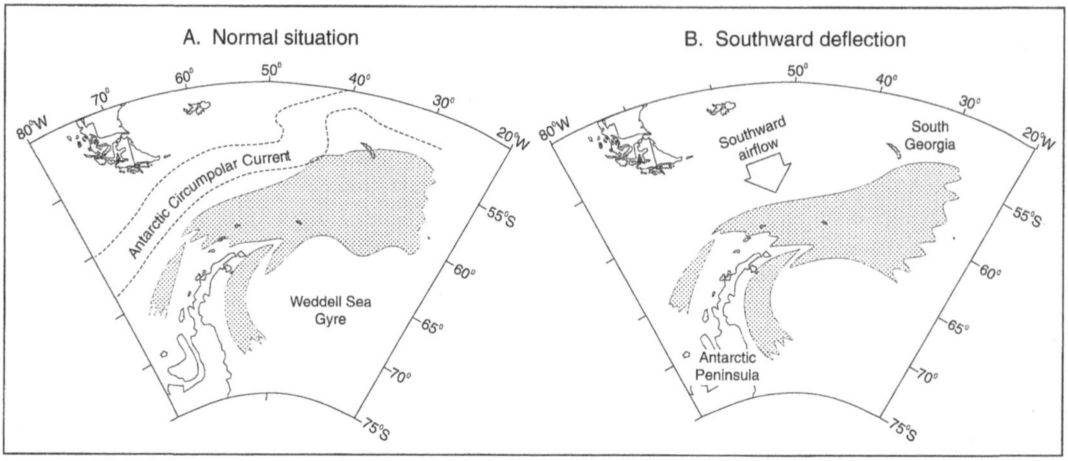

Fig 5.21 (A) The normal position of the area of high krill abundance (shaded), where krill from the Weddell Sea Gyre and Bransfield Strait are entrained into the Antarctic Circumpolar Current. (B) The possible effect of enhanced southward airflow, which may displace the krill-rich current south of South Georgia and its large numbers of krill-dependent species (source: after Priddle *et al.*, 1988).

changes may have occurred in its spatial distribution and availability, particularly around South Georgia and the Scotia Sea. Alterations in the geography of krill will impact on all those species of seals and birds that rely on local access to krill, especially during the breeding season (see Figure 4.25). For the king penguins and fur seals of South Georgia the consequences will be profound, as there are few alternative oceanic islands nearby. However, if food resources are available, climate warming may allow successful southward expansion of such species to the northern part of the Antarctic Peninsula, a process that may already have begun as far as the Antarctic fur seal is concerned (see Chapter 4). In general, it seems that climate warming and changes in atmospheric circulation patterns would mean shifts in all those systems that are atmospherically forced, and these shifts will force changes in the marine biogeography of the Southern Ocean.

Any increase in water temperature might also impact on how far north the sea ice might extend. However, since the overall extent and variability of the sea ice is controlled mainly by currents and winds rather than by *in situ* temperature-controlled freezing and melting (Allison, 1992), this effect may be limited or difficult to detect. Ice formation and retreat processes are of great biological significance in the Southern Ocean, because the sea ice provides the stable conditions in which phytoplankton blooms can become established (Priddle *et al.*, 1992). Gloerson and Campbell (1991) have linked changes in extent and thickness of Arctic sea ice to global warming, but as yet no links have been established between warming and Antarctic sea ice extent. However, higher temperatures may lead to later and thinner sea ice formation and in time may result in a decrease in total ice extent, together with an alteration in the timing of its advance and retreat. This will impact on the triggering of phytoplankton blooms and the productivity of the Southern Ocean in the seasonal sea ice zone, and on this basis most observers predict a decline in primary production (Priddle *et al.*, 1992; Spindler and Dieckmann, 1994). However, a reduced sea ice extent also means more light reaching the water column and, possibly, enhanced phytoplankton growth (Marchant, 1992). The extent to which this will offset any reduction in the inoculating power of a reduced sea ice community is unknown.

Murphy and King (1997) point out that, irrespective of whether the decline in Antarctic summer sea ice extent inferred by de la Mare (1997) is naturally induced or not, the shift occurred at the same time as the final reduction of whale numbers. This means that irreversible shifts in the wider ecosystem may already have occurred that now make the recovery of whale populations to pre-exploitation levels unrealistic. Shifts may have occurred in the pattern of primary productivity, krill recruitment and distribution of higher predators that now make the management of marine systems a short-term task where change is abrupt (Murphy and King, 1997). There is also evidence of significant interannual variability in seawater temperature, winter sea ice, phytoplankton production and nutrient utilisation, as demonstrated by Murphy *et al.*, (1995) for shallow water at Signy Island. Clarke and Leakey (1996) suggest that subdecadal oceanographic variability driven by atmospheric processes may well carry through to primary production. If proven, such a relationship would be of profound ecological significance, for the resultant variability would propagate through the marine food web. Variation in the extent of sea ice may also affect the balance of species in the MIZ. For example, Loeb *et al.* (1997) suggest that a decreased frequency of winters with extensive sea ice development may result in reduced krill availability but may favour increased salp availability: extensive sea ice favours early seasonal krill spawning but inhibits spring salp blooms, and poor sea ice development promotes extensive salp blooms and poor krill spawning. If these effects are real then a long-term warming trend in the Antarctic Peninsula area could have profound effects on the dominance of krill and salp populations and thus on krill-dependent predators (*ibid.*).

With less ice, organisms would also be exposed to increased UV irradiance, and those that have photorepair or photoprotective mechanisms would be favoured to an unknown extent. It is also clear that simple and direct relationships are unlikely. For example, a reduction in sea ice extent will probably be manifest first in reductions in ice concentration and an increased frequency of unfrozen leads (Simmonds, 1992) (see Table 5.1). Yet this will serve to increase precipitation and snowfall onto the sea ice surface. The attendant cooling associated with thermal and albedo effects may lead to local expansion and thickening, especially in areas such as the Weddell and Ross Seas (Priddle *et al.*, 1992). If this occurs then the primary productivity model of Arrigo *et al.* (1997) suggests that enhanced snowfall as a result of higher temperatures may lead to increased loading of the sea ice, infiltration and nutrient supply. This may lead to enhanced primary productivity of the sea ice community.

Reductions in sea ice extent may also affect the efficiency of the phytoplanktonic fixing of atmospheric CO_2 (see Chapter 4) in at least two ways. First, reduction in sea ice extent may result in the southward extension of species more common in warmer oceans, and the introduction into the Southern Ocean of blooms associated with, for example, the coccolithophorids may alter nutrient dynamics and reduce CO_2 drawdown from the atmosphere into the ocean and thus enhance warming (Priddle *et al.*, 1992). Second, the two dominant components of MIZ phytoplankton, the diatoms and the alga *Phyaeocystis pouchetii*, differ markedly in their response to UV radiation. The marine diatoms are more affected by UV-B radiation because they have relatively low concentrations of UV-B-absorbing compounds, yet the colonial stage of *Phyaeocystis* in the Southern Ocean has five–ten times higher concentrations of these compounds than recorded in *Phyaeocystis* from other oceans (Marchant, 1992). *Phyaeocystis* growth may then be favoured by a restricted sea ice cover, but because its nutritional value is less than that of the diatoms on which krill and copepods selectively graze, there emerges the possibility of nutrient limitation to the zooplankton as well as a reduction in vertical carbon flux as a result of reduced diatom abundance (*ibid.*). Evidence from warmer oceans suggests that when diatoms are replaced by other species of phytoplankton, fish productivity is markedly reduced (Barber and Chavez, 1983). If climate change also resulted in a reduction in wind forcing, both diatom abundance and the vertical flux of carbon would fall even further because diatoms have high sinking rates and require turbulent mixing to remain in the photic zone. However, there may be some degree of negative feedback since *Phyaeocystis* is the main producer of dimethyl sulphide (DMS) in Antarctic waters, contributing up to 10% of the total global DMS flux to the atmosphere. DMS forms the basis for the sulphate aerosols that serve as cloud condensation nuclei over the sea (Charleson *et al.*, 1987). The resultant increase in cloud cover caused by enhanced *Phyaeocystis* abundance may increase the sea surface albedo and may help to moderate any climate warming.

It is also important to recall that the development of sea ice in autumn and winter around Antarctica is directly associated with the production of cold and dense Antarctic Bottom Water (Gordon, 1988) (see Chapter 4). Reductions in the amount of sea ice as a result of warming would reduce the production of Antarctic Bottom Water and influence the delivery of mineral nutrients such as phosphate, nitrate and silicate to the ecosystems of the Southern Ocean. Part of the cold nutrient-rich water also flows northwards, where it mixes with the waters of lower latitudes, contributing greatly to their productivity. Changes to the production of this water may well force changes in the marine productivity of oceans beyond the immediate Southern Ocean: in the Atlantic, Antarctic Bottom

Water finds its way well north of the equator! The release of low-salinity water from the melting of sea ice in spring also allows the development of a shallow layer of surface water whose presence is crucial to the spring phytoplankton bloom (Marchant, 1992). Reduction in the amount of sea ice would impact on the physical extent, timing and location of this important layer of water and may effect as yet unknown impacts on an area widely regarded as being the most productive in the Southern Ocean.

Reduction in the amount of sea ice would also have a profound impact on the species that use it as a breeding platform and that find their food in the adjacent sea. Generally speaking, if the amount of food remains unchanged as a result of reduced sea ice cover, the occurrence of more open water and leads would benefit those seals and penguins that breed on the continent or in the seasonal sea ice zone in that they would require to expend less energy as they moved between breeding and feeding sites. Crabeater seals that now pup on the outer margins of the sea ice would simply be further south at pupping time than at present (*ibid.*). However, for those species that inhabit the sub-Antarctic islands or northern parts of the Antarctic Peninsula, foraging close to the sea ice edge might be more remote and require more energy expenditure to access food (see Figure 4.25). However, a fundamental problem remains that it is not known if less ice means more prey or less prey (Croxall, 1992). There is some evidence to link the scarcity of krill in some years to the alterations in ocean circulation patterns resulting from El Niño–Southern Oscillation (ENSO) events (*ibid.*). In the year following two ENSO events, 1984 and 1987, krill populations were much reduced and this impacted on the breeding success of krill predators. However, other reductions in breeding performance in krill-eating species are unrelated to ENSO events (*ibid.*).

In spite of this uncertainty, various population fluctuations in ice-breeding seals and penguins as a result of past changes in sea ice distributions give some clues about the possible future effects of climate-induced reductions in sea ice. All of the ice-breeding seals show four or five year cycles in their age structures, which have been related to the strength of ENSO events via greater sea ice cover and thus a reduction in reproductive performance. For penguins, extensive sea ice cover increases the time and effort expended in reaching the breeding site at the start of the season in order to commence breeding or, during the season, to feed chicks (*ibid.*). In East Antarctica, the years of highest and lowest breeding success of Adélie penguins and petrels is directly related to the abundance or scarcity of food and the early and late breakout of sea ice, respectively (Whitehead *et al.*, 1990). However, for the emperor penguin, early break-up of the sea ice before the chicks can fend for themselves can lead to breeding failure and falling numbers, as has occurred in Terre Adélie since 1976 (Jouventin and Weimerskirch, 1991). Both emperor and Adélie penguins, together with snow petrels, are closely dependent on the sea ice all year round but, although they are sensitive to its changes, the impact of less sea ice in any one year is different for each species.

The way in which the sea ice environment may change in the future will impact in differing ways on predators depending on whether they or their prey are directly associated with sea ice or are remote from it. Indirect effects on the nature and extent of the sea ice habitat for feeding or breeding are likely to be important; for example, the occurrence, size and persistence of polynyas and ice-free leads are vitally important to the ice-breeding seals, which need juxtaposed water and ice for feeding, mating and suckling. Reductions in sea ice cover should favour the ice-breeding seals. However, reductions in the amount of sea ice may fundamentally affect the seeding of the spring phytoplankton blooms in the MIZ. Any reduction in phytoplankton biomass will impact on krill and other grazers and on all those species that feed on zooplankton. What appears certain is that the impact of change in the physical systems and biological systems will vary between different regions in the Antarctic.

5.4.4 Sensitivity of Antarctic environments to change

There are many references to the fragility and sensitivity of Antarctic systems to change and of their susceptibility to disturbance (Smith, 1990; Adamson and Adamson, 1992; Block, 1994). Generally speaking, biological systems in the Antarctic are subject to extreme environmental conditions and display relatively simple characteristics, with few species and short food webs with few links (Block, 1994). Where few species are involved then the decline of one of these undoubtedly has a greater effect in the ecosystem than otherwise and, together with a low population density and slow growth, these ecosystems are likely to be more sensitive to enhanced stress than their lower-latitude counterparts. However, there is also evidence of rapid recovery in the face of large perturbations in ecosystem stability. For example, the

Southern Ocean has arguably adjusted well to the wholesale removal of some seal species and most of the great whales. This ability to adjust to large perturbations can be argued to demonstrate an inherent resilience rather than fragility (Dunbar, 1973)! Nevertheless, two points are important. First, there is great variability in the degree of sensitivity to change in both the terrestrial and marine ecosystems. Second, the limited scale of impact of past human activity on the continent may now be changing rapidly with the onset of human-driven climate change. The changes that might be involved as a result of the latter scenario have been outlined above, but it is possible to speculate on the degree of sensitivity or resilience displayed by Antarctic ecosystems to such changes.

Terrestrial ecosystem sensitivity to change

Antarctic vegetation is at the geographical and physiological limit of plant growth and must therefore be sensitive to any additional stress, especially arising from changes in climate (Adamson and Adamson, 1992) and human activity. Most of the vegetation of continental Antarctica is composed of lower plants such as the mosses, which occupy moist sites, and the lichens, which occupy dry, exposed sites. The terrestrial ecosystem is finely tuned and is sensitive to a mosaic of fluctuating microenvironmental conditions, and the patchy stands of vegetation are particularly vulnerable to even minimal disturbance (Bonner, 1984; Walton, 1987c). As a result of slow rates of growth, colonisation and community development, the sensitivity of such environments is revealed when disruption leads to protracted recovery, particularly in dry habitats (Smith, 1993a and b). Many of the adaptations that favour the survival of Antarctic species may place them at a disadvantage when conditions change. The storage of biomass in root systems to avoid extremes of temperature and wind leads to restricted growth of above-ground foliage: a disadvantage under climate warming, when immigrants may produce large amounts of above-ground foliage and shade out dwarf plants. The biogeographical distribution of many species seems to be more limited by the lack of moisture operating through the water balance of organisms than by temperature alone (Kennedy, 1993). For example, the maximum populations of terrestrial invertebrates relate to periods of moisture availability rather than to periods of warmth, so there is great sensitivity to even small-scale changes in the length of melt season. Long cold winters with late-lying snow banks reduce the productivity of terrestrial ecosystems and produce habitats that are essentially fragile and easily disrupted. In addition, having been geographically isolated for a considerable period of time, the terrestrial biota of Antarctica are genetically impoverished, with species that are slow to reproduce sexually, and this adds to their sensitivity as indicators of climate change. The extremely simple structure of Antarctic plant and animal communities also means that the direct or indirect effects of climate change should be readily observed (Adamson and Adamson, 1992).

Since the sensitivity displayed by terrestrial ecosystems is a function of existence in, and adaptation to, an extreme environment, it is more likely that climate change will reduce rather than increase fragility, because most habitats will benefit initially from the predicted increases in coastal precipitation, enhanced summer temperatures and more and earlier snow melt. All of these changes serve to reduce the influence of the two major constraints on plant growth and animal activity: limited moisture availability and low temperature. However, many invertebrates have flexible survival strategies and are thus better equipped to cope with climate change than plants: they can move to a more favourable microclimate, for example. Some locations may suffer a deterioration in conditions, but on average most existing habitats will gain. Some plants are already well-adapted to higher UV-B irradiation by the production of photoprotective pigments, yet others are likely to suffer negative impacts as a result of enhanced tissue damage and lowered productivity. Competition from immigrant species or shifts in dominance of existing species present additional stresses. Local immigration of Antarctic species into adjacent communities has always occurred, but what may now be changing is the rate of expansion and the opportunity for exotic species to immigrate. Overall, the evidence points to Antarctic terrestrial ecosystems being relatively sensitive to change and easily disturbed, especially the plant communities. The animal communities may be more robust and many species have the ability to adopt flexible survival strategies when conditions demand.

Marine ecosystem sensitivity to change

Unlike the terrestrial environment, the Southern Ocean is not subject to the extremes of temperature seen on the continent. Nevertheless, it is subject to conditions that lead to freezing of its topmost layer over winter, and many of its species share similar ecological and physiological adaptations in response to extreme cold and seasonality. Cold-adapted species such as fish and benthic fauna have physiologies that are well-adjusted to their environment and have, for example, invested in reproductive strategies that work best where food is limited on account of the cold: slow growth, long life, low fecundity, large and few eggs, and brooding of young. However, these characteristics do not suit them to dispersal into new, areas, so they may prove sensitive to future changes in water temperature for two reasons: because of competition from immigrant species; and because they are intolerant of changes in water temperature itself. Many of the fish populations have also been shown to be susceptible to overfishing as a result of slow rates of growth and recruitment, so their ecology, and the ecology of other fish and benthic species, may be more fragile than appears. However, in spite of the above sensitivity to environmental and predation conditions shown by fish and benthos, the overriding impression of the ecosystem of the Southern Ocean is that it is capable of absorbing substantial change. This results from the remarkable resilience shown by the marine ecosystem in adjusting to the massive predation of some species of seals and great whales that has occurred over the last 200 years or so. After a short interval, the populations of fur seals in the sub-Antarctic began to recover rapidly and ultimately exceed the original numbers. The removal of the great whales, largely in the first half of this century, led to a krill 'surplus', which resulted in increases in the numbers of krill eaters. However, even while exploitation of some whale species was progressing, the remaining whales had responded to change by reduced age at maturity and increased pregnancy levels (see Figure 5.10). Some species of seals show a similar response, and together with the increased breeding success of the bird populations, the Southern Ocean displays more characteristics of a robust ecosystem than a fragile one.

However, it is important to remember that this example of recovery and resilience is related to predation of species close to the top of the food web of the Southern Ocean. What remains unclear is the sensitivity of the marine ecosystem to changes that might directly affect the phytoplankton and zooplankton species that support much of the food web. At these trophic levels, there is likely to be greater sensitivity to change however introduced. For example, southward shifts of the Polar Frontal Zone may have led to low krill years and poor breeding success of krill eaters around South Georgia. Since the annual march of the seasonal sea ice zone is probably responsible for much of the inoculation of the spring phytoplankton blooms in this part of the Southern Ocean, it follows that primary productivity is likely to be sensitive to any changes in the nature, timing and extent of sea ice.

One of the most likely outcomes of any change in climate is that the current distribution of the sea ice will change, probably becoming more discontinuous with more open-water leads and polynyas as well as reduction in extent and duration. The effect that this might have on primary production can only be guessed at, but overall it may result in a reduction in productivity. Perhaps even more significant would be the effect of perturbations in the population of phytoplankton grazers such as copepods and krill. Any major reduction in krill on account of phytoplankton failure, changes to ocean currents or overfishing of krill itself has the potential to impact seriously on the entire ecosystem. Questions of fragility and resilience are clearly related to the nature, extent and magnitude of the change to which the system is subjected, and at present levels of knowledge it is still unclear whether a reduction in sea ice extent might lead to more or less krill for predators to eat. Overall, the past evidence suggests that the ecosystems of the Southern Ocean have reacted robustly to change, especially the removal of the top predators. However, the marine ecosystem is likely to respond to any future reduction in the amount of phytoplankton and zooplankton in a much more sensitive way.

5.4.5 Summary

It may be that adaptation to the cold condition of the continent and Southern Ocean has taken the Antarctic fauna up an evolutionary cul-de-sac (Clarke and Crame, 1992). However, geological evidence suggests that faunas in the past have coped with climate change and it is possible that environmental change, although perhaps detrimental in the short term, is actually a stimulus to evolution in the long term. A similar argument can be made for Antarctic flora. Notwithstanding this, the likely changes contingent upon climate change are likely to be significant. Overall, it may be that in a warmer world terrestrial ecosystems will be relieved of the two major limiting factors to growth: availability of moisture and low temperature. If this is the case then such ecosystems will be subject to increased productivity and expanding populations of native plants and animals but also to enhanced competition from immigrant species. The effect on the marine ecosystem is more difficult to predict but will focus on changes in the productivity of the seasonal sea ice zone and its effects on phytoplankton productivity and krill numbers and distribution. Whether changes in sea ice distribution lead to reductions in food and hence in the breeding success of seals and penguins is unknown. The potential effects of enhanced UV-B exposure on Antarctic marine and terrestrial organisms may lead to more sensitive species being replaced by more tolerant ones. However, little is known about the UV photobiology of Antarctic organisms under field conditions or of the longer-term ecological consequences (Karentz, 1991; 1992). Although the dire consequences of some predictions have not materialised, it is unclear what the cumulative effects of enhanced UV-B exposure will be, over what timescale they may become apparent, and how any effects may be transferred through the food web.

5.5 Conclusion

The extremes of the Antarctic environment have resulted in organism adaptations to allow life at low temperatures within a strongly seasonal regime via physiological, ecological and behavioural adaptation. For example, Antarctic plants can withstand freezing and have developed survival strategies by avoiding severe conditions and adopting ground-hugging habits. Antarctic animals are also physiologically adapted to cope with the effects of low temperatures, and the poikilotherms, which exist close to the environmental temperature, adopt strategies such as slow growth. The homeotherms use thermoregulation mechanisms that allow them to remain active throughout the year. They are less controlled by their immediate habitat and avoid the severest of conditions by migration and selection of advantageous breeding sites. Such adaptive flexibility provides them with an advantage in the face of any fundamental environmental change.

Two types of indicator have been used to examine the potential impacts of climate change on physical systems: direct and indirect. Examining the direct indicators of climate change shows evidence of warming in the Antarctic Peninsula and in the islands of the Scotia Sea but inconclusive evidence for warming of the continent, possibly on account of sparse and inadequate data coverage. This is at odds with the predictions of warming from computer models of various types, most of which indicate some CO_2-driven warming in the future with associated increased precipitation in the sea ice zone and along the continental coast. The best estimates indicate a warming of 1 °C over the MIZ for a 1% a^{-1} CO_2 increase over 100 years. These temperature increases may lead to some enhanced melting of the polar ice sheets, but the amount of increase is unlikely to affect the long-lived stability of the East Antarctic ice sheet substantially. A more important effect, resulting from enhanced snowfall in coastal areas, may be local ice sheet thickening. The extent of sea ice surrounding the continent is also important since the seasonal girdle of ice enhances winter cooling. The ice shelves that flow from the continent into the sea are thought to buttress the ice sheet itself and with climate warming it is possible that they might disintegrate and contribute to ice sheet collapse. The ensuing sea-level rise might further threaten the stability of the West Antarctic ice sheet. However, while local warming may be responsible for the collapse of some ice shelves in the Antarctic Peninsula area, the stability of the large southerly ice shelves is not yet in question.

Any changes in the physical systems will also affect the biological systems. The two major constraints on plant growth and animal activity in the Antarctic are limited moisture availability and low temperature, and since the organisms survive at the very limits of life, the terrestrial ecosystems are fragile and slow to recover from disruption. However, it is likely that climate change will reduce rather than increase fragility, because most plants will benefit initially from the predicted increases in coastal precipitation, enhanced summer temperatures and more and earlier snow melt. The flexible survival strategies of invertebrates should allow them to cope better with climate change than plants. The ecosystem of the Southern Ocean has shown in the past that it is capable of absorbing substantial change through the removal of the top predators and, in this respect, is resilient rather than fragile. However, with climate change it is likely to be more sensitive to changes to the zooplankton and phytoplankton that occupy crucial positions in the food web. It should be noted that substantial environmental change has occurred throughout the recent geological past and that, in general, Antarctic ecosystems have successfully altered to cope with the change. Although possibly detrimental in the short term, climate change may be a stimulus to adaptation and evolution in the long term.

Further reading

Department of the Arts, Sport, the Environment and Territories (1992) *Impact of Climate Change on Antarctica-Austalia*. Australian Government Publishing Service, Canberra.

Drewry, D.J., Laws, R.M. and Pyle, J.A. (eds) (1992) Antarctica and Environmental Change. *Philosophical Transactions of the Royal Society of London*, 338(1285), 199–334.

Everson, I. (1987) Life in a cold environment. In Walton, D.W.H. (ed.) *Antarctic Science*. Cambridge University Press, Cambridge. 71–138.

Harris, C.M. and Stonehouse, B. (eds) (1991) *Antarctica and Global Environmental Change*. Belhaven Press, London. 198pp.

Hempel, G. (ed.) (1994) *Antarctic Science: Global Concerns*. Springer-Verlag, Berlin. 287pp.

Houghton, J.T., Meira Filho, L.G., Callander, B.A., Harris, N., Kattenberg, N. and Maskell, K. (1996) *Climate Change 1995. The Science of Climate Change*. Cambridge University Press, Cambridge. 572pp.

SCAR (1993) *The Role of the Antarctic in Global Change*. Scientific Committee on Antarctic Research, Cambridge. 54pp.

PART 2 Antarctic exploration, exploitation and human impact

CHAPTER 6

The human dimension: exploration, exploitation, politics and science

Objectives

Key themes covered in this chapter:

- the changing perceptions of Antarctica and its resources
- the changing motivation for human interest in the Antarctic
- the role of science in facilitating international political agreement
- the Antarctic Treaty
- contemporary issues and tensions: science, politics, resources, conservation and environmental management

6.1 Introduction

There is a considerable body of literature on the discovery, exploration and development of the Antarctic (e.g. Mill, 1905; Markham, 1921; Hayes, 1932; Christie, 1951; Kirwan, 1959; Bertrand, 1971; Quartermain, 1967; *Reader's Digest*, 1985; Headland, 1989; Fogg, 1992). Human interest in the region has encompassed geographical and scientific discovery, adventure, commercial exploitation of resources, political strategy and, most recently, tourism, environmental management and conservation. Perceptions of the Antarctic and its potential resources have therefore varied according to external imperatives involving the interplay of human curiosity, economics, science, strategic interests, politics and conservation. The aim of this chapter is to highlight the themes and geography of the evolving human interest in the resources of Antarctica, the Southern Ocean and the sub-Antarctic islands.

The isolation of Antarctica, combined with its size and physical geography (see Chapters 2, 3 and 4), has exerted fundamental constraints on the discovery, exploration and exploitation of the resources. A series of snapshots (Figure 6.1) reveals the changing pattern of perception of the geography of Antarctica, the last continent to be discovered. The existence of the continent, although suspected, remained a matter of speculation until the eighteenth century. The pattern of discovery that followed focused on two main approaches and core areas, predicated both by physical geography and human motives. The first, via the South Atlantic, offered accessibility to the Antarctic Peninsula and offshore islands and the living resources of the adjacent seas. The second, via New Zealand, to the Ross Sea provided the shortest route to the compelling goal of the South Pole. The interior of the continent remained unexplored until the beginning of the twentieth century, and even then large areas of its coastline were largely unknown. It was not until the advent of aircraft, and more recently satellite imagery, that the geography of the more remote coastal areas and much of the interior was determined. Several phases can be recognised in the awareness, exploration and exploitation of Antarctica (Gould, 1978; Mickleburgh, 1987; Headland, 1989), each broadly characterised by a particular combination of motives and perceptions in the geographical imagination (Dodds, 1997). Although these phases overlap to some extent, they nevertheless provide a useful temporal framework.

6.2 Phase 1: from myth to reality (developments pre-1895)

The first phase involved a combination of exploration, geographical discovery, economic activity and science. In most but not all cases, scientific investigations were secondary concerns and involved mapping, survey and description of the natural environment.

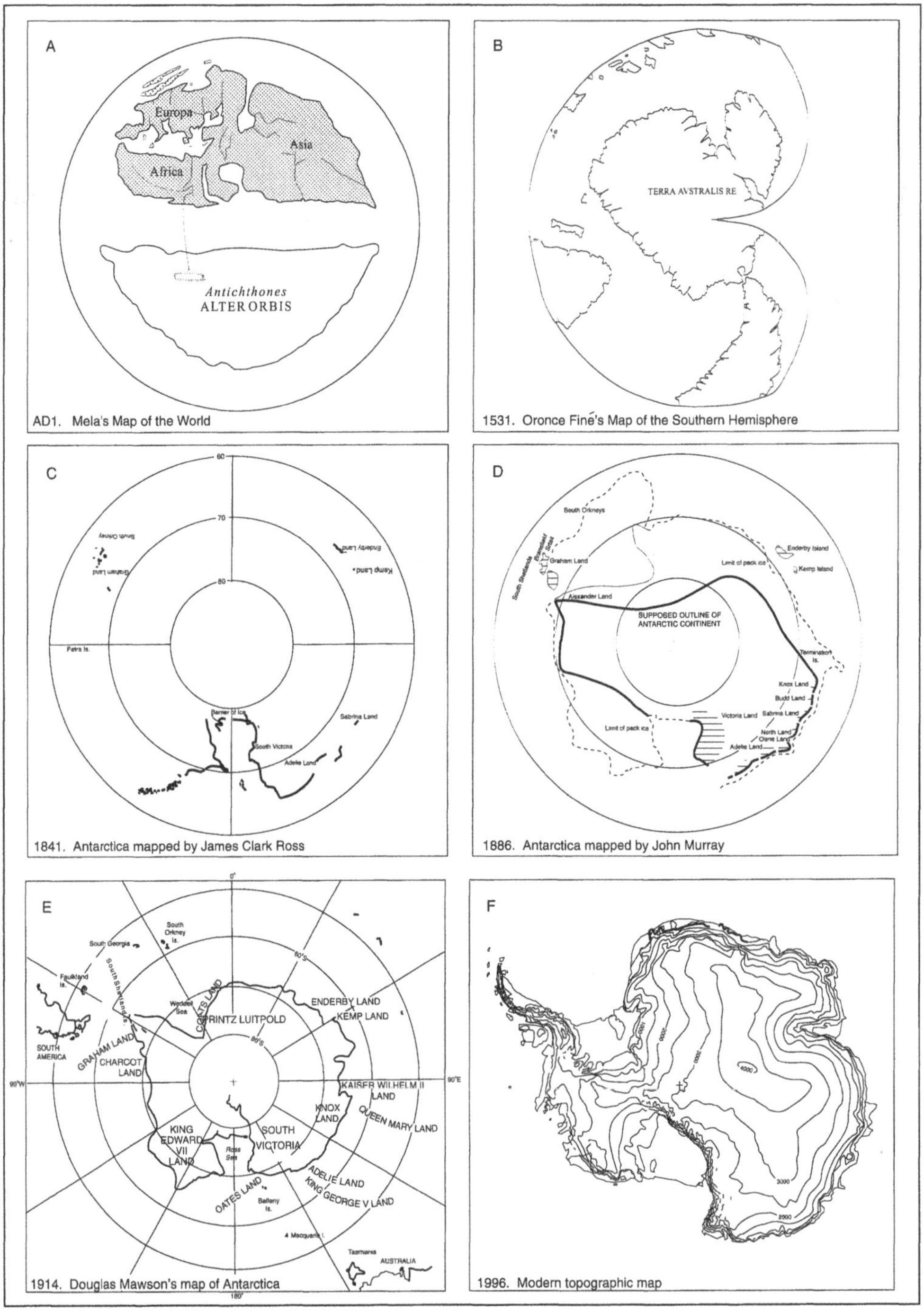

Fig 6.1 The changing geography of Antarctica: (A) in AD1 by Mela (after Cameron, 1974); (B) in 1531 by Oronce Finé depicting a large southern continent, 'Terra Australis', which was copied by later cartographers, including Mercator (after King, 1969); (C) in 1841 by James Clark Ross including the discoveries of sealers (after Ross, 1847; Sugden, 1982); (D) in 1886 by John Murray (after Murray, 1886; Sugden, 1982), the dashed line represents the limits of pack ice; (E) in 1914 by Douglas Mawson (after Mawson, 1914); and (F) a modern map of Antarctica (after Drewry, 1991).

The concept of a great southern continent dates back to the time of the early Greek philosophers. In the sixth century BC Pythagoras postulated that the Earth was spherical. His later followers, including Aristotle, proposed from the point of view of symmmetry that a large continental land mass must exist in the Southern Hemisphere to balance the northern continents (*cf.* Figure 6.1A). Ptolemy in *c.* AD 150 envisaged the probable existence of a large southern continent, '*terra incognita*', linked to Africa and Asia. The idea of a vast habitable land, '*terra australis incognita*', persisted in sixteenth-century world maps (Tooley, 1985) (Figure 6.1B). Despite the discoveries made on southern voyages by Portuguese, Dutch, English, French and Spanish navigators that Africa and South America were not linked to a great southern continent, the idea lingered on into the eighteenth century.

According to Rarotongan legend, in the seventh century a Polynesian called Ui-te-Rangiora voyaged deep into the Southern Ocean and encountered huge icebergs and snow (Quartermain, 1967). It was some time, however, before the first formal discoveries of land were made, namely the sub-Antarctic islands of the Southern Ocean: South Georgia (1675), Bouvet Island (1739) and Îles Kerguelen (1772). The second voyage of James Cook in 1772–1775 represents an important historical landmark in several respects. First, in circumnavigating the Southern Ocean, Cook disproved the existence of a large habitable continent extending north of 60° S. He crossed the Antarctic Circle three times, reaching 71° 10' S in the Bellingshausen Sea. Although he did not sight Antarctica itself, Cook described an icy, inhospitable environment and effectively dispelled earlier hopes of an inhabited southern continent with new opportunities for trade:

> The risk one runs in exploreing a coast in these unknown and Icy Seas, is so very great, that I can be bold to say, that no man will ever venture farther than I have done and that the lands which may lie to the South will never be explored. Thick fogs, Snow storms, Intense Cold and every other thing that can render Navigation dangerous one has to encounter and these difficulties are greatly heightned by the enexpressable horrid aspect of the Country, a Country doomed by Nature never once to feel the warmth of the Sun's rays, but to lie for ever buried under everlasting snow and ice . . . . It would have been rashness in me to have risked all which had been done in the Voyage, in finding out and exploaring a Coast which even when done would have answered no end whatever, or been of the least use either to Navigation or Geography or indeed any other Science.
>
> (Beaglehole, 1961, pp.637–638)

Cook's expedition furnished the first scientific information about the Southern Ocean (Rubin, 1982), but he envisaged no benefits to humankind arising from the area:

> Lands doomed by nature to everlasting frigidness and never once to feel the warmth of the Sun's rays, whose horrible and savage aspect I have no words to describe; such are the lands we have descovered, what may we expect those to be which lie more to the South, for we may reasonably suppose that we have seen the best as lying most to the North, whoever has resolution and perseverance to clear up this point by proceeding farther than I have done, I shall not envy him the honour of the discovery but I will be bold to say that the world will not be benefited by it.
>
> (Beaglehole, 1961, p.646).

Ironically, however, it was Cook's report of plentiful seals on South Georgia that opened the way for the first phase of commercial exploitation of Antarctic resources (see Chapter 7). The sealing industry ushered in a period of intense activity (Christie, 1951; Bertrand, 1971; Headland, 1984), with British sealers at South Georgia in 1778, and American vessels there in the early 1790s. Initially, fur seals were the target and when lucrative markets for seal skins opened in China in 1785 (Mitterling, 1959), massive and ultimately unsustainable slaughter of fur seals spread across the islands of the Southern Ocean. Elephant seals were also taken for their oil. In 1819, the South Shetlands Islands were discovered by William Smith, providing a new, but short-lived focus for the sealing industry as British and American vessels sought to exploit this open-access resource (Christie, 1951; Bertrand, 1971; Jones, 1975, 1985a, and b; Smith and Simpson, 1987). While the fur seal industry declined rapidly after the mid-1820s, elephant seals continued to be taken for their oil, particularly by American vessels at Îles Crozet, Îles Kerguelen and Heard Island (Bertrand, 1971; Richards, 1992). Although fur sealing revived in the 1870s at South Georgia and the South Shetland Islands, this was short-lived.

These early sealing voyages made significant contributions to the knowledge of the geography of the approaches to Antarctica and the continental margins (Mitterling, 1959; Bertrand, 1971; Headland, 1989), although many discoveries were probably not reported because intense commercial competition encouraged secrecy among the sealing captains. The coast of the Antarctic Peninsula was explored by James Weddell, Edward Bransfield, Nathaniel Palmer and George Powell. Powell and Palmer also discovered the South Orkney Islands in 1821, and Weddell pushed south to 74° 15' in the sea that now carries his name (1822–1824). John Biscoe circumnavigated the continent (1830–1832), sighting Enderby Land, Adelaide Island and the Biscoe Islands. James Eights, an American naturalist, made observations on the geology and natural history of the South Shetland Islands as part of a sealing cruise. In terms of historical landmarks, sealing ship masters were probably the first to set foot on the continent in 1820–1821 (Headland, 1994b; 1996).

During the first half of the nineteenth century, four major naval expeditions with government funding had geographical discovery and science as their principal objectives. A Russian naval expedition led by Bellingshausen circumnavigated the continent between 1819 and 1821 and is acknowledged with the first sighting in 1820, at the Fimbul Ice Shelf in Dronning Maud Land. Dumont D'Urville led a French naval expedition to the Atlantic and Pacific Ocean sectors, and discovered Terre Adélie in 1840. The United States Exploring Expedition (1838–1842), under Charles Wilkes, voyaged 2400 km along the edge of the East Antarctic pack ice, confirming the existence of an Antarctic continent (Bertrand, 1971). The British naval expedition of James Clark Ross explored the Ross Sea area in 1841, discovering the Ross Ice Shelf and Victoria Land (Figure 6.1C). These early expeditions produced important scientific observations on meteorology, natural history and terrestrial magnetism (Fogg, 1992). However, by the mid-nineteenth century, knowledge of Antarctica was still confined to a few small coastal areas separated by great distances: the Antarctic Peninsula, South Shetland Islands, South Orkney Islands, South Georgia, Enderby Land, Kemp Coast (Kemp in 1833), Balleny Islands and Sabrina Coast (Balleny in 1839) and the Ross Sea (Figures 6.1 C and 6.1 D).

During the latter part of the nineteenth century, important scientific activities included the oceanographic cruises of the *Challenger* expedition in the Southern Ocean (1872–1876) and the German investigations on South Georgia that formed part of the First International Polar Year in 1882–1883. Eduard Dallmann aboard the steam whaler *Grönland*, the first such powered vessel to operate in Antarctic waters, explored the northern part of the Antarctic Peninsula (1873–1874). Several exploratory voyages were also made to assess the potential for whaling: by C.A. Larsen with *Jason* and other ships in the Weddell Sea area (1892–1893); a Dundee fleet, including *Balaena*, also in the Weddell Sea area (1892–1893); and L. Kristensen and H.J. Bull aboard *Antarctic* in the Ross Sea (1894–1895).

6.3 Phase 2: the heroic era (c. 1895–1915)

The 'heroic era', between *c.*1895 and 1915 (Hayes, 1932), saw a remarkable development of geographical exploration and scientific study in Antarctica. This activity reflected several factors, including growth of the natural sciences, wider public awareness and media interest in exploration, geography and popular science, the interest of industrial and commercial entrepreneurs prepared to support expeditions, and an imperialistic interest in the last unclaimed continent. In 1895, illuminating the lack of knowledge about Antarctica, the International Geographical Congress in London adopted a resolution to encourage Antarctic exploration, and in the next two decades fifteen major national expeditions visited the region (Table 6.1). These expeditions differed from earlier ventures in that their funding came from individuals, public subscriptions and scientific societies, and sometimes governments, and their objectives often included the establishment of land bases from which to explore the interior (Figure 6.2). The geographical pattern of this activity focused on the Ross Sea area, George V Land and the Antarctic Peninsula–Weddell Sea area (see Figure 6.1E). The first wintering south of the Antarctic Circle was made in 1898 by the Belgian Antarctic Expedition aboard *Belgica*, trapped by ice in the Bellingshausen Sea, and at the first land station at Cape Adare in 1899 by Borchgrevink and the *Southern Cross* expedition. National prestige was an important motivation during this period, and the attainment of the South Pole a powerful incentive. In many cases, comprehensive scientific programmes ran in parallel with, or provided supporting justification for, geographical exploration (Walton and Bonner, 1985b; Fogg, 1992). The undertakings and achievements of Scott, Shackleton and Nordenskjöld are well-known, but significant contributions were also made by the expeditions of Gerlache, Charcot, Bruce, Drygalski and Mawson. Bruce's work, for example, added greatly to knowledge of the oceanography and biology of the Southern Ocean (Speak, 1992).

Table 6.1 Major expeditions during the 'heroic era'

Expedition	Ship	Dates	Leader
Belgian Antarctic Expedition	*Belgica*	1897–99	de Gerlache
British Antarctic Expedition	*Southern Cross*	1898–1900	Borchgrevink
German South Polar Expedition	*Gauss*	1901–03	von Drygalski
Swedish South Polar Expedition	*Antarctic*	1901–04	Nordenskjöld
British National Antarctic Expedition	*Discovery*	1901–04	Scott
Scottish National Antarctic Expedition	*Scotia*	1902–04	Bruce
French Antarctic Expedition	*Français*	1903–05	Charcot
British Antarctic Expedition	*Nimrod*	1907–09	Shackleton
French Antarctic Expedition	*Pourquoi Pas?*	1908–10	Charcot
Norwegian Antarctic Expedition	*Fram*	1910–12	Amundsen
Japanese Antarctic Expedition	*Kainan Maru*	1910–12	Shirase
British Antarctic Expedition	*Terra Nova*	1910–13	Scott
German South Polar Expedition	*Deutschland*	1911–12	Filchner
Australasian Antarctic Expedition	*Aurora*	1911–14	Mawson
Imperial Trans-Antarctic Expedition	*Endurance*	1914–16	Shackleton

Fig 6.2 The first land bases for exploration of the interior of the Antarctic continent were established at the end of the nineteenth century and at the beginning of the twentieth century. Photo of Shackleton's hut, built at Cape Royds in 1908. The photo shows the hut in 1967 after restoration work. It is now protected as a historic site (source: photograph by B.L. Mason, US Navy, reproduced by permission of Canterbury Museum, Ref. No. 5701, Christchurch, New Zealand).

The heroic era also saw the rapid development of the second main phase of commercial exploitation of Antarctic resources. The first whaling station was established on South Georgia in 1904, and the industry rapidly grew in economic importance thereafter (see Chapter 7); whale oil was an important strategic resource during the First World War. This period also saw consolidated claims of sovereignty when Britain established the Falkland Islands Dependencies, motivated at least in part by a desire to control the lucrative and strategic whaling industry.

6.4 Phase 3: the mechanical age (1916–1958)

The decades after the First World War saw the growth of technical and mechanical aids in the pursuit of Antarctic exploration, science and economic exploitation. Three principal foci of human interest emerge from this period: whaling, exploration and science, accompanied by increased intrusion of politics.

Principal among the technical developments were the use of aircraft, motor transport, radios and aerial photography (Bertrand, 1971). Aircraft, in particular, made a significant contribution to the exploration of both the coast and the continental interior, especially on the expeditions of Wilkins, Byrd, Mawson, Christensen, Ritscher, Ellsworth, Rymill and Ronne. Geographically, major gaps in the coastline were completed, notably between the Weddell Sea and 80° E, while further exploratory effort continued in the Antarctic Peninsula and Ross Sea areas. The British Grahamland Expedition (1934–1937) discovered King George VI Sound and demonstrated that the Antarctic Peninsula was not an archipelago as had previously been suggested. International collaboration developed, for example with the British–Australian–New Zealand Expedition (1929–1931), between the Falkland Islands Dependencies Survey and the Ronne Antarctic Research Expedition (1947–1948), and with the Norwegian–British–Swedish Antarctic Expedition (1949–1952). Ship-based activities also continued. On the one hand, the whaling effort advanced remorselessly as technology progressed (see Chapter 7): this allowed the exploitation of new pelagic catching grounds and reduced the need for land bases, and also facilitated new geographical discoveries, notably of Dronning Maud Land in 1930–1931 (Headland, 1993a). On the other hand, major investigations of marine biology and hydrography were instigated in 1925 by the Discovery Committee and financed through a tax on whale oil production (Figure 6.3). Expeditions were funded by the governments of Germany (1938–1939) and the United States (1939– 1941), and there was greater government involvement in scientific research. Two American expeditions, in 1946–1947 and 1947–1948, epitomised the application of mechanical technology to exploration and scientific discovery. The former, 'Operation Highjump', is the largest expedition ever mounted in Antarctica, involving the deployment of over 4700 men, thirteen ships and numerous aircraft (Bertrand, 1971). It achieved important results in meteorology, aerial reconnaissance (about 60% of the coast was photographed) and field operation techniques.

During this period, political and strategic motives came to the fore as several countries declared territorial claims in Antarctica. By the end of the 1920s, claims had been declared by or on behalf of Britain, France and New Zealand, followed in the next decade or so by Australia, Norway, Argentina and Chile (Section 6.5). Whereas 'effective occupation' required to underpin sovereignty claims had previously been maintained by occasional visits, by the 1940s permanent occupation was deemed essential. Britain established the first permanently manned bases on the Antarctic Peninsula at Port Lockroy and Hope Bay (Figure 6.4) as part of 'Operation Tabarin' (1944–

Fig 6.3 Scott's former ship, *Discovery*, was refitted in the 1920s as a research vessel for studies of whale and marine biology and oceanography. Directed by the Discovery Committee and initially funded by royalties from the whaling industry, this work resulted in major scientific contributions that were continued by a more modern vessel, *Discovery II*, which operated from 1929 to 1951. *Discovery* has now returned to the town where it was built and forms the focal point of a polar education visitor centre at Discovery Point, Dundee, Scotland (source: photograph courtesy of Discovery Point, Dundee).

Fig 6.4 The British bases at Port Lockroy (1944) and Hope Bay (1945) were established as part of 'Operation Tabarin' and later operated by the Falkland Islands Dependencies Survey (FIDS) and the British Antarctic Survey. Although both were later abandoned, Port Lockroy has recently been restored and is now a tourist attraction (source: photograph by I.M. Lamb of Hope Bay under construction in 1945, British Antarctic Survey Archives Ref. No. AD6/19/1(D165/26), reproduced with permission of the British Antarctic Survey).

Fig 6.5 Sledging party at the northern end of the Antarctic Peninsula in 1955. Dog teams provided the backbone of field travel by survey and geology parties during the 1950s and 1960s. Under the terms of the Protocol on Environmental Protection, the last dogs were removed from Antarctica in 1994 (source: photograph by N. Leppard, British Antarctic Survey Archives Ref: no. AD6/19/2 (D378/19). Crown copyright material in the Public Record Office is reproduced by permission of the Controller of Her Majesty's Stationery office.

1945) (Fuchs, 1982). Undertaken for strategic reasons to strengthen Britain's claims to sovereignty, following the earlier declaration of territorial claims in the same area by Argentina and Chile, this initiative also denied access to German naval raiders (Beck, 1986b; Dodds, 1994). Subsequently in 1947 Argentina and Chile established bases in the same area. All three countries now had overlapping territorial claims, which gave rise to the so-called 'Antarctic problem' (Christie, 1951). After 1945, further British bases were established and operated by the Falkland Islands Dependencies Survey, later renamed the British Antarctic Survey in 1962. Such bases served the dual function of underpinning territorial claims and supporting scientific activity, principally in meteorology, surveying and geology (Figure 6.5). In practice, scientific activity and

Table 6.2 Stations of SCAR Nations operating in the Antarctic, Winter 1997. Stations are numbered clockwise from the Greenwich Meridian. See Figure 6.6 (p182).

1	Amundsen-Scott	United States	22	Vernadsky	Ukraine
2	Maiti	India	23	Palmer	United States
3	Novolazarevskaya	Russia	24	Captain Arturo Prat	Chile
4	*Marion Island	South Africa	25	†Great Wall	China
5	Syowa	Japan	26	†Presidente Eduardo Frei	Chile
6	Dome Fuji	Japan	27	†Bellingshausen	Russia
7	Molodezhnaya	Russia	28	†Artigas	Uruguay
8	*Alfred Faure, Îles Crozet	France	29	†King Sejong	Korea
9	Mawson	Australia	30	†Jubany	Argentina
10	*Port aux Français, Îles Kerguelen	France	31	†Arctowski	Poland
11	Zhongshan	China	32	†Comandante Ferraz	Brazil
12	*Martin de Viviès, Îles Amsterdam	France	33	General Bernado O'Higgins	Chile
13	Davis	Australia	34	Esperanza	Argentina
14	Mirny	Russia	35	Marambio	Argentina
15	Casey	Australia	36	Orcadas	Argentina
16	Dumont d'Urville	France	37	*Bird Island	United Kingdom
17	*Macquarie Island	Australia	38	Belgrano II	Argentina
18	McMurdo	United States	39	Halley	United Kingdom
19	Scott Base	New Zealand	40	*Gough Island	South Africa
20	Rothera	United Kingdom	41	Neumayer	Germany
21	San Martin	Argentina	42	SANAE	South Africa

*Stations north of 60° †Stations on King George Island

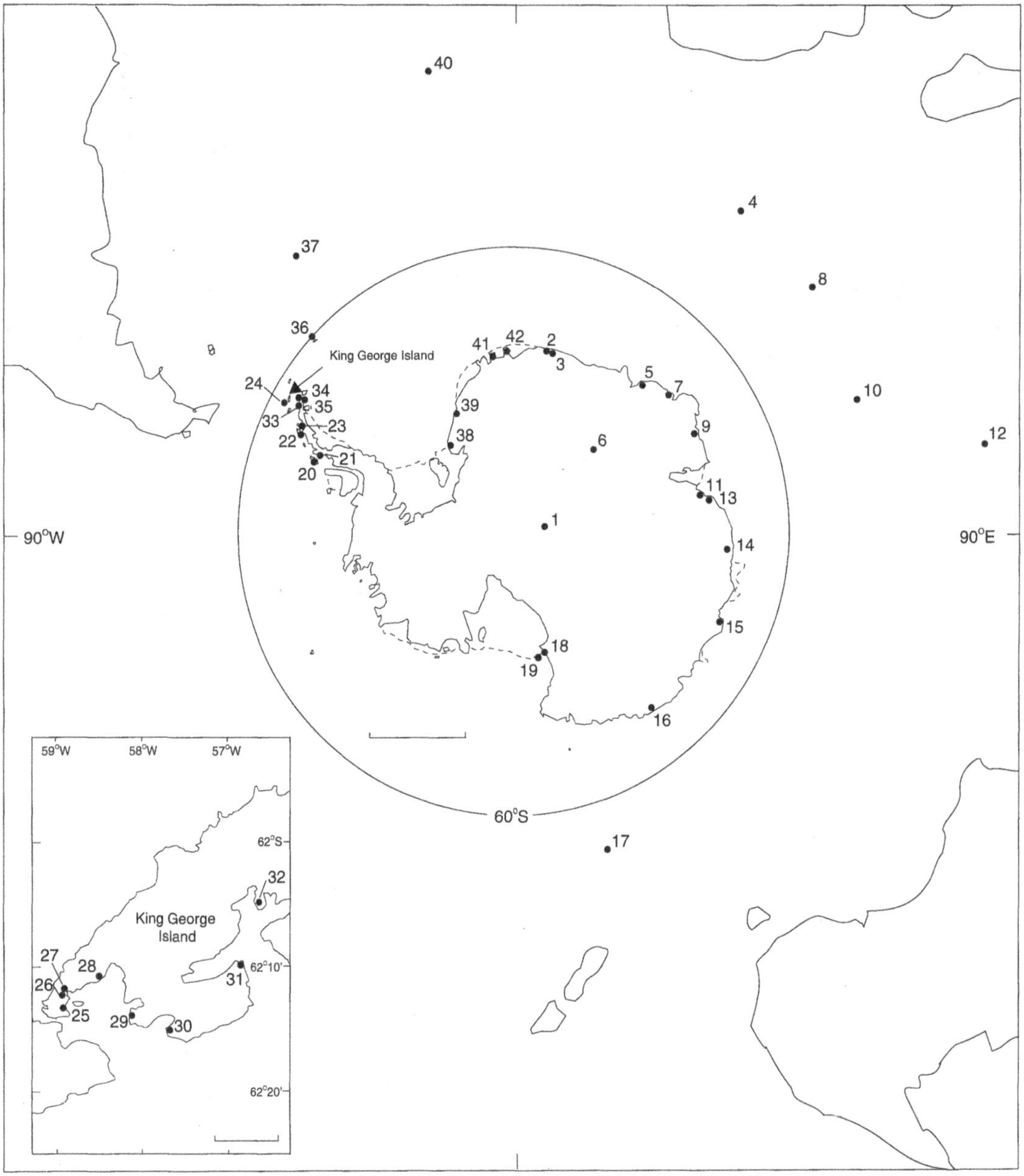

Fig 6.6 Stations of SCAR nations operating in the Antarctic during winter 1997 (source: modified from SCAR Bulletin 127, 1997). Two additional stations were open at King Edward Point, South Georgia (UK), and Václav Vojtech, Nelson Island (Czech Republic). See Table 6.2 on p. 181 for the stations corresponding to the numbers.

exploration became a means of maintaining and advancing sovereignty claims as part of a wider political strategy (Indreeide, 1990; Dodds, 1994). The spatial implication can be seen in the relatively high density of bases now established in the most easily accessible area, the Antarctic Peninsula (Figure 6.6). Other early post-war developments were the siting of Australian bases on Macquarie Island, Heard Island and at Mawson in Mac Robertson Land, and a French base on the coast of Terre Adélie. The political dimension was explicit in the objectives of 'Operation Highjump', one of which was to 'consolidate and extend the basis for United States claims in the Antarctic, if such should subsequently be made' (Bertrand, 1971). The mechanical age was epitomised by the Commonwealth Trans-Antarctic Expedition (1955–1958), which used mechanised transport for a journey in the 'heroic' tradition (the first crossing of Antarctica), combined with scientific investigations.

The mechanical age culminated in the International Geophysical Year (IGY) in 1957–1958, which represents an important landmark not only in Antarctic science and geography, but also in Antarctic politics and the international management of the continent. Antarctica was selected by the International Council of Scientific Unions (ICSU) as a focus for intensive scientific study during the IGY on account of its interest, both in a regional and global context, for auroral studies, geomagnetism, cosmic rays, ionospherics, glaciology, seismology, gravity and meteorology (*cf.* Fogg, 1992). Scientific programmes and logistics were coordinated and the distribution of scientific stations agreed by a special committee to extend the previously small geographical spread of activity (Figure 6.7). In addition, supporting programmes in geology, biology and medical research were developed. In total, twelve countries were involved, 55 stations occupied (including 47 south of 60° S) on the continent and islands of the Southern Ocean (King, 1969), and over 5000 personnel deployed. There was a surge in the number of wintering stations and personnel (Beltramino, 1993; Headland, 1993b) (Figure 6.8). Overall, the IGY greatly expanded

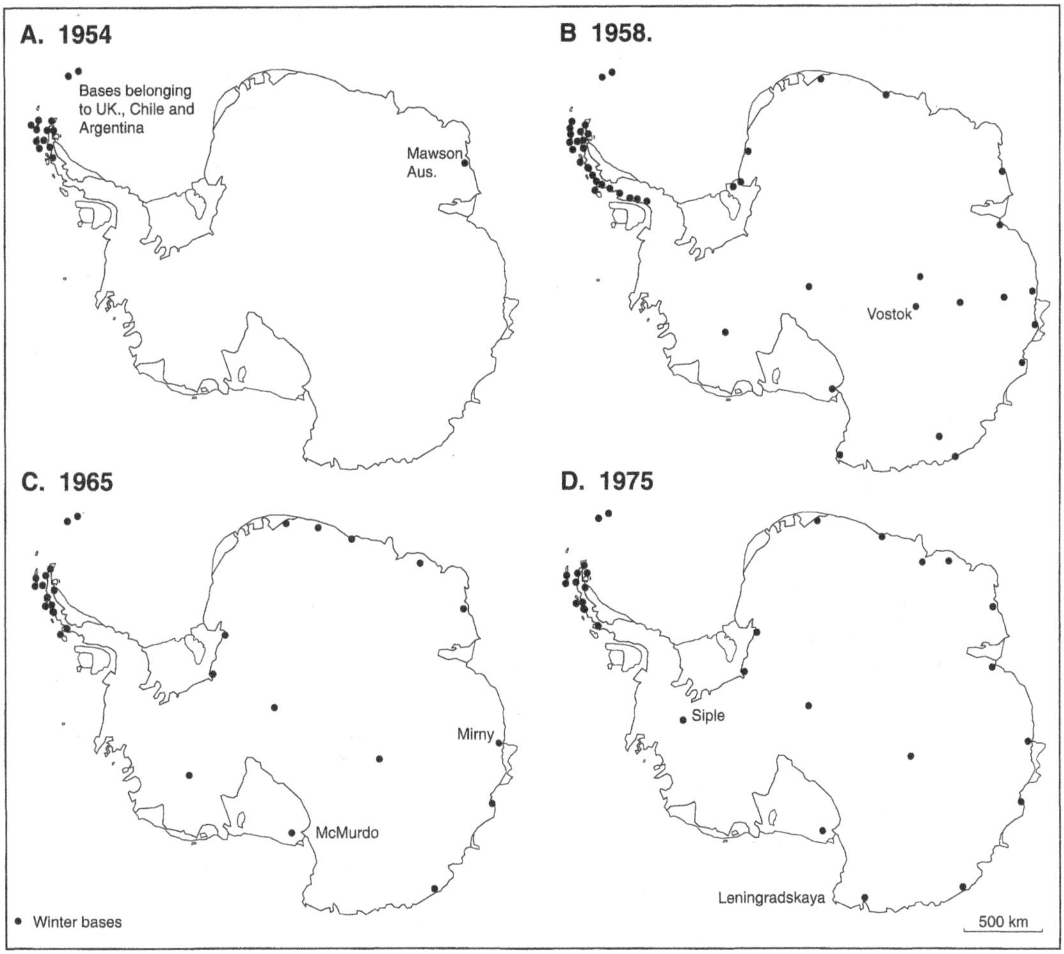

Fig 6.7 The changing spatial pattern of winter stations operating in Antarctica: (A) before the IGY (1954); (B) during the IGY (1958); (C) post-IGY (1965); and (D) in 1975 (source: Sugden, 1982).

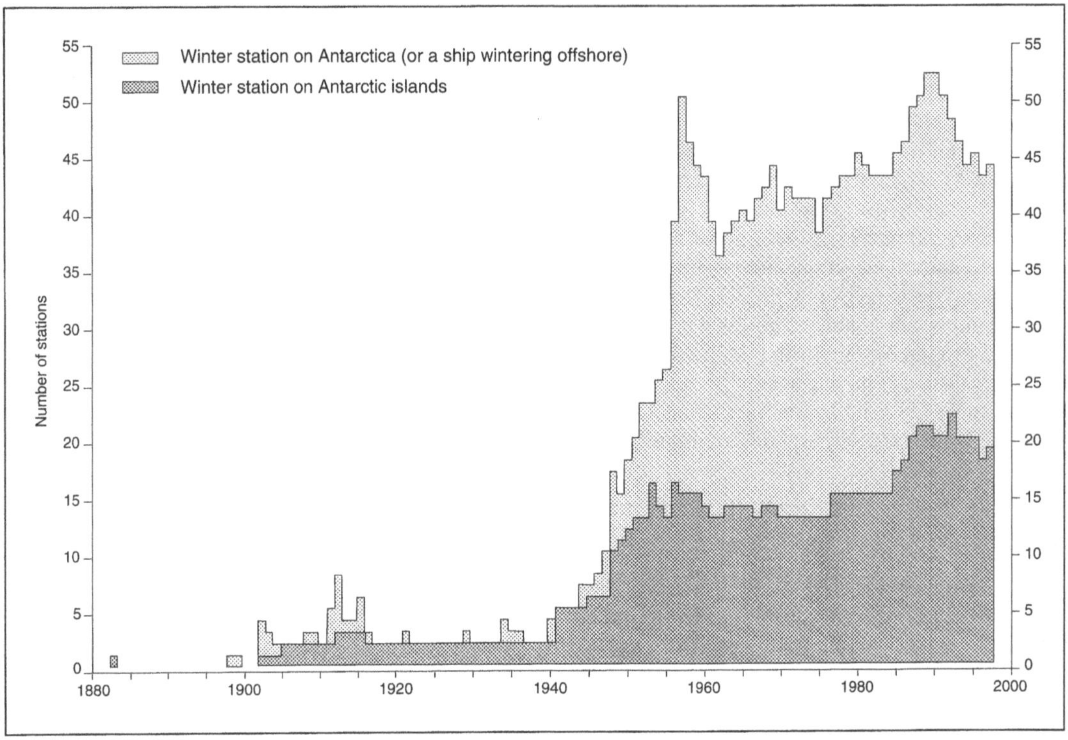

Fig 6.8 The growth of winter stations operating in the Antarctic (source: R.K. Headland, Scott Polar Research Institute).

geographical knowledge and scientific achievement in Antarctica (Sullivan, 1961; Eklund and Beckman, 1963; Forbes, 1967) and significantly raised the profile of Antarctic science. Moreover, it also set a pattern for future international cooperation in both science and politics (see below).

In the area of resource exploitation, the mechanical age saw the expansion of the whaling industry away from coastal stations into the open ocean through the use of factory ships. Exploitation was driven by economic forces and although a succession of regulatory measures were adopted, they failed to ensure sustainable harvesting, and the industry declined in the 1960s (see Chapter 7).

6.5 Background to the Antarctic Treaty

The key agreement that provides the fundamental legal and political framework for Antarctic affairs is the Antarctic Treaty, which was signed in 1959. Events leading to the Antarctic Treaty were greatly influenced by two factors: (1) the geopolitical situation involving conflicting sovereignty claims over Antarctica and the background of East–West tensions during the Cold War; and (2) the success of the IGY in promoting international scientific cooperation in Antarctica. Following its ratification in 1961, the treaty is generally agreed to have proved remarkably successful in fostering international cooperation in Antarctica and in ensuring its peaceful use for scientific purposes. The Antarctic Treaty System (ATS) is a descriptive term that encompasses the Antarctic Treaty itself and a series of additional instruments: the Agreed Measures for the Conservation of Antarctic Fauna and Flora (1964), the Convention for the Conservation of Antarctic Seals (1972), the Convention on the Conservation of Antarctic Marine Living Resources (1980) and the Protocol on Environmental Protection (1991) (see Chapter 9). Negotiations for a minerals regime, the Convention on the Regulation of Antarctic Minerals Resources (CRAMRA), were concluded in 1988, but this has not come into force and was overtaken by the Protocol on Environmental Protection (see Chapter 9). The system is therefore an evolutionary one, developed both through the measures adopted in these instruments and through recommendations agreed at regular meetings of the Consultative Parties to the

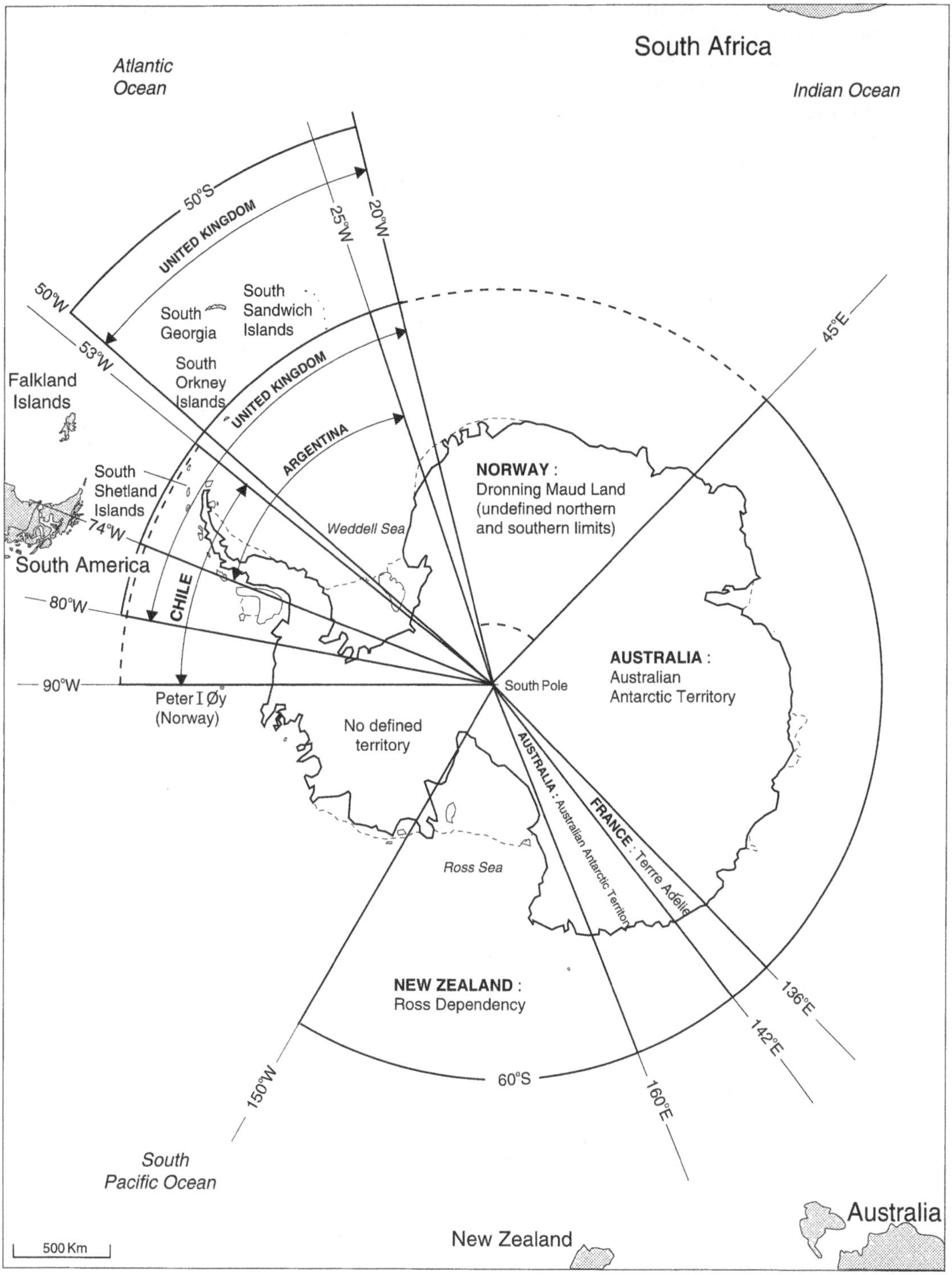

Fig 6.9 Territorial claims in Antarctica. The following claims partly overlap south of 60°S: United Kingdom, 20° W to 80° W; Argentina 25° W to 74° W; Chile 53°W to 90°W. Between 50°S and 60°S, South Georgia and South Sandwich Islands, claimed by the United Kingdom, were part of Falkland Islands Dependencies until 1985. (source: R.K. Headland, Scott Polar Research Institute).

Antarctic Treaty. The Scientific Committee on Antarctic Research (SCAR) provides independent scientific support to the treaty parties.

Historically, the Antarctic Treaty was developed against a background of geopolitical tensions arising from a variety of territorial claims (Hanessian, 1965; Sollie, 1984). The UK first delimited a claim in 1908, subsequently elaborated in 1917, to the territories defined as the Falkland Islands Dependencies. Subsequently, claims were made by France (1924), the UK on behalf of New Zealand (1923) and Australia (1933), Norway (1939), Chile (1940) and Argentina (1943). The claims of the UK, Argentina and Chile all overlap to some extent (Figure 6.9). At least 15% of the continent remains without defined territories, mainly Marie Byrd Land. The foundations for these claims variously include discovery, contiguity, occupation, inherited rights, geological affinity, proximity, sectoral rights, succession to Spanish rights under the Papal Bull of 1493 and other administrative or symbolic acts (Auburn, 1982; Quigg, 1983; Sahurie, 1992). Other states, including the USA and Russia, have not recognised these claims, but have reserved their own rights to make claims. The motivation behind such national interests in Antarctica includes the legacy of the imperial tradition (Beck, 1983), the perceived potential for resource exploitation in the long term, strategic interests, geopolitical interests (Child, 1988) and scientific investigation. States have expressed their legal positions through a variety of measures, claimant states by enacting legislation applicable to their claimed territories, establishing administrative functions such as post offices, issuing maps, protesting about infringements of their claimed territories by other national expeditions, and establishing settlements (e.g. by Argentina at Esperanza and Chile on King George Island). Non-claimant states have not recognised such acts and have continued to conduct their activities regardless. In the period before 1959, disputes therefore existed between claimant and non-claimant states and between those with overlapping claims (Christie, 1951; Maquieira, 1986; Orrego Vicuña, 1986; Beck, 1986b, 1987a), and there were fears about further escalation of the sovereignty issue; for example, in 1952 there was a clash at Hope Bay when Argentines fired on Falkland Islands Dependencies Survey personnel landing to rebuild the British base, which had been destroyed earlier by a fire.

A further source of tension was superpower rivalry and the potential extension of the Cold War to Antarctica (Beck, 1986b, 1990b); for example, the USA considered it strategically important to pre-empt any Soviet claims or threats that might arise from the military use of the region. The possible development of superpower military activities and nuclear testing was also a matter of immediate strategic and environmental concern to some states adjacent to Antarctica, notably Australia and New Zealand; for example, during the IGY the USSR had stations in the sector claimed by Australia. During the post-war years, various solutions to the sovereignty problem centred around United Nations involvement, but encountered opposition from claimant states (Quigg, 1983; Orrego Vicuña, 1988; Peterson, 1988). The Antarctic Treaty however, set aside the sovereignty question. It offered a form of internationalisation among those countries with direct interests in the area and received the support of both claimant and non-claimant states.

The Antarctic Treaty has its roots partly in the IGY, the results of which were outstandingly successful both in scientific terms and in furthering international cooperation. This was achieved despite the political tensions between the superpowers and between the individual states claiming sovereignty over parts of Antarctica. Effectively a 'gentleman's agreement' was reached to set aside political and territorial arguments to allow scientific activity to proceed unimpeded. At the time, considerable interest was expressed in maintaining the momentum of the IGY on scientific, strategic and political grounds. The outcome was a proposal by the USA in 1958 to address the problem of territorial claims and to set aside the continent for science. Following informal discussions (*cf.* Beck, 1985b) and a Treaty Conference, the Antarctic Treaty was signed in December 1959 and came into force in June 1961 after ratification by the original twelve signatories. Undoubtedly, the success of the IGY helped to create the necessary political atmosphere for commencing negotiations. The basic principles of the IGY were embodied in the treaty, including independence of scientific research, free exchange of information and suspension of territorial claims. Although the treaty is overtly founded on the fruits of scientific cooperation, its development was significantly motivated by political interests to avoid international conflict and chaos in the region (Heap, 1983; Laws, 1987; Østreng, 1989).

6.6 The Antarctic Treaty: objectives and scope

The objectives of the treaty were formulated to ensure that Antarctica is used for peaceful purposes and were founded on the value of science and international cooperation in scientific investigations. The twin tenets of peaceful use and international harmony thus accord with the charter of the United Nations. The treaty comprises 14 articles (Appendix 1). These prohibit activities of a military nature, although allowing the use of military personnel for scientific and other peaceful purposes (Article I). Nuclear explosions and disposal of radioactive waste are also prohibited (Article V). Article IV, the crux of the treaty, addresses the sovereignty issue, stating that nothing in the treaty shall be interpreted as affecting existing claims or the basis of any claim by other states; nor is the position of states compromised where they choose to recognise, or refuse to recognise, the claims of others. Further, while the treaty remains in force, no activities undertaken shall affect claims, and no new claims or enlargement of existing claims shall be asserted. In effect, this means that nothing in the treaty, or anything done under it, can be used either to support or counter sovereignty claims. Other provisions facilitate scientific activity (Article II) and open exchange of information and personnel (Article III). Provisions are made for inspection of a state's facilities by other parties to the treaty (Article VII). The provisions of the treaty apply to the area south of 60° S, including ice shelves, but without prejudice to rights under international law relating to the high seas (Article VI). Article IX establishes the mechanism for the operation of the treaty and its development, including the recommendation of measures for the preservation and conservation of living resources, while Article XI sets out procedures for settling disputes. Although the treaty included a provision for review at any time after 30 years (1991) if so requested by any Consultative Party (ATCP) (Article XII), there is no limit to its duration. Such a review has not been conducted.

The treaty was signed by the original twelve nations that participated in the IGY; they achieved ATCP status on ratification and may keep it indefinitely. The treaty is open to all countries for accession through acceptance of its provisions, but only those with significant research interests qualify for ATCP status (Article IX). In December 1997, in addition to the original signatories, there are fourteen other ATCPs and seventeen acceding states (Table 6.3), making a total of 43 adherent states and representing over three-quarters of the world's population. Essentially, the treaty involves a functional basis for participation in decision making through the requirement for ATCP status, rather than a political or ideological one (Scully, 1986), and provides a restricted form of internationalisation. Since 1983, non-ATCPs have been invited to attend Consultative Meetings but with non-voting status and this has helped to deflect some of the criticism about the closed nature and exclusivity of decision making under the treaty.

Consultative Meetings provide the mechanism by which the treaty is operated. The ATCPs adopt recommendations, which require unanimous approval. These then become provisions, which complement the original treaty articles. Recommendations adopted have related to matters such as telecommunications, tourism, non-governmental expeditions, transport and logistics, designation of protected areas, human impacts on the environment, and exchange of information (Heap, 1994). From 1995, recommendations have been made in three categories: measures, decisions and resolutions. Measures contain provisions intended to be legally binding when adopted by all ATCPs; decisions are made on internal organisational matters; resolutions are hortatory. In addition, a number of Special Consultative Meetings have been held to discuss particular issues, e.g. relating to CRAMRA and the Protocol on Environmental Protection.

The treaty consists of obligations, but there are no sanctions to enforce observance. The principal gains of the ATCPs are to keep their options open for the future and to retain their influence in the region. The treaty offers states a means of cooperating despite fundamental differences. The difficult question of sovereignty was tackled in a practical manner, since any other solution might have deepened the existing political discords. Article IV was formulated to protect the different positions regarding sovereignty and has been described as the 'cornerstone' of the treaty (Watts, 1986a). It follows a so-called 'bifocal' approach, allowing claimant and non-claimant states to interpret its meaning differently, and so was fundamental to the support of both. Although it is one of the strengths of the treaty, Article IV has not resolved the sovereignty problem, and territorial claims still exist; indeed the sovereignty issue remains an important underlying factor in Antarctic politics (Beck, 1991b). Article IV does not prevent the contracting parties from asserting their sovereignty rights, but there is a recognition that it is not in their political interest to do so and thereby risk reopening old conflicts (Watts, 1986a). However, during the last two decades there have been fears that this situation could change on account of speculation over the potential mineral resources of the region (Chapter 2) and their ownership. This was not a key issue when the treaty was developed and was therefore not resolved at the time. The recent prohibition on mining for at least 50 years included in the Protocol on Environmental Protection has helped to defuse the situation (see Chapter 9).

Table 6.3 Antarctic Treaty member states

Antarctic Treaty made 1 December 1959: came into force 23 June 1961. (The treaty has no limit on its duration. It may be reviewed at the request of a Consultative Party.)

Contracting parties, in chronological order		
United Kingdom	***31 May 1960***	
South Africa	***21 June 1960***	
Belgium	***26 July 1960***	
Japan	***4 August 1960***	
United States of America	***18 August 1960***	
Norway	***24 August 1960***	
France	***16 September 1960***	
New Zealand	***1 November 1960***	
Russia[3]	***2 November 1960***	
Poland	**8 June 1961**	(29 July 1977)
Argentina	***23 June 1961***	
Australia	***23 June 1961***	
Chile	***23 June 1961***	
Czech Republic[4]	14 June 1962	
Slovak Republic[4]	14 June 1962	
Denmark	20 May 1965	
Netherlands	**30 March 1967**	(19 November 1990)
Romania	15 September 1971	
Germany (DDR)[1]	**19 November 1974**	(5 October 1987)
Brazil	**16 May 1975**	(12 September 1983)
Bulgaria	11 September 1978	
Germany (DBR)[1]	**5 February 1979**	(3 March 1981)
Uruguay	**11 January 1980**	(7 October 1985)
Papua New Guinea[2]	16 March 1981	
Italy	**18 March 1981**	(5 October 1987)
Peru	**10 April 1981**	(9 October 1989)
Spain	**31 March 1982**	(21 September 1988)
China, People's Republic	**8 June 1983**	(7 October 1985)
India	**19 August 1983**	(12 September 1983)
Hungary	27 January 1984	
Sweden	**24 April 1984**	(21 September 1988)
Finland	**15 May 1984**	(9 October 1989)
Cuba	16 August 1984	
South Korea	**28 November 1986**	(9 October 1989)
Greece	8 January 1987	
North Korea	21 January 1987	
Austria	25 August 1987	
Ecuador	**15 September 1987**	(19 November 1990)
Canada	4 May 1988	
Colombia	31 January 1989	
Switzerland	15 November 1990	
Guatemala	31 July 1991	
Ukraine	28 October 1992	
Turkey	24 January 1996	

Original signatories: 12 states, which signed the treaty on 1 December 1959, are italicised. The dates are those of the deposition of the instruments of ratification, approval or acceptance of the treaty.
Consultative parties of the treaty:
26 states, comprising 12 orginal signatories and 14 others that achieved this status after becoming actively involved in Antarctic research (with dates in parentheses). A total of 43 states are adherent to the treaty.

[1] The two German states unified from 3 October 1990.
[2] Succeeded to the treaty after becoming independent of Australia.
[3] Formerly the Soviet Union, represented by Russia from December 1991.
[4] Succeeded to the treaty as part of Czechoslovakia, which separated into two republics from 1 January 1993.

Source: R.K. Headland, Scott Polar Research Institute.

It is generally agreed, at least among the adherent states, that the Antarctic Treaty has worked well (Parsons, 1987; Beeby, 1991), although there have been external challenges and criticisms of its legitimacy (see below). According to Scully (1984), this success reflects a recognition by the claimant and non-claimant states that common standards and objectives are possible without the resolution of sovereignty. Indeed, the sovereignty issue may have served to strengthen the Antarctic Treaty System (Beck, 1991b), notably because of the potential for conflict if the system disintegrated. Thus while this issue may be set aside, in practice the need to meet the requirements of claimant and non-claimant states, especially in negotiations over marine living resources and minerals resources, has required much effort. This is not uncommon in international negotiations, where the first priority of governments is to protect their national interests. While this highlights the lack of a truly international approach to the management of Antarctica, it nevertheless reflects the realities of international politics.

The Antarctic Treaty has developed in two main ways. First, there has been a significant increase in political interest in Antarctica (see below), reflected in the growing numbers of acceding states. Second, in response to changing needs and conditions as well as to international debate on Antarctica, the treaty has evolved through a system of further agreements and instruments addressing conservation, resource management and environmental management (Bush 1990; Heap, 1994; see Chapter 9). With the signing of the Protocol on Environmental Protection, conservation has become firmly secured, together with peace and science, as a central objective of the treaty.

6.7 Phase 4: the Antarctic Treaty period (1959–present)

The period since 1959 has seen Antarctica come to occupy a prominent position in world affairs (e.g. Beck, 1990c), and it may no longer be considered 'a pole apart' as described by Quigg (1983). Antarctica has emerged from the geopolitics of the Cold War period, with its emphasis on territorial claims, military power and strategic interests, to a new geopolitics in which there has been a focus on peace, cooperation and environmental protection (Chaturvedi, 1996). While this has not been without its problems as new challenges have arisen, the treaty and its subsequent development (see Chapter 9) have, nevertheless, provided a stable political framework for the international management of Antarctica by those states with established territorial or scientific interests in the region. Since the mid-1970s, there has been considerable interest in Antarctic resources: the exploitation of fish and krill in the Southern Ocean; speculation about the potential minerals wealth of both the continent and its continental shelves; and the emergence of tourism. Some countries, with no tradition of interest in Antarctica, have mounted political pressure to have the area recognised as part of the common heritage of mankind and administered under the United Nations in order for them to gain a share of the potential resources. The debate raised fundamental questions about how Antarctica should be managed, by whom and for what purpose. The growth in global environmental awareness and concern about conservation has also had a major impact on Antarctic affairs. The environmental value of the region and the effects of human activities have become issues of international concern, in part through realisation of the unique scientific value of this largely unspoiled continent and its role in global processes, and in part through the efforts of environmental non-governmental organisations (NGOs) and bodies such as the IUCN and the World Commission on Environment and Development. The NGOs view Antarctica as a conservation symbol and wish to see the continent declared a 'world park'. The high media profile of these organisations and their growing political influence within the treaty states contributed to the present moratorium on mining activities and the rapid development of the Protocol on Environmental Protection (see Chapter 9), reflecting a significant change in emphasis to environmental protection. There has also been a significant growth of interest in Antarctic affairs by international lawyers and political analysts, concerned with the legal, environmental and political implications of resource exploitation and management (e.g. Auburn, 1982; Beck, 1986b; Joyner and Chopra, 1988; Peterson, 1988; Jørgensen-Dahl and Østreng, 1991; Watts, 1992; Sahurie, 1992; Francioni and Scovazzi, 1996). Such interest, and the numerous reports that have been produced, give the impression of 'a continent surrounded by advice' (Beck, 1988b, p.285).

Scientific interest in Antarctica has also advanced substantially following the stimulus of the IGY. In particular, the global significance of Antarctica has become apparent in a number of key environmental spheres, including climate history, global climate processes and global warming, sea-level change, and springtime thinning of stratospheric ozone and the effects of enhanced UV radiation on biological systems (see Chapters 2–5). Integrated research via large international programmes is accepted as the only way to address these issues. The discovery of the ozone hole, in particular, focused much attention on the region. Other developments have also stimulated applied research to underpin advice on the management of marine living resources and environmental problems, a trend that is likely to continue through the requirements for environmental impact assessments and monitoring under the Protocol on Environmental Protection.

The following sections examine these themes in more detail, including the links between politics, resources, science and environmental management.

6.7.1 Antarctica and international politics

During the 1960s, Antarctica remained largely apart from world politics. However, in the late 1970s its re-emergence followed from growing speculation about resources, interest in regulating minerals activities and the later negotiations for a minerals convention, and the growth of global environmental politics and the interest of environmental NGOs in Antarctic issues. This period also saw a large increase in accessions to the Antarctic Treaty: between 1961 and 1976 only six additional states acceded, but between 1977 and 1991 the increase was 22 (Figure 6.10). This trend has been viewed as a direct response to the interest in resource exploitation (Laws, 1987).

A parallel development involved the discussion of Antarctica at meetings of the United Nations and other international organisations, including the Non-Aligned Movement, the Organisation of Eastern Caribbean States, the Caribbean Community, the Islamic Summit Conference, the League of Arab States, the Organisation of African Unity, the South Pacific Forum and the European Parliament (Beck, 1988b, 1989a and b). This wider interest stemmed from the desire of a group of developing nations to share in the perceived resources of the continent and its seas (e.g. Chaturvedi, 1990). These nations, notably Malaysia and Pakistan (which now has an Antarctic research programme but is not a party to the treaty), have been strong critics of the legitimacy of the ATS, which they regard as being run by a select club of countries, and they have challenged both the right of a few nations alone to make decisions regarding the management of all activities in Antarctica, and their lack of accountability (Zain, 1986, 1987; Haron, 1986, 1988, 1991; Koroma, 1988). Instead, they have strongly advocated that Antarctica should be regarded as the common heritage of mankind and that its resources should form part of the global commons, to be shared equitably as

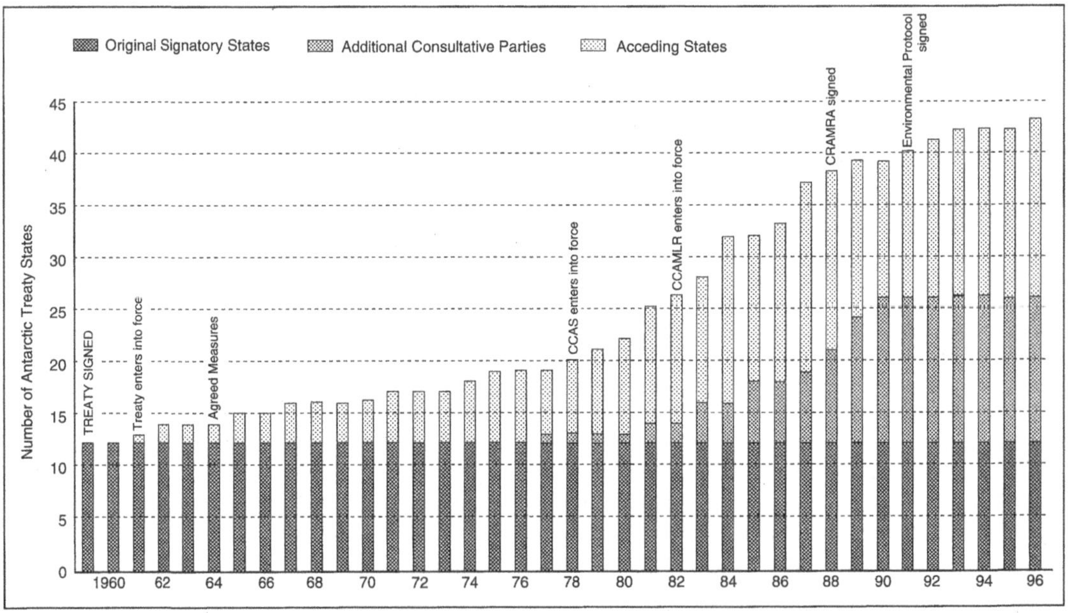

Fig 6.10 The growth in the number of Antarctic Treaty nations (source: *Antarctic Journal of the United States*, 31(1), 1996).

part of a wider internationalisation of the region through the involvement of the UN; scientific activity and practical expertise should not be an exclusive criterion for participation in the decision-making process. The background to this argument lies in the New International Economic Order and the common heritage principles embodied in the UN Convention on the Law of the Sea (UNCLOS) (1982) and the internationalisation of the deep seabed, which provide for benefits from exploitation to be shared among the whole international community, including disadvantaged nations. The apparent secrecy of the minerals regime negotiations (1982–1988) was a particular focus of criticism.

The thrust of the developing nations argument has been that Antarctica should be internationalised through a new agreement, negotiated through the UN, to extend the benefits under the existing system to the wider international community. Although the possibility of UN involvement in Antarctica was first raised in the late 1940s and again in the late 1950s, support was generally not forthcoming from nations directly involved in the region. However, on the initiative of Malaysia, the issue was first tabled at the UN in 1983 (Beck, 1984). Since then, Antarctica has been the subject of a factual study by the UN (Beck, 1985a) and was debated annually until 1994, then biennially (Beck, 1986a, 1987b, 1988a, 1989a, 1990a, 1991a, 1992, 1994b, 1994c; Wolfrum, 1991). Discussions have been strongly polarised between treaty and non-treaty countries, and resolutions adopted have called for an enhanced role for the UN in Antarctic affairs, wider participation in negotiations on the minerals regime and a moratorium on minerals regime discussions. The UN resolutions also attracted support from countries critical of South Africa's participation in the Antarctic Treaty. The treaty parties, however, have continued to ignore such resolutions and have strongly defended the ATS and its achievements (e.g. Scully, 1986; Woolcott, 1986; Orrego Vicuña, 1987). Several developing states (e.g. Brazil and India) acceded to the treaty during the minerals negotiations, weakening the case of the opponents, and a minerals convention was eventually concluded despite the UN resolutions. However, the minerals potential was never likely to live up to its exaggerated expectations on any realistic timescale (see Chapter 7) and, even without the minerals ban now in force, it is probable that for the forseeable future 'Antarctica's contribution as a commons would be meagre in solving basic problems of underdevelopment' (Sahurie, 1992, p.578).

In 1989 and 1990 the critics broadened their attack, reflecting growing environmental concerns about Antarctica (Beck, 1990a, 1991a, 1992, 1993). Further UN resolutions focused on Antarctica's role in the global environment, the need for comprehensive environmental protection, a ban on mining, the applicability of UNCLOS to the Southern Ocean, creation of a world park, adverse impacts from overcrowding of scientific stations and the establishment of a UN international station. In spite of a shift in emphasis from seeking a share of mineral resources to concern for environmental issues, Antarctica did not feature prominently at the UN Conference on Environment and Development in Rio de Janeiro in 1992. However, Agenda 21 reaffirmed the importance of the region for global environmental research and exhorted that the results should be made more freely available (Beck, 1993). Overall, reflecting the strongly polarised views, the UN has had only a minimal influence on Antarctic affairs (Beck, 1991a, 1994c), but the debates have served to raise awareness of the issues and indirectly may have helped to encourage the evolution of the ATS and more openness in the conduct of its affairs (S. Harris, 1991). At the UN in 1994, consensus was seemingly restored as the critics acknowledged progress in a number of areas, particularly concerning the Protocol on Environmental Protection; whether or not this consensus is maintained is likely to depend on the progress in implementation of the protocol (Beck, 1995).

The Falklands War in 1982 between the UK and Argentina also focused international attention on the region and its resources. From a British, as well as an Argentinian, viewpoint, the Falkland Islands have been regarded as a 'gateway' to the Antarctic (Child, 1990; Beck, 1994a). Although fears were expressed about the possible extension of the conflict to Antarctica, the treaty continued to function unhindered (Beeby, 1991). From a British perspective, government interest in the area had been in decline (Beck, 1983), but the war had the effect of reversing this trend. In the aftermath, the updated report of the Shackleton Committee (Shackleton, 1982) noted the need to view the future of the Falkland Islands in a wider Antarctic perspective and highlighted the greater resource potential of the South Georgia seas. In November 1982, the British government announced a 60% increase in funding for the British Antarctic Survey, which increased the scientific research effort as well as strengthening Britain's strategic presence in the region.

6.7.2 Science and politics

Antarctica has been described as a 'continent for science' (Lewis, 1973) and an 'international laboratory for science', with scientific knowledge viewed as its greatest export (Indreeide, 1990). Antarctic science in its own right came of age during the IGY and since then has developed within the framework of the Antarctic Treaty and under the coordination of SCAR, a permanent sub-committee of the ICSU (see Chapter 9). This period has seen the application of new technology, including satellite remote sensing, deep ice coring and radio echo-sounding of the ice sheet, supported by aircraft and ships (Figures 6.11 and 6.12). Research has become increasingly multidisciplinary and problem-oriented. Significant advances have been made, particularly in major interdisciplinary fields involving multinational cooperative research on the upper atmosphere, ice sheet dynamics and palaeoclimates, geological history and marine biology. An important development has been the recognition of the global relevance of Antarctic science and the significance of Antarctica in the functioning of the Earth's atmosphere–geosphere– biosphere systems (Budd, 1986; Roots, 1986; Drewry, 1988, 1993; SCAR, 1989, 1993a; Walton and Morris, 1990; Laws, 1990) (see Chapters 2–5). The results of this work have had significant implications for policy decisions at a global level; for example, the 1987 Montreal Protocol (see Chapter 8) followed the discovery of a springtime depletion of ozone in the Antarctic stratosphere.

Much of the achievement of the IGY reflected a spirit of international cooperation between scientists and was fundamental to the subsequent development and success of the Antarctic Treaty (Indreeide, 1990). The scientific imperative offered a solution to fears over sovereignty conflicts and the possibility of an East–West clash in Antarctica as an extension of the Cold War (Østreng, 1989). That the solution effectively involved an evasion of the sovereignty problem has probably contributed to its success and to the development of science during the period of the

A

B

Fig 6.11 (A) Field party travelling overland by skidoo in the Antarctic Peninsula (source: authors). (B) Such field parties are supported by modern ski-equipped aircraft such as this De Havilland Twin Otter of the British Antarctic Survey (source: photograph courtesy of A. Alsop).

Fig 6.12 RRS *James Clark Ross* is a modern ice-strengthened ship that came into service with the British Antarctic Survey in 1990. It is designed to support advanced marine research as well as to transport cargo and personnel (source: photograph courtesy of A. Alsop).

Antarctic Treaty. Under the terms of the treaty, scientific activity became a condition for achieving political status and maintaining a strategic interest in the region through participation in Consultative Meetings, given the difficulties in justifying the substantial expense of otherwise maintaining a presence; as such, science has been described as a 'currency of credibility' (Herr and Hall, 1989; Quilty, 1990). As science and politics have become closely linked (Beck, 1986b; Elzinga and Bohlin, 1989), scientific cooperation is now a clear political objective, the success of which can be used by the treaty nations to justify their continued administration of Antarctica in the face of criticism from non-treaty states and environmental NGOs (Østreng, 1989). In helping to provide the public justification for the policies of the treaty states, science thus has an instrumental value as well as an intrinsic value. The cost to science has involved some loss in academic freedom, and with the increasing regulation of activities, the displacement of scientific priorities by political and legal issues at the forefront in Antarctic affairs (Laws, 1987; Herr and Hall, 1989). Equally, however, it is doubtful that the amount of research that has been funded would have been achieved were it not for underlying political, strategic and economic motives (Indreeide, 1990).

The patent use of science for political ends, both to acquire a voice in Antarctic affairs and to maintain a strategic presence in the light of the resource potential or for geopolitical reasons, also carries negative implications. The requirement for states seeking ATCP status to demonstrate substantial scientific interest has been equated with the operation of scientific stations. As a growing number of nations have acceded to the treaty, concern has been expressed about duplication of effort and the questionable scientific value of some research programmes (Drewry, 1988, 1989; Anon, 1990c). This applies particularly to King George Island in the South Shetlands Islands, where the relative ease of access has led to overcrowding of bases (see Chapter 8) in support of national political objectives, (i.e. to achieve ATCP status and therefore a voice in Antarctic affairs, or to underpin tacitly future sovereignty claims through demonstration of effective occupation). Research into globally important problems requires more effective international planning, coordination and funding, and a more scientifically effective approach might involve, for example, the sharing of facilities within multinational programmes and a wider geographical spread of research activity. A significant development has been the accession of the Netherlands to ATCP status in 1990, although it had no plans to construct a station (Pannatier, 1994). Instead the Netherlands has developed a range of programmes in cooperation with other nations, including the sharing of facilities at Poland's Arctowski station on King George Island.

6.7.3 Science and environmental management

A further development during the Antarctic Treaty period has seen the growing involvement of science in an advisory and monitoring capacity for environmental and resource management and conservation (Heap, 1988; Østreng, 1989; Walton and Morris, 1990; Davis, 1992; Elzinga, 1993a). This trend has its roots in the attempts to regulate whaling via the International Whaling Commission. In this case the scientific advice, albeit based on incomplete data, was largely disregarded in the face of economic and political pressures. Science has been closely integrated in the development of the various instruments of the ATS through the work of SCAR (see Chapter 9). The Scientific Committee of CCAMLR has a key role in ensuring that conservation measures are based on the best scientific advice available and has been assisted in its work by the major programmes organised by SCAR to improve understanding of key components in the Southern Ocean marine ecosystem (El-Sayed, 1994). This has not been without difficulties, however, since the scientific uncertainties have allowed nations with fisheries interests to resist regulation of resource exploitation (see Chapters 7 and 9). SCAR has also played an important part in Antarctic conservation through its work with the IUCN, for example in relation to the Antarctic protected area system and conservation of sub-Antarctic islands (see Chapter 9). Science will also have a key input into the implementation of the measures in the Protocol on Environmental Protection, for example in respect of environmental assessment and monitoring.

SCAR has an essential role in ensuring greater international coordination of science, involving wider participation at different levels (Drewry, 1988). It also has a proactive role in the area of environmental management: to ensure that scientific activities are compatible with environmental impact requirements; to provide the information base for evaluating and monitoring wider impacts; and to guide the development of future regulations and environmental management principles (Heap, 1991a; Walton, 1991). Antarctic policymaking is no longer the preserve of the diplomats and scientists with experience in the area (Drewry, 1994); it now includes a wider political involvement (Davis, 1990), increasingly influenced by the environmental movement. Environmental concerns are now high on the Antarctic political agenda, so scientists, as the major user group on the continent, have a more overt responsibility and requirement to demonstrate good stewardship. Under the Protocol on Environmental Protection, environmental effects of human activities, including scientific investigations, must now be evaluated (see Chapter 9); existing and abandoned facilities are being cleaned up and improved practices implemented, for example for waste disposal procedures, to meet the new environmental standards. Thus science and its logistical support will cost more. The demand for environmental controls has been strongly driven by environmental NGOs, producing a defensive reaction from the scientific community to the overstatement of the case by the NGOs. The scientific concerns have centred on unnecessarily restrictive measures for environmental protection, increased bureaucracy, the time required to complete environmental assessments and monitor impacts, and the possible constraints that could be placed on scientific freedom (Laws, 1991, 1994a; Elzinga, 1993a; National Research Council, 1993). Furthermore, there is a worry that funding may be diverted from basic science to support clean-up of past mistakes and to undertake environmental monitoring (Drewry, 1993; Laws, 1994a). There has been a basic difference of approach between the scientists and the environmental NGOs. The former are concerned to prevent unnecessary and bureaucratic restrictions on their freedom to pursue their investigations; the latter are seeking to impose their goals, irrespective of the facts. Fortunately, there is an atmosphere of improving relations, and provided that sensible procedures are adopted for implementation of the measures in the protocol, then science and good stewardship can progress interactively, with benefits to both (National Research Council, 1993): good stewardship requires good science to underpin management decisions and policy development, but it also ensures the protection of a resource of global significance for scientific study. While in practice there are still uncertainties to be resolved (see Chapter 10), environmental management now has a much greater priority than in 1959.

6.7.4 Resource development, environmental management and conservation

From a resources viewpoint, the treaty period has seen several important trends, including the decline of the whaling industry, growing exploitation of the finfish and krill resources of the Southern Ocean, interest in mineral resources and development of tourism (see Chapters 7 and 8). On the management side, increasingly strict regulatory measures have been developed by the treaty parties under CCAMLR, CRAMRA and the Protocol on Environmental Protection (see Chapter 9), and latterly by the International Whaling Commission (see Chapter 7). The potential resources of Antarctica were the subject of much speculation during the 1970s and 1980s, with unsubstantiated and exaggerated claims made about the mineral wealth of the region (Table 6.4) Despite much interest in minerals resources, in part fuelled by the oil crisis of 1973 and by the prediction of substantial oil and gas reserves in the Antarctic, the potential is largely unfounded, and there is now a ban on minerals-related activities under the Protocol on Environmental Protection. In contrast, the fishing industry has already had a significant impact on certain stocks, but the predicted potential of the krill fishery, despite highly optimistic forecasts, has not been realised. Given the delay in adopting specific management measures for the latter, its slow development has been fortunate from an ecological perspective. Recent years have also seen a significant expansion in Antarctic tourism (see Chapter 8), coinciding with a demand for new frontiers in the industry and also with the growth of 'ecotourism'.

Much concern has been expressed about the political and environmental implications of resource

Table 6.4 Speculation about Antarctic resources

'Antarctica is mankind's last remaining treasure-house, other than the deep sea': statement made at UN meeting in 1988 (quoted in Beck 1989a, p. 329)
'there could be another Middle East in Antarctica' (quoted in Tinker, 1979)
'vast hoard of fuel believed to be buried in this frozen wasteland' (Spivak, 1974)
'I have heard that the South Pole is made of gold and I want my piece of it' Dr Mahathir, Malaysian Prime Minister, reported in *The Guardian*, 26 November 1988

exploitation, including the issues of ownership and the future of the Antarctic Treaty (Mitchell, 1977; Peterson, 1980; Westermayer, 1982; Luard, 1984; Zorn, 1984; Shapley, 1985). The various claims and concerns have also served to raise both the political profile and public awareness of Antarctica, leading to external challenges to the ATS from two directions. First, there have been the moves at the UN to have the region and its resources recognised as the common heritage of mankind. Second, the environmental NGOs, concerned about the possible impacts of mineral exploitation and the environmental impacts of scientific and logistic activities, have actively pursued their goal to have the continent declared a 'world park'. Although this has been resisted by the treaty parties, the NGOs have nevertheless exerted considerable influence both during the debate over the exploitation of minerals resources and in ensuring the adoption of high standards of environmental management and awareness in the Protocol on Environmental Protection (see Chapter 9). The minerals issue has therefore played a significant part not only in widening global interest in Antarctica but also ironically in its protection.

Recent years have therefore seen a significant change in attitudes to environmental management. Much attention has been focused on the cumulative effects of scientific stations and associated logistics activities, which have left a localised but significant legacy in the form of terrestrial and marine pollution at a number of localities around the periphery of the continent. In many areas, this legacy is now being addressed through clean-up programmes, and the Protocol on Environmental Protection now provides a legal framework for improved standards for waste management and local pollution control. However, global pollution potentially poses a greater and more widespread threat to the Antarctic environment, particularly through climate warming from greenhouse gas emissions and stratospheric ozone depletion.

The Antarctic Treaty period has therefore seen a more complex interplay between resource development, environmental management, conservation and politics. In the 1970s and 1980s, the focus was on resource management, whereas in the late 1980s and the 1990s it has shifted to comprehensive environmental protection as Antarctica rose to prominence in the global environmental agenda. The ATS has nevertheless proved flexible in the face of new pressures arising from the broadening interest in Antarctic resources, the growth in environmental awareness and informed concern about human impacts at a global scale (see Chapter 9). Although conservation was addressed only briefly in Article IX in the original treaty, its importance was recognised in numerous recommendations and in the development of the Agreed Measures and later in CCAS, CCAMLR and in the negotiations leading to CRAMRA. Most recently, the Protocol on Environmental Protection provides for the first time an overall framework for mandatory environmental protection and management. Under the protocol, Antarctica becomes a 'natural reserve, devoted to peace and science'. This carries a commitment to the comprehensive protection of the environment and ecosystems of Antarctica, including their value not only for scientific research of global significance but also for their wilderness and aesthetic values.

6.8 Conclusion

The changing geography of Antarctica has not merely involved improving knowledge about the physical outline of the continent and its ice cover, as depicted in Figure 6.1, but has encompassed the development of a geography of resources, politics and environmental management. While Antarctica remains unique for its scientific, ecological and environmental values, it is no longer a forgotten continent, of interest to a few hardy explorers, commercial entrepreneurs or scientists. It is now perceived as a global concern in terms of its political management, its influence on the dynamics of our planet's physical and biological systems and our understanding of how they have evolved, and its significance as a global conservation symbol as the last unspoiled corner of the world. As the motivation for human interest has changed, so have the constituents of Antarctica, who now include not only scientists, politicians and those who seek to exploit its resources, but also a much wider spectrum of environmentally conscious people across the globe. National interests in Antarctica are driven by a changing interplay of legal, political, strategic, economic, scientific and environmental motives (Beck, 1996). Scientific activity, nevertheless, remains the prime reason for a permanent human presence in Antarctica, and international scientific cooperation played a central role in preparing the ground for the Antarctic Treaty. However, as science has become bound up with politics, resource management and environmental management, tensions have arisen concerning scientific freedom and the use of science for political or strategic ends.

The environment of Antarctica is now a prime focus of interest in terms of its value for scientific investigation, nature conservation and tourism. Environmental management is a key issue, not only from the point of view of impacts on ecosystems and the landscape, but also in terms of sustainable exploitation of marine living resources. There has thus been a fundamental shift in perception of Antarctica from James Cook's original vision of a geographically remote, hostile and useless environment to one in which that environment itself is now considered to be threatened by human activities and so require protection. This contrasts sharply with attitudes in the past, as reflected in the poor record of environmental management in Antarctica and the depletion of its marine resources, catalogued in the following two chapters.

Further reading

Beck, P.J. (1986) *The International Politics of Antarctica.* Croom Helm, London.

Chaturvedi, S. (1996) *The Polar Regions: A Political Geography*. John Wiley, Chichester.

Fogg, G.E. (1992) *A History of Antarctic Science.* Cambridge University Press, Cambridge.

Hempel, G. (ed.) (1994) *Antarctic Science. Global Concerns.* Springer-Verlag, Berlin.

Reader's Digest (1985) *Antarctica. Great Stories from the Frozen Continent.* Reader's Digest. London, Sydney, etc.

Simpson-Housley, P. (1992) *Antarctica: Exploration, Perception and Metaphor*. Routledge, London.

Walton, D.W.H. (ed.) (1987) *Antarctic Science.* Cambridge University Press, Cambridge.

Walton, D.W.H. and Morris, E.M. (1990) Science, environment and resources in Antarctica. *Applied Geography,* 10, 265–286.

CHAPTER 7 Environmental impacts: Antarctic resource exploitation

Objectives

Key themes and questions covered in this chapter:

- patterns of exploitation of marine living resources – seals, whales, finfish, krill
- regulating the exploitation of open-access resources: can sustainable management be achieved?
- the ecosystem approach to resource management – an ideal or an attainable goal?
- cultural attitudes and the management of non-endangered species
- mineral resources and environmental issues

7.1 Introduction

Environmental impacts on the terrestrial and marine ecosystems of Antarctica and on the atmosphere principally arise from the exploitation of marine living resources; from human activities associated with scientific research programmes and the supporting logistics, transport and station facilities; from tourism; and from the global dispersal of pollutants from sources external to Antarctica, including the effects of global warming and stratospheric ozone depletion. The potential threats from the exploitation of non-living resources have been effectively removed for the forseeable future through the ban on minerals activities included in the Protocol on Environmental Protection. The aim of this chapter is to review and assess the past, present and anticipated human activities associated with resource exploitation in Antarctica, together with their environmental impacts. In Chapter 8 the impacts on the Antarctic environment arising from human occupation and visitation and from various forms of pollution are examined. These two chapters provide a perspective from which to assess the developing environmental management framework of the ATS and the alternative proposals put forward by the environmental NGOs (Chapter 9).

The past exploitation of marine living resources is important in understanding the nature of the environmental debate about Antarctic resources. In a largely chronological sequence, the development of the sealing industry in the late eighteenth and early nineteenth centuries was followed by whaling during the early–mid twentieth century and finfish and krill fisheries during the later twentieth century. The development and impact of these industries, and the depletion of the resource base through overfishing (except in the case of krill), provides important lessons for regulating the development of sustainable exploitation of current and future marine resources. Although minerals activities are now effectively banned for the forseeable future, the minerals issue and the attendant questions of ownership, regulation and potential environmental impacts dominated Antarctic affairs during the mid–late 1980s. As well as presenting a serious challenge to the ATS, the minerals issue highlighted the importance of environmental management and paved the way for the Protocol on Environmental Protection.

7.2 Seals

7.2.1 Introduction

The sealing industry represents the first phase in the exploitation of Antarctic resources (Figure 7.1) and occasioned the first major human impacts on the Antarctic marine ecosystem. Given the geographical distribution of the target seal species, these impacts were spatially focused on the sub-Antarctic islands and the islands on the periphery of the Antarctic Peninsula.

7.2.2 Development and impact of the industry

Early in the seventeenth century, sealers from New England and Europe had extended their activities to the beaches of the South Atlantic in search of blubber for oil from elephant seals and the valuable skins of the fur seal. Both of these species provided a vulnerable target since they breed onshore in large, discrete colonies, returning to the same sites each spring where the females feed the pups for about three months in territories closely guarded by the bulls. When attacked, the bulls attempt to defend their territories rather than flee into the sea, and the females remain with their pups. As the beaches of southern South America and the Falkland Islands were cleared, the sealers spread their efforts further south. Following Cook's visit to South Georgia in 1775 and the publication of his observations of great numbers of seals there, sealers first arrived at the island in the late eighteenth century to find the beaches densely populated by Antarctic fur seals and elephant seals (Christie, 1951; Bonner, 1968; Headland, 1984).

By the early 1820s, 1.2 million fur seal skins in total may have been taken during the course of the industry at the island (Weddell, 1825). In 1801 alone, seventeen vessels removed 122,000 fur seal skins (Christie, 1951), and 91 vessels were at the island in 1820–21, with as many as 3000 men employed

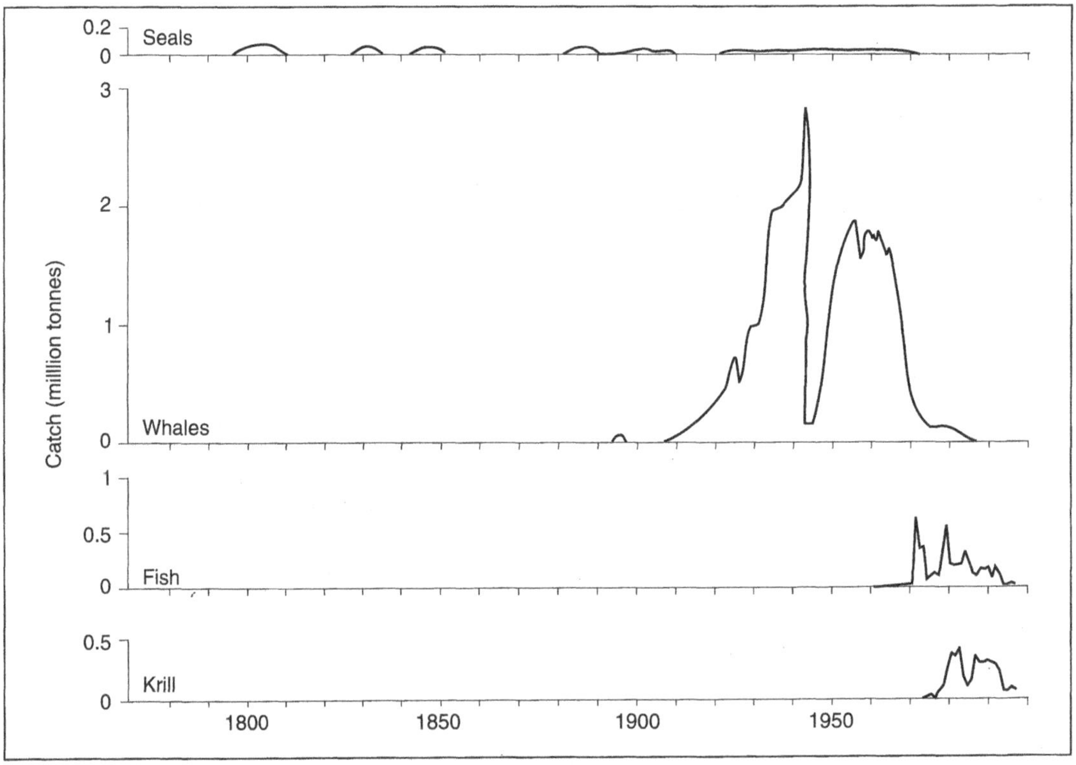

Fig 7.1 Exploitation of Antarctic marine living resources. The figure shows the successive phases of harvesting of seals, whales, finfish and krill (source: after Laws, 1989; CCAMLR Statistical Bulletins).

onshore. Graphic descriptions of the methods employed in the slaughter are given by Christie (1951) and Bonner and Laws (1964), based on the accounts of James Weddell, Robert Fildes and Edmund Fanning. Thereafter, with the resource depleted, the sealers moved on in search of new sources of fur seals (Figure 7.2). Following the voyage of William Smith on the brig *Williams* in 1819, word soon spread of the discovery of new land and sealing beaches in the South Shetland Islands. During the following season about 30 American and 15–20 British ships were active around the islands (Christie, 1951), with as many as a quarter of a million seals taken (Bonner, 1968). In the succeeding years, the sealers virtually eliminated the fur seal stock, although there was a brief revival in the 1870s (Smith and Simpson, 1987). Further stocks in the South Orkney Islands and the island groups in the Indian and Pacific ocean sectors were similarly depleted (Bonner and Laws, 1964). The fur seal population of Macquarie Island was eliminated ten years after the discovery of the island in 1810, with as many as 193,300 seals slaughtered (Shaughnessy *et al.*, 1988). Similarly at Îles Crozet and the Prince Edward Islands, fur seals were severely depleted between 1805 and 1810 (Richards, 1992).

As fur seal stocks declined, the emphasis shifted to the exploitation of elephant seals for blubber-oil used for lighting, lubrication and in the preparation of leather (see Figure 7.2). South Georgia, Îles Kerguelen and Heard Island provided particularly rich harvests (Bonner and Laws, 1964). At Macquarie Island, the elephant seals were exploited between 1810 and 1919 (Cumpston, 1968) and it is estimated that the initial population of about 93,000–110,000 was reduced by 70% between 1820 and 1830, when harvesting ceased (Hindell and Burton, 1988). When restarted in 1875, it appears to have been pursued at a much lower and relatively sustainable level. At South Georgia, elephant sealing continued for much of the nineteenth century (Headland, 1984), and Murphy (1948) gives an excellent account of one of the last old-style sealing expeditions there on the brig *Daisy*. After 1909, the industry was regulated until it closed in 1964 (see below), making a significant contribution to oil production at the Grytviken whaling station (Dickinson, 1993).

The large stocks of the four other Antarctic seal species, particularly the crabeaters, have not been exploited to any extent (Bonner, 1985a), probably because of their inaccessibility as creatures of the sea ice. In 1964, a Norwegian expedition carried out exploratory commercial sealing in the area between the South Shetland Islands and the South Orkney Islands, catching over 850 adults, principally of crabeater seals (Øritsland, 1970). In 1986/87 a Soviet expedition to the Antarctic took 4802 seals (Bonner, 1990). As the most abundant of the seal species, the crabeaters represent a vast natural resource, but harvesting in the future is unlikely to be either commercially viable or morally acceptable (Bonner, 1985a).

The sealing industry, together with whaling, also had a considerable, but unmeasured, impact on the Southern Ocean ecosystem (Laws, 1977, 1985) in

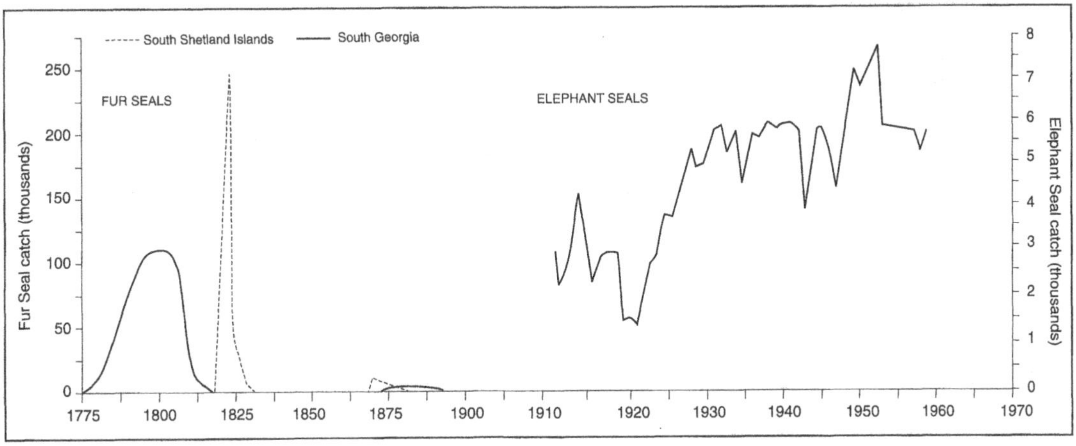

Fig 7.2 Catches of Antarctic fur seals at South Georgia and the South Shetland Islands and of elephant seals at South Georgia. The latter shows only the period when the industry was managed during the twentieth century (sources: data from Bonner and Laws, 1964; Holdgate, 1984).

terms of reduced predation. The increases in krill-dependent species (including chinstrap and Adélie penguins) and increases in fish- and squid-dependent species (e.g. king penguins) observed over the twentieth century (Croxall *et al.*, 1988; Croxall, 1992; see Chapter 4), and attributed to reduced levels of competition by whales (e.g. Laws, 1985), were almost certainly preceded by similar effects as a result of sealing.

On many islands, such as South Georgia (Headland, 1984), harvesting of penguins was pursued alongside sealing activities both for oil and for fuel for boiling down seal blubber. On Macquarie Island, up to 150,000 penguins were taken each year at the end of the nineteenth century. On Heard Island, populations were also significantly reduced (Cumpston, 1968; Rounsevell and Copson, 1982). Penguin eggs have also been collected by sealers and whalers, as have those of other seabirds (Croxall *et al.*, 1984). Overall, the impacts on penguins seem to have been locally significant (Croxall, 1987).

7.2.3 Regulation of sealing

By 1909, the less intensively exploited elephant seals of South Georgia had recovered sufficiently to allow further harvesting under licence by the Falkland Islands Dependencies administration (Bonner and Laws, 1964; Headland, 1984; Bonner, 1985a; Dickinson, 1989; Laws, 1994b). Management restrictions allowed an annual quota of up to 6000 adult bulls to be taken in any one year from three of the four sealing areas into which the island was divided. In addition, a close season and sealing reserves were established. During the late 1940s, the industry was unable to sustain higher quotas, and a revised management plan based on biological principles was adopted. Although sealing ended in 1964 with the closure of the whaling industry on the island, its management in this latter period is regarded as highly successful, representing a sustainable use of a renewable natural resource (Bonner, 1985a; Laws, 1994b) (Figure 7.3). In total, about a quarter of a million seals were taken during this phase (see Figure 7.2), producing about 75,000 tonnes of oil (Bonner and Laws, 1964).

In the 1960s, concern grew about possible future exploitation of seals. The Agreed Measures for the Conservation of Antarctic Flora and Fauna did not protect seals at sea or on the sea ice, and a separate convention, the Convention for the Conservation of Antarctic Seals, was adopted in 1972 (came into force in 1978). This covers the area south of 60°S and includes protection for Ross, elephant and fur seals, catch limits for Weddell, crabeater and leopard seals and a range of additional conservation measures (see Chapter 9).

Fig 7.3 Sealing at South Georgia continued until the early 1960s under a management plan that allowed for sustainable exploitation of elephant seals. Flensing an elephant seal, South Georgia, 1958 (source: photograph by N. Bonner, reproduced by permission of J. Bonner).

7.2.4 Recovery of seal populations

The rate and extent to which seal stocks have recovered from exploitation has been variable. Elephant seals had recovered sufficiently on South Georgia by the early twentieth century to permit further controlled exploitation, and currently the population appears to be stable (McCann and Rothery, 1988; Laws, 1994b; Boyd *et al.*, 1996). At Macquarie Island and island groups in the Indian Ocean, numbers are declining (Pascal, 1985; Burton, 1986; Hindell and Burton, 1988; Wilkinson and Bester, 1988; Guinet *et al.*, 1992), but whether this is due to fisheries competition, environmental changes or inherent population dynamics remains unresolved (Pascal, 1985; Hindell and Burton, 1987; Vergani and Stanganelli, 1990; Guinet *et al.*, 1992; Hindell *et al.*, 1994). The recovery of the fur seal population (Figure 7.4) at South Georgia has been slower, but nevertheless remarkable (Payne, 1977; Boyd, 1993). Following the last kills in 1907, no further seals were recorded there until 1915 (Bonner, 1968). By 1956, a small breeding colony at Bird Island was producing 3500 pups (Bonner, 1985a). Between 1958 and 1972, the annual rate of population increase was 16.8%, and in 1975/76 an estimated 90,000 pups were born on South Georgia (Payne, 1977). Currently, the population on South Georgia is about 2.05 million (Arnould and Croxall, 1995), and by the end of the present century it may be near to regaining its former abundance. In addition, breeding colonies have become established elsewhere from South Georgia stock; the west coast of the Antarctic Peninsula is now frequented by fur seals and they are expanding on the South Shetland Islands, South Orkney Islands and South Sandwich Islands. One factor that may have favoured the recovery of the fur seals is the increased availability of krill arising from the great reduction in whale stocks (see Chapter 4). The possibility of expanded krill harvesting in the future could have significant implications for the sustainability of the recovery of the fur seals (Doidge and Croxall, 1985). Recolonisation has also occurred on Macquarie Island (Shaughnessy *et al.*, 1988), Heard Island (Shaughnessy and Goldsworthy, 1990) and the Prince Edward Islands (Wilkinson and Bester, 1990), although at much lower levels. At Îles Crozet, rapid growth comparable with that at South Georgia has been recorded over the last decade (Guinet *et al.*, 1994).

Fig 7.4 Fur seal breeding beach at South Georgia. Fur seal numbers have now recovered from exploitation, resulting in high densities and erosion of tussock grass and peat (source: authors).

The recovery of the fur seal population on South Georgia has resulted in very high colony densities and this has produced extensive destruction of the tussock grass, together with peat and soil erosion at Bird Island (Bonner, 1985b). In addition, there is a serious risk to the habitat of ground-nesting and burrowing birds (the endemic pintail, *Anas georgica*, and the pipit, *Anthus antarcticus*) and invertebrates, although there appear to be no significant adverse effects on the breeding success of wandering albatrosses (Croxall *et al.*, 1990c). Similar vegetation disturbance has been reported on Heard Island (Scott, 1990), and on Signy Island in the South Orkney group, the fragile terrestrial vegetation has been severely damaged and eutrophication of freshwater lakes has occurred as a result of the large summer influx of seals from breeding colonies on South Georgia (Smith, 1988a). The long-term impact of such changes raises difficult management questions (*ibid.*; Hodgson *et al.*, 1997); for example, should a protected species of seal be allowed to increase in numbers at the expense of the diversity of the terrestrial and freshwater ecosystems, or should they be managed through carefully controlled exclusion fencing or culling? Adding to the dilemma is the possibility that the existing vegetation patterns may have developed only as a result of reduced seal numbers through the sealing activities in the early nineteenth century, and although pre-exploitation numbers may have been high, the densities may have been lower. Thus recovery of the fur seal population appears not to have been accompanied by a return to pre-exploitation conditions. It is clear that while components of an ecosystem may recover from human perturbation, there may be significant differences from the original undisturbed condition (Bonner, 1985b).

7.2.5 Summary

The sealing industry was the first commercial exploitation of Antarctic marine resources and the first major perturbation of the marine ecosystem arising from human activities. Indirectly, the industry contributed to geographical exploration of the coasts of the sub-Antarctic islands and the Antarctic Peninsula as sealing captains searched for new untapped resources (see Chapter 6). The pattern of exploitation set unfortunate precedents since, in the absence of regulation, intense competition for short-term profits produced unsustainable levels of harvesting. As individual beaches and islands were successively cleared of seals, the sealers moved on to the next location, repeating the process in 'boom and bust' fashion, until the industry collapsed when the resource became depleted. With time the seal populations have recovered significantly, assisted in the case of the fur seals by reduced whale numbers and decreased predator competition. The overall, long-term effects of sealing on the ecosystem, however, are difficult to disentangle from those of other human activities, particularly whaling.

7.3 Whales

7.3.1 Introduction

The development and ultimately the collapse of the whaling industry in the Southern Ocean is a story of overexploitation of a natural resource in the pursuit of non-sustainable returns. The history of the industry is well-documented in a number of sources (e.g. Brown, 1963; Tønnessen, 1970; Gulland, 1974, 1976; McHugh, 1974; Jackson, 1978; Tønnessen and Johnsen, 1982; Birnie, 1985). In contrast to the sealing phase of marine resource exploitation, the whaling industry indirectly made significant contributions to scientific knowledge not only about whales but also about the wider marine ecosystem and oceanography of the Southern Ocean, most notably through the work of the Discovery Investigations. The Antarctic fishery was based on catching whales at their feeding areas during the southern summer, when the whales rapidly develop a thick layer of blubber. As a result, the geographical development of the industry was intimately related to (1) the spatial distribution of krill in the Southern Ocean; (2) the location of suitable sites for shore stations and floating factories; and (3) the development of handling and processing techniques that allowed the industry to operate latterly over a wider geographical area. The industry developed initially in krill-rich coastal areas of the Scotia Sea from shore-based stations and floating factories. It later spread to the high seas of the Southern Ocean. As technology improved, driven by the relentless pursuit of profit to sustain an intensely competitive and over-capitalised industry, the main species of the great whales were progressively overfished. Three main phases occurred in the development of Antarctic whaling: 1904 to the mid-1920s; the mid-1920s to 1939; and post-1939.

7.3.2 Development of the industry

1904–mid-1920s

The discovery of large numbers of whales in Antarctic waters was reported by a number of early explorers, including Bouvet, Cook, Ross and Weddell, but it was some time before economic conditions and technological developments allowed their exploitation. Until the end of the nineteenth century, whaling was largely a Northern Hemisphere activity, based principally on the right whales; these swam at relatively slow speeds and continued to float when killed, in contrast to the faster rorquals. However, by the beginning of the twentieth century several factors contributed to the development of the Antarctic fishery. The decline of the northern whale stocks was paralleled by technological developments that allowed pursuit of rorquals, including the steam-driven catcher boat with powerful winches, a bow-mounted gun that fired harpoons with attached lines and exploding grenades, an accumulator to take the strain on the line, and an air pump to inflate the dead whales to prevent sinking. It was then possible to haul whales to the surface and tow them to shore stations or factory ships. Developments in industrial processing, notably hydrogenation, provided important new markets for whale oil in the production of soap and margarine, particularly in the years after the First World War.

The biogeography of whales in the Southern Ocean is driven by the availability of krill and, strategically placed within easy reach of dense krill concentrations in the Scotia Sea (see Chapter 4), South Georgia was the initial focus of the Antarctic whaling effort. For the whalers, South Georgia also provided secure harbours, fresh water and a relatively benign environment in the midst of a hostile ocean at a time when only parts of the continental coastline were known and knowledge of sea ice conditions was sketchy. The Compañia Argentina de Pesca S.A. established the first shore-based station at Grytviken on South Georgia in 1904 (Walton, 1982b). Initial catches of humpback whales from the coastal waters of the island were sufficient to warrant the establishment of floating factories in several fjords and further shore stations at Leith Harbour (1909), Ocean Harbour (1909), Husvik (1910), Stromness (1912) and Prince Olav Harbour (1917) (Headland, 1984). Whaling extended to the South Shetland Islands in 1906, when the first Norwegian factory ship, *Admiralen*, and two catchers were located there, together with a Chilean factory ship; the construction of a shore station on Deception Island followed in 1910/11. Whaling also developed at the South Orkney Islands and to a lesser extent the South Sandwich Islands in 1911/12.

The industry in the Falkland Islands Dependencies developed rapidly, from one shore station in 1904 and a single catcher taking 195 whales to six shore stations, 21 floating factories and 62 catchers taking 10,670 whales in 1912/13. By the second decade of the twentieth century, catches from the Dependencies exceeded those from the rest of the world and expansion was temporarily halted by a ceiling imposed on licences by the British Colonial Office. This coincided with the decline of the humpback and a switch to blue and fin whale catching (Figure 7.5). The South Orkney Islands fishery proved significant in several respects (Tønnessen and Johnsen, 1982): it showed that it was possible to operate in a sea ice environment and much was learned about the relationships between ice, krill and whale stocks.

Pelagic whaling mid-1920s–1939

From the end of the 1920s, whaling on a global scale had become firmly focused on the Antarctic. Hitherto, the fishery had been based on shore stations (Figure 7.6) or factory ships anchored in sheltered harbours and with catching restricted to a geographical range near the processing plants. In 1925/26 the successful deployment of the first factory ship with a stern slipway, the *Lancing*, showed that vessels could operate on the high seas independent of shore bases (Figure 7.7). The development of pelagic fishing also meant that factory ships could follow the catchers into the whaling grounds. Whaling expanded geographically to include the krill-rich waters of the Ross Sea, along the edge of the pack ice and the areas more remote from the sub-Antarctic islands. On the open seas, the operators avoided both payment of royalties and restrictions on catches, the resource became common property, and the spatial distribution of activity became truly circumpolar (see Figure 4.22). As technology developed, factory ship sizes increased, with improved processing plant and increased efficiency, and more powerful catchers equipped with radios were deployed. The scale of investment was huge; for example, the total capital outlay of the Southern Whaling and Sealing Company fleet in the late 1920s was nearly £1 million (Jackson, 1978). Production increased sharply between 1927 and 1931 as the Atlantic and Ross Sea areas were intensively fished. A record catch of over 40,000 whales in 1930/31 produced over 600,000 tonnes of oil from six shore stations, 41 floating factories and 232 catchers. The result was overproduction of oil and collapse of the market. By agreement during the following season, the

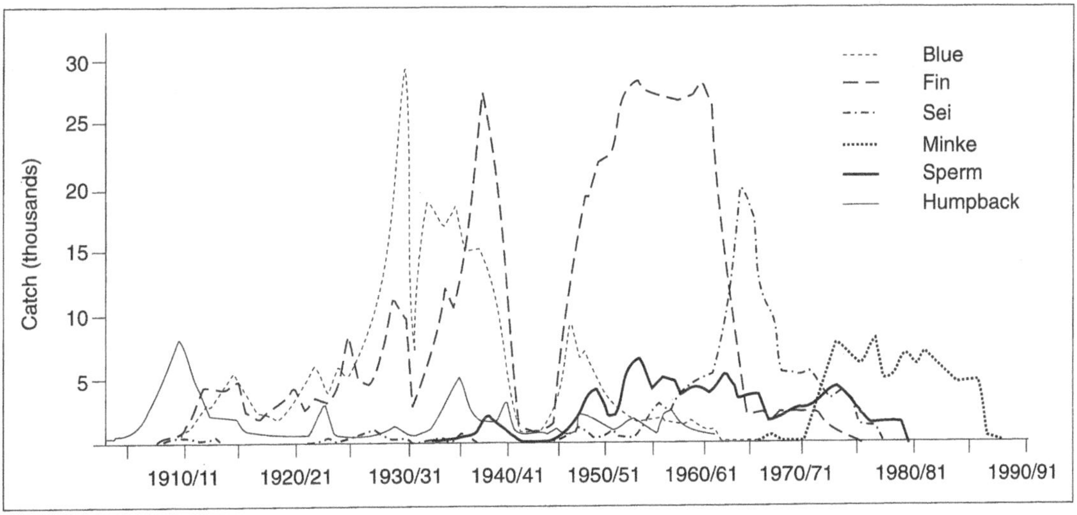

Fig 7.5 Annual catches of whales in the Antarctic, showing the catching effort that was successively focused on humpback, blue, fin, sei and minke whales as each target species was overfished (source: redrawn from Laws, 1989).

Fig 7.6 The whaling station at Leith Harbour, South Georgia, around 1950. The station opened in 1909 and operated almost continuously until 1965, when whaling ended at the island. Berthed are the factory ships *Southern Garden* and probably, *Polar Chief* (source: photograph courtesy of Christian Salvesen plc).

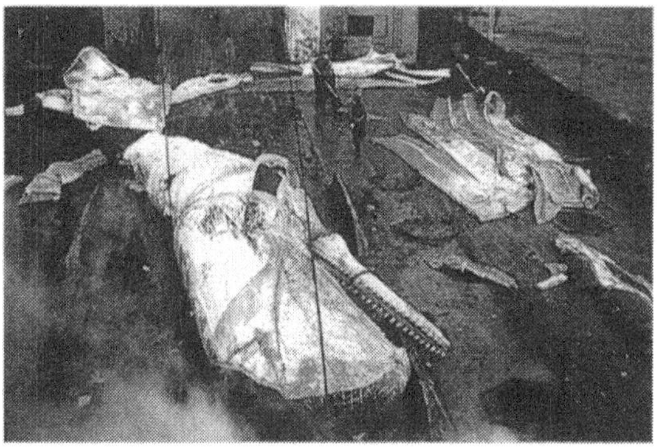

Fig 7.7 Flensing a sperm whale on the plan deck of the floating factory ship *Southern Venturer* in 1960. The deployment of such ships enabled the geographical expansion of the industry (source: photographed by W. Greenfield and reproduced courtesy of Christian Salvesen plc).

Norwegian fleets stayed in harbour and only one British pelagic fleet deployed. By this time (about 1927) the South Shetlands fishery was practically finished (the shore station on Deception Island closed after 1931), and substantially reduced stocks around South Georgia led to the closure of four shore stations. Greater efficiency could be gained by the pelagic operation; in season 1930/31 the five shore stations at South Georgia produced 188,000 barrels of oil, compared with 199,100 barrels from one floating factory (*Kosmos*) alone (Basberg, 1993). Only Grytviken operated continuously, although Leith Harbour and Husvik later reopened (Headland, 1984). A second boom in whaling occurred between 1934 and 1939 as the world market for whale oil rose after the Depression. Previously, the industry had been dominated by Norwegian and British operators, but other countries now entered the scene, notably Japan in 1933/34 and Germany in 1936/37. In 1938/39, 34 factory ships operating with 270 catchers took over 36,000 whales.

1939–present

Whaling declined sharply during the Second World War (see Figure 7.5) but resumed later with fleets deployed from Britain, Norway, Japan, the USSR, South Africa and the Netherlands. There was no evidence that stocks had recovered during the intervening years (Brown, 1963) but, encouraged by the relatively high demand and prices for the oil, catches in the post-war years were high (see Figure 7.5). At the peak of the fishery in 1961/62, 21 factory ships and three shore stations were operating. Catches of blue whales and then fin whales successively collapsed through overfishing and were replaced in turn by smaller sei and minke whales. The Norwegian dominance of the industry was ultimately replaced by Japan and the Soviet Union.

The last of the shore stations on South Georgia (see Figure 7.6) closed on 15 December 1965, the end of South Georgia whaling. The last of Britain's pelagic fleet was sold to Japan in 1963, followed by the Netherlands' fleet in 1964. Norway withdrew its last factory ship in 1968, leaving Japan and USSR as the only nations whaling in the Antarctic. By the 1985/86 season, the International Whaling Commission, a regulatory body set up in 1946, had introduced zero catch quotas for all commercial whaling to allow stocks to be reassessed. This moratorium remains in force (see below). Both Japan and the USSR initially continued commercial whaling 'under objection'. However, all commercial whaling in the Antarctic ceased at the end of the 1986/87 season, although some whaling for 'scientific' purposes has continued.

7.3.3 Whale products

The oil derived from most whale species is a fat formerly used in the production of soap, margarine and lard compound (Roberts, 1939; Tønnessen and Johnsen, 1982). Sperm whale oil is a wax and cannot be hydrogenated; its principal uses have been for industrial waxes and lubricants and in the cosmetics industry. Both oils could be used for lighting and lubrication purposes. At end of the nineteenth century, whale oil was replaced by mineral oil for lighting and lubrication, previously its most important uses. The basis for market expansion after 1905 hinged on world demand for animal fats and developments in industrial processing. At a time of rising consumption of soap and margarine, the relatively low price of whale oil compared with animal fat encouraged its wider use. Following improvements in the hydrogenation process in 1929, whale oil was extensively used in margarine. Glycerine, a by-product of soap manufacture, became a major product of whale oil during the First World War. Additional products included meat meal for animal feeds and bone meal for fertiliser. The best grades of oil were obtained from the blubber and bones. Initially only the blubber was used, and the remainder left to rot. However, regulations were later introduced to ensure more efficient use of the resource (see Section 7.3.5).

After the Second World War, there was a growing emphasis on by-products, notably the production of meat for human and animal (including pet food) consumption, at the expense of oil production. By the mid-1960s, the value of by-products exceeded 50% of the value of total production in the Norwegian industry (Basberg, 1993). The shift from oil to meat production coincided with the decline of the larger whale species and the rise of sei fishing; sei have less blubber and more valuable meat than the larger species (Gulland, 1976). In the 1980s, the high value of the meat was able to sustain the Japanese industry where others failed (Clark and Lamberson, 1982). Overall, during the main phase of pelagic whaling (1925–1965) the industry was one of the most productive and economically valuable of the oceanic fisheries (Clark and Lamberson, 1982).

7.3.4 Impact of the industry

The direct impact of the industry can be seen in the large and rapid declines in the whale stocks (see Figure 4.23). It has been estimated that the biomass of baleen and sperm whales has been reduced from about 45.7×10^6 to about 9.6×10^6 t, about 21% of its pre-exploitation level; and the numbers from about 1.05×10^6 to about 0.58×10^6, about 55% of their pre-exploitation levels (Berkman, 1992; Croxall, 1992) (see Table 4.2), although these global figures mask massive reductions in humpback whales (97%), blue whales (95%) and fin whales (21%). The target species changed through time as each became sucessively depleted (see Figures 4.23 and 7.5). Initially, shore-based whaling concentrated on the humpbacks, which were easily caught in the coastal waters near the stations and floating factories. As numbers fell, catchers extended their field of operations, seeking out the largest whales available, the blues, and then the next largest, the fins, during the boom years after the mid-1920s. As their numbers also declined, attention focused successively on the smaller sei and minke whales. Sperm whale catches have also formed a small but significant part of the Antarctic industry, particularly after the Second World War (see Figure 7.5). As the species harvested changed, so did the locations of the principal activity: blue whales were generally caught furthest south near the ice, with fin and sei successively further north. Although no species has become extinct, there is no evidence of significant recovery of the larger whales despite the introduction of various conservation measures (see Section 7.3.5). Only minke whales, exploited at the end of commercial whaling, are relatively abundant.

Various other effects have been attributed to the large decrease in whale numbers and the potential availability to other species of about 1.5×10^8 t a^{-1} of krill formerly consumed by the whales (Laws, 1977, 1985; Gambell, 1985; Croxall, 1992; see also Chapter 4). These include an increase in pregnancy rates of fin and sei whales, and apparent reductions in the age of sexual maturity of fin, sei and minke whales. Fur seals have increased on South Georgia (see Section 7.2.4), and populations of crabeater seals and chinstrap, Adélie and macaroni penguins have increased at annual rates of 2–10% since the mid-1940s (Croxall and Prince, 1979; Croxall *et al.*, 1981, 1984, 1988; Laws, 1985; Jouventin and Weimerskirch, 1990; Croxall, 1992; see also Chapter 4). These indirect effects have a clear spatial distribution, focused predominantly on the Scotia Sea since this is where the greatest overlap in the distributions of whales, seals, seabirds and krill occurs. Although not all the 'krill surplus' need necessarily have been taken up, and other factors such as climate change and variability in sea ice conditions may partly explain the increased populations (Taylor and Wilson, 1990; Taylor *et al.*, 1990; Croxall, 1992; Fraser *et al.*, 1992; Trathan *et al.*, 1996), it is evident that there has been a major perturbation in the balance of species in the Southern Ocean ecosystem and that the present balance differs significantly from the pre-exploitation situation. Under these circumstances, it is unclear to what extent the whale stocks can recover under current and future management (Gambell, 1985).

The whaling industry, through the location of shore stations and pollution of inshore waters, also had additional localised impacts at South Georgia, Deception Island and Signy Island (see Chapter 8). Such stations, however, are now of interest for their industrial archaeology (Basberg *et al.*, 1996) and form an attraction for tourism (see Chapter 8).

7.3.5 Regulation and management of whaling

Concern about overcatching in the late 1920s and early 1930s was motivated by economic factors, including overproduction of oil, and by fears that the Antarctic fishery would follow the northern fishery into uncontrolled harvesting and collapse of stocks. In response, a range of regulatory measures to conserve and manage the stocks has been introduced over the years (Ruud, 1956; Brown, 1963; Gulland, 1974; Tønnessen and Johnsen, 1982; Birnie, 1985; Gulland, 1990), the relative lack of success of which has depended on two factors. First, clear stock assessments have not been available early enough to provide incontrovertible evidence that stocks were seriously threatened and therefore to allow the implementation of effective measures (Gulland, 1976). Second, the burden of proof has rested with the regulators rather than the industry.

Pre-1940

During the period 1904–1925, whaling was largely conducted from British-claimed territory, which was administered by the Falkland Islands government for the British Colonial Office. A system of leases for shore-based stations and licences for mobile floating factories was established and a tax levied on the oil produced. Economically and politically, this provided a source of revenue and a means of regulating the industry and asserting British claims to the Dependencies. Whaling in this early period was very wasteful and only about 50% of the total oil content of the whale was utilised (Tønnessen and Johnsen, 1982).

> Companies literally wallowed in whales, selecting only the best blubber and allowing the carcass, with 60 per cent of its content, to drift away. A total of 2,300 whales were caught [in the 1907/08 season], producing only 61,000 barrels of oil, representing 26.5 barrels on an average from a large fat, whale – an all-time low in the utilisation of the raw material and a record in waste. Some 65,000/70,000 barrels of oil . . . were simply thrown away.
>
> (*ibid.*, p. 184).

To reduce the waste, leases granted to the last four companies in South Georgia (1909–1911) required whales to be processed in their entirety. Additional measures introduced to preserve stocks included restricting the number of station leases at South Georgia to nine; catching calves or females with calves banned in 1912; catching of humpbacks banned in 1918.

The 1929 Norwegian Whaling Act provided a framework for subsequent international regulations. Its principal stipulations included the protection of right whales and females with calves; the setting of minimum size limits for blue and fin whales; full utilisation of carcasses; appointment of inspectors to factory ships; and reporting of catch and production statistics to a new body, the Bureau of International Whaling Statistics. During the 1932–1934 seasons, recognising that it was to their advantage to control the production of oil not only to improve the price, but also to ensure the future of the industry, the main British and Norwegian companies reached voluntary agreements on season limits, catch quotas (estimated in terms of the blue whale unit (BWU)), oil production and minimum lengths for species caught. These measures proved far more effective than other regulations. However, as markets improved following the Depression, other countries, not bound by restrictions on catching, commenced whaling in the Antarctic. By the mid-1930s, with twelve countries participating in Antarctic pelagic whaling, the need for global restrictions was recognised, and in 1937 an International Agreement for the Regulation of Whaling was signed by nine whaling countries. This essentially incorporated the earlier measures, but no restrictions were placed on catch numbers. The measures were unsuccessful in achieving a sustainable harvest and, in the following season, the greatest number of whales ever was taken – over 46,000. By the end of the 1930s, various international conferences and agreements had produced a range of conservation measures, as much motivated by a need to maintain oil prices as to preserve stocks: protecting right and humpback whales (no longer a commercial catch anyway) and females with calves, setting size limits, restricting factory ships to areas south of 40° S, establishing a sanctuary in the Pacific area, setting a fixed season, appointment of government inspectors on factory ships, licensing of vessels and collection of catch statistics. Despite these measures, the industry continued to be driven by intense competition between companies in order to maximise profits because of the discontinuation of voluntary catch quotas. Overcapitalisation of the industry resulted, and wasteful methods continued (Ruud, 1956; Jackson, 1978, p.215). Even under regulation the numbers of whales killed in the 1930s far exceeded the catches in the 1920s. In fact the quotas set in 1931 were above world demand, and market conditions were more effective in regulating the industry than were conservation measures.

Post-1940 developments and the International Whaling Commission

In 1944, the Allied whaling nations agreed a catch limit of 16,000 BWU for the first two post-war seasons, well below the catches of 25,000 BWU in 1937/38. In 1946, fifteen nations signed the International Convention for the Regulation of Whaling 'to provide for the proper conservation of whale stocks, and thus make possible the orderly development of the whaling industry'. As part of the convention, an executive body, the International Whaling Commission (IWC), first met in 1949. Its terms of reference recognised that whales formed a valuable resource that could be harvested under proper management, and the commission was assigned responsibility for developing and applying regulatory measures; its performance has been critically reviewed in a number of sources (e.g. McHugh, 1974; Gambell, 1977; Birnie, 1985; Holt, 1985a and b; Andresen, 1989; Cherfas, 1989). However, the commission was given no powers to enforce regulation, to restrict the numbers of factory ships or shore stations, or to allocate quotas to them. Faced with a powerful, highly capitalised and competitive industry, conservation measures, which would have reduced profits, could not be agreed. Catches fell substantially below quotas in 1962/63. However, catch limits were not reduced downwards to take account of rapidly depleting stocks, and the gap between catches and sustainable yields increased. However, increasing competition occurred as fleets raced to obtain the largest share possible before the total quota was reached and the season closed. Consequently, there was significant wastage in the processing of carcasses since the imperative was to produce as much oil as possible: processing of other products was avoided.

The IWC attempted to provide a biological basis for management but in its early years lacked quantitative assessment of stocks, and doubts about stock levels and the precise impacts of catches were exploited by the operators to maintain high quota levels in the 1950s. In order to resolve this uncertainty, the IWC established an independent committee of scientists to assess the stocks, their sustainable yields and conservation measures, using mathematical models of population dynamics (Allen, 1980).

The recommendations of this committee included total protection for blue and humpback whales, a large reduction in the catch of fin whales and regulation based on quotas for individual species (Chapman *et al.*, 1964). At first the industry was not prepared to accept such constraints, but at a special meeting of the IWC in 1965 it was agreed to reduce quotas progressively to below the sustainable yields; this involved a sharp drop to 4500 BWU in 1965/66. In 1967/68 the total quota was reduced to 3200 BWU. Eventually, in 1972/73, the BWU system was abandoned and quotas set for each species, allowing stocks to be managed according to their individual requirements.

Overall, post-war whaling suffered from a lack of strategic planning, with emphasis on short-term economic factors rather than longer-term sustainable development. Regulatory measures were inadequate and based on quotas that were much too high. Also, overinvestment in the industry led to increased expedition numbers and a declining relative share of a fixed total catch. Individual companies maintained their share by increasing the number and efficiency of their catchers and the capacity of their factories, and resisting quota reductions. During the boom years, the length of season required to catch the quota declined from 122 days to obtain 318,082 tonnes of oil in 1946/47 to only 58 days to obtain 334,641 tonnes in 1955/56 (Jackson, 1978). It is also apparent that monitoring and enforcement of conservation measures were inadequate. For example, it has emerged that, over a period of 30 years from the late 1940s, the former Soviet Union not only deliberately under-reported its catches (by as much as 48% worldwide), but also that its fleets harvested protected species and ignored other conservation measures (Zemsky *et al.*, 1995).

The IWC was established to regulate the optimum utilisation of whales and to provide the necessary management underpinning for conservation of the resource and thereby ensure the 'orderly development' of the industry. In reality, it was driven by members with a strong interest in commercial exploitation, and most decisions were based on short-term economic considerations rather than the requirements for longer-term sustainable harvesting (McHugh, 1974). In particular, the IWC failed to take appropriate measures to ensure the latter, for example to set individual quotas and revise these downwards in the light of declining abundance, and to set quotas for individual species and stocks (Gulland, 1974, 1990). Moreover, the BWU formed an inappropriate unit of quota, since it allowed the most profitable whales to be taken regardless of scarcity. The IWC has been further constrained by a provision that allows members to object to any majority decision within 90 days; a state doing so is then not bound by the decision concerned. The measures adopted by the IWC were therefore inadequate to prevent the depletion of successive whale stocks. By the time the blue whale was protected in 1967, there were only 1000 left. Similarly, when fin whales were protected in 1976, and sei in 1978, the Russian, Japanese and Norwegian fleets had virtually achieved their commercial destruction.

Developments since 1970

In the early 1970s, reflecting not only the poor record of the IWC but also a moral argument against whaling, the rapidly growing conservation movement began to express wider concern about the activities of the whaling industry. At the United Nations Conference on the Human Environment held in Stockholm in 1972, a resolution was passed calling for a ten-year moratorium on commercial whaling. In 1974, the IWC revised its management policy to recognise stocks that required complete protection until they could recover and those that could be exploited at different levels in different management units (Gambell, 1977). However, it continued to be hampered by a lack of accurate quantitative data on stocks, and some stocks continued to be depleted. A view also emerged that single species management was inappropriate and that management should be based on ecosystem concepts. Paradoxically, the public campaign to halt whaling made the industry less willing to make short-term concessions and encouraged maximisation of catches (Gulland, 1976).

The period 1972–1984 saw a fundamental change in the composition of the IWC and the evolution of its role from a management one to a protectionist one. Between the late 1970s and 1983 its membership grew from 15 to 41 nations as non-whaling nations with a growing interest in common heritage resources joined the IWC. Environmental NGOs provided encouragement to developing countries to participate and to adopt the environmentalist policy to halt whaling (Andresen, 1989). The organisation thus changed in character from what has been described as a 'whalers club' to one with a much stronger conservation imperative.

In 1982, the nations opposed to whaling achieved the necessary three-quarters voting majority for the IWC to introduce zero catch levels in the 1985/86 season. These were to last for a period of up to ten years to allow the development of a 'revised management procedure' (RMP) to provide a sound basis for setting sustainable catch limits. Provision was made for revision of these limits after assessment of the effects on whole stocks by 1990. However, what is in effect a moratorium remains in place and although the RMP was accepted by the IWC in 1994, other aspects (e.g. enforcement measures) have yet to be agreed. Implementation of the moratorium has undoubtedly been strengthened by support from the USA, which has fisheries statutes providing for economic sanctions against countries weakening the effectiveness of international conservation legislation (*ibid.*; Cherfas, 1989). The last two nations whaling in the Antarctic, Japan and the former Soviet Union, had ceased commercial operations by the end of the 1986/87 season.

More recently, attention has focused on the question of catching for 'scientific purposes' under the IWC regulations (Harwood, 1990). Particularly contentious was a proposal by Japan for a twelve-year research programme involving an annual catch of 825 minke and 50 sperm whales (Nagasaki, 1990). Reservations about the scientific value of the proposal were widespread (de la Mare, 1990) since the development of non-lethal sampling and analytical techniques (Hoelzel and Amos, 1988) and computer modelling of population dynamics (de la Mare, 1990) offer alternative approaches. Environmentalists argued that it simply represented a guise for continued commercial whaling. Nevertheless, Japan was catching around 300 minke whales a year up to 1995 for scientific purposes, and the figure rose to 625 in 1996.

In addition to the moratorium, a proposal by France in 1992 for an Antarctic Whale Sanctuary in the Southern Ocean, in which all whaling would be banned, was overwhelmingly approved at the 1994 meeting of the IWC. The sanctuary will be reviewed every ten years, and although it does not explicitly exclude scientific whaling, the rationale for the latter is now difficult to sustain and it seems likely that pressure will continue for Japan to discontinue this practice. However, Japan and pro-whaling groups are challenging the legality of the sanctuary on the grounds that it is contary to the provisions of the 1946 Convention (Burke, 1997).

Since 1972, environmental groups have argued that scientific knowledge is not adequate to determine sustainable yields or set catch quotas. The environmentalists have also successfuly employed a moral argument that whales are a special species and should not be harvested on a commercial basis (e.g. Barstow, 1990). Indeed, in the high-profile campaign in the 1970s the whale became a global symbol for environmental protection. The crux of the debate is whether whaling is morally wrong or whether whales can be regarded as a legitimate resource that can be managed and harvested in a sustainable fashion (Frost, 1979; Aron, 1988). The unfortunate history of the industry and its mismanagement gives doubtful confidence in the latter approach, although the scientific basis is now greatly improved through the RMP. Ironically, just as the science is better able to underpin management, the ethical and moral arguments against whaling have proved more compelling (Butterworth, 1992).

7.3.6 Summary

The biogeography of krill in the Southern Ocean and its influence on the distribution of whales largely defined the natural boundary conditions for the development of whaling. The fishery followed a clear progression from slow whales close to easily accessed island factories to larger, more distant whale species. Technological developments played an important role, enabling expansion of whaling into the pelagic realm and the processing of whales at sea. Such was the efficiency of the industry that successive species were hunted to the edge of commercial and biological viability.

In terms of economic models, the depletion of the resource by the whaling companies can be rationalised as achieving profit maximisation (Clark and Lamberson, 1982), and there was little economic incentive to conserve whale stocks. Ultimately, the lesson is 'that no amount of good faith and international agreement can prevent the exhaustion of wasting resources in international waters so long as they are exploited under a system of free competition and when profit is the sole determinant of the level of activity' (Jackson, 1978, pp.257–258).

Management of the industry singularly failed to ensure sustainable use of whales, reflecting the supreme difficulties in effectively regulating an open-access resource. During the first 20 years or so after its establishment, the IWC oversaw the overfishing and depletion of nearly all the world's whale populations. It was unable to safeguard the resource and to secure the orderly development of the industry. Instead, the industry progressed through boom and collapse, driven by economics and the goal of profit maximisation. In the face of scientific uncertainties about the impact of the industry on the state of stocks, the burden of proof lay with the regulators, and the IWC had no power to implement precautionary restraints to ensure long-term sustainable harvesting and to prevent the decline of the resource. The whale has now become a symbol of humanity's inability to manage a common natural resource in a responsible and sustainable manner.

From a conservation viewpoint, the battle to save and protect the remaining stocks of the major species had largely been won by the mid-1970s, although in the final analysis it was economics and not conservation management that 'saved' the whales. However, it will be many years before the depleted stocks recover: the minke is the only whale now potentially harvestable. The issue then is not so much about saving whales from extinction as about the sustainable management of migratory species, and the ecosystem of which they are part, to allow recovery of stocks (cf. Holt, 1993). If stocks become sufficient to allow sustainable exploitation under the RMP, as in the case of the minke whale at present, and providing that any wider impacts on the marine ecosystem are considered acceptable, then continuation of the moratorium will be as much a matter of ethical arguments and cultural attribution of special status to whales, as considerations sustainability since it is questionable whether commercial whaling meets any compelling human needs. It is clear that in the unlikely event of commercial whaling ever being resumed in the Antarctic, the sustainable basis will be subject to intense scrutiny and operations hampered by the anti-whaling activities and campaigning of environmental NGOs.

7.4 Finfish

7.4.1 Introduction

Finfish play an important role in the Antarctic marine food web, both as predators of krill and as prey for higher-level predators (e.g. Kock, 1992; Hureau, 1994; see also Chapter 4). During the late 1960s and early 1970s, the unrestricted international waters of the Southern Ocean, with their seemingly high levels of primary and secondary production (see Chapter 4), became the focus of attention of distant-water fishing fleets, particularly from the former USSR and the Eastern Bloc countries. Initially, the potential krill resource in the Scotia Sea was the principal attraction for the fishing fleets. However, the problems of processing and marketing krill products hindered the expansion of that industry (see Section 7.5). At the same time, it was apparent that there was a substantial finfish resource in the Southern Ocean and finfish were then pursued more intensively than krill, particularly since harvesting, processing and marketing could proceed through the use of existing methods (Everson, 1978).

7.4.2 Development of the industry

Development of the Antarctic fishing industry has been outlined in a number of sources (Kaczynski, 1984; Kock, 1985, 1991, 1992, 1994; Kock *et al.*, 1985; Hureau and Slosarczyk, 1990; Kock and Köster, 1990). Commercial exploitation dates from the start of the twentieth century, when whalers encountered large stocks of marbled notothenia (*Notothenia rossii*) in the waters near South Georgia. Although barrels of salted fish were exported to Buenos Aires, the distance from markets and the greater returns from whaling effectively precluded any significant development of the fishery, and attempts by Argentinian, Norwegian and Japanese companies in the 1930s, 1950s and 1960s also failed (Dickinson, 1985). The geography of fisheries development is closely linked to both the oceanography and biogeography of the Southern Ocean. Many of the most abundant fish species graze the plentiful krill found along the edge of the sea ice and in the gyres of the Antarctic Circumpolar Current (see Chapter 4), particularly where it is joined by the Antarctic Coastal Current in the Scotia Sea. Others graze the abundant benthic fauna found closer inshore on the shelf areas adjacent to the Antarctic Peninsula, the islands of the Scotia Sea and Îles Kerguelen. On both counts, biogeographical factors indicate that the Scotia Sea and Îles Kerguelen offer the greatest fisheries potential. Exploratory fishing by the Soviet Union in 1961 in the Scotia Sea developed into large-scale fishing in 1969–1971 around South Georgia and Îles Kerguelen. These areas have remained the most important fishing grounds, although fisheries have also subsequently developed closer to the Antarctic continent, particularly around the South Orkney Islands, the South Shetland Islands, the northern part of the Antarctic Peninsula and the Ob and Lena Banks (Kock *et al.*, 1985; Kock, 1991, 1992) (Figure 7.8). Until 1990, commercial fishing was undertaken almost exclusively by Eastern Bloc countries. Up to 1992/93, 2.08 million tonnes had been caught in the Atlantic sector and 1.74 million tonnes in the Indian Ocean sector, with 83% and 94% of these totals, respectively, being caught around South Georgia and Îles Kerguelen (Kock, 1994). Of the total catch in these sectors, about 90% was taken by the USSR, principally of the Nototheniidae family of demersal (bottom-dwelling) fish (Everson, 1978; Kock, 1985) (Figure 7.9); there are no significant stocks of pelagic fish. Until 1981/82, harvesting was undertaken mainly by bottom trawling up to depths of 500 m. Subsequently, mid-water trawls towed close to the bottom have been used for mackerel icefish, *Champsocephalus gunnari*, and yellowfin notothenia, *Patagonotothen guntheri* (Kock, 1992).

During the initial stages of the Atlantic fishery, large harvests of *N. rossii* (1969/70 and 1970/71) and *C. gunnari* (1976/77) were taken from South Georgia waters. The focus then moved to *C. gunnari* around the South Orkney Islands and the South Shetland Islands (1977/78 and 1978/79), *N. rossii* around the South Shetland Islands and Antarctic Peninsula area (1979/80) and *P. guntheri* around Shag Rocks at South Georgia (from 1978/79). More recently, catches have extended to the Patagonian toothfish, *Dissostichus eleginoides*, and the lantern fish, *Electrona carlsbergi*, from the continental slope of South Georgia and the open ocean, respectively (Figure 7.10). In the Indian Ocean sector, initial attention also focused on *N. rossii*, grey notothenia (*Lepidonotothen squamifrons*) and *C. gunnari* in the Kerguelen area, and then on *L. squamifrons* on the Ob and Lena Banks (1977/78). In the same sector, exploratory fishing has been conducted along the coast of the Antarctic continent, but the Pacific sector appears to have little potential for a commercial fishery (*ibid.*).

Peak catches (Figure 7.11) reflect several factors, including increased fishing effort and abundant year class recruitment to exploited stocks, but principally the exploitation of new fishing grounds and stocks (Hureau and Slosarczyk, 1990). These peak year catches have proved to be unsustainable and reflect the exploitation of accumulated stocks (Kock *et al.*, 1985). Since 1992, fishing activity has declined in the South Atlantic area due to restrictions under CCAMLR on

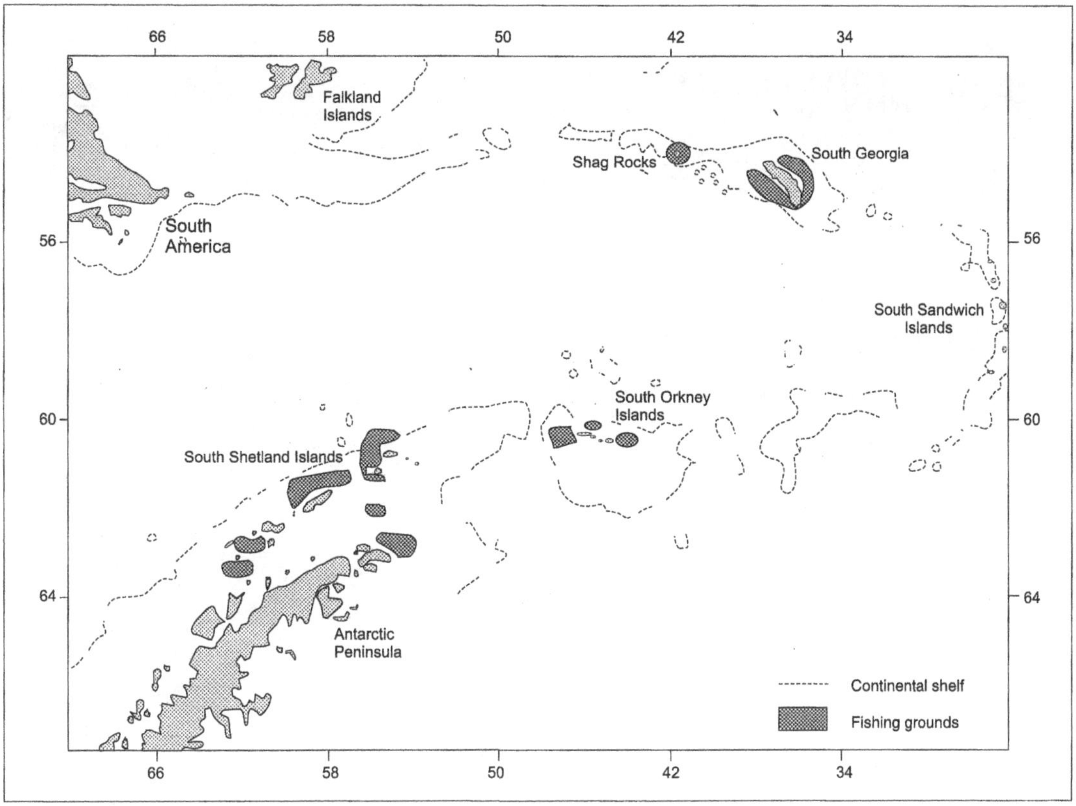

Fig 7.8 Shelf areas and fishing grounds in the Atlantic sector of the Southern Ocean (source: redrawn after Kock, 1992).

target species and a reduction in the number of vessels from the former Soviet Union. In 1995/96 the only significant catches were of *D. eleginoides* from the South Georgia and Kerguelen areas, totalling 8700 tonnes, the majority of the catch being taken by Chile and France. However, unreported catches in the Îles Crozet and Prince Edward Islands areas may have been as large or larger than the total reported catch. There has also been significant unregulated and illegal fishing for *D. eleginoides* around South Georgia, Îles Kerguelen, and Heard and McDonald Islands (Lugten, 1997); the former by vessels from states not parties to CCAMLR, the latter by vessels from CCAMLR states that have reflagged in non-signatory states. Although several successful prosecutions have occurred, in the case of illegal fishing this has not acted as a deterrent, and the British government reported to the commission in 1995 that around South Georgia 'catches from illegal fishing now exceed those taken legitimately' (cited in Pearce, 1996).

7.4.3 Impacts of the industry

In total, twelve species have been commercially fished, and twelve of the thirteen stocks that can be assessed are now depleted (Kock and Köster, 1990; Kock, 1991, 1992). Some species have been seriously overfished; for example the population of *N. rossii* may have been reduced to less than 10% of pre-harvesting levels (Basson and Beddington, 1991). In the first two years (1969/70 and 1970/71), about 500,000 tonnes of *N. rossii* were caught around South Georgia, then catches declined and stocks have not recovered. For the Atlantic sector, Kock and Köster (1990) estimated that the present stock size of *N. rossii* is now less than 5% of its pre-exploitation level; recruitment has also decreased, but this may in part reflect predation by increased numbers of fur seals on South Georgia (Kock and Köster, 1990). The stocks of *C. gunnari* in the South Orkney Islands–Antarctic Peninsula area

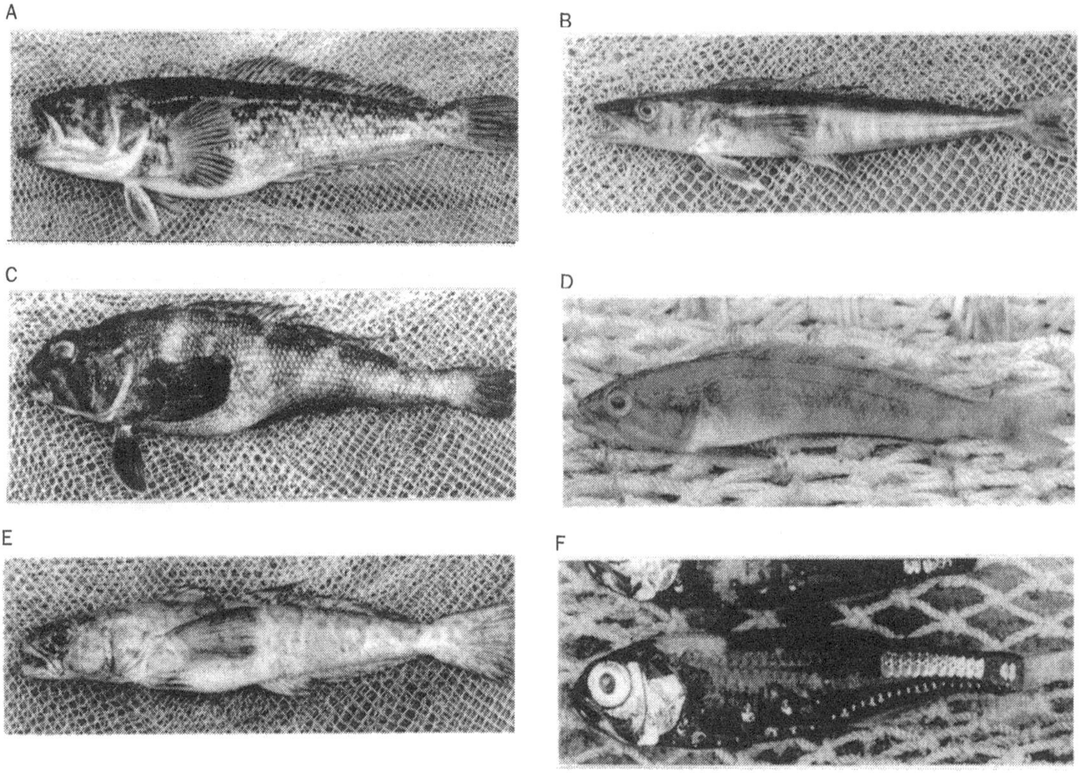

Fig 7.9 Commercially exploited fish species of the Southern Ocean. (A) *Notothenia rossii*; (B) *Champsocephalus gunnari*; (C) *Lepidonotothen squamifrons*; (D) *Patagonotothen guntheri*; (E) *Dissostichus eleginoides*; (F) *Electrona carlsbergi* (source: photographs courtesy of K.-H. Kock).

and *L. squamifrons* in the Îles Kerguelen area are also considered to be heavily depleted (Duhamel and Hureau, 1990; Kock and Köster, 1990; Kock, 1991, 1992), and in some years total allowable catches have not been achieved (Kock, 1991). Slow growth rates, late achievement of sexual maturity and low reproduction rates make stocks unsuitable for sustainable high-yield harvesting (see Chapter 4). Recovery is slow after overexploitation.

Adverse, indirect impacts of fishing have also arisen (Croxall *et al.*, 1984; Jouventin *et al.*, 1984; Croxall, 1987; Kock, 1992). First, the industry may act as a competitor with seabirds and seals, the principal fish predators. South Georgia seabirds and seals consume about 1.6 million tonnes of fish annually, but this is the least important component of their diet (Croxall *et al.*, 1985; see also Chapter 4). Moreover, assessing potential competition is fraught with uncertainties (Croxall, 1987), and there is no direct evidence to relate changes in predator abundance or distribution to reduced fish stocks (Kock, 1992). Thus although commercial fishing has been implicated in the decline in elephant seal numbers on islands in the Indian Ocean sector (Pascal, 1985), other environmental factors may be more important (Hindell and Burton, 1987; McCann and Rothery, 1988; Guinet *et al.*, 1992).

Second, seals may suffer enhanced rates of mortality arising from entanglement in discarded or lost fishing nets, lines and debris, and from ingestion of plastics (see Chapter 8). For example, extrapolation from studies at Bird Island to the total population of fur seals on South Georgia suggests that the total number of seals entangled per year is about 5000 (possible maximum figure of 15,000), with a mortality rate of about 38% (Croxall *et al.*, 1990b; Arnould and Croxall, 1995). Only 12% of entangled seals have been females, so that the impact on the South Georgia population of 2.05 million is small, but nevertheless unnecessary.

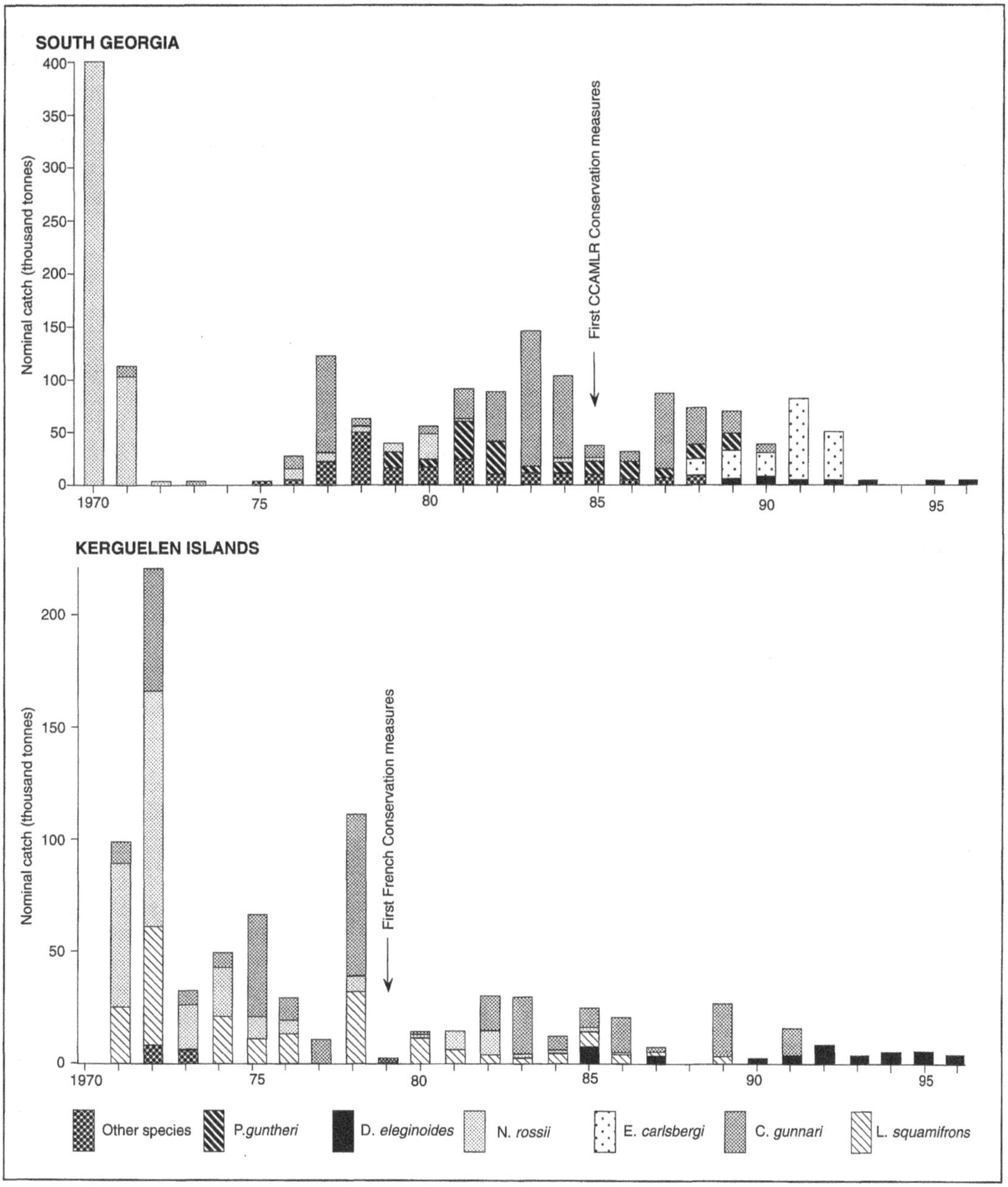

Fig 7.10 Annual catches of finfish by species in the area of South Georgia and Îles Kerguelen, 1970–1996. The fishing years are split across two calendar years, and the years on the figure refer to the second year in each case (e.g. 1993 refers to the 1992/93 fishing year) (source: after Kock, 1994; CCAMLR Statistical Bulletins).

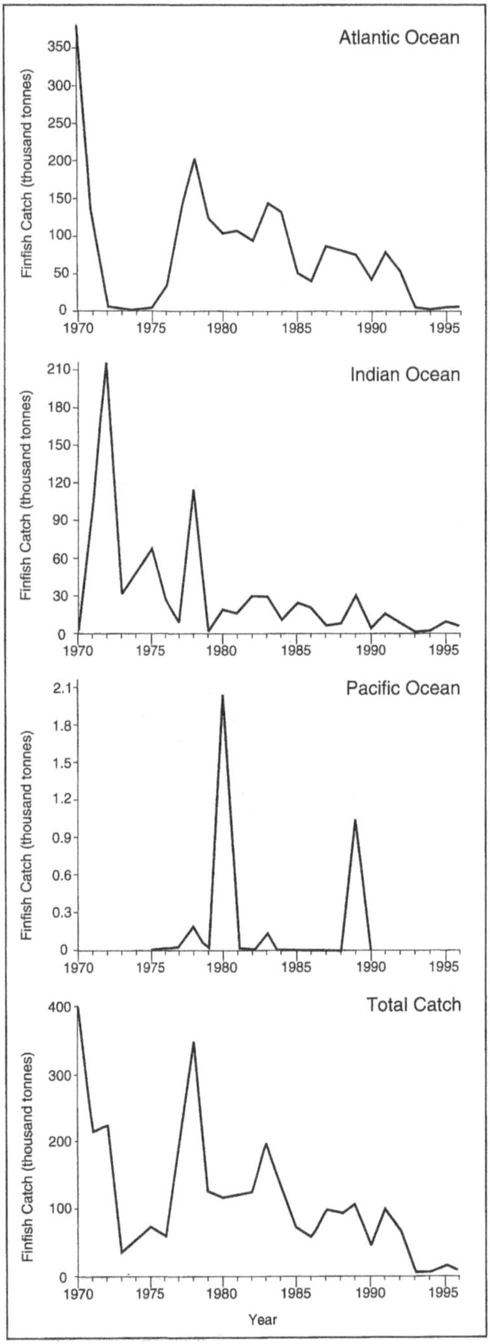

Fig 7.11 Annual catches of finfish in the Atlantic, Indian and Pacific Ocean areas and total annual catches for the Southern Ocean, 1970–1996. The fishing years are split across two calendar years, and the years on the figure refer to the second year in each case (e.g. 1993 refers to the 1992/93 fishing year) (source: data updated from Miller, 1991; Nicol, 1991; CCAMLR Statistical Bulletins).

A third area of concern is the high incidental mortality of seabirds, principally albatrosses, associated with the rapid growth of longline fishing for tuna in the Southern Hemisphere since the 1960s and the development of longline fisheries for *D. eleginoides* in 1988/89 on the edge of the shelf and in the deep sea around South Georgia and Îles Kerguelen (Bergin, 1997; Weimerskirch *et al.*, 1997). Incidental mortality arising from entanglement in baited longline hooks has been imputed as a major cause of the decrease in the wandering albatross populations at Îles Crozet (Weimerskirch and Jouventin, 1987; Jouventin and Weimerskirch, 1991) and Macquarie Island (de la Mare and Kerry, 1994), and the black-browed albatross population at Îles Kerguelen (Weimerskirch *et al.*, 1987; Jouventin and Weimerskirch, 1991). At Bird Island, South Georgia, annual rates of incidental mortality in the wandering albatross population attributed to the longline tuna-fishing industry at higher latitudes may exceed 2–3% of adults and 15% of juveniles (Croxall *et al.*, 1990a). High juvenile mortality in black-browed and particularly grey-headed albatrosses at South Georgia has also been attributed to incidental mortality associated with fishing activity (Prince *et al.*, 1994). Overall, Brothers (1991) made a conservative estimate that 44,000 albatrosses of different species were killed in the Southern Hemisphere oceans each year. However, a major source of the impacts, the tuna longline fisheries, lie outside the area of jurisdiction of CCAMLR. On the positive side, discarded material from fishing vessels may provide additional food for albatrosses and petrels (Croxall *et al.*, 1988; Jouventin and Weimerskirch, 1991).

Finally, bottom trawling, used for catching most of the finfish, may cause significant disturbance to benthic communities, although the effects are largely unknown (Kock, 1992, 1994). Bottom trawling is now prohibited as a precautionary measure to protect benthic communities and by-catch species.

7.4.4 Regulation of the industry

The fishery as a whole has been regulated under CCAMLR since 1982 (see Chapter 9), and since 1991 regulatory measures have been in place for all commercial fisheries. Several states have enacted additional measures for the waters around their sovereign territories, thus providing the option to implement more stringent controls if required. The UK established a 200-mile maritime zone around South

Fig 7.12 Squid-jigging fishing vessel in the South Atlantic (source: photograph by P. Rodhouse, courtesy of BAS).

Georgia and the South Sandwich Islands in 1993. Australia has a 200-nautical-mile exclusive economic zone around its territories. In the Îles Kerguelen area, France declared a 200-mile exclusive economic zone in 1978 and the fishery is regulated by licensing; France also has controls around Îles Crozet (Duhamel and Hureau, 1990; Hureau and Slosarczyk, 1990; Kock, 1992).

Measures adopted under CCAMLR have included prohibition of the fishing of certain species, specification of minimum net mesh sizes to protect immature fish, the setting of total allowable catches (TACs) for stocks in particular areas, closure of fisheries in specific areas, and the establishment of closed seasons. Thus directed fishing of *N. rossii* has been prohibited since 1986 and TACs have been set for other species. All finfishing was prohibited in the South Orkney Islands and Antarctic Peninsula area in 1990/91, and the *P. guntheri* fishery was closed at South Georgia in 1990/91. Further, a 12-nautical-mile prohibition zone has been established around South Georgia since 1985 to protect the spawning grounds of *C. gunnari* and other species. For the 1996/97 season, directed fishing was prohibited for all species except *C. gunnari*, *D. eleginoides* and *E. carlsbergi*.

In 1989/90, a system of inspection of fishing vessels was agreed to ensure compliance with conservation measures. However, the costs of operating such a system are substantial and enforcement is a matter for flag states. Although legal action has been taken in a number of cases by national authorities, and also by the South Georgia government in respect of illegal fishing within its waters, violations have continued (Pearce, 1996; Lugten, 1997).

Further regulations approved in 1994 (including modifications to fishing techniques, e.g. use of streamers, weights, setting longlines only at night and with minimum use of lights) should help to reduce incidental seabird mortality from longline fishing operations in the area south of the Polar Frontal Zone, although this is dependent on effective implementation (Dalziell and de Poorter, 1993), as well as on the success of measures adopted to tackle the problem outside the CCAMLR area (Bergin, 1997). Measures to control disposal of waste at sea, under MARPOL and CCAMLR, appear to have contributed to a reduction in the entanglement of fur seals in packaging bands at South Georgia (Arnould and Croxall, 1995).

Significant overfishing and depletion of many stocks had already occured before CCAMLR came into force, and initially management action was taken retrospectively. CCAMLR in its early years also faced a number of difficulties (see Chapter 9), but over time the measures have been strengthened and various precautionary measures adopted, including the requirement for prior notification of new fisheries, such as for squid around South Georgia (Figure 7.12). Nevertheless, the burden of proof lies with the regulators and there remain deficiencies in the information needed for stock assessment and management (Miller, 1991; Kock, 1992). As a result, in the early years of CCAMLR, it was difficult to reach consensus on more stringent measures required for the recovery of the depleted stocks and their sustainable management (Miller, 1991; Kock 1992). It is easy to advocate adoption of the 'precautionary principle' under such circumstances, although implementation was opposed by the vested interests of the fishing state members of CCAMLR. Improvements in data and developments in stock assessment modelling incorporating risk assessment and the presentation of options have, however, strengthened the scientific advice and the case for precautionary measures. In contrast, the recent illegal catching of *D. eleginoides* not only highlights the problems of enforcement of CCAMLR measures and ensuring that states whose vessels are involved take appropriate action, but also threatens the effectiveness of the conservation measures in force. The issue of reflagging is addressed in the FAO Agreement to Promote the Compliance with International Conservation and Management Measures by Fishing Vessels on the High Seas (Compliance Agreement), but so far few states have ratified it. This is part of the wider problem, not unique to the Antarctic, that although international fishing laws and regional agreements, such as CCAMLR, exist, they are frequently either not enforced by states or ignored by fishermen.

7.4.5 Summary

The development of the finfish industry has followed a similar pattern to the exploitation of other marine resources, involving discovery, exploitation and depletion of each new stock. The spatial impacts of fishing have been felt most in the Scotia Sea and around Îles Kerguelen, where target species were abundant. Commercial fishing expanded rapidly in advance of assessment of impacts on stocks and the development of a substantive basis for sustainable management and conservation measures. Consequently, stocks were rapidly overexploited and the future of the industry now depends on the success or otherwise of the conservation measures implemented by CCAMLR and the rate of recovery of stocks. Even with recovery of stocks, Kock (1992) considered that annual sustainable yields are unlikely, ultimately, to exceed 100,000 tonnes. Also, further uncertainty about the progress of the industry is related to future economic conditions in the former Soviet bloc countries, formerly the principle fishing nations, and the viability of their distant-water fleets.

7.5 Krill

7.5.1 Introduction

Krill (see Figure 4.14) is a central species in the food web of the Southern Ocean ecosystem (Knox, 1994) (see Chapter 4). Realisation of its potential as a resource, rich in protein, dates from the 1960s, although the existence of large stocks had been known for some time. The commercial development of the krill fishery coincided with the decline of Antarctic whaling and the establishment of 200-mile wide exclusive economic zones in other parts of the world. As distant-water fishing fleets were forced to seek new areas, attention focused on the krill stocks of the Southern Ocean. Considerable claims were made for the potential of the resource, arising from its food value and the supposed 'surplus' created by the fall in consumption by the depleted whale stocks. Estimates based on the reduction in whale numbers suggested an annual 'surplus' in the region of 150×10^6 t (Gulland, 1970; Mackintosh, 1970; Knox, 1984; Laws, 1985). Although such estimates relied on rather sweeping assumptions about biological factors and the food web, the potentially significant size of the resource was readily apparent. Early projections were made that the potential annual harvest could be as much as double the existing world fish production (about 70×10^6 t in the late 1970s) (Moiseev, 1970; Golosov, 1984).

The composition of krill (25% lipid, 49% protein and 2.5% chitin by dry matter) gives it a high nutritional value as a food source rich in protein. The content of vitamins A and B is high, as is that of essential amino acids such as arginine, lysine, leucine and phenylalanine; minerals such as potassium, iron and manganese and unsaturated fatty acids are also present (Suzuki and Shibata, 1990). The nutritive value of krill is less than that of whole egg protein but slightly higher than milk protein (casein) (*ibid.*). Krill has therefore been viewed as a potential source of food for the developing world (Mitchell and Sandbrook, 1980).

7.5.2 Development of the industry

The growth of the krill fishery has been charted in various publications (Everson, 1977; McElroy, 1984;

Sahrhage, 1989). The USSR commenced exploratory fishing for krill in 1961/62 and was followed by Japan in 1972/73. Subsequently, other countries including Bulgaria, West Germany, Poland, Chile, Korea and Taiwan have been involved. Large-scale commercial fishing took off in 1976/77 (at a time when finfish stocks were becoming overfished). Catches, principally by the USSR and Japan, rose to a peak of 528,000 tonnes in 1981/82, representing about 16% of the world's marine crustacean harvest by weight (Nicol, 1991), but subsequently declined (see Figure 7.13) due to problems in processing and marketing the products. However, catches again increased and stood at about 395,000 tonnes in 1988/89, with the Soviet Union taking 76% and Japan 20%. At the peak in 1982, the Soviet Union took 93% of the catch. Due to reductions in the Russian and Ukrainian fishery, catches fell to 83,961 tonnes in 1993/94, with Japan taking about 74%. Catches rose again to 118,712 tonnes in 1994/95 due to increased activity by the Ukraine, but fell back to 95,039 tonnes in 1995/96. The krill fishery is currently the principal economic activity in the Southern Ocean and is the largest single-species crustacean fishery in the world.

The spatial distribution of the krill fishery is dictated principally by the occurrence of krill 'hot spots', notably in the zone of mixing between the Antarctic Circumpolar Current and the Weddell Sea Gyre in the Scotia Sea and adjacent to the Ross Sea (Balleny Islands and Bellingshausen Sea) (see Figure 4.15). Other areas where krill are abundant in the Antarctic Coastal Current occur in the Indian Ocean (off Enderby Land and Prydz Bay and on the Kerguelen–Gaussberg Ridge) and in the upwelling zone of the Antarctic Divergence (see Chapter 4). Of these areas, the Scotia Sea and the South Atlantic, in particular, are also favoured by suitable navigation conditions, greater reliability of catches arising from bigger and more extensive concentrations of krill, and a longer operational season (Makarov *et al.*, 1994); this area is now the main focus of the industry (Figure 7.13). For these reasons, the Scotia Sea is central to the spatial distribution of the krill fishery, which is located mainly on the continental shelf or close to the shelf break. However, the spatial pattern also varies seasonally with the northern limit of the sea ice. In the South Atlantic, winter fishing is located around South Georgia, but in summer moves south to the South Orkney and South Shetland islands (Everson and Goss, 1991). The krill fishery is undertaken by large factory trawlers equipped to process the catch at sea.

Various estimates exist for krill biomass and annual production (Everson, 1977; El-Sayed, 1985; Ross and Quetin, 1988; Miller and Hampton, 1989). Standing stocks are probably several hundred million tonnes, and annual production is of a similar magnitude (see Chapter 4). The current catch therefore represents less than 0.5% of the estimated krill biomass. Despite the optimistic expectations, the full potential of the krill fishery has not been realised for a combination of technical, environmental and economic factors. Problems in processing krill arise from their small size, their weakness to mechanical stresses, their rapid decomposition after catching and difficulties in separating the meat from the shell (Sahrhage, 1989; Nicol, 1994). Enzymes in krill rapidly break down the body after death, so that krill for human consumption must be processed within 3 hours of catching, and within 10 hours for animal feed. At present, only half the catch is sold for human consumption, the rest as fish meal (Nicol, 1989). Four main types of product have been marketed: whole krill, peeled tail meat, mince and meal (Grantham, 1977; McElroy, 1980; Nicol, 1989; Suzuki and Shibata, 1990). Krill have been frozen and sold whole in Japan for human consumption, and in the USSR whole frozen krill and krill meal have been used as animal feed. Krill paste for human consumption was not a success in the USSR due to poor quality and limited storage life. The use of minced krill as a protein source in processed foods has been more successful, as has the sale of canned tail meat in the USSR and Chile. High concentrations of fluoride found in krill are concentrated in the exoskeleton and are not a problem if the shell and meat are separated quickly after catching (Nicol, 1989). However, krill for animal foodstuffs may contain four times permitted EC levels of fluoride and must be mixed with other foods for poultry and pigs (*ibid.*); krill meal is not normally fed to cattle. Krill is more suitable for farmed fish, since the fluoride accumulates in the bones rather than the flesh, or for animals that are farmed for their fur.

Additional logistical and operating problems and costs arise from remoteness from home ports, support services, other fishing grounds and markets (Everson 1978; McElroy, 1984; Kaczynski, 1984) (Table 7.1). Although safe anchorages exist near to the fishing grounds (e.g. South Georgia, Deception Island), there are no established harbours with facilities for undertaking major repairs. Fishing fleets therefore comprise a mix of trawlers, mother ships,

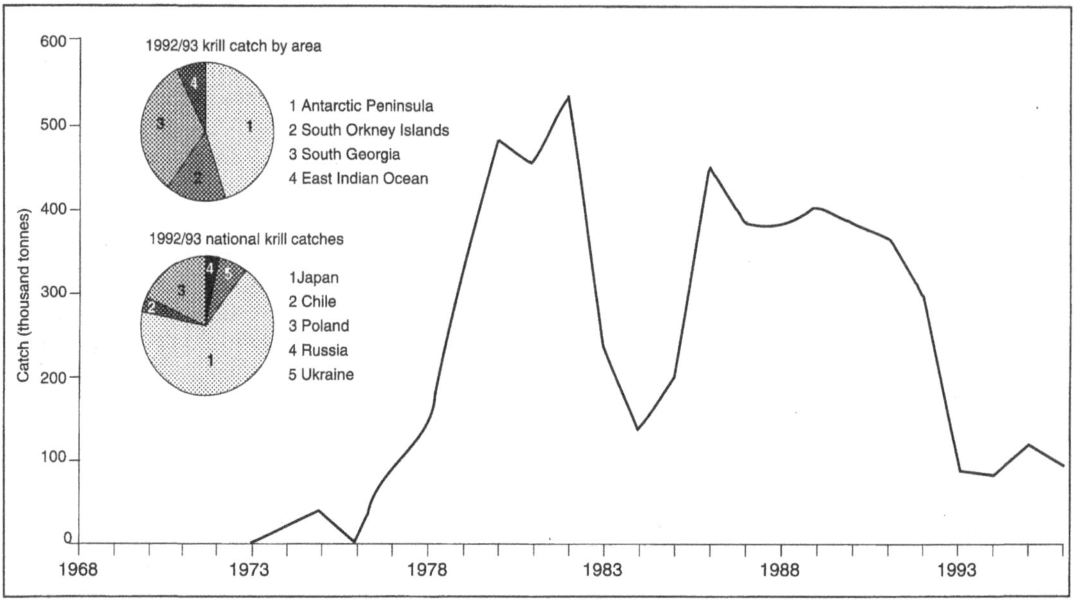

Fig 7.13 Annual catches of Krill in the Southern Ocean, 1968–1996. Insets show (A) the percentage of the catch from CCAMLR statistical areas in 1992/93; and (B) the percentage of the catch by country in 1992/93. The fishing years are split across two calendar years, and the years on the figure refer to the second year in each case (e.g. 1993 refers to the 1992/93 fishing year) (source: after Nicol, 1994; CCAMLR *Statistical Bulletins*).

Table 7.1 Approximate distances (in nautical miles) between various Southern Ocean fishing grounds and nearest ports

	Punta Arenas/Ushvaia	River Plate	Cape Town	Otago	Fremantle
Deception Island (South Shetlands)	500	1690	3517	4493	6546
South Georgia	1080	1420	2590	5208	5796
Îles Kerguelen	4519	5036	2500	3761	2270
Macquarie Island	4220	5590	3758	700	2359
Cape Adare	4380	5743	6289	1530	1205

Source: Everson, 1978

tankers, tugs and research vessels (Everson, 1978). Weather and ice conditions provide a harsh working environment. Further, spatial and temporal variations in krill abundance (El-Sayed, 1985; Hempel, 1985; Priddle *et al.*, 1988; Everson, 1992) introduce uncertainty of production; for example, krill abundance was low around South Georgia in 1977/78, 1983/84, 1990/91 and 1993/94.

Economic analysis by McElroy (1980) suggests that the optimal market for development of the industry is a high-value and moderate-volume product necessary to achieve adequate return on investment, bearing in mind the added operating and processing costs, and the short season. Since krill products are relatively costly to produce, and krill has no special intrinsic property on which a large new market could be developed, its competitiveness with other products remains a critical factor. High-value, low-volume products, aimed at the lower end of the luxury market, seem likely to continue to be the principal attraction, at least in the short term. The great potential envisaged for krill as a cheap source of protein for the developing world has not materialised, and expansion has slowed in the last few years, probably since supply exceeds demand for existing products.

Future expansion in the industry is likely to depend on technological and market developments, particularly of the valuable, high-quality products necessary to justify operating costs. One approach is to develop more acceptable products in the processed food area (e.g. in burgers, stews and sauces) and the improved marketing of tail meat as a substitute for shrimps (Nicol, 1989). According to Nicol, the FAO has estimated that krill could attract up to 10% of the market for edible shrimps, equivalent to 60,000 tonnes of tail meat per year. Potentially more important are several chemical by-products for use in the medical and biotechnological industries (*ibid.*). Krill has a relatively high content of enzymes such as endopeptidases and exopeptidases, which have a medical role in breaking down the dead tissues that occur in wounds and ulcers. Chitin in the shell of krill, and its derivative chitosan, also have a number of potential medical applications in slowing bleeding and improving the healing of wounds. Another property of these substances is that they are biodegradable and therefore could make a significant contribution to materials for the packaging industry or for chemicals or drugs that work on slow release. The development of biodegradable plastics based on chitin and chitosan is currently under investigation in Japan. Further expansion of krill harvesting probably awaits a technological breakthrough in the processing of such chemical substances and developments in the pharmaceutical and biochemical industries. Future developments might also be improved by the establishment of shore bases in countries around the Southern Ocean and by the exploitation of other resources during the remainder of the sub-Antarctic year (Sahrhage, 1989).

7.5.3 Impacts of the industry

Concerns about the potential ecological impact of the krill fishery have been voiced for some time (El-Sayed and McWhinnie, 1979; Croxall *et al.*, 1984; Knox, 1984). Based on the experience of the whaling industry, the need for management and further research was clearly recognised by biologists before intensive harvesting developed (Gulland, 1970; Holdgate 1970). So far no adverse effects on krill stocks have been identified (Kock, 1994), although methods used to measure krill abundance have poor resolution.

Since krill occupy an important position in the Antarctic food web, the potential impacts of the krill fishery extend to the entire ecosystem. A particular worry has been that catch levels might affect krill predator species (Knox, 1984; Laws, 1985). Key considerations are the huge biomass of krill, its circumpolar distribution and role as a major component in the diet of fish, birds and marine mammals, particularly crabeater seals, chinstrap penguins, macaroni penguins and some petrels and fur seals (Croxall *et al.*, 1985; Croxall, 1987; see Chapter 4). In many cases, krill predators have increased in numbers (Croxall *et al.*, 1988), so a krill 'surplus' may not exist (Nicol, 1994), although assessment of impacts has been hampered by the lack of basic information needed to model predator–prey relationships (Croxall, 1987; Miller, 1991). However, based as it is largely on a high-price, low-volume market, the krill fishery at present appears to have had a small ecological impact on a global scale, particularly when current catch levels are compared with predator requirements and the apparent overall size of the resource (Croxall, 1987). However, any longer-term development of high-volume products and a heavily capitalised industry could have far-reaching consequences.

It is still unclear whether krill form a single or several discrete populations. Since major krill concentrations and fisheries are located close to large predator colonies of penguins and seals at, for example, South Georgia (Hunt *et al.*, 1992) and the South Shetland Islands (Kock, 1994) (see Figure 4.15), significant impacts could arise if large catches were to be taken from a small area or from a single stock (Everson, 1977; Miller, 1991); this might be reflected in declining populations of dependent species, increased competition between krill consumers and changes in species abundance and composition. Intensive fishing in areas where predators are restricted by foraging range from their breeding sites could also have a significant impact (Everson, 1977). Everson and Goss (1991) noted that competition between predators and the fishery coincided during the summer months in the South Orkney and South Shetland areas. This concern may be partly allayed by the results of a study of the Japanese krill fishery in the waters around the South Shetland Islands. Here Ichii *et al.* (1994) found an insignificant coincidence between the krill fishing areas and penguin foraging areas and that the catch levels were at least one order of magnitude smaller than the krill biomass in the intensively fished areas. However, it is also important to consider the effects of natural variations in krill abundance, as around South Georgia in 1993/94, when breeding success of dependent species was poor due to a scarcity of krill. In seasons when krill are locally scarce, a large fishing effort could seriously exacerbate the food shortage for breeding seabird and seal colonies.

A large krill fishery might also impede recovery of whale stocks (see Section 7.3). A further effect concerns the by-catches of finfish while trawling for krill. This imposes extra stress on small populations or stocks that are already depleted (*cf.* Williams, 1985; Kock, 1992) and is thought to be significant around South Georgia (Everson *et al.*, 1992). Incidental seabird mortality is also associated with mid-water trawling methods for krill (Kock, 1994; see also Section 7.4.3).

7.5.4 Regulation of the industry

Concern about the potential impact of the krill fishery on the Southern Ocean ecosystem and the need for regulation played an important part in the development of CCAMLR and the emphasis on an ecosystem management approach (see Chapter 9). CCAMLR came into force in 1982 but was negotiated when krill fishing was in its early, albeit rapidly growing, phase. Its adoption provided for the first time an opportunity to regulate a major Antarctic fishery before depletion of the target species. CCAMLR provides a framework for the development of a sustainable fishing regime: harvesting should not deplete the resource, and stocks of dependent species in the ecosystem should not be adversely affected. Despite the acknowledged importance of krill, it was 1991 before the first regulatory measure for krill was introduced. The apparently slow progress in this area reflects a number of factors (Miller 1991; Nicol, 1991, 1992; Croxall *et al.*, 1992): the modest size of catches in relation to the resource potential; the pressing need to focus initially on the finfish industry, where stocks were already depleted; a lack of biological data on which to establish conservation measures (*cf.* Ross and Quetin 1986; Everson, 1988, 1992; Priddle *et al.*, 1988; Miller and Hampton, 1989; Nicol and de la Mare, 1993; Makarov *et al.*, 1994); difficulty in achieving consensus support for precautionary measures; and a lack of monitoring programmes to identify impacts on dependent species and to separate natural annual variations in krill abundance from fishing impacts.

Regulatory measures are now in place for the principal krill fishery areas in the Atlantic and Indian Oceans. The annual catch limits are 1.5×10^6 for the South Atlantic, 450,000 tonnes for the Prydz Bay and western Indian Ocean area and 775,000 tonnes for the eastern Indian Ocean. However, in terms of wider ecosystem impacts, the location and timing of the catch may be as important as the total catch, so that in view of the sensitivity of predator species on adjacent islands in the Scotia Sea area, some form of geographical zoning to protect the food resource in inshore waters during the breeding season may be advisable. Thus precautionary limits will be set for specific subareas in the South Atlantic, such as South Georgia, if a threshold catch of 620,000 tonnes is reached in any fishing season. The krill industry is also required to comply with other measures such as those in place for the protection of CCAMLR Ecosystem Monitoring Programme (CEMP) sites (see Chapter 9).

In view of the difficulties in developing an ecosystem approach, it has been advocated that management initially be focused on selected areas where there are already some existing data on fisheries and predator–prey relationships (*cf.* Butterworth, 1986; Everson, 1988; Miller and Hampton 1989). The most appropriate areas are considered to be South Georgia, the South Orkney Islands and the Antarctic Peninsula (Croxall *et al.*, 1985). New approaches using computer modelling of the fishery and the wider ecosystem (Nicol and de la Mare, 1993) will allow different management options to be tested and their risks assessed. An important component of management involves monitoring of indicator species that are sensitive to variations in krill abundance, in order to detect early impacts (Everson, 1988; Croxall *et al.*, 1988). The problem remains, however, to distinguish statistically significant impacts arising from commercial harvesting from large natural fluctuations in the absence of long-term monitoring data (Croxall, 1988). The CCAMLR Ecosystem Monitoring Programme (CEMP) has been developed to address such questions (see Chapter 9).

7.5.5 Summary

Central to the functioning of the Southern Ocean ecosystem, krill also represents an attractive resource as one of the last, large, open-access fisheries. However, for CCAMLR to provide a politically realistic management framework in which conservation includes rational use, much more scientific information is needed on population age structure, natural mortality rates, abundance and distribution, and predator–prey relationships to develop biologically based management models. It is a paradox that some of the information needed will not be available until the fishery itself expands (Nicol, 1990). However encouraging that precautionary catch limits have now been set, it is critical that they be implemented and enforced and that catches and impacts be monitored. So far, the large stock size, low market value and high operating costs have saved the krill stocks from depletion, more than any directed conservation action. However, as most of the world's commercial species are now fished to capacity or overfished, greater attention may return to krill. If the economic viability of the fishery improves, then any failure to implement and enforce the precautionary management regime could have far-reaching consequences for the Southern Ocean ecosystem.

7.6 Other fisheries

7.6.1 Cephalopods

Cephalopods, principally squid, are a common occurrence in the stomach contents of whales, seals and seabirds, which together consume 34×10^6 t annually (Clarke, 1985). According to Clarke, the stocks in the Southern Ocean are large, but unquantified, and have probably increased both directly as a result of the commercial exploitation of their main predator, the sperm whale, and indirectly through the reduction in baleen whale numbers and therefore in the competition for krill, which squid eat (Rodhouse, 1990). As yet, there is little information about their distribution and biology and they have not been fished commercially, although there has been some exploratory fishing in the Atlantic sector of the Southern Ocean (Figure 7.12). Nevertheless, they are a resource of some potential (Everson, 1978; Rodhouse, 1989a, 1990). Squid are a high-value product, and with large fisheries established in waters around the Falkland Islands and New Zealand (Rodhouse, 1989b), it may only be a matter of time before the stocks of the Southern Ocean are exploited, although, as with whaling and finfish, the impacts are likely to centre on the South Georgia area. Rodhouse (1990) identified several species around South Georgia with commercial potential, notably *Martialia hyadesi*, but considered the stocks to be particularly vulnerable to overfishing since they are liable to large fluctuations (see Chapter 4). The indirect impacts on cephalod predators there, particularly seabirds and elephant seals, would also be severe. Much more information on squid biology and stocks is required for planning the sustainable management of any future fishery and to assess its likely impact on the food chain of the Southern Ocean (Rodhouse, 1989a).

In an exploratory fishery, a Korean vessel took 52 tonnes of *M. hyadesi* around South Georgia in 1995/96. Subsequently, CCAMLR was notified of a new fishery for this species for 1996/97 by the Republic of Korea and the UK.

7.6.2 Crabs

In 1991/92, a US company began an experimental fishery for the crab *Paralomis spinosissima*, at South Georgia. In 1992/93, 299 tonnes were taken and 497 tonnes in 1995/96. Various conservation measures are in place, but the fishery did not continue in 1996/97 .

7.7 Minerals

7.7.1 Introduction

The first mineral resources were recorded by Frank Wild, who discovered coal measures close to the upper Beardmore Glacier during Shackleton's 1907–1909 expedition (David and Priestley, 1914). In the 1970s and 1980s, reflecting concern about world mineral reserves, especially oil and gas, attention became focused on the potential mineral wealth of Antarctica. Given the limited information available on Antarctic minerals, this attention was based on speculation rather than evidence. Nevertheless, when viewed in the context of Gondwanaland and earlier supercontinents, the wide distribution of mineral resources within other parts of these supercontinents suggests that similar resources might also occur in Antarctica (Chapter 2) (Craddock, 1990). However these are not yet proven, far less known to be economically exploitable in the face of hazardous environmental conditions, logistical and technological problems, and major political and environmental concerns (e.g. Larminie, 1987, 1991; Parsons, 1987; Behrendt, 1990; Willan *et al.*, 1990; Anderson, 1991; Rowley *et al.*, 1991). Moreover, under the Protocol on Environmental Protection (see Chapter 9) mineral exploration and exploitation activities are prohibited. Although this may be modified or amended at any time by the Consultative Parties and reviewed after 50 years, it is improbable that minerals exploitation will take place in Antarctica in the foreseeable future. However, the issue formed a major political theme and source of debate in Antarctic affairs during the 1980s.

7.7.2 Distribution of mineral occurrences

Metallic and non-metallic mineral occurrences have been widely reviewed in relation to the geological framework of Antarctica (e.g. Holdgate and Tinker,

1979; Zumberge, 1979; Thomson and Swithinbank, 1985; Rowley *et al.*, 1983, 1991; Office of Technology Assessment, 1989; St John, 1990; Splettstoesser and Dreschhoff, 1990; Willan *et al.*, 1990) (see Chapter 2), and the speculated localities of Rowley *et al.* (1983) are summarised in Figure 2.8. Apart from coal and iron, most minerals are present in small and uneconomic deposits, and the mineral resources of Antarctica can be classified as speculative and hypothetical (Office of Technology Assessment, 1989; Willan *et al.*, 1990).

Coal occurs in a number of locations in the Transantarctic Mountains as part of the Beacon Supergroup (Splettstoesser, 1985; Rose and McElroy, 1987; Coates *et al.*, 1990), and in the Prince Charles Mountains (see Chapter 2). Although the coal-bearing rocks are relatively abundant, they are of little potential resource significance because of the problems of exploitation compared with existing reserves elsewhere in the world. The largest deposits of banded iron formation occur in the Prince Charles Mountains, particularly at Mount Ruker, but the relatively low iron content and their distance from the coast make their possible exploitation uneconomic (Tingey, 1990). Copper mineralisation is common on the Antarctic Peninsula, and many other mineral occurrences have been recorded there (Rowley and Pride, 1982; Pride *et al.*, 1990; Willan *et al.*, 1990). However, re-examination has failed to reveal evidence for significant geochemical anomalies or mineralisation processes (Willan *et al.*, 1990) (see Chapter 2).

Geological assessments emphasise that current knowledge is confined to records of mineral occurrence. Detailed exploration and drilling will be required before the extent and grade of mineralisation can be established and whether any potentially economic deposits exist (Crockett and Clarkson, 1987; Rowley *et al.*, 1991). Since the existing ice-free area for exploration is limited to less than 2% of the continent, statistical comparison with the density of mineral occurrences in other parts of Gondwanaland suggests that the number of accessible deposits likely to be discovered is low (Wright and Williams, 1974).

One area that has been the source of much interest is the Dufek intrusion, where a variety of mineral occurrences, notably platinum and chromium group elements, cobalt, nickel, vanadium, copper and iron, have been recognised (Ford, 1990). De Wit (1985) and de Wit and Kruger (1990) considered that platinum could be exploited economically and described the operation of a hypothetical mine based on comparisons with other polar areas (de Wit, 1985). Others have questioned the costings (Office of Technology Assessment, 1989; Willan, 1989; Beike, 1990; Rowley *et al.*, 1991), since existing knowledge is based only on reconnaissance scientific surveys (Ford and Himmelberg, 1991). Assessment of the full potential thus awaits directed mineral exploration.

Technical comparisons with Arctic mining operations suggest that mining could be undertaken in more accessible coastal locations in the Antarctic if economic deposits were discovered. However, the problems of remoteness, logistics and environmental hazards would impose high investment and operating costs, which combined with the high levels of risk would make such a venture uneconomic under current market conditions, even for high-value minerals such as platinum (Office of Technology Assessment, 1989; Beike, 1990). Deposits of higher potential elsewhere in the world are more likely to be exploited before commercial interest extends to the Antarctic. Should market conditions change, and if the ban on minerals activities was ever relaxed, then minerals of high unit value could become a target for exploration in areas such as the Dufek intrusion, although more accessible areas such as the Antarctic Peninsula might be more attractive exploration propositions. Since world resources of most major non-fuel minerals have increased over the past 50 years, there appears to be no foreseeable shortage (Hodges, 1995) that might trigger demand for exploitation in Antarctica.

7.7.3 Hydrocarbons

The prospects for commercial exploration for offshore hydrocarbons are considered better than for onshore minerals. The continental margins and adjacent shelves of Antarctica (see Figure 2.9) offer potential opportunities for the existence of oil and gas fields in sedimentary basins (J.B. Anderson, 1991; Behrendt, 1991). Elliott (1988) reported the occurrence of hydrocarbon residues in a core from McMurdo Sound, and hydrocarbons have also been reported from surface sediments in the Bransfield Strait (Whiticar *et al.*, 1985), but such observations tend to generate speculation; they are not indicative of deposits. In the absence of exploration, estimates of potential reserves have been based principally on comparisons with other parts of the world. Of 21 onshore and offshore sedimentary basins identified in Antarctica, St John (1986) considered that ten were viable as exploration targets, and he estimated the total potential of these to be 203 billion barrels of oil equivalent. Such estimates are widely regarded as not only highly speculative, but probably meaningless (Crockett and Clarkson, 1987; Larminie, 1987; Behrendt, 1990). The main basins containing thick sequences of sedimentary rocks that may provide potential oil sources were reviewed in Chapter 2. However, more detailed surveys and drilling would be necessary to allow evaluation of possible resources. The prospects for onshore exploration appear minimal.

The estimated economics of Antarctic oil extraction are not encouraging (Garrett, 1984; Behrendt, 1991; Beike, 1992). Garrett estimated investment costs of $100,000 per daily barrel, compared with $30,000 for the Chukchi Sea off Alaska in 1983 figures. In any event, only the largest fields would be potentially exploitable: an assessment of the development of hypothetical resources in the Ross Sea suggested that only a 4000 million barrel field would be profitable at an oil price double that in 1989 (Office of Technology Assessment, 1989). According to the assessment, commercial oil development 'could be feasible in the next century but only if several optimistic assumptions prove out. These include technological advances; sustained, relatively high oil prices; and a reduction in excess OPEC production capacity that currently depresses the world oil market. It will also require the presence and discovery of large oil deposits, an expeditious process for resolving environmental and operating policies, and sensible and measured taxation of Antarctic oil revenues. OTA concludes that if any one of these assumptions do not hold, oil development in the Antarctic will not occur' (p. 170). World oil prices are currently relatively low, and although the market is vulnerable to political instabilities in producer areas, known or anticipated reserves are relatively high.

Outside the Antarctic Treaty area, the sedimentary basins around the Falkland Islands are now generating active exploration interest. In the long term, if oil is confirmed, this area may offer better prospects in terms of operating conditions than areas close to the Antarctic continent. It also lies outwith the area covered by the Protocol on Environmental Protection ban on minerals exploration and exploitation.

7.7.4 Other possible resources

Other possible exploitable resources that have been considered include mineral nodules and icebergs. The former occur on the floor of the Southern Ocean and contain iron and manganese, with some copper, nickel and cobalt (Frakes and Moreton, 1990) (Figure 2.10). However, they have a lower metal content than deposits at lower latitudes and have not been surveyed systematically. On the basis of present knowledge, these deposits have little commercial prospects.

Antarctic icebergs have been viewed as a possible source of fresh water for locations in the Southern Hemisphere, California and the Middle East (Weeks and Campbell, 1973; Weeks, 1980; Wadhams, 1990). Wadhams outlined a possible iceberg utilisation scheme, concluding that a combination of water use and generation of electric power through the conversion of ocean thermal energy probably offered the best commercial prospects. However, uncertainties and problems remain regarding the stability of towed bergs, insulation and protection from melting and wave erosion, and processing at destination points.

7.7.5 Constraints and problems related to minerals exploitation

A range of environmental, technical and political factors add to the risks and costs of exploiting Antarctic mineral resources (Pontecorvo, 1982; Mitchell, 1983; Keys, 1984; Parsons, 1987; J.B. Anderson, 1991; Behrendt, 1991; Larminie, 1991; Rowley *et al.*, 1991; Beike, 1992). Major problems to be overcome in exploration, development and exploitation are associated with environmental hazards that include sea ice, icebergs, rapid changes in wind, sea and sea ice conditions, strong katabatic winds in coastal areas, low temperatures, adverse climate and short working season (*cf.* Chapter 2). In the past 15 years, two ships have been sunk in the Ross Sea, and there have been shipping accidents elsewhere in Antarctica (see Chapter 8). Hydrocarbon exploitation offshore would face a contintental shelf with an average depth of 500 m: shelves elsewhere in the world average about 200 m deep. Further, icebergs are known to have left deep plough marks in the shelf down to 350 m, although the deepest may be related to low sea levels in the past (Deacon, 1984). Extensive areas of the sea floor, including the shelf, are susceptible to mass movements (Reid and Anderson, 1990; Wright *et al.*, 1993). Other problems may arise from gas hydrates and neotectonic hazards, including volcanism and faulting (Behrendt, 1990).

Most reviews emphasise the logistical and technical problems of operating in the Antarctic (Crockett and Clarkson, 1987; Rose and McElroy, 1987; Office of Technology Assessment, 1989). These stem from remoteness, lack of fresh water supply, absence of existing infrastructure and internal communications, the environmental sensitivity of the region, and the need for medical, environmental management and emergency procedures. There is a lack of good harbours and links with inland sites. Transport of oil and ores would require ice-strengthened tankers (environmental risks) or long-distance pipelines to ice-free harbours (risks from icebergs). Technology is already available for exploratory drilling, at least during the short summer, season, but production from artificial gravel islands, as utilised in the Arctic, does not appear feasible, because of the greater water depths. Thus while some aspects of Arctic technology would be transferable, production would require significant technological advances (e.g. Parsons, 1987; Office of Technology Assessment, 1989), perhaps in the form of sea-floor well heads and the use of submarine tankers (Crockett and Clarkson, 1987). While experience in the Arctic has shown that the industry is capable of operating in hostile environments (Roots, 1983; Croasdale, 1986; Lyons, 1991), the operating and logistical problems arising from greater remoteness and climatic severity in the Antarctic would be more difficult than in the Arctic, with consequent implications for costs and environmental management.

Political factors include both developments under the ATS (Chapter 9) and wider international developments such as political instability in producer countries. Nations with limited resources or resources vulnerable to disruption may view the Antarctic as a potential source of strategic supplies. For some countries, securing access in the long term to such resources is probably a stronger motive than short-term profits. Conversely, existing producer nations may wish to restrain Antarctic developments, which would provide competition or reduce world prices. From a political viewpoint, sovereignty issues, particularly in areas of overlapping territorial claims, would probably emerge if mineral exploitation were to be permitted (Sollie, 1983). Ultimately, the main constraint to exploitation, at least in the short to medium term, is economics and competition with resources on the inhabited continents. Contrary to predictions in the 1960s and 1970s, there have been no minerals shortages, and world mineral reserves have increased (Hodges, 1995).

7.7.6 Environmental impacts

Environmental considerations have been one of the major factors in the debate about Antarctic minerals exploitation. Concerns over potential impacts and the need for effective safeguards have been well-publicised, particularly by a range of influental environmental NGOs opposed to minerals development (e.g. Kimball, 1984). The impacts of potentially damaging activities would arise principally during the exploitation phases of development (Holdgate and Tinker, 1979; Zumberge 1979; Crockett and Clarkson, 1987; Keys, 1984; Rutford, 1986; Parker and Angino, 1990). During the exploration and prospecting phases, impacts would be broadly analagous to those of scientific research.

Ore mining operations would involve stockpiling, tailings/waste disposal, ore separation and concentration facilities, the production of dust and industrial effluents, and the risk of heavy metal and chemical pollution. Severe, long-term (hundreds of years) effects would probably occur at a local scale (<100 km^2), with moderate to low effects over a wider area (>1000 km^2). One assessment of the impacts of hydrocarbon development is illustrated in Table 7.2. Both mining and hydrocarbon development would also require the construction of onshore support facilities and infrastructure, including buildings, living quarters, roads, harbours, fuel storage tanks, hard runways, power generation and transmission networks, and water and sewage treatment works. The potential for pollution would arise from fuel combustion, waste disposal, fuel and other chemical spillages, and from increased shipping traffic, with increased probability of accidents. Probable impacts would include terrain modification, competition with wildlife for limited ice-free areas, displacement of wildlife, and localised pollution of terrestrial and nearshore ecosystems. Terrestrial impacts could be severe and of wider significance if located near sensitive ecosystems, wildlife colonies or protected areas. They are also likely to be relatively long-lived due to slow rates of biodegradation and ecosystem recovery (see Chapter 8).

Threats to marine ecosystems, arising from oil spills (from blowouts or tanker operations) would potentially be extensive and severe, depending on the location and timing of incidents and weather conditions; for example, coastal accidents during the breeding season could have far-reaching impacts if crude oil were to be involved. There are also major gaps in knowledge concerning degradation of hydrocarbons in Antarctic sea waters and prediction of the impacts and fate of particular pollutants and the rates of ecosystem recovery.

In sum, there would be both local and regional impacts of commercial minerals activities. The impacts of onshore mining would probably be severe but local in scale; those of hydrocarbon

Table 7.2 Summary and assessment of potential environmental impacts associated with hydrocarbon exploitation activities in Antarctica

	Survey phase		Exploration drilling phase			Development and exploitation phase					
	Ship-related activity	Seismic sounding	Ship-related activity	Drilling mud & cutting losses	Oil leaks, blowouts	Ship-related activity	Small spills & blowouts	Major spills & blowouts	Land base terminal impacts	Gas flares & evaporative losses	Construction dredging impact
Atmosphere											
Quality	0	0	0	0	1/2/3	0	1/2/3	1/3/3	0	1/3/3	1/3/3
Ocean (open)											
Quality	0	0	0	0	0	0	0	1/3/3	0	0	0
Local water chemistry	1/1/1	0	1/1/1	3/2/1	3/3/2	1/1/1	3/3/2	3/3/2	0	0	–
Phytoplankton, primary productivity	0	0	0	0	1/3/2	0	1/3/2	1/3/2	–	0	–
Zooplankton, particularly krill	0	0	0	0	1/3/2	0	1/3/2	1/3/2	–	0	–
Higher predators, including birds and seals	0	1/1/1	0	0	3/3/2	0	3/3/2	3/3/2	–	0	–
Continental shelf seas											
Quality	1/1/1	0	1/1/1	0	0	0	0	1/3/3	0	0	0
Local water chemistry	3/2/1	0	3/2/1	3/2/1	3/3/2	1/3/1	3/3/2	3/3/2	3/3/1	0	3/3/1
Phytoplankton, primary productivity	0	0	0	0	1/3/2	0	1/3/2	2/3/2	0	0	0
Zooplankton, particularly krill	0	0	0	0	1/3/2	0	1/3/2	2/3/2	0	0	0
Higher predators, including birds and seals	0	1/1/1	0	0	3/3/2	0	3/3/2	3/3/2	3/3/1	3/1/1	3/3/1
Benthos	0	0	0	3/2/1	3/3/1	0	3/3/1	3/3/2	3/3/1	0	3/3/1
Shoreline	0	3/3/1	0	0	3/3/1	0	3/3/1	3/3/3	3/3/1	0	3/3/1
Ice											
Pack	0	0	0	0	2/1/1	0	2/1/1	2/1/1	–	0	0
Shelf	–	0	–	3/2/1	3/3/2	–	3/3/2	3/3/2	–	0	0
Land surface											
Ice and snow	–	0	–	0	0	–	0	3/2/1	1/3/2	0	2/3/1
Exposed terrain	–	3/3/1	–	3/3/1	3/3/1	–	3/3/1	3/3/1	3/3/1	0	3/3/1
Freshwater bodies	–	3/3/1	–	3/3/1	3/3/1	–	3/3/1	3/3/1	3/3/1	0	3/3/1

Entries as Intensity/Duration/Extent.
Intensity range: 1 (negligible), 2 (moderate), 3 (severe). Duration range: 1 (one week), 2 (one week to one year), 3 (greater than one year). Extent range: 1 (<100 km^2), 2 (100–10,000 km^2), 3 (>10,000 km^2). Shoreline range: 1 (10 km), 2 (10–1000 km), 3 (> 1000 km). 0 indicates very slight but no measurable impact; – indicates no impact. This assessment is based on several assumptions, including the use of conventional technology; blowouts could be controlled before the onset of winter; an increased potential for all impacts to increase in the development and exploitation phase; impacts on lakes could be greater than indicated because of their small areal extent; impacts on atmospheric quality are estimated to be transient during any land-based development and exploitation.

Source: Rutford, 1986.

exploitation could range from locally to regionally severe, in the case of a major blowout or accident to a crude oil tanker.

7.7.7 Regulation of potential mineral exploitation activities

The environmental issues associated with minerals exploitation were much debated during the discussions leading up to the formulation of CRAMRA (see Chapter 9). CRAMRA attempted to address these concerns through a series of measures that indicate the constraints likely to be applied should minerals exploitation ever be considered. CRAMRA specifies that no mineral resource activity should take place in Antarctica unless a set of specific conditions are met. It provides procedures for the assessment of possible environmental impacts and for the regulation of approved activities in compliance with environmental safeguards. However, CRAMRA was criticised by environmental organisations over its failure to consider the question of liability for environmental damage from minerals activities. CRAMRA has not come into force and was overtaken by the adoption of the Protocol on Environmental Protection (see Chapter 9). Under the protocol, any activity, other than scientific research, relating to mineral resources is banned indefinitely unless the protocol is amended. Such amendment to lift or modify the ban is only possible with the agreement of all the Consultative Parties or at a review conference, which may be held 50 years after the protocol comes into force, provided that (1) it is supported by a majority of the parties at the time, including three-quarters of the current parties, and (2) there is in place a binding legal regime for determining whether minerals activities are acceptable and, if so, for their regulation. However, if any amendment is not ratified within three years, then any country could opt to withdraw from the protocol, and after two years of declaring its intention to withdraw, could unilaterally begin such activities.

Recent analysis using economic value concepts applied to development and environmental protection (Cullen, 1994) shows that with present uncertainties and constraints, the economic benefits forgone by the world community under the ban appear to be close to zero for the foreseeable future, particularly when set against the value of environmental protection benefits, and this situation seems likely to continue. However, if major technological changes were to reduce potential costs, or if shortages of supply from alternative sources of scarce minerals were to arise, then the 'option value' of mineral development could increase. According to this analysis, the approach adopted under the protocol is sensible, both in current economic terms and in keeping future options open (see Chapter 9).

7.7.8 Summary

Assessments of the minerals resources of Antarctica suggest that only higher-grade ores or the very largest oilfields, if either exist, are likely ever to be viable (Crockett and Clarkson, 1987; Office of Technology Assessment, 1989). The ultimate constraints will probably be economic and political, rather than environmental. On the economic side, these will include world markets (demand, supply and prices), the rate of depletion of existing reserves, the rate of discovery of new reserves elsewhere, the rise in demand from developing countries, advances in technology that will allow improved recovery from existing resources, and the development of alternative energy sources. Changes in policy may see greater use of renewable resources, recycling and energy efficiency savings. Politically, the stability of producer areas, as in the Middle East and the countries of the former Soviet Union, will affect supply and pricing and therefore the potential attractiveness of Antarctic resources.

The opportunity cost of a minerals moratorium appears to be close to zero for the foreseeable future. However, given that scarcity of certain high-value minerals could occur, then in economic terms there is sense in not foreclosing options. From a political viewpoint, the successful implementation of the Protocol on Environmental Protection and the continued high-profile interventions of environmental NGOs should ensure that any minerals exploitation is delayed into the forseeable future. For the present, there is no great sacrifice in agreeing a moratorium. Ultimately it may be questionable whether a moratorium would survive in the face of serious global economic pressure.

7.8 Conclusion

The pattern of exploitation of open-access marine resources (seals, whales, finfish) has followed a predictable sequence of events: discovery of a new resource; a rush to exploit it; overcapitalisation of fishing fleets; overfishing of the target species; collapse of the fishery; then exploitation of a new target species. The geographical impacts have been felt most in the more biologically productive areas such as the Scotia Sea. While from a perspective of conservation and sustainable management the lessons from the past are apparent, they were rarely put into practice in the face of short-term greed. Where management has been implemented, it has generally involved too little too late when stocks were already depleted. The current moratorium on whaling owes more to cultural attitudes and public pressure than to arguments about sustainable management. As with open-access fisheries elsewhere in the world, the historical lesson is whether sustainable management can operate through a process of consensus and voluntary compliance while the onus of proof to demonstrate significant impacts remains with the regulators: the absence of a strong authority and a shared commitment to sustainable exploitation is only likely to delay the inevitable depletion of the target resource. Against this background, the early years of CCAMLR were not encouraging, particularly as many stocks of finfish were already overfished before the convention came into force. However, given the political difficulties and the necessity of achieving consensus between fishing and non-fishing interests, significant progress has been made in developing a range of regulatory and precautionary measures, although the problem of ensuring compliance remains. Whether CCAMLR can ultimately prevent a repeat of the 'tragedy of the commons' in the case of the krill fishery remains to be seen. Fortunately, so far the main constraint has been the low market demand, allowing time to develop a more secure basis for a precautionary approach and a programme to monitor impacts on dependent species.

Management of the living resources of the Southern Ocean, and of krill especially, presents particular challenges. Traditional single-species fisheries management is inappropriate, but the lack of basic data and the complexities of modelling species interactions make the application of an ecosystem approach to management problematic and seemingly unattainable in the near future (Butterworth, 1986; Miller and Hampton, 1989; Kock, 1994). Two fundamental considerations need to be resolved (Miller, 1991): the implementation of measures necessary to underpin sustainable management and exploitation as opposed to 'boom and bust'; and the recognition of the wider ecological impacts of resource exploi-tation and the extent to which changes in harvested species produce detectable changes in other components of the ecosystem. Both are being addressed by CCAMLR.

It is clear that 'a vast uncontrolled experiment' has been carried out in the Southern Ocean (Laws, 1985) in the relentless pursuit of open-access resources and that significant changes have occurred in species composition as a result of sealing, whaling and fishing. Whether such changes are reversible is uncertain. In any case, it is probably impractical to manage the ecosystem to this end. The cautionary note sounded by Gambell (1985) regarding the ecosystem of the Southern Ocean is quite salutary: 'even the most cautious disturbance of the present relationships may have far-reaching and possibly irreversible effects' (p. 240).

To what extent have the lessons of the past, the 'boom and bust' exploitation, been learned? Certainly the pattern of overexploitation of finfish in the 1970s and 1980s is not encouraging. Nor is comparison with existing open-access fisheries elsewhere in the world, which are generally in serious decline (FAO, 1995). Undoubtedly, the same economic pressures also underlie current activity in the Southern Ocean, as they did in the past, and as other world fisheries decline, renewed pressure on Antarctic resources seems inevitable in the future. However, other circumstances have also changed. Regulatory mechanisms are now stronger than at any time in the past and a precedent has been set in the establishment of precautionary catch limits for krill harvesting. Scientific knowledge of the Southern Ocean ecosystem is improving, as are modelling techniques for fisheries management. There is now increased public awareness of environmental issues as a result of politically influential campaigning by NGOs. Sustainable management is now an internationally recognised guiding principle for the exploitation of renewable resources, and governments may be more willing to impose sanctions on states contravening international conservation agreements. The threat of such sanctions may be the only guarantee ultimately of ensuring enforcement of conservation measures. The current recession in the krill fishery offers the members of CCAMLR a unique opportunity, possibly the last, to pre-empt the next cycle of overexploitation of Antarctic marine living resources.

Potential exploitation of non-marine living resources has proved an important catalyst, both in focusing attention on environmental impacts in Antarctica and in accelerating the development of comprehensive measures for environmental protection. The ban on minerals activities has taken the heat out of an environmentally contentious and politically divisive issue, at least for the forseeable future.

Further reading

Bonner, W.N. and Walton, D.W.H. (eds) (1987) *Key Environments. Antarctica*. Pergamon Press, Oxford.

Knox, G.A. (1994) *The Biology of the Southern Ocean*. Cambridge University Press. Cambridge.

Kock, K.-H. (1992) *Antarctic Fish and Fisheries*. Cambridge University Press, Cambridge.

Miller, D.G.M. (1991) Exploitation of Antarctic marine living resources: a brief history and a possible approach to managing the krill fishery. *South African Journal of Marine Science*, 10, 321–329.

Mitchell, B. and Sandbrook, R. (1980) *The Management of the Southern Ocean*. International Institute for Environment and Development, London.

Splettstoesser, J.F. and Dreschhoff, G.A.M. (eds) (1990) *Mineral Resources Potential of Antarctica*. American Geophysical Union, Antarctic Research Series, 51, Washington DC.

CHAPTER 8

Environmental impacts: human impacts on the environment

Objectives

Key themes and questions covered in this chapter:

- legacy of past environmental standards
- changing attitudes to environmental stewardship
- impact of chronic marine pollution and marine pollution incidents
- impact of global air pollution and global warming
- significance of human impacts at local and continental scales
- environmental clean-up
- to what extent should introduced species be managed?
- is greater regulation of tourism required?

8.1 Introduction

The impacts of human activity on the global environment have received increasing attention in recent decades. Antarctica has not escaped human impacts, resulting both directly from activities on and around the continent and indirectly from effects initiated elsewhere. Direct, but small-scale, impacts on the terrestrial and marine environments arise from the presence of scientific stations and support activities, and from tourism, as well as from the introduction of non-native species of plants and animals, principally on the sub-Antarctic islands. Indirect, but continental-scale, effects arise from atmospheric pollution (including CFCs responsible for stratospheric ozone depletion and greenhouse gases for global warming), sourced from lower latitudes and carried by atmospheric circulation to the polar region, which may influence the natural operation of physical and biological systems in the Antarctic.

The sensitivity of Antarctic ecosystems to human activities is reflected in a number of factors (see Chapter 5):

- the relatively limited extent of ice-free areas at a continental scale;
- the sensitivity of permafrost soils to disturbance;
- the low species diversity, slow rates of growth and impoverished nutrient status of terrestrial ecosystems;
- slow rates of biological and chemical processes and therefore slow rates of recovery from impacts;
- the concentration of primary production during the short spring season and in particular geographical localities, e.g. the MIZ, the sub-Antarctic islands and the terrestrial ice-free areas of the continent;
- the geographical concentrations of breeding areas and feeding grounds so that major pollution incidents or localised over-harvesting of prey species may have disproportionate impacts on populations of several species;
- the occurrence of large natural perturbations, particularly in the marine ecosystem (e.g. in krill reproduction), which could exacerbate impacts (or equally mitigate them) depending on timing;
- the vulnerability to competition from introduced species;
- the rapid transfer of pollutants along short food chains to higher species;
- the accumulation of pollutants in the fat stores of mammals.

This chapter reviews the legacy of human effects on the Antarctic environment and the measures adopted to mitigate the impacts.

8.2 Marine pollution

8.2.1 Introduction

Pollution of the Antarctic marine environment may arise from localised sources (e.g. stations and ships) or from the long-distance transfer of chemical pollutants in the atmosphere, which are then scavenged and incorporated into sea water and marine food chains. It may involve chronic contamination, usually of a limited area, or it may be associated with major pollution incidents. Shipborne transport, in particular, has always been a feature of human activity in the Antarctic: it is the principal means of access and support for scientific activities and more recently for tourist visits, and for commercial fisheries in the Southern Ocean. In areas such as the Antarctic Peninsula, shipping activity associated with tourism has increased notably in recent years (see Section 8.5).

8.2.2 Chronic marine pollution

Localised chronic, long-term pollution in the nearshore marine environment is associated principally with human activities at major scientific stations (e.g. Kennicutt *et al.*, 1995). At McMurdo, the largest station in Antarctica and the logistic centre of the US Antarctic Program, significant localised pollution has been measured in Winter Quarters Bay. This area has a relatively long history of human activity dating back to Scott's expedition in 1902. It has been used as a docking site for cargo ships and icebreakers, and the former main rubbish dump for the station was located on its shores. Macerated sewage is also discharged in the bay (Howington *et al.*, 1992). Sediments in the bay contain moderately high levels of polychlorinated biphenyls (PCBs) (100–1400 ng g^{-1} dry weight) and polychlorinated terphenyls (PCTs) (30–1200 ng g^{-1} dry weight) (Risebrough *et al.*, 1990), and high levels of petroleum hydrocarbons (up to 4500 ppm) and heavy metals (Lenihan *et al.*, 1990). The most heavily polluted sediment is confined to an inner area of about 0.1 km^2 within the bay by a submarine sill, although an area five–ten times this size has been modified by human activity. The levels of contamination are locally as high as in polluted harbours elsewhere in the world (Lenihan *et al.*, 1990). The impact of the contamination (Figure 8.1) is reflected in the low abundance of benthic species, a significantly altered community structure and behavioural changes (*ibid;* Lenihan, 1992). However, conditions improve rapidly outside the bay (Lenihan *et al.*, 1990; Risebrough *et al.*, 1990; Lenihan and Oliver, 1995). Overall, the area affected is small, but recovery will require many decades (Lenihan *et al.*, 1990; Lenihan and Oliver, 1995).

Arthur Harbor, near Palmer Station on the Antarctic Peninsula, provides a second example of intense localised pollution of the nearshore marine environment. Since the 1940s, Arthur Harbor has been the location for scientific research, first by a British base (Base N), then by the US Palmer Station. It has been a focus of ship activity (for logistics and more recently tourism), and waste has also been dumped and incinerated locally. High concentrations of hydrocarbon contamination recorded in sediments and limpets derive from diesel fuel spills from ships and runoff from onshore spills and incineration sites (Kennicutt *et al.*, 1992a and b). Polycyclic aromatic hydrocarbon (PAH) contamination from activities at Palmer Station has also been recorded in fish (McDonald *et al.*, 1992). PCB residues detected in snow near the station probably derive from waste incineration (Risebrough *et al.*, 1976), but this has now ceased and all solid wastes are removed by ship.

Localised pollution has also been measured in Factory Cove at Signy Island in the South Orkney Islands (Cripps, 1992c), the site of a British base since 1947; previously, whaling operations were conducted variously from factory ships and a small shore station during the 1920s (Marr, 1935). Hydrocarbon concentrations in sea water were slightly higher than in the open ocean; those in sediments on the floor of the cove showed elevated levels. However, the extent of the contamination is confined to an area within several hundred metres of the station.

Several bays on the coast of South Georgia have a history of pollution arising from the whaling industry. At Grytviken, where a whaling station operated from 1904 to 1965, heavy pollution of the local marine enviroment through the discharge of large amounts of whale refuse and fuel oil is recorded in sediment cores (Platt and Mackie, 1979). However, by the early 1970s the macrobenthic community showed no apparent adverse effects (Platt, 1978; Mackie *et al.*, 1978), although later sampling of benthic invertebrates showed significantly higher concentrations of PAHs than samples from a pristine environment at Signy Island (Clarke and Law, 1981). Other South Georgia fjords with derelict whaling stations also reveal localised pollution of the shoreline and nearshore marine environment, including oiling

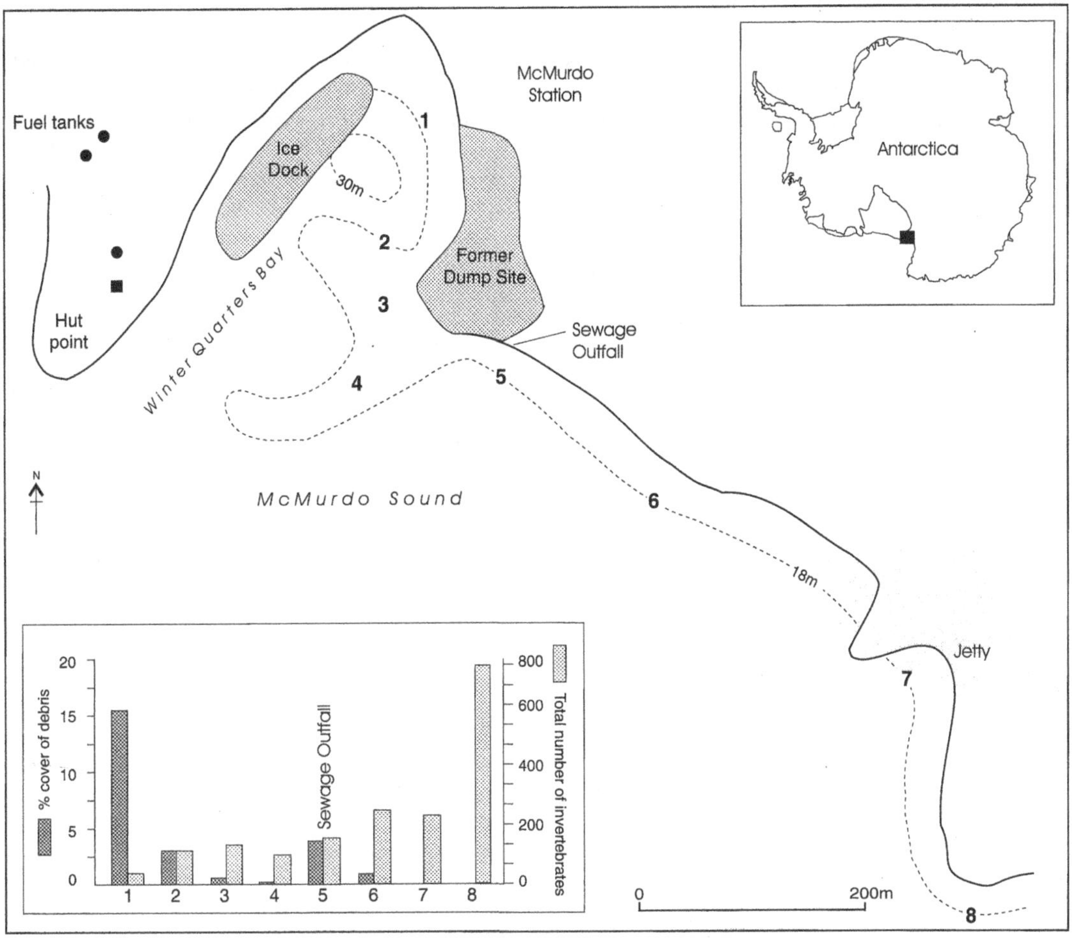

Fig 8.1 The human impact on Winter Quarters Bay, McMurdo Sound, as reflected in the percentage cover of human debris on the floor of the bay and the numbers of infaunal invertebrates (means of six samples per location, each covering 0.0075 m^2) (source: data from Lenihan *et al.*, 1990).

of seals and penguins (Cripps, 1989; Cripps and Priddle, 1991). The wider marine ecosystem of the Southern Ocean, however, is relatively uncontaminated by anthropogenic hydrocarbons (Cripps, 1990, 1992a and b; Cripps and Priddle, 1991).

Examples of chronic, but more diffuse, pollution include the presence of organochlorine compounds at higher levels than expected in the Antarctic marine food chain, although at lower concentrations than in more northern latitudes: DDT and PCBs have been detected in seals, penguins, seabirds, fish species and plankton and ascribed to pollution from global or Southern Hemisphere sources outside Antarctica (e.g. Risebrough, 1977; Subramanian *et al.*, 1986; Joiris and Overloop, 1991; Focardi *et al.*, 1992; Larsson *et al.*, 1992). Joiris and Overloop (1991) reported contamination levels of organochlorine residues in plankton comparable with those in the North Sea but seven times lower in terms of sea-water volume. Also, species that breed in the Antarctic and migrate north of the Polar Frontal Zone have been observed to contain higher levels of organochlorine contaminants than non-migratory species (Luke *et al.*, 1989).

Concentrations of heavy metals in Antarctic sea waters are also relatively low (Flegal *et al.*, 1993) compared with measurements in more northern latitudes, and this is reflected in the generally low levels of accumulation from anthropogenic sources in marine organisms (e.g. Honda *et al.*, 1987; Berkman and Nigro, 1992; Kureishy *et al.*, 1993; Thompson *et al.*, 1993). However, relatively high natural background levels of trace metals have been recorded in some species of crustaceans and seals (e.g. Petri and Zauke, 1993; Malcolm *et al.*, 1994; Szefer *et al.*, 1994), suggesting a need for caution in assessing baseline levels and wider global comparisons.

Another form of chronic marine pollution increasingly documented in recent years is the accumulation of plastic waste and fishing debris from commercial fisheries (Gregory *et al.*, 1984b; Gregory, 1987; Ryan, 1987; Slip and Burton, 1991; Walker *et al.*, 1997) (see Chapter 7). The movement of debris in the Southern Ocean is influenced by the northeastward Antarctic Circumpolar Current (see Chapter 4) and concentrates at the northern limit of the Polar Frontal Zone. Plastic pollution is thus greatest in the sub-Antarctic in the region of the Polar Frontal Zone and declines on each side of it. At higher latitudes, levels appear to be low, but with local concentrations associated with scientific stations and areas of logistic operations (Gregory *et al.*, 1984b). Impacts on marine life include the entanglement of fur seals in the debris (Croxall *et al.*, 1990b; Arnould and Croxall, 1995; see Chapter 7). Ingestion of small particles by Antarctic seabirds has also been noted, although higher levels in more migratory species suggest derivation from more northern latitudes (van Franeker and Bell, 1988), where the impact on the mortality of marine life is well-documented (Laist, 1987). Gregory *et al.* (1984b) concluded that although the amount of debris was small in the New Zealand sector of Antarctica, it represented a problem that could not be easily dismissed.

8.2.3 Major marine pollution incidents

There have been several well-publicised major pollution incidents. On 3 December 1987, the Australian re-supply ship *Nella Dan* ran aground on Macquarie Island (Pople *et al.*, 1990), and 264,000 litres of light marine diesel and 6000 litres of oil were released into the sea over a two-week period before the ship was scuttled in deep water. For the first day or so, the oil was washed ashore in an area on the eastern side of the island but then was driven offshore by the prevailing westerly winds. In the immediate aftermath of the spill, thousands of dead marine invertebrates were washed ashore along a 2 km stretch of coast, but penguins and seals were largely unaffected. One year after the incident, the impact remained evident in the reduced densities of marine invertebrates in the lower littoral and sublittoral zones, compared with the upper littoral zone (*ibid*; Simpson *et al.*, 1995). The long-term effects of the spill and the recovery time of the contaminated area remain uncertain.

A more serious oil pollution incident occurred on 28 January 1989, when the *Bahia Paraiso*, an Argentine Navy re-supply ship, ran aground about 2 km from the US Palmer Station (Anon, 1989a) in an area of islands of significant biological importance for seabirds, seals and intertidal and subtidal communities. The ship had over 300 passengers, including 81 tourists, and crew on board and was carrying diesel fuel arctic (DFA) (a blend of diesel and jet fuel), jet fuel gasoline and compressed gas cylinders, in total over 1,100,000 litres. More than 680,000 litres, principally DFA, are estimated to have been released, producing slicks that covered 100 km^2 of the sea surface and contaminating Arthur Harbor and Biscoe Bay (Kennicutt *et al.*, 1990, 1991a and b). In a joint clean-up operation between the NSF and the Argentine and Chilean navies, surface oil was collected by skimming and much of the remaining fuel was successfully removed from the ship (Anon, 1989a; Penhale, 1989).

Although diesel fuel is not as persistent as crude oil, its effects are more toxic. Algae and limpet populations in the intertidal zone downwind of the wreck were seriously affected, and 50% of the limpets may have perished in a *c.* 2 km zone around the wreck (Kennicutt *et al.*, 1990). About 300 seabirds, principally Adélie penguins and blue-eyed shags, died in the three weeks after the event (Kennicutt *et al.*, 1990), although more deaths went unrecorded at sea. The spill occurred during the principal period when parent birds forage for food at sea, and most of the *c.* 30,000 penguins in the area are thought to have come into contact with the oil (Penhale, 1989). Skua mortality may also have been exacerbated through parental neglect of chicks, but this indirect effect was difficult to establish (*cf.* Eppley and Rubega, 1989, 1990; Barinaga, 1990; Trivelpiece *et al.*, 1990a; Eppley, 1992). Effects on fish and marine mammals appear to have been negligible, the latter being absent from the area at the time of the spill.

The short-term effects of the spill were restricted to within a few kilometres of the wreck. Two months after the spill, there were no detectable effects on subtidal macrofaunal assemblages (Hyland *et al.*, 1994), and after two years, little contamination was detected in intertidal limpets or subtidal sediments, although limpet populations had only partially recovered (Kennicutt and Sweet, 1992). At the time, fears were expressed about a major environmental disaster affecting krill and penguins (Anon, 1989c), but several factors mitigated the impact, including the high volatility of the fuel spilled, the natural flushing of the high-energy wave environment, the exposed nature of the mainly shingle or boulder beaches, and the lack of fine sediment to hold the oil.

Observations made one year after the spill (Kennicutt *et al.*, 1991a) revealed pockets of contamination in the vicinity of the wreck. Subtidal and beach sediments, intertidal algae, limpets and sediments showed only low levels of contamination. Freshwater ponds on several islands continued to show elevated levels of hydrocarbons transported by seabirds. Even after two years, persistent leakage from the wreck resulted in downwind beaches and limpet communities being freshly contaminated with PAHs, but the biological impact is difficult to assess (Kennicutt and Sweet, 1992). Elsewhere, little contamination from the original spill persisted after two years. Subsequently, further clean-up was undertaken in December 1992–January 1993, when all the remaining fuel and oil were removed from the wreck in a joint Dutch–Argentine operation costing an estimated US$4 million (Anon, 1993). It seems likely, therefore, that any long-term effects will be minor. Kennicutt and Sweet (1992) concluded that, apart from containment at source, spills of comparable size and volatility in high-energy environments may be best managed by allowing natural processes to run their course. Green *et al.* (1992) reached a similar conclusion from an experimental study near Davis Station.

Other incidents reported include the grounding of a Russian tanker on Grand Terre, Îles Kerguelen, producing a small spill (600 tonnes) of petroleum (Jouventin *et al.*, 1984). Hundreds of southern rockhopper penguins were reported killed after a small oil spill at the station on Marion Island in 1980 (Williams, 1984). In 1989, the Peruvian vessel *Humboldt* ran aground in Marion Cove at King George Island in the South Shetland Islands, producing a small fuel spill (Anon, 1989b).

8.2.4 Assessment of environmental impacts

Although the review above has highlighted contamination of the Antarctic marine environment arising from both chronic pollution and major oil spills, this needs to be placed in a wider Antarctic and global context. The first point to be stressed is that general levels of pollution of Antarctic seas and ecosystems arising from anthropogenic sources are acknowledged to be low by world standards (e.g. Cripps and Priddle, 1991). The Antarctic environment therefore provides valuable baselines against which to compare and evaluate global changes in pollution. Second, even where higher levels of pollution occur, they are of very limited spatial extent relative to the size of the continent and are usually near to sites where there has been chronic low-level contamination. These sites are related to current and former research stations and abandoned whaling stations, which are geographically clustered in the Antarctic Peninsula, the sub-Antarctic islands and the Ross Sea.

Much attention has focused on the *Bahia Paraiso* incident because it underscores the current vulnerability of the Antarctic Peninsula and Scotia Sea, a biologically important area already a focus of activity for both scientific purposes and tourist cruises. The effects of the *Bahia Paraiso* fuel spill were significantly mitigated by physical processes and severe weather conditions. This allayed the initial fears of a major ecological disaster and highlights the problem of predicting impacts. Any increase in ship activity as a result of tourism, resource exploitation or scientific activity enhances the risk of pollution incidents that may be more difficult to mitigate. The limited information available suggests that hydrocarbon-oxidising bacteria are present in low density compared with temperate regions (Kennicutt *et al.*, 1990), and rates of biodegradation of hydrocarbons appear to be low except in the sub-Antarctic (Delille and Vaillant, 1990; Green *et al.*, 1992; Karl, 1992).

It is acknowledged that relatively little is known about the fate of oil pollution and its potential impacts on Antarctic ecosystems, and there is a need for further assessment of current background levels of pollution at both a local and a regional scale (e.g. Cripps and Priddle, 1991) and for studies of the long-term impacts arising from past contamination. One of the difficulties in establishing longer-term effects is that not enough is known about the natural dynamics and variability of the populations and ecosystems in response to other environmental stresses or variations in food supplies. It is also unclear to what extent the findings from the extensive studies undertaken on the effects of contamination on Arctic ecosystems, notably in association with the oil exploitation on the North Slope of Alaska, can be applied (Kennicutt *et al.*, 1991a; Cripps and Priddle, 1991). Similarly, there is little information on the impacts of pollution on marine species. It is known that penguins are particularly vulnerable to oil spills as they swim between feeding grounds and breeding colonies (Randall *et al.*, 1980; Croxall 1987; Culik *et al*., 1991). Oil contamination reduces the insulating and waterproofing properties of their feathers and hence their natural ability to continue feeding and to survive under extreme environmental conditions (Culik *et al*., 1991). An experimental study has also shown that krill, a key component of the marine ecosystem, are highly susceptible to toxic effects of oil pollution (Ogrodowczyk, 1981).

Recent fuel spills from stations and ships in the Antarctic are nearly all of refined fuels (e.g. diesel and petrol). Such spills have an initial, but short-lived, toxic effect, particularly on marine invertebrates. However, recovery is rapid. Heavy oil, which has much more long-lasting effects, is not used by virtually all Antarctic operators, including tour operators. With the prohibition on all minerals activities for 50 years under the Protocol on Environmental Protection, the threat of spills of heavy crude oil has been removed in the interim. The seriousness of such a threat should not be underestimated, and a major pollution incident involving heavy oil released into low-energy bays at the onset of the winter freeze-up could have had long-lasting and devastating consequences on breeding beaches packed with seals and penguins during the spring. The *Exxon Valdez* spill in Alaska in 1989 provides a salutary lesson and may be more indicative of the potential effects of a major spill of crude oil hydrocarbons should oil production ever be permitted in Antarctica. This incident involved 41 million litres of heavy oil and contamination of 2000 km of rocky shorelines in Prince William Sound and the Gulf of Alaska. Here, the initial impact was severe, and debate still continues about the longer-term effects (Pain, 1993; Wiens, 1996).

8.2.5 Pollution controls

Various pollution control measures are in place under both international legislation and the Antarctic Treaty. The International Convention for the Prevention of Pollution from Ships (MARPOL, 1973/1978) provides measures for the control of marine pollution. Under Annexes I, II and V of the convention, the sea south of 60° S is designated as a Special Area and the discharge of oil, oily wastes, plastics and other garbage has been prohibited since 1992. The provisions under Annex IV (Prevention of Marine Pollution) of the Protocol on Environmental Protection follow closely those of MARPOL. As with all such measures, however, the problem is to ensure strict enforcement and monitoring on the high seas.

In the past, the measures adopted under the Antarctic Treaty (Heap, 1994) have been far from comprehensive and there are questions on how stringently they have been applied, with attention drawn by the environmental lobby to unsatisfactory operational procedures at a number of bases (e.g. Manheim, 1988; Greenpeace, 1988, 1990, 1991, 1992). However, there have now been significant improvements in waste disposal (see below), and most national operators and tour ships now make the effort to ensure that waste disposal is done properly in compliance with the measures set out in the protocol, notably under Annex I (Environmental Impact Assessment), Annex III (Waste Disposal and Waste Management) and Annex IV (Prevention of Marine Pollution) (see Chapter 9). However, legislation cannot cover major accidents such as the *Bahia Paraiso* incident. Ironically, this may have served a useful purpose both in highlighting the potential risks of marine traffic growth in the Antarctic Peninsula area, and in providing an impetus to develop international management responses, for example through the Council of Managers of National Antarctic Programmes (COMNAP) (Chapter 9).

8.2.6 Summary

Pollutants are widely present in the Antarctic marine environment but at significantly lower levels than elsewhere in the world. Local areas of severe pollution, principally associated with past waste management practices now recognised to have been inadequate, are very limited in extent in a continental context. However, there is still little understanding of the fate of such pollutants, their toxic effect and the cumulative impacts on marine biota. Environmental awareness is now greatly increased among the treaty states, and significant improvements in pollution prevention and control have been instituted, or are planned, under the requirements of the Protocol on Environmental Protection. The principal concerns for the future are twofold. The first is to ensure adequate monitoring and compliance with the protocol measures. The second involves the increased levels of marine traffic, particularly associated with the tourist industry, and the increased potential for further major shipping accidents. The geographical pattern of pollution risk is significant, since the areas most affected by human activity in the Antarctic overlap with the limited extent of ice-free ground essential to the survival and breeding success of most of the major Antarctic wildlife species.

8.3 Atmospheric pollution

8.3.1 Introduction

The significance of the Antarctic for studies of global atmospheric pollution has already been noted (see Chapter 1) through the preservation of trace gases and airborne particulates, including heavy metals and organic pollutants, in snow and ice layers (Wolff and Peel, 1985; Peel, 1989; Wolff, 1990). The source of most global pollutants is in mid-latitudes, particularly in the Northern Hemisphere. However, as interhemispheric exchange is weak, the principal sources of Antarctic pollutants include industrial centres in the Southern Hemisphere superimposed on a more general atmospheric loading from the Northern Hemisphere (Wolff, 1992). It is important to distinguish between particulates and gases (Wolff, 1990). Gases are relatively well mixed and have longer residence times in the atmosphere and show similar concentrations and trends in the Antarctic to elsewhere in the world, for example CO_2 (Robinson *et al.*, 1988). Particulates and reactive gases have a shorter residence time and are effectively scavenged by precipitation in the circumpolar cyclonic zone (Shaw, 1989). However, winter development of the polar vortex (see Chapter 2) produces wind patterns over Antarctica that impede meridional exchange and the ability of particulates to reach the interior of the continent, although some particulates and diatoms are known to reach Antarctica from air subsiding into the vortex. When the vortex breaks down in summer, winds carrying particulates can penetrate into the interior. In spite of this source, much of the continent and its remote interior is affected by the impact of global pollutants to a far more limited degree than other continents. The principal value of Antarctica in this respect therefore lies in the near pristine environment, which allows snow and ice accumulation to provide an ideal archive of data on past and current trends in global climate and pollution (Wolff, 1990).

8.3.2 Global sources

Measurements have been made of several global pollutants in Antarctic snow and ice (Peel, 1989; Wolff, 1990, 1992). These principally comprise (1) a carbonaceous component of soot (elemental carbon) and organic compounds derived from the incomplete combustion of fossil fuels, incineration processes and the use of chemicals such as pesticides; and (2) an inorganic component of various chemical elements including heavy metals derived from industrial activity.

Soot pollution appears to be confined to the vicinity of scientific stations (e.g. Warren and Clarke, 1990), and there is no effect comparable to that of Arctic haze. However, there is also some evidence that soot may be transported to coastal areas of Antarctica from external sources under particular meteorological conditions (Murphey and Hogan, 1992).

The discovery of DDT compounds and subsequently PCBs in Antarctic biota and snow (e.g. Sladen *et al.*, 1966; Peel, 1975; Risebrough *et al.*, 1976) in the 1960s and early 1970s had a significant effect in raising environmental awareness that global pollutants had reached even the remotest parts of the world. Although some contamination could be ascribed to scientific stations (see below), the principal inputs were derived from global, mainly Southern Hemisphere, sources (Risebrough *et al.*, 1976; Risebrough 1977). Manufactured organochlorine compounds have now been widely detected in Antarctic air and snow (Tanabe *et al.*, 1983; Focardi *et al.*, 1991; Larsson *et al.*, 1992; Bidleman *et al.*, 1993), marine biota (see Section 8.2.2), mosses and lichens (Focardi *et al.*, 1991) and lake sediments (Sarkar *et al.*, 1994). However, concentrations of organochlorine compounds in the Antarctic atmosphere appear to have declined sharply during the last decade (Larsson *et al.*, 1992; Bidleman *et al.*, 1993).

The detection of heavy metals in snow has been beset by measurement problems, particularly concerning contamination (Peel, 1989). The data for lead (Wolff, 1990, 1992; Boutron *et al.*, 1994) suggest that present concentrations (up to 11 ng kg^{-1}) are five to ten times pre-industrial values, but the magnitude of the increase is uncertain and may be influenced by emissions from scientific stations or field operations (see below). The background levels of lead pollution in Antarctic snow reflect predominantly Southern Hemisphere sources (Rosman *et al.*, 1994; Wolff and Suttie, 1994). From a snowpit study in Coats Land, Wolff and Suttie (1994) showed that concentrations there peaked at 12 ng kg^{-1} around 1980 and then declined to about 5.5 ng kg^{-1} in the 1980s. For comparison, lead levels in Greenland snow in the 1960s were over 100 times pre-industrial levels but appear to have decreased in the last decade (Wolff and Peel, 1988; Boutron *et al.*, 1991). There have been no apparent trends discerned in other heavy metals, or in nitrate and sulphate concentrations, in Antarctic snow associated with global pollution (Wolff,

1990, 1992; Görlach and Boutron, 1992). In contrast, in Greenland nitrate levels have doubled and sulphate levels trebled during the last century (Neftel *et al.*, 1985; Finkel *et al.*, 1986; Mayewski *et al.*, 1986), reflecting the proximity to industrial centres in the Northern Hemisphere. The average atmospheric concentration of mercury (0.55 ng m^{-3}) measured at Ross Island during 1987–1989 is the lowest yet recorded in the world (de Mora *et al.*, 1993).

Radioactive fallout from atmospheric nuclear testing, which began in the 1950s, is a further source of pollution (Jouzel *et al.*, 1979; Pourchet *et al.*, 1983; Wolff, 1990, 1992); at the South Pole the peak level of tritium in 1966 was 2000TU (tritium units) compared with a natural background value of 32TU. However, this has had some scientific benefits in that radionuclide profiles and measurements of total beta-radioactivity have proved useful in dating snow cores (e.g. Peel and Clausen, 1982) and in meteorological studies (Pourchet *et al.*, 1983). The Antarctic Treaty prohibits nuclear explosions and disposal of nuclear wastes in Antarctica.

The global emission of greenhouse and ozone-depleting gases, from sources external to Antarctica, represents a further form of indirect atmospheric pollution that may impact significantly on the region. The possible impacts and the uncertainties surrounding the predictions are examined in some detail in Chapter 5.

8.3.3 Local sources

Pollutant levels are also enhanced by human activity on the continent, particularly from the use of fuel and, in the past, from waste burning at scientific stations (Figure 8.2), but also more widely spread through field camps and vehicle and aircraft operations (Boutron and Patterson, 1987; Boutron and Wolff, 1989; SCAR Panel of Experts on Waste Disposal, 1989; Rosman *et al.*, 1994). It has been estimated that the total particulate emissions on or adjacent to land amount to less than 200t a^{-1}, most of which is blown out to sea by the prevailing winds (SCAR Panel of Experts on Waste Disposal, 1989). Analysis of inventories of emissions to the atmosphere from human activities in Antarctica (Boutron and Wolff, 1989; Suttie and Wolff, 1993) suggests that pollution by most species (sulphur, cadmium, copper and zinc) is limited and localised. Boutron and Wolff (1989) noted that lead from aviation fuel could be widely dispersed and that the atmospheric loading of lead from local, as opposed to global, emissions could be an order of magnitude

Fig 8.2 Open burning of wastes, seen here at McMurdo Station in 1978, has been a source of local pollution in the past. Many states, including the USA, have now ceased this practice. Under the Protocol on Environmental Protection, all open burning of waste will be phased out by 1998/99 (source: photograph courtesy of I.B. Campbell).

higher than other species and account for 20% of the widely dispersed total annual fallout of lead south of 60° S. Although it was recommended by the ATCPs in 1976 that leaded fuels should not be used, no measures to achieve this have been implemented. However, virtually all national programmes now operate aircraft that use unleaded fuel. Only the Russian programme was recently using older aircraft running on unleaded fuel, but these have now ceased operating due to chronic funding problems.

Generally, the impacts of atmospheric pollution from local sources are likely to be confined to the vicinity of stations (e.g. Warren and Clarke, 1990; Suttie and Wolff, 1993; Rosman *et al.*, 1994; Stenberg *et al.*, in press) and therefore to be locally concentrated specifically where there is a high density of bases and human activity, as in the Antarctic Peninsula and the Ross Sea area. Given that most stations are sited on the coast and that prevailing katabatic winds are offshore, the bulk of local atmospheric pollution may be scavenged by precipitation over the Southern Ocean. The spread of most emissions inland should therefore be limited; for example, Suttie and Wolff (1993) were unable to detect any heavy metal contamination of the snow surface above background levels beyond a distance of 10 km inland (upwind) from Halley Research Station (UK). Similarly, most emissions from ships are dispersed over the ocean.

Low levels of organochlorine pollutants have been found in terrestrial moss and lichen species both in areas exposed to human activity and in more remote locations (Bacci *et al.*, 1986; Focardi *et al.*, 1991). It is apparent that some localised occurrences of organochlorine contamination can be sourced to waste burning formerly carried out at particular scientific stations (Riseborough *et al.*, 1976, 1990; Monod *et al.*, 1992).

8.3.4 Regulations to control global atmospheric pollution

Predictions of enhanced global warming as a result of the emission of anthropogenically produced greenhouse gases (principally carbon dioxide, methane, chlorofluorocarbons and nitrous oxide) are now firmly established through the consensus provided by the Intergovernmental Panel on Climate Change (IPCC) (Houghton *et al.*, 1996), although the magnitude, timing and regional variations are uncertain. If CO_2 emissions were held at their present levels, the concentration in the atmosphere would reach nearly twice the pre-industrial value by the end of the twenty-first century, and the average global temperature might rise by 2.5 °C. The United Nations Framework Convention on Climate Change (1992) aims to achieve

> stabilisation of greenhouse gas concentrations in the atmosphere at a level that would prevent dangerous anthropogenic interference with the climate system. Such a level should be achieved within a time frame sufficient to allow ecosystems to adapt naturally to climate change, to ensure that food production is not threatened and to enable economic development to proceed in a sustainable manner.
>
> (from Johnson, 1993, p.62)

The convention requires the parties to take 'precautionary measures to anticipate, prevent or minimise the causes of climate change and mitigate its adverse effects. Where there are threats of serious or irreversible damage, lack of full scientific certainty should not be used as a reason for postponing such measures' (*ibid*, p.62). However, although a non-binding agreement was made to reduce methane and CO_2 emissions to 1990 levels by the year 2000, most industrialised nations now seem unlikely to meet this target (Masood, 1996). At a meeting of the signatories to the convention in Kyoto in 1997 a legally binding protocol was negotiated with targets and a timetable of reductions in emissions of greenhouse gases. The outcome included an overall target for industrialised countries to reduce emissions by 5.2% below 1990 levels by between 2008 and 2012. However, such a target falls far short of the deep cuts necessary to stabilise greenhouse gases in the atmosphere (Houghton *et al.*, 1996). To stabilise $C0_2$ at around twice the pre-industrial level by 2150 would require emissions to be reduced to about 50% of their present values.

Given the formidable political and economic obstacles to be overcome in agreeing targets for reductions (Anon, 1995; Clayton, 1995; Pearce, 1997), such cuts were improbable. Even if emissions ceased immediately, global warming would still continue for a century or more, since the greenhouse gases have relatively long residence times in the atmosphere (Houghton *et al.*, 1996). Because of the lag between greenhouse gas build-up and the climate response, there is already an inbuilt mechanism for future global warming. The clear implication is that further global warming is highly probable, but the magnitude and specific effects in Antarctica are currently uncertain (see Chapter 5), although some effects may already be apparent on the Antarctic Peninsula (see Chapters 3 and 5). In the longer term, much may depend on the negotiation of amendments to the Kyoto Protocol to achieve further cuts in emission of long-lived greenhouse gases, as well as on the extent to which any such cuts are actually implemented, in the face of incontrovertible evidence of climate change comparable to the discovery of the ozone 'hole'.

Other forms of atmospheric pollution are regulated by a variety of international treaties and domestic legislation (e.g. Elsom, 1992). These include, for example, the 1979 Geneva Convention on Long Range Transboundary Air Pollution, a framework convention that has been developed in subsequent protocols, and various national air quality standards. While such measures will help to reduce Northern Hemisphere loadings of a range of pollutants, for Antarctica the principal concerns are likely to arise from expanding economic and industrial development and biomass burning in the developing countries of the Southern Hemisphere. Pollution from these sources in Antarctica looks set to increase at an unknown rate and with unknown impacts. However, given the relative isolation of the continent and the partial blocking effect of the polar vortex, significant impacts are likely to be less than would otherwise be expected.

8.3.5 Regulations to control ozone depletion

The publication in 1985 of measurements from Halley Research Station demonstrating a startling and unexpected thinning (by as much as 40%) of the ozone layer over Antarctica (Farman *et al.*, 1985), and subsequently confirmed by satellite data (Stolarski *et al.*, 1986), focused attention on the role of chlorine and other chemicals from anthropogenic sources in the destruction of ozone in the Antarctic atmosphere during the winter and spring (see Chapters 2 and 5). The resulting concern prompted a rapid political response in accelerating an international agreement, the Montreal Protocol on Substances that Deplete the Ozone Layer (signed in 1987 and came into force in 1989), to limit the production of ozone-depleting chemicals.

The Montreal Protocol originally provided for a freeze at 1986 levels on the production of chlorofluorocarbons (CFCs) and halons (the principal, but not the only, ozone-destroying gases) to be followed by cuts of up to 50% in emissions by 1999. However, under amendments to the protocol in 1990 and 1992, the process was accelerated to phase out production of CFCs, halons and some other ozone-depleting substances in industrialised countries by 1996, except where no substitutes are available. However, production in developing countries can continue until 2010. Although the problem has been recognised, governments have been unwilling, in the face of political and economic pressure, to agree more stringent controls or to implement some of the requirements for phasing out CFCs and other gases before replacements are available (Johnston, 1992; O'Riordan, 1995). For example, consumption of hydrochlorofluorocarbons (HCFCs), a substitute developed for CFCs but which also contain ozone-destroying chlorine atoms, will not be phased out in industrialised countries until 2030 (2040 for developing countries). Similarly, production and consumption of methyl bromide, which may be responsible for about 10% of ozone destruction, will not be phased out in developing countries until 2015, and various exceptions are to be allowed. For industrialised countries the deadline is 2005.

As a result of the controls introduced, the stratospheric concentrations of chlorine and bromine are predicted to stabilise by the turn of the century, and then to fall over the next 50 years (Elkins *et al.*, 1993; van der Leun *et al.*, 1995; Montzka *et al.*, 1996), assuming adherence to the amended targets in the Montreal Protocol. Although the ozone hole will therefore begin to heal slowly, the abundance of anthropogenic chlorine in the stratosphere will remain until the middle of next century at a level above that at which the Antarctic ozone 'hole' forms (Jones and Shanklin, 1995). This nevertheless represents an improvement on earlier predictions of a recovery period lasting about a century, before the revisions to the Montreal Protocol (Rowland, 1990; Benedick, 1991). In the long term, much will depend on the actual rate of reduction of global emissions of ozone-destroying chemicals and the effectiveness of implementation of the Montreal Protocol, an area heavily circumscribed by political and economic factors (Johnston, 1992; O'Riordan, 1995; Passacantando and Carothers, 1995). In particular, there is concern about non-compliance and illegal trade in CFCs (Brack, 1996). Further uncertainty arises from the recent cooling of the stratosphere, which, if it persists, may delay recovery of global ozone levels.

8.3.6 Pollution controls under the Antarctic Treaty

The ATCPs formally addressed local atmospheric pollution from waste disposal in 1989 and subsequently in the Protocol on Environmental Protection. Under the latter, the open burning of wastes is to be phased out by the end of 1998/99. This, and improved waste management methods including the removal of plastic materials where practical and the use of incinerators designed to reduce harmful emissions to the maximum extent practicable, should help to reduce further emissions at scientific stations, although the use of fossil fuels will continue to affect local areas. The appropriateness and effectiveness of waste incineration in Antarctica has been questioned by NGOs such as Greenpeace, and SCAR has now recommended that as far as possible all incineration should be stopped. This policy has been adopted by some states such as the UK, USA, Germany and South Africa, which now remove all their wastes except sewage and domestic wastes.

8.3.7 Summary

Excluding the indirect effects of the greenhouse and ozone-destroying gases, it appears that the impact of global atmospheric pollution in the Antarctic is limited, but that there are significant uncertainties regarding the reliability of many earlier reported measurements. In spite of the overall low levels of atmospheric pollution, some land areas experience locally enhanced levels as a result of the density of scientific stations and activity in the Antarctic Peninsula, the islands of the Scotia Sea, the Ross Sea area and the ice-free oases of East Antarctica. It is clear that much work remains to be done to establish current levels and trends in pollutant concentrations and their sources. This is particularly important both for global process studies and, at a local scale, for establishing baselines for monitoring the impacts of human activities and the effectiveness of measures adopted under the Protocol on Environmental Protection. Whereas pollution from local sources is now being much more stringently controlled than in the past, under the measures in the protocol, contaminants from global sources not only have a much wider distribution but their control is also more difficult to regulate.

By comparison, the potential effects of global warming and springtime depletion of stratospheric ozone are of considerably greater significance not only because of the regional-scale impacts on Antarctic systems but also for the wider feedbacks on global climate, sea levels and biogeochemical cycling (see Chapter 5). In the long term, much will depend on the rate of reduction of global emissions of greenhouse gases and ozone-destroying gases and therefore on the effectiveness of the measures adopted under the wider international agreements of the UN Framework Convention on Climate Change and the Montreal Protocol.

8.4 Scientific stations and support facilities

8.4.1 Introduction

Human impacts in Antarctica have occurred in parallel with the early exploration of the region and the development and progress of scientific investigations (see Chapter 6). Over the first half of the twentieth century, expedition numbers were small, technology was relatively limited and overall impacts at a continental scale were minimal (Kriwoken, 1991). Beginning with the IGY in 1958/59, however, there was expansion of national scientific programmes, with accompanying scientific stations and supporting infrastructure, and a significant increase in the geographical extent of activity (see Chapter 6). In 1994, seventeen countries operated 36 year-round stations with over 3000 summer personnel, but the number of stations has been higher in the recent past (see Chapter 6). Some of these stations, constructed during the IGY or earlier, are now obsolete or dilapidated, requiring replacement by more modern facilities, while others have been abandoned or removed. On a continental scale, the impact of scientific stations and support activities is minimal. However, the local impacts on terrestrial and nearshore marine environments have increased significantly in some areas over the past 40 years. The restricted extent of accessible sites, together with a spatial distribution of activity focused on the ice-free areas of the Antarctic Peninsula, the islands of the Scotia Sea, the Ross Sea area and a few continental oases, has served to magnify the local impacts. Many bases are located in, or close to, sensitive marine, terrestrial and inland water environments, so the impacts are of potentially wider significance, even where the activity is on a small scale (Heap and Holdgate, 1986; Walton, 1987d; Smith 1990).

This section examines the local environmental impacts of scientific stations and support activities and their effects on habitats and wildlife.

8.4.2 General environmental impacts of stations

Scientific research and supporting logistics have so far been the major activities on the Antarctic continent, and its remote and relatively undisturbed and unpolluted environment will continue to be of fundamental value for scientific studies. However, such activities inevitably impact on the environment to varying degrees. Benninghoff and Bonner (1985) noted that the impacts might arise deliberately, incidentally or accidentally, and they identified those likely to produce significant effects (Table 8.1) and the resulting impacts (Table 8.2).

Concern about the detrimental impacts of human activities on highly sensitive Antarctic terrestrial environments and ecosystems began to be expressed during the 1960s and early 1970s (Parker, 1972). Before then, environmental considerations were generally secondary, if addressed at all, in Antarctic operations. Waste disposal was haphazard, with rubbish left to accumulate or dumped into the sea. Disused bases were simply abandoned and left to deteriorate; accumulated rubbish was not removed and there was no assessment of their possible historical heritage value. Environmental impact assessments (EIAs) were only occasionally carried out for major projects, and then voluntarily (*cf.* Parker and Howard, 1977; Parker, 1978). Waste disposal practice and standards differed between national operators, often failing to comply with Antarctic Treaty recommended guidelines, and variously involving landfill dumps, dumping offshore onto sea ice, discharging sewage and effluents into either the sea or snowpack, burning and incineration of combustibles, and removing certain hazardous materials from the continent (retrograding) (Barnes *et al.*, 1988). By modern standards, there is therefore a significant legacy of poor environmental management in Antarctica. The most severe effects are associated with bases in three locations: the Ross Sea area, the Antarctic Peninsula and the ice-free areas of East Antarctica.

Cameron (1972) described many instances of poor waste management in the past at the US base at McMurdo Sound (Figures 8.3 and 8.4), the largest in Antarctica with a summer population of *c.*1200, and at remote field camps in the Dry Valleys, where waste and microbial contaminants were dispersed (*cf.* Cameron *et al.*, 1977). At that time, an essentially short-term view was taken of waste management (Reed and Sletten, 1989). For example, at McMurdo, waste and scrap were

Table 8.2 Impacts on terrestrial environments

habitat destruction/modification
destruction/removal/modification of biota, fossils, ventifacts, etc.
modification of vital rates of biota (disturbance to production and/or growth)
modification of distribution of biota
introduction of alien biota
pollution by:

- biocides and noxious substances
- nutrients (eutrophication)
- radionuclides
- inert materials
- electromagnetic radiation
- noise

modification of thermal balance of environment
aesthetic intrusion

Source: Benninghoff and Bonner, 1985.

Table 8.1 Types of activity likely to have a significant impact on the Antarctic environment.

Logistic/support activities

1 Establishment of bases, stations, airstrips, etc., or the extension of existing facilities (by, say, >10%).
2 Increases in personnel or personnel movements, aircraft flights, etc. (by, say, >10%).
3 Major changes in amount (by, say, >10%) or type of power generation, or fuel consumption.
4 Any operation affecting areas valued mainly for their sterile or pristine nature, e.g. dry valleys, remote ice cap areas.
5 Any operation involving the closing of a station that has been active for several years.

Scientific activities

1 Interference with or modifications of endangered or unique systems, communities or populations.
2 Operations that might adversely affect SPAs or SSSIs.
3 Introduction of alien biota with the potential to multiply or disperse.
4 Any operation affecting areas valued mainly for their sterile or pristine nature, e.g. dry valleys, remote ice cap areas.
5 Application of biologically active substances that have the potential to spread so as to cause perceptible effects outside their area of application.
6 Operations that might perceptibly impede the recovery of any endangered, threatened or depleted populations.
7 Experiments deliberately designed to create adverse changes in populations or communities (perturbation experiments) that extend over areas of >100 m^2 or, possibly, even less, particularly if unique systems are involved.
8 Operations that will adversely affect populations for which long time-series of data have been (or are being) collected to establish the status of the population.
9 Introduction of radionuclides into the environment, where their subsequent recovery and removal cannot reasonably be assured.
10 Drilling operations involving the use of drilling fluids other than water and/or possible escape or vertical movement of subterranean fluids.
11 Marine seismic surveys involving the use of explosive charges.

Source: Benninghoff and Bonner, 1985.

placed on the adjacent sea ice and then left to float offshore to deeper water during the annual break-up, a practice also followed at other bases on the Antarctic Peninsula. This was not always successful and scrap piled up on the shore where the ice grounded, (see Figure 8.3). The practice terminated in 1987, and the USA and other states now retrograde all solid and hazardous waste from Antarctica. However, there still remains a major clean-up requirement for old dump sites, which have simply been bulldozed over at several bases, McMurdo included (see Figure 9.6).

Fig 8.3 Poor waste disposal practices in the past led to the accumulation of rubbish at older Antarctic stations, such as in this photograph of McMurdo Station in 1960. Although such debris has now been removed from many stations, including McMurdo, several bases still have a poor waste disposal record (source: photograph courtesy of Charles Swithinbank).

Destabilisation of soils has also occurred locally in the McMurdo Sound area, where removal of snow cover or disturbance of the ground surface associated with construction activities has allowed permafrost to melt (Campbell and Claridge, 1987; Campbell *et al.*, 1994). The release of water from thawing of the permafrost beneath scraped soil surfaces has caused soil shrinkage, slumping and the formation of salt efflorescences. Inevitably, some disturbance of soils is also caused by fieldwork activity and may be relatively long-lasting (Campbell and Claridge, 1987). Deeper imprints such as vehicle tracks showed little recovery after 30 years at Marble Point on the western side of McMurdo Sound, but shallow footprints appear to be more readily obliterated where freeze–thaw processes are active in the soil (Campbell *et al.*, 1993). Low levels of localised heavy metal contamination have also been recorded in soils disturbed by human activity in the McMurdo Sound area (Claridge *et al.*, 1995).

The Antarctic Peninsula has also been a focus of much human activity and numerous instances of locally adverse impacts have been noted (Lipps, 1979; Greenpeace, 1990, 1991). The South Shetland Islands, in particular, not only have a long historical association with sealing and whaling, but also in the last few decades have become the centre of greatly increased pressure associated with the establishment of scientific stations (Headland and Keage, 1985; Harris, 1991a).

Fig 8.4 McMurdo Station, Ross Island, is the largest scientific station in Antarctica and has the appearance of a small town (source: photographed in 1983, courtesy of Ed Stump).

In view of their relative accessibility to South America by sea and air and the relatively large extent of ice-free ground, they have proved an attractive destination for national expeditions and scientific programmes. King George Island, and Fildes Peninsula in particular, now has the greatest concentration of multinational activities in Antarctica (Figure 8.5A and B). Ten scientific stations are currently operational, and there are various field huts and refuges (Harris, 1991a). Scientific and tourist travel and logistic support is greatly assisted by a hard rock runway constructed by Chile at Marsh Station in 1979/80, which can land wheeled Hercules C-130 aircraft throughout the year. As well as scientific facilities and accommodation, the Chilean settlement, Villa Las Estrellas, includes an 80-bed hostel, hospital, school and separate accommodation for the fifteen – eighteen families who live there (*ibid*). The Russian Bellingshausen Station was built on a Specially Protected Area (SPA), designated in 1966 for outstanding ecological interest (see Chapter 9). After this occurred, three other countries, Chile, Uruguay and China, followed suit to make Fildes Peninsula one of the most damaged and polluted areas in Antarctica. A network of roads now links these bases and large areas of the lush cover of moss have been damaged or destroyed. Due to the pressure on the area, the SPA status was revoked in 1975, and a site of special scientific interest (SSSI) (in two parts) designated to protect geological features.

Biologically, King George Island is important for its diverse plant and animal life; geomorphologically, it is noted for periglacial patterned ground features and suites of raised shorelines (see Chapter 3). The impacts associated with the various activities on the environment of the island, which includes several SSSIs (*ibid*) are summarised in Table 8.3. While there is no quantitative assessment of either the individual activities or their total impact, it is clear that existing management procedures under the ATS have completely failed to provide adequate environmental safeguards. Harris (1991b) proposed a set of new management approaches involving multinational cooperation to help to rectify the situation.

The ice-free oases of East Antarctica have also been subject to localised disturbance from construction of bases. In the Schirmacher Oasis in Dronning Maud Land, the German (former German Democratic Republic) Georg Forster and Russian

A

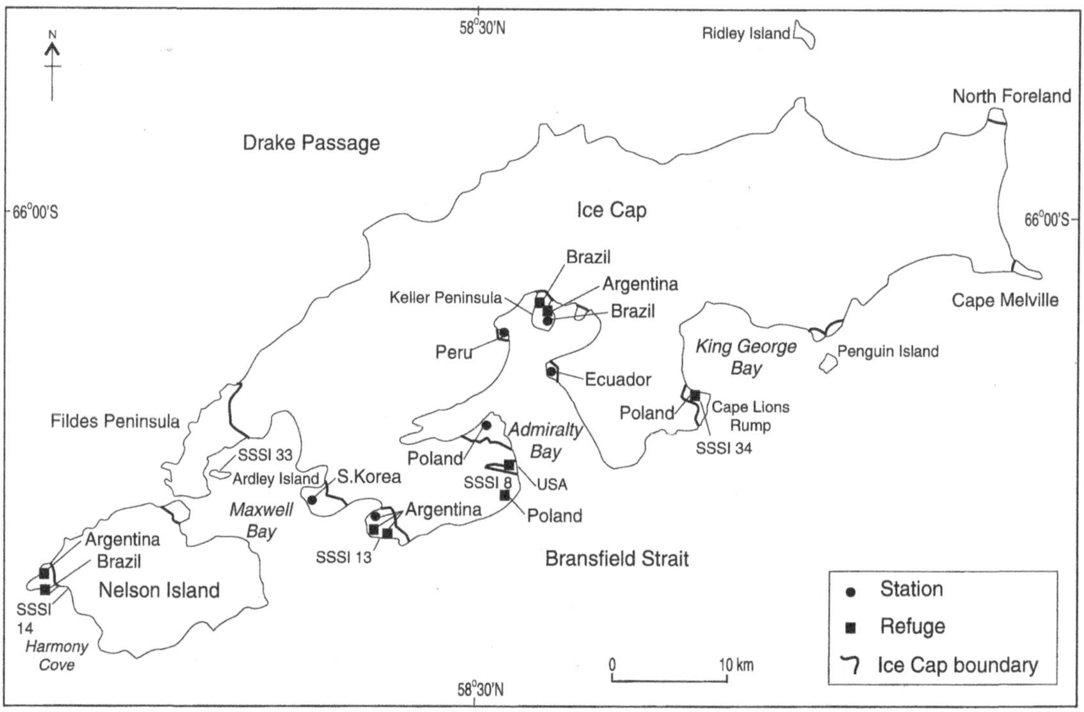

Fig 8.5 (A) Locations of stations and protected areas on King George Island and Nelson Island in the South Shetland Islands. (B) Distribution of human activities and protected areas on Fildes Peninsula, King George Island. A former UK base on Keller Peninsula has been removed (source: modified from Harris, 1991a).

B

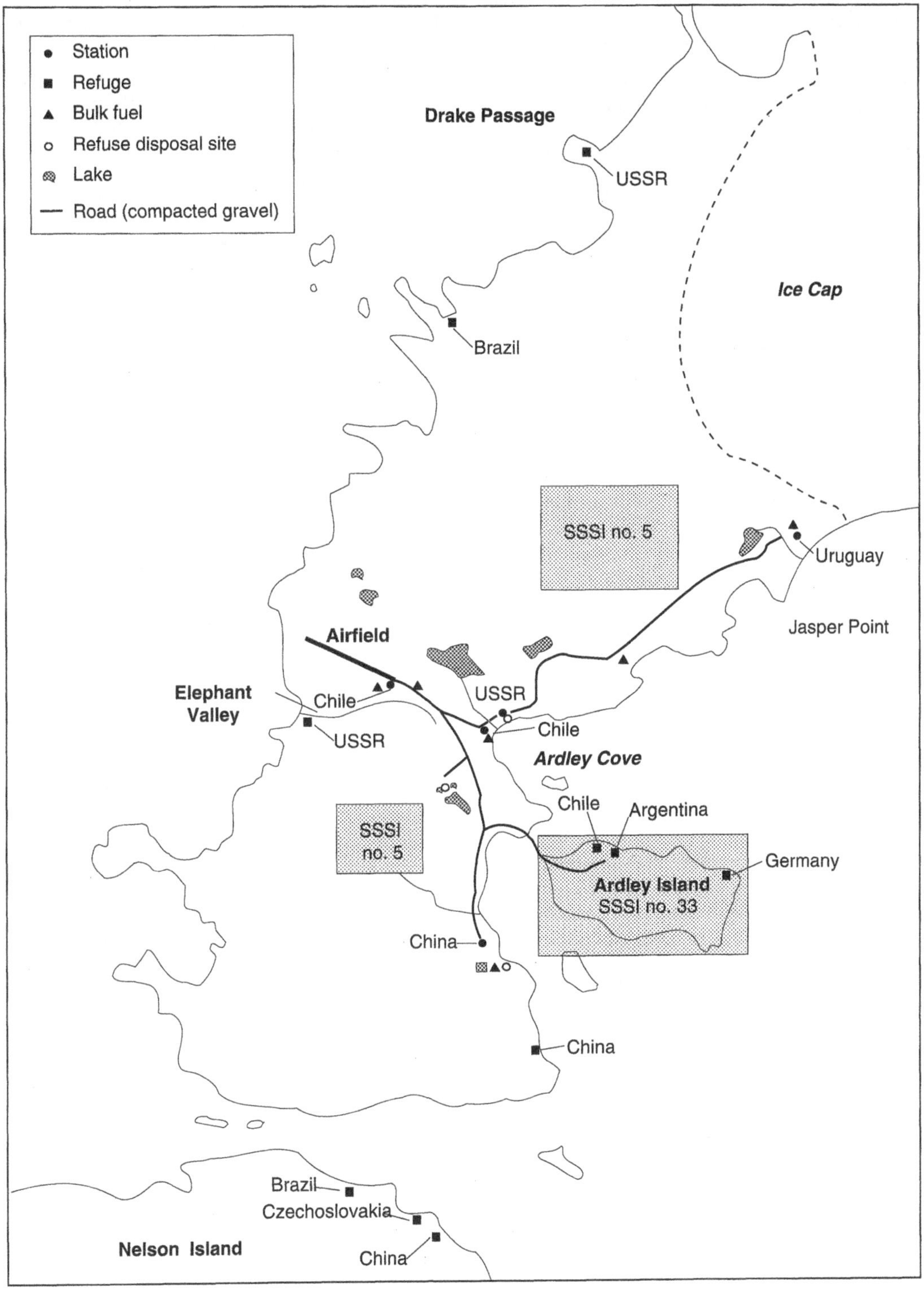

Fig 8.5 Continued

Table 8.3 Summary of human activities and impacts on King George Island

Activity	Effect
Aircraft activity	wildlife disturbance
Ship movements	marine pollution
Overland vehicle movements	damage to soils and vegetation
Fuel leakage	soil contamination
Waste disposal	litter; chemical contamination; air pollution from incineration
Fuel emissions	air pollution and fallout
Scientific research	abandoned experimental equipment; painted rocks, etc.; disturbance of wildlife
Introduced species	potential impact on ecosystems
Tourism	disturbance of wildlife, terrestrial vegetation, patterned ground

Source: Harris, 1991a.

Novolazarevskaya stations are located within 1 km of each other and the Indian station, Maitri, is nearby. The Russian infrastructure includes earlier stations dating from the late-1950s and occupies an area of 4-5 km^2. Much scrap had accumulated in the area between the Russian and German stations, and the state of the area and waste management practices at Novolazarevskaya were heavily criticised by a Swedish inspection team in 1994. However, the area has now been cleaned up following a joint project between Russia and Germany (see section 8.4.5). In the Larsemann Hills, Australia established a summer base in 1986, to be followed by year-round stations by the USSR and China, all within 3 km of each other. There is also a Russian airstrip on the ice. A number of environmental impacts associated with the construction and operation of the stations have been recorded (Burgess *et al.*, 1992; Ellis-Evans, 1996): erosion caused by poorly constructed vehicle tracks, minor hydrocarbon spills, local hydrological changes, disturbance of the temperature and nutrient status of a small freshwater lake accompanied by changes in its microbial community structure, accumulation of rubbish and dispersal of unsecured items in the strong katabatic winds (Figure 8.6). In particular, waste disposal procedures appear to have been lax, particularly around the Chinese and Russian stations. Overall, the impacts appear to be confined to an area of 10 km^2, but this nevertheless represents 5% of the ice-free area of the Larsemann Hills with impacts that are still occurring despite the supposedly greater environmental awareness of the 1990s.

The impacts associated with Casey Station at Newcombe Bay have been examined by Kriwoken (1991). A site on the north side of the bay was originally occupied by the American Wilkes Station during the IGY. This was later taken over by Australia while Casey Station was constructed on the south side of the bay. Casey Station was replaced in 1986/87 by a new facility constructed nearby. According to Kriwoken, the cumulative impact of these developments on the ice-free area has been significant, irreversibly altering an area of 9.5 ha, particularly as the activity has been dispersed across a relatively large area, with 2 km of road now servicing the site. Biologically, the ice-free area is important for its moss and lichen communities, and includes two SSSIs. One of these sites has been disturbed by the laying of cables, and Adamson and Seppelt (1990) noted further disturbance and devegetation over a distance of several hundred metres to the east of the new station outside the SSSI. Significant airborne alkaline pollution by cement dust has also occurred downwind from a concrete-batching platform used for five summers during the rebuilding work. Lichens were severely affected, becoming bleached and physically damaged, although mosses were apparently unaffected (Adamson *et al.*, 1994). Although the standards of environmental planning and management required at the time under Australian legislation and the Antarctic Treaty were not met, the Australian Antarctic Division now operates a comprehensive environmental management programme.

8.4.3 Fuel spills at stations

There have been several recorded instances of terrestrial pollution by hydrocarbons. In the vicinity of Palmer Station, Kennicutt *et al.* (1992a) found locally high concentrations of hydrocarbons in soil samples arising from fuel spills and leakage. Even higher levels were found at Old Palmer Station and at the former site of British Base N and its associated incineration site. At two other US stations major fuel leaks have occurred (Wilkniss, 1990). At Williams Field, the air facility for McMurdo, fuel storage bladders were found in 1989 to have leaked 260,000 litres. The spilled fuel was contained within berms or the snowpack, and 100,000 litres were recovered by direct pumping, and 11,000 litres were extracted from the contaminated snowpack. The fate of the remaining fuel is not documented. There have also been numerous spills from fuel storage tanks and pipelines at McMurdo Station (Tumeo and Larson, 1994). At the Amundsen–Scott South Pole Station, 150,000 litres of fuel were lost into the snowpack from an undetected leak in a fuel line during the winter of 1989. Recovery of the lost fuel was not attempted and the impact of the spill was considered negligible.

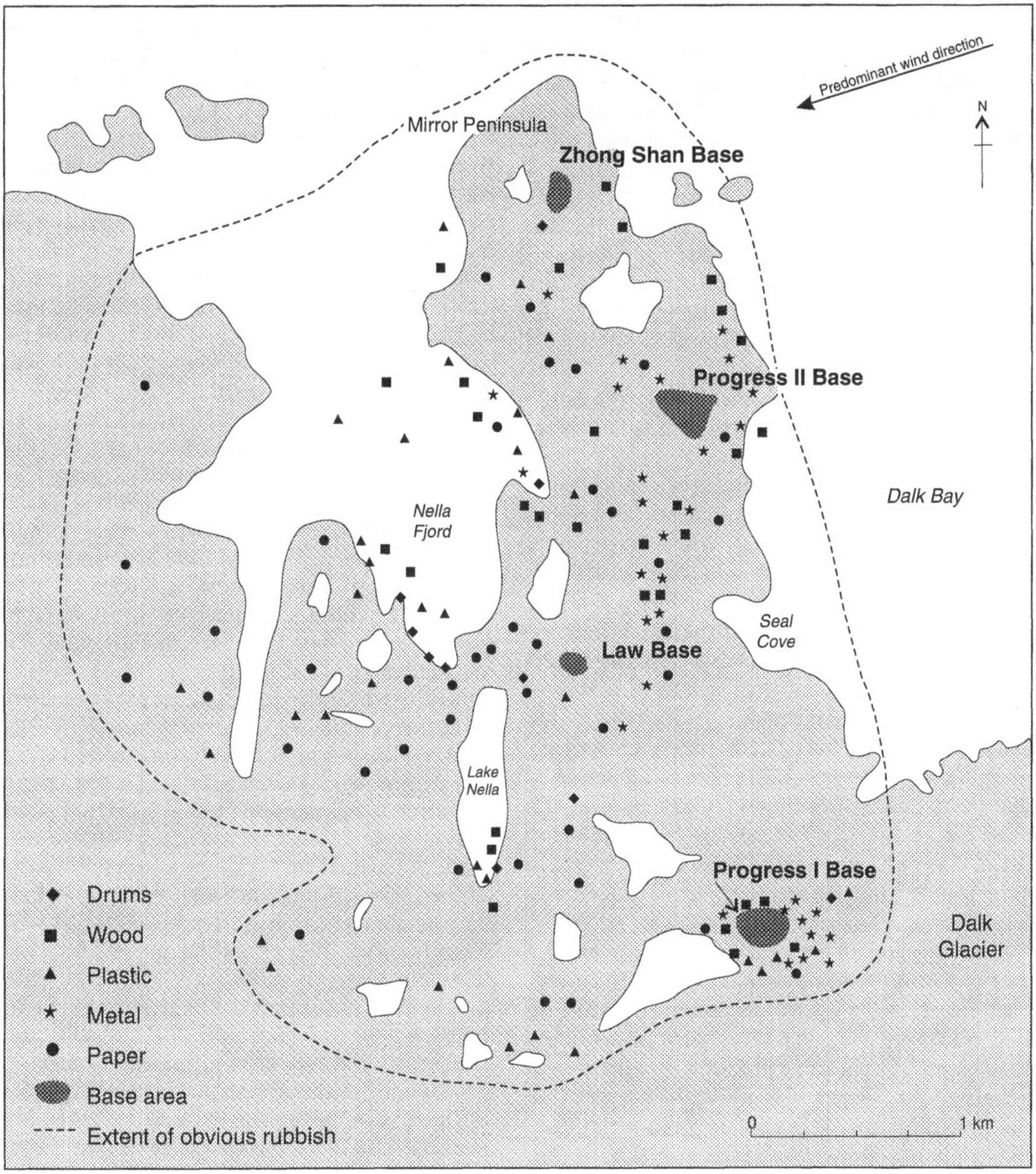

Fig 8.6 Dispersal of rubbish in the Larsemann Hills, south of Mac Robertson Land, East Antarctica (source: redrawn from Burgess *et al.*, 1992).

Extensive fuel contamination has been noted near the Chilean station, Marsh, on Fildes Peninsula. Between 1987 and 1991, a 20,000-litre rubber fuel bladder drained into a small valley used by nesting birds and elephant seals (Szabo, 1994). Contamination was extensive since there was no containment system around the fuel storage area.

At Australia's Casey Station, about 90,000 litres of light fuel oil leaked from a storage tank in June 1990 but was contained in a natural rock depression. Up to half the spill of the volatile fuel may have been lost through evaporation, and the environmental impact is considered to have been minimal: there was no vegetation at the site and no wildlife present at the time of year (Anon, 1990a). A subsequent spill of 190,000 litres occurred in January 1991 (Anon, 1991). In July 1994, a spill of 80,000 litres of fuel oil occurred at Argentina's Marambio Station (Pearce, 1994) but the impacts are undocumented. Until the problem was addressed recently (Smith, 1995a), the derelict whaling stations on South Georgia continued to be a source of pollution, including fuel oil leakages (Headland, 1984; Cripps, 1989; Cripps and Priddle, 1991). Poor fuel management and the high number of local spills at a variety of stations continues to be a source of great concern to both the Antarctic Treaty parties and the NGOs (Greenpeace, 1994; US Inspection Team, 1995).

There have been few studies of the potential impacts of fuel spills on Antarctic terrestrial ecosystems and their likely recovery rates. Low levels of hydrocarbon-utilising bacteria have been recorded in mineral soils in Antarctica (Konlechner, 1985; Kerry, 1990, 1993; Green *et al.*, 1992). While the potential exists for microbial degradation in some soils, this is likely to be confined to the short summer period and to proceed at a slow rate under the harsh environmental conditions. Some pollutants are likely to persist for up to a century (Konlechner, 1985), and longer in subsurface soils, where biodegradation and volatilisation are inhibited (Kerry, 1993). In an experimental study of beach sediments contaminated by light fuel oil near Davis Station, Green *et al.* (1992) found that most of the fuel was lost by volatilisation rather than biodegradation. These findings are in agreement with the similarly low rates of biodegradation in marine oil spills (see Section 8.2.4).

There are currently no local sources of radioactive contamination in Antarctica. However, a nuclear reactor was used to generate power at McMurdo Station between 1962 and 1972, when it was closed down because of operating problems. It was removed completely by the end of the 1975/76 summer, and retrograded together with contaminated rockfill and 12,000 tons of soil (Wilkes and Mann, 1978).

8.4.4 Human impacts on wildlife

The ice-free coastal areas of Antarctica play a disproportionate role, both as a habitat for plant and animal life and as summer breeding sites for birds and seals. Since these areas are also the focus of human activity and provide the most suitable sites for the location of scientific stations, competition arises between humans and wildlife.

There is well-documented disturbance to Adélie penguin colonies arising from human activity and aircraft operations, leading to reduced hatching and fledging success and therefore to decline in breeding populations (Sladen and Leresche, 1970; Thomson, 1977; Culik *et al.*, 1990; Wilson *et al.*, 1990, 1991; Cooper *et al.*, 1994; Woehler *et al.*, 1994; Giese, 1996). Chronic disturbance due to the presence of humans may produce stress and increased heart rates at distances as great as 30 m, even when there is no apparent visible reaction (Culik *et al.*, 1990); small colonies appear to be more sensitive to disturbance than large ones (Giese, 1996). Detailed studies by Wilson *et al.* (1991) showed that the reactions of nesting birds depended on the progress in the breeding cycle; the adult birds deserted their nests more readily when large chicks were present rather than smaller chicks and eggs. Birds returning from foraging deviated by 70 m due to the presence of a single person 20 m away from a routine penguin pathway. Wilson *et al.* also demonstrated that aircraft activity is highly stressful, even at distances greater than 1 km. Helicopter activity over a period of three days discouraged foraging penguins from returning to their nests, caused a 15% decrease in bird numbers at the colonies studied and contributed to a nest mortality of 8%. Larger helicopters have been observed to cause panic reactions in adults from a distance of 1.5 km (Culik *et al.*, 1990). In one notable incident in 1990, the mass death of 7000 king penguins through stampeding and asphyxiation on Macquarie Island was attributed to disturbance from a passing aircraft (Rounsevell and Binns, 1991).

The Adélie penguin rookery at Cape Royds also suffered severe disturbance and reduction in breeding pair numbers arising from uncontrolled visitor pressure after the establishment of McMurdo Station in 1956 (Thomson, 1977). However, following the introduction of visitor restrictions in 1963 jointly by the US and New Zealand authorities, numbers of breeding pairs increased in response to reduced levels of disturbance, but it has taken twelve – fourteen years to regain the original population numbers. Elsewhere, the respective effects of different human activities on penguin colonies are more difficult to disentangle; for example, a recent 10–20% decline in Adélie and chinstrap penguin populations at Admiralty Bay on King George Island, South Shetland Islands, has been variously attributed to overfishing of krill (C. Anderson, 1991) and human disturbance (Culik and Wilson, 1991).

Examples of large-scale impacts and direct competition for space are provided by the establishment of the joint US–New Zealand base at Cape Hallett and the airstrip at the French base at Point Geologie, both of which impinged on Adélie penguin breeding colonies. During construction of the base at Cape Hallett in 1956, 7500 penguins were moved, and the associated infrastructure ultimately covered an area of 4.4 ha out of a total area of 24.3 ha previously occupied by nesting birds (Wilson *et al.*, 1990). Additional disturbance to the penguins arose from snowdrift accumulation downwind of large objects, and scientific observations including extensive banding experiments. The base became disused in 1973 and was removed between 1984 and 1988, with some of the large equipment dumped on the sea ice. Penguin numbers declined from 62,900 breeding pairs in the first census in 1959 to 37,000 pairs in 1968, but aerial surveys have shown that the population has now recovered, reaching 66,319 pairs in 1987 (Figure 8.7). Although Adélie penguin colonies may experience natural long-term fluctuations (Trivelpiece *et al.*, 1990b; Taylor *et al.*, 1990; Wilson, 1990), the marked decline at Cape Hallett is ascribed directly to human disturbance and loss of habitat.

Human activity may also have been a contributory factor in the local decline in penguin numbers at the Point Geologie Archipelago in Terre Adélie, where Adélie penguin numbers have increased on all the islands except the largest, on which the French station Dumont d'Urville is located (Jouventin and Weimerskirch, 1990). The emperor penguin colony there also decreased in size by about 60% to 2300 breeding pairs between 1962 and 1989. From an environmental perspective, the construction of an airstrip, beginning in 1983, was particulary controversial in view of its impact on breeding penguins, petrels and skuas. Because of limited level ice-free ground, the project involved construction of a 1100 m long airstrip on a causeway linking three islands (Engler *et al.*, 1990). Environmental organisations and scientific observers questioned the approval for such a project within the framework of the Antarctic Treaty and the ATCP's approach to environmental management and impact assessment in general (*cf.* Auburn, 1984; Barnes, 1986; Joyner, 1986; May, 1988; Suter, 1991). The extent to which seabird numbers will recover from the disturbance is uncertain; for example, it was reported in 1992 that only 1800 out of 3200 displaced pairs of Adélie penguins had returned to new nesting sites nearby (MacKenzie, 1992). In January 1994, the runway was seriously damaged by a seismic sea wave, and the French government decided not to undertake repairs both because of the costs and to prevent further disruption to the bird colonies (Patel, 1994).

Both direct and indirect human impacts on skuas have been reviewed by Hemmings (1990). Increases in local populations have been favoured by the presence of station garbage as a food source and by introductions of prey species on sub-Antarctic islands. Reductions have also occurred in response to direct killing, destruction or modification of habitats, as at Pointe Geologie, and reduction of prey availability. However, the kleptoparasitic and opportunist behaviour of skuas makes them adaptable to human presence (Young, 1990). On the Antarctic Peninsula, skuas have proved resilient to increased levels of human activity in the vicinity of the UK research station at Rothera Point (Shears, 1995).

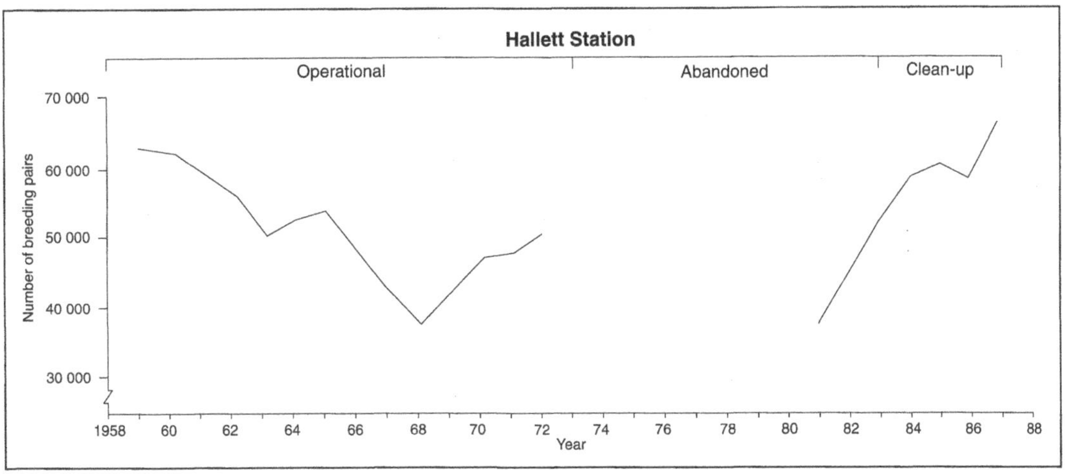

Fig 8.7 Decline in numbers of breeding pairs of Adélie penguins at Cape Hallett, 1959–1968, and their subsequent recovery mainly between 1981 and 1987 in relation to changing human activities in the area (source: after Wilson *et al.*, 1990).

8.4.5 Regulation and controls

In the past ten years there has been a fundamental change in the attitudes of the ATCPs to waste disposal and environmental management in Antarctica. This has occurred through SCAR and particularly through the activities of the environmental NGOs, which have raised wider awareness of the issues and helped to accelerate improvements in working practices. Concern has been expressed by the NGOs about delays in the implementation of approved recommendations, variations in the levels of compliance between different states and the lack of enforcement measures. It is clear that the procedures established in the past under the treaty have proved inadequate and that the treaty parties failed to criticise non-compliance with existing measures in each other's operations or ensure enforcement (Auburn, 1982). On the other hand, while recognising the need for improvements, there is justifiable concern among national operators and scientists about the costs of clean-up of poor practice in the past, the costs of adopting new measures with limited cost–benefit returns and the diversion of funding from scientific research.

The ATCPs have agreed various recommendations concerning environmental conduct, both of a general nature and in relation to specific activities, culminating in a series of rules and guidelines adopted at the XVth Consultative Meeting in 1989 (Heap, 1994). Based in part on a report by SCAR (Benninghoff and Bonner, 1985), guidelines were produced for EIA in 1987 to include provision for 'initial environmental evaluation' to determine the impact of a proposed activity. Where significant impacts were indicated, 'comprehensive environmental evaluation' was required, and this procedure has been adopted in the Protocol on Environmental Protection (see Chapter 9) and will apply, for example, to the siting of new stations; such procedures will be carried out by the ATCPs for their own operations. Tough regulations concerning waste disposal procedures (SCAR Panel of Experts on Waste Disposal, 1989) have been established in the protocol, which thus provides the basis for significantly improved standards of environmental management. Implementation of such standards, however, will incur clear financial, administrative and practical costs for national Antarctic operators, and it remains to be seen whether they will be complied with uniformly by all ATCPs.

Annex III of the protocol sets out measures for waste management and reduction of waste production and disposal to minimise impacts on the environment. These include:

- obligations to ensure that waste management is properly considered in planning and carrying out activities in Antarctica, covering issues such as storage, disposal, removal, minimisation, recycling and audit;
- wastes to be returned to country of origin as far as possible;
- old and current waste disposal sites to be cleaned up by their generator, but with the provision that wastes are not to be removed if that would cause greater impacts than if they were left in place;
- specific wastes to be removed if generated after the protocol comes into force, including materials such as batteries, fuel, fuel drums, toxic materials and plastics;
- combustible wastes to be incinerated, subject to emission standards to be determined, and residues to be removed from the treaty area; open burning to be phased out by 1998/99;
- restrictions on waste disposal on ice-free areas and into freshwater systems;
- wastes generated at field camps to be removed to base stations or ships for appropriate disposal;
- conditions for the discharge of sewage and liquid domestic wastes into the sea;
- prohibition of substances such as PCBs, pesticides and polystyrene packaging;
- preparation of waste management plans, exchange of information, and appointment of waste management officials to develop and monitor plans.

The protocol also strengthens the requirements for environmental monitoring of human activities (Walton and Shears, 1994). As a result of these developments, pollution control and waste management procedures have been much improved, and considerable accumulated scrap and other material has been removed from the continent (e.g. Anon, 1990b), but it remains abundantly clear that large amounts still remain. Environmental organisations have helped to focus wider attention on whether nations have delayed the implementation of measures or even acted on them at all (e.g. Greenpeace, 1994; Szabo, 1994). The activities of the National Science Foundation (NSF), which under the 1978 US Antarctic Conservation Act has responsibility for waste management and pollution control at US stations, have been the target of considerable criticism by environmental groups (Manheim, 1988, 1992); for example, the USA burned waste in an open landfill pit at McMurdo until 1991 with no monitoring or control of emissions. As a result of such pressure and the requirements to comply with the new regulations, waste management at US stations is now controlled under the Antarctic

Conservation Act (National Science Foundation, 1989, 1991; Draggan, 1992; Draggan and Wilkniss, 1992) and in 1990 the NSF initiated a five-year clean-up operation costing $30 million. In addition to improved waste management measures, including the closure of the landfill site, cessation of open burning and incineration at US bases and removal of most waste from the continent, the USA has also undertaken the clean-up of the its former East Base on Stonnington Island, last used by Finn Ronne in 1947/48, and now an historic monument (Parfit, 1993), and the removal of Old Palmer Station and the disused British base on Anvers Island.

Although impacts are closely related to numbers of people working in Antarctica, and the US programme has by far the largest number there, this does not necessarily mean that the numbers of critical reports of US operations is a reflection of their activities having created more problems. Rather it is also the case that US openness has allowed problems to be identified. The same legacy of past environmental mismanagement applies to most other nations operating in the Antarctic, and these nations could well adopt similar openness in both their activities and responses. Greenpeace inspections of active and disused bases have revealed many problems and shortcomings both in relation to current waste management activities and the deteriorating condition of abandoned bases (Greenpeace, 1988, 1990, 1991, 1992, 1994; Szabo, 1994). While improvements have undoubtedly been made and the requirement to comply with the provisions of the Protocol on Environmental Protection should lead to further progress, there is still concern that levels of compliance have been variable between different states and that over-reliance is placed on self-policing policies (Greenpeace, 1994; Szabo, 1994). It is clear that bodies like Greenpeace have played a valuable role not only in documenting the extent of the problems and raising wider awareness but also in forcing the pace of development for higher standards of environmental management, including a practical demonstration of what can be achieved at World Park base (see Chapter 9).

Environmental management procedures now adopted within the British Antarctic Survey (BAS) provide an illustration of the kind of changes required of all nations in the Antarctic (British Antarctic Survey, 1991). An environmental officer and a working group on environmental management and conservation, ensure implementation of the provisions of the protocol at all BAS bases and on board all ships. EIA procedures are now applied to all scientific and logistic projects. At Rothera Point, key environmental indicators are being measured as part of monitoring the impact of a recently constructed airstrip (Figure 8.8). Waste management

Fig 8.8 The Rothera research station of the British Antarctic Survey is situated on the Antarctic Peninsula and is now supported by a crushed rock airstrip accessible by Dash-7 aircraft from the Falkland Islands, allowing easier access for scientists (source: aerial photograph courtesy of A. Alsop).

procedures are detailed in a Waste Management Handbook (Shears, 1993), and there is an annual waste management plan and audit of disposal. All waste other than sewage and domestic waste is now removed from Antarctica, and hazardous waste is returned to the UK for safe disposal. Incineration is not undertaken at UK Antarctic bases. A survey and clean-up programme of the eighteen abandoned British bases and field huts has been undertaken (Shears and Hall, 1992) (Figures 8.9 and 8.10). Four

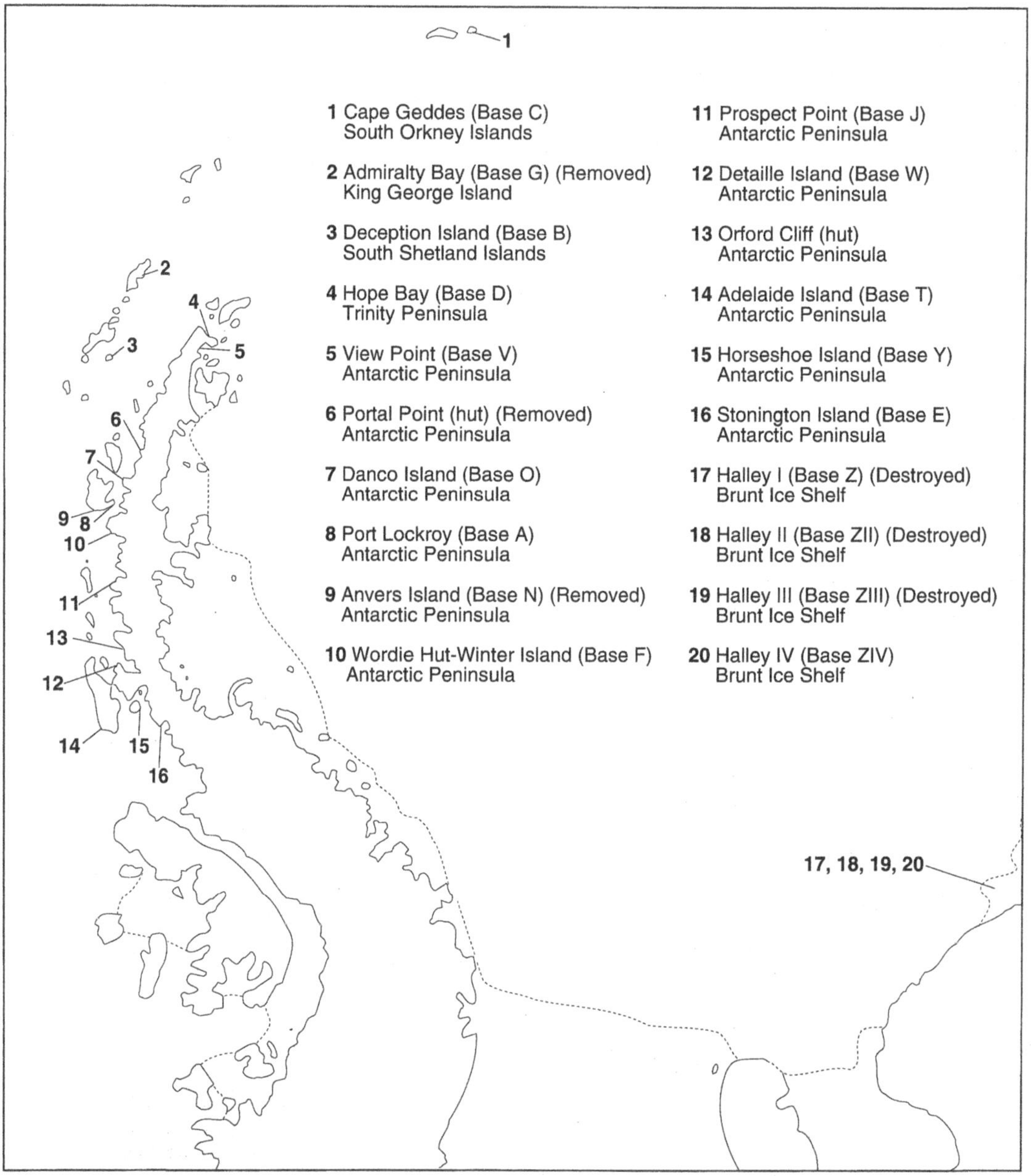

Fig 8.9 Abandoned British bases and field huts in the Antarctic Treaty area. Four have been declared as historic monuments: Port Lockroy, Wordie House at Faraday, Horseshoe Island including the adjacent refuge of Blaiklock Hut, and Stonington Island (source: Shears, 1993).

Fig 8.10 Clean-up of the abandoned British base at Deception Island in the South Shetland Islands involved the removal of accessible waste material and scrap. This work forms part of a wider programme undertaken by the British Antarctic Survey to clean up or remove abandoned British bases and field huts (source: photograph by J. Hall, courtesy of BAS).

abandoned bases of recognised historic significance have been designated as Historic Sites and Monuments (see Chapter 9). They are being cleaned up and conserved and are becoming tourist attractions (see Section 8.5); others of limited value will be removed. Measures have also been taken to comply with the protocol requirements for the prevention of marine pollution from ship discharges, the conservation of Antarctic flora and fauna and the management of protected areas. The education and training of staff to improve environmental awareness is also undertaken. Similarly, the United States has conducted an environmental assessment of its entire Antarctic programme (National Science Foundation, 1991), and an independent environmental audit has been completed for the New Zealand Antarctic Programme (Royds Consulting Ltd, 1994; Smith *et al.*, 1994) to assess the level of compliance with the new standards set out in the Protocol on Environmental Protection. Proposals for construction of an airstrip at Rothera Point, where there are no bird or seal colonies, were accompanied by the first comprehensive environmental evaluation (CEE), which concluded that impacts would be localised and relatively insignificant (BAS, 1989). The runway became operational in the 1991/92 summer. CEEs have now been undertaken by New Zealand, France and South Africa for projects such as new station construction and scientific drilling (Shears, 1994). France's design for the new inland Dome C Station requires all wastes to be retrograded. Another welcome development has involved international cooperation to clean up heavily impacted areas, for example between Russia and Germany for the area around Novolazarevskaya and Georg Forster stations in the Schirmacher Oasis in Dronning Maud Land.

In spite of the above examples of responsible EIA procedures, since 1987 several major projects have been undertaken without EIA; for example, the Maldonado base of Ecuador was built in 1989 close to a giant petrel nesting area without an EIA and has been occupied only once or twice for a few weeks since its construction. It remains that there are no mechanisms to force a country to halt activities detrimental to the environment: each country is both 'poacher' and 'gamekeeper'. EIAs would be more meaningful if they were conducted by truly independent agencies rather than by the states or operators that stand to gain from the activity. The dangers of such self-policing might be better avoided if other nations followed the New Zealand lead in using independent environmental audits.

While recognising the great need to correct the worst mistakes of the past and to improve standards of environmental management, there is equally a legitimate concern that an over-rigorous environmental agenda may not be cost-effective and may divert funding and resources away from scientific research (Walton, 1990b). It will thus be necessary to prioritise between the remediation of what may be termed 'aesthetic eyesores' (Bonner, 1994) and the removal of materials that represent genuine environmental hazards. In addition, the cost-benefits, financial as well as environmental, of removing non-toxic wastes from remote areas or toxic wastes from dump sites (with the attendant risk of remobilising the pollutants) and nearshore sediments require careful evaluation as part of evolving, medium-term national strategies for environmental rehabilitation. Unfortunately, such legitimate concerns and doubts are likely to be eagerly embraced by those nations whose activities fall well short of modern environmental standards and whose political will and technical expertise are not up to the task.

8.4.6 Summary

Awareness of environmental stewardship has only developed significantly in the past few decades and this has meant that terrestrial human impacts in the past pursuit of exploration and scientific research have been locally significant in a few areas of

Antarctica, producing a legacy of polluted 'hot spots'. However, in assessing the environmental effects of human activity, it is important not to lose sight of the vast size of the continent and that large areas of ice sheet and ice shelf are relatively robust in their capacity to absorb human impacts. Thus, while the local impact of individual scientific stations might be significant over an average area of 2–3 km^2 in terms of physical damage and 80 km^2 in terms of fallout of pollutants, this 'footprint' is unimportant on the scale of the whole continent (Laws, 1991; Aust and Shears, 1996). However, such a stance ignores the geographical perspective developed above that highlights the sometimes strong degree of overlap between sites used for human activity and those essential to the survival of plants and animals. Stations are usually sited, and sometimes crowded, on regionally scarce ice-free areas that provide not only important habitats and wildlife breeding grounds, but also the near 'pristine' conditions fundamental to many scientific investigations. The impact of human activity on local areas therefore assumes a significance disproportionate to the size of area and should not be dismissed or disregarded.

Attitudes and working practices have changed significantly in the past few years, and the Protocol on Environmental Protection now provides a secure basis for more responsible environmental management and the mitigation of the impacts that necessarily accompany the human presence in Antarctica. Yet the protocol has not solved the wide differences in environmental awareness and attitudes that exist between the Antarctic nations. Until these differences are addressed, then effective implementation of environmental controls may be compromised. This is a major concern since Antarctica is no longer a forgotten frontier where human activity can continue in disregard of the environment. However, results of global significance have been emerging from the research conducted there and some impacts on the environment are part of the necessary price to pay, but the paradox is that such work requires an unpolluted and undisturbed environment. Scientists therefore have an important role to ensure that the impact of their work is both minimal and sustainable in the sense that it does not compromise the scientific programmes of the future. Furthermore, the science that is done must also be worth the cost in environmental terms. It must therefore be of global or regional significance.

8.5 Tourism

8.5.1 Introduction

Despite its remoteness and its popular perception as an inhospitable and forbidding continent, Antarctica's unquestionable potential for tourism, increasingly regarded as one of its prime resources, is readily apparent from the many coffee-table picture books, television programmes and popular articles depicting the spectacular scenery and wildlife of the continent. To these principal attractions can be added the historical interests associated with the heroic age of Antarctic exploration and the scope for mountaineering and adventure travel. Antarctica also has appeal as an 'ultimate' tourist destination as well as an intellectual appeal: the boundaries of tourism and education have increasingly converged, notably through increased awareness of global environmental issues. Most operators of smaller cruise ships (up to 140 passengers) follow the ecotourism approach established by Lars-Eric Linblad in the 1960s (Stonehouse, 1994) (Figure 8.11). This combines an 'expedition' atmosphere with lecture programmes and briefings by specialists and an emphasis on environmentally sensitive behaviour and shore landings supervised by experienced staff; indeed some cruises have been described as having the 'feel of a university field trip' (Cardozo and Hirsch, 1989). Although various measures have been instituted or proposed to minimise the environmental impacts of tourism, and despite the environmentally conscious attitude adopted by most tour operators (Stonehouse, 1992a; Swithinbank, 1993a), there is a widely expressed view that a comprehensive appraisal of the growing industry, its impacts and regulation is now timely (IUCN, 1991). The recently adopted Protocol on Environmental Protection should provide the framework for such a review.

8.5.2 Development of Antarctic tourism

Tourism in the Antarctic is principally based on ship-borne cruises with shore landings, and to a much lesser extent on air overflights and landings, and has a relatively short history (Reich, 1980; Enzenbacher 1992a, 1993, 1994a; Headland, 1994a; Stonehouse, 1994). The first tourist flights began in 1956 (to the Antarctic Peninsula from Chile) and the first seaborne cruises in 1958. Since then, it is estimated that almost 70,000 tourists have visited Antarctica up to 1995/96. The

Fig 8.11 Brochure for expedition-style ecotourism, with cruises usually departing from Ushuaia in Argentina to cruise the Antarctic Peninsula and Scotia Sea (source: by courtesy of Quark Expeditions, Inc.).

data (Figure 8.12) show annual fluctuations in numbers of visitors, reaching a total of over 9000 in 1995/96. Such numbers now significantly exceed the total of scientific and logistic staff engaged in national programmes (estimated at 4000: Enzenbacher, 1993), although most tourists only visit Antarctica for a short time (e.g. a two-week cruise). Tourism is now probably the second largest commercial activity in Antarctica after fishing (Herr, 1989). Several categories of tourist activity have been recognised (Reich, 1980): shipborne cruises, tourist flights, private expeditions, goodwill, VIP and media visits, and off-duty visits by station staff. Ship capacities have ranged from about 1250 to less than 50 passengers, but most carry between 100 and 250 (Reich, 1980; Quigg, 1983; Enzenbacher, 1992a). In 1996/97, 104 cruises were made by a total of 13 ships (NSF data).

Airborne tourism has had a much smaller corner of the market (see Figure 8.12) (less than 6% since 1980/81; Enzenbacher, 1993), reflecting the relatively few airfields (only four hard runways), the lack of land-based passenger facilities and accommodation, and the very high costs. In addition, the USA has not allowed proposals for commercial flights to McMurdo, nor has the UK to Rothera. Tourist overflights of Antarctica were initiated in the 1970s and undertaken by Quantas and Air New Zealand. However, in 1979, when an Air New Zealand DC10 crashed on the slopes of Mount Erebus with the loss of all 257 passengers and crew on board (*cf.* Auburn, 1983; Suter, 1991), overflights ended for safety reasons and have only recently resumed in 1994/95 from Australia (Headland and Keage, 1995). The main developments in airborne tourism have now switched to the South Shetland Islands (Boswall, 1986) and inland parts of the continent (Swithinbank, 1993a). In the former case, the construction of a hard airstrip at Teniente Rodolfo Marsh Station on King George Island in 1979/80 has allowed regular flights from Chile from 1982 onwards, with accommodation provided in the first Antarctic hotel, 'Estrella Polar'. In the latter case, interest has focused principally on the Ellsworth Mountains, which includes Vinson Massif (the highest mountain in Antarctica). To access this area, a private company, Adventure Network International, operates a summer camp at Patriot Hills and uses wheeled aircraft (C-130) on a bare ice runway in support of private expeditions (Swithinbank, 1988b, 1993b, 1996). While costs remain high, numbers of visitors to the interior are likely to remain low (Swithinbank, 1993a). Helicopters are also used by some ships, increasing access to more remote and pristine areas.

Information on private expeditions (both scientific and non-scientific), a category that includes both high-profile climbing and adventure expeditions as well as less publicised visits by small boats, is more limited (e.g. Boczek, 1988). Up to 1980, Reich (1980) estimated a total figure of only 100–200 visitors in this category, but this figure has increased by an unknown amount since then. Goodwill and VIP visits have featured in the programmes of most nations involved in the Antarctic, and media representatives and film makers have also been frequent visitors in small numbers. Finally, off-duty visits are made by scientific and other personnel participating in the various national Antarctic programmes; their numbers are unquantified and their impacts may be considered as part of those associated with station activities.

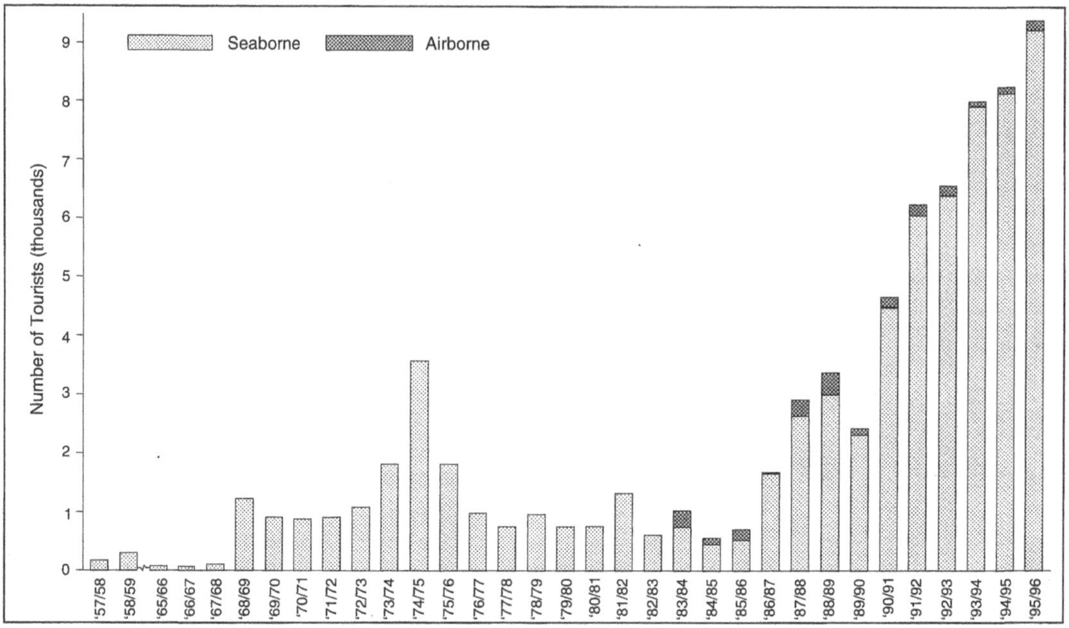

Fig 8.12 Estimated annual numbers of tourist visitors to Antarctica, 1957–1996 (sources: updated from National Research Council, 1993; Enzenbacher, 1994a; B Stonehouse, personal communication based on NSF data).

8.5.3 Impacts of tourism

Most of the current tourist activity is spatially focused on two areas – principally the Antarctic Peninsula (including the South Shetland Islands) and South Georgia – and to a much lesser extent the Ross Sea area (including McMurdo Sound, Cape Adare, and Commonwealth Bay) and East Antarctica (sometimes including islands such as Macquarie, Heard and Kerguelen). The reasons for this pattern are summarised in Table 8.4 and are related to access and interest (Figure 8.13). Within these areas, however,

Table 8.4 Tourist potential of Antarctica by geographical sector

	South America	New Zealand	Australia
Distance from Antarctica	1000 km	3000 km	>3000 km
Antarctic climate	mild, cloudy	very cold	very cold and windy
Islands *en route*	South Georgia, Falkland Islands, South Sandwich Islands , South Orkney Islands, South Shetland Islands,	Snares Islands, Auckland Islands, Campbell Island Macquarie Island, Balleny Islands	Heard Island, Îles Kerguelen, Îles Crozet
Sea access in summer	open	Ross Sea pack ice	offshore pack ice
Wildlife	several species of penguins, birds and seals on islands and Antarctic coasts	several species of penguins, birds and seals on islands and Antarctic coasts	several species of penguins, birds and seals on islands and Antarctic coasts
Historic sites	Charcot, Nordenskjöld Shackleton, Rymill, FIDS, whaling stations	Borchgrevink, Scott, Shackleton, Byrd	Mawson
Scientific bases	many	few	few

Source: Wace, 1990.

Fig 8.13 Tourists on the bow of the *Kapitan Khlebenikov* enjoy the sunshine and spectacular scenery of the Lemaire Channel, Antarctic Peninsula (source: photograph by S. Nicholini, courtesy of Quark Expeditions, Inc.).

itineraries are broadly similar and landfall destinations especially show a spatial preference for a relatively small number of locations, set amidst spectacular scenery, centred on occupied or unoccupied bases, wildlife colonies, former whaling stations and historical sites; for example, over 220 landing sites are cited but fewer than 50 are used regularly (B. Stonehouse, personal communication based on NSF data). During the 1995/96 season, six sites each had a total of over 3000 visitors, and two, Half Moon Island and Whalers Bay, each had over 5000 visitors. During the 1996/97 season, the restored British base at Port Lockroy, now a historic monument manned during the summer, attracted 4292 visitors from 65 ship and nineteen yacht visits (J.R. Shears, personal communication). Some scientific stations have also experienced heavy visitor pressure, particularly on the South Shetland Islands and the Antarctic Peninsula, with attendant concerns about disruption of scientific programmes and station routines. In response, the US authorities have imposed restrictions on the visits of tourists to Palmer Station on the Antarctic Peninsula, and in 1993 appointed a ranger from the US National Parks Service to handle visiting groups (Mervis, 1993a). The British Antarctic Survey strictly regulates the number of cruise ship visits per year to its stations (two – four according to the station) (Enzenbacher, 1994b). Historic era huts are a further focus of attraction, resulting in problems of visitor management as well as long-term preservation of the structures and associated artefacts, which are showing significant deterioration (Hughes, 1994; Hughes and Davis, 1995; Harrowfield, 1995).

As early as the 1970s, concern existed about the environmental impact of tourism (Roberts, 1978). In its more extreme forms, such concern is associated with a general opposition to all forms of commercial enterprise in the Antarctic, but there have also been more reasoned arguments against the development of tourism and private expeditions (*cf.* Nicholson, 1986; Boczek, 1988; Beck, 1990d). These focal locations are frequently in the most sensitive and fragile terrestrial environments in the few ice-free areas occurring at the periphery of the continent as well as being the main breeding sites for wildlife (Figure 8.14) (see Chapter 3). There are therefore concerns about the actual and potential direct environmental impacts (Enzenbacher, 1992b; Hall and McArthur, 1993), including harassment of wildlife (particularly during the breeding seasons of birds and mammals, when most visits take place), introduction and spread of bird or other animal diseases, encroachment on Specially Protected Areas and Sites of Special Scientific Interest (now Antarctic Specially Protected Areas and Antarctic Specially Managed Areas), disturbance of historical sites, disposal of waste from cruise ships and environmental effects of shipping accidents. Visits to scientific stations may disrupt routines and research work, and impose on resources. There is a lack of coordinated search and rescue facilities in the Antarctic and there is an implicit humanitarian obligation on the various national operators to assist in the event of an emergency. In effect, the latter are providing a form of hidden

Fig 8.14 Wildlife colonies are a major tourist attraction, but too many visitors may disturb breeding birds. In recognition of this, many tour operators are members of the International Association of Antarctic Tour Operators (IAATO), one of whose objectives is to promote and practise safe and environmentally responsible travel to the Antarctic (source: photograph by S. Nicholini, courtesy of Quark Expeditions, Inc.).

subsidy to the tour operators (Herr, 1989). Similarly, there is an implied obligation on these same organisations to undertake costly clean-up operations following shipping accidents, drawing on limited physical and human resources (see above).

Conversely, on the positive side, many visitors to the Antarctic are sensitive to environmental issues (perhaps more so than some ships' crews or government expedition support staff), gain a greater awareness of the global importance of Antarctica and are potentially influential advocates able to strengthen political support not only for conservation but also for scientific research in the Antarctic (Tangley, 1988; IUCN, 1991). In addition, tourist operations may be able to assist research activities, for example through providing transport for field parties (Wace, 1990). Equally, tourism may be used to support national programme logistics, and countries such as Argentina, Chile and France are already exploiting this source of revenue; for example, the new French research vessel has berths available for tourists who want to visit the French sub-Antarctic islands.

It is clear that Antarctic tourism is now an established and growing activity. Although the visitor totals are still relatively small, tourism is spatially focused in environmentally sensitive areas that are accessible and have special interests of much wider significance. Impacts may therefore be locally high, and there are now valid concerns about the further growth of tourism and the longer-term impacts. However, in spite of reports of disturbance to wildlife and vegetation (e.g. Smith, 1988b; Codling, 1982; Wace, 1990; Enzenbacher, 1992b; Sitwell, 1993), to date there has been little systematic monitoring of the environmental impacts of tourist activities but some research in progress represents a significant step forward in acquiring appropriate data on which to base future management strategies (Stonehouse, 1992a, 1993; Enzenbacher, 1992b).

According to Stonehouse (1992a), 'in a period during which several expedition and survey ships have been damaged or lost, some with considerable environmental impact, no ship dedicated to tourism has been lost, suffered serious damage, or made a serious impact on the environment' (p.214). In a similar vein, the IUCN (1991) reported that 'experience to date suggests that, in general, tourist operations have been conducted in a responsible manner and undesirable impacts have not been severe, especially compared to environmental impacts of scientific and associated logistical activity' (p.56). However, by way of qualification, the IUCN report also adds that the Mount Erebus air disaster and the shipping accidents noted in Section 8.2.3 illustrate the type of impacts that could arise from unregulated development of tourist activities. More worrying for most Antarctic operators is the possibility of a major accident involving a tour ship. Any loss of a tourist vessel could be potentially serious in view of the difficulty in mounting a major rescue attempt.

8.5.4 Regulation of tourism

Various measures to regulate tourism have been implemented under the Antarctic Treaty since 1966 and more recently by the tour operators in furtherance of a treaty recommendation in 1994 (Nicholson, 1986; Beck, 1990d; Heap, 1994; Stonehouse, 1992a; Enzenbacher, 1995). These include the preparation of guidelines for visitor behaviour and procedures to be followed by operators, involving requirements to circulate advance information about tourist expeditions, to conduct EIAs and to prevent pollution. The recognition of Areas of Special Tourist Interest was proposed in 1975. However, none were identified, and this designation has now been overtaken by the agreement on two types of protected area under Annex v of the Protocol on Environmental Protection (see Chapter 9). The protocol covers all activity in Antarctica, including tourism.

In 1975, the ATCPs acknowledged that tourism was a legitimate activity in Antarctica, but translating the recommendations for its regulation into an accepted code of practice has been difficult (Beck, 1990d; Stonehouse, 1990). It remains the responsibility of individual treaty members to ensure that their nationals comply with the various recommendations relating to tourism and conservation (see Chapter 9). However, some countries, including the UK, Australia and the USA, have incorporated regulations governing the conduct of visitors to the Antarctic into their own legislative measures. The USA, New Zealand and Australia have also employed observers aboard Antarctic cruise ships operated by tour companies based in, or operating from, those countries.

Major tour operators have been at the forefront in the development of an environmentally responsible industry, an approach promoted since 1991 through the International Association of Antarctic Tour Operators (IAATO) (Enzenbacher, 1992a; Splettstoesser and Folks, 1994). IAATO recognises that 'environmentally conscious tourism to Antarctica will benefit the continent's future preservation' (Stonehouse, 1992a, p.215). To this end, IAATO members work within the framework of the Antarctic Treaty, including the Protocol on Environmental Protection, and they have implemented the most comprehensive set of guidelines so far for both shipborne tours and aircraft-supported tours to the continental interior (Stonehouse, 1992b; Splettstoesser and Folks, 1994). IAATO included twelve members in 1993, which were responsible for 70% of all tourists in 1992/93 (Enzenbacher, 1994a), and membership has since increased to sixteen in 1995. In effect, management of a large part of current tourist activities is self-regulated by the industry, although IAATO now has formal links with the ATS as an invited attendee at Consultative Meetings.

Various codes of conduct have been produced for tour operators and visitors to the Antarctic. In addition to those produced by IAATO, SCAR published a brochure, *A Visitor's Guide to the Antarctic and its Environment* (1980), and the treaty parties recommended the adoption of two new codes in 1994: *Guidance for Visitors to the Antarctic* and *Guidance for Those Organising and Conducting Tourism and Non-Governmental Activities in the Antarctic*.

The Protocol on Environmental Protection (see Chapter 9) provides the most comprehensive set of measures so far for protection of the Antarctic environment. All tourists and tour operators from ATCP states will be covered by their appropriate domestic enabling legislation, and all tourist activities will require EIA. The issue of liability will also be covered in the Liability Annex currently being negotiated by the treaty parties. However, there are still long-standing inadequacies relating to aspects of compliance, enforcement, jurisdiction and insurance that remain to be addressed satisfactorily (Auburn, 1982; Nicholson, 1986; Boczek, 1988; Beck, 1990d; Manheim, 1990; Enzenbacher, 1995). There is also the difficulty of regulating the activities of non-ATCPs and non-IAATO members. For example, in the 1992/93 season, of the twelve cruise ships operating, only five were registered in states party to the Antarctic Treaty, the remainder being registered in the Bahamas, Liberia and Panama. In addition, 30% of tourists are carried by operators outside IAATO and therefore currently outwith their self-regulating framework and stringent operational procedures.

Currently, there are conflicting views about how to address these issues and whether a separate tourism annex to the Protocol on Environmental Protection is necessary (Enzenbacher, 1992b, 1993; Herr, 1993; Vidas, 1996). The view of many ATCPs and the industry is that the measures in the protocol, including the requirement to prepare EIAs, already cover tourist activities and that implementation of the protocol should take priority over the development of new measures. Certainly, the measures in the protocol and the visitor and operator guidelines that are now in place will encourage responsible tourism and help to minimise environmental impacts. However, while a welcome step, such developments do not address the complex jurisdictional questions relating to liability, enforcement and compliance by third parties outside the treaty states and IAATO. Nor are they a substitute for a comprehensive tourist management framework (Davis, 1995). Such a framework, developed in partnership between the treaty parties, IAATO and the NGOs, could usefully address a range of issues (*cf.* IUCN, 1991). First, there is a need for guidance and standards for EIA for tourist activities. Second, as well as the requirement for tour operators to carry out EIAs for their activities, there is a need for management plans and independent monitoring at the most frequently used sites to help to assess the cumulative impacts and the acceptable types and levels of use. Third, there is the coordination of visits to the more popular sites to ensure that impacts remain within acceptable levels. Fourth, there is the establishment of a network of robust and carefully chosen sites to spread the pressure of tourist visits and provide a range of alternative attractions. In practical terms, such an approach would control the spread of tourist impacts and would facilitate monitoring and enforcement, which in any case may only be realistic at such sites. A balance would need to be found between sites that have sufficient tourist appeal (e.g. with wildlife and scenic interests) and where environmental impacts can be contained and monitored. To avoid conflict between tourism and scientific activities and to prevent adverse environmental impacts, appropriate planning and management, such as land use zoning, could progress towards a system of integrated environmental management that also incorporates protected areas and other uses of land and water. Implementation of such zoning would require cooperation between the ATCPs, particularly in the Antarctic Peninsula area, where there are sensitivities concerning overlapping sovereignty claims and territorial control. If such an approach were adopted, IAATO would continue to have a key role in ensuring that the high standards of operation generally demonstrated by its members were maintained.

A tourist management framework would also have to consider jurisdictional issues and the regulation of operators or ships outwith the sphere of IAATO and the treaty parties, although these matters are not likely to be easily resolved. Similarly, the regulation of private 'adventure' expeditions is also problematic. These include mountaineering and overland journeys and visits by yachts. Although expeditions from countries party to the Protocol on Environmental Protection are governed by its measures as far as environmental protection are concerned, enforcement is clearly very difficult. Also, adventure expeditions can, and do, get into difficulties and expect national operators to come to their rescue (e.g. Anon, 1994). While there is a clear requirement for such expeditions to be self-supporting, this is more difficult to achieve in practice and raises difficult questions about licensing and liability (e.g. Johnston, 1997).

In the sub-Antarctic island groups, where sovereignty is well-established, various controls on tourist activity have been implemented (e.g. Hall and Wouters, 1994; Wouters and Hall, 1995a). Management plans have been produced for the New Zealand island groups, which include setting quotas for visitor numbers and levying fees from visitors to support management work (Sanson, 1994; Wouters and Hall, 1995b). At Macquarie Island, a similar system is operated by Australia as part of the management plan for the island (Hall and McArthur, 1993). Tourism is permitted where it does not conflict with the overall objectives of nature conservation and is carefully managed; for example visits are ship-based, walkways and viewing platforms are provided in appropriate areas, and strict conditions aim to minimise environmental impacts (Hall and McArthur, 1993; Stephenson, 1993). On South Georgia, Britain encourages tour operators to restrict their visits to designated areas at Grytviken and the Bay of Isles, but in practice this is often ignored (Smith, 1995a).

8.5.5 Future developments

Although visitor numbers may now have peaked, and fewer tourists were expected to go south in 1996/97 than in recent years, tourism is likely to remain a significant activity in Antarctica. Possible future developments may see the exploitation of new opportunities. For example, utilising hard airstrips accessible from South America and the Falkland Islands could allow tour operators to offer combined fly–cruise tours in the Antarctic Peninsula area, as well as shorter trips, to a larger market (Swithinbank, 1993a). Another area of development may arise from increased access to the continental interior by direct flights from South America, South Africa, Australia and New Zealand using large, wheeled aircraft to land on blue-ice runways (Mellor and Swithinbank, 1989; Swithinbank, 1991, 1993a). This would inevitably raise questions about air traffic control safety and the environmental impact on pristine areas in the interior. Developments in other sectors appear less certain unless air access is facilitated, for example in Australian Antarctic Territory (AAT) (Herr, 1989). However, an Australian parliamentary report advised the government against granting approval for airstrips and land-based facilities for tourists until a full conservation strategy had been developed for AAT (House of Representatives Standing Committee on Environment, Recreation and the Arts, 1989; Wace, 1990).

Development proposals by an Australian company in 1989 point to perceived commercial opportunities in Antarctica. The plan, which did not proceed, included a £100 million conference centre, hotel and hospital complex in the Vestfold Hills oasis near Australia's Davis Station, served by a runway capable of taking jumbo jets and up to 16,000 tourists per year (House of Representatives Standing Committee on Environment, Recreation and the Arts, 1989; Anon, 1989d).

Increased demand for adventure activities involving camping and trekking may see onshore resource centres and lodges developed in areas that are able to absorb large numbers of visitors from larger ships (Stonehouse, 1994). Proposals for tourist 'camps' on King George Island have been produced by Marine Expeditions, but they did not proceed. Such developments would carry far greater risk to the environment than existing ship-based operations. Whether such developments are appropriate in Antarctica is questionable, but the issue may ultimately need to be addressed. If such developments do proceed, then it will be necessary to consider the options for more formal wilderness management (*cf.* Hendee *et al.*, 1990) possibly through local parks, although this would raise further difficult questions concerning funding, administration and sovereignty in areas of overlapping claims, as well as the practicalities of operations currently beyond the scope of the Protocol on Environmental Protection (Stonehouse and Crosbie, 1995).

There is a continuing search for new and different sites for tourists to visit; for example there have been helicopter tours into the Dry Valleys (Enzenbacher, 1993) and Russian icebreakers used as cruise ships are now capable of reaching previously inaccessible areas. A key concern here is that the sensitivity and scientific value of such areas may not be known or adequately protected (*cf.* Chapter 9). Also of concern is the deployment of larger cruise ships (up to 400 passengers) and a weakening of the original Linblad ethos of environmentally sensitive Antarctic tourism (Stonehouse and Crosbie, 1995). Another form of development may see the leasing of accommodation or station facilities to tour operators (Sayers, 1993), which might prove attractive to some nations starved of support funding. Interest in the novelty aspect of Antarctica is also likely to continue: for example, the world's southernmost marathon was run in 1995 by 150 participants on Fildes Peninsula in the South Shetland Islands. However, it is highly questionable whether mass participation events or 'stunts' are appropriate in this environment. Some states (e.g. Chile) assist tourist activities through the use of airstrips, and future developments may see further provision of tourist facilities in tacit pursuance of territorial claims.

8.5.6 Summary

In the wider context of human impacts in Antarctica, those of tourist activities appear to be small in relation to the past effects of scientific expeditions and commercial activities such as fishing. Generally, smaller ships adhering to the requirements of the Protocol on Environmental Protection and putting ashore small numbers of people well-supervised and following recognised codes of conduct, as in the traditional Linblad approach, probably have minimal impact. While the tourist presence continues to be transient and essentially ship-based and travel in the Antarctic remains hazardous, this situation seems likely to continue. However, the long-term, cumulative effects of repeated visits at the more frequently used sites are more difficult to judge, particularly where they are geographically focused on specific areas of the Antarctic Peninsula and the Ross Sea. In these situations, tourism represents a further human stress on the biota, and until existing environmental impacts and their cumulative effects are comprehensively assessed, and carrying capacities of particular areas are established, management should follow precautionary principles; for example, new areas should not be visited before EIA is carried out. There are clear implications here for scientific research, since information is needed to underpin this work, as well as the development of site management plans and the development of a broader management framework for tourism.

With increased public interest in environmental issues and heightened awareness of Antarctica, its attraction as an ecotourism destination seems likely to continue. It is fortunate that governments, scientists and environmentalists regard tourism as a valid activity, since any ban would be difficult if not impossible to enforce. Moreover, educational and environmentally conscious tourism has an important role to play in fostering wider public support for protection of the Antarctic environment and the compliance with agreed standards. The development of an appropriate management strategy is needed to address the concerns of all parties and to achieve an industry that is sustainable. Currently, the industry is self-regulating, and it is in the long-term commercial interest of IAATO to ensure minimal environmental impacts and to promote good practice in environmentally conscious tourism. However, there is also a need to ensure that regulations apply to all tour operators. In view of the global importance of the Antarctic environment and its sensitivity to disturbance, it seems clear that the present mix of official recommendations and voluntary codes of practice is unsatisfactory. A more formal management strategy developed in partnership between the treaty parties and IAATO, with appropriate scientific underpinning and monitoring, and within the existing framework of the Protocol on Environmental Protection, remains a priority.

8.6 Introduced species

8.6.1 Introduction

The known impacts of introduced species relate mostly to the sub-Antarctic islands. At least fifteen species of mammal have been introduced, both intentionally and accidentally, to these islands, which are the only Antarctic terrestrial ecosystems capable of sustaining mammals (Holdgate and Wace, 1961; Bonner, 1984a; Clark and Dingwall, 1985; Leader-Williams, 1985, 1988; Walton, 1987d; Chapuis *et al.*, 1994). Many accompanied the sealers and whalers in the nineteenth and early twentieth centuries, and in the 1950s more were brought by the expeditions that manned scientific bases. The planned introductions have included domestic animals (Holdgate and Wace, 1961) as sources of food and companionship, and reindeer for sport on South Georgia (Leader-Williams, 1988) (see Figure 3.31); unplanned introductions have been principally of rats and mice. Several of the deliberate introductions have failed and at present only eight species of mammal are known to be established on the various islands within or close to the Polar

Table 8.5 Human impacts on the sub-Antartic islands

Island	Jurisdiction	Total area (km^2)	Protected area (km^2)	Whaling stations (no. of years)	Scientific stations[1]	No. of established introduced vascular species	No. of established introduced mammal species	Damage by introduced herbivores[3]
South Georgia	UK[2]	3756	24.5	60	P	22	3 Norway rat, house mouse, reindeer	3
South Sandwich Islands	UK[2]	618	–	–	T	0	0	0
Bouvet Island	Norway	54	54	–	T	0	0	0
Marion Island	S.Africa	298	–	–	P	14	2 house mouse, cat	1
Prince Edward Island	S.Africa	47	–	–	–	2	0	0
Îles Crozet	France	345	225	–	P	27	4 house mouse, rabbit, black rat, cat	1
Îles Kerguelen	France	7215	80	3	P	10	7 black rat, house mouse, rabbit, sheep, cat reindeer, mouflon	3
Heard Island	Australia	380	380	–	T	0	0	0
McDonald Island	Australia	2.6	2.6	–	–	0	0	0
Macquarie Island	Australia	127.9	127.9	–	P	4	4 black rat, house mouse rabbit, cat	3

[1] P: permanent; T: temporary
[2] Also claimed by Argentina
[3] Degree of impact (including non-established introductions) 0 – negligible; 1 – slight; 2 – moderate; 3 – locally severe.

Sources: Leader-Williams, 1985; Walton, 1987d; Dingwall and Lucas, 1992; Smith, 1995b.

Front (Table 8.5); only the Heard and Macdonald Island group, the South Sandwich Islands and Bouvet Island are free of introduced mammals.

Numerous plant and insect species have also been introduced (Bonner and Lewis Smith, 1985; Clark and Dingwall, 1985; Walton, 1987d; Cooper and Condy, 1988; Smith, 1996). Various field experiments to grow non-indigenous plants have fortunately had a negligible impact (Smith, 1996). On the Antarctic continent, non-native bacteria and fungi have been introduced, usually unintentionally, to almost sterile and pristine areas (e.g. Cameron *et al.*, 1977; Vincent, 1988).

8.6.2 Impacts of introduced species

The impact of these introductions on the native floras and faunas of the sub-Antarctic islands has been substantial (Holdgate and Wace, 1961; Holdgate, 1970; Leader-Williams, 1985) (Table 8.6), reflecting the simplicity of the terrestrial ecosystems and their limited capacity for readjustment. In many cases, the introduced mammals readily adapted to the climatic conditions, and populations expanded under conditions of abundant food availability, lack of competition from native species and lack of predators; subsequently numbers have fallen in balance with the food supply. The impacts include predation on native species, vegetation modification through grazing pressure and competition with, and in some cases elimination of, native species.

Among the principal impacts are the grazing of herbivores (sheep, rabbits and reindeer on various islands) and predation by rats and cats. Tussock grasslands and herbfields have been overgrazed, resulting in soil erosion, the spread of alien grasses and loss of habitat for burrow- and ground-nesting birds and invertebrates. Cats and rats have locally reduced or eliminated burrow- and ground-nesting birds, for example on Macquarie Island (Rounsevell and Brothers, 1984), Îles Kerguelen and Îles Crozet (Jouventin *et al.*, 1984; Jouventin and Weimerskirch, 1991) and rats likewise on South Georgia (Croxall *et al.*, 1984). Initially, the cats were introduced to control rodents, but the birds have often proved easier prey! Thus on Marion Island, five cats introduced as pets in 1949 had expanded to about 3400 by 1977 and were estimated to be killing 450,000 burrowing petrels per annum (Bloomer and Bester, 1992). The extinction there by 1965 of the common diving petrel has been attributed to cat predation (Williams, 1984). Introduced species may also have indirect effects; for example, on Marion Island, predation of mice on insect larvae, which play an important role in the decomposition of plant litter, is thought to have a significant impact on nutrient mineralisation and recycling processes (Crafford, 1990). On South Georgia, the introduced reindeer have had a significant impact on the vegetation of their restricted range areas in the form of overgrazing of tussock grass, macrolichens and *Acaena magellanica*, invasion of grazed areas by the introduced species *Poa annua*, and trampling damage (Bonner, 1984a).

The introduction and dispersion of microorganisms through human activity may have led to changes in species composition or diversity, disruption of competitive interactions between species, and alteration of microbial gene pools; habitats may also have been damaged or destroyed and nutrient status altered (White, 1995). However, the scale of any such effects is unknown (Vincent, 1988) but is likely to be small and localised. Climate change may ultimately have greater and more widespread effects. Such change could enhance the viability, growth and competitiveness of introduced species, and enhanced rates of decomposition of organic matter could significantly alter the nutrient supply within the unique soil and freshwater ecosystems of continental Antarctica.

Concerns that human activity may inadvertently spread animal diseases to Antartica have arisen following the recent discovery of an avian pathogen, infectious bursal disease virus, in emperor penguin chicks and adult Adélie penguins at sites near Mawson Station (Gardner *et al.*, 1997). Although no adverse effects have been detected in the penguins, the highly contagious virus may make the chicks susceptible to infections and retard their growth. The virus could have been transported to Antarctica in poultry products and transferred via scavengers such as skuas. Since the virus can be carried in contaminated clothing and vehicles, its spread in Antarctica could be facilitated by the movement of scientists and tourists.

Table 8.6 Impacts of introduced mammals on the flora and fauna of sub-Antarctic islands.

Affected species	Status	Extinct	Locally eliminated or reduced	Locally spreading or beneficiary	Island group	Animal responsible	Reason
Tussock grassland	Native						
Poa flabellata			*		S. Georgia	Reindeer	Grazing
Poa foliosa			*		Macquarie	Rabbits	Grazing
Poa cookii			*		Kerguelen	Rabbits	Grazing
Heath	Native						
Acaena magellanica			*		S. Georgia	Reindeer	Grazing
			*		Crozet	Rabbits	Grazing
					Kerguelen	Reindeer	Grazing
			*		Kerguelen	Rabbits	Dispersion of seeds
Azorella selago			*		Kerguelen	Rabbits	Grazing
Herbfield	Native						
Stilbocarpa polaris			*		Macquarie	Rabbits	Grazing
Pringlea antiscorbutica			*		Kerguelen	Rabbits	Grazing
Lichens	Native		*		S. Georgia	Reindeer	Grazing
			*		Kerguelen	Reindeer	Grazing
Poa annua	Introduced			*	S. Georgia	Reindeer	Grazing-tolerant
				*	Macquarie	Rabbits	Grazing-tolerant
				*	Kerguelen	Rabbits	Grazing-tolerant
Parakeet *Cyanoramphus novae-zealandiae*	Endemic	1880–1891			Macquarie	Cats/Wekas	Predation
Rail *Rallus philippensis*	Endemic	1894			Macquarie	Cats/Wekas	Predation
Pipit *Anthus antarcticus*	Endemic		*		S. Georgia	Rats	Predation
Burrow-nesting prions	Native		*		S. Georgia	Rats	Predation
and petrels			*		Macquarie	Cats	Predation
(Most species)			*		Kerguelen	Cats	Predation
			*		Marion	Cats	Predation
			*		Crozet	Cats	Predation
Skuas	Native			*	Kerguelen	Rabbits	Additional prey
Catharacta lonnbergi				*	Macquarie	Rabbits	Additional prey
Wekas *Gallirallus australis*	Introduced			*	Macquarie	Rabbits	Prey

Source: Leader-Williams, 1988

8.6.3 Management Controls

In the sub-Antarctic islands, management control has involved a number of measures, including the periodic shooting of cats on Îles Kerguelen and the introduction of myxomatosis on Macquarie Island and Îles Kerguelen, but attempts so far have had only limited success (Leader-Williams, 1985). On Marion Island, various measures, including introduction of a viral disease, feline panleucopaenia, have been employed successfully to eradicate cats (Bloomer and Bester, 1992). On other islands, measures to control or eradicate introduced mammals have been implemented or are planned (*cf.* Dingwall, 1995). However, even where eradication of introduced species is practical, it will be costly and will require assessment of priorities and benefits. Although work reviewed by Leader-Williams (1988) suggests at least partial recovery of some ecosystems when introduced mammals are removed, some ecosystems will have altered irrevocably (Holdgate and Wace, 1961; Bonner, 1984a; Leader-Williams, 1985). In view of the growing recognition of the conservation significance of the sub-Antarctic islands (Bonner and Lewis Smith, 1985; Clark and Dingwall, 1985; Smith and Smith, 1987), further evaluation and more concerted management efforts to protect refugias for native floras and faunas are urgently required (Jouventin *et al.*, 1984; Walton, 1987d; see Chapter 9).

In most cases, management controls can be justified on scientific grounds. However, as Leader-Williams *et al.* (1989) have argued, this is not the case for reindeer on South Georgia, where the impacts, although significant, are localised and the survival of native species and communities is not threatened. Indeed, on the positive side, the introductions have provided a unique scientific opportunity to study population expansions and dynamics (Leader-Williams, 1988). Further, the grazing pressure and the destruction of tussock grass areas may have exerted a control on rat populations.The present ranges of the reindeer herds are restricted by tidewater glaciers. However, if climate warming and glacier recession continue at South Georgia (cf. Gordon and Timmis, 1992), then the herds may be able to extend their range further and increase the spatial extent of the damage to the island's vegetation. Ultimately, therefore, management of the reindeer will involve value judgements to be made in light of an overall management strategy for South Georgia.

In Antarctica, provisions to preclude the introduction of non-indigenous species were included in the Agreed Measures, although it appears that these were not always implemented (Smith, 1996). Annex II of the Protocol on Environmental Protection extends these controls. The prohibition on introductions of non-native plants and animals is continued, and permits may be issued to allow the import only of domestic plants and laboratory plants and animals. Measures are also included to prevent the introduction of micro-organisms via importation of live poultry and non-sterile soil and to regulate the disposal of poultry waste. In addition, under Annex II, all dogs were removed from Antarctica in 1994.

8.6.4 Summary

The sub-Antarctic islands have had a long history of contact with human activity, through sealing and whaling and more recently through the siting of scientific stations. The level of human modification, however, has been variable (e.g. Clark and Dingwall, 1985) and this applies to the impacts of introduced species as well as to other activities. What is required is an overall conservation management strategy for these islands (see Chapter 9), which would include an assessment of the desirability and practicality of exterminating introduced species as part of the wider management objectives for each island. Although the management ideal may be to remove all introduced species, this may cause unacceptable disturbances, since some ecosystems will have adjusted to new conditions. It is also unlikely that the pre-introduction balance of species and community structures could be regained. In spite of these *caveats*, however, it seems just as important to assess what can be done to restore ecosystems to an approximation of their former states by removal of introduced species as it does to restore sites by the removal of human debris and abandoned scientific bases that serve no further purpose.

8.7 Conclusion

Direct human impacts on the Antarctic environment have developed in parallel with the expansion of scientific activities and the establishment of temporary and permanent stations, and more recently with other human activities such as tourism. Cumulative impacts have been greatest on those ice-free margins of the continent that have proved most accessible and that have a long history of occupation. Locally, the impacts have been high in terms of poor waste disposal, pollution, wildlife disturbance and soil disturbance, reflecting a low priority accorded to careful environmental management in the past. While such impacts are spatially limited when viewed at the scale of the continent, they are nevertheless more significant when seen in the context of the more limited extent of the ice-free areas. Measures to regulate impacts have been ignored or applied inconsistently in the past, according to perceived national interests. Impacts of the tourist industry are small and localised, but growing cumulatively, and probably the greatest concern is over the extension of the human 'footprint' to the more pristine parts of the continent. There is also concern about increased levels of marine traffic associated with tourism and the potential for major incidents in biologically vulnerable areas such as the Antarctic Peninsula. Human impacts on some sub-Antarctic islands have been severe, particularly where introduced species have competed with the natural floras and faunas.

Recently, however, there has been a marked change in attitudes to environmental management, encouraged by a growing global consciousness both of the sensitivity of the Antarctic environment and the need for sensitive environmental stewardship, as well as of environmental issues generally. This change is reflected in the measures now set out in the Protocol on Environmental Protection (see Chapter 9). Scientists, as the main users of Antarctica, now not only concern themselves with the environmental impact of their research and that of the necessary logistical support, but also contribute to improve wider public understanding and justification of their work. The environmental effects of human activities can no longer be disregarded or left to others to clean up afterwards. It is therefore important that the measures in the protocol are fully implemented and enforced, as legally required, by all states and other parties operating in Antarctica.

Indirect effects from most forms of global pollution have so far been minor, reflecting the relative isolation of Antarctica, but they are nevertheless measurable and widespread. In the longer term, however, such pollution may present the greatest threat to Antarctica since it is much more difficult to control than localised pollution, which is now subject to stringent controls through the measures in the Protocol on Environmental Protection. Of particular concern are the potential effects of global warming from global emissions of greenhouse gases and the springtime depletion of the ozone layer over Antarctica arising from the production and release of CFCs and other ozone-destroying gases in the industrialised and developing world. As yet, the effects on Antarctic ecosystems remain unclear, but could be very significant. At a continental scale, global warming may ultimately produce more significant and wide-ranging changes than other forms of localised pollution or environmental disturbance, in the form of changes in climate, species distributions, ice-sheet behaviour and sea-level rise (see Chapter 5).

Further reading

Aust, A. and Shears, J. (1996) Liability for environmental damage in Antarctica. *Review of European Community & International Environmental Law*, 5, 312–320.

de Poorter, M. and Dalziell, J.C. (eds) (1997) *Cumulative Environmental Impacts: Minimisation and Management*. Proceedings of the IUCN Workshop on Cumulative Impacts in Antarctica, Washington DC, 18–21 September 1996, IUCN, Gland.

Hall, C.M. and Johnston, M.E. (eds) (1995) *Polar Tourism. Tourism in the Arctic and Antarctic Regions*. Wiley, Chichester.

Kennicutt II, M.C. and Champ, M.A. (eds) (1992) Environmental awareness in Antarctica: history, problems and future solutions. *Marine Protection Bulletin*, 25 (9–12).

May, J. (1988) *The Greenpeace Book of Antarctica. A New View of the Seventh Continent*. Dorling Kindersley, London.

Szabo, M. (1994) *State of the Ice: An Overview of Human Impacts in Antarctica*. Greenpeace International, Amsterdam.

PART 3

Past and future environmental management of the Antarctic

CHAPTER 9 The Antarctic Treaty System: a developing framework for environmental management

Objectives

Key themes and questions covered in this chapter:

- development of environmental protection and resource management under the Antarctic Treaty System
- the Antarctic protected areas system
- CCAMLR and problems of regulating common access fisheries
- CRAMRA: is mining in Antarctica compatible with environmental protection?
- the Protocol on Environmental Protection: a framework for environmental stewardship
- the role of NGOs as a force for change
- the common heritage and world park approaches: alternatives to the Antarctic Treaty?
- the sub-Antarctic islands: management issues and priorities

9.1 Introduction

During the period of development of the ATS since 1961, measures for conservation and environmental management have been added principally in an *ad hoc* fashion. Although the treaty itself did not incorporate specific conservation measures, it was recognised that conservation would require to be addressed in furthering its principles and objectives. Under the ATS, conservation and environmental management have been developed through a series of legal instruments and treaty recommendations (Figure 9.1), which establish general principles and address specific activities and threats (Table 9.1; see also Chapters 7 and 8) (Holdgate, 1970, 1987; Roberts, 1977; Bonner, 1984b, 1990; Boczek, 1986; Heap and Holdgate, 1986; Heap, 1994; Harris and Meadows, 1992). This chapter reviews these measures.

9.2 The Antarctic Treaty System: Environmental Protection and Resource Management

The legal instruments now in force comprise the Agreed Measures on the Conservation of Antarctic Fauna and Flora (the Agreed Measures), the Convention for the Conservation of Antarctic Seals (CCAS) and the Convention on the Conservation of Antarctic Marine Living Resources (CCAMLR); these are reviewed below in Sections 9.3–9.5. Strict controls to minimise the environmental impacts of minerals activities were included in the Convention on the Regulation of Antarctic Mineral Resource Activities (CRAMRA), and although this convention failed to be ratified, a number of its measures have been incorporated in the Protocol on Environmental

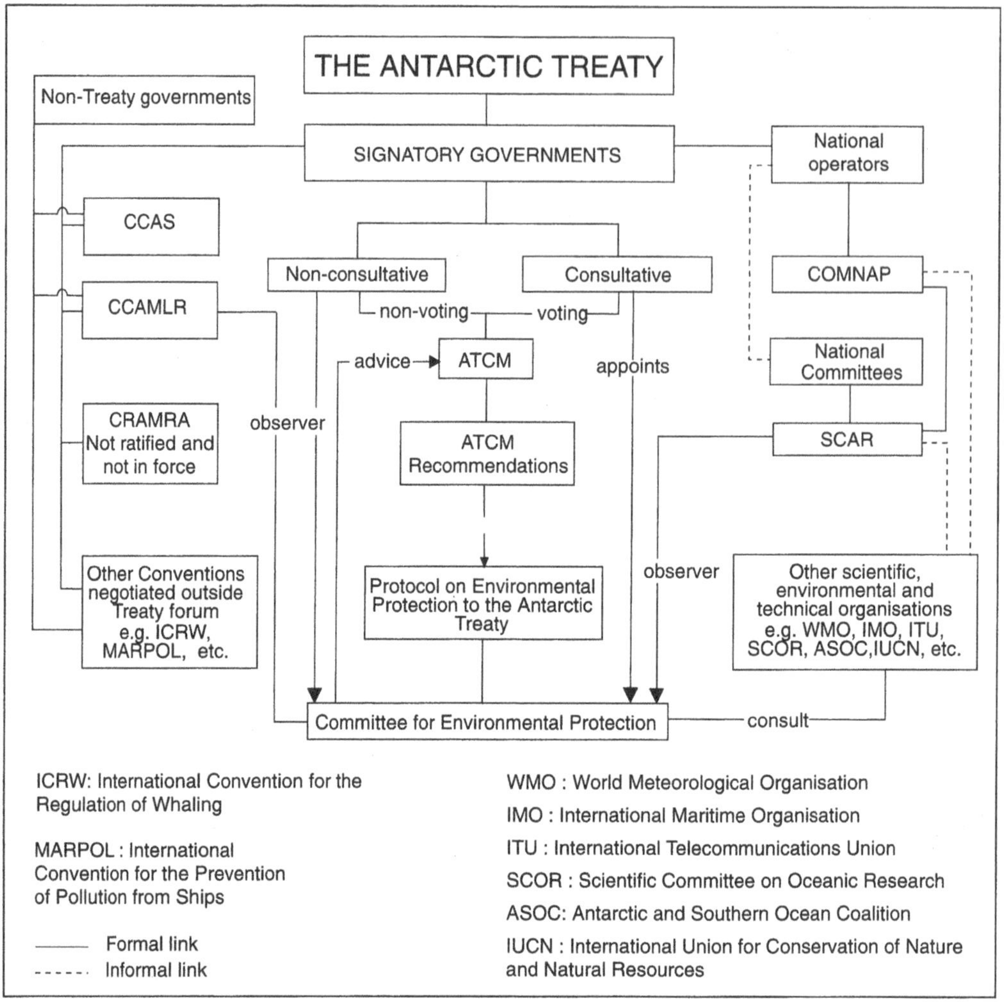

Fig 9.1 Summary of the main instruments and institutions for environmental protection and management in Antarctica under the Antarctic Treaty System (source: Harris and Meadows, 1992).

Table 9.1 Actual and perceived threats to the Antarctic environment

Activities and impacts	Scale of impact
commercial exploitation of living resources (impacts on both target species and dependent species in the food chain, principally whales, finfish, krill, seals, squids)	regional
pollution from local sources (impacts on terrestrial and marine ecosystems, involving both chronic contamination and major pollution incidents)	local
introduction of alien species (impacts on terrestrial ecosystems)	local–regional
tourism (local pollution and disturbance of wildlife and protected areas)	local
commercial minerals activities, including offshore oil and onshore minerals (potential impacts on terrestrial and marine ecosystems)	local–regional
scientific expeditions and logistical support activities (impacts on terrestrial and marine ecosystems)	local
pollution from global sources, including effects of global warming from greenhouse gas emissions and stratospheric ozone depletion (impacts on terrestrial and marine ecosystems)	regional

Protection to the Antarctic Treaty adopted in 1991, the provisions of which are reviewed in Section 9.7. Some activities in the Southern Ocean are also regulated by other international agreements, for example marine pollution (International Convention for the Prevention of Pollution from Ships – MARPOL 73/78), whaling (the International Convention for the Regulation of Whaling – see Chapter 7) and exploitation of the seabed (Law of the Sea Convention). Other international conventions (e.g. Ramsar Convention, Bonn Convention, World Heritage Convention) do not apply in Antarctica, because sovereignty is not recognised under the Antarctic Treaty. However, they do apply in the case of sub-Antarctic islands where sovereignty has been established, if the sovereign states have ratified the conventions.

The treaty parties have also adopted numerous recommendations relating to environmental management. These cover general provisions for protection of the Antarctic environment, environmental impact assessment, siting of stations, aspects of environmental conduct (waste management, waste disposal, marine pollution, oil contamination, disposal of nuclear waste, use of radio-isotopes and scientific drilling), environmental monitoring, tourism and non-governmental activities and use of Antarctic ice (Heap, 1994). The recommendations encourage good environmental practice and require, if appropriate, to be implemented through national legislation (e.g. Andersen and Rudolph, 1989). Many have subsequently been consolidated into the mandatory framework of the Protocol on Environmental Protection. Although the recommendations must be formally approved by the governments of all the negotiating parties before they become legally binding, in general most countries have undertaken implementation as soon as possible. In 1995, new categories of recommendation were introduced (see Chapter 6).

The approach adopted so far has been one of 'prevention rather than cure' (Heap, 1994) and this is reflected particularly in the Agreed Measures, CCAS, CCAMLR and CRAMRA, which were negotiated in advance of either significant impacts or commencement of potentially damaging activities (except in the case of finfish exploitation). In the 1980s, however, protection of the Antarctic environment increasingly became a global concern. This arose from growing public debate on the negotiations for an Antarctic minerals regime, the adoption of Antarctica as a 'big issue' for global campaigning by environmental NGOs and increasing awareness of the importance of Antarctica in global environmental processes and change (see Chapter 1). The environment also became a political means through which to challenge the ATS, notably by certain developing nations (see Chapter 6). Further, there was growing concern expressed by NGOs about the environmental impact of scientific activities, lapses in good conduct, the effectiveness of existing measures and variations in the level of compliance, delays in approval of recommendations by treaty parties, and the lack of enforcement mechanisms (Heap and Holdgate, 1986; Bonner and Angel, 1987; Barnes, 1988). Although the ATCPs were discussing measures for comprehensive environmental protection, progress was slow, in part reflecting the preoccupation with CRAMRA in the 1980s. Consequently, there was pressure from a range of different sources, including the NGOs and the scientific community through SCAR, for a more strategic approach to environmental protection in Antarctica (Heap and Holdgate, 1986; Burgess, 1990; Triggs, 1990; IUCN, 1991; Walton, 1991). This pressure, allied with the stalling of CRAMRA in 1989–1990, forced the pace towards negotiation of an environmental instrument. Some countries favoured a Conservation Convention, others an Environmental Protocol. The advantages of the protocol are that it is supplementary to the treaty itself, and therefore enhances it, and through a framework format with annexes can be amended relatively easily to meet changing environmental requirements. In contrast, a free-standing convention would have been more difficult to amend and would have involved an erosion of the status of the ATCM by removing from it the environmental agenda.

9.3 Agreed Measures for the Conservation of Antarctic Fauna and Flora

Recognition in the Antarctic Treaty of the need for the protection of the living resources of Antarctica led to the adoption of the Agreed Measures for the Conservation of Antarctic Fauna and Flora in 1964 (Anderson, 1968; Heap, 1994). The Agreed Measures apply to the area south of 60° S, including all ice shelves, although, as with the treaty itself, there is derogation to high seas rights, which include sealing, fishing and whaling as well as freedom of passage. Under this instrument, the Treaty Area is considered a Special Conservation Area and, within that area, the Agreed Measures provide for the protection of native mammals and birds, the minimization of harmful interference to native fauna, the designation of Specially Protected Areas (SPAs), and the protection of Ross and fur seals. The importation of non-indigenous species is regulated to prevent the accidental introduction of parasites and diseases. Under certain conditions, permits may be issued for specific activities involving the killing or capturing of species for scientific research or entry into SPAs. The Agreed Measures have been reviewed at subsequent treaty meetings and further recommendations adopted. These relate to the designation of SPAs (additions and deletions) and review of selection criteria. In addition, the need was recognised for the designation of Sites of Special Scientific Interest (SSSIs) to protect scientific research interests. The Agreed Measures provided an early framework for Antarctic conservation but relied on voluntary compliance (see Chapter 8), reflecting sensitivities about jurisdiction and sovereignty (Bush, 1990). Many of the measures in the Agreed Measures are now subsumed or developed in the the Annexes to the Protocol on Environmental Protection.

9.3.1 Protected areas

Many of the limited ice-free areas of the continent and adjacent islands are of outstanding importance, both for their ecological systems and as breeding grounds for birds and seals, and are extremely sensitive to disturbance through human activities (see Figure 9.2a, Chapters 3 and 8). Under the Agreed Measures, provision was made for the establishment of SPAs – areas of outstanding scientific interest that require protection to preserve their unique ecological systems. Access is prohibited except by permit for 'a compelling scientific purpose that cannot be served elsewhere'. The first fifteen SPAs were designated in 1966, and from 1989 all SPAs have been required to have management plans. In 1975, a second category of protected area, SSSIs, was introduced in recognition of the distinction between the need for virtually complete protection in some areas (SPAs) and the requirements to protect a resource for scientific research, including non-biological interests (Heap, 1994). SSSIs may be designated to protect scientific research where there is 'a demonstrable risk of interference' with the investigations or where an exceptionally important site requires long-term protection (Figure 9.2b). Each SSSI is designated for a specific period of time, normally ten years, after which it is reviewed. SSSIs also have management plans, which may allow other non-harmful scientific activities. However, in contrast to SPAs, which have mandatory requirements, permits are not required for access to SSSIs and compliance with management plans is voluntary. Hence the legal protection of SSSIs has been relatively weak. Currently, there are a total of 20 SPAs covering an area of about 183 km^2 and 35 SSSIs covering an area of about 2684 km^2 (Figures 9.3 and 9.4; Tables 9.2, 9.3 and 9.4). Marine SSSIs were introduced in 1987; currently there are five. SPAs were initially established in an unsystematic manner, but subsequently criteria were developed by SCAR for the selection of particular categories of site (Table 9.5). Usher and Edwards (1986b) stressed that representativeness, followed by uniqueness, should be the pre-eminent criteria for the selection of protected areas in Antarctica.

There has been widespread criticism of the system of protected areas (Bonner and Lewis Smith, 1985; Keage, 1986; Boczek, 1986; Bonner and Angel, 1987; Broady, 1987; Dingwall and Lucas, 1992; Smith, 1994b). In terms of the classification of Antarctic ecosystems, it is apparent that SPAs for terrestrial coastal ecosystems predominate and since most are located on the Antarctic Peninsula or offshore islands, the spatial distribution is biased and not fully representative of Antarctic ecosystems. This reflects a lack of knowledge of remoter areas together with a need to protect threatened sites. However, as yet there has been no systematic inventory and application of site assessment criteria to the full range of Antarctic terrestrial, freshwater and marine ecosystems. Emphasis has been placed on unique areas and protection for scientific research. In addition, it has been noted that in some cases the original scientific reasons for designation have not been validated (Bonner and Smith, 1985), nor have sites been monitored to ensure that conservation objectives are being met. No fully marine SPAs have been designated and there is a low representation of nearshore ecosystems and freshwater ecosystems; geological and geomorphological interests have also been neglected.

Lucas and Dingwall (1987) noted further criticisms. Given the size and significance of Antarctica, there are relatively few protected areas and these are relatively limited in extent and lack buffer zones; the total land area protected by SPAs, SSSIs and CEMP Sites is about 790 km^2, equivalent to approximately 0.007% of the continental area of Antarctica (Foreign and Commonwealth Office, 1997). The scope of protected areas is limited to those selected for scientific purposes, neglecting other categories recognised by IUCN (*cf.*

Fig 9.2a The Dry Valleys of Victoria Land are unique for their extreme polar desert ecosystems, physical features and aesthetic value. They have been the focus of much scientific and visitor activity but are highly sensitive to disturbance and slow to recover. Although strict regulations for human activity are now in place, this has not always been the case. The photo shows Taylor Valley, with Taylor Glacier (bottom centre) and Lake Bonney. US Navy trimetrogon aerial photo no. 250 (TMA 540 F31), Antarctic Map and Photography Library. Data available from US Geological Survey, EROS Data Center, Sioux Falls, SD, USA.

Fig 9.2b Barwick Valley SSSI (No. 3) is one of the least disturbed areas of the Dry Valleys. The ice-covered, saline lakes are Lake Vashka (centre) and Hourglass Lake (foreground). Only small parts of the Dry Valleys are formally protected, but an integrated environmental management plan has been proposed for the entire area (Vincent, 1996) (source: photograph courtesy of Trevor Chinn).

Table 9.2 Categories of protected areas with existing site designations.

Category	Date introduced	Number and objectives	Total area (km²)
Specially Protected Area (SPA)	1966	20 to protect areas of outstanding scientific interest to preserve their unique ecological systems	184.0
Site of Special Scientific Interest (SSSI)	1972	35 to prevent interference with scientific investigations and to protect sites of exceptional scientific interest from human interference	2684
Historic Site and Monument	1972	72 to preserve and protect historic monuments	–
Seal Reserve	1972	3 to protect seal breeding areas and sites of long-term research	215,217
Tomb	1980	1 to ensure that the site of the 1979 air crash on Mount Erebus is left undisturbed	–
CEMP Site	1992	2 to safeguard sites contributing to the CCAMLR Ecosystem Monitoring Programme (CEMP)	4.4
Specially Reserved Area (SRA)	1989	1 proposal 480 to protect areas of outstanding proposed geological, glaciological, geomorphological, aesthetic, scenic or wilderness value	
Multiple-Use Planning Area (MPA)	1989	1 proposal 1535 to assist in coordinating human activities in areas where there are risks of material interference or cumulative environmental impacts	
Area of Special Tourist Interest	1975	0 areas designated for tourist visits	

Source: Foreign and Commonwealth Office, London, 1997.

Table 9.3 Specially Protected Areas in Antarctica.

No.	SPA	No.	SPA
1	Taylor Rookery	17	Litchfield Island
2	Rookery Islands	18	Northern Coronation Island
3	Ardery Island, Odbert Island	19	Lagotellerie Island
4	Sabrina Island	20	'New College Valley'
5	Beaufort Island	21	Avian Island
7	Cape Hallett	22	'Cryptogam Ridge'
8	Dion Islands	23	Forlidas Pond, Davis Valley Ponds
9	Green Island	24	'Pointe Geologie Archipelago'
13	Moe Island		
14	Lynch Island		
15	Southern Powell Island and adjacent islands		
16	Coppermine Peninsula		

Note: SPAs 6, 10, 11 and 12 now redesignated as SSSIs 4, 6, 32 and 5, respectively.

Source: Foreign and Commonwealth Office, London, 1997.

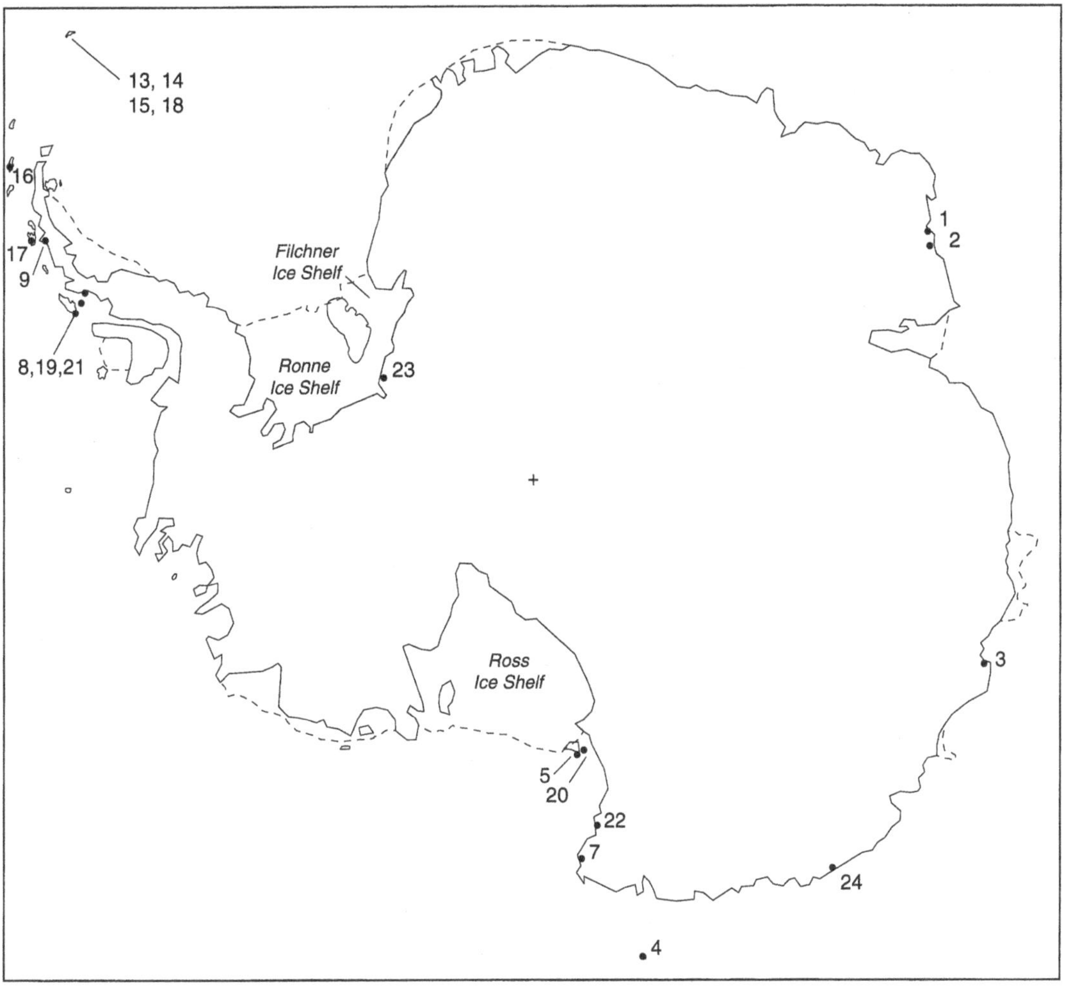

Fig 9.3 Distribution of protected areas: SPAs. The sites are listed in Table 9.3. (source: Foreign and Commonwealth Office, 1997).

McNeely and Miller, 1983), for example multiple-use areas where management objectives allow resources to be used in a sustainable way; spectacular landforms; and landscapes of outstanding scenic or aesthetic value. Management plans are inadequate, both in terms of basic content and in setting out clear objectives and policies (Kriwoken and Keage, 1989) also advocated the establishment of a management committee for protected areas. Most of these points have now been addressed in the Protocol on Environmental Protection (see below) but it remains to be seen how quickly changes are implemented in practice. In the past, the concentration of protected areas in the more accessible localities is understandable since there has not been the same need, as in other parts of the world, to address the conflicts between human pressures and nature conservation over large areas of the continent away from the main foci of scientific and tourist activities. However, as the geographical range of these activites expands, there is now a strong case for a more comprehensive system of protected areas.

SPAs have generally been successful in reducing adverse human impacts, but notable failures have occurred, usually concerning the location of scientific stations either close to, or within, SPAs. For example, both the USSR and Chile built bases within Fildes Peninsula SPA, and although the Agreed Measures were not then in force, both nations had

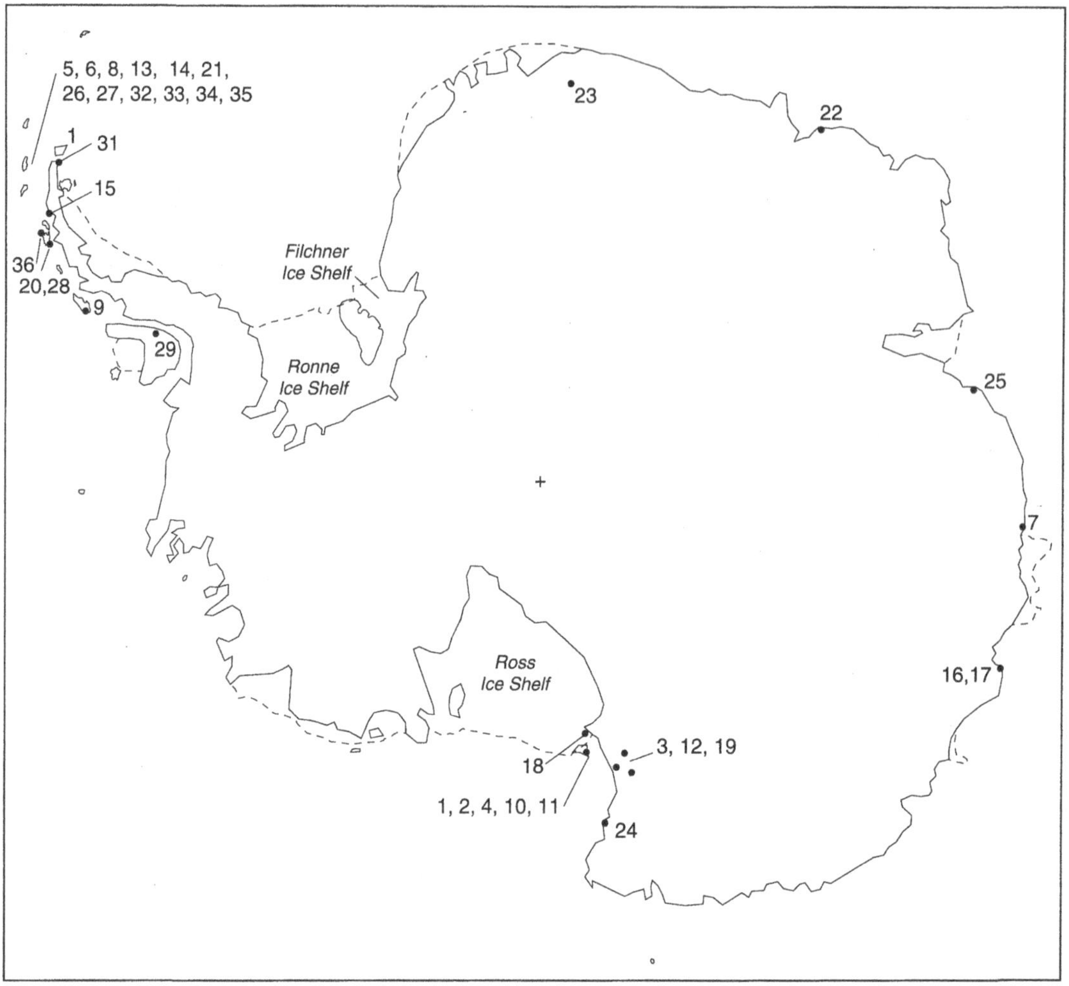

Fig 9.4 Distribution of protected areas: SSSIs. The sites are listed in Table 9.4. (source: Foreign and Commonwealth Office, 1997).

been involved in the preparation of the SPA. The location of these bases was neither in the spirit of the Agreed Measures nor demonstrated a commitment to them. Further management controls were not applied and extensive disturbance and localised pollution of the terrestrial and freshwater ecosystems ensued (Bonner and Smith, 1985; Harris, 1991a). The SPA was later reduced in size and finally deleted in 1975 as pressures on the area increased (see Chapter 8). This example starkly demonstrates the limited priority placed on conservation interests in the face of the political imperative to establish stations in an accessible region. Several other SPAs are close to stations, for example 1 and 2 (Mawson), 3 (Casey) and 24 (Dumont d'Urville), but this may be seen as a positive measure in focusing protection on areas where there are identifiable risks of human impacts, providing that management controls are strictly applied.

The importance of historic sites and monuments has also been recognised, with 72 sites identified as worthy of protection (Figure 9.5 and Table 9.6). These include huts and relics of expeditions from the heroic era, abandoned stations, rock shelters, cairns, graves and commemorative plaques and monuments recording significant events (Headland, 1994c). Again, however, there has been no systematic overview of historic sites and some of those included are of questionable historic interest, while some

Table 9.4 Sites of Special Scientific Interest in Antarctica.

No.	SSSI	No.	SSSI
1	Cape Royds	20	Biscoe Point
2	Arrival Heights	21	Parts of Deception Island
3	Barwick Valley	22	'Yukidori Valley'
4	Cape Crozier	23	Svarthamaren
5	Fildes Peninsula	24	Summit of Mount Melbourne
6	Byers Peninsula	25	Marine Plain
7	Haswell Island	26	'Chile Bay' (Discovery Bay)
8	Western shore of Admiralty Bay	27	Port Foster
9	Rothera Point	28	South Bay
10	Caughley Beach	29	Ablation Point – Ganymede Heights
11	'Tramway Ridge'	31	Mount Flora
12	Canada Glacier	32	Cape Shirreff
13	Potter Peninsula	33	Ardley Island
14	Harmony Cove	34	Lions Rump
15	Cierva Point and offshore islands	35	Western Bransfield Strait
16	North-eastern Bailey Peninsula	36	Eastern Dallmann Bay
17	Clark Peninsula		
18	North-western White Island		Note: SSSI No. 30 now redesignated as SPA No. 21.
19	Linnaeus Terrace		

Source: Foreign and Commonwealth Office, London, 1997.

notable omissions include relics of the sealing era on the South Shetland Islands (Smith and Simpson, 1987). Although the set of guidelines (see Table 9.5) to be used in drawing up proposals, produced in 1995, is a welcome step, there is still a need for a wider strategy. Many sites are now deteriorating through age and exposure to the elements (Harrowfield, 1995) and there is no clear policy for conservation, restoration or removal (Hughes and Davis, 1995). Moreover, there is no clear area of designation, and the immediate surroundings of the monuments are usually excluded, so that adjacent developments can devalue their aesthetic quality (Keage, 1986; Figure 9.6). Some of the huts in the Ross Sea sector have been the subject of restoration measures undertaken by the Antarctic Heritage Trust based in New Zealand (Harrowfield, 1990; Ritchie, 1990), and, following a survey of its abandoned bases, the UK has recommended four on the Antarctic Peninsula for designation (see Chapter 8).

Other categories of protected area include three Seal Reserves established under CCAS and two sites under the CCAMLR Ecosystem Monitoring Programme (CEMP) (see Section 9.5). Both sites are in the South Shetland Islands, at Seal Islands, and at Cape Shirreff and the San Telmo Islands, Livingston Island. The latter are designated by the CCAMLR Commission to safeguard monitoring programmes for bird and seal parameters at specified sites, and their status is mandatory. The designation of Areas of Special Tourist Interest was proposed in 1975, but none has been identified.

Two further categories of protected area were recommended in 1989: Specially Reserved Areas (SRAs) of outstanding value for their geological, glaciological, geomorphological, aesthetic, scenic or wilderness qualities; and Multiple-use Planning Areas (MPAs), where there is a risk of mutual interference from activities or cumulative environmental impacts. However, the recommendations have not subsequently been approved. Thus although two sites have been proposed, North of Dufek Massif (SRA) and Southwest Anvers Island (MPA), neither these sites nor the designations have any formal status.

Under Annex v of the Protocol on Environmental Protection the system of protected areas has been restructured and now comprises Antarctic Specially Protected Areas (ASPAs) (incorporating the former SPAs and SSSIs) and Antarctic Specially Managed Areas (ASMAs) (see Section 9.7). There is an understanding that in due course the draft MPA will become an ASMA, on submission of a revised management plan, and the draft SRA, an ASPA. Sites or monuments of recognised historic value will continue to be listed as Historic Sites and Monuments. The Protocol provides a good basis for addressing many of the deficiencies of the existing system for protected areas noted above and for the development of better management planning and monitoring to ensure that the conservation objectives are being met. For example, Annex v extends mandatory provisions to SSSIs

Table 9.5 Site selection categories for SPAs, ASPAs, ASMAs and Historic Sites and Monuments.

SPAs	ASPAs	ASMAs	Historic Sites and Monuments
areas that should be kept inviolate so that in future they may be used for purposes of comparison with localities that have been disturbed by humans	areas kept inviolate from human interference so that future comparisons may be possible with localities affected by human activities	any area, including any marine area, where activities are being conducted or may in the future be conducted may be designated an ASMA to assist in the planning and coordination of activities, avoid possible conflicts, improve cooperation between parties or minimise environmental impacts	a particular event of importance in the history of science or exploration occurred at the place
representative examples of the major Antarctic land and freshwater systems	representative examples of major terrestrial, including glacial and aquatic, ecosystems and marine ecosystems	ASMAs may include areas where activities pose risks of mutual interference or cumulative environmental impacts, and sites or monuments of recognised historic value	a particular association with a person who played an important role in the history of science or exploration of Antarctica
areas with unique complexes of species; areas that contain specially interesting breeding colonies of birds or mammals	areas with important or unusual assemblages of species, including major colonies of breeding native birds or mammals		a particular association with a notable feat of endurance or achievement
areas that are the type locality or only known habitat of any plant or invertebrate species	the type locality or only known habitat of any species		representative of, or forms part of, some wide-ranging activity that has been important in the development of Antarctica
	areas of particular interest to ongoing or planned scientific research		particular technical or architectural value in its materials, design or method of construction
	examples of outstanding geological, glaciological or geomorphological features		the potential, through study, to reveal information or has the potential to educate people about significant human activities in Antarctica
	areas of outstanding aesthetic and wilderness value		symbolic or commemorative value for people of many nations
	sites or monuments of recognised historic value		
	such other areas as may be appropriate to protect outstanding environmental, scientific, historic, aesthetic or wilderness values		

Sources: Bonner and Smith, 1985; SCAR Bulletin 110, 1993; SCAR Bulletin 121, 1996.

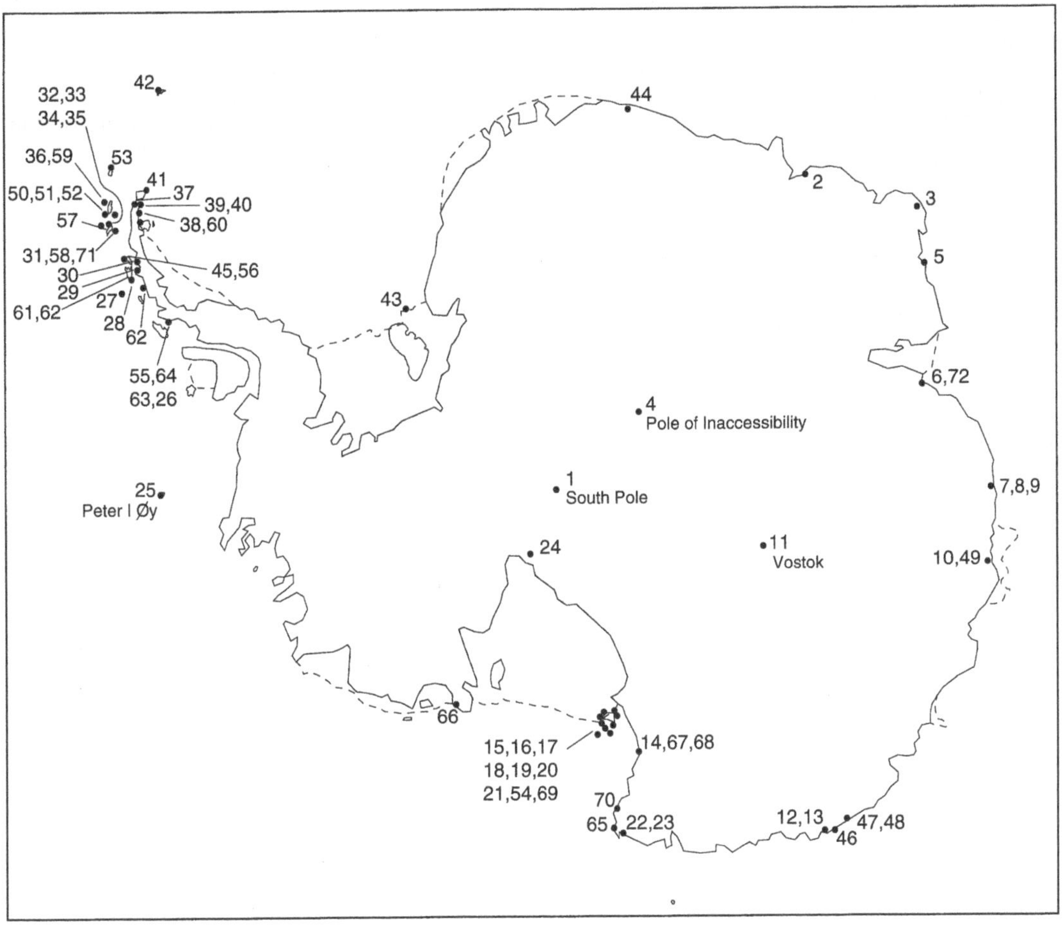

Fig 9.5 Distribution of Historic Sites and Monuments (source: Foreign and Commonwealth Office, 1997). (See Table 9.6.)

through converting them to ASPAs and extends the list of categories of site covered by such provisions, emphasing site selection within a systematic environmental geographical framework (see Table 9.3). Permits will be required for all visits to ASPAs, providing more stringent control on access than exists, for example for most protected sites in the UK. Specific recommendations for practical implementation of the measures in Annex v have been prepared jointly by SCAR and IUCN (Smith *et al.*, 1994).

International designations of protected areas (e.g. World Heritage sites, Ramsar sites, IUCN categories) cannot be applied in Antarctica because of the particular legal and political circumstances concerning the question of sovereignty; such sites can only be proposed by a government within its territory. It has been argued, however, that appropriate international recognition, in the form of parallel designations based on comparable criteria and assessment procedures, is desirable, both to strengthen the existing system and to integrate Antarctica into the global conservation framework (IUCN, 1991; Holdgate, 1994). In the light of other priorities, particularly to improve existing site coverage and to develop detailed management plans, this form of internationalisation is unlikely to happen quickly.

9.4 Convention for the Conservation of Antarctic Seals (CCAS)

Historically, stocks of fur seals and elephant seals were significantly depleted through exploitation on

Table 9.6 Historic Sites and Monuments in Antarctica.

No.	HSM	No.	HSM
1	flag mast, South Pole	41	hut and grave, Paulet Island
2	rock cairn and plaques, Ongul Island	42	huts, magnetic observatory and graveyard, Scotia Bay
3	rock cairn and plaque, Proclamation Island	43	cross, 'Piedrabuena Bay', Filchner Ice Front
4	bust and plaque, 'Pole of Inaccessibility'	44	plaque, Nivlisen Ice Front
5	rock cairn and plaque, Cape Bruce	45	plaque, Metchnikoff Point
6	rock cairn and canister, Walkabout Rocks	46	building and installations, Port-Martin
7	stone and plaque, Mabus Point	47	building, Île des Pétrels
8	monument sledge and plaque, Mabus Point	48	cross, Île des Pétrels
9	cemetery, Buromsky Island	49	pillar, Bunger Hills
10	observatory, Bunger Hills	50	plaque, Fildes Peninsula
11	tractor and plaque, Vostok Station	51	grave and cross, Admiralty Bay
12	cross and plaque, Cape Dension	52	monolith, Fildes Peninsula
13	hut, Cape Denison	53	monolith and plaques, Elephant Island
14	ice cave, Inexpressible Island	54	bust, Ross Island
15	hut, Cape Royds	55	buildings and artefacts, Stonington Island
16	hut, Cape Evans	56	remains of hut and environs, Waterboat Point
17	cross, Cape Evans	57	plaque, 'Yankee Bay'
18	hut, Hut Point	58	cairn and plaque, Whalers Bay
19	cross, Hut Point	59	cairn, Half Moon Beach
20	cross, Observation Hill	60	cairn and plaque, 'Penguins Bay'
21	hut, Cape Crozier	61	'Base A', Port Lockroy
22	hut, Cape Adare	62	'Base F' (Wordie House), Winter Island
23	grave, Cape Adare	63	'Base Y', Horseshoe Island
24	rock cairn, Mount Betty	64	'Base E', Stonington Island
25	hut and plaque, Framnësodden	65	message post, Foyn Island
26	installations, Barry Island	66	cairn, Scott Nunataks
27	cairn and plaque, Megalestris Hill	67	rock shelter, 'Granite House', Cape Geology
28	cairn, pillar and plaque, Port Charcot	68	depot, Hell's Gate Moraine
29	lighthouse, Lambda Island	69	message post, Cape Crozier
30	shelter, Paradise Harbour	70	message post, Cape Wadworth
31	plaque, Whalers Bay	71	whaling station, Whalers Bay
32	monolith, Greenwich Island	72	cairn, Tryne Islands
33	shelter, cross and plaque, Greenwich Island		
34	bust, Greenwich Island		
35	cross and statue, Greenwich Island		
36	plaque, Potter Cove		
37	statue, Trinity Peninsula		
38	hut, Snow Hill Island		
39	hut, Hope Bay		
40	bust, grotto, statue, flag mast, graveyard and stele, Hope Bay		

Source: Foreign and Commonwealth Office, London, 1997.

the sub-Antarctic islands and Antarctic Peninsula (see Chapter 7). Although these have now substantially recovered, there was concern that uncontrolled sealing in the future might target the potentially large resource of previously unexploited southern species. Indeed, an exploratory Norwegian commercial sealing expedition was mounted in 1964. Concern was such that precautionary measures were negotiated in advance of any large-scale commercial harvesting since, under the Agreed Measures, which have a derogation to high seas rights, seals were only protected on land or ice shelves and not in the sea or on sea ice. The Convention for the Conservation of Antarctic Seals was adopted in 1972 and came into force in 1978; it applies to the seas south of 60° S. The objectives of the convention are to achieve protection, scientific study and rational use of Antarctic seals, with any harvesting regulated on a scientific basis so as to maintain the balance of the marine ecosystem. It provides legal protection for Ross, southern elephant and southern fur seals and annual catch limits for crabeater, leopard and Weddell seals; but pelagic sealing of these latter species is prohibited. Permits may be issued to allow 'limited quantities' of any species to be taken to provide 'indispensable food for men or dogs', or for research or museum specimens.

Fig 9.6 Scott's Discovery Expedition hut at Winter Quarters Bay, Ross Island, constructed in 1902, is now protected as a historic site. The sequence of photos shows the changing environmental impact between 1902 and 1989 (after Lyons, 1993). (A) *Discovery* at Hut Point, 1902 (photograph from the Quartermain Collection, Ref. No. 1702, courtesy of Canterbury Museum, Christchurch, New Zealand). (B) Hut Point in 1960 with McMurdo Station in the background. Rubbish has been bulldozed into the sea on the far side of the bay. Note the sunken ship behind Scott's hut, which is derelict (source: photograph courtesy of Charles Swithinbank). (C) Hut Point in 1962. Rubbish is still being used as landfill. The wrecked ship has been removed (source: photograph courtesy of Charles Swithinbank). (D) Hut Point in 1989. The landfill site has been covered up and rubbish removed. Scott's hut is now cordoned off and restored (source: photograph courtesy of Charles Swithinbank). (See Figure 8.3.)

Additional provisions establish closed seasons, sealing zones and seal reserves, and reporting requirements on catches. The three seal reserves established are at the South Orkney Islands, in the Ross Sea and at Cape Hallett, their purpose being to protect all seals from harvesting and to maintain their role in the marine ecosystem.

Although feared at the time of negotiation of CCAS, commercial sealing has not resumed, and measures adopted under the convention provide the basis for regulation and monitoring, if required. In this respect, CCAS was the first international conservation agreement to be adopted before the start of exploitation of the resource it was designed to protect. CCAS is now somewhat dated in comparison with the measures developed under CCAMLR (see Chapter 7 and Section 9.5), and it has not been integrated into the ecosystem approach adopted in the latter convention. Realistically, however, it is improbable that commercial sealing will ever again take place in the Antarctic on account of opposition from the public and environmental NGOs, as well as the low economic returns.

9.5 Convention on the Conservation of Antarctic Marine Living Resources (CCAMLR)

The principal motivation behind CCAMLR arose from the development of exploratory commercial harvesting of krill in the Southern Ocean in the late 1960s/early 1970s, a fishery that was then predicted to have an immense potential (see Chapter 7). As an open-access resource, krill was considered attractive to distant-water fleets excluded from other fishing grounds through the extension of national fisheries limits to 200 miles under the Law of the Sea Convention. The USSR had already operated a significant fishery in the Southern Ocean for finfish (stocks of some species of fish were already depleted) and there were fears concerning the impacts of large-scale, unregulated harvesting of a species central to the Antarctic marine ecosystem. In addition, the FAO and the United Nations Development Program had proposed a food development programme for the developing world based on utilising Antarctic marine resources (Mitchell and Kimball, 1979). Although not specifically mentioned in the convention, it is acknowledged that regulation of the krill fishery was central to its development and the adoption of an ecosystem approach (Edwards and Heap, 1981). However, the ATCPs had additional political motives (Auburn, 1982; Lagoni, 1984), including avoidance of sovereignty claims driven by national desire to gain exclusive control of marine living resources, and a concern to retain authority over the region in the face of external interest by bodies such as the FAO.

The objectives of the convention, which came into force in 1982, are to ensure conservation and rational use of Antarctic marine living resources based on sound scientific knowledge (Edwards and Heap, 1981; Auburn, 1982; Barnes, 1982a; Gardam, 1985; Howard, 1989; Heap, 1994). The convention sets out three conservation principles, which should be read as a whole:

1 prevention of decrease in the size of any harvested population to levels below those which ensure its stable recruitment. For this purpose its size should not be allowed to fall below a level close to that which ensures the greatest net annual increment;
2 maintenance of the ecological relationships between harvested, dependent and related populations of Antarctic marine living resources and the restoration of depleted populations to the levels defined in sub-paragraph (1) above; and
3 prevention of changes or minimization of the risk of changes in the marine ecosystem which are not potentially reversible over two or three decades, taking into account the state of available knowledge of the direct and indirect impact of harvesting, the effect of the introduction of alien species, the effects of associated activities on the marine ecosystem and of the effects of environmental changes, with the aim of making possible the sustained conservation of Antarctic marine living resources.

(Heap, 1994).

A Commission, supported by an advisory Scientific Committee, meets annually to develop and set regulatory measures.

CCAMLR followed CCAS in providing a basis for a precautionary approach to regulating exploitation in advance of a major krill fishery. CCAMLR's ecosystem approach means that measures adopted to protect or regulate the exploitation of one species must involve consideration of the impact on other species and the integrity of the ecosystem as a whole (Edwards and Heap, 1981). This approach differs from most other fisheries treaties, which are concerned with protecting individual species. It carries the implication that in order to minimise wider impacts, acceptable harvesting levels of one species are likely to be lower, especially for krill (*cf.* Basson and Beddington, 1991).

The convention applies to all types of marine life, including finfish, molluscs, crustaceans, mammals and birds in the area south of the Antarctic Polar Front (Figure 9.7) and so includes areas outside the provisions of the Antarctic Treaty, notably around some of the sub-Antarctic islands. However, sovereign states within CCAMLR can opt to exclude their coastal waters from the provisions of specific conservation measures and to establish their own regulations. The convention is open to accession by any state interested in research on, or harvesting of, the marine living resources within the area covered by the convention; currently there are 23 members, including the original fifteen signatories, and a further six states party to the convention but not members of the Commission. The political framework of the convention thus allows the contracting parties to maintain a central role in the management of the Southern Ocean, but it cannot prevent the involvement of other states or NGOs in the high seas areas of the Southern Ocean. Thus non-CCAMLR parties can, and do, fish in the CCAMLR area (see Chapter 7), although in the view of the CCAMLR parties they do so in contravention of the objectives and principles of the convention and outside of its management regime.

Conservation measures that the Commission may adopt include setting precautionary species catch limits, on a regional basis as appropriate; designation of protected species and protection areas; establishment of open and closed seasons; regulation of

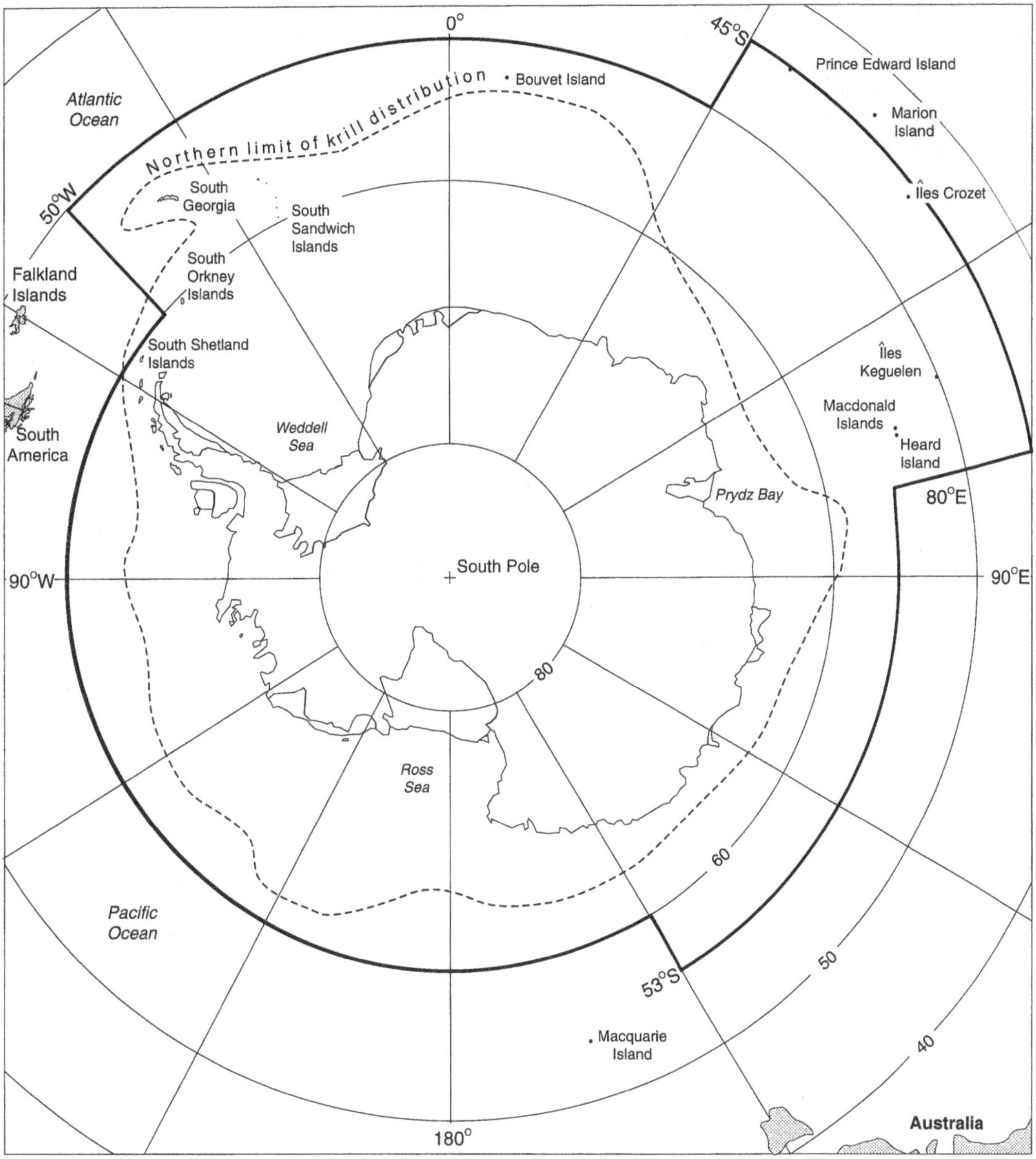

Fig 9.7 Boundary of the area covered by CCLAMR.

catching effort and methods of harvesting; and the requirement for the prior notification of new fisheries. Decision making is based on a consensus approach but, crucially, any member may subsequently opt out of conservation measures to which it is opposed. Enforcement and compliance is a matter for individual contracting parties, in effect protecting sovereignty rights.

As with any fisheries management regime, the requirement for adequate scientific knowledge is fundamental to the working of the convention, but despite the results from programmes such as BIOMASS (see Chapter 4), there has been a lack of data on the dynamics of the marine ecosystem and the productivity of harvested and dependent species to underpin decision making. The development of management models of sustainable harvesting for krill, let alone for multiple species, remains elusive, so single- rather than multiple-species approaches have been adopted. Equally, there is still a lack of comprehensive scientific data to allow assessment of impacts of harvesting target species on the wider ecosystem. Monitoring under the CEMP of selected indicator species that are most sensitive to resource changes is therefore critical in this respect. Obtaining sufficient commercial data on catch and effort statistics from some of the fishing states was also problematic in the early years of CCAMLR, yet these states also argued that there were insufficient data to justify stringent conservation measures (Howard, 1989). Nevertheless, with agreement on reporting procedures, the flow of information has now improved and with it the scientific basis of advice from the Scientific Committee. The adoption in 1992 of a scientific observation scheme should further assist in this respect. Also, improvements in modelling techniques have allowed risk assessment of different management options.

The objectives of CCAMLR are laudable, but their implementation has been problematic (Bonner, 1987) and widely criticised, particularly during the initial years (e.g. Frank, 1983; Howard, 1989; Nicol, 1991, 1992; Kock, 1992, 1994). Underlying problems have been associated with weaknesses within the convention, the slow progress in adopting measures due to the resistance of the fishing states and data inadequacies. The measures adopted have been viewed as reactive and too late to protect the many fish stocks that were already sufficiently depleted as to be non-economic when CCAMLR came into force (Howard, 1989). CCAMLR was slow to adopt measures to enable the recovery of the depleted stocks (Kock, 1994) and to regulate and monitor the krill fishery, although the harvest was already at substantial levels when the convention came into force (Nicol, 1991). Another area of weakness has been the delay in implementing measures for inspection and enforcement (Frank, 1983; Howard, 1989; Kock, 1994), and the lack of independent monitoring. The combination of consensus voting and objection procedures has effectively allowed the fishing states to block the faster development of precautionary conservation measures.

The reality is that CCAMLR represents a compromise between different competing interests – between fishing and non-fishing states, and between claimant and non-claimant states concerning territorial claims offshore. The adoption of a consensus approach was therefore necessary to protect the interests of all the parties involved (Orrego Vicu˜na, 1991), and slow progress has been inevitable when conflicting interests between fishing and conservation have had to be balanced, with the burden of proof resting with the non-fishing states. This could have been mitigated if conservative interim measures had been adopted until data were available to allow more considered scientific assessments (Heap, 1988), but instead commercial fishing was unrestricted until the need for regulation was agreed. Overall, CCAMLR has faced real difficulties in translating its general objectives into an integrated framework of practical management measures. Ultimately, the fundamental problem remains that of regulating the exploitation of a resource subject to open, competitive access and driven by powerful economic motives. In other situations, effective fisheries management has been achieved only where unilateral authoritative action has been taken, usually by one state exercising sovereignty over the resource (Parsons, 1987). There is thus a need for greater political will to make CCAMLR work (Heap and Holdgate, 1986), but with resources at stake, this has proved difficult.

Nevertheless, in spite of the practical and political difficulties, there has been some optimism about the achievements of CCAMLR in comparison with other international fisheries conventions and also on account of recent moves towards a more precautionary management policy (Gulland, 1988a; Heap, 1991b; Miller, 1991; Orrego Vicuña, 1991; Croxall *et al.*, 1992; Kock, 1994; Stokke, 1996). Up to 1990, most of the conservation measures (*cf.* Miller, 1991; Kock, 1992) reflected the pressure on finfish stocks (see Chapter 7). Since then, precautionary catch limits have been set for krill and there are restrictions to ensure that harvesting is not concentrated in any one area. Progress has also been made in establishing a catch reporting system for krill and other fisheries, and the development of a management policy and a programme (CEMP) for monitoring dependent species to investigate whether fisheries' impacts on the ecosystem can be distinguished from natural changes (Agnew, 1997). As part of CEMP, studies of various predator, prey and environmental indicators are in progress at 15 sites around the Antarctic, with three areas selected for intense study (South Gerorgia, the South Shetland Islands and Prydz Bay). Two of the sites on the South Shetland Islands are protected as CEMP sites (see Section 9.3.1). Since 1993, all commercial fisheries within the waters covered by CCAMLR have been regulated by appropriate conservation measures. In addition, all new and exploratory fisheries now require prior notification and assessment before approval, while significant research fishing programmes (taking more than 50 tonnes) require endorsement from the CCAMLR Commission. Additional measures are in place to prevent incidental mortality of seabirds in longline fisheries and disposal of plastic packaging at sea (see Chapter 7).

The future challenge is to translate further the principles of sustainable management into practical measures with effective regulation. The recent measures relating to krill are therefore encouraging. However, until precautionary action is accepted as a general management principle and the fishing states are persuaded of its value, the difficulties with CCAMLR will remain (Croxall *et al.*, 1992; Kock, 1992), as with all fisheries conventions. Conversely, had CCAMLR not been in place, then the overfishing in the late 1960s and 1970s would undoubtedly have continued, with further stocks severely depleted in the absence of any other regulation in the Southern Ocean. Thus, despite its imperfections and lack of enforcement measures, CCAMLR deserves recognition for bringing order and regulation to fisheries management. The main concerns for the future are the recent growth of unregulated and illegal fishing and the need to ensure compliance with CCAMLR measures. However, without sanctions the latter may prove difficult, particularly where it infringes on the sovereignty issue.

9.6 Convention on the Regulation of Antarctic Mineral Resource Activities (CRAMRA)

The Antarctic Treaty did not address the question of mineral resources and their possible exploitation. As the economic potential of Antarctica became subject to increasing speculation in the 1970s, the issue raised a number of difficult questions for the ATCPs that appeared less pressing than the urgent need to agree measures for the conservation of marine living resources since experimental krill harvesting had commenced. Although commercial mining activity was not imminent, there had been exploratory inquiries from several oil companies in the early 1970s (Auburn 1982), and the ATCPs were concerned to adopt a precautionary approach, particularly since the difficulties of negotiating a minerals regime would have been greatly enhanced after any exploitable deposits were found. Moreover, unregulated mining activity would have carried potentially serious environmental risks (Zumberge, 1979). Any realistic prospect of mineral resources would also have raised legal and sovereignty issues and possibly threatened the stability of the ATS; for example the Dufek Intrusion, speculated to have significant resource potential, lies within the territorial claims of the UK, Argentina and Chile. In addition, non-treaty parties and developing states, perceiving the possibility of great Antarctic mineral riches, were keen to acquire common heritage access, drawing parallels with the UN Law of the Sea Convention and the mineral resources of the deep seabed (see Chapter 6). This political concern principally underlay the urgency to develop a minerals regime (Beeby, 1983; Sollie, 1983).

Accordingly, it was agreed in 1981 to pursue a comprehensive minerals regime within the ATS, and although the discussions concluded in 1988 with the adoption of CRAMRA, the convention has not been ratified and has not come into force. Nevertheless the negotiation of CRAMRA provided important groundwork for the Protocol on Environmental Protection (see Section 9.7). Much has been written on the background to CRAMRA (e.g. Beeby, 1983; Mitchell, 1983; Kimball, 1984; Zorn, 1984; Watts, 1987; Triggs, 1987c; Heap, 1990b; Andersen, 1991), and comprehensive accounts are given by Orrego Vicuña (1988) and Wolfrum (1991). CRAMRA sought to establish a regulatory framework for minerals prospecting, exploration and development activities (Heap, 1990b). Although it had been argued that adoption of CRAMRA would actually increase the probability that mining would eventually take place, the convention adopted a presumption against mining except under strict conditions. It thus differed from CCAMLR, under which activities were permitted initially until the need for regulation was agreed. CRAMRA represented a carefully negotiated framework that sought to accommodate (1) the requirements of the claimant states in relation to the non-claimant states, (2) the interests of the wider international community and (3) the need to achieve a balance between 'establishing a practical and workable regulatory system for possible minerals development in Antarctica' that does not 'undermine other uses of Antarctica nor significantly alter its relatively pristine environment' (Kimball, 1990). To achieve this, the convention set out the principles, standards and conditions under which minerals activities might be permitted. Institutions to be set up under CRAMRA included an Antarctic Mineral Resource Commission, an advisory Scientific, Technical and

Environmental Committee, Regulatory Committees for each area opened to exploitation, and an Arbitral Tribunal. Under CRAMRA, before an area could be opened for exploration and possible development, the consensus of all members of the Commission was required. Unless sufficient information was available on environmental impacts, and if these impacts were significantly adverse, the activities would not be permitted. Furthermore, technology and procedures must be available to ensure compliance with a strict environmental code; monitoring must be undertaken to identify adverse effects; and there must be a capacity to respond effectively to accidents that impact adversely on the environment. Provision for liability for environmental damage arising from any minerals activities was also in the convention, but was incomplete. From the perspective of the developing countries, although CRAMRA did not incorporate the principle of common heritage as such, it did recognise the need to take broader interests into account and to encourage international participation by such countries. There have been several reviews of how the convention might operate in practice (e.g. Triggs, 1988; Beck, 1989c; Joyner, 1989, 1991; Office of Technology Assessment, 1989; Scully and Kimball, 1989; Auburn, 1990; Kimball, 1990; Wolfrum, 1991).

Overall, CRAMRA represented an important component of the ATS, filling a major gap in the conservation legislation (Auburn, 1990). Potentially, it provided strong controls to safeguard the Antarctic environment (Kimball, 1991; Joyner, 1991; Wolfrum, 1991). Nevertheless, criticisms of both the details and the general approach and purpose were widespread (e.g. Triggs, 1990; Barnes, 1991; IUCN, 1991; Joyner, 1991). For example, uncertainties concerned compliance with, and enforcement of, the regulations, and the meaning of terms such as 'significant' in relation to environmental impact was also open to interpretation. Environmental NGOs expressed concern that limitations on liability and provisions for environmental protection were not sufficiently strict (Environmental Policy and Law, 1988). They vigorously opposed CRAMRA on the grounds that mining was not compatible with comprehensive environmental protection and advocated instead the declaration of Antarctica as a World Park. Most damaging were criticisms that CRAMRA actually encouraged minerals activities in Antarctica, since a legal regime would have provided stability for minerals operators, compared with additional risks from uncontrolled exploitation. Nevertheless, according to Joyner (1991), the overarching purpose of CRAMRA was to act as a failsafe mechanism, not as an encouragement to mining activities. There was also a serious political concern that permits to undertake mining could be granted within the territory of a claimant state without its approval, thus presenting a challenge to sovereignty claims.

CRAMRA was adopted in 1988 but, in 1989, both Australia and France declined to sign the convention, advocating that no mining should ever take place in Antarctica and that an agreement on comprehensive environmental protection should be sought, under which Antarctica would be recognised as a 'natural reserve, land of science'. Countries including Belgium, Italy and New Zealand agreed, but others, including Britain, Japan and the USA, argued against a permanent ban on mining, although recognising the need for comprehensive environmental protection. Although environmental concerns were cited as the basis of Australia's rejection of CRAMRA, other factors were fears about loss of sovereignty (Beck, 1991b), economic motives (notably the absence of royalty concessions), the wide public mood of opposition to mining on environmental grounds (Beck, 1990c; Blay and Tsamenyi, 1990), the pressure of the environmental NGOs, and the belief that there might be votes in a 'green' policy towards Antarctica. However, the failure of Australia and France to sign the convention would not have ultimately prevented CRAMRA from coming into force, since both states could have subsequently acceded to the convention. The fate of CRAMRA was sealed by the pronouncement by New Zealand in 1990 that, although it had signed CRAMRA, it would not ratify the convention. Since the convention requires ratification by sixteen states, including all the claimants, to come into force, New Zealand's statement effectively shelved CRAMRA.

Two events in 1989, the sinking of the *Bahio Paraiso* near Anvers Island and the *Exxon Valdez* oil spill in Alaska, also significantly enhanced public concerns about the environmental risks of minerals activities in Antarctica and strengthened the NGOs' campaign. At the outset of the CRAMRA negotiations, a mining ban was considered unrealistic by the ATCPs and it was therefore expedient to assume that mining was not incompatible with environmental protection and to proceed on that basis (Triggs, 1990). However, the strong challenge to this assumption has shifted the balance in favour of the primacy of environmental protection.

Although never implemented, CRAMRA provides a potential regime if minerals activities are ever allowed in the future. It also provided an important basis (e.g. in terms of definitions) for the negotiations on comprehensive protection of the Antarctic environment that culminated in the Protocol on Environmental Protection. The environmental measures incorporated within CRAMRA were the most comprehensive that had been developed within the ATS (Kimball, 1991) but were equally applicable to other existing activities. Thus a number of the crucial measures in CRAMRA have been incorporated or extended in the protocol to cover other human activities (Wolfrum, 1991; Joyner, 1996a).

9.7 Protocol on Environmental Protection to the Antarctic Treaty

The threat to CRAMRA coming into force galvanised the treaty parties, led by Australia and France, to develop measures for the comprehensive protection of the Antarctic environment (Burgess, 1990; Scully, 1991; Blay, 1992; Francioni, 1993). NGOs continued to assert strongly the case for a world park, with increasing public support in some countries. There was also underlying concern that if the mining issue remained unresolved, and in the absence of a comprehensive environmental regime, strong pressure would be voiced at the 1992 UN Conference on Environment and Development (Rio Summit) for a solution to be developed under UNEP (Kämmerer, 1992). Politically, it was imperative for the treaty parties to find a solution within the framework of the ATS, since wider internationalisation, as under a UN framework, would weaken both the ATS and the interests of the claimant states. In October 1991, the Consultative Parties adopted the Protocol on Environmental Protection to the Antarctic Treaty (SCAR, 1993b). The protocol, which comprises 27 Articles and five Annexes (Appendix 2), represents a significant landmark in the development of the Antarctic Treaty, establishing for the first time a comprehensive and legally binding system for protection of the Antarctic environment and ecosystems. Its major strengths are in drawing together the various existing measures and codes of procedure into a single framework, in filling gaps in the existing measures, in setting a clear environmental code for all types of human activity in Antarctica and in instituting mandatory environmental impact assessment and monitoring procedures for most activities. Moreover, it incorporates an indefinite ban on all minerals activities, other than for scientific research. This may be reviewed after 50 years, or before if there is a consensus. The protocol is a framework regime and can therefore be developed in the future through the system of annexes. All decisions relating to the protocol will be made by consensus at Antarctic Treaty Consultative Meetings.

The protocol designates Antarctica as 'a natural reserve, devoted to peace and science'. It establishes principles for the planning and conduct of all activities in Antarctica by states party to the agreement, including scientific research programmes as well as non-government activities such as tourism, to limit adverse impacts on the environment and ecosystems, and sets out procedures for the impacts of these activities to be assessed before they take place. This protection extends to 'the intrinsic value of Antarctica, including its wilderness and aesthetic values'. The need to preserve the value of Antarctica for scientific research, particularly relating to the global environment, is also recognised. There is a requirement for monitoring to provide early warning of adverse impacts. Exchange of information and cooperation in planning and conducting activities are emphasised, for example in relation to siting new stations to avoid cumulative impacts arising from overconcentration in particular areas. In future, states seeking Consultative Party status will be required to have ratified the protocol.

A Committee for Environmental Protection (CEP) will be established to provide advice and recommendations to ATCMs regarding the implementation of the protocol and its operation. Compulsory measures for the settlement of disputes include the establishment of an Arbitral Tribunal and recourse to the International Court of Justice. Rules covering liability for damage arising from activities in Antarctica will be incorporated in a future annex (*cf.* Wilder, 1993; Blay and Green, 1995). Amendment of the protocol may be done at any time with the agreement of all the contracting parties, and its operation may be reviewed after a period of 50 years at the request of any of the contracting parties. However, any modification or amendment would require majority support, including three-quarters of the Consultative Parties at the time the protocol was adopted. Under such a review, any decision to allow mining would additionally require a legally binding regulatory regime to be in place. The protocol thus allows a decision on whether mining should ever occur to be postponed and it does not foreclose options for future generations possibly living under different economic circumstances and resource constraints (Redgwell, 1992). At the insistence of the USA, the protocol contains a 'walkaway' clause whereby any nation could withdraw from the protocol and unilaterally commence mining activities after five years (see Chapter 7).

Detailed provisions include mandatory procedures for environmental impact assessment: an initial environmental evaluation (IEE), which is applicable when an activity is likely to have only a minor or transitory impact, and a comprehensive environmental evaluation (CEE) where an activity is considered likely to have a significant impact or if this is indicated by an IEE. Strict regulations for waste disposal and management at stations and field camps are specified; for example, specific types of waste must be removed and open burning of waste is to be phased out. Regulations to prevent marine pollution, similar (and in part stronger) to those under the International Convention for the Prevention of Pollution from Ships (MARPOL 73/78) (*cf.* Rothwell and Kaye, 1994; Joyner, 1996b), govern discharge of oil, liquids harmful to the marine environment, garbage and sewage. The enforcement of the protocol will subsume elements of the ATS, including the Agreed Measures. Protection of the native fauna and flora includes designation of 'Specially Protected Species' of native mammals, birds and plants. All dogs had to be removed from Antarctica by April 1994. Area Protection and Management establishes two categories of protected area, which may include marine areas: Antarctic Specially Protected Areas (ASPAs) and Antarctic Specially Managed Areas (ASMAs). ASPAs provide mandatory protection for 'outstanding environmental, scientific, historic, aesthetic, or wilderness values, any combination of those values, or ongoing or planned scientific research' and include existing SPAs and SSSIs (see Section 9.3.1). ASMAs may be designated to assist in the planning and coordination of activities, to improve cooperation or minimise environmental impacts and include areas subject to cumulative environmental impacts and sites of historic importance. Both ASPAs and ASMAs require management plans. Entry into ASPAs is prohibited, except by permit; ASMAS operate under a code of conduct specified in the management plan.

The Protocol entered into force in January 1998 after ratification by all 26 states that were ATCPs at the time it was adopted, although a number have yet to implement domestic legislation. The UK introduced domestic legislation in 1994 (Antarctic Act, 1994). There were strong political pressures to ratify the Protocol since its failure to enter into force would have led to a major loss of credibility in the ATS and in the ability of the ATCPs to ensure environmental safeguards for Antarctica. Although the ATCPs had earlier agreed to treat the provisions of the Protocol as being in force to the extent possible, it is apparent that the response was variable (Greenpeace, 1994; see Chapter 8). The priority thus remains to ensure that the measures in the Protocol are full implemented as now legally required. The one country, Pakistan, which is not a party to the treaty, and which has an Antarctic research programme, has indicated its support for the Protocol but is not bound by it.

Overall, the protocol is a compromise between national interests and wider concerns for effective environmental protection in Antarctica (Wolfrum, 1991). In contrast to the provisions in CRAMRA, it will be the responsibility of individual governments to ensure that the environmental assessment and other compliance procedures are followed, a weakness that has been criticised (Foreman, 1992; Welch, 1992). This perpetuates a principal defect in the past: the lack of an independent, centralised regulatory authority to ensure compliance with standards (Joyner, 1986; Barnes, 1988, 1991). The CEP has no more than an advisory function. It will have no authority to enforce the measures in the protocol, and attempts to create a body with enforcement powers, such as an Environmental Protection Agency, were perceived as a threat to the national interests of claimant states and therefore opposed (Wolfrum, 1991). There is thus no means to ensure that all breaches of the protocol will be addressed, for example if parties are reluctant to contest such breaches on political or other grounds through the formal dispute arbitration mechanism (Wilder, 1995).

Other criticisms of the protocol concern the slow progress on the issue of liability (van Bennekom, 1992) and the lack of clear definition of standards to be applied in making environmental evaluations. Lyons (1993) highlighted the lack of guidelines for determining when an EIA is required and at what level, and the need for independent assessments to consider alternatives to the proposed activity. EIA should not be merely a 'rubber stamping' exercise for a course of action that has already been decided by individual nations. Also, while CEEs must be circulated among the contracting parties for comment, neither the measures that they contain nor the comments received are binding on the nation proposing the development concerned. Many states lack the resources to undertake rigorous EIA procedures and subsequent monitoring, and there is ambiguity as to whether activities with more than a minor or transitory impact can be prohibited in the absence of enforcement mechanisms. Ultimately, public credibility and

NGO pressure will be important considerations, particularly in the absence of any international enforcement mechanism under the protocol. A further aspect of concern relates to the number of qualifying statements in the provisions, for example in Annexes III and IV. While these reflect some of the practical difficulties in operating in Antarctica, they could provide potential loopholes for poor practices. Also, many practical questions regarding implementation of the protocol remain to be answered; in particular how to achieve good environmental stewardship yet avoid imposing unnecessary restrictions and administrative burdens on science programmes (Rothwell, 1992; National Research Council, 1993). There have also been calls for a specific annex on tourism (see Chapter 8).

What emerges from the protocol is that in spite of increasing attention being paid to environmental issues in Antarctica and the development of more stringent measures, there is still questioning by the environmental NGOs not only about whether the measures go far enough to protect the Antarctic environment but also about the true level of commitment to their enforcement. In part this reflects the different approaches of the NGOs and the ATCPs, the former demanding absolute protection as their central policy objective, the latter seeking to balance a range of objectives as well as being faced with real practical and economic difficulties in implementation. However, the measures for environmental protection are now in place and the onus is on contracting parties to enforce them. The switch to an environmental agenda has also successfully deflected attention away from potentially divisive sovereignty and minerals ownership issues and criticism of the ATS. It has also deferred decisions on minerals activity to a date 50 years hence, when the global minerals situation could be different. In the intervening years, as much will depend on public opinion, awareness and the policies of environmental NGOs as on the positions of the treaty state governments. Should the protocol lapse, then mineral resources would return to centre stage in Antarctic politics.

9.8 The role of non-governmental organisations and other bodies

Non-governmental organisations (NGOs) have taken an increasingly important interest in Antarctic affairs, particularly since the mid-1980s (Barnes, 1986, 1991; Kimball, 1988; Suter, 1991; Herr, 1996). This stems from the growth in interest in the exploitation of Antarctic resources and concern for the environmental impacts on the last unspoiled continent on Earth. As exploitation of finfish and krill developed and negotiations for CCAMLR and later CRAMRA progressed, NGOs such as the Antarctic and Southern Ocean Coalition (ASOC) (an alliance of environmental organisations from over 40 countries) and Greenpeace International became increasingly active and politically influential. The World Conservation Union (IUCN), which includes membership from governments, NGOs, research institutions and conservation agencies, has also taken an active interest in conservation in Antarctica. The negotiations on CRAMRA, in particular, focused attention on the possible impacts of mining and raised strong popular demands for a ban on mining activities to preserve the wilderness qualities of Antarctica. A compelling argument was that if the existing system was unable to regulate present activities, it would be even less appropriate to deal with the regulation of a powerful multinational minerals industry. ASOC, Greenpeace and the International Institute for Environment and Development (IIED) exerted considerable pressure to improve the environmental safeguards in CRAMRA (Auburn, 1990), and the resulting convention significantly reflects their efforts, although not to the full extent that they sought. Lobbying from the NGOs also played an important role in the decisions by New Zealand, Australia, France and other countries to stall CRAMRA and support a ban on mining. Subsequently, ASOC and IUCN were invited to participate as invited experts at the Special Consultative Meeting that led to the Protocol on Environmental Protection, while other NGOs were represented on national delegations. NGOs were thus able to exert a significant influence in the negotiations. ASOC, IUCN and IAATO are among bodies currently invited to attend ATCMs as invited experts.

NGOs have contributed in several respects to Antarctic affairs (Kimball, 1988; Barnes, 1991). They have significantly raised international public awareness of Antarctic issues through high-profile propaganda campaigns in the media and through publications (e.g. Barnes, 1982b; Brewster, 1982; May, 1988). They have also influenced the development of policy at national and international levels, combining political lobbying with the provision of supporting scientific information and publications, particularly in the USA (Kimball, 1988).

The NGOs, notably Greenpeace (Figure 9.8), have also played an important role in monitoring the compliance of national expeditions with environmental safeguards and have widely publicised poor practice in waste management and fuel handling (Greenpeace, 1994; Szabo, 1994). During the construction of the French airfield at Pointe Geologie (see Chapter 8), they even conducted lobbying on the spot. In the USA, the Environmental Defense Fund has been highly critical of waste management practices in the US Antarctic Program, administered by the National Science Foundation (NSF) (Manheim, 1988), and has helped to accelerate changes; for example, it took successful legal action against the NSF to ensure compliance with the National Environmental Protection Act (Mervis, 1993b). The NGOs have strongly advocated the establishment of an independent Antarctic Environmental Protection Agency (Barnes, 1986, 1991), but this has been resisted by the ATCPs as infringing on their authority. The NGOs also attach symbolic significance to the wilderness values of Antarctica as the last great unspoiled continent and have campaigned for the declaration of Antarctica as a world park (see Section 9.10.2), involving no minerals developments and the strict management of marine resources, tourism, scientific research and supporting logistics. As part of a successful high-profile campaign, Greenpeace established a base in Antarctica in 1987 called 'World Park' to publicise Antarctic environmental issues. The base was dismantled and completely removed in 1992, and although the project demonstrated that bases could be operated with minimal environmental impact, it achieved little scientifically to justify its 'footprint' on the continent. More recently, in 1994/95 Greenpeace visited ten stations to check on compliance with the protocol. The NGOs, notably ASOC, have also lobbied hard for more effective regulation of Southern Ocean fisheries under CCAMLR.

Fig 9.8 The Greenpeace vessel, *Gondwana*, berthed at Lyttleton Harbour, New Zealand, before sailing to Antarctica, 1990 (source: authors).

In the absence of an official environmental regulatory authority within the ATS, the NGOs act as an important independent 'watchdog' monitoring implementation of and compliance with the Protocol on Environmental Protection and CCLAMR, in the assessment of EIAs and in providing advice on environmental management to the treaty parties. However, this role also carries a responsibility for the NGOs to maintain a high level of scientific credibility rather than promote the alarmist speculation or inaccurate interpretations that have sometimes marred past campaigns.

In collaboration with SCAR, IUCN has played an important proactive role in the development of an Antarctic conservation strategy, based on the principles of sustainable management set out in the World Conservation Strategy (IUCN, 1980; Heap and Holdgate, 1986). *A Strategy for Antarctic Conservation* (IUCN, 1991) sets out a framework and objectives for long-term conservation in Antarctica (Appendix 3), some of which are incorporated within the Protocol on Environmental Protection. IUCN and SCAR have also produced recommendations for developing the Antarctic protected areas system (section 9.3.1) and for enhancing the conservation of sub-Antarctic Islands (section 9.11). Other developments have seen the formation of IAATO (see Chapter 8) to self-regulate the tourist industry.

In the longer term, it remains to be seen whether the NGOs will maintain the same level of commitment to Antarctic conservation. With the protocol achieved, involvement in the Antarctic may have less attraction for some campaiging NGOs, both on account of the cost and their need for the type of high-profile issues that generate funding from the public. Greenpeace, however, proposes not only to maintain its 'watchdog' role over activities in Antarctica, but also to campaign to raise wider awareness of the impacts of activities elsewhere in the world (e.g. climate change and stratospheric ozone depletion) on Antarctica. It also plans to work with the scientific community to publicise the value of Antarctic research.

9.9 Scientific Committee on Antarctic Research

The Scientific Committee on Antarctic Research (SCAR) is a scientific committee of ICSU, 'charged with initiation, promotion and coordination of scientific activity in the Antarctic, with a view to framing and reviewing scientific programmes of circumpolar scope and significance' (SCAR, 1987). Membership is open to countries with substantial and continuing research interest in the area. SCAR operates independently of the Consultative Parties but performs an advisory role to them, providing independent and authoritative scientific and technical advice and information (Zumberge, 1986; Rinaldi, 1991). It has played an important role in the development of the Agreed Measures, CCAS, CCAMLR, CRAMRA and the Protocol on Environmental Protection, as well as providing advice on the impact of minerals exploitation, waste disposal, environmental impact assessment and designation of SPAs and SSSIs (Zumberge, 1979; Benninghoff and Bonner, 1985; Bonner and Lewis Smith, 1985; Rutford, 1986; SCAR Panel of Experts on Waste Disposal, 1989). SCAR also has an important function in setting the research agenda and in promoting cooperation in international scientific programmes (Fifield, 1987; SCAR, 1989); for example, the BIOMASS programme has contributed significantly to knowledge about Antarctic marine living resources and thereby assisted in the development of resource management measures. Currently, SCAR is coordinating six major programmes (see Chapter 1), and it cooperates with other international bodies to ensure that Antarctic science is included in global research programmes such as the World Ocean Circulation Experiment.

The Council of Managers of National Antarctic Programs (COMNAP), an independent body federated to SCAR, has prepared guidelines on a range of issues including waste management, preparation of oil spill contingency plans and implementation of EIAs. It also has a key coordinating role, for example in the development of a regional oil spill contingency plan for the South Shetland Islands.

As science becomes increasingly involved with environmental issues and human impacts in Antarctica, SCAR is well-placed to advise on the appropriateness and effectiveness of regulatory measures. The president of SCAR is also invited to attend meetings of the CEP. However, there is concern about the financial costs to SCAR of a significantly expanded role in providing detailed environmental advice at the expense of focusing on primary scientific research (Laws, 1994a). Also, as science and politics become even more closely intertwined, a further challenge to SCAR is to maintain its scientific integrity and independence and to avoid being drawn too closely into policy issues and decision making (Zumberge, 1986; Heap, 1988). Some might argue that this is already in train, since political reality increasingly demands research on the big issues, such as the ozone 'hole', which can best be addressed by large directed research programmes. But these programmes tend towards rigidity and conservatism, which might ill serve Antarctic science in the long run. It is an uncomfortable fact that funding for some 'blue skies' research (of the kind that originally discovered the Antarctic ozone 'hole') is increasingly scarce and is being diverted towards the directed research programmes that help to provide answers to the applied problems of the day.

9.10 Alternative approaches to environmental protection in Antarctica

Two general approaches have been advocated for the development of environmental protection in Antarctica. The first of these, as outlined in the preceding sections, has involved building on the existing Antarctic Treaty System (Bonner, 1987; Holdgate, 1987). The second has involved calls for a common heritage or world park alternative (*cf.* Østreng, 1991).

9.10.1 Common heritage of mankind

In the 1980s, a number of developing nations advocated the adoption of a common heritage approach to Antarctic resources as negotiations proceeded towards an Antarctic minerals regime (Francioni, 1987; Triggs, 1987b; Chopra, 1988; Orrego Vicuña, 1988; Herber, 1991; Jacobsson, 1994). This approach, similar to a proposal by Balch (1910), involved replacing the ATS, by an alternative regime and the participation of all nations, regardless of their active involvement in the area, under the United Nations (see Chapter 6). It included the rejection of national ownership of territory or resources, and the safeguard of common resources for future sustainable use and common benefit (Triggs, 1984; Orrego Vicuña, 1988). The common heritage approach as initially advocated in the 1980s therefore implied controlled development of resources and equitable distribution of benefits but ignored the fact that Antarctica is subject to a number of longstanding territorial claims. Whether these would take precedence in international law against political demands that Antarctica be managed in the interests of the wider international community is a matter of legal argument (*cf.* Triggs, 1984, 1987b; Francioni, 1987). Further, it was politically unrealistic to anticipate that claimant states might renounce their claims for a common heritage approach, or that the ATCPs would accede to the replacement of the existing system.

Since the late 1980s, the thrust of the common heritage argument has shifted focus from resource to environmental issues and this was confirmed in a 1989 UN resolution calling for a world park or nature reserve for the benefit of all humanity (Suter, 1991). This shift, together with the adoption of the Protocol on Environmental Protection and the increased participation and openness at ATCMs, is evidence of an apparent convergence in current approaches. Thus development of the existing regime, which may itself be regarded as a form of common heritage management, may best ensure the future management of Antarctica for the whole of humankind (Francioni, 1987), and this is now recognised by former critics (Haron, 1991). This approach is also more realistic and pragmatic than application of the similar concept of 'public heritage' advocated by Suter (1991). Under the latter, environmental protection would have absolute priority, and Antarctica and the surrounding seas would be held in public trust for present and future generations with no exploitation of resources other than for science and tourism. A more sceptical view is that Antarctica has not been 'saved' and that the convergence of approaches is more apparent than real (*cf.* Beck, 1995) and concealed behind a protocol that postpones the real decision on the future for 50 years.

9.10.2 World park

Most environmental organisations advocate the protection of Antarctica in perpetuity under the umbrella of a 'world park' designation (*cf.* Barnes, 1982a, 1991; Mosley, 1984; Rigg, 1990; Rothwell, 1990a, 1990b). The idea received impetus at the Second World Conference on National Parks in 1972, organised by the IUCN, which recommended that Antarctica and its surrounding seas be established as the first world park under the auspices of the UN. New Zealand took this proposal to the ATCM in 1975, but it received little support. Nevertheless, it was widely promoted by other campaigning environmental groups as an alternative to a mining regime (Barnes, 1982a; Mosley, 1984). The concept also found favour with the governments of Australia and France, which used terms such as 'wilderness reserve' and 'nature reserve, land of science', in order to preserve the option of sovereignty claims since 'world park' has connotations of the common heritage approach (Beck, 1991b). World park designation would specifically exclude minerals exploitation, involve a precautionary approach to all Antarctic activities and use regulation to ensure common standards and enforcement. The arguments advanced for world park designation invoke the wilderness concept, protecting Antarctica for its own intrinsic value and its unique scientific opportunities (e.g. Mosley, 1984; May, 1988; Barnes, 1991; Herber, 1992; Phillips, 1992). From an economic perspective, it is also arguable that the international common good, as reflected in these values, is best served by a world park arrangement rather than through the benefits that might accrue from minerals exploitation (Herber, 1992). The objectives under a world park designation remain unclear but would probably include some or all of the following, as expressed by various NGOs and individuals:

- Antarctica would remain a continent for peace;
- the main objective would be to establish priority for comprehensive nature conservation and wilderness values in relation to all human activities;
- science would remain the principal reason for human activity on the continent but would be limited to those aspects either unique to Antarctica or vital in understanding or monitoring global processes and changes, although the environmental impacts of such activity would be evaluated;
- likewise tourism would continue but under regulation to minimise environmental impacts;
- harvesting of marine resources would continue under the conservation principles espoused in CCAMLR but under greater regulation;
- mineral development activities would be banned;
- an independent Environmental Protection Agency would be established to set and monitor environmental standards for all human activities in Antarctica;
- the protected area system would be developed.

The case against a world park has been argued forcibly by Law (1990) on the grounds that existing human impacts are negligible. To this may be added a fear that such an approach would destroy the benefits that have accrued under the ATS. There is also concern among some scientists that the environmental controls would be restrictive and unnecessarily bureaucratic (see above and Chapter 6), although this view is not shared by all scientists (*cf.* Patel and Mayer, 1991). It is also unclear how a world park approach would be funded and whether the existing treaty states would continue to support scientific activities at the same levels as at present, particularly if strategic motives for a presence on the continent were to be removed (see Chapter 6).

9.11 Sub-Antarctic islands

The islands north of 60° S lie outwith the area of the Antarctic Treaty. They are among the most heavily modified areas of the Antarctic region (Clark and Dingwall, 1985): their marine living resources have been exploited for some 200 years (see Chapter 7), and many have experienced significant ecological changes as a result of the introduction of non-indigenous species (see Chapter 8). Nevertheless, some islands and parts of islands remain essentially unmodified and are therefore of outstanding scientific importance – the South Sandwich Islands, Bouvet Island, Prince Edward Island, Heard Island, the Macdonald Islands and the more remote of Îles Crozet and Îles Kerguelen. The islands are covered by national legislation, and the extent and type of protection vary (Bonner and Lewis Smith, 1985; Clark and Dingwall, 1985; Dingwall, 1995). In some cases, conservation management broadly parallels the measures for protected areas specified under the Antarctic Treaty, as in the case of South Georgia (Headland, 1984; Smith, 1995a). Macquarie Island is designated as a nature reserve (Biosphere Reserve). Marion Island and Prince Edward Island are effectively managed as nature reserves although not statutorily designated as such (Cooper and Condy, 1988). Heard Island and the Macdonald Islands are notable for the absence of introduced plants and animals (Keage, 1982) and are protected under Australian legislation, which includes control of access. These islands, together with Macquarie Island, are also being considered for World Heritage designation. There are no designated marine reserves, although in many cases conservation measures, including controls on fishing activity, extend to territorial waters. For the islands that lie south of the Polar Frontal Zone, exploitation of their marine resources may be regulated under CCAMLR (see Section 9.5).

Several management issues and priorities have been identified (Clark and Dingwall, 1985; Smith and Smith, 1987; Walton, 1987d; IUCN, 1991; Dingwall and Lucas, 1992; Dingwall, 1995), the most

Fig 9.9 The former whaling station at Grytviken, South Georgia, 1975. The station is now a tourist attraction and includes a recently opened whaling museum (source: authors).

pressing of which are to prevent the introduction of exotic species to relatively undisturbed ecosystems and to control or eradicate introduced species elsewhere (see Chapter 8). Other impacts are localised around scientific stations and the derelict whaling stations on South Georgia (Smith, 1995b). Realisation of the potential archaeological and tourist value of the latter requires long-term maintenance and management, and the establishment of a whaling museum at Grytviken is a welcome development (Figure 9.9). Indirect effects are associated with the exploitation of fish and krill offshore and in more distant waters and include possible competition for food resources, as well as incidental seabird mortality and entanglement of fur seals in fishing debris (see Chapter 7). Increased tourism may also bring extra pressure on sensitive environments (e.g. Scott and Kirkpatrick, 1994; Wouters and Hall, 1995b) but access is controlled at most islands (Dingwall and Lucas, 1992) (see Chapter 8). More generally, there is a need for a more rationalised and standardised system of protective legislation and enforcement, extension of the network of protected areas, the inclusion of historic sites (e.g. associated with the sealing and whaling eras: Headland, 1984; Townrow, 1988) and development of appropriate management plans and policies, including the implementation of EIA (*cf.* Dingwall, 1995). Many of the measures adopted in the Protocol on Environmental Protection could be usefully adopted and applied to the sub-Antarctic islands without infringing on issues of individual sovereignty.

9.12 Conclusion

The Antarctic Treaty was originally framed to address political issues and to ensure that scientific research could be conducted in a demilitarised continent free from sovereignty disputes or Cold War antagonism. In this respect, it has been spectacularly successful. During the period of its existence, however, there have been significant changes in the wider global geopolitical framework, new challenges have emerged from external sources, and resources and environmental protection have become high priorities. Critics have successfully challenged the exclusive nature of the Antarctic Treaty 'club' and sought wider representation and participation in the administration of Antarctic affairs. Initially, this was motivated by perceptions of Antarctica's resource potential and the desire of developing countries to share in the benefits of exploitation. This was underpinned by developments in political thinking, including the New International Economic Order and the principle of the Common Heritage of Mankind embodied in the the Law of the Sea Convention. At the same time, the environmental NGOs have mounted highly effective campaigns to obtain greater levels of protection for the Antarctic environment (latterly supported by developing countries at the UN), questioning both the adequacy of existing measures and the commitment of the treaty parties to their effective enforcement, and forcing the treaty parties to be more accountable in environmental matters.

The positive response of the treaty parties to the environmental agenda, through the development of the Protocol on Environmental Protection, has increased the legitimacy of the ATS in the eyes of its critics, whereas CRAMRA failed the test of international acceptance. The development of the protocol, which recognises Antarctica as a natural reserve for peace and science, and which sets out a formal framework for environmental protection, has gone some way towards achieving an apparent convergence of views. Broadly, the treaty parties and their critics agree that Antarctica should be managed for all humankind in a way that maintains its unique environment, its value for science and its freedom from military conflict. A natural reserve with strong environmental protection within the ATS is the only practical way forward at present that accommodates the interests of all parties. Most of the objectives, if not the terminology, of the world park concept could be developed under the ATS as part of the progress towards comprehensive environmental protection. Such an approach is more realistic in terms of avoiding the wider internationalisation issue, in recognising the past success of the ATS and in leaving open the sovereignty question. Replacement of the ATS by a common heritage or world park arrangement in which all claims to sovereignty are renounced is not seen as a particularly realistic option (Falk, 1991). Indeed, 'the quest for a substitute regime universally acceptable to the world community is likely to be illusory and to divert effort from the actions urgently needed now' (Holdgate, 1987, p.141). The key issue therefore remains one of how to spread the benefits and perfect the management mechanisms so as to command widespread international support for the existing system (Zumberge and Kimball, 1986), as well as to ensure that the ideals of a world park and a common heritage approach are permanently safeguarded through effective implementation of the Protocol on Environmental Protection and its further development as necessary. In the short term, the priorities are to ensure the implementation of the Protocol on Environmental Protection, to conclude negotiations for a liability annex to the protocol and to deal with the management of tourism.

Further reading

Cook, G. (ed.) (1990) *The Future of Antarctica. Exploitation Versus Preservation*. Manchester University Press, Manchester.

Cross, M. (1991) Antarctica: exploration or exploitation. *New Scientist*, 130, 29–32.

Elliott, L.M. (1994) *International Environmental Politics. Protecting the Antarctic*. Macmillan, Basingstoke.

Harris, C.M. and Meadows, J. (1992) Environmental management in Antarctica. Instruments and institutions. *Marine Pollution Bulletin*, 25, 239–249.

Herr, R.A., Hall, H.R. and Haward, M.G. (eds) (1990) *Antarctica's Future: Continuity or Change?* Australian Institute of International Affairs, Hobart.

IUCN (1991) *A Strategy for Antarctic Conservation*. IUCN, Gland.

Jørgensen-Dahl, A. and Østreng, W. (eds) (1991) *The Antarctic Treaty System in World Politics*. Macmillan, Basingstoke.

Parsons, A. (1987) *Antarctica: the Next Decade*. Cambridge University Press, Cambridge.

Patel, J. and Mayer, S. (eds) (1991) *Antarctica. The Scientists' Case for a World Park*. Greenpeace, London.

Stokke, O.S. and Vidas, D. (eds)(1996) *Governing the Antarctic. The Effectiveness and Legitimacy of the Antarctic Treaty System*. Cambridge University Press, Cambridge.

Suter, K. (1991) *Antarctica. Private Property or Public Heritage?* Zed Books, London.

Verhoeven, J., Sands, P. and Bruce, M. (eds) (1992) *The Antarctic Environment and International Law*. Graham & Trotman, London.

CHAPTER 10 Conclusion: challenges and opportunities

Objectives

Key themes and questions covered in this chapter:

- assessment of resource exploitation and human effects on the environment of Antarctica;
- assessment of the success of the Antarctic Treaty;
- implementation of the measures in the Protocol on Environmental Protection;
- future development of the Antarctic Treaty System;
- future directions in science, resource management and environmental issues.

10.1 Introduction

In 1987, the influential Brundtland Report set out clear aims for the management of Antarctica: 'During the forthcoming period of change, the challenge is to ensure that Antarctica is managed in the interests of all humankind, in a manner that conserves its unique environment, preserves its value for scientific research, and retains its character as a demilitarized, non-nuclear zone of peace' (World Commission on Environment and Development, 1987, p. 279). These objectives provide a useful perspective from which to review the past management of the Antarctic and its resources. This chapter first looks back to examine the extent to which these objectives have been attained over the period of human involvement with the Antarctic, particularly during the time of the Antarctic Treaty, and it then looks ahead to future developments.

Throughout the short history of direct human involvement, interest in the Antarctic and its geography has been driven not only by a basic quest for discovery, exploration and scientific knowledge, but also by a motivation to exploit its resources, and particularly the living resources of the Southern Ocean. By virtue of its size, isolation, hostile climate, awesome scenery and abundant wildlife, the region has held a special fascination in the human imagination, with perceptions that range from a useless frozen waste to a source of untapped riches. With modern scientific investigation has come a greater appreciation of the role of the Antarctic in the planet's oceanography, atmospheric circulation and climate, and geological systems. With interest in open-access resources has come unsustainable exploitation of marine species, disruption of the Southern Ocean ecosystem and enhanced political intervention in Antarctic affairs. With modern communications has come a wider awareness of the beauty and natural heritage value of Antarctica, the sensitivity of the environment, the tourist potential, the human impacts and the need for the highest standards of environmental management. Antarctica is thus no longer the 'pole apart' of some earlier writers; it is now firmly on the world's geographical stage. Such a situation has arisen from a combination of several factors outlined in earlier chapters:

- the recognition of the role and significance of the Antarctic for understanding processes of global change, both at present (stratospheric ozone depletion, climate warming and sea-level change) and in the past (record of climate change in ice sheet cores and geological history);
- the exploitation, and in many cases depletion, of marine living resources and speculation about the potential of minerals resources;

- the 'politicisation' of Antarctica, initially through sovereignty claims and latterly through the desire of developing countries to share in the apparent wealth of perceived 'common' resources in line with the new international economic order, accompanied by demands for a more open and accountable management regime expressed in forums such as the UN;
- the development of a legal framework for the international management of Antarctica and human activities there;
- the growth of global environmental awareness and the recognition of the symbolic value of Antarctica as the last unspoiled wilderness on Earth, fostered by the high-profile media activities of the environmental NGOs campaigning for the declaration of Antarctica as a world park and for improved standards of environmental management by national operators;
- the growth of Antarctic tourism.

The following section considers some of the wider questions and forces that lie behind these changes, which have both shaped the geography of Antarctica and given rise to the fundamental issues that now face the scientists, environmental managers, conservationists, politicians and resource developers discussed in the subsequent section. It also provides a basis for an assessment of the human interaction with the environment and resources of Antarctica and the success of the Antarctic Treaty System.

10.2 Retrospect

10.2.1 Resource issues

The high seas resources of the Southern Ocean are not owned by any particular state and access is open to those with the desire and technology to exploit them. The inevitable consequence has been a cyclic pattern of exploitation and depletion of a succession of marine living resources which began long before the Antarctic Treaty. There have been major perturbations of the Southern Ocean ecosystem, particularly through sealing and whaling, that have affected not only the target species but also a wide range of competitor and prey species. However, the full effects of sealing and whaling exploitation via interspecific responses in the Antarctic food web can only be guessed. Where the need for regulation was recognised and implemented, as through the IWC, this was constrained by the conflicting objectives of the exploiters to maximise their return on capital outlay and of those concerned with longer-term conservation through sustainable management. With the exception of krill and some finfish, the result has been a dismal failure to adopt and implement a sustainable management regime before the target species were depleted. In the case of whaling, this pattern was favoured by uncertainty over scientific assessments of stocks, market forces, the precedents of exploitation largely uncontrolled except by the operators, the recourse to opt-outs from conservation measures, a lack of political sanctions (or a will to apply them), and a lack of strong and organised public pressure on politicians to take effective action. Market forces ultimately exerted a far greater influence on the patterns of resource exploitation than did ideas of sustainable management. It was only in the 1970s that circumstances changed, whaling became an important environmental issue, and public opinion was mobilised against exploitation by carefully organised campaigning by environmental organisations. Politicians were also prepared to employ sanctions against those states breaking international agreement to halt commercial whaling. Regulation was then effective, albeit at a time when the industry was scarcely commercially viable. Although scientific assessment of stocks shows that catching of minke whales could be resumed on a sustainable basis, the whaling ban continues. Thus cultural attitudes and public sentiment have become more influential than scientific assessment in the management of certain non-endangered species.

The exploitation of finfish has strong parallels with whaling, in the form of depletion of successive target species. Although much has been made of the potential krill resource, overexploitation of which could have far-reaching consequences on the Southern Ocean ecosystem, this potential has not yet been realised. However, through CCAMLR, the means are now in place for the sustainable management of krill, finfish and other marine living resources. CCAMLR came into force after the finfishery was already well-established and stocks of several species depleted. Not surprisingly, early progress with CCAMLR encountered problems similar to those faced by the IWC, with the onus of proof to justify controls resting with the regulators rather than the industry. CCAMLR also set over-ambitious objectives, at least in the short term, in that inadequate information was available to develop an ecosystem approach to multi-species fisheries management. It is now clear that the political will was not there either, particularly among the fishing states, to support not only the ideals of CCAMLR but also to enforce the regulatory measures adopted. However, in recent years CCAMLR has brought order and regulation to fisheries management in the Southern Ocean, and there is now a presumption, at least in intent, that new fisheries should be regulated from their commencement rather than retrospectively. It is therefore encouraging to see a range of precautionary measures now in force, including catch limits for krill, although the latter should be seen against a background of difficult market conditions and reduced catches by the former Soviet bloc countries. Should more intensive fishing for krill resume, including the involvement of new participants such as China, where there is possible interest (Zou, 1993), this could severely test the strength of CCAMLR. Despite the recent progress under CCAMLR, doubts remain about the levels of enforcement and the extent of illegal and unregulated fishing, and therefore about the ultimate effectiveness of international regulation of open-access resources in the presence of strong economic interests. This is not unique in respect of international fisheries management and echoes the pessimistic thesis of Johnston (1992) that current international environmental controls are weak and that until national sovereignty is ceded to a body with full international control then the 'tragedy of the commons' will continue to be repeated while powerful self-interests remain unrestrained.

The question of exploitation of potential minerals resources has generated much argument, particularly concerning ownership and the possible environmental impacts. Over the space of less than ten years a profound shift in attitudes occurred from the expectation that exploitation would one day occur, albeit with internationally agreed regulations to mitigate the environmental impacts, to a situation in which there is now a ban on minerals activities in the Protocol on Environmental Protection. Again this shift in attitude reflects in part the mobilisation of popular opinion by the environmental NGOs. Equally, however, following the shelving of CRAMRA, and given both the prime concern to maintain the stability of the ATS in the face of external criticism (see Chapter 6) and that minerals activities were not imminent, there was considerable political gain for the ATCPs from the agreement of a minerals ban and a strong environmental protection regime. Although the ban on minerals activity is indefinite, it is reviewable after 50 years, reflecting a political compromise between the conservationists' ideal of an unconditional ban and the interests of those states that might one day be potential exploiters. In effect, the solution is probably a realistic one from several viewpoints. On the one hand, it may be considered inappropriate to foreclose the option for minerals resource activities for future generations. Equally, from a pragmatic perspective, while exploitation is unlikely in the forseeable future and any Antarctic minerals would be resources of the last resort, it may not be realistic to assume that a permanent ban on minerals activity would survive serious future world shortages, should they arise. On the other hand, future technological advances may allow not only recovery of smaller and lower-grade resources elsewhere in the world, but also enhanced recycling. Even under conditions of political upheaval in current producer areas, would it really be more economic to exploit Antarctica, bearing in mind that no exploration has yet been done there, than to recover smaller and lower-grade resources elsewhere or to increase recycling? In addition to economic and political factors, the other unknown is how cultural attitudes to Antarctica will develop. Will it continue to remain a largely unblemished environmental symbol in the global consciousness, or will it one day come to be seen as a last reservoir of valuable resources in an overcrowded and resource-depleted planet?

A further question concerns whether Antarctica and its resources are part of the common heritage of mankind as proposed by some developing nations. Although under regulation by CCAMLR, marine living resources in the high seas are open access, and are therefore theoretically accessible to exploitation by developing nations. As far as minerals are concerned, such nations are least well placed to participate in any exploitation, but this is currently an academic point while the minerals ban remains in place. Should any minerals exploitation ever take place, there is arguably a moral case that such nations should share in any benefits as much as any of the current claimant states in Antarctica. Indeed, the whole question of revenues and sovereignty claims would be likely to prove a source of intense political and legal dispute, particularly were mining to proceed in areas of overlapping territorial claims. Thus from a political, as well as an environmental perspective, the present mining ban may be regarded as the best possible outcome.

10.2.2 Environmental management issues

Human activity in Antarctica has left a major legacy in the form of disturbance to the Southern Ocean ecosystem. There have also been significant, but highly localised, impacts on the terrestrial environment. To some extent the latter are understandable during the decades up to the 1960s, given the different environmental ethic and standards prevalent at the time, and the harsh operating conditions, 'frontier spirit' and limited technology available to earlier expeditions. However, in spite of changes in attitudes to environmental stewardship since the 1970s, many of the national operators were slow to adopt in practice an environmental ethic appropriate to the 'specialness' of Antarctica and therefore to apply in practice the high standards that should be commensurate with this recognition. In the last few years, however, the situation has improved significantly and under the Protocol on Environmental Protection almost all activities will require EIA.

The principal impacts have been in the form of localised chronic marine pollution, occasional major marine pollution incidents associated with shipping accidents, disturbance of soil and vegetation, disturbance of wildlife, and poor management of waste disposal. Although the effects are limited at a contintental scale, they assume a disproportionate significance because of the limited extent of ice-free areas and the competition with wildlife at such sites. Such biologically vulnerable areas coincide with marine traffic routes and station locations, as well as tourist destinations, notably along the Antarctic Peninsula and in parts of the Ross Sea area. In the sub-Antarctic islands, human impacts have been significant, particularly through the effects of introduced species on the natural ecosystems. From an environmental perspective, the ban on minerals developments is a significant step forward, since the start of mining would have raised significantly the potential for further local and regional environmental impacts, despite the precautionary measures in CRAMRA. So far, global atmospheric pollution, although detectable in Antarctica, has had limited direct effect on the continent and surrounding oceans. Potentially much more serious at a continental scale are the indirect effects of stratospheric ozone depletion and climate warming through the emission of greenhouse gases.

Notwithstanding the statements of good intent in the formulation of conventions, regulations and guidelines, just how good has been the record of the ATCPs in terms of practical conservation and environmental management? The answer is disappointing, insofar as an enduring 'footprint', albeit localised, has been left on most of the ice-free areas. There have been clear contraventions of treaty recommendations and SCAR advice regarding waste management. More seriously, there has been scant regard for environmental considerations by states establishing stations in areas such as the South Shetland Islands. The protected areas system has been shown to be inadequate not only in terms of coverage but also in terms of the ease with which it has been infringed. In many cases, the spirit of the ATS rules and conservation philosophy has hardly been honoured, most notably in the case of the Point Geologie airstrip, but also in terms of routine station management and operational practice. Effectively, there has been a failure to ensure compliance, hence the calls by NGOs for a formal inspectorate to work with the CEP (Sabella, 1992). The criticism of stations is not about their presence *per se* in Antarctica, but rather about their concentration for political and strategic gain in restricted areas and the poor standards of environmental stewardship in the past. If anything, there is an argument for more stations but sited according to wider scientific requirements and with appropriate environmental impact assessment. Overall, there has been a reticence to apply principles of environmental management developed elsewhere, for example in land-use planning and zoning or the setting of limits of acceptable change. Generally, short-term operational needs, or strategic requirements for the siting of new stations at least cost, have dictated priorities at the expense of scientific or environmental considerations. Remedial or mitigating measures or improvements in operational procedures have often only followed publicity, exposure and criticism by NGOs. The effects of tourist activity have generally been low but have not been monitored. Particular criticism may be directed at the more affluent and developed states for not taking a voluntary lead in applying improved standards of environmental management and utilising the professional expertise readily available to them. While the national operators may plead they were prohibited on the grounds of cost, they may be considered to have failed to have pressed the case sufficiently strongly to their governments. At the same time, governments have backed away from the economic and moral responsibilities of a sustainable Antarctic presence and the implications of environmental agreements; for example, the designation of SPAs on paper is not enough – they need to be managed appropriately and regulations enforced.

The record of the ATCPs in the past has therefore been better in terms of formulating measures than in their implementation. However, even with regard to the former, progress on the development of comprehensive environmental management was slow and only accelerated following the breakdown of CRAMRA. The failure has been in the implementation of the measures adopted and the tolerance of breaches, both in letter and in spirit, by other member states.

This leads to the question of whether the mistakes of the past should be cleared up, by whom, and who bears the cost. Strictly, it should be a case of the 'polluter pays'. However, many states facing domestic economic strictures are unlikely to be willing or in a position to achieve much, and international cooperation is essential; for example, as demonstrated between Germany and Russia for the clean-up of the areas around Georg Forster and Novolazarevskaya stations. There also needs to be a cost–benefit analysis, not only in financial terms, but also in environmental terms. It may not be beneficial to clean up if the disturbance or pollution is spread over a wider area.

Any assessment of the environmental impacts must include an evaluation of the benefits arising from human activities. Can the impacts adjacent to scientific stations be justified in terms of the scientific output? Overall, the answer is probably a qualified yes. However, in areas such as the South Shetland Islands and the Antarctic Peninsula many stations are located primarily for political and strategic purposes rather than scientific ones. In such instances, it is questionable that the impacts can be justified to society at large in terms of the quality of scientific output, but only in terms of national political benefits. The impacts on Fildes Peninsula have been particularly severe.

A further question concerns the resilience of Antarctic ecosystems and their potential for recovery. Some ecosystems, particularly in the terrestrial realm, are highly sensitive and very slow to recover, taking perhaps hundreds of years. Some marine species and ecosystems, however, may be relatively robust over a period of decades or tens of decades, as seen in the recovery of elephant seal and fur seal populations at South Georgia and penguin populations at Cape Hallet, as well as in the recovery of the Arthur Harbor area from the *Bahia Paraiso* oil spill. Such species may be adapted to large natural perturbations anyway (see Chapter 4), although the effects of human activity could exert a significant additional stress. Furthermore, impending climate change could produce far more significant changes in species distributions and diversity than localised human impacts (see Chapter 5). This is not to excuse the human impacts but rather to place them in a wider perspective. Overall, the impacts on the more robust marine environment, from science and logistics activities as opposed to resource exploitation, while they may be long-term as in the case of accumulation of pollutants in sediments, are spatially localised. Impacts on the terrestrial environment are generally more significant because of the greater sensitivity of such environments, the greater persistence of the impacts and the limited extent of the ice-free areas.

Although the above discussion presents a rather pessimistic retrospective view, this should be placed in the context of the scale of the impacts, the improvements in practice in the last few years, and the volume of scientific output (see Section 10.2.4). While spatially focused, the impacts are tiny on the scale of the continent. However, the NGOs view even small impacts with great concern and question whether any impact is justified, so that to some extent the argument is really about the symbolic value of Antarctica. Nevertheless, there has been poor practice in the past when viewed from today's standards and expectations of environmental management, as well as a sufficient degree of non-compliance with existing measures to call into question the commitment of some ATCPs (Elliott, 1994). Recent developments have seen significant changes in attitude and in operational procedures by most national operators, in compliance with the Protocol on Environmental Protection, and they have rightly tightened up on practice to the extent that scientists are now concerned about over-restrictive, bureaucratic conditions and diversion of funding away from science. In terms of environmental legislation, Antarctica is no worse and probably now has more stringent controls than many other areas of the world. Nevertheless, although establishing a mandatory framework for environmental management, the protocol falls short of providing for an independent monitoring and enforcement authority, responsibility instead remaining with individual governments as a concession to the sensitivities of the ATCPs about sovereignty and jurisdiction issues. The overall high quality of scientific output has been one of the major achievements under the Antarctic Treaty and offsets, but does not excuse, the environmental impacts. However, where stations have been established for political or strategic gain, this has resulted in the duplication of scientific effort or output of more limited value. Under these latter circumstances, the environmental impacts are even less justifiable.

10.2.3 Political issues and the ATS

The Antarctic Treaty has proved a remarkably enduring and stable platform for the conduct of Antarctic affairs and has survived internal tensions, arising for example over territorial claims, and external pressures, arising for example from the Cold War, the Falklands War and challenges to its legitimacy from developing countries at the UN. The treaty was only the beginning of the process, and it has proved flexible in responding to meet the new challenges, in terms of accommodating the sovereignty issue, in the development of additional measures and conventions addressing resource management and in expanding its membership. This has not, however, been without paradoxes. On the one hand, the ATCPs have been far-sighted in the development of CCAS, CCAMLR and CRAMRA, which may be considered considerable achievements in the face of the need to achieve consensus. On the other hand, they have proved slow to respond to the more practical aspects of implementation and enforcement of the measures in these conventions as well as to the growth of global environmental consciousness. Singularly, the ATCPs failed to recognise the influence, publicity skills and successful track record of NGOs over big campaigning issues and the growing political influence of the green movement in a number of member states, which helped to force the pace of change. While it may be that sensitivities over jurisdiction and sovereignty have militated against institutional structures to enforce implementation of environmental measures, and inertia in the consensus-negotiating process delayed progress towards comprehensive environmental protection, it is apparent that the ATCPs were overtaken by the speed and force of external developments. The NGOs have had a clear, unobstructed view of their goal, whereas the ATCPs have had to negotiate a complex course around issues of sovereignty, jurisdiction, and economic and political interests. The political reality was that the stability of the regime was more important than enforcement of the rules, since disputes about violations might have destabilised the system – in effect there was a reluctance to rock the boat (Elliott, 1994). To the credit of the ATCPs, however, is their rapid response in adopting the Protocol on Environmental Protection.

The answer to the question of how should Antarctica be managed – by those with the practical experience and active interests in the region, or through a wider international arrangement under the UN – appears to have been addressed by the overall achievements of the ATS. It is difficult to imagine a wider system administered under the UN that would have achieved so much, particularly through the involvement of states with limited experience in the region: consensus would have been much more difficult to achieve. While at times the ATS created the impression of an exclusive 'club', promulgating an attitude of 'we know what is best', this attitude has changed for a more open system of membership and participation in Antarctic affairs. Motivated principally by an interest in minerals resources, the strength of this external challenge has now been weakened following the ban on minerals activities for the foreseeable future. The focus is now on environmental protection, and the Protocol on Environmental Protection has widened the legitimacy of the ATS.

The other main challenge has arisen from the environmental NGOs, which have strongly advocated the declaration of Antarctica as a world park. In the face of the political realities, this objective, although admirable, appears unattainable. However, by responding in a positive manner to this challenge, the ATCPs have significantly narrowed the gap between the existing system and that proposed by the environmental movement, but at the same time preserving the stability of the ATS and its achievements. Given the political realities, it is probably more realistic for the environmental movement to work towards change and improved standards of environmental management and enforcement within the existing regime.

10.2.4 Science

The last three decades have seen significant developments in Antarctic science. As outlined in earlier chapters, major programmes of work in a range of fields – geology/geophysics, glaciology, marine biology and atmospheric sciences – have yielded results of global importance and revealed the fundamental role of Antarctica in modulating many global processes, but also as a source of unique archives of palaeoenvironmental data that span a range of timescales. Science, having advanced from largely observation-based studies of local and regional phenomena to encompass theories and models of global change, thus continues to be the prime focus and justification for human activity in Antarctica.

Science has also played a significant role in facilitating international political agreement in Antarctica, as well as underpinning the development of environmental and resource management. The requirements for science as demonstrated in the IGY set the whole framework for the Antarctic Treaty. Science has further helped to advance national interests in Antarctica, and for some countries the symbolic value of science has been more important than the results in terms of establishing a presence on the continent. This is exemplified in the rush to establish stations and to qualify for ATCP status during the 1980s in order to ensure a place at the negotiating table for CRAMRA and involvement in any future minerals resource exploitation. In addition to this political dimension, science has become increasingly involved in issues of environmental and marine resource management. SCAR has contributed significantly to the development of provisions for environmental protection, and in some instances has advised more stringent measures than were adopted. SCAR also played a central part in the coordination of research on the marine ecosystem through the BIOMASS programme, which has helped to provide some of the necessary understanding for managing the exploitation of marine living resources through CCAMLR. Science has also been fundamental in the subsequent programmes developed through the Scientific Committee of CCAMLR, notably on ecosystem monitoring and in the development of increasingly authoritative advice on fisheries management. The scope of science has thus broadened, with research increasingly directed at the questions that politicians and environmental managers are asking – Is the ice sheet stable? How will sea level change? What are the implications of the ozone hole and enhanced levels of UV irradiance for species and ecosystems? What limits should be set for marine resource harvesting? What are the effects of the fisheries and how can they be monitored? What are the impacts of human activities? Overall, in the past, this linking of science with politics, resources and environmental management has been to the benefit of advancing scientific knowledge about Antarctica and its physical and biological systems, both in terms of the priority accorded to science as a keystone of the Antarctic Treaty and in terms of funding support. However, more difficult issues may lie ahead (see Section 10.3.3).

10.2.5 Assessment of the achievements of the Antarctic Treaty System

Overall, the Antarctic Treaty System has many significant achievements (Table 10.1). It has ensured the

Table 10.1 Summary of achievements and shortfalls of the Antarctic Treaty System

Achievements

- maintenance of Antarctica as a zone of peace, free from military and nuclear activity;
- Setting aside the sovereignty issue and territorial claims;
- Strengthening of the position of the ATS through evolution in the face of external challenges and to meet new demands, as reflected in the development of treaty recommendations, CCAS, CCAMLR, CRAMRA and the Protocol on Environmental Protection;
- changing political attitudes and moves towards a more open regime, involving growth in participation in the ATS;
- development of a broad consensus that the ATS is an effective mechanism, as demonstrated by the achievements in its first 30 years, notably the provision of a remarkably stable framework for international cooperation between states;
- recognition of the need for greater responsibility for environmental stewardship and improved standards, in line with changing attitudes to environmental management;
- adoption of the Protocol on Environmental Protection, which has reinforced conservation and environmental management as a cornerstone of the ATS, together with peace and science;
- a ban on minerals activities;
- convergence of NGO and ATCP objectives for the use and management of Antarctica;
- greater management control of the exploitation of marine living resources under CCAMLR;
- a focus on globally important scientific research;
- enhanced public awareness of Antarctica and its role in global processes, accompanied by concern for its conservation;
- a start made by many states to implement the measures in the Protocol on Environmental Protection and to clear up the waste accumulated from the past.

Shortfalls

- failure to prevent overexploitation of finfish resources;
- slow response in practice to adopt improved standards of environmental management and sustainable resource management measures;
- failure to enforce implementation of environmental measures;
- use of science for political/strategic ends, resulting in duplication of effort;
- local proliferation of bases built on overcrowded sites;
- unmonitored expansion of the tourist industry.

preservation of Antarctica for scientific research, maintained the region as a zone of peace and prevented conflict over territorial claims. Despite the views of some nations, it is also arguable that Antarctica under the ATS is managed in the interests of most of humankind; the expanded composition of the treaty parties now represents 70% of the world's population. The ATS is an evolving system open for accession by all nations and is supported by a diverse range of countries of widely differing persuasions. To this extent, Antarctica is already internationalised and all countries have an opportunity to participate throughout the whole of Antarctica. There have been substantial gains of world importance to scientific knowledge. The ATS has encouraged and facilitated scientific research, including international programmes, and the results have been made available to all nations. The ATS has promoted the principles of the UN Charter through international cooperation, demilitarisation and establishment of working links with UN bodies (e.g. WMO). While the responsibilities, obligations and costs have been met by the ATCPs, the principal benefits in the form of scientific results have been made widely available.

The primary purpose of the Antarctic Treaty was to secure peace in the region and to protect scientific activities, against a background of sovereignty claims and superpower tensions, and therefore did not address environmental management and resource exploitation in a significant way. The ATS, therefore, has evolved in response to changing global priorities and challenges, accompanied by a greatly expanded level of membership. It has been sufficiently flexible not only to withstand the external challenges from the Non-Aligned Movement, the UN and the NGOs but also to accommodate the needs of the ATCPs themselves (internal challenges), for example during the CRAMRA negotiations and in dealing with its subsequent stalling. The achievements of the treaty are remarkable, particularly in terms of establishing a framework for international cooperation, given the range and mix of nations involved. The ATCPs have effectively isolated Antarctica from wider political issues. Thus, the treaty has successfully survived confrontations between member parties elsewhere, for example during the Cold War and the Falklands War. North Korea and South Korea both attend ATCMs; and South Africa continued to participate in ATCMs despite international condemnation of the apartheid regime and exclusion from most other international forums.

As far as conservation of Antarctica's unique environment is concerned, the past few years have seen significant positive developments in respect of resource and environmental management compared with the shortfalls in earlier decades, when the achievements were more on paper than in practice (see Table 10.1). Although significant improvements have now been made, and the Protocol on Environmental Protection provides the strongest framework so far for environmental stewardship, many practical and political questions and uncertainties relating to its implementation have yet to be resolved, as outlined below. Equally, there have been reservations concerning the effectiveness of resource management, particularly of marine living resources, although again significant improvements have been made with the adoption of precautionary measures under CCAMLR.

Broadly, therefore, the challenge posed in the Brundtland Report has been addressed and at least three of the four aims attained: management of Antarctica for all humankind, preservation of its value for science and maintenance of peace. As far as the fourth aim is concerned, although there has been significant progress on conservation and resource management, particularly in the formulation of conventions and recommendations, criticism can be levelled at the slow rate of development, implementation and enforcement of practical measures. While acknowledging the political realities required to achieve consensus, the failure, until recently, of individual states with more developed awareness and expertise to take a stronger lead in environmental matters is a more serious indictment. Unfortunately, the philosophy of 'out of sight, out of mind' prevailed in the past and it was the need to meet external legitimacy and wider accommodation that eventually forced the pace of change. The challenge now is to convert the good intent in the protocol into effect.

10.3 Prospect

10.3.1 Implementation of the Protocol on Environmental Protection

Is the Protocol on Environmental Protection an idealistic or a practical framework for environmental stewardship? The protocol sets out a vision for the future as well as specific measures. While the ATCPs agreed at the XVIth ATCM in 1991 that the measures should be applied as far as possible in the interim before the protocol became legally binding, it is clear that not all states did so, that EIA procedures were not being widely followed and that some personnel in the field had only limited awareness of the requirements (Greenpeace, 1994). Now that the protocol has entered into force, its implementation is a high priority, but the practicalities are also a matter of concern. For example, how will different approaches to enabling legislation in different countries be accommodated (Walton and Thomson, 1994)? How will the money be found for proactive environmental management, including the costs of surveys, monitoring, preparation of management plans and clean-up of past impacts, particularly at a time when there is pressure on governments to cut back Antarctic budgets (e.g. Mervis, 1995)? Will the money be found for the required environmental improvements, management programmes and environmental monitoring at the expense of scientific research? How will those states facing severe domestic economic difficulties respond? How will different perceptions among national operators of when comprehensive environmental evaluations are required be resolved in practice? It is relatively easy to insist that standards must be applied consistently and that there must be rigorous environmental assessment of individual and cumulative impacts, but how will this work in practice in the absence of independent EIA, and how strongly will effective enforcement be applied? Will the CEP take a strong line, or will it be constrained by requirements for consensus or become politicised? Will it interact effectively with SCAR? Furthermore, the complex issue of liability is still to be resolved. Thus, while the implementation of the Protocol on Environmental Protection represents a major step forward, it is merely the first stage in a longer-term process towards achieving a practical regime for environmental stewardship.

Despite increasing attention being paid to environmental issues in Antarctica under the protocol, there are still concerns as to whether the measures go far enough to protect the Antarctic environment and also about the degree of commitment to its enforcement. In the past, breaches in environmental regulations have either been overlooked or justified as a necessary part of the conduct of scientific activities in a remote and hostile environment. Relying on the goodwill of the ATCPs alone may no longer be adequate in the light of increased levels of activity, the global interest in the Antarctic environment and the adoption of the most stringent set of environmental protection measures to date under the protocol. On the one hand, there is the view of many scientists, who advocate that activities should be allowed where the benefits outweigh the impacts, providing the impacts are minimised; they fear excessive, and in their view unnecessary, constraints on their activities at the expense of rather limited environmental gains and consequently favour a policy of counselling rather than policing in the implementation of the protocol (Heywood, 1993). The environmental NGOs, on the other hand, advocate stronger controls under an independent inspectorate or Environmental Protection Agency (e.g. Barnes, 1988), and that no activity should be allowed unless it is demonstrated to have no significant impact on the environment. However, it is not clear how such an agency would be funded, how it would interact with the CEP or how its authority would be established. In view of the hostility of the ATCPs, it is doubtful that an independent agency could function effectively, and the current approach will prevail but under the watchful eye of the environmental NGOs and the UN. The pragmatic way ahead would be to give the protocol time to bed down and then to conduct an independent evaluation of the effectiveness of its measures and the success of their implementation. In a sense, the ATCPs are on probation and it is in their long-term interests to ensure that the protocol works in such a way that ensures a balance between comprehensive environmental protection and minimisation of environmental impacts arising from those legitimate activities where the scientific benefits outweigh the impacts. At the same time, it must not impose unnecessarily restrictive constraints that may discourage high quality scientific research on issues of global relevance. The balance will be difficult to achieve, but the traditional approach of negotiation and persuasion may offer more flexibility than a rigid policing system. The concern, however, is that if this approach fails, then the back-up enforcement procedures are inadequate. One solution that merits examination is an extension of the protocol to grant powers to the CEP in this respect (Wilder, 1995). Much will also depend on the environmental NGOs maintaining a close interest and alerting infringements in a responsible manner. The need now is not for more legislation, but rather for effective implementation and enforcement of existing measures and clear apportionment of responsibility for liability. The real test will be whether the ATCPs have the political will to make the protocol work.

10.3.2 Development of the Antarctic Treaty System

The shift in focus from CRAMRA and minerals exploitation to environmental protection has successfully deflected the challenge to the legitimacy of the ATS by the developing countries. The development of the Protocol on Environmental Protection has also found favour among the environmental critics. The debate appears now to be less over the objectives for Antarctica (science, natural reserve, peace), and more over the mechanism for implementing a strategy for environmental management. Should this remain in the preserve of the ATCPs or be entrusted to some international organisation such as the UN or an international Environmental Protection Agency? Under the existing political realities, the common heritage or world park approaches are unacceptable to the ATCPs, and currently the convergence of opinion amongst the advocates of each approach appears to be to work towards common objectives within the existing system. Thus a future under the ATS seems to be the best way forward, especially as science will continue to be the prime overt motive for human presence on the continent. The focus of political energies should now be on implementation of the protocol, in ensuring the enforcement of its measures and, where appropriate, overseeing its development.

A fundamental question concerns whether the ideals of the world park and common heritage concepts are safeguarded by the present system, given that both the sovereignty and minerals issues are deferred. There are three perspectives to this. One, it would be wrong, perhaps impossible, to foreclose for future generations the opportunity to exploit Antarctic minerals, and that the present approach is right to leave them the option. Two, the cynics would argue that the present approach is based on political and strategic considerations. Since minerals extraction is currently uneconomic, there is nothing to be lost and all to be gained by moving with the tide of environmental opinion, but ultimately to retain the option for minerals exploitation if and when the conditions are right. Three, there is the realist's view that, given the conflicting interests, the protocol is a significant step forward and represents a balance between the idealist's view and what is pragmatic. There is still much to be done, but by ensuring that the protocol is implemented and its measures enforced, attitudes may be progressively and widely changed. While there may be dangers in this pragmatic approach, including further adverse environmental impacts, it appears to offer the only realistic way forward given the improbability that the ATCPs will relinquish their political and strategic interests.

While the ending of the Cold War may motivate some governments to accept a reduced presence in Antarctica, compounded by the ban on minerals activities, for many countries the underlying strategic motives of sovereignty, geopolitics and mineral resources will continue to ensure a presence, particularly as the date for the optional review of the minerals ban approaches. It is in the national interests of most consumer countries to have a stake in Antarctica while minerals activity is still ultimately a possibility. If mining were to be banned permanently, this might diminish the motivation for an active presence. However, the option for review in 50 years time still holds out the long-term possibility that minerals exploitation could one day take place and therefore is likely to maintain the interest of such countries. A parallel may be drawn with the motivation for many new states to become involved in the ATS while discussions on CRAMRA were being pursued, to ensure that their national interests were represented in any regime that was negotiated. It will be interesting to see if current levels of scientific research funding and activity continue in the twenty-first century as a means of enhancing potential claims on possible future resources and their exploitation.

A further development is likely to include a wider recognition of the expertise of the environmental NGOs, particularly those such as ASOC and Greenpeace, as well as bodies like the IUCN, in contributing constructively to the discussion of environmental matters and the practical implementation of the measures in the Protocol on Environmental Protection. Thus the future should see a more integrated role for some NGOs as they become more accepted by the establishment (Herr and Davis, 1992) and are offered the opportunity to help to develop solutions. In particular, they offer (1) a source of expertise in environmental management; (2) a means of providing independent appraisal of standards of performance; and (3) expertise in raising public awareness and influencing public opinion. Equally, having forced the pace of change, the more campaigning NGOs now have a responsibility to adopt a more measured approach than perhaps has been taken in the past. They must increasingly face the challenge of both contributing through the established negotiating processes in which the scientific evidence and facts are carefully evaluated, and in maintaining an independent watch on practical implementation. Whether this role will sit easily with the NGOs' campaigning style and the emphasis on issues with a high public profile that generate both public awareness and funding, remains to be seen. It is also probable that the UN will continue to provide a forum for wider discussion of Antarctic affairs, with critics ready to focus attention on any lapses in the performance of the ATCPs in implementing the protocol.

A further area that will require attention concerns the relationship between the ATS and the Law of the Sea Convention. This applies not only in relation to the resources of the deep sea-bed south of 60° S, but also in relation to the potential assertion of claims to the continental shelf and EEZs by the claimant states. If asserted, such claims are likely to reawaken the sovereignty issue; if not, this may be regarded as a weakening of claims.

Developing the analysis of Falk (1991), the future success of the ATS, as viewed from the perspective of the 1990s, will be judged on the achievement in relation to several key aspects:

- the continued achievement of peace;
- improved credibility in environmental management;
- implementation and enforcement of the measures in the Protocol on Environmental Protection;
- greater openness and accountability by the ATCPs and the fostering of wide international support for the ATS through increased participation at all levels;
- the amount and quality of scientific output relating to global issues;
- the sustainable management of marine living resources;
- the management of tourism;
- the response of the ATS to further changes and new challenges.

10.3.3 Role of science

Antarctic science faces several challenges. If the scientific value of Antarctica is to be fully realised, then some impact on the environment is inevitable. The challenge is therefore twofold: first, to minimise that impact, to ensure that the research is properly co-ordinated, that it does not duplicate other work and that it cannot be done as effectively at alternative locations where environmental impacts would be less; second, to ensure that the quality of scientific output is sufficiently high to justify such impacts as do occur. While the emphasis should continue to be on globally important research undertaken in a way that leaves minimal impact on the environment, there is concern that the quality and output of this science will deteriorate at the expense of environmental considerations (Heap, 1989). On the other side of the equation, scientists need to inform better the politicians and policy makers about the value and implications of their work, as well as the practical issues of which they have direct experience. Scientists have tended to ignore this aspect, particularly as the political climate in the past allowed them simply to get on with their work. It is a question of finding a new balance between cutting-edge research and providing the information and analysis to underpin environmental management and to inform the policy makers (Scully and Kimball, 1989). However, the increasing demands on science are unlikely to be matched by increased funding.

From a political perspective, some research has undoubtedly been used as a justification for a strategic presence in Antarctica rather than being based purely on its intrinsic quality and the importance of the questions addressed. To quote Shapley (1985), 'politics – and not science – is the real reason behind the treaty powers' willingness to foot the bill for an Antarctic presence' (p. 227). However, with the ending of the Cold War and the competition between the USA and the USSR to maintain a strategic presence, scientists are likely to be called on increasingly to justify their expensive presence on the continent in scientific terms alone. For example, this is already happening with a review of the US Antarctic programme initiated by the Senate committee that oversees the NSF's budget, under pressure from Congress to cut costs (Kiernan, 1996). Furthermore, with the ban on minerals activities, a further plank in the argument for a strategic presence and support of science has at least in the medium term been removed. For those nations with long-standing territorial claims, however, the political and strategic imperative is likely to remain, although the high costs of maintaining a presence, as well as the reduced prospects of possible minerals gains, may see some scaling down of activity or refocusing of scientific priorities, as in the case of the closure of the UK base at Faraday and the switch of Signy Island from year-round to summer-only operations. It is also possible that research on environmental management may be used by some states to justify a presence to monitor the impacts of research that is not necessarily needed.

Some retraction of scientific activity also seems likely due to escalating costs, particularly in the case of those countries facing domestic economic difficulties; for example, several stations of the former USSR have been closed, and the inland station Vostok is now used only in the summer. An extreme measure of costs may be gauged from the planned rebuilding of the only year-round inland station, Scott–Amundsen Station at the South Pole, estimated at up to

$180 million. A scaled-down version costing *c.* $120 million has been proposed, but to meet part of these costs, a House of Representatives review panel has recommended that the research budget be cut by $20 million over a period of five years (Mervis, 1997). Except where there are ulterior strategic motives (see above), science will be focused on issues of global significance and on underpinning environmental management. Other possibilities may see the establishment of European programmes (e.g. the European Project for Ice Coring in Antarctica) or greater sharing of stations and programmes to spread costs, as currently occurs between the Scandinavian nations. The recent handover to the Ukraine of the redundant British base at Faraday on the Antarctic Peninsula may be a harbinger of further transfers. Britain is relieved of the obligation to remove the station and clean up the site, while the Ukraine has gained a functional station, albeit an ageing one.

The future is likely to see science even more closely interwoven with political, economic and environmental threads (Drewry, 1988, 1993; Davis, 1990; Walton and Morris, 1990). This is already leading to more 'directed' science, with resources deployed in underpinning the needs of marine resource management, environmental monitoring and impact assessments (*cf.* Keage and Quilty, 1988). These needs should not be underestimated, and the requirement for baseline information and follow-up monitoring is likely to be substantial if the measures in the protocol are to be strictly enforced. The fear that funds will be diverted away from cutting-edge science is a very real one: the concern is that politicians, having agreed to the protocol, must ensure that funding is provided for its implementation. There are also concerns that less wealthy nations will lack the resources to undertake the required baseline studies and monitoring of human impacts. It will also be a significant task to set monitoring standards, to coordinate monitoring programmes and to assess the results to inform management. Where will such developments leave scientific freedom and 'blue skies' research? There is certainly concern among scientists about reordering of research priorities to favour environmentally driven research, which may give a higher political payoff in the short term but reduce scientific freedom (*cf.* Elzinga, 1993b; Drewry, 1994).

Drewry (1988, 1993) identified several needs to assure the future of science: more critical evaluation of programme quality, cost-effectiveness and responsiveness to international issues and forces that drive scientific research. From an applied perspective, science must be flexible and responsive to the needs of the politicians and policy makers so that the scientific underpinning is in place to ensure future developments in Antarctica are both sustainable and sensitive to the unique environment and the global significance of the area and its surrounding oceans. From a scientific perspective, there must equally be a place for visionary 'blue skies' research. A key role for SCAR is to develop the necessary vision for the global value of such work and to convince the politicians that Antarctic science competes on the best possible footing with other demands for science funding. Achieving this balance does not necessarily mean a loss of scientific freedom; rather it is a matter of presentation so that the wider value of programmes and results is communicated to the policy makers and diplomats involved in Antarctic Treaty meetings, as well as more generally to a growing lay audience interested in the Antarctic.

As part of the bargain for working in the Antarctic, scientists have a responsibility to ensure that their investigations are conducted in a manner that is sensitive to the environment, an approach, for example, that is now part of BAS policy in the UK. From the perspective of NGOs, scientific activities may no longer be regarded as justifiable, regardless of their impact, just because they carry the magic label 'science' (Anon. 1990d). The value of the science should be weighed against its impact, including the logistic support. Thus if a station or research programme is not adding significantly to science, then any significant environmental impacts must be seriously questioned. Conversely, where the contribution to science is substantial, either in terms of addressing global issues or work that cannot be done elsewhere, then some significant but controlled impacts at a local scale may have to be part of the bargain for society (Bonner, 1994). Thus a degree of common sense will be required to ensure that, providing precautions are taken to minimise impacts, such research is not rendered prohibitively expensive or impractical.

Antarctica may no longer continue to be exclusively a continent for science, but the scientists, through SCAR, must ensure that there is a sound scientific basis for present and future environmental legislation (Heap, 1991a) and that the regulations and standards that are adopted are practical and based on the best scientific advice. There is therefore some concern about the weakening of the role of SCAR within the ATS and its lack of a vision for the future (*cf.* Elzinga, 1993b). Ultimately, the strength of the environmental movement and its political influence may offer an opportunity for science to regain and establish a stronger position within the ATS through offering new opportunities to forge links between science, environmental management and policy. For science, the balance between 'blue skies', directed and routine science within an overall programme is achieveable. While there is a trend towards the internationalisation of science addressing multidisciplinary questions of wider global relevance, routine measurements are also needed to compile time-series data. Without this, it may be difficult to make new discoveries, as for example the detection of stratospheric ozone depletion, or to test new models.

It is still an open question whether the ideal of Antarctica as a continent for science with truly international programmes, as opposed to a continent where science serves political or strategic purposes (Elzinga, 1993a), can be achieved. However, within the wider geopolitical frame it may become less important for some countries to maintain their scale of presence in Antarctica with the easing of the Cold War and, at least in the short term, with reduced interest in minerals and hydrocarbons. If this is borne out, science will need to be justified even more because it is good science rather than because it is located in Antarctica. For other countries, however, the strategic motive for a scientific presence in Antarctica will remain.

10.3.4 Responses to global change

As revealed in the geological records from ocean floor and ice sheet cores, change over different timescales has been a fundamental feature of Antarctic physical and biological systems in the past. Such change will continue in the future as part of the natural evolution of the planet. However, there is growing concern about the role of anthropogenic factors in forcing the pace and direction of such change, particularly through the effects of global warming and depletion of stratospheric ozone. The concern is not only about the impacts on the continent, but also about the wider global feedbacks between the Antarctica and the world's climatic and oceanic systems.

Currently, predictions of future climate change in the Antarctic are beset with uncertainty because of the complexity of modelling the links between the atmosphere, oceans and sea ice (see Chapter 5). Recent results have modified earlier predictions of amplified global warming in the Antarctic region (Cattle *et al.*, 1992), but even under more modest warming there are likely to be significant changes in sea ice extent and concentration, and increased precipitation over coastal areas. Much interest and speculation centre on the possible responses of the West and East Antarctic ice sheets. There have been predictions of catastrophic collapse of the former, possibly due to the weakening of ice shelf buttressing, and the rise in global sea level that would follow (see Chapters 3 and 5). However, major uncertainties concern the role of increased precipitation and increased snow accumulation on the ice sheets, which could be accompanied by a lowering of sea level. The stability of the East Antarctic ice sheet is also of concern because of the significantly larger potential contribution to global sea-level rise if melting were to occur. There are conflicting interpretations of the evidence from the geological record for its long-term stability, but the concensus is that the ice sheet has been more or less stable for the last 14 Ma. The response times of both ice sheets are likely to range from hundreds to thousands of years, and scenarios of catastrophic collapse and melt are unlikely. Possible changes in sea ice extent and variability are equally difficult to predict, but the effects are potentially far-reaching, both in terms of climate feedbacks and impacts on the biosphere (see Chapter 5).

Whether or not the significant changes observed recently in the ice shelves at the northern end of the Antarctic Peninsula (see Chapter 2) are a harbinger of more widespread changes to come is difficult to assess, because while such changes may be a response to global warming, the large ice shelves further south are less close to climatic thresholds for break-up. The immediate threat to the stability of the West Antarctic ice sheet appears to be minimal, since the northern ice shelves buttress relatively small amounts of land-based ice, so the effect of local break-up and ice draw-down on sea level may not be too great. What is less certain is whether the demise of the northern ice shelves represents the start of a pattern that will ultimately be repeated further south with much wider global impacts, if significant global warming occurs. The break-up of the Antarctic Peninsula ice shelves may therefore represent an early warning and a message of urgency for political decisions at a global scale concerning the stricter control of greenhouse gas emissions than the modest reductions agreed at Kyoto in December 1997.

Biosphere responses to global warming are likely to include changes in terrestrial plant distributions, production, microbial activity and nutrient cycling. In the marine ecosystem, changes may be anticipated in species distributions and numbers, feeding grounds and breeding patterns. Such changes may also be compounded by the effects of stratospheric ozone depletion, although these are circumscribed by greater uncertainty about species and commmunity responses and predictions that the ozone 'hole' will disappear by the middle of the twenty-first century. The potential is nevertheless significant for changes in biogeochemical cycling and feedback on global climate. The effects arising from these changes may have far greater impact on the ecosystems and landscapes of the Antarctic than the localised human activities on the continent. Much will depend on the sensitivity of phytoplankton and zooplankton, as well as that of their consumers, such as krill, for while the Southern Ocean ecosystem has responded robustly to the removal of top predators, the same may not be the case if the productivity of those species at the centre of the food web is reduced significantly.

Geographically, the effects of any changes in physical and biological systems are likely to be evident initially in the sub-Antarctic islands and the Antarctic Peninsula. As described in Chapter 5, such changes may already be happening in the form of glacier recession and ice shelf break-up, reduced sea ice extent and expansion of species breeding ranges and plant colonisation.

10.3.5 Resources

The issue of resource exploitation, in its various forms, is set to continue as a significant theme in Antarctic affairs. One concern relates to the exploitation of the marine living resources of the Southern Ocean. In the past, their exploitation has produced the 'predictable results of un-managed exploitation of a common property, open access resource' (Gulland, 1987, p. 117). This is part of a much wider problem relating to all resources, namely the failure to follow a precautionary approach when faced with scientific uncertainties. It remains to be seen whether CCAMLR will ultimately be any more successful in achieving sustainable management of the krill fishery or the remaining stocks of finfish if faced with pressures to intensify these fisheries as resources elsewhere in the world decline. As with the Protocol on Environmental Protection, the major uncertainties concern the effectiveness of implementation of the measures adopted under CCAMLR. Present indications are encouraging insofar as precautionary measures are in place, but the real test as far as the krill fishery is concerned may yet be to come, since levels of exploitation to date have been relatively low in relation to the size of the resource.

A second area of activity with a potentially significant impact on Antarctica is the development of tourism. Implementation of the measures set out in the protocol is required to ensure that the the benefits that arise in terms of raising Antarctic awareness continue to outweigh the impacts. Practical management questions will need to be addressed as part of a comprehensive tourist management framework; for example, should tourism be focused on specific locations where the impacts can be managed and monitored, or should it be dispersed to spread the impact? Future development of the industry might see the establishment of seasonal shore-based facilities and expansion of overland activities. While any such development would be subject to EIA under the protocol, it would undoubtedly raise the threshold for environmental disturbance in those accessible areas that are already of greatest vulnerability. Related development might see cooperation between science and tourism; for example, the leasing of disused bases or those temporarily closed might provide a source of funding for scientific research or environmental restoration. From a legal perspective there are concerns as to whether tour operators outside the control of states party to the treaty or IAATO will comply with the measures adopted under the protocol and the extent of any liability that they may have for environmental impacts.

The issue of minerals exploitation is postponed, rather than settled, since the ban may be reviewed in 50 years time, or earlier if there is consensus between the ATCPs. If particular minerals become globally scarce, is it realistic to expect the ban on Antarctic mining to continue? Currently it would be uneconomic to exploit Antarctic minerals, even if they were known to exist, so there is little to be lost and much to be gained in political terms, both globally and domestically, by the ATCPs in deferring to the environmental concerns, not least the securing of the continuation of the ATS because of the fears that minerals exploitation would re-open existing claims to sovereignty. Ultimately, in the global search for resources, Antarctica may again appear on the agenda, reopening the tensions between resource exploitation and environmental protection. The fundamental uncertainty is whether mining or hydrocarbon development and the resulting economic benefits, with appropriate but certainly not failsafe environmental safeguards, will ever be judged acceptable by world public opinion, or whether cultural attitudes about the environmental value of Antarctica will change.

10.3.6 Human impacts and environmental management

How pristine is the Antarctic? First, there have been large impacts in the past on the marine ecosystem through exploitation of marine living resources. These are likely to continue in the future with the exploitation particularly of krill, fish and squid. Second, human impact on Antarctica needs to be seen on the scale of the whole continent (see Chapter 9, Table 9.1). It is relatively restricted in spatial terms, and the Antarctic continent still remains the least disturbed terrestrial biogeographic realm on the planet (Hannah *et al.*, 1994). However, the impacts have been high locally. While still small at a continental scale, they assume greater significance in terms of the much more limited extent of the ice-free areas. In reality, however, the greatest threats in the future may arise from human activities elsewhere, for example from any significant global warming and stratospheric ozone depletion. Considerable uncertainties exist in predicting the impacts on Antarctic physical and biological systems and their interactions, but any such impacts could have far-reaching consequences at a global scale because of the feedback between Antarctica and global climatic and oceanic systems. Impacts arising from these wider global changes, together with the threat of overfishing of krill in the Southern Ocean, are likely to be the prime concerns. This is not to deny or excuse local environmental impacts, but rather to advocate a balanced overview of priorities.

At a local scale, environmental management should improve significantly through implementation of the measures in the Protocol on Environmental Protection, and many states are already doing so. However, there are doubts as to the degree of commitment that will ultimately be shown for such implementation. Particular concerns have been expressed about concentrations of bases in certain areas. Is this good or bad? Is it better to concentrate or disperse the effects? Arguably, it is better from an environmental viewpoint to concentrate the impacts and to focus management on these areas, employing management zonation to ensure optimal use of particular areas, be it for science, tourism, habitat or ecosystem protection, or even minerals exploitation should it eventually occur; the failure in the past has been to deliver such integrated management. Conversely, from a scientific viewpoint it may make more sense to have stations more widely dispersed. Developments in relatively pristine areas (including tourist activities) must, however, be subject to the most stringent environmental assessment. Above all, precautionary principles need to be applied where activities are suspected of having a detrimental effect on the environment or where there is a lack of scientific information to make an informed assessment of the risks. An understanding of the natural processes is essential for proper hazard assessment and the setting of clearly defined critical loads and monitoring targets. Important value judgements will be required to assess whether the benefit of an activity outweighs its environmental impacts.

Environment protection from human disturbance should also be enhanced by the development of a more comprehensive system of protected areas. While it is unlikely that additional types of designation (e.g. wilderness parks) will be implemented in view of the likely problems over issues such as sovereignty, administration and funding, the management approaches and principles developed elsewhere should provide valuable practical guidance for the management of tourists and visitors. However, it remains to be seen how such management can best be instituted within the ATS. Where not already applied, then at least similar standards to those specified in the Protocol on Environmental Protection should be adopted in the sub-Antarctic islands. There is a strong argument for a greater level of coordination in approaches to environmental management between the different island groups and a clear identification of management priorities that would lead to an integrated management framework or strategy. There are difficult questions to be addressed concerning the extent to which introduced species can or should be removed and the extent to which disturbed ecosystems should be managed to allow recovery or be left undisturbed to attain new equilibria.

Finally, political developments outside the Antarctic arena will have a significant bearing on the impacts of global change in the Antarctic, in respect of significant global warming and stratospheric ozone depletion. These concern progress on negotiating amendments to the Kyoto Protocol for further reductions in emissions of greenhouse gases and the enforcement of compliance with the measures in the Montreal Protocol and its amendments to phase out ozone-depleting chemicals. Ultimately, the success or otherwise of such global environmental agreements may have a greater impact on protection of the Antarctic environment and ecosystems at the continental scale.

10.4 Conclusion

The eighteenth-century view of Captain Cook, that Antarctica is an icy wilderness with no possible use (see Chapter 6) is a far cry from the recognition of the Antarctic region as an integral, and in some respects a keystone part of the Earth as a whole system. That Antarctica remains the least spoiled continent on Earth and that activities there must be carried out with proper regard for the environmental consequences is also undisputed. Extending the analogy of Gould (1978) that the stages in Antarctic development parallel Man's intellectual development (see Chapter 6), the Antarctic may be considered now to have entered an 'age of global environmental consciousness', reflecting both the scientific awareness of its role in global change and a wider concern about global environmental issues and human impacts on the more pristine parts of the planet. With the ending of the Cold War and the ban on minerals activities, there has been a weakening of the political and strategic arguments for some countries to maintain at existing levels an increasingly expensive permanent presence on the continent. The case for such a presence will increasingly fall on the value of scientific programmes, on the quality and global significance of the science and a favourable assessment of its environmental impacts. The principal challenges and opportunities in the future may be summarised as follows:

- implementation and further development, if necessary, of the Protocol on Environmental Protection, beyond the statement of policy aims and regulations to more problematic tasks of monitoring compliance with, and ensuring enforcement of, the measures;
- achievement of a proper balance and integration between science on the one hand and environmental programmes and management on the other;
- improved understanding of the role of the Antarctic in global change;
- sustainable management of the living resources of the Southern Ocean;
- persuading governments of the risks and implications of global pollution from human activities elsewhere in the world;
- development of coordinated and cost–effective approaches to environmental management and regulation, involving SCAR, the environmental NGOs, IUCN and the CEP to provide the best scientific advice possible;
- improvement of information databases and information management to underpin the science and monitoring essential for sound environmental management and regulation of activities;
- development of the protected areas system and development of integrated management plans for the focal areas of human activity, including tourism;
- development of a coordinated environmental management programme for the sub-Antarctic islands.

In a positive vein, these challenges may be viewed as opportunities to build on the already considerable achievements of the ATS. It is particularly encouraging, therefore, to see the recent convergence of the NGO and ATCP approaches and the apparent restoration of consensus at the UN. The Antarctic is no longer the largely exclusive realm of scientists but now has a much wider constituency with additional diverse political, economic and conservation objectives. Furthermore, to quote from Beck (1990e), 'the region's issues concerning ownership, resource management and conservation are merely Antarctic manifestations of global problems' (p. 128). Not only can politicians, scientists, resource managers and conservationists concerned with the Antarctic learn from experience elsewhere, but the successes, and indeed failures, of the ATS and the management of Antarctic resources have clear lessons for the rest of the world.

The future lies in maintaining the primacy of the consensus that the Antarctic is a region for peace, science and natural heritage and that it should be managed for all humankind. As recognised in the Brundtland Report, commmunication and cooperation are fundamental:

> To focus on longer-term strategies to preserve and build on the achievements of the existing Treaty System, nations must create the means to foster dialogue among politicians, scientists, environmentalists, and industries from countries within and outside it.
>
> (World Commission on Environment and Development, 1987, p.286).

The scientific work set in motion by the IGY has clearly shown the Antarctic region to be vital to the health of the planet. That it remains pristine is perhaps our safest guarantee towards ensuring that this health is maintained into the future, irrespective of changes elsewhere. It follows that the task must surely be to keep the human impact on Antarctica and its ocean to an absolute minimum, and at the same time to utilise its scientific wealth in pursuit of the vital records of the planet's systems and to cherish its landscapes and wildlife as a source of aesthetic inspiration.

Further reading

Barnes, J.N. (1988) Legal aspects of environmental protection in Antarctica. In Joyner, C.C. and Chopra, S.K. (eds) *The Antarctic Legal Regime*. Martinus Nijhoff Publishers, Dordrecht, 241–268.

Barnes, J.N. (1991) Protection of the environment in Antarctica: are present regimes enough? In Jørgensen-Dahl, A. and Østreng, W. (eds) *The Antarctic Treaty System in World Politics*. Macmillan, Basingstoke.

Drewry, D.J. (1994) Conflicts of interest in the use of Antarctica. In Hempel, G. (ed.) *Antarctic Science. Global Concerns*. Springer-Verlag, Berlin, 12–30.

Jackson, A. (ed.) (1996) *On the Antarctic Horizon. Proceedings of the International Symposium on the Future of the Antarctic Treaty System*, Ushuaia, Argentina, 20–24 March 1995. Australian Antarctic Foundation, Hobart.

Stokke, O.S. and Vidas, D. (eds) (1996). *Governing the Antarctic. The Effectiveness and Legitimacy of the Antarctic Treaty System*. Cambridge University Press, Cambridge.

Walton, D.W.H. and Morris, E.M. (1990) Science, environment and resources in Antarctica. *Applied Geography*, 10, 265–286.

APPENDIX 1 Text of the Antarctic Treaty

The Governments of Argentina, Australia, Belgium, Chile, the French Republic, Japan, New Zealand, Norway, the Union of South Africa, the Union of Soviet Socialist Republics, the United Kingdom of Great Britain and Northern Ireland, and the United States of America,

Recognizing that it is in the interest of all mankind that Antarctica shall continue for ever to be used exclusively for peaceful purposes and shall not become the scene or object of international discord;

Acknowledging the substantial contributions to scientific knowledge resulting from international co-operation in scientific investigation in Antarctica;

Convinced that the establishment of a firm foundation for the continuation and development of such co-operation on the basis of freedom of scientific investigation in Antarctica as applied during the International Geophysical Year accords with the interests of science and the progress of all mankind;

Convinced also that a treaty ensuring the use of Antarctica for peaceful purposes only and the continuance of international harmony in Antarctica will further the purposes and principles embodied in the Charter of the United Nations;

Have agreed as follows:

Article I

1. Antarctica shall be used for peaceful purposes only. There shall be prohibited, *inter alia*, any measure of a military nature, such as the establishment of military bases and fortifications, the carrying out of military manoeuvres, as well as the testing of any type of weapon.

2. The present Treaty shall not prevent the use of military personnel or equipment for scientific research or for any other peaceful purpose.

Article II

Freedom of scientific investigation in Antarctica and co-operation toward that end, as applied during the International Geophysical Year, shall continue, subject to the provisions of the present Treaty.

Article III

1. In order to promote international co-operation in scientific investigation in Antarctica, as provided for in Article II of the present Treaty, the Contracting Parties agree that, to the greatest extent feasible and practicable:

(a) information regarding plans for scientific programs in Antarctica shall be exchanged to permit maximum economy of and efficiency of operations;

(b) scientific personnel shall be exchanged in Antarctica between expeditions and stations;

(c) scientific observations and results from Antarctica shall be exchanged and made freely available.

2. In implementing this Article, every encouragement shall be given to the establishment of co-operative working relations with those Specialized Agencies of the United Nations and other technical organizations having a scientific or technical interest in Antarctica.

Article IV

1. Nothing contained in the present Treaty shall be interpreted as:

(a) a renunciation by any Contracting Party of previously asserted rights of or claims to territorial sovereignty in Antarctica;

(b) a renunciation or diminution by any Contracting Party of any basis of claim to territorial sovereignty in Antarctica which it may have whether as a result of its activities or those of its nationals in Antarctica, or otherwise;

(c) prejudicing the position of any Contracting Party as regards its recognition or non-recognition of any other State's rights of or claim or basis of claim to territorial sovereignty in Antarctica.

2. No acts or activities taking place while the present Treaty is in force shall constitute a basis for asserting, supporting or denying a claim to territorial sovereignty in Antarctica or create any rights of sovereignty in Antarctica. No new claim, or enlargement of an existing claim, to territorial sovereignty in Antarctica shall be asserted while the present Treaty is in force.

Article V

1. Any nuclear explosions in Antarctica and the disposal there of radioactive waste material shall be prohibited.

2. In the event of the conclusion of international agreements concerning the use of nuclear energy, including nuclear explosions and the disposal of radioactive waste material, to which all of the Contracting Parties whose representatives are entitled to participate in the meetings provided for under Article IX are parties, the rules established under such agreements shall apply in Antarctica.

Article VI

The provisions of the present Treaty shall apply to the area south of 60° South Latitude, including all ice shelves, but nothing in the present Treaty shall prejudice or in any way affect the rights, or the exercise of the rights, of any State under international law with regard to the high seas within that area.

Article VII

1. In order to promote the objectives and ensure the observance of the provisions of the present Treaty, each Contracting Party whose representatives are entitled to participate in the meetings referred to in Article IX of the Treaty shall have the right to designate observers to carry out any inspection provided for by the present Article. Observers shall be nationals of the Contracting Parties which designate them. The names of observers shall be communicated to every other Contracting Party having the right to designate observers, and like notice shall be given of the termination of their appointment.

2. Each observer designated in accordance with the provisions of paragraph 1 of this Article shall have complete freedom of access at any time to any or all areas of Antarctica.

3. All areas of Antarctica, including all stations, installations and equipment within those areas, and all ships and aircraft at points of discharging or embarking cargoes or personnel in Antarctica, shall be open at all times to inspection by any observers designated in accordance with paragraph 1 of this Article.

4. Aerial observation may be carried out at any time over any or all areas of Antarctica by any of the Contracting Parties having the right to designate observers.

5. Each Contracting Party shall, at the time when the present Treaty enters into force for it, inform the other Contracting Parties, and thereafter shall give them notice in advance, of

(a) all expeditions to and within Antarctica, on the part of its ships or nationals, and all expeditions to Antarctica organized in or proceeding from its territory;

(b) all stations in Antarctica occupied by its nationals; and

(c) any military personnel or equipment intended to be introduced by it into Antarctica subject to the conditions prescribed in paragraph 2 of Article I of the present Treaty.

Article VIII

1. In order to facilitate the exercise of their functions under the present Treaty, and without prejudice to the respective positions of the Contracting Parties relating to jurisdiction over all other persons in Antarctica, observers designated under paragraph 1 of Article VII and scientific personnel exchanged under sub-paragraph 1(b) of Article III of the Treaty, and members of the staffs accompanying any such persons, shall be subject only to the jurisdiction of the Contracting Party of which they are nationals in respect of all acts or omissions occurring while they are in Antarctica for the purpose of exercising their functions.

2. Without prejudice to the provisions of paragraph 1 of this Article, and pending the adoption of measures in pursuance of subparagraph 1(e) of Article IX, the Contracting Parties concerned in any case of dispute with regard to the exercise of jurisdiction in Antarctica shall immediately consult together with a view to reaching a mutually acceptable solution.

Article IX

1. Representatives of the Contracting Parties named in the preamble to the present Treaty shall meet at the City of Canberra within two months after the

date of entry into force of the Treaty, and thereafter at suitable intervals and places, for the purpose of exchanging information, consulting together on matters of common interest pertaining to Antarctica, and formulating and considering, and recommending to their Governments, measures in furtherance of the principles and objectives of the Treaty, including measures regarding:

(a) use of Antarctica for peaceful purposes only;

(b) facilitation of scientific research in Antarctica;

(c) facilitation of international scientific co-operation in Antarctica;

(d) facilitation of the exercise of the rights of inspection provided for in Article VII of the Treaty;

(e) questions relating to the exercise of jurisdiction in Antarctica;

(f) preservation and conservation of living resources in Antarctica.

2. Each Contracting Party which has become a party to the present Treaty by accession under Article XIII shall be entitled to appoint representatives to participate in the meetings referred to in paragraph 1 of the present Article, during such times as that Contracting Party demonstrates its interest in Antarctica by conducting substantial research activity there, such as the establishment of a scientific station or the despatch of a scientific expedition.

3. Reports from the observers referred to in Article VII of the present Treaty shall be transmitted to the representatives of the Contracting Parties participating in the meetings referred to in paragraph 1 of the present Article.

4. The measures referred to in paragraph 1 of this Article shall become effective when approved by all the Contracting Parties whose representatives were entitled to participate in the meetings held to consider those measures.

5. Any or all of the rights established in the present Treaty may be exercised as from the date of entry into force of the Treaty whether or not any measures facilitating the exercise of such rights have been proposed, considered or approved as provided in this Article.

Article X

Each of the Contracting Parties undertakes to exert appropriate efforts, consistent with the Charter of the United Nations, to the end that no one engages in any activity in Antarctica contrary to the principles or purposes of the present Treaty.

Article XI

1. If any dispute arises between two or more of the Contracting Parties concerning the interpretation or application of the present Treaty, those Contracting Parties shall consult among themselves with a view to having the dispute resolved by negotiation, inquiry, mediation, conciliation, arbitration, judicial settlement or other peaceful means of their own choice.

2. Any dispute of this character not so resolved shall, with the consent, in each case, of all parties to the dispute, be referred to the International Court of Justice for settlement; but failure to reach agreement on reference to the International Court shall not absolve parties to the dispute from the responsibility of continuing to seek to resolve it by any of the various peaceful means referred to in paragraph 1 of this Article.

Article XII

1. (a) The present Treaty may be modified or amended at any time by unanimous agreement of the Contracting Parties whose representatives are entitled to participate in the meetings provided for under Article IX. Any such modification or amendment shall enter into force when the depositary Government has received notice from all such Contracting Parties that they have ratified it.

(b) Such modification or amendment shall thereafter enter into force as to any other Contracting Party when notice of ratification by it has been received by the depositary Government. Any such Contracting Party from which no notice of ratification is received within a period of two years from the date of entry into force of the modification or amendment in accordance with the provision of subparagraph 1(a) of this Article shall be deemed to have withdrawn from the present Treaty on the date of the expiration of such period.

2. (a) If after the expiration of thirty years from the date of entry into force of the present Treaty, any of the Contracting Parties whose representatives are entitled to participate in the meetings provided for under Article IX so requests by a communication addressed to the depositary Government, a Conference of all the Contracting Parties shall be held as soon as practicable to review the operation of the Treaty.

(b) Any modification or amendment to the present Treaty which is approved at such a Conference by a majority of the Contracting Parties there represented, including a majority of those whose representatives are entitled to participate in the meetings provided for under Article IX, shall be communicated by the depositary Government to all Contracting Parties immediately after the termination of the Conference and shall enter into force in accordance with the provisions of paragraph 1 of the present Article

(c) If any such modification or amendment has not entered into force in accordance with the provisions of subparagraph 1(a) of this Article within a period of two years after the date of its communication to all the Contracting Parties, any Contracting Party may at any time after the expiration of that period give notice to the depositary Government of its withdrawal from the present Treaty; and such withdrawal shall take effect two years after the receipt of the notice by the depositary Government.

Article XIII

1. The present Treaty shall be subject to ratification by the signatory States. It shall be open for accession by any State which is a Member of the United Nations, or by any other State which may be invited to accede to the Treaty with the consent of all the Contracting Parties whose representatives are entitled to participate in the meetings provided for under Article IX of the Treaty.

2. Ratification of or accession to the present Treaty shall be effected by each State in accordance with its constitutional processes.

3. Instruments of ratification and instruments of accession shall be deposited with the Government of the United States of America, hereby designated as the depositary Government.

4. The depositary Government shall inform all signatory and acceding States of the date of each deposit of an instrument of ratification or accession, and the date of entry into force of the Treaty and of any modification or amendment thereto.

5. Upon the deposit of instruments of ratification by all the signatory States, the present Treaty shall enter into force for those States and for States which have deposited instruments of accession. Thereafter the Treaty shall enter into force for any acceding State upon the deposit of its instruments of accession.

6. The present Treaty shall be registered by the depositary Government pursuant to Article 102 of the Charter of the United Nations.

Article XIV

The present Treaty, done in the English, French, Russian and Spanish languages, each version being equally authentic, shall be deposited in the archives of the Government of the United States of America, which shall transmit duly certified copies thereof to the Governments of the signatory and acceding States.

APPENDIX 2 Protocol on Environmental Protection to the Antarctic Treaty

PREAMBLE

The States Parties to this Protocol to the Antarctic Treaty, hereinafter referred to as the Parties,

Convinced of the need to enhance the protection of the Antarctic environment and dependent and associated ecosystems;

Convinced of the need to strengthen the Antarctic Treaty system so as to ensure that Antarctica shall continue forever to be used exclusively for peaceful purposes and shall not become the scene or object of international discord;

Bearing in mind the special legal and political status of Antarctica and the special responsibility of the Antarctic Treaty Consultative Parties to ensure that all activities in Antarctica are consistent with the purposes and principles of the Antarctic Treaty;

Recalling the designation of Antarctica as a Special Conservation Area and other measures adopted under the Antarctic Treaty system to protect the Antarctic environment and dependent and associated ecosystems;

Acknowledging further the unique opportunities Antarctica offers for scientific monitoring of and research on processes of global as well as regional importance;

Reaffirming the conservation principles of the Convention on the Conservation of Antarctic Marine Living Resources;

Convinced that the development of a comprehensive regime for the protection of the Antarctic environment and dependent and associated ecosystems is in the interest of mankind as a whole;

Desiring to supplement the Antarctic Treaty to this end;

Have agreed as follows:

Article 1
DEFINITIONS

For the purposes of this Protocol:

(a) "The Antarctic Treaty" means the Antarctic Treaty done at Washington on 1 December 1959;

(b) "Antarctic Treaty area" means the area to which the provisions of the Antarctic Treaty apply in accordance with Article VI of that Treaty;

(c) "Antarctic Treaty Consultative Meetings" means the meetings referred to in Article IX of the Antarctic Treaty;

(d) "Antarctic Treaty Consultative Parties" means the Contracting Parties to the Antarctic Treaty entitled to appoint representatives to participate in the meetings referred to in Article IX of that Treaty;

(e) "Antarctic Treaty system" means the Antarctic Treaty, the measures in effect under that Treaty, its associated separate international instruments in force and the measures in effect under those instruments;

(f) "Arbitral Tribunal" means the Arbitral Tribunal established in accordance with the Schedule to this Protocol, which forms an integral part thereof;

(g) "Committee" means the Committee for Environmental Protection established in accordance with Article 11.

Article 2
OBJECTIVE AND DESIGNATION

The Parties commit themselves to the comprehensive protection of the Antarctic environment and dependent and associated ecosystems and hereby designate Antarctica as a natural reserve, devoted to peace and science.

Article 3

Environmental Principles

1. The protection of the Antarctic environment and dependent and associated ecosystems and the intrinsic value of Antarctica, including its wilderness and aesthetic values and its value as an area for the conduct of scientific research, in particular research essential to understanding the global environment, shall be fundamental considerations in the planning and conduct of all activities in the Antarctic Treaty area.
2. To this end:
   (a) activities in the Antarctic Treaty area shall be planned and conducted so as to limit adverse impacts on the Antarctic environment and dependent and associated ecosystems;
   (b) activities in the Antarctic Treaty area shall be planned and conducted so as to avoid:
      (i) adverse effects on climate or weather patterns;
      (ii) significant adverse effects on air or water quality;
      (iii) significant changes in the atmospheric, terrestrial (including aquatic), glacial or marine environments;
      (iv) detrimental changes in the distribution, abundance or productivity of species or populations of species of fauna and flora;
      (v) further jeopardy to endangered or threatened species or populations of such species; or
      (vi) degradation of, or substantial risk to, areas of biological, scientific, historic, aesthetic or wilderness significance;
   (c) activities in the Antarctic Treaty area shall be planned and conducted on the basis of information sufficient to allow prior assessments of, and informed judgements about, their possible impacts on the Antarctic environment and dependent and associated ecosystems and on the value of Antarctica for the conduct of scientific research; such judgements shall take account of:
      (i) the scope of the activity, including its area, duration and intensity;
      (ii) the cumulative impacts of the activity, both by itself and in combination with other activities in the Antarctic Treaty area;
      (iii) whether the activity will detrimentally affect any other activity in the Antarctic Treaty area;
      (iv) whether technology and procedures are available to provide for environmentally safe operations;
      (v) whether there exists the capacity to monitor key environmental parameters and ecosystem components so as to identify and provide early warning of any adverse effects of the activity and to provide for such modification of operating procedures as may be necessary in the light of the results of monitoring or increased knowledge of the Antarctic environment and dependent and associated ecosystems; and
      (vi) whether there exists the capacity to respond promptly and effectively to accidents, particularly those with potential environmental effects;
   (d) regular and effective monitoring shall take place to allow assessment of the impacts of ongoing activities, including the verification of predicted impacts;
   (e) regular and effective monitoring shall take place to facilitate early detection of the possible unforeseen effects of activities carried on both within and outside the Antarctic Treaty area on the Antarctic environment and dependent and associated ecosystems.
3. Activities shall be planned and conducted in the Antarctic Treaty area so as to accord priority to scientific research and to preserve the value of Antarctica as an area for the conduct of such research, including research essential to understanding the global environment.
4. Activities undertaken in the Antarctic Treaty area pursuant to scientific research programmes, tourism and all other governmental and non-governmental activities in the Antarctic Treaty area for which advance notice is required in accordance with Article VII (5) of the Antarctic Treaty, including associated logistic support activities, shall:
   (a) take place in a manner consistent with the principles in this Article; and
   (b) be modified, suspended or cancelled if they result in or threaten to result in impacts upon the Antarctic environment or dependent or associated ecosystems inconsistent with those principles.

Article 4

Relationship with the Other Components of the Antarctic Treaty System

1. This Protocol shall supplement the Antarctic Treaty and shall neither modify nor amend that Treaty.

2. Nothing in this Protocol shall derogate from the rights and obligations of the Parties to this Protocol under the other international instruments in force within the Antarctic Treaty system.

Article 5

Consistency with the Other Components of the Antarctic Treaty System

The Parties shall consult and co-operate with the Contracting Parties to the other international instruments in force within the Antarctic Treaty system and their respective institutions with a view to ensuring the achievement of the objectives and principles of this Protocol and avoiding any interference with the achievement of the objectives and principles of those instruments or any inconsistency between the implementation of those instruments and of this Protocol.

Article 6

Co-operation

1. The Parties shall co-operate in the planning and conduct of activities in the Antarctic Treaty area. To this end, each Party shall endeavour to:
   (a) promote co-operative programmes of scientific, technical and educational value, concerning the protection of the Antarctic environment and dependent and associated ecosystems;
   (b) provide appropriate assistance to other Parties in the preparation of environmental impact assessments;
   (c) provide to other Parties upon request information relevant to any potential environmental risk and assistance to minimize the effects of accidents which may damage the Antarctic environment or dependent and associated ecosystems;
   (d) consult with other Parties with regard to the choice of sites for prospective stations and other facilities so as to avoid the cumulative impacts caused by their excessive concentration in any location;
   (e) where appropriate, undertake joint expeditions and share the use of stations and other facilities; and
   (f) carry out such steps as may be agreed upon at Antarctic Treaty Consultative Meetings.
2. Each Party undertakes, to the extent possible, to share information that may be helpful to other Parties in planning and conducting their activities in the Antarctic Treaty area, with a view to the protection of the Antarctic environment and dependent and associated ecosystems.
3. The Parties shall co-operate with those Parties which may exercise jurisdiction in areas adjacent to the Antarctic Treaty area with a view to ensuring that activities in the Antarctic Treaty area do not have adverse environmental impacts on those areas.

Article 7

Prohibition of Mineral Resource Activities

Any activity relating to mineral resources, other than scientific research, shall be prohibited.

Article 8

Environmental Impact Assessment

1. Proposed activities referred to in paragraph 2 below shall be subject to the procedures set out in Annex I for prior assessment of the impacts of those activities on the Antarctic environment or on dependent or associated ecosystems according to whether those activities are identified as having:
   (a) less than a minor or transitory impact;
   (b) a minor transitory impact; or
   (c) more than a minor or transitory impact.
2. Each Party shall ensure that the assessment procedures set out in Annex I are applied in the planning processes leading to decisions about any activities undertaken in the Antarctic Treaty area pursuant to scientific research programmes, tourism and all other governmental and non-governmental activities in the Antarctic Treaty area for which advance notice is required under Article VII (5) of the Antarctic Treaty, including associated logistic support activities.
3. The assessment procedures set out in Annex I shall apply to any change in an activity whether the change arises from an increase or decrease in the intensity of an existing activity, from the addition of an activity, the decommissioning of a facility, or otherwise.
4. Where activities are planned jointly by more than one Party, the Parties involved shall nominate one of their number to coordinate the implementation of the environmental impact assessment procedures set out in Annex I.

Article 9

Annexes

1. The Annexes to this Protocol shall form an integral part thereof.
2. Annexes, additional to Annexes I–IV, may be adopted and become effective in accordance with Article IX of the Antarctic Treaty.

3. Amendments and modifications to Annexes may be adopted and become effective in accordance with Article IX of the Antarctic Treaty, provided that any Annex may itself make provision for amendments and modifications to become effective on an accelerated basis.
4. Annexes and any amendments and modifications thereto which have become effective in accordance with paragraphs 2 and 3 above shall, unless an Annex itself provides otherwise in respect of the entry into effect of any amendment or modification thereto, become effective for a Contracting Party to the Antarctic Treaty which is not an Antarctic Treaty Consultative Party, or which was not an Antarctic Treaty Consultative Party at the time of the adoption, when notice of approval of that Contracting Party has been received by the Depositary.
5. Annexes shall, except to the extent that an Annex provides otherwise, be subject to the procedures for dispute settlement set out in Articles 18 to 20.

Article 10
ANTARCTIC TREATY CONSULTATIVE MEETINGS

1. Antarctic Treaty Consultative Meetings shall, drawing upon the best scientific and technical advice available:
   (a) define, in accordance with the provisions of this Protocol, the general policy for the comprehensive protection of the Antarctic environment and dependent and associated ecosystems; and
   (b) adopt measures under Article IX of the Antarctic Treaty for the implementation of this Protocol.
2. Antarctic Treaty Consultative Meetings shall review the work of the Committee and shall draw fully upon its advice and recommendations in carrying out the tasks referred to in paragraph 1 above, as well as upon the advice of the Scientific Committee on Antarctic Research.

Article 11
COMMITTEE FOR ENVIRONMENTAL PROTECTION

1. There is hereby established the Committee for Environmental Protection.
2. Each Party shall be entitled to be a member of the Committee and to appoint a representative who may be accompanied by experts and advisers.
3. Observer status in the Committee shall be open to any Contracting Party to the Antarctic Treaty which is not a Party to this Protocol.
4. The Committee shall invite the President of the Scientific Committee on Antarctic Research and the Chairman of the Scientific Committee for the Conservation of Antarctic Marine Living Resources to participate as observers at its sessions. The Committee may also, with the approval of the Antarctic Treaty Consultative Meeting, invite such other relevant scientific, environmental and technical organisations which can contribute to its work to participate as observers at its sessions.
5. The Committee shall present a report on each of its sessions to the Antarctic Treaty Consultative Meeting. The report shall cover all matters considered at the session and shall reflect the views expressed. The report shall be circulated to the Parties and to observers attending the session, and shall thereupon be made publicly available.
6. The Committee shall adopt its rules of procedure which shall be subject to approval by the Antarctic Treaty Consultative Meeting.

Article 12
FUNCTIONS OF THE COMMITTEE

1. The functions of the Committee shall be to provide advice and formulate recommendations to the Parties in connection with the implementation of this Protocol, including the operation of its Annexes, for consideration at Antarctic Treaty Consultative Meetings, and to perform such other functions as may be referred to it by the Antarctic Treaty Consultative Meetings. In particular, it shall provide advice on:
   (a) the effectiveness of measures taken pursuant to this Protocol;
   (b) the need to update, strengthen or otherwise improve such measures;
   (c) the need for additional measures, including the need for additional Annexes, where appropriate;
   (d) the application and implementation of the environmental impact assessment procedures set out in Article 8 and Annex I;
   (e) means of minimising or mitigating environmental impacts of activities in the Antarctic Treaty area;
   (f) procedures for situations requiring urgent action, including response action in environmental emergencies;
   (g) the operation and further elaboration of the Antarctic Protected Area system;
   (h) inspection procedures, including formats for inspection reports and checklists for the conduct of inspections;

(i) the collection, archiving, exchange and evaluation of information related to environmental protection;
(j) the state of the Antarctic environment; and
(k) the need for scientific research, including environmental monitoring, related to the implementation of this Protocol.

2. In carrying out its functions, the Committee shall, as appropriate, consult with the Scientific Committee on Antarctic Research, the Scientific Committee for the Conservation of Antarctic Marine Living Resources and other relevant scientific, environmental and technical organizations.

Article 13

COMPLIANCE WITH THIS PROTOCOL

1. Each Party shall take appropriate measures within its competence, including the adoption of laws and regulations, administrative actions and enforcement measures, to ensure compliance with this Protocol.
2. Each Party shall exert appropriate efforts, consistent with the Charter of the United Nations, to the end that no one engages in any activity contrary to this Protocol.
3. Each Party shall notify all other Parties of the measures it takes pursuant to paragraphs 1 and 2 above.
4. Each Party shall draw the attention of all other Parties to any activity which in its opinion affects the implementation of the objectives and principles of this Protocol.
5. The Antarctic Treaty Consultative Meetings shall draw the attention of any State which is not a Party to this Protocol to any activity undertaken by that State, its agencies, instrumentalities, natural or juridical persons, ships, aircraft or other means of transport which affects the implementation of the objectives and principles of this Protocol.

Article 14

INSPECTION

1. In order to promote the protection of the Antarctic environment and dependent and associated ecosystems, and to ensure compliance with this Protocol, the Antarctic Treaty Consultative Parties shall arrange, individually or collectively, for inspections by observers to be made in accordance with Article VII of the Antarctic Treaty.
2. Observers are:
   (a) observers designated by any Antarctic Treaty Consultative Party who shall be nationals of that Party; and
   (b) any observers designated at Antarctic Treaty Consultative Meetings to carry out inspections under procedures to be established by an Antarctic Treaty Consultative Meeting.
3. Parties shall co-operate fully with observers undertaking inspections, and shall ensure that during inspections, observers are given access to all parts of stations, installations, equipment, ships and aircraft open to inspection under Article VII (3) of the Antarctic Treaty, as well as to all records maintained thereon which are called for pursuant to this Protocol.
4. Reports of inspections shall be sent to the Parties whose stations, installations, equipment, ships or aircraft are covered by the reports. After those Parties have been given the opportunity to comment, the reports and any comments thereon shall be circulated to all the Parties and to the Committee, considered at the next Antarctic Treaty Consultative Meeting, and thereafter made publicly available.

Article 15

EMERGENCY RESPONSE ACTION

1. In order to respond to environmental emergencies in the Antarctic Treaty area, each Party agrees to:
   (a) provide for prompt and effective response action to such emergencies which might arise in the performance of scientific research programmes, tourism and all other governmental and non-governmental activities in the Antarctic Treaty area for which advance notice is required under Article VII (15) of the Antarctic Treaty, including associated logistic support activities; and
   (b) establish contingency plans for response to incidents with potential adverse effects on the Antarctic environment or dependent and associated ecosystems.
2. To this end, the Parties shall:
   (a) co-operate in the formulation and implementation of such contingency plans; and
   (b) establish procedures for immediate notification of, and co-operative response to, environmental emergencies.
3. In the implementation of this Article, the Parties shall draw upon the advice of the appropriate international organisations.

Article 16

LIABILITY

Consistent with the objectives of this Protocol for the comprehensive protection of the Antarctic environment and dependent and associated ecosystems, the

Parties undertake to elaborate rules and procedures relating to liability for damage arising from activities taking place in the Antarctic Treaty area and covered by this Protocol. Those rules and procedures shall be included in one or more Annexes to be adopted in accordance with Article 9 (2).

Article 17

Annual Report by Parties

1. Each Party shall report annually on the steps taken to implement this Protocol. Such reports shall include notifications made in accordance with Article 13 (3), contingency plans established in accordance with Article 15 and any other notifications and information called for pursuant to this Protocol for which there is no other provision concerning the circulation and exchange of information.
2. Reports made in accordance with paragraph 1 above shall be circulated to all Parties and to the Committee, considered at the next Antarctic Treaty Consultative Meeting, and made publicly available.

Article 18

Dispute Settlement

If a dispute arises concerning the interpretation or application of this Protocol, the parties to the dispute shall, at the request of any one of them, consult among themselves as soon as possible with a view to having the dispute resolved by negotiation, inquiry, mediation, conciliation, arbitration, judicial settlement or other peaceful means to which the parties to the dispute agree.

Article 19

Choice of Dispute Settlement Procedure

1. Each Party, when signing, ratifying, accepting, approving or acceding to this Protocol, or at any time thereafter, may choose, by written declaration, one or both of the following means for the settlement of disputes concerning the interpretation or application of Articles 7, 8 and 15 and, except to the extent that an Annex provides otherwise, the provisions of any Annex and, insofar as it relates to these Articles and provisions, Article 13:
   (a) the International Court of Justice;
   (b) the Arbitral Tribunal.
2. A declaration made under paragraph 1 above shall not affect the operation of Article 18 and Article 20 (2).
3. A Party which has not made a declaration under paragraph 1 above or in respect of which a declaration is no longer in force shall be deemed to have accepted the competence of the Arbitral Tribunal.
4. If the parties to a dispute have accepted the same means for the settlement of a dispute, the dispute may be submitted only to that procedure, unless the parties otherwise agree.
5. If the parties to a dispute have not accepted the same means for the settlement of a dispute, or if they have both accepted both means, the dispute may be submitted only to the Arbitral Tribunal, unless the parties otherwise agree.
6. A declaration made under paragraph 1 above shall remain in force until it expires in accordance with its terms or until three months after written notice of revocation has been deposited with the Depositary.
7. A new declaration, a notice of revocation or the expiry of a declaration shall not in any way affect proceedings pending before the International Court of Justice or the Arbitral Tribunal, unless the parties to the dispute otherwise agree.
8. Declarations and notices referred to in this Article shall be deposited with the Depositary who shall transmit copies thereof to all Parties.

Article 20

Dispute Settlement Procedure

1. If the parties to a dispute concerning the interpretation or application of Articles 7, 8 or 15 or, except to the extent that an Annex provides otherwise, the provisions of any Annex or, insofar as it relates to these Articles and provisions, Article 13, have not agreed on a means for resolving it within 12 months of the request for consultation pursuant to Article 18, the dispute shall be referred, at the request of any party to the dispute, for settlement in accordance with the procedure determined by Article 19 (4) and (5).
2. The Arbitral Tribunal shall not be competent to decide or rule upon any matter within the scope of Article IV of the Antarctic Treaty. In addition, nothing in this Protocol shall be interpreted as conferring competence or jurisdiction on the International Court of Justice or any other tribunal established for the purpose of settling disputes between Parties to decide or otherwise rule upon any matter within the scope of Article IV of the Antarctic Treaty.

Article 21

Signature

This Protocol shall be open for signature at Madrid on the 4th of October 1991 and thereafter at Washington until the 3rd of October 1992 by any State which is a Contracting Party to the Antarctic Treaty.

Article 22

Ratification, Acceptance, Approval or Accession

1. This Protocol is subject to ratification, acceptance of approval by signatory States.
2. After the 3rd of October 1992 this Protocol shall be open for accession by any State which is a Contracting Party to the Antarctic Treaty.
3. Instruments of ratification, acceptance, approval or accession shall be deposited with the Government of the United States of America, hereby designated as the Depositary.
4. After the date on which this Protocol has entered into force, the Antarctic Treaty Consultative Parties shall not act upon a notification regarding the entitlement of a Contracting Party to the Antarctic Treaty to appoint representatives to participate in Antarctic Treaty Consultative Meetings in accordance with Article IX (2) of the Antarctic Treaty unless that Contracting Party has first ratified, accepted, approved or acceded to this Protocol.

Article 23

Entry into Force

1. This Protocol shall enter into force on the thirtieth day following the date of deposit of instruments of ratification, acceptance, approval or accession by all States which are Antarctic Treaty Consultative Parties at the date on which this Protocol is adopted.
2. For each Contracting Party to the Antarctic Treaty which, subsequent to the date of entry into force of this Protocol, deposits an instrument of ratification, acceptance, approval or accession, this Protocol shall enter into force on the thirtieth day following such deposit.

Article 24

Reservations

Reservations to this Protocol shall not be permitted.

Article 25

Modification or Amendment

1. Without prejudice to the provisions of Article 9, this Protocol may be modified or amended at any time in accordance with the procedures set forth in Article XII (1) (a) and (b) of the Antarctic Treaty.
2. If, after the expiration of 50 years from the date of entry into force of this Protocol, any of the Antarctic Treaty Consultative Parties so requests by a communication addressed to the Depositary, a conference shall be held as soon as practicable to review the operation of this Protocol.
3. A modification or amendment proposed at any Review Conference called pursuant to paragraph 2 above shall be adopted by a majority of the Parties, including 3/4 of the States which are Antarctic Treaty Consultative Parties at the time of adoption of this Protocol.
4. A modification or amendment adopted pursuant to paragraph 3 above shall enter into force upon ratification, acceptance, approval or accession by 3/4 of the Antarctic Treaty Consultative Parties, including ratification, acceptance, approval or accession by all States which are Antarctic Treaty Consultative Parties at the time of adoption of this Protocol.
5. (a) With respect to Article 7, the prohibition on Antarctic mineral resource activities contained therein shall continue unless there is in force a binding legal regime on Antarctic mineral resource activities that includes an agreed means for determining whether, and, if so, under which conditions, any such activities would be acceptable. This regime shall fully safeguard the interests of all States referred to in Article IV of the Antarctic Treaty and apply the principles thereof. Therefore, if a modification or amendment to Article 7 is proposed at a Review Conference referred to in paragraph 2 above, it shall include such a binding legal regime.
   (b) If any such modification or amendment has not entered into force within 3 years of the date of its adoption, any Party may at any time thereafter notify to the Depositary of its withdrawal from this Protocol, and such withdrawal shall take effect 2 years after receipt of the notification by the Depositary.

Article 26

Notifications by the Depositary

The Depositary shall notify all Contracting Parties to the Antarctic Treaty of the following:

(a) signatures of this Protocol and the deposit of instruments of ratification, acceptance, approval or accession;
(b) the date of entry into force of this Protocol and any additional Annex thereto;
(c) the date of entry into force of any amendment or modification to this Protocol;
(d) the deposit of declarations and notices pursuant to Article 19; and
(e) any notification received pursuant to Article 25 (5) (b).

Article 27

AUTHENTIC TEXTS AND REGISTRATION WITH THE UNITED Nations

1. This Protocol, done in the English, French, Russian and Spanish languages, each version being equally authentic, shall be deposited in the archives of the Government of the United States of America, which shall transmit duly certified copies thereof to all Contracting Parties to the Antarctic Treaty.
2. This Protocol shall be registered by the Depositary pursuant to Article 102 of the charter of the United Nations.

SCHEDULE TO THE PROTOCOL

ARBITRATION

Article 1

1. The Arbitral Tribunal shall be constituted and shall function in accordance with the Protocol, including this Schedule.
2. The Secretary referred to in this Schedule is the Secretary General of the Permanent Court of Arbitration.

Article 2

1. Each Party shall be entitled to designate up to three Arbitrators, at least one of whom shall be designated within three months of the entry into force of the Protocol for that Party. Each Arbitrator shall be experienced in Antarctic affairs, have thorough knowledge of international law and enjoy the highest reputation for fairness, competence and integrity. The names of the persons so designated shall constitute the list of Arbitrators. Each Party shall at all times maintain the name of at least one Arbitrator on the list.
2. Subject to paragraph 3 below, an Arbitrator designated by a Party shall remain on the list for a period of five years and shall be eligible for redesignation by that Party for additional five year periods.
3. A Party which designated an Arbitrator may withdraw the name of that Arbitrator from the list. If an Arbitrator dies or if a Party for any reason withdraws from the list the name of an Arbitrator designated by it, the Party which designated the Arbitrator in question shall notify the Secretary promptly. An Arbitrator whose name is withdrawn from the list shall continue to serve on any Arbitral Tribunal to which that Arbitrator has been appointed until the completion of proceedings before the Arbitral Tribunal.
4. The Secretary shall ensure that an up-to-date list is maintained of the Arbitrators designated pursuant to this Article.

Article 3

1. The Arbitral Tribunal shall be composed of three Arbitrators who shall be appointed as follows:
   (a) The party to the dispute commencing the proceedings shall appoint one Arbitrator, who may be its national, from the list referred to in Article 2. This appointment shall be included in the notification referred to in Article 4.
   (b) Within 40 days of the receipt of that notification, the other party to the dispute shall appoint the second Arbitrator, who may be its national, from the list referred to in Article 2.
   (c) Within 60 days of the appointment of the second Arbitrator, the parties to the dispute shall appoint by agreement the third Arbitrator from the list referred to in Article 2. The third Arbitrator shall not be either a national of a party to the dispute, or a person designated for the list referred to in Article 2 by a party to the dispute, or of the same nationality as either of the first two Arbitrators. The third Arbitrator shall be the Chairperson of the Arbitral Tribunal.
   (d) If the second Arbitrator has not been appointed within the prescribed period, or if the parties to the dispute have not reached agreement within the prescribed period on the appointment of the third Arbitrator, the Arbitrator or Arbitrators shall be appointed, at the request of any party to the dispute and within 30 days of the receipt of such request, by the President of the International Court of Justice from the list referred to in Article 2 and subject to the conditions prescribed in subparagraphs (b) and (c) above. In performing the functions accorded him or her in this subparagraph, the President of the Court shall consult the parties to the dispute.
   (e) If the President of the International Court of Justice is unable to perform the functions accorded him or her in subparagraph (d) above or is a national of a party to the dispute, the functions shall be performed by the Vice-President of the Court, except that if the Vice-President is unable to perform the functions or is a national of a party to the dispute the functions shall be performed by the next most senior member of the Court

who is available and is not a national of a party to the dispute.

2. Any vacancy shall be filled in the manner prescribed for the initial appointment.
3. In any dispute involving more than two Parties, those Parties having the same interest shall appoint one Arbitrator by agreement within the period specified in paragraph 1 (b) above.

Article 4

The party to the dispute commencing proceedings shall so notify the other party or parties to the dispute and the Secretary in writing. Such notification shall include a statement of the claim and the grounds on which it is based. The notification shall be transmitted by the Secretary to all Parties.

Article 5

1. Unless the parties to the dispute agree otherwise, arbitration shall take place at The Hague, where the records of the Arbitral Tribunal shall be kept. The Arbitral Tribunal shall adopt its own rules of procedure. Such rules shall ensure that each party to the dispute has a full opportunity to be heard and to present its case and shall also ensure that the proceedings are conducted expeditiously.
2. The Arbitral Tribunal may hear and decide counterclaims arising out of the dispute.

Article 6

1. The Arbitral Tribunal, where it considers that *prima facie* it has jurisdiction under the Protocol, may:
   (a) at the request of any party to a dispute, indicate such provisional measures as it considers necessary to preserve the respective rights of the parties to the dispute;
   (b) prescribe any provisional measures which it considers appropriate under the circumstances to prevent serious harm to the Antarctic environment or dependent or associated ecosystems.
2. The parties to the dispute shall comply promptly with any provisional measures prescribed under paragraph 1 (b) above pending an award under Article 10.
3. Notwithstanding the time period in Article 20 of the Protocol, a party to a dispute may at any time, by notification to the other party or parties to the dispute and to the Secretary in accordance with Article 4, request that the Arbitral Tribunal be constituted as a matter of exceptional urgency to indicate or prescribe emergency provisional measures in accordance with this Article. In such case, the Arbitral Tribunal shall be constituted as soon as possible in accordance with Article 3, except that the time periods in Article 3 (1) (b), (c) and (d) shall be reduced to 14 days in each case. The Arbitral Tribunal shall decide upon the request for emergency provisional measures within two months of the appointment of its Chairperson.
4. Following a decision by the Arbitral Tribunal upon a request for emergency provisional measures in accordance with paragraph 3 above, settlement of the dispute shall proceed in accordance with Articles 18, 19 and 20 of the Protocol.

Article 7

Any Party which believes it has a legal interest, whether general or individual, which may be substantially affected by the award of an Arbitral Tribunal, may, unless the Arbitral Tribunal decides otherwise, intervene in the proceedings.

Article 8

The parties to the dispute shall facilitate the work of the Arbitral Tribunal and, in particular, in accordance with their law and using all means at their disposal, shall provide it with all relevant documents and information, and enable it, when necessary, to call witnesses or experts and receive their evidence.

Article 9

If one of the parties to the dispute does not appear before the Arbitral Tribunal or fails to defend its case, any other party to the dispute may request the Arbitral Tribunal to continue the proceedings and make its award.

Article 10

1. The Arbitral Tribunal shall, on the basis of the provisions of the Protocol and other applicable rules and principles of international law that are not incompatible with such provisions, decide such disputes as are submitted to it.
2. The Arbitral Tribunal may decide, *ex aequo et bono*, a dispute submitted to it, if the parties to the dispute so agree.

Article 11

1. Before making its award, the Arbitral Tribunal shall satisfy itself that it has competence in respect of the dispute and that the claim or counterclaim is well founded in fact and law.
2. The award shall be accompanied by a statement of reasons for the decision and shall be communicated to the Secretary who shall transmit it to all Parties.

3. The award shall be final and binding on the parties to the dispute and on any Party which intervened in the proceedings and shall be complied with without delay. The Arbitral Tribunal shall interpret the award at the request of a party to the dispute or of any intervening Party.
4. The award shall have no binding force except in respect of that particular case.
5. Unless the Arbitral Tribunal decides otherwise, the expenses of the Arbitral Tribunal, including the remuneration of the Arbitrators, shall be borne by the parties to the dispute in equal shares.

Article 12

All decisions of the Arbitral Tribunal, including those referred to in Articles 5, 6 and 11, shall be made by a majority of the Arbitrators who may not abstain from voting.

Article 13

1. This Schedule may be amended or modified by a measure adopted in accordance with Article IX (1) of the Antarctic Treaty. Unless the measure specifies otherwise, the amendment or modification shall be deemed to have been approved, and shall become effective, one year after the close of the Antarctic Treaty Consultative Meeting at which it was adopted, unless one or more of the Antarctic Treaty Consultative Parties notifies the Depositary, within that time period, that it wishes an extension of that period or that it is unable to approve the measure.
2. Any amendment or modification of this Schedule which becomes effective in accordance with paragraph 1 above shall thereafter become effective as to any other Party when notice of approval by it has been received by the Depositary.

ANNEX I
TO THE PROTOCOL ON ENVIRONMENTAL PROTECTION TO THE ANTARCTIC TREATY
ENVIRONMENTAL IMPACT ASSESSMENT

Article 1

Preliminary Stage

1. The environmental impacts of proposed activities referred to in Article 8 of the Protocol shall, before their commencement, be considered in accordance with appropriate national procedures.
2. If an activity is determined as having less than a minor or transitory impact, the activity may proceed forthwith.

Article 2

Initial Environmental Evaluation

1. Unless it has been determined that an activity will have less than a minor or transitory impact, or unless a Comprehensive Environmental Evaluation is being prepared in accordance with Article 3, an Initial Environmental Evaluation shall be prepared. It shall contain sufficient detail to assess whether a proposed activity may have more than a minor or transitory impact and shall include:
   (a) a description of the proposed activity, including its purpose, location, duration and intensity; and
   (b) consideration of alternatives to the proposed activity and any impacts that the activity may have, including consideration of cumulative impacts in the light of existing and known planned activities.
2. If an Initial Environmental Evaluation indicates that a proposed activity is likely to have no more than a minor or transitory impact, the activity may proceed, provided that appropriate procedures, which may include monitoring, are put in place to assess and verify the impact of the activity.

Article 3

Comprehensive Environmental Evaluation

1. If an Initial Environmental Evaluation indicates or if it is otherwise determined that a proposed activity is likely to have more than a minor or transitory impact, a Comprehensive Environmental Evaluation shall be prepared.
2. A Comprehensive Environmental Evaluation shall include:
   (a) a description of the proposed activity including its purpose, location, duration and intensity, and possible alternatives to the activity, including the alternative of not proceeding, and the consequences of those alternatives;
   (b) a description of the initial environmental reference state with which predicted changes are to be compared and a prediction of the future environmental reference state in the absence of the proposed activity;
   (c) a description of the methods and data used to forecast the impacts of the proposed activity;

(d) estimation of the nature, extent, duration and intensity of the likely direct impacts of the proposed activity;
(e) consideration of possible indirect or second order impacts of the proposed activity;
(f) consideration of cumulative impacts of the proposed activity in the light of existing activities and other known planned activities;
(g) identification of measures, including monitoring programmes, that could be taken to minimise or mitigate impacts of the proposed activity and to detect unforeseen impacts and that could provide early warning of any adverse effects of the activity as well as to deal promptly and effectively with accidents;
(h) identification of unavoidable impacts of the proposed activity;
(i) consideration of the effects of the proposed activity on the conduct of scientific research and on other existing uses and values;
(j) an identification of gaps in knowledge and uncertainties encountered in compiling the information required under this paragraph;
(k) a non-technical summary of the information provided under this paragraph; and
(l) the name and address of the person or organization which prepared the Comprehensive Environmental Evaluation and the address to which comments thereon should be directed.

3. The draft Comprehensive Environmental Evaluation shall be made publicly available and shall be circulated to all Parties, which shall also make it publicly available, for comment. A period of 90 days shall be allowed for the receipt of comments.
4. The draft Comprehensive Environmental Evaluation shall be forwarded to the Committee at the same time as it is circulated to the Parties, and at least 120 days before the next Antarctic Treaty Consultative Meeting, for consideration as appropriate.
5. No final decision shall be taken to proceed with the proposed activity in the Antarctic Treaty area unless there has been an opportunity for consideration of the draft Comprehensive Environmental Evaluation by the Antarctic Treaty Consultative Meeting on the advice of the Committee, provided that no decision to proceed with a proposed activity shall be delayed through the operation of this paragraph for longer than 15 months from the date of circulation of the draft Comprehensive Environmental Evaluation.
6. A final Comprehensive Environmental Evaluation shall address and shall include or summarise comments received on the draft Comprehensive Environmental Evaluation. The final Comprehensive Environmental Evaluation, notice of any decisions relating thereto, and any evaluation of the significance of the predicted impacts in relation to the advantages of the proposed activity, shall be circulated to all Parties, which shall also make them publicly available, at least 60 days before the commencement of the proposed activity in the Antarctic Treaty area.

Article 4
Decisions to be Based on Comprehensive Environmental Evaluations

Any decision on whether a proposed activity, to which Article 3 applies, should proceed, and, if so, whether in its original or in a modified form, shall be based on the Comprehensive Environmental Evaluation as well as other relevant considerations.

Article 5
Monitoring

1. Procedures shall be put in place, including appropriate monitoring of key environmental indicators, to assess and verify the impact of any activity that proceeds following the completion of a Comprehensive Environmental Evaluation.
2. The procedures referred to in paragraph 1 above and in Article 2 (2) shall be designed to provide a regular and verifiable record of the impacts of the activity in order, *inter alia*, to:
   (a) enable assessments to be made of the extent to which such impacts are consistent with the Protocol; and
   (b) provide information useful for minimising or mitigating impacts, and, where appropriate, information on the need for suspension, cancellation or modification of the activity.

Article 6
Circulation of Information

1. The following information shall be circulated to the Parties, forwarded to the Committee and made publicly available:
   (a) a description of the procedures referred to in Article 1;
   (b) an annual list of any Initial Environmental Evaluations prepared in accordance with Article 2 and any decisions taken in consequence thereof;

(c) significant information obtained, and any action taken in consequence thereof, from procedures put in place in accordance with Articles 2 (2) and 5; and
(d) information referred to in Article 3 (6).
2. Any Initial Environmental Evaluation prepared in accordance with Article 2 shall be made available on request.

Article 7

Cases of Emergency

1. This Annex shall not apply in cases of emergency relating to the safety of human life or of ships, aircraft or equipment and facilities of high value, or the protection of the environment, which require an activity to be undertaken without completion of the procedures set out in this Annex.
2. Notice of activities undertaken in cases of emergency, which would otherwise have required preparation of a Comprehensive Environmental Evaluation, shall be circulated immediately to all Parties and to the Committee and a full explanation of the activities carried out shall be provided within 90 days of those activities.

Article 8

Amendment or Modification

1. This Annex may be amended or modified by a measure adopted in accordance with Article IX (1) of the Antarctic Treaty. Unless the measure specifies otherwise, the amendment or modification shall be deemed to have been approved, and shall become effective, one year after the close of the Antarctic Treaty Consultative Meeting at which it was adopted, unless one or more of the Antarctic Treaty Consultative Parties notifies the Depositary, within that period, that it wishes an extension of that period or that it is unable to approve the measure.
2. Any amendment or modification of this Annex which becomes effective in accordance with paragraph 1 above shall thereafter become effective as to any other Party when notice of approval by it has been received by the Depositary.

ANNEX II
TO THE PROTOCOL ON ENVIRONMENTAL PROTECTION TO THE ANTARCTIC TREATY
CONSERVATION OF ANTARCTIC FAUNA AND FLORA

Article 1

Definitions

For the purposes of this Annex:

(a) "native mammal" means any member of any species belonging to the Class Mammalia, indigenous to the Antarctic Treaty area or occurring there seasonally through natural migrations;
(b) "native bird" means any member, at any stage of its life cycle (including eggs), of any species of the Class Aves indigenous to the Antarctic Treaty area or occurring there seasonally through natural migrations;
(c) "native plant" means any terrestrial or freshwater vegetation, including bryophytes, lichens, fungi and algae, at any stage of its life cycle (including seeds, and other propagules), indigenous to the Antarctic Treaty area;
(d) "native invertebrate" means any terrestrial or freshwater invertebrate, at any stage of its life cycle, indigenous to the Antarctic Treaty area;
(e) "appropriate authority" means any person or agency authorized by a Party to issue permits under this Annex;
(f) "permit" means a formal permission in writing issued by an appropriate authority;
(g) "take" or "taking" means to kill, injure, capture, handle or molest, a native mammal or bird, or to remove or damage such quantities of native plants that their local distribution or abundance would be significantly affected;
(h) "harmful interference" means:
   (i) flying or landing helicopters or other aircraft in a manner that disturbs concentrations of birds and seals;
   (ii) using vehicles or vessels, including hovercraft and small boats, in a manner that disturbs concentrations of birds and seals;
   (iii) using explosives or firearms in a manner that disturbs concentrations of birds and seals;
   (iv) wilfully disturbing breeding or moulting birds or concentrations of birds and seals by persons on foot;
   (v) significantly damaging concentrations of native terrestrial plants by landing aircraft, driving vehicles, or walking on them, or by other means; and

(vi) any activity that results in the significant adverse modification of habitats of any species or population of native mammal, bird, plant or invertebrate.

(i) "International Convention for the Regulation of Whaling" means the Convention done at Washington on 2 December 1946.

Article 2

CASES OF EMERGENCY

1. This Annex shall not apply in cases of emergency relating to the safety of human life or of ships, aircraft, or equipment and facilities of high value, or the protection of the environment.
2. Notice of activities undertaken in cases of emergency shall be circulated immediately to all Parties and to the Committee.

Article 3

PROTECTION OF NATIVE FAUNA AND FLORA

1. Taking or harmful interference shall be prohibited, except in accordance with a permit.
2. Such permits shall specify the authorized activity, including when, where and by whom it is to be conducted and shall be issued only in the following circumstances:
   (a) to provide specimens for scientific study or scientific information;
   (b) to provide specimens for museums, herbaria, zoological and botanical gardens, or other educational or cultural institutions or uses; and
   (c) to provide for unavoidable consequences of scientific activities not otherwise authorized under sub-paragraphs (a) or (b) above, or of the construction and operation of scientific support facilities.
3. The issue of such permits shall be limited so as to ensure that:
   (a) no more native mammals, birds, or plants are taken than are strictly necessary to meet the purposes set forth in paragraph 2 above;
   (b) only small numbers of native mammals or birds are killed and in no case more native mammals or birds are killed from local populations than can, in combination with other permitted takings, normally be replaced by natural reproduction in the following season; and
   (c) the diversity of species, as well as the habitats essential to their existence, and the balance of the ecological systems existing within the Antarctic Treaty are maintained.
4. Any species of native mammals, birds and plants listed in Appendix A to this Annex shall be designated "Specially Protected Species", and shall be accorded special protection by the Parties.
5. A permit shall not be issued to take a Specially Protected Species unless the taking:
   (a) is for a compelling scientific purpose;
   (b) will not jeopardize the survival or recovery of that species or local population; and
   (c) uses non-lethal techniques where appropriate.
6. All taking of native mammals and birds shall be done in the manner that involves the least degree of pain and suffering practicable.

Article 4

INTRODUCTION OF NON-NATIVE SPECIES, PARASITES AND DISEASES

1. No species of animal or plant not native to the Antarctic Treaty area shall be introduced onto land or ice shelves, or into water in the Antarctic Treaty area except in accordance with a permit.
2. Dogs shall not be introduced onto land or ice shelves and dogs currently in those areas shall be removed by April 1, 1994.
3. Permits under paragraph 1 above shall be issued to allow the importation only of the animals and plants listed in Appendix B to this Annex and shall specify the species, numbers and, if appropriate, age and sex and precautions to be taken to prevent escape or contact with native fauna and flora.
4. Any plant or animal for which a permit has been issued in accordance with paragraphs 1 and 3 above, shall, prior to expiration of the permit, be removed from the Antarctic Treaty area or be disposed of by incineration or equally effective means that eliminates risk to native fauna or flora. The permit shall specify this obligation. Any other plant or animal introduced into the Antarctic Treaty area not native to that area, including any progeny, shall be removed or disposed of, by incineration or by equally effective means, so as to be rendered sterile, unless it is determined that they pose no risk to native flora or fauna.
5. Nothing in this Article shall apply to the importation of food into the Antarctic Treaty area provided that no live animals are imported for this purpose and all plants and animal parts and products are kept under carefully controlled conditions and disposed of in accordance with Annex III to the Protocol and Appendix C to this Annex.
6. Each Party shall require that precautions, including those listed in Appendix C to this Annex, be taken to prevent the introduction of micro-organisms (e.g., viruses, bacteria, parasites, yeasts, fungi) not present in the native fauna and flora.

Article 5
INFORMATION

Each Party shall prepare and make available information setting forth, in particular, prohibited activities and providing lists of Specially Protected Species and relevant Protected Areas to all those persons present in or intending to enter the Antarctic Treaty area with a view to ensuring that such persons understand and observe the provisions of this Annex.

Article 6
EXCHANGE OF INFORMATION

1. The Parties shall make arrangements for:
   (a) collecting and exchanging records (including records of permits) and statistics concerning the numbers or quantities of each species of native mammal, bird or plant taken annually in the Antarctic Treaty area;
   (b) obtaining and exchanging information as to the status of native mammals, birds, plants, and invertebrates in the Antarctic Treaty area, and the extent to which any species or population needs protection;
   (c) establishing a common form in which this information shall be submitted by Parties in accordance with paragraph 2 below.
2. Each Party shall inform the other Parties as well as the Committee before the end of November of each year of any step taken pursuant to paragraph 1 above and of the number and nature of permits issued under this Annex in the preceding period of 1st July to 30th June.

Article 7
RELATIONSHIP WITH OTHER AGREEMENTS OUTSIDE THE ANTARCTIC TREATY SYSTEM

Nothing in this Annex shall derogate from the rights and obligations of Parties under the International Convention for the Regulation of Whaling.

Article 8
REVIEW

The Parties shall keep under continuing review measures for the conservation of Antarctic fauna and flora, taking into account any recommendations from the Committee.

Article 9
AMENDMENT OR MODIFICATION

1. This Annex may be amended or modified by a measure adopted in accordance with Article IX (1) of the Antarctic Treaty. Unless the measure specifies otherwise, the amendment or modification shall be deemed to have been approved, and shall become effective, one year after the close of the Antarctic Treaty Consultative Meeting at which it was adopted, unless one or more of the Antarctic Treaty Consultative Parties notifies the Depositary, within that time period, that it wishes an extension of that period or that it is unable to approve the measure.
2. Any amendment or modification of this Annex which becomes effective in accordance with paragraph 1 above shall thereafter become effective as to any other Party when notice of approval by it has been received by the Depositary.

APPENDICES TO THE ANNEX

APPENDIX A:
SPECIALLY PROTECTED SPECIES

All species of the genus *Arctocephalus*, Fur Seals.
Ommatophoca rossii, Ross Seal.

APPENDIX B:
IMPORTATION OF ANIMALS AND PLANTS

The following animals and plants may be imported into the Antarctic Treaty area in accordance with permits issued under Article 4 of this Annex:
(a) domestic plants; and
(b) laboratory animals and plants including viruses, bacteria, yeasts and fungi.

APPENDIX C:
PRECAUTIONS TO PREVENT INTRODUCTION OF MICRO-ORGANISMS

1. Poultry. No live poultry or other living birds shall be brought into the Antarctic Treaty area. Before dressed poultry is packaged for shipment to the Antarctic Treaty area, it shall be inspected for evidence of disease, such as Newcastle's Disease, tuberculosis, and yeast infection. Any poultry or parts not consumed shall be removed from the Antarctic Treaty area or disposed of by incineration or equivalent means that eliminates risks to native flora and fauna.
2. The importation of non-sterile soil shall be avoided to the maximum extent practicable.

ANNEX III
TO THE PROTOCOL ON ENVIRONMENTAL PROTECTION TO THE ANTARCTIC TREATY
WASTE DISPOSAL AND WASTE MANAGEMENT

Article 1
GENERAL OBLIGATIONS

1. This Annex shall apply to activities undertaken in the Antarctic Treaty area pursuant to scientific research programmes, tourism and all other governmental and non-governmental activities in the Antarctic Treaty area for which advance notice is required under Article VII (5) of the Antarctic Treaty, including associated logistic support activities.
2. The amount of wastes produced or disposed of in the Antarctic Treaty area shall be reduced as far as practicable so as to minimise impact on the Antarctic environment and to minimise interference with the natural values of Antarctica, with scientific research and with other uses of Antarctica which are consistent with the Antarctic Treaty.
3. Waste storage, disposal and removal from the Antarctic Treaty area, as well as recycling and source reduction, shall be essential considerations in the planning and conduct of activities in the Antarctic Treaty area.
4. Wastes removed from the Antarctic Treaty area shall, to the maximum extent practicable, be returned to the country from which the activities generating the waste were organized or to any other country in which arrangements have been made for the disposal of such wastes in accordance with relevant international agreements.
5. Past and present waste disposal sites on land and abandoned work sites of Antarctic activities shall be cleaned up by the generator of such wastes and the user of such sites. This obligation shall not be interpreted as requiring:
   (a) the removal of any structure designated as a historic site or monument; or
   (b) the removal of any structure or waste material in circumstances where the removal by any practical option would result in greater adverse environmental impact than leaving the structure or waste material in its existing location.

Article 2
WASTE DISPOSAL BY REMOVAL FROM THE ANTARCTIC TREATY AREA

1. The following wastes, if generated after entry into force of this Annex, shall be removed from the Antarctic Treaty area by the generator of such wastes:
   (a) radio-active materials;
   (b) electrical batteries;
   (c) fuel, both liquid and solid;
   (d) wastes containing harmful levels of heavy metals or acutely toxic or harmful persistent compounds;
   (e) poly-vinyl chloride (PVC), polyurethane foam, polystyrene foam, rubber and lubricating oils, treated timbers and other products which contain additives that could produce harmful emissions if incinerated;
   (f) all other plastic wastes, except low density polyethylene containers (such as bags for storing wastes), provided that such containers shall be incinerated in accordance with Article 3 (1);
   (g) fuel drums; and
   (h) other solid, non-combustible wastes;

   provided that the obligation to remove drums and solid non-combustible wastes contained in subparagraphs (g) and (h) above shall not apply in circumstances where the removal of such wastes by any practical option would result in greater adverse environmental impact than leaving them in their existing locations.
2. Liquid wastes which are not covered by paragraph 1 above and sewage and domestic liquid wastes, shall, to the maximum extent practicable, be removed from the Antarctic Treaty area by the generator of such wastes.
3. The following wastes shall be removed from the Antarctic Treaty area by the generator of such wastes, unless incinerated, autoclaved or otherwise treated to be made sterile:
   (a) residues of carcasses of imported animals;
   (b) laboratory culture of micro-organisms and plant pathogens; and
   (c) introduced avian products.

Article 3
WASTE DISPOSAL BY INCINERATION

1. Subject to paragraph 2 below, combustible wastes, other than those referred to in Article 2 (1), which are not removed from the Antarctic Treaty area shall be burnt in incinerators which to the

maximum extent practicable reduce harmful emissions. Any emission standards and equipment guidelines which may be recommended by, inter alia, the Committee and the Scientific Committee on Antarctic Research shall be taken into account. The solid residue of such incineration shall be removed from the Antarctic Treaty area.

2. All open burning of wastes shall be phased out as soon as practicable, but no later than the end of the 1998/1999 season. Pending the completion of such phase-out, when it is necessary to dispose of wastes by open burning, allowance shall be made for the wind direction and speed and the type of wastes to be burnt to limit particulate deposition and to avoid such deposition over areas of special biological, scientific, historic, aesthetic or wilderness significance including, in particular, areas accorded protection under the Antarctic Treaty.

Article 4

Other Waste Disposal on Land

1. Wastes not removed or disposed of in accordance with Articles 2 and 3 shall not be disposed of onto ice-free areas or into fresh water systems.
2. Sewage, domestic liquid wastes and other liquid wastes not removed from the Antarctic Treaty area in accordance with Article 2, shall, to the maximum extent practicable, not be disposed of onto sea ice, ice shelves or the grounded ice-sheet, provided that such wastes which are generated by stations located inland on ice shelves or on the grounded ice-sheet may be disposed of in deep ice pits where such disposal is the only practicable option. Such pits shall not be located on known ice-flow lines which terminate at ice-free areas or in areas of high ablation.
3. Wastes generated at field camps shall, to the maximum extent practicable, be removed by the generator of such wastes to supporting stations or ships for disposal in accordance with this Annex.

Article 5

Disposal of Waste in the Sea

1. Sewage and domestic liquid wastes may be discharged directly into the sea, taking into account the assimilative capacity of the receiving marine environment and provided that:
   (a) such discharge is located, wherever practicable, where conditions exist for initial dilution and rapid dispersal; and
   (b) large quantities of such wastes (generated in a station where the average weekly occupancy over the austral summer is approximately 30 individuals or more) shall be treated at least by maceration.
2. The by-product of sewage treatment by the Rotary Biological Contacter process or similar processes may be disposed of into the sea provided that such disposal does not adversely affect the local environment, and provided also that any such disposal at sea shall be in accordance with Annex IV to the Protocol.

Article 6

Storage of Waste

All wastes to be removed from the Antarctic Treaty area, or otherwise disposed of, shall be stored in such a way as to prevent their dispersal into the environment.

Article 7

Prohibited Products

No polychlorinated biphenyls (PCBs), non-sterile soil, polystyrene beads, chips or similar forms of packaging, or pesticides (other than those required for scientific, medical or hygiene purposes) shall be introduced onto land or ice shelves or into water in the Antarctic Treaty area.

Article 8

Waste Management Planning

1. Each Party which itself conducts activities in the Antarctic Treaty area shall, in respect of those activities, establish a waste disposal classification system as a basis for recording wastes and to facilitate studies aimed at evaluating the environmental impacts of scientific activity and associated logistic support. To that end, wastes produced shall be classified as:
   (a) sewage and domestic liquid wastes (Group 1);
   (b) other liquid wastes and chemicals, including fuels and lubricants (Group 2);
   (c) solids to be combusted (Group 3);
   (d) other solid wastes (Group 4); and
   (e) radioactive material (Group 5).
2. In order to reduce further the impact of waste on the Antarctic environment, each such Party shall prepare and annually review and update its waste management plans (including waste reduction, storage and disposal), specifying for each fixed site, for field camps generally, and for each ship (other than small boats that are part of the operations of fixed sites or of ships and taking into account existing management plans for ships):

(a) programmes for cleaning up existing waste disposal sites and abandoned work sites;
(b) current and planned waste management arrangements, including final disposal;
(c) current and planned arrangements for analysing the environmental effects of waste and waste management; and
(d) other efforts to minimise any environmental effects of wastes and waste management.

3. Each such Party shall, as far as is practicable, also prepare an inventory of locations of past activities (such as traverses, field depots, field bases, crashed aircraft) before the information is lost, so that such locations can be taken into account in planning future scientific programmes (such as snow chemistry, pollutants in lichens or ice core drilling).

Article 9
CIRCULATION AND REVIEW OF WASTE MANAGEMENT PLANS

1. The waste management plans prepared in accordance with Article 8, reports on their implementation, and the inventories referred to in Article 8 (3), shall be included in the annual exchanges of information in accordance with Articles III and VII of the Antarctic Treaty and related Recommendations under Article IX of the Antarctic Treaty.
2. Each Party shall send copies of its waste management plans, and reports on their implementation and review, to the Committee.
3. The Committee may review waste management plans and reports thereon and may offer comments, including suggestions for minimising impacts and modifications and improvement to the plans, for the consideration of the Parties.
4. The Parties may exchange information and provide advice on, inter alia, available low waste technologies, reconversion of existing installations, special requirements for effluents, and appropriate disposal and discharge methods.

Article 10
MANAGEMENT PRACTICES

Each Party shall:

(a) designate a waste management official to develop and monitor waste management plans; in the field, this responsibility shall be delegated to an appropriate person at each site;
(b) ensure that members of its expeditions receive training designed to limit the impact of its operations on the Antarctic environment and to inform them of requirements of this Annex; and
(c) discourage the use of poly-vinyl chloride (PVC) products and ensure that its expeditions to the Antarctic Treaty are advised of any PVC products they may introduce into that area in order that these products may be removed subsequently in accordance with this Annex.

Article 11
REVIEW

This Annex shall be subject to regular review in order to ensure that it is updated to reflect improvement in waste disposal technology and procedures and to ensure thereby maximum protection of the Antarctic environment.

Article 12
CASES OF EMERGENCY

1. This Annex shall not apply in cases of emergency relating to the safety of human life or of ships, aircraft or equipment and facilities of high value or the protection of the environment.
2. Notice of activities undertaken in cases of emergency shall be circulated immediately to all Parties and to the Committee.

Article 13
AMENDMENT OR MODIFICATION

1. This Annex may be amended or modified by a measure adopted in accordance with Article IX (1) of the Antarctic Treaty. Unless the measure specifies otherwise, the amendment or modification shall be deemed to have been approved, and shall become effective, one year after the close of the Antarctic Treaty Consultative Meeting at which it was adopted, unless one or more of the Antarctic Treaty Consultative Parties notifies the Depositary, within that time period, that it wishes an extension of that period or that it is unable to approve the amendment.
2. Any amendment or modification of this Annex which becomes effective in accordance with paragraph 1 above shall thereafter become effective as to any other Party when notice of approval by it has been received by the Depositary.

ANNEX IV
TO THE PROTOCOL ON ENVIRONMENTAL PROTECTION TO THE ANTARCTIC TREATY
PREVENTION OF MARINE POLLUTION

Article 1
DEFINITIONS

For the purposes of this Annex:

(a) "discharge" means any release howsoever caused from a ship and includes any escape, disposal, spilling, leaking, pumping, emitting or emptying;

(b) "garbage" means all kinds of victual, domestic and operational waste excluding fresh fish and parts thereof, generated during the normal operation of the ship, except those substances which are covered by Articles 3 and 4;

(c) "MARPOL 73/78" means the International Convention for the Prevention of Pollution from Ships, 1973, as amended by the Protocol of 1978 relating thereto and by any other amendment in force thereafter;

(d) "noxious liquid substance" means any noxious liquid substance as defined in Annex II of MARPOL 73/78;

(e) "oil" means petroleum in any form including crude oil, fuel oil, sludge, oil refuse and refined oil products (other than petrochemicals which are subject to the provisions of Article 4);

(f) "oily mixture" means a mixture with any oil content; and

(g) "ship" means a vessel of any type whatsoever operating in the marine environment and includes hydrofoil boats, air-cushion vehicles, submersibles, floating craft and fixed or floating platforms.

Article 2
APPLICATION

This Annex applies, with respect to each Party, to ships entitled to fly its flag and to any other ship engaged in or supporting its Antarctic operations, while operating in the Antarctic Treaty area.

Article 3
DISCHARGE OF OIL

1. Any discharge into the sea of oil or oily mixture shall be prohibited, except in cases permitted under Annex I of MARPOL 73/78. While operating in the Antarctic Treaty area, ships shall retain on board all sludge, dirty ballast, tank washing waters and other oily residues and mixtures which may not be discharged into the sea. Ships shall discharge these residues only outside the Antarctic Treaty area, at reception facilities or as otherwise permitted under Annex I of MARPOL 73/78.
2. This Article shall not apply to:
   (a) the discharge into the sea of oil or oily mixture resulting from damage to a ship or its equipment:
      (i) provided that all reasonable precautions have been taken after the occurrence of the damage or discovery of the discharge for the purpose of preventing or minimising the discharge; and
      (ii) except if the owner or the Master acted either with intent to cause damage, or recklessly and with the knowledge that damage would probably result; or
   (b) the discharge into the sea of substances containing oil which are being used for the purpose of combating specific pollution incidents in order to minimise the damage from pollution.

Article 4
DISCHARGE OF NOXIOUS LIQUID SUBSTANCES

The discharge into the sea of any noxious liquid substance, and any other chemical or other substances, in quantities or concentrations that are harmful to the marine environment, shall be prohibited.

Article 5
DISPOSAL OF GARBAGE

1. The disposal into the sea of all plastics, including but not limited to synthetic ropes, synthetic fishing nets, and plastic garbage bags, shall be prohibited.
2. The disposal into the sea of all other garbage, including paper products, rags, glass, metal, bottles, crockery, incineration ash, dunnage, lining and packing materials, shall be prohibited.
3. The disposal into the sea of food wastes may be permitted when they have been passed through a comminuter or grinder, provided that such disposal shall, except in cases permitted under Annex V of MARPOL 73/78, be made as far as practicable from land and ice shelves but in any case not less than 12 nautical miles from the nearest land or ice

shelf. Such comminuted or ground food wastes shall be capable of passing through a screen with openings no greater than 25 millimetres.

4. When a substance or material covered by this article is mixed with other such substance or material for discharge or disposal, having different disposal or discharge requirements, the most stringent disposal or discharge requirements shall apply.
5. The provisions of paragraphs 1 and 2 above shall not apply to:
   (a) the escape of garbage resulting from damage to a ship or its equipment provided all reasonable precautions have been taken, before and after the occurrence of the damage, for the purpose of preventing or minimising the escape; or
   (b) the accidental loss of synthetic fishing nets, provided all reasonable precautions have been taken to prevent such loss.
6. The Parties shall, where appropriate, require the use of garbage record books.

Article 6

Discharge of Sewage

1. Except where it would unduly impair Antarctic operations:
   (a) each Party shall eliminate all discharge into the sea of untreated sewage ("sewage" being defined in Annex IV of MARPOL 73/78) within 12 nautical miles of land or ice shelves;
   (b) beyond such distance, sewage stored in a holding tank shall not be discharged instantaneously but at a moderate rate and, where practicable, while the ship is *en route* at a speed of no less than 4 knots.

   This paragraph does not apply to ships certified to carry not more than 10 persons.
2. The Parties shall, where appropriate, require the use of sewage record books.

Article 7

Cases of Emergency

1. Articles 3, 4, 5 and 6 of this Annex shall not apply in cases of emergency relating to the safety of a ship and those on board or saving life at sea.
2. Notice of activities undertaken in cases of emergency shall be circulated immediately to all Parties and to the Committee.

Article 8

Effect on Dependent and Associated Ecosystems

In implementing the provisions of this Annex, due consideration shall be given to the need to avoid detrimental effects on dependent and associated ecosystems, outside the Antarctic Treaty area.

Article 9

Ship Retention Capacity and Reception Facilities

1. Each Party shall undertake to ensure that all ships entitled to fly its flag and any other ship engaged in or supporting its Antarctic operations, before entering the Antarctic Treaty area, are fitted with a tank or tanks of sufficient capacity on board for the retention of all sludge, dirty ballast, tank washing water and other oil residues and mixtures, and have sufficient capacity on board for the retention of garbage, while operating in the Antarctic Treaty area and have concluded arrangements to discharge such oily residues and garbage at a reception facility after leaving that area. Ships shall also have sufficient capacity on board for the retention of noxious liquid substances.
2. Each Party at whose ports ships depart *en route* to or arrive from the Antarctic Treaty area undertakes to ensure that as soon as practicable adequate facilities are provided for the reception of all sludge, dirty ballast, tank washing water, other oily residues and mixtures, and garbage from ships, without causing undue delay, and according to the needs of the ships using them.
3. Parties operating ships which depart to or arrive from the Antarctic Treaty area at ports of other Parties shall consult with those Parties with a view to ensuring that the establishment of port reception facilities does not place an inequitable burden on Parties adjacent to the Antarctic Treaty area.

Article 10

Design, Construction, Manning and Equipment of Ships

In the design, construction, manning and equipment of ships engaged in or supporting Antarctic operations, each Party shall take into account the objectives of this Annex.

Article 11

Sovereign Immunity

1. This Annex shall not apply to any warship, naval auxiliary or other ship owned or operated by a State and used, for the time being, only on government non-commercial service. However, each Party shall ensure by the adoption of appropriate measures not impairing the operations or operational capabilities of such ships owned or

operated by it, that such ships act in a manner consistent, so far as is reasonable and practicable, with this Annex.

2. In applying paragraph 1 above, each Party shall take into account the importance of protecting the Antarctic environment.
3. Each Party shall inform the other Parties of how it implements this provision.
4. The dispute settlement procedure set out in Articles 18 to 20 of the Protocol shall not apply to this Article.

Article 12

Preventive Measures and Emergency Preparedness and Response

1. In order to respond more effectively to marine pollution emergencies or the threat thereof in the Antarctic Treaty area, the Parties, in accordance with Article 15 of the Protocol, shall develop contingency plans for marine pollution response in the Antarctic Treaty area, including contingency plans for ships (other than small boats that are part of the operations of fixed sites or of ships) operating in the Antarctic Treaty area, particularly ships carrying oil as cargo, and for oil spills, originating from coastal installations, which enter into the marine environment. To this end they shall:
   (a) co-operate in the formulation and implementation of such plans; and
   (b) draw on the advice of the Committee, the International Maritime Organization and other international organizations.
2. The Parties shall also establish procedures for cooperative response to pollution emergencies and shall take appropriate response actions in accordance with such procedures.

Article 13

Review

The Parties shall keep under continuous review the provisions of this Annex and other measures to prevent, reduce and respond to pollution of the Antarctic marine environment, including any amendments and new regulations adopted under MARPOL 73/78, with a view to achieving the objectives of this Annex.

Article 14

Relationship with MARPOL 73/78

With respect to those Parties which are also Parties to MARPOL 73/78, nothing in this Annex shall derogate from the specific rights and obligations thereunder.

Article 15

Amendment or Modification

1. This Annex may be amended or modified by a measure adopted in accordance with Article IX (1) of the Antarctic Treaty. Unless the measure specifies otherwise, the amendment or modification shall be deemed to have been approved, and shall become effective, one year after the close of the Antarctic Treaty Consultative Meeting at which it was adopted, unless one or more of the Antarctic Treaty Consultative Parties notifies the Depositary, within that time period, that it wishes an extension of that period or that it is unable to approve the measure.
2. Any amendment or modification of this Annex which becomes effective in accordance with paragraph 1 above shall thereafter become effective as to any other Party when notice of approval by it has been received by the Depositary.

ANNEX V
TO THE PROTOCOL ON ENVIRONMENTAL PROTECTION TO THE ANTARCTIC TREATY
AREA PROTECTION AND MANAGEMENT

Article 1

Definitions

For the purposes of this Annex:

a) "appropriate authority" means any person or agency authorised by a Party to issue permits under this Annex;
b) "permit" means a formal permission in writing issued by an appropriate authority;
c) "Management Plan" means a plan to manage the activities and protect the special value or values in an Antarctic Specially Protected Area or an Antarctic Specially Managed Area.

Article 2

Objectives

For the purposes set out in this Annex, any area, including any marine area, may be designated as an Antarctic Specially Protected Area or an Antarctic

Specially Managed Area. Activities in those Areas shall be prohibited, restricted or managed in accordance with Management Plans adopted under the provisions of this Annex.

Article 3

ANTARCTIC SPECIALLY PROTECTED AREAS

1. Any area, including any marine area, may be designated as an Antarctic Specially Protected Area to protect outstanding environmental, scientific, historic, aesthetic or wilderness values, any combination of those values, or ongoing or planned scientific research.
2. Parties shall seek to identify, within a systematic environmental–geographical framework, and to include in the series of Antarctic Specially Protected Areas:
   (a) areas kept inviolate from human interference so that future comparisons may be possible with localities that have been affected by human activities;
   (b) representative examples of major terrestrial, including glacial and aquatic, ecosystems and marine ecosystems;
   (c) areas with important or unusual assemblages of species, including major colonies of breeding native birds or mammals;
   (d) the type locality or only known habitat of any species;
   (e) areas of particular interest to ongoing or planned scientific research;
   (f) examples of outstanding geological, glaciological or geomorphological features;
   (g) areas or outstanding aesthetic and wilderness value;
   (h) sites or monuments of recognised historic value; and
   (i) such other areas as may be appropriate to protect the values set out in paragraph 1 above.
3. Specially Protected Areas and Sites of Special Scientific Interest designated as such by past Antarctic Treaty Consultative Meetings are hereby designated as Antarctic Specially Protected Areas and shall be renamed and renumbered accordingly.
4. Entry into an Antarctic Specially Protected Area shall be prohibited except in accordance with a permit issued under Article 7.

Article 4

ANTARCTIC SPECIALLY MANAGED AREAS

1. Any area, including any marine area, where activities are being conducted or may in the future be conducted, may be designated as an Antarctic Specially Managed Area to assist in the planning and co-ordination of activities, avoid possible conflicts, improve co-operation between Parties or minimise environmental impacts.
2. Antarctic Specially Managed Areas may include:
   (a) areas where activities pose risks of mutual interference or cumulative environmental impacts; and
   (b) sites or monuments of recognised historic value.
3. Entry into an Antarctic Specially Managed Area shall not require a permit.
4. Notwithstanding paragraph 3 above, an Antarctic Specially Managed Area may contain one or more Antarctic Specially Protected Areas, entry into which shall be prohibited except in accordance with a permit issued under Article 7.

Article 5

MANAGEMENT PLANS

1. Any Party, the Committee, the Scientific Committee on Antarctic Research or the Commission for the Conservation of Antarctic Marine Living Resources may propose an area for designation as an Antarctic Specially Protected Area or an Antarctic Specially Managed Area by submitting a proposed Management Plan to the Antarctic Treaty Consultative Meeting.
2. The area proposed for designation shall be of sufficient size to protect the values for which the special protection or management is required.
3. Proposed Management Plans shall include, as appropriate:
   (a) a description of the value or values for which special protection or management is required;
   (b) a statement of the aims and objectives of the Management Plan for the protection or management of those values;
   (c) management activities which are to be undertaken to protect the values for which special protection or management is required;
   (d) a period of designation, if any;
   (e) a description of the area, including:
      (i) the geographical co-ordinates, boundary markers and natural features that delineate the area;
      (ii) access to the area by land, sea or air including marine approaches and anchorages, pedestrian and vehicular routes within the area, and aircraft routes and landing areas;

(iii) the location of structures, including scientific stations, research or refuge facilities, both within the area and near to it; and
(iv) the location in or near the area of other Antarctic Specially Protected Areas or Antarctic Specially Managed Areas designated under this Annex, or other protected areas designated in accordance with measures adopted under other components of the Antarctic Treaty System;

(f) the identification of zones within the area, in which activities are to be prohibited, restricted or managed for the purpose of achieving the aims and objectives referred to in subparagraph (b) above;
(g) maps and photographs that show clearly the boundary of the area in relation to surrounding features and key features within the area;
(h) supporting documentation;
(i) in respect of an area proposed for designation as an Antarctic Specially Protected Area, a clear description of the conditions under which permits may be granted by the appropriate authority regarding:
(i) access to and movement within or over the area;
(ii) activities which are or may be conducted within the area, including restrictions on time and place;
(iii) the installation, modification, or removal of structures;
(iv) the location of field camps;
(v) restrictions on materials and organisms which may be brought into the area;
(vi) the taking of or harmful interference with native flora and fauna;
(vii) the collection or removal of anything not brought into the area by the permit holder;
(viii) the disposal of waste;
(ix) measures that may be necessary to ensure that the aims and objectives of the Management Plan can continue to be met; and
(x) requirements for reports to be made to the appropriate authority regarding visits to the area;
(j) in respect of an area proposed for designation as an Antarctic Specially Managed Area, a code of conduct regarding:
(i) access to and movement within or over the area;
(ii) activities which are or may be conducted within the area, including restrictions on time and place;
(iii) the installation, modification, or removal of structures;
(iv) the location of field camps;
(v) the taking of or harmful interference with native flora and fauna;
(vi) the collection or removal of anything not brought into the area by the visitor;
(vii) the disposal of waste; and
(viii) any requirements for reports to be made to the appropriate authority regarding visits to the area; and
(k) provisions relating to the circumstances in which Parties should seek to exchange information in advance of activities which they propose to conduct.

Article 6
DESIGNATION PROCEDURES

1. Proposed Management Plans shall be forwarded to the Committee, the Scientific Committee on Antarctic Research and, as appropriate, to the Commission for the Conservation of Antarctic Marine Living Resources. In formulating its advice to the Antarctic Treaty Consultative Meeting, the Committee shall take into account any comments provided by the Scientific Committee on Antarctic Research and, as appropriate, by the Commission for the Conservation of Antarctic Marine Living Resources. Thereafter Management Plans may be approved by the Antarctic Treaty Consultative Parties by a measure adopted at an Antarctic Treaty Consultative Meeting in accordance with Article IX 1) of the Antarctic Treaty. Unless the measure specifies otherwise, the Plan shall be deemed to have been approved 90 days after the close of the Antarctic Treaty Consultative Meeting at which it was adopted, unless one or more of the Consultative Parties notifies the Depositary, within that time period, that it wishes an extension of that period or is unable to approve the measure.
2. Having regard to the provisions of Articles 4 and 5 of the Protocol, no marine area shall be designated as an Antarctic Specially Protected Area or an Antarctic Specially Managed Area without the prior approval of the Commission for the Conservation of Antarctic Marine Living Resources.

3. Designation of an Antarctic Specially Protected Area or an Antarctic Specially Managed Area shall be for an indefinite period unless the Management Plan provides otherwise. A review of a Management Plan shall be initiated at least every five years. The Plan shall be updated as necessary.
4. Management Plans may be amended or revoked in accordance with paragraph 1 above.
5. Upon approval Management Plans shall be circulated promptly by the Depositary to all Parties. The Depositary shall maintain a record of all currently approved Management Plans.

Article 7

Permits

1. Each Party shall appoint an appropriate authority to issue permits to enter and engage in activities within an Antarctic Specially Protected Area in accordance with the requirements of the Management Plan relating to that Area. The permit shall be accompanied by the relevant sections of the Management Plan and shall specify the extent and location of the Area, the authorised activities and when, where and by whom the activities are authorised and any other conditions imposed by the Management Plan.
2. In the case of a Specially Protected Area designated as such by past Antarctic Treaty Consultative Meeting which does not have a Management Plan, the appropriate authority may issue a permit for a compelling scientific purpose which cannot be served elsewhere and which will not jeopardise the natural ecological system in that Area.
3. Each Party shall require a permit-holder to carry a copy of the permit while in the Antarctic Specially Protected Area concerned.

Article 8

Historic Sites and Monuments

1. Sites or monuments of recognised historic value which have been designated as Antarctic Specially Protected Areas or Antarctic Specially Managed Areas, or which are located within such Areas, shall be listed as Historic Sites and Monuments.
2. Any Party may propose a site or monument of recognised historic value which has not been designated as an Antarctic Specially Protected Area or an Antarctic Specially Managed Area, or which is not located within such an Area, for listing as a Historic Site or Monument. The proposal for listing may be approved by the Antarctic Treaty Consultative Parties by a measure adopted at an Antarctic Treaty Consultative Meeting in accordance with Article IX (1) of the Antarctic Treaty. Unless the measure specifies otherwise, the proposal shall be deemed to have been approved 90 days after the close of the Antarctic Treaty Consultative Meeting at which it was adopted, unless one or more of the Consultative Parties notifies the Depositary, within that time period, that it wishes an extension of that period or is unable to approve the measure.
3. Existing Historic Sites and Monuments which have been listed as such by previous Antarctic Treaty Consultative Meetings shall be included in the list of Historic Sites and Monuments under this Article.
4. Listed Historic Sites and Monuments shall not be damaged, removed or destroyed.
5. The list of Historic Sites and Monuments may be amended in accordance with paragraph 2 above. The Depositary shall maintain a list of current Historic Sites and Monuments.

Article 9

Information and Publicity

1. With a view to ensuring that all persons visiting or proposing to visit Antarctica understand and observe the provisions of this Annex, each Party shall make available information setting forth, in particular:
   (a) the location of Antarctic Specially Protected Areas and Antarctic Specially Managed Areas;
   (b) listing and maps of those Areas;
   (c) the Management Plans, including listings of prohibitions relevant to each Area;
   (d) the location of Historic Sites and Monuments and any relevant prohibition or restriction.
2. Each Party shall ensure that the location and, if possible, the limits of Antarctic Specially Protected Areas, Antarctic Specially Managed Areas and Historic Sites and Monuments are shown on its topographic maps, hydrographic charts and in other relevant publications.
3. Parties shall co-operate to ensure that, where appropriate, the boundaries of Antarctic Specially Protected Areas, Antarctic Specially Managed Areas and Historic Sites and Monuments are suitably marked on the site.

Article 10

Exchange of Information

1. The Parties shall make arrangements for:
   (a) collecting and exchanging records, including records of permits and reports of visits,

including inspection visits, to Antarctic Specially Protected Areas and reports of inspection visits to Antarctic Specially Managed Areas;

(b) obtaining and exchanging information on any significant change or damage to any Antarctic Specially Managed Area, Antarctic Specially Protected Area or Historic Site or Monument; and

(c) establishing common forms in which records and information shall be submitted by Parties in accordance with paragraph 2 below.

2. Each Party shall inform the other Parties and the Committee before the end of November of each year of the number and nature of permits issued under this Annex in the preceding period of 1st July to 30th June.

3. Each Party conducting, funding or authorising research or other activities in Antarctic Specially Protected Areas or Antarctic Specially Managed Areas shall maintain a record of such activities and in the annual exchange of information in accordance with the Treaty shall provide summary descriptions of the activities conducted by persons subject to its jurisdiction in such areas in the preceding year.

4. Each Party shall inform the other Parties and the Committee before the end of November each year of measures it has taken to implement this Annex, including any site inspections and any steps it has taken to address instances of activities in contravention of the provisions of the approved Management Plan for an Antarctic Specially Protected Area or Antarctic Specially Managed Area.

Article 11

CASES OF EMERGENCY

1. The restrictions laid down and authorized by this Annex shall not apply in cases of emergency involving safety of human life or of ships, aircraft or equipment and facilities of high value or the protection of the environment.
2. Notice of activities undertaken in cases of emergency shall be circulated immediately to all Parties and to the Committee.

Article 12

AMENDMENT OR MODIFICATION

1. This Annex may be amended or modified by a measure adopted in accordance with Article IX (1) of the Antarctic Treaty. Unless the measure specifies otherwise, the amendment or modification shall be deemed to have been approved, and shall become effective, one year after the close of the Antarctic Treaty Consultative Meeting at which it was adopted, unless one or more of the Antarctic Treaty Consultative Parties notifies the Depositary, within that time period, that it wishes an extension of that period or that it is unable to approve the measure.
2. Any amendment or modification of this Annex which becomes effective in accordance with paragraph 1 above shall thereafter become effective as to any other Party when notice of approval by it has been received by the Depositary.

APPENDIX 3

IUCN: Summary of actions needed to implement an Antarctic conservation strategy

1. Strengthening conservation principles, policies and goals

- conservation of the Antarctic environment should be a principal objective of international policy and should receive appropriate emphasis
- the existing ATS should be extended to incorporate comprehensive environmental protection principles and legally binding conservation measures
- full recognition should be given to wilderness values and the importance of the region for global environmental research
- measures should be based on the precautionary principle

2. Strengthening legal instruments and enforcement measures

- need for new instruments to secure comprehensive environmental protection and setting out clear principles for regulation of human activities
- transparency and accessibility are essential
- stronger legal obligations and strengthening of measures to ensure compliance

3. Strengthening Antarctic institutions

- establishment of a permanent treaty secretariat
- establishment of an appropriate kind of advisory committee to advise on environmental protection requirements

4. Strengthening of public information and education

- public awareness campaign should form part of comprehensive environmental protection strategy
- improved training programmes emphasising environmentally responsible conduct for all Antarctic visitors

5. Strengthening Antarctic science and its management

- development of a long-term scientific strategy by SCAR according to certain guidelines, and including monitoring environmental variables of global importance
- stations should be located where there are opportunities for valuable research and to avoid clustering, which gives rise to duplication of effort and environmental damage; location of stations in protected areas should be prohibited
- countries undertaking good-quality programmes should be accorded Consultative Party status without the need to establish a station
- SCAR should assist in the development of quality scientific programmes and increased international cooperation
- further development of guidelines to minimise the impact of scientific activities
- more applied research to monitor the impacts of human activity, including resource development,

and translation of the results into effective management action
- development of monitoring programmes relating to global environmental change and their integration with programmes of other international organisations

6. Strengthening conservation measures in Antarctic logistics and establishment of stations

- ensure that existing provisions relating to operation of vehicles, aircraft and ships are applied and where necessary reviewed and strengthened; development of contingency plans for emergency action, e.g. in relation to oil spills
- careful planning to ensure minimal impacts both during the operational life of stations and if and when they become disused
- preparation of operational rules for stations that could be incorporated in management plans for the area
- improved waste management standards and procedures to be pursued
- use of the most appropriate technology
- strict application of EIA procedures and adherence to their recommendations

7. Strengthening of Antarctic protected areas

- systematic assessment of key sites based on an improved biogeographical framework
- broad participation in the preparation of comprehensive and legally binding management plans for all protected areas
- according protection to outstanding landscapes and historical sites under the World Heritage Convention
- integration of the management of protected areas and areas used for human activity, e.g. through multiple-use planning areas or application of biosphere reserve management principles
- safeguarding additional sites on sub-Antarctic islands
- limiting impacts of introduced species through increased control and eradication programmes

8. Strengthening conservation measures for managing Antarctic tourism

- comprehensive review of tourism issues and development of management guidelines through the interaction of all appropriate parties
- recognition of identified areas for tourist visits, coupled with monitoring of impacts

9. Strengthening the conservation of marine living resources

- agreement on objectives of a long-term conservation management plan for the resources
- elaboration of management procedures to determine the controls on exploitation, based on agreed criteria for sustainable use
- adoption of a comprehensive management plan for the rehabilitation of depleted stocks
- establishment of links between CCAS, the IWC and CCAMLR to develop long-term management plans for seals and whales within the CCAMLR framework

10. Strengthening measures to protect Antarctica against minerals exploitation

- mineral resource activities should be excluded from Antarctica under a binding legal agreement
- if a future agreement is reached to permit minerals activities, there must be prior establishment of a legal regulatory regime with environmental provisions as least as strong as those in CRAMRA

Source: IUCN (1991).

References

Abrams, R.W. (1985) Energy and food requirements of pelagic aerial seabirds in different regions of the African sector of the Indian Ocean. In Siegfried, W.R., Condy, P.R. and Laws, R.M. (eds) *Antarctic Nutrient Cycles and Food Webs*. Springer-Verlag, Berlin, 466–472.

Adamson, E. and Seppelt, R.D. (1990) A comparison of airborne alkaline pollution damage in selected lichens and mosses at Casey Station, Wilkes Land, Antarctica. In Kerry, K.R. and Hempel, G. (eds) *Antarctic Ecosystems. Ecological Change and Conservation*. Springer-Verlag, Berlin, 347-353.

Adamson, E., Adamson, H. and Seppelt, R. (1994) Cement dust contamination of *Ceratodon purpureus* at Casey, East Antarctica: damage and capacity for recovery. *Journal of Bryology*, 18, 127–137.

Adamson, H. and Adamson, E. (1992) Possible effects of global climate change on Antarctic terrestrial vegetation. In Department of the Arts, Sport, the Environment and Territories. *Impact of Climate Change on Antarctica – Australia*. Australian Government Publishing Service, Canberra, 52–62.

Adamson, H., Wilson, M., Selkirk, P. and Seppelt, R. (1988) Photoinhibition in Antarctic mosses. *Polarforschung*, 58, 103–111.

Agnew, D.J. (1997) The CCAMLR Ecosystem Monitoring Programme. *Antarctic Science*, 9, 235–242.

Agrímsson, H. (1989) Developments leading to the 1982 decision of the International Whaling Commission for a zero catch quota 1986–90. In Andresen, S. and Østreng, W. (eds) *International Resource Management: the Role of Science and Politics*. Belhaven Press, London, 221–231.

Aguaya, A. (1978) The present status of the Antarctic fur seal *Arctocephalus gazella* at the South Shetland Islands. *Polar Record*, 19(119), 167–176.

Ahmadjian, V. (1970) Adaptations of Antarctic terrestrial plants. In Holdgate, M.W. (ed.) *Antarctic Ecology*. Vol. 2. Academic Press, London, 801–811.

Allen, K.R. (1980) *Conservation and Management of Whales*. Butterworths, London.

Alley, R.B. and Whillans, I.M. (1991) Changes in the West Antarctic Ice Sheet. *Science,* 254, 959-963.

Alley, R.B., Blankenship, D.D., Rooney, S.T. and Bentley, C.R. (1989) Sedimentation beneath ice shelves – the view from Ice Stream B. *Marine Geology*, 85, 101–120.

Allison, I.F. (1992) The Antarctic cryosphere: evidence of the impacts of change and strategies for detection. In Department of the Arts, Sport, the Environment and Territories. *Impact of Climate Change on Antarctica – Australia.* Australian Government Publishing Service, Canberra, 62–66.

Allison, I.F. and Keage. P.L. (1986) Recent changes in the glaciers of Heard Island. *Polar Record*, 23, 255–271.

Amos, A.F. (1984) Distribution of krill (*Euphausia superba*) and the hydrography of the Southern Ocean: large scale processes. *Journal of Crustacean Biology*, 4(1), 306–329.

Ancel, A., Visser, H., Handrich,Y., Masman, D. and Le Maho,Y. (1997) Energy saving in huddling penguins. *Nature*, 385, 304–305.

Andersen, R.M. and Rudolph, L. (1989) On solid international ground in Antarctica: a U.S. strategy for regulating environmental impact on the continent. *Stanford Journal of International Law*, 26, 93–151.

Andersen, R.T. (1991) Negotiating a new regime: how CRAMRA came into existence. In Jørgensen-Dahl, A. and Østreng, W. (eds), *The Antarctic Treaty System in World Politics*. Macmillan, Basingstoke, 94–109.

Anderson, C. (1991) Penguins losing the struggle? *Nature*, 350, 294.

Anderson, D. (1968) The conservation of wildlife under the Antarctic Treaty. *Polar Record*, 14, 25–32.

Anderson, J.B. (1990) Geology and hydrocarbon potential of the Antarctic continental margin. In Splettstoesser, J.F. and Dreschhoff, G.A.M. (eds) *Mineral Resources of Antarctica*. American Geophysical Union, Antarctic Research Series 51, Washington DC, 175–201.

Anderson, J.B. (1991) The Antarctic continental shelf: results from marine geological and geophysical investigations. In Tingey, R.J. (ed.) *The Geology of Antarctica*. Clarendon Press, Oxford, 285–334.

Anderson, J.B., Bartek, L. and Thomas, M.A., (1991) Seismic and sedimentologic record of glacial events on the Antarctic Peninsula shelf. In Thomson, M.R.A., Crame, J.A. and Thomson, J.W. (eds) 1991. *Geological Evolution of Antarctica*. Cambridge University Press, Cambridge, 687–691.

Anderson, J.B., Brake, C.F. and Myers, N.C. (1984) Sedimentation of the Ross Sea Continental Shelf, Antarctica. *Marine Geology*, 57, 295–333.

Anderson, J.J. (1965) Bedrock geology of Antarctica: a summary of exploration, 1831–1962. In Hadley, T.B. (ed.) *Geology and Palaeontology of the Antarctic*. American Geophysical Union, Antarctic Research Series 6, Washington DC, 1–70.

Anderson, R.T. (1991) Negotiating a new regime: how CRAMRA came into existence. In Jørgensen-Dahl, A. and Østreng, W. (eds) *The Antarctic Treaty System in World Politics*. Macmillan, Basingstoke, 94–109.

Andresen, S. (1989) Science and politics in the international management of whales. *Marine Policy*, 13, 99–117.

Andrews, J.T. (1975) *Glacier Systems*. Duxbury, Massachusetts.

Anon (1989a) Argentine ship sinks near Palmer Station. *Antarctic Journal of the United States*, 24(2), 3–9.

Anon (1989b) Peru's first station is on King George Island. *Antarctic*, 11(11), 420–421.

Anon (1989c) Oil slick could scar Antarctica for a century. *New Scientist*, 121(1651), 31.

Anon (1989d) Architect plans holiday town in Antarctica. *New Scientist*, 122(1665), 22.

Anon (1990a) Midwinter fuel spill contained. *ANARE News*, 62, 13.

Anon (1990b) NSF reports focus on improving USAP environmental practices. *Antarctic Journal of the United States*, 25(2), 1–2.

Anon (1990c) Antarctic wilderness. *Nature*, 348, 267–268.

Anon (1990d) A scientific perspective. *Eco*, 77(4), 2.

Anon (1991) Fuel spill at Casey, news release from the Division, 23 January '91. *Aurora*, 10(3), 32.

Anon (1993) Last of the oil removed from the *Bahia Pariso*. *Antarctic*, 12(11–12), 400–401.

Anon (1994) Private Norwegian expedition ends in tragedy: USAP dispatches SAR team to retrieve party of four. *Antarctic Journal of the United States*, 29(2), 11–12.

Anon (1995) Global warming and cooling enthusiasm. *The Economist*, 335 (7908), 67–68.

Armstrong, R.L. (1978) Late Cenozoic McMurdo Volcanic Group and dry valley glacial history, Victoria Land, Antarctica. *New Zealand Journal of Geology and Geophysics*, 21, 685–698.

Armstrong, T.E., Roberts, B. and Swithinbank, C.W.M. (1973) *Illustrated Glossary of Snow and Ice*, 2nd edn. Scott Polar Research Institute, Cambridge.

Arnould, J.P.Y. and Croxall, J.P. (1995) Trends in entanglement of Antarctic fur seals (*Arctocephalus gazella*) in man-made debris at South Georgia. *Marine Pollution Bulletin*, 30, 707–712.

Aron, W. (1988) The commons revisited: thoughts on marine mammal management. *Coastal Management*, 16, 99–110.

Arrigo, K.R., Worthen, D.L., Lizotte, M.P., Dixon, P. and Dieckmann, G. (1997) Primary production in Antarctic sea ice. *Science*, 276, 394–397.

ASOC (1990) The convention on Antarctic conservation. ASOC Information Paper ANT/SCM/NGO INF Paper 1990-1. Antarctic and Southern Ocean Coalition, Washington DC.

Atkinson, R.J., Matthews, W.A., Newman, P.A. and Plumb, R.A. (1989) Evidence of the mid-latitude impact of Antarctic ozone depletion. *Nature*, 340, 290–294.

Auburn, F.M. (1982) *Antarctic Law and Politics*. C. Hurst and Co., London.

Auburn, F.M. (1983) The Royal Commission on the Mount Erebus air disaster. *Polar Record*, 21, 359–367.

Auburn, F.M. (1984) The Antarctic minerals regime: sovereignty, exploration, institutions and environment. In Harris, S. (ed.) *Australia's Antarctic*

Policy Options. CRES Monograph II, Centre for Resource and Environmental Studies, Australian National University, Canberra, 271–300.

Auburn, F.M. (1990) Convention on the Regulation of Antarctic Mineral Resource Activities. In Splettstoesser, J.F. and Dreschhoff, G.A.M. (eds) *Mineral Resources Potential of Antarctica*. American Geophysical Union, Antarctic Research Series 51, Washington DC, 259–271.

Aust, A. and Shears, J. (1996) Liability for environmental damage in Antarctica. *Review of European Community & International Environmental Law*, 5, 312–320.

Australian Government Printing Service. (1992) *Impact of Climate Change, Antarctica*. Canberra, 78pp.

Bacci, E., Calamari, D., Gaggi, C., Fanelli, R., Focardi, S. and Morosini, M. (1986) Chlorinated hydrocarbons in lichen and moss samples from the Antarctic Peninsula. *Chemosphere*, 15, 747–754.

Balch, T.W. (1910) The Arctic and Antarctic regions and the Law of Nations. *American Journal of International Law*, 4, 265–275.

Barber, R.T. and Chavez, F.P. (1983) Biological consequences of El Niño. *Science*, 222, 1203–1210

Barinaga, M.(1990) Eco-quandary: what killed the skuas? *Science*, 20 July, Vol. 249, 243.

Barker, P.F. and Burrell, J. (1977) The opening of the Drake Passage. *Marine Geology*, 25, 15–34.

Barker, P.F. and Griffiths, D.H. (1972) The evolution of the Scotia Ridge and Scotia Sea. *Philosophical Transactions of the Royal Society of London*, A271, 151–183.

Barker, P.F. and Hill, I.A. (1981) Back–arc extension of the Scotia Sea. *Philosophical Transactions of the Royal Society of London*, A300, 249–262.

Barker, P.F., Dalziel, I.W.D. and Storey, B.C. (1991) Tectonic development of the Scotia Arc Region. In Tingey, R.J. (ed.) *The Geology of Antarctica*. Clarendon Press, Oxford. 215–248.

Barker, P.F., Kennett, J.P. and scientific party. (1988) Weddell Sea palaeooceanography: preliminary results of ODP Leg 113. *Palaeogeography, Palaeoclimatology and Palaeoecology*, 67, 75–102.

Barnes, J.N. (1982a) The emerging Convention on the Conservation of Antarctic Marine Living Resources: an attempt to meet the new realities of resource exploitation in the Southern Ocean. In Charney, J. (ed.) *The New Nationalism and the Use of Common Spaces. Issues in Marine Pollution and the Exploitation of Antarctica*. Allanheld, Osmun Publishers, Totowa, New Jersey, 239–286.

Barnes, J.N. (1982b) *Let's Save Antarctica*. Greenhouse Publications, Richmond, Victoria.

Barnes, J.N. (1986) The future of Antarctica. Environmental issues and the role of NGOs. In Wolfrum, R. (ed.) *Antarctic Challenge II. Conflicting Interests, Co–operation, Environmental Protection, Economic Development*. Duncker & Humblot, Berlin, 413–445.

Barnes, J.N. (1988) Legal aspects of environmental protection in Antarctica. In Joyner, C.C. and Chopra, S.K. (eds) *The Antarctic Legal Regime*. Martinus Nijhoff Publishers, Dordrecht, 241–268.

Barnes, J.N. (1991) Protection of the environment in Antarctica: are present regimes enough? In Jørgensen–Dahl, A. and Østreng, W. (eds) *The Antarctic Treaty System in World Politics*. Macmillan, Basingstoke, 186–228.

Barnes, J.N., Lipperman, P.J. and Rigg, K. (1988) Waste management in Antarctica. In Wolfrum, R. (Ed.), *Antarctic Challenge III. Conflicting Interests, Co-operation, Environmental Protection, Economic Development*. Duncker & Humblot, Berlin, 491–529.

Barnola, J.M., Raynaud, D., Korotkevitch, Y.S. and Lorius, C. (1987) Vostok ice core provides 160,000–year record of atmospheric CO_2. *Nature*, 329, 408–414.

Barrett, P.J. (ed.) (1986) Antarctic Cenozoic history from the MSSTS–1 drillhole, McMurdo Sound. *DSIR Bulletin*, 237, Science Information Service, Wellington.

Barrett, P.J. (1991a) Antarctica and global climatic change: a geological perspective. In Harris, C.M. and Stonehouse, B. (eds) *Antarctica and Global Climatic Change*. Belhaven Press, London, 35–50.

Barrett, P.J. (1991b) The Devonian to Triassic Beacon Supergroup of the Transantarctic Mountains and correlatives in other parts of Antarctica. In Tingey, R.J. (ed.) The *Geology of Antarctica*. Clarendon Press, Oxford, 120–152.

Barrett, P.J. and Hambrey, M.J. (1992) Plio-Pleistocene sedimentation in Ferrar Fiord, Antarctica, *Sedimentology*, 39, 109–123.

Barrett, P.J., Adams, C.J., McIntosh, W.C., Swisher III, C.C. and Wilson, G.S. (1992) Geochronological evidence supporting Antarctic deglaciation three million years ago. *Nature*, 359, 816–818.

Barrett, P.J., Hambrey, M.J. and Robinson, P.R. (1991) Cenozoic glacial and tectonic history from CIROS–1, McMurdo Sound. In Thomson, M.R.A., Crame, J.A. and Thomson, J.W. (eds) *Geological Evolution of Antarctica*. Cambridge University Press, Cambridge, 651–656.

Barron, J. and Leg 119 Shipboard Scientific party (1988) Early glaciation of Antarctica. *Nature*, 333, 303–304.

Barrow, C.J. (1978) Postglacial pollen diagrams from South Georgia (sub-Antarctic) and West Falkland Island (South Atlantic) *Journal of Biogeography*, 5, 251–274.

Barry, R.G. and Chorley, R.J. (1992) *Atmosphere, Weather and Climate*, 6th edn, Methuen, London.

Barsch, V.D. and Stablein, G. (1984) Frostdynamik und permafrost in eisfreien gebieten der Antarktischen Halbinsel. *Polarforschung*, 54(2), 111–119.

Barstow, R. (1990) Beyond whale species survival – peaceful coexistence and mutual enrichment as a basis for human–cetacean relations. *Mammal Review*, 20, 65–73.

Barton, C.M. (1965) The geology of the South Shetland Islands – III. The stratigraphy of King George Island. *British Antarctic Survey Scientific Report*, 44.

BAS (1987) British Antarctic Survey Annual Report for 1986/87. British Antarctic Survey, Cambridge, 85.

BAS (1989) *Proposed Construction of a Crushed Rock Airstrip at Rothera Point, Adelaide Island, British Antarctic Territory. Final Comprehensive Environmental Evaluation.* Natural Environment Research Council, Swindon.

BAS (1991) Environmental management. *BAS Environmental Information Leaflets*, Nos 1–5.

BAS (1993) *Antarctica – a Topographic Database.* Sheet BAS (Misc) 7. British Antarctic Survey, Cambridge.

Basberg, B.L. (1993) Survival against all odds? Shore station whaling at South Georgia in the pelagic era, 1925–1960. In Basberg, B.L. Ringstad, J.E. and Wexelsen, E. (eds) *Whaling and History. Perspectives on the Evolution of the Industry*. Kommandør Chr. Christensens Hvalfangstmuseum Publikasjon 29, Sandefjord, 157–167.

Basberg, B.L., Naevestad, D. and Rossnes, G. (1996) Industrial archaeology at South Georgia: methods and results. *Polar Record*, 32, 51–66.

Basson, M. and Beddington, J.R. (1991) CCAMLR: the practical implications of an eco-system approach. In Jørgensen–Dahl, A. and Østreng, W. (eds) *The Antarctic Treaty System in World Politics*. Macmillan, Basingstoke, 54–69.

Beaglehole, J.C. (1961) *The Journals of Captain James Cook on his Voyages of Discovery. II. The Voyage of the Resolution and Adventure 1772–1775*. Hakluyt Society, Cambridge.

Bechervaise, J. (1962) *The Far South*. Angus and Robertson, London.

Beck, P.J. (1983) Britain's Antarctic dimensions. *International Affairs*, 59, 429–444.

Beck, P.J. (1984) The United Nations and Antarctica. *Polar Record*, 22, 137–144.

Beck, P.J. (1985a) The United Nations' study on Antarctica 1984. *Polar Record*, 22, 499–504.

Beck, P.J. (1985b) Preparatory meeting for the Antarctic Treaty 1958–59. *Polar Record*, 22, 653–664.

Beck, P.J. (1986a) Antarctica at the United Nations, 1985: the end of consensus? *Polar Record*, 23, 159–166.

Beck, P.J. (1986b) *The International Politics of Antarctica*. Croom Helm, London.

Beck, P.J. (1987a) A cold war. Britain, Argentina and Antarctica. *History Today*, 37, 16–23.

Beck, P.J. (1987b) The United Nations and Antarctica, 1986. *Polar Record*, 23, 683–690.

Beck, P.J. (1988a) Another sterile annual ritual? The United Nations and Antarctica 1987. *Polar Record*, 24, 207–212.

Beck, P.J. (1988b) A continent surrounded by advice: recent reports on Antarctica. *Polar Record*, 24, 285–291.

Beck, P.J. (1989a) Antarctica at the UN 1988: seeking a bridge of understanding. *Polar Record*, 25, 329–334.

Beck, P.J. (1989b) A new polar factor in international relations. *The World Today*, 45, 65–68.

Beck, P.J. (1989c) Convention on the Regulation of Antarctic Mineral Resource Activities: a major addition to the Antarctic Treaty System. *Polar Record*, 25, 19–32.

Beck, P.J. (1990a) The UN goes green on Antarctica: the 1989 session. *Polar Record*, 26, 323–325.

Beck, P.J. (1990b) Antarctic as a zone of peace: a strategic irrelevance?: A historical and contemporary survey. In Herr, R.A., Hall, H.R. and Haward, M.G. (eds) *Antarctica's Future: Continuity or Change?* Australian Institute of International Affairs, Hobart, 193–224.

Beck, P.J. (1990c) Antarctica enters the 1990s: an overview. *Applied Geography*, 10, 247–263.

Beck, P.J. (1990d) Regulating one of the last tourism frontiers: Antarctica. *Applied Geography*, 10, 343–56.

Beck, P.J. (1990e) International relations in Antarctica: Argentina, Chile and the great powers. In Morris, M.A. (ed.) *Great Power Relations in Argentina, Chile and Antarctica*. Macmillan, Basingstoke, 101–130.

Beck, P.J. (1991a) Antarctica,Viña del Mar and the 1990 UN debate. *Polar Record*, 27, 211–216.

Beck, P.J. (1991b) The Antarctic resource conventions implemented: consequences for the sovereignty issue. In Jørgensen–Dahl, A. and Østreng, W. (eds)

The Antarctic Treaty System in World Politics. Macmillan, Basingstoke, 229–276.

Beck, P.J. (1992) The 1991 UN session: the environmental protocol fails to satisfy the Antarctic Treaty System's critics. *Polar Record*, 28, 307–314.

Beck, P.J. (1993) The United Nations and Antarctica, 1992: still searching for that elusive convergence of view. *Polar Record*, 29, 313–320.

Beck, P.J. (1994a) Looking at the Falkland Islands from Antarctica: the broader regional perspective. *Polar Record*, 30, 167–180.

Beck, P.J. (1994b) The United Nations and Antarctica 1993: continuing controversy about the UN's role in Antarctica. *Polar Record*, 30, 257–264.

Beck, P.J. (1994c) Asia, Antarctica and the United Nations. In Herr, R.A. and Davis, B.W. (eds) *Asia in Antarctica.* Centre for Resource and Environmental Studies, Australian National University, Canberra, 5–39.

Beck, P.J. (1995) The United Nations and Antarctica, 1994: the restoration of consensus? *Polar Record*, 31, 419–424.

Beck, P.J. (1996) Identifying national interests in Antarctica: the case of Canada. *Polar Record*, 32, 335–346.

Beddington, J.R. and May, R.M. (1982) The harvesting of interacting species in a natural ecosystem. *Scientific American*, 247(5), 42–49.

Beeby, C.D. (1983) An overview of the problems which should be addressed in the preparation of a regime governing the mineral resources of Antarctica. In Orrego Vicuña, F. (ed.) *Antarctic Resources Policy: Scientific, Legal and Political Issues.* Cambridge University Press, Cambridge, 191–198.

Beeby, C.D. (1991) The Antarctic Treaty System: goals, performance and impact. In Jørgensen–Dahl, A. and Østreng, W. (eds) *The Antarctic Treaty System in World Politics.* Basingstoke, Macmillan, 4–21.

Behrendt, J.C. (1983) Are there petroleum resources in Antarctica? In Behrendt, J.C. (ed) *Petroleum and Mineral Resources of Antarctica.* United States Geological Survey, Circular No. 909, 3–24.

Behrendt, J.C. (1990) Recent geophysical and geological research in Antarctica related to the assessment of petroleum resources and potential environmental hazards to their development. In Splettstoesser, J.F. and Dreschhoff, G.A.M. (eds) *Mineral Resources Potential of Antarctica.* American Geophysical Union, Antarctic Research Series 51, Washington DC, 163–174.

Behrendt, J.C. (1991) Scientific studies relevant to the question of Antarctica's petroleum resource potential. In Tingey, R.J. (ed.) *The Geology of Antarctica.* Clarendon Press, Oxford, 588–616.

Behrenfeld, M.J., Lee II, H. and Small, L.F. (1994) Interactions between nutritional status and long–term responses to ultraviolet–B radiation stress in a marine diatom. *Marine Biology*, 118, 523–530.

Beike, D. (1990) An engineering economic evaluation of mining in Antarctica: a case study of platinum. In Splettstoesser, J.F. and Dreschhoff, G.A.M. (eds) *Mineral Resources Potential of Antarctica.* American Geophysical Union, Antarctic Research Series 51, Washington DC, 53–67.

Beike, D. (1992) An engineering–economic evaluation of offshore oil and gas development in the Ross Sea, Antarctica. In Thompson, B.J. (ed.) *Research Reports of the Link Energy Fellows* Vol. 7. University of Rochester Press, Rochester, 1–61.

Beltramino, J.C.M. (1993) *The Structure and Dynamics of Antarctic Population.* Vantage Press, New York.

Benedick, R.E. (1991) *Ozone Diplomacy.* Harvard University Press, Cambridge, Mass.

Bengtson, J.L., Ferm, L.M., Härkönen, T.J. and Stewart, B.S. (1990) Abundance of Antarctic fur seals in the South Shetland Islands, Antarctica, during the 1986/87 austral summer. In Kerry, K.R. and Hempel, G. (eds) *Antarctic Ecosystems. Ecological Change and Conservation.* Springer–Verlag, Berlin, 265–270.

Benninghoff, W.S. and Bonner, W.N. (1985) *Man's Impact on the Antarctic Environment: A Procedure for Evaluating Impacts for Scientific and Logistic Activities.* SCAR, Cambridge.

Benson, C.S. (1962) The role of air pollution in arctic planning and development. *Polar Record*, 14, 783–790.

Bentley, C.R. (1996) Water kept liquid by warmth from within. *Nature*. 381, 645.

Bentley, C.R. and Giovinetto, M.B. (1991) Mass balance of Antarctica and sea level change. In Weller, G., Wilson, J., and Severin, B. (eds) *Role of Polar Regions in Global Change.* Proceedings of the International Conference, University of Alaska. Geophysical Institute and Center for Global Change and Arctic System Research, University of Alaska, Fairbanks; 489–494.

Bergin, A. (1997) Albatross and longlining – managing seabird bycatch. *Marine Policy*, 21, 63–72.

Berkman, P.A. (1992) The Antarctic marine ecosystem and humankind. *Reviews in Aquatic Sciences*, 6, 295–333.

Berkman, P.A. and Nigro, M. (1992) Trace metal concentrations in scallops around Antarctica: extending the mussel watch programme to the Southern Ocean. *Marine Pollution Bulletin*, 24, 322–323.

Bertrand, K.J. (1971) *Americans in Antarctica, 1775–1948*. American Geographical Society, Special Publication No. 39, New York.

Best, P.B. (1993) Increased rates in severely depleted stocks of baleen whales. I.C.E.S. *Journal of Marine Science*, 50, 169–186.

Bidigare, R.R. (1989) Potential effects of UV–B radiation on marine organisms of the Southern Ocean: distributions of phytoplankton and krill during austral spring. *Photochemistry and Photobiology*, 50, 469–477.

Bidleman, T.F., Walla, M.D., Roura, R., Carr, E. and Schmidt, S. (1993) Organochlorine pesticides in the atmosphere of the Southern Ocean and Antarctica, January–March, 1990. *Marine Pollution Bulletin*, 26, 258–262.

Bilder, R.B. (1982) The present legal and political situation in Antarctica. In Charney, J.I. (ed.) *The New Nationalism and the Use of Common Spaces. Issues in Marine Pollution and the Exploitation of Antarctica*. Allanhead, Osmun Publishers, Totowa, New Jersey, 167–205.

Birkenmajer, K. (1988) Geochronology of Tertiary glaciations on King George Island, West Antarctica. *Bulletin of the Polish Academy of Sciences. Earth Sciences*, 37, 27–48.

Birnie, P. (1985) *International Regulation of Whaling: from Conservation of Whaling to Conservation of Whales and the Regulation of Whale–Watching*, 2 vols. Oceana Publications Inc., New York, London.

Birnie, R.V. and Gordon, J.E. (1980) Drainage systems associated with snow melt, South Shetland Islands, Antarctica. *Geografiska Annaler*, 62A, 57–62.

Birnie, R.V. and Thom, G. (1982) Preliminary observations on two rock glaciers in South Georgia, Falkland Islands Dependencies. *Journal of Glaciology*, 28, 377–386.

Black, R.F. and Berg, T.E. (1964) Glacier fluctuations recorded by patterned ground, Victoria Land. In Adie, R.J. (ed.) *Antarctic Geology*, North Holland Publishing Co., Amsterdam, 107–122.

Blay, S. and Green, J. (1995) The development of a liability annex to the Madrid Protocol. *Environmental Policy and Law*, 25, 24–37.

Blay, S.K.N. (1992) New trends in the protection of the Antarctic environment: the 1991 Madrid Protocol. *The American Journal of International Law*, 86, 377–399.

Blay, S.K.N. and Tsamenyi, B.M. (1990) Australia and the Convention for the Regulation of Antarctic Mineral Resource Activities (CRAMRA) *Polar Record*, 26, 195–202.

Bleasel, J. (1986) An Australian perspective. In Miller, T.B. (ed.) *Australia, Britain and Antarctica*. Australian Studies Centre, Institute of Commonwealth Studies, University of London, London, 37–50.

Block, W. (1980) Survival strategies in polar terrestrial arthropods. *Biological Journal of the Linnean Society*, 14, 29–38.

Block, W. (1990) Cold tolerance of insects and other arthropods. *Philosophical Transactions of the Royal Society of London*, B326, 613–633.

Block, W. (1994) Terrestrial ecosystems: Antarctica. *Polar Biology*, 14, 293–300.

Block, W. and Convey, P. (1995) The biology, life cycle and ecophysiology of the Antarctic mite *Alaskozetes antarcticus*. *Journal of the Zoological Society of London*, 236, 431–449.

Bloom, A.L. (1983) Sea level and coastal changes. In Wright Jr, H.E. (ed.) *Late Quaternary Environments of the United States*. University of Minnesota Press, Minneapolis, 42–51.

Bloomer, J.P. and Bester, M.N. (1992) Control of feral cats on sub–Antarctic Marion Island, Indian Ocean. *Biological Conservation*, 60, 211–219.

Bockheim, J.G. (1990) Soil development rates in the Transantarctic Mountains. *Geoderma*, 47, 59–77.

Bockheim, J.G. (1995) Permafrost distribution in the Southern Circumpolar Region and its relation to the environment: a review and recommendations for further research. *Permafrost and Periglacial Processes*, 6, 27–45.

Boczek, B.A. (1986) Specially protected areas as an instrument for the conservation of Antarctic nature. In Wolfrum R. (ed.) *Antarctic Challenges II: Conflicting Interests, Co–operation, Environmental Protection, Economic Development*. Duncker & Humblot, Berlin, 65–101.

Boczek, B.A. (1988) The legal status of visitors, including tourists, and non–governmental expeditions in Antarctica. In Wolfrum, R. (ed.) *Antarctic Challenges III: Conflicting Interests, Co–operation, Environmental Protection, Economic Development*. Duncker & Humblot, Berlin, 455–490.

Bonner, W.N. (1968) The fur seal of South Georgia. *British Antarctic Survey Scientific Reports*, No. 56.

Bonner, W.N. (1984a) Introduced mammals. In Laws, R.M. (ed.) *Antarctic Ecology*, Vol. 1. Academic Press, London, 237–278.

Bonner, W.N. (1984b) Conservation and the Antarctic. In Laws, R.M. (ed.) *Antarctic Ecology*, Vol. 2. Academic Press, London, 822–850.

Bonner, W.N. (1985a) Birds and mammals – Antarctic seals. In Bonner, W.N. and Walton,

D.W.H. (eds) *Key Environments. Antarctica.* Pergamon Press, Oxford, 202–222.

Bonner, W.N. (1985b) Impact of fur seals on the terrestrial environment at South Georgia. In Siegfried, W.R., Condy, P.R. and Laws, R.M. (eds) *Antarctic Nutrient Cycles and Food Webs.* Springer–Verlag, Berlin, 641–646.

Bonner, W.N. (1987) Recent developments in Antarctic conservation. In Triggs, G.D. (ed.) *The Antarctic Treaty Regime. Law, Environment and Resources.* Cambridge University Press, Cambridge, 143–149.

Bonner, W.N. (1989) Environmental assessment in the Antarctic. *Ambio*, 18, 83–89.

Bonner, W.N. (1990) International agreements and the conservation of Antarctic systems. In Kerry, K.R. and Hempel, G. (eds) *Antarctic Ecosystems. Ecological Change and Conservation.* Springer-Verlag, Berlin, 386–393.

Bonner, W.N. (1994) Environmental protection and science in the Antarctic. In Hempel, G. (ed.) *Antarctic Science. Global Concerns.* Springer-Verlag, Berlin, 6–11.

Bonner, W.N. and Angel, M.V. (1987) Conservation and the Antarctic environment: the Working Group Reports of the joint IUCN/SCAR Symposium on the Scientific Requirements for Antarctic Conservation. *Environment International*, 13, 137–144.

Bonner, W.N. and Laws, R.M. (1964) Seals and sealing. In Priestley, R., Adie, R.J. and Robin, G. de Q. (eds) *Antarctic Research.* Butterworths, London, 163–190.

Bonner, W.N. and Smith, R.I.L. (1985) *Conservation Areas in the Antarctic.* SCAR, Cambridge.

Bonner, W.N. and Walton, D.W.H. (1985) Marine habitats – introduction. In Bonner, W.N. and Walton, D.W.H. (eds) *Key Environments: Antarctica.* Pergamon Press, Oxford, 133–134.

Boswall, J. (1986) Airborne tourism 1982–1984: a recent Antarctic development. *Polar Record*, 23, 187–191.

Boutron, C.F. and Patterson, C.C. (1987) Relative levels of natural and anthropogenic lead in recent Antarctic snow. *Journal of Geophysical Research*, 92, 8454–8464.

Boutron, C.F. and Wolff, E.W. (1989) Heavy metal and sulphur emissions to the atmosphere from human activities in Antarctica. *Atmospheric Environment*, 23, 1669–1675.

Boutron, C.F., Candelone, J–P. and Hong, S. (1994) Past and recent changes in the large-scale tropospheric cycles of lead and other heavy metals as documented in Antarctic and Greenland snow and ice: a review. *Geochimica et Cosmochimica Acta*, 58, 3217–3225.

Boutron, C.F., Görlach, U., Candelone, J.–P., Bolshov, M.A. and Delmas, R.J. (1991) Decrease in anthropogenic lead, cadmium and zinc in Greenland snows since the late 1960s. *Nature*, 353, 153–156.

Boyd, I.L. (1993) Pup production and distribution of breeding Antarctic fur seals (*Arctocephalus gazella*) at South Georgia. *Antarctic Science*, 5, 17–24.

Boyd, I.L., Walker, T.R. and Poncet, J. (1996) Status of southern elephant seals at South Georgia. *Antarctic Science*, 8, 237–244.

Brack, D. (1996) *International Trade and the Montreal Protocol.* Earthscan Publications Ltd, London.

Bradley, R.S. (1985) *Quaternary Palaeoclimatology.* Allen & Unwin. Boston.

Bradshaw, J.D. (1991) Cretaceous dispersion of Gondwana: continental and oceanic spreading in the south–west Pacific – Antarctic sector. In Thomson, M.R.A., Crame, J.A. and Thomson, J.W. (eds) *Geological Evolution of Antarctica.* Cambridge University Press, Cambridge, 581–586.

Brasseur, G. and Hitchmann, M.H. (1988) Stratospheric response to trace gas perturbations: changes in ozone and temperature distributions. *Science*, 240, 629–634.

Brennan, K. (1983) Criteria for access to the resources of Antarctica: alternatives, procedure and experience applicable. In Orrego Vicuña, F. (Ed.), *Antarctic Resources Policy: Scientific, Legal and Political Issues.* Cambridge University Press, Cambridge, 217–227.

Brewster, B. (1982) *Antarctica: Wilderness at Risk.* A.H. & A.W. Reed Ltd, Wellington.

Briggs, D.J. and Smithson, P.A. (1985) *Fundamentals of Physical Geography.* Hutchinson, London.

Broady, P.A. (1987) Protection of terrestrial plants and animals in the Ross Sea region, Antarctica. *New Zealand Antarctic Record*, 8 (1), 18–41.

Bromwich, D.H. and Kurtz, D.D. (1982) Experiences of Scott's Northern Party: evidence for a relationship between winter katabatic winds and the Terra Nova Bay polynya. *Polar Record*, 21, 137–146.

Brothers, N. (1991) Albatross mortality and associated bait loss in the Japanese longline fishery in the Southern Ocean. *Biological Conservation*, 55, 255–268.

Brown, A.D. (1991) The design of CRAMRA: how appropriate for the protection of the environment?

In Jørgensen-Dahl, A. and Østreng, W. (eds) *The Antarctic Treaty System in World Politics*. Macmillan, Basingstoke, 110–119.

Brown, S.G. (1963) A review of Antarctic whaling. *Polar Record*, 11, 555–566.

Bruno, L.A., Baur, H., Graf, T., Schluchter, C. and Wieler, R. (1997) The dating of Sirius group tillites in the Antarctic dry valleys with cosmogenic He–3 and Ne–21. *Earth and Planetary Science Letters*, 147(1–4), 37–54.

Budd, W.F. (1982) The role of Antarctica in Southern Hemisphere weather and climate. *Austalian Meterological Magazine*, 30, 265–272.

Budd, W.F. (1986) The Antarctic Treaty as a scientific mechanism (post-IGY) – contributions of Antarctic scientific research. In Polar Research Board, *Antarctic Treaty System. An Assessment*. National Academy Press, Washington DC, 103–151.

Budd, W.F., McInnes, B.J., Jenssen, D. and Smith, I.N. (1987) Modelling the response of the West Antarctic Ice Sheet to a climatic warming. In van der Veen, C.J. and Oerlemans, J. (eds) *Dynamics of the West Antarctic Ice Sheet*. Dordrecht, D. Riedel, 321–358.

Burgess, J. (1990) Comprehensive environmental protection of the Antarctic: new approaches for new times. In Cook, G. (ed.) *The Future of Antarctica. Exploitation Versus Protection*. Manchester University Press, Manchester, 53–67.

Burgess, J.S., Spate, A.P. and Norman, F.I. (1992) Environmental impacts of station development in the Larsemann Hills, Princess Elizabeth Land, Antarctica. *Journal of Environmental Management*, 36, 287–299.

Burke, W.T. (1997) Legal aspects of the IWC decision on the Southern Ocean Sanctury. *Ocean Development and International Law*, 28, 313–27.

Burmester, H.C. (1989) Liability for damage from Antarctic mineral resource activities. *Virginia Journal of International Law*, 29, 621–660.

Burton, H. (1986) A substantial decline in numbers of the southern elephant seal at Heard Island. *Tasmanian Naturalist*, 86, 4–8.

Burton, H.R. (1981) Chemisry, physics and evolution of Antarctic saline lakes: a review. *Hydrobiologia*, 82, 339–62.

Bush, W.M. (1990) The Antarctic Treaty System: a framework for evolution. The concept of a system. In Herr, R.A., Hall, H.R. and Haward, M.G. (eds) *Antarctica's Future: Continuity or Change*? Australian Institute of International Affairs, Hobart, 119–179.

Butterworth, D.S. (1986) Antarctic marine ecosystem management. *Polar Record*, 23, 37–47.

Butterworth, D.S. (1992) Science and sentimentality. *Nature*, 357, 532–534.

Cailleaux, A. (1968) Periglacial of McMurdo Strait (Antarctica) *Biuletyn Peryglacjalny*, 17, 57–90.

Caldwell, M., Teramura, A.H., Tevini, M., Bornman, J.F., Björn, L.O. and Kulandaivelu, G. (1995) Effects of increased solar ultraviolet radiation on terrestrial plants. *Ambio*, 24, 166–173.

Callaghan, T.V. and Lewis, M.C. (1971) The growth of *Phleum alpinum* L. in contrasting habitats at a sub-Antarctic station. *New Phytologist*, 70, 1143–1154.

Callaghan, T.V., Sonesson, M. and Sømme, L. (1992) Responses of terrestrial plants and invertebrates to environmental change at high latitudes. *Philosophical Transactions of the Royal Society of London Series B*, 338, 279–288.

Cameron, I. (1974) *Antarctica: the Last Continent*. Cassell, London.

Cameron, R.E. (1972) Pollution and conservation of the Antarctic terrestrial ecosystem. In Parker, B.C. (ed.) *Proceedings of the Colloquium on Conservation Problems in Antarctica*. Virginia Polytechnic Institute and State University, Blacksburg, Virginia, 267–306.

Cameron, R.E., Honour, R.C. and Morelli, F.A. (1977) Environmental impact studies of Antarctic sites. In Llano, G.A. (ed.) *Adaptations Within Antarctic Ecosystems*. Proceedings of the Third SCAR Symposium on Antarctic Biology. Smithsonian Institution, Washington DC, 1157–1176.

Cameron, R.L. (1964) Glaciological studies at Wilkes Station, Budd Coast, Antarctica in Mellor, M. (ed.) *Antarctic Snow and Ice Studies*. American Geophysical Union, Antarctic Research Series 2, Washington DC, 1–36.

Campbell, I.B. and Claridge G.G.C. (1987) *Antarctica: Soils, Weathering Processes and Environment*. Elsevier, Amsterdam.

Campbell, I.B. and Claridge, G.G.C. (1988) Landscape evolution in Antarctica. *Earth Science Reviews*, 25, 345–353.

Campbell, I.B., Balks, M.R. and Claridge, G.G.C. (1993) A simple visual technique for estimating the impact of fieldwork on the terrestrial environment in the ice-free areas of Antarctica. *Polar Record*, 29, 321–328.

Campbell, I.B., Claridge, G.G.C. and Balks, M.R. (1994) The effect of human activities on moisture content of soils and underlying permafrost from the McMurdo Sound region, Antarctica. *Antarctic Science*, 6, 307–314.

Cardozo, Y. and Hirsch, B. (1989) Antarctic tourism '89. *Sea Frontiers*, 35, 282–291.

Cassidy, W.A. (1991) Meteorites from Antarctica. In Tingey, R.J. (ed.) *The Geology of Antarctica*. Clarendon Press, Oxford, 653–666.

Cattle, H. (1991) Global climate models and Antarctic climatic change. In Harris, C.M. and Stonehouse, B. (eds) *Antarctica and Global Climatic Change*. Belhaven Press, London, 21–34.

Cattle, H. and Roberts, D.L. (1988) The performance of atmospheric general models in polar regions. *World Climate Programme and World Climate Research Programme. Sea Ice and Climate*. Report of the third session of the JSC Working Group on Sea Ice and Climate. Geneva, WMO, 1A–22A.

Cattle, H., Jenkins, G. and Vaughan, D. (1997) Sea level and Antarctic ice shelf disintegration. *Globe*, No. 36, UK Global Environmental Research Office, Swindon, 4.

Cattle, H., Murphy, J.M and Senior, C.A. (1992) The response of Antarctic climate in general circulation model experiments with transiently increasing carbon dioxide concentrations. *Philosophical Transactions of the Royal Society of London, B* 338, 209–218.

Chambers, M.J.G. (1966) Investigations of patterned ground at Signy Island, South Orkney Islands: II; Temperature regimes in the active layer. *British Antarctic Survey Bulletin*, 10, 71–83.

Chapman, D.G., Allen, K.R. and Holt, S.J. (1964) Reports of the Committee of Three Scientists on the special investigation of the Antarctic whale stocks. *Report of the International Commission on Whaling*, 14, 32–106.

Chapuis, J.L., Bousses, P. and Barnaud, G. (1994) Alien mammals, impact and management in the French subantarctic islands. *Biological Conservation*, 67, 97–104.

Charleson, R.L., Lovelock, J.E., Andreae, M.O., Warren, S.G. (1987) Oceanic phytoplankton, atmospheric sulfur, cloud albedo and climate. *Nature*, 326, 655–661.

Charney, J.I. (1982) Future strategies for an Antarctic mineral resource regime – can the environment be protected? In Charney, J.I. (ed.) *The New Nationalism and the Use of Common Spaces: Issues in Marine Pollution and the Exploitation of Antarctica*. Allanheld, Osman Publishers, Totowa, New Jersey, 206–238.

Chaturvedi, S. (1990) *Dawning of Antarctica. A Geopolitical Analysis*. Segment Books, New Dehli.

Chaturvedi, S. (1996) *The Polar Regions: A Political Geography*. John Wiley, Chichester.

Cherfas, J. (1989) *The Hunting of the Whale. A Tragedy That Must End*. Penguin, London.

Child, J. (1988) *Antarctica and South American Geopolitics: Frozen Lebensraum*. Praeger, New York.

Child, J. (1990) 'Latin *lebensraum*': the geopolitics of Ibero–American Antarctica. *Applied Geography*, 10, 287–305.

Chinn, T. J.H. (1990) The Dry Valleys. In Hatherton, T. (ed.) *Antarctica: the Ross Sea region*. DSIR Publishing, Department of Scientific and Industrial Research, Wellington, New Zealand, 137–154.

Chopra, S.K. (1988) Antarctica as a commons regime: a conceptual framework for cooperation and coexistence. In Joyner, C.C. and Chopra, S.K. (eds) *The Antarctic Legal Regime*. Martinus Nijhoff Publishers, Dordrecht, 163–186.

Chown, S.L. and Smith, V.R. (1993) Climate change and the short-term impact of feral house mice at the sub-Antarctic Prince Edward Islands. *Oecologia*, 96, 508–516.

Christie, E.W.H. (1951) *The Antarctic Problem*. Allen & Unwin, London.

Clapperton, C.M. (1971) Geomorphology of the Stromness Bay–Cumberland Bay Area, South Georgia. *British Antarctic Survey Scientific Reports*, No. 70.

Clapperton, C.M. and Sugden, D.E. (1982) Late Quaternary glacial history of George VI Sound area, West Antarctica. *Quaternary Research*, 18, 243–267.

Clapperton, C.M. and Sugden, D.E. (1983) Geomorphology of the Ablation Point Massif, Alexander Island, Antarctica. *Boreas*, 12, 125–135.

Clapperton, C.M., Sugden, D.E., Birnie, R.V. Hansom, J.D. and Thom, G. (1978) Glacier fluctuations in South Georgia and comparison with other island groups in the Scotia Sea. In van Zinderen Bakker, E.M. (ed.) *Antarctic Glacial History and World Palaeoenvironments*. A.A. Balkema, Rotterdam. 95–104.

Clapperton, C.M., Sugden, D.E., Birnie, J. and Wilson, M.J. (1989) Late-Glacial and Holocene glacier fluctuations and environmental change on South Georgia, Southern Ocean. *Quaternary Research*, 31, 210–228.

Claridge, G.G.C. and Campbell, I.B. (1985) Physical geography – soils. In Bonner, W.N. and Walton, D.W.H. (eds) *Key Environments: Antarctica*. Pergamon Press, Oxford, 62–70.

Claridge, G.G.C., Campbell, I.B., Powell, H.K.J., Amin, Z.H. and Balks, M.R. (1995) Heavy metal contamination in some soils of the McMurdo Sound region, Antarctica. *Antarctic Science*, 7, 9–14.

Clark, C.W. and Lamberson, R. (1982) An economic history and analysis of pelagic whaling. *Marine Policy*, 6, 103–120.

Clark, M.R. and Dingwall, P.R. (1985) *Conservation of Islands in the Southern Ocean. A Review of the Protected Areas of Insulantarctica*. IUCN, Gland.

Clark, P.U. (1995) Fast glacier flow over soft beds. *Science*, 267, 43–44.

Clarke, A. (1980) A reappraisal of the concept of metabolic cold adaptation in polar marine environments. *Biological Journal of the Linnean Society of London*, 14, 77–92.

Clarke, A. and Leakey, R.J.G. (1996) The seasonal cycle of phytoplankton, macronutrients, and the microbial community in a nearshore Antarctic marine ecosystem. *Limnology and Oceanography*, 41(6), 1281–1294.

Clarke, A. and Crame, J.A. (1992) The Southern Ocean benthic fauna and climate change: a historical perspective. *Philosophical Transactions of the Royal Society of London*, B 338, 299–309.

Clarke, A. and Law, R. (1981) Aliphatic and aromatic hydrocarbons in benthic invertebrates from two sites in Antarctica. *Marine Pollution Bulletin*, 12, 10–14.

Clarke, M.R. (1985) Marine habitats – Antarctic cephalopods. In Bonner, W.N. and Walton, D.W.H. (eds) *Key environments: Antarctica*. Pergamon Press, Oxford, 193–200.

Clayton, K.M. (1995) The threat of global warming. In O'Riordan T. (ed.) *Environmental Science for Environmental Management*. Longman Scientific & Technical, Harlow, 110–130.

Coates, D.A., Stricker, G.D. and Landis, E.R. (1990) Coal geology, coal quality and coal resources in Permian rocks of the Beacon Supergroup, Transantarctic Mountains, Antarctica. In Splettstoesser, J.F. and Dreschhoff, G.A.M. (eds) *Mineral Resources Potential of Antarctica*. American Geophysical Union, Antarctic Research Series 51, Washington DC 133–162.

Codling, R.J. (1982) Sea-borne tourism in the Antarctic: an evaluation. *Polar Record*, 21, 3–9.

Coles, P. (1991) Engineering feat or blight? *Nature*, 350, 301.

Colhoun, E.A. (1991) Geological evidence for changes in the East Antarctic ice sheet (60°–120° E) during the last glaciation. *Polar Record*, 27, 345–355.

Collins, N.J. and Callaghan, T.V. (1980) Predicted patterns of photosynthetic production in maritime Antarctic mosses. *Annals of Botany*, 45, 601–620.

Comiso, J.C. and Gordon, A.L. (1987) Recurring polynyas over the Cosmonaut Sea and the Maud Rise. *Journal of Geophysical Research*. 92, 2819–2833.

Conroy, J.W.H. (1975) Recent increases in penguin populations in the Antarctic and SubAntarctic. In Stonehouse, B. (ed.) *The Biology of Penguins*. London, Macmillan, 321–336.

Cooper, A.K., Davey, F.J. and Hinz, K. (1990) Geology and hydrocarbon potential of the Ross Sea, Antarctica. In St John, B. (ed.) *Antarctica as an Exploration Frontier – Hydrocarbon Potential, Geology, and Hazards*. American Association of Petroleum Geologists, AAPG Studies in Geology, 31, Tulsa, Oklahoma, 47–67.

Cooper, A.K., Davey, F.J., and Behrendt, J.C. (1987) Seismic stratigraphy and structure of the Victoria Land basin western Ross Sea, Antarctica. In Cooper, A.K. and Davey, F.J. (eds) *The Antarctic Continental Margin: Geology and Geophysics of the Western Ross Sea*. Earth Science Series, Vol. 5B, Circum–Pacific Council for Energy and Mineral Resources, Houston, Texas. 27–76.

Cooper, J. and Condy, P.R. (1988) Environmental conservation at the sub–Antarctic Prince Edward Islands: a review and recommendations. *Environmental Conservation*, 15, 317–326.

Cooper, J., Avenant, N.L. and Lafite, P.W. (1994) Air drops and king penguins: a potential conservation problem at sub-Antarctic Marion Island. *Polar Record*, 30, 277–282.

Craddock, C. (1990) The mineral resources of Gondwanaland. In Splettstoesser, J.F. and Dreschhoff, G.A.M. (eds) *Mineral Resources Potential of Antarctica*. American Geophysical Union, Antarctic Research Series 51, Washington DC, 1–6.

Craddock, C. (1982) Antarctica and Gondwanaland. In Craddock, C. (ed.) *Antarctic Geoscience*. University of Wisconsin Press, Madison, 3–13.

Crafford, J.E. (1990) The role of feral house mice in ecosystem functioning on Marion Island. In Kerry, K.R. and Hempel, G. (eds) *Antarctic Ecosystems. Ecological Change and Conservation*. Springer-Verlag, Berlin, 359–364.

Cripps, G.C. (1989) Problems in the identification of anthropogenic hydrocarbons against natural background levels in the Antarctic. *Antarctic Science*, 1, 307–312.

Cripps, G.C. (1990) Hydrocarbons in the seawater and pelagic organisms of the Southern Ocean. *Polar Biology*, 10, 393–402.

Cripps, G.C. (1992a) Baseline levels of hydrocarbons in seawater of the Southern Ocean. Natural variability and regional patterns. *Marine Pollution Bulletin*, 24, 109–114.

Cripps, G.C. (1992b) Natural and anthropogenic hydrocarbons in the Antarctic marine environment. *Marine Pollution Bulletin*, 25, 266–273.

Cripps, G.C. (1992c) The extent of hydrocarbon contamination in the marine environment from a research station in the Antarctic. *Marine Pollution Bulletin*, 25, 288–292.

Cripps, G.C. and Priddle, J. (1991) Hydrocarbons in the Antarctic marine environment. *Antarctic Science*, 3, 233–250.

Croasdale, K.R. (1986) Arctic offshore technology and its relevance to the Antarctic. In Polar Research Board, *Antarctic Treaty System. An Assessment*. National Academy Press, Washington DC, 245–263.

Crockett, R.N. and Clarkson, P.D. (1987) The exploitation of Antarctic minerals. *Environment International*, 13, 121–132.

Cross, M. (1991) Antarctica: exploration or exploitation. *New Scientist*, 130 (1774), 29–32.

Croxall, J.P. (1984) Seabirds. In Laws, R.M. (ed.) *Antarctic Ecology*, Vol. 2. Academic Press, London, 533–616.

Croxall, J.P. (1987) The status and conservation of Antarctic seals and seabirds: a review. *Environment International*, 13, 55–70.

Croxall, J.P. (1992) Southern Ocean environmental changes: effects on seabird, seal and whale populations. *Philosophical Transactions of the Royal Society of London*, B338, 319–328.

Croxall, J.P. (1997) Emperor ecology in the Antarctic winter. *Trends in Ecology and Evolution*, 12(9), 333–334.

Croxall, J.P. and Prince, P.A. (1979) Antarctic seabird and seal monitoring studies. *Polar Record*, 19, 573–595.

Croxall, J.P. Everson, I. and Miller, D.G.M. (1992) Management of the Antarctic krill fishery. *Polar Record*, 28, 64–66.

Croxall, J.P., McCann, T.S., Prince, P.A. and Rothery, P. (1988) Reproductive performance of seabirds and seals at South Georgia and Signy Island, South Orkney Islands, 1976–1987: implications for Southern Ocean monitoring studies. In Sahrhage, D. (ed.) *Antarctic Ocean and Resources Variability*. Springer–Verlag, Berlin, 261–285.

Croxall, J.P., Pickering, S.P.C. and Rothery, P. (1990c) Influence of the increasing fur seal population on wandering albatrosses *Diomedea exulans* breeding on Bird Island, South Georgia. In Kerry, K.R. and Hempel, G. (eds) *Antarctic Ecosystems: Ecological Change and Conservation*. Springer–Verlag, Berlin, 237–240.

Croxall, J.P., Prince, P.A., Hunter, I., McInnes, S.J. and Copestake, P. (1984) The seabirds of the Antarctic Peninsula, islands of the Scotia Sea, and Antarctic continent between 80°W and 20°W: their status and conservation. In Croxall, J.P., Evans, P.G.H. and Schreiber, R.W. (eds) *Status and Conservation of the World's Seabirds*. International Council for Bird Preservation, ICBP Technical Publication No. 2, Cambridge, 637–666.

Croxall, J.P., Prince, P.A. and Ricketts, C. (1985) Relationships between prey life-cycles and the extent, nature and timing of seal and seabird predation in the Scotia Sea. In Siegfried, W.R., Condy, P.R. and Laws, R.M. (eds) *Antarctic Nutrient Cycles and Food Webs*. Springer–Verlag, Berlin, 516–533.

Croxall, J.P., Rodwell, S.G. and Boyd, I.L. (1990b) Entanglement in man–made debris of Antarctic fur seals at Bird Island, South Georgia. *Marine Mammal Science*, 6, 221–233.

Croxall, J.P., Rootes, D.M. and Price, R.A. (1981) Increases in penguin populations at Signy Island, South Orkney Islands. *British Antarctic Survey Bulletin*, 54, 47–56.

Croxall, J.P., Rothery, P., Pickering, S.P.C. and Prince, P.A. (1990a) Reproductive performance, recruitment and survival of wandering albatrosses *Diomedea exulans* at Bird Island, South Georgia. *Journal of Animal Ecology*, 59, 775–796.

Culik, B., Adelung, D. and Woakes, A.J. (1990) The effect of disturbance on the heart rate and behaviour of Adélie penguins (*Pygoscelis adeliae*) during the breeding season. In Kerry, K.R. and Hempel, G. (eds) *Antarctic Ecosystems. Ecological Change and Conservation*. Springer-Verlag, Berlin, 177–182.

Culik, B. and Wilson, R. (1991) Penguins crowded out? *Nature*, 351, 340.

Culik, B.M., Wilson, R.P., Woakes A.T. and Sanudo, F.W. (1991) Oil pollution of Antarctic penguins: effects on energy metabolism and physiology. *Marine Pollution Bulletin*, 22, 388–391.

Cullen, J.J., Neale, P.J. and Lesser, M.P. (1992) Biological weighting function for the inhibition of phytoplankton photosynthesis by ultraviolet radiation. *Science*, 258, 646–650.

Cullen, R. (1994) Antarctic minerals and conservation. *Ecological Economics*, 10, 143–155.

Cumpston, J.S. (1968) Macquarie Island. *ANARE Scientific Reports Series* A(1), No. 93. Antarctic Division, Department of External Affairs, Melbourne.

Daly, K.L. (1990) Overwintering development, growth and feeding of larval *Euphausia superba* in the Antarctic marginal zone. *Limnology and Oceanography*, 35, 1564–1576.

Dalziel, I.W.D. (1983) The evolution of the Scotia arc: a review. In Oliver, R.L., James, P.R. and Jago, J.B. (eds) *Antarctic Earth Science*. Cambridge University Press, Cambridge, 283–288.

Dalziell, J. and de Poorter, M. (1993) Seabird mortality in longline fisheries around South Georgia. *Polar Record*, 29, 143–145.

Davenport, J. (1989) Feeding, oxygen uptake, ventilation rate and shell growth in the Antarctic protobranch bivalve mollusc *Yoldia eightsi* (Courthouy) In Heywood, R.B. (ed.) *University Research in Antarctica*. Proceedings of the British Antarctic Survey Special Topic Award Scheme Symposium, 9–10 Nov. 1988. British Antarctic Survey, Cambridge, 57–65.

Davey, F.J. (1985) The Antarctic margin and its possible hydrocarbon potential. *Tectonophysics*, 114, 443–470.

David, T.W.E. and Priestley, R.E. (1914) *British Antarctic Expedition 1907–9. Reports on the Scientific Investigations. Geology, Vol. 1. Glaciology, Physiography, Stratigraphy, and Tectonic Geology of South Victoria Land*. Heinemann, London.

Davidson, A.T. and Marchant, H.J. (1994) The impact of ultraviolet radiation on *Phaeocystis* and selected species of Antarctic marine diatoms. In Weiler, C.S. and Penhale, P.A. (eds) *Ultraviolet Radiation in Antarctica: Measurements and Biological Effects*. American Geophysical Union, Antarctic Research Series 62, Washington DC, 187–205.

Davidson, A.T., Bramich, D., Marchant, H.J. and McMinn, A. (1994) Effects of UV-B irradiation on growth and survival of Antarctic marine diatoms. *Marine Biology*, 119, 507–515.

Davis, B. (1990) Science and politics in Antarctic and Southern Oceans policy: a critical assessment. In Herr, R.A., Hall, H.R. and Howard, M.G. (eds) *Antarctica's Future: Continuity or Change?* Australian Institute of International Affairs, Hobart, 39–45.

Davis, B.W. (1992) Focussing an Antarctic research programme: the Australian experience. *Polar Record*, 28, 51–56.

Davis, P.B. (1995) Antarctic visitor behaviour: are guidelines enough? *Polar Record*, 31, 327–334.

Davis, R.A., (1981) Structure and function of two Antarctic terrestrial moss communities. *Ecological Monographs*, 51, 149–168.

de Baar, H.J.W., Buma, A.G.J., Nolting, R.F., Cadee, G. C., Jacques, G. and Treguer, P.J. (1990) On iron limitation of the Southern Ocean: experimental observations in the Weddell and Scotia Seas. *Marine Ecology Progress Series*, 65, 105–122.

de Baar, H.J.W., de Jong, J.T.M., Bakker, D.C.E., Loscher, B.M. and Smetachek, V. (1995) Importance of iron for plankton blooms and carbon dioxide drawdown in the Southern Ocean. *Nature*, 373, 412–415.

de. Broyer, C. (1977) Analysis of gigantism and dwarfness of Antarctic and sub–Antarctic Gammaridean amphipoda. In Llano, G.A. (ed.) *Adaptations within Antarctic Ecosytems*. Proceedings of the 3rd SCAR Symposium on Antarctic Biology. Smithsonien Institute, Washington DC 327–334.

de la Mare, W. (1990) Problems of 'scientific' whaling. *Nature*, 345, 771.

de la Mare, W.K. and Kerry, K.R. (1994) Population dynamics of the wandering albatross (*Diomedea exulans*) on Macquarie Island and the effects of mortality from longline fishing. *Polar Biology*, 14, 231–241.

de la Mare, W. (1997) Abrupt mid-twentieth-century decline in Antarctic sea-ice extent from whaling records. *Nature*, 389, 57–60.

de Mora, S.J., Patterson, J.E. and Bibby, D.M. (1993) Baseline atmospheric mercury studies at Ross Island, Antarctica. *Antarctic Science*, 5, 323–326.

dc Poorter, M. and Dalziell, J.C. (1997) *Cumulative Environmental Impacts in Antarctica. Minimisation and Management*. Proceedings of the IUCN Workshop on cumulative impacts in Antarctica, Washington DC, 18–21 September, 1996. IUCN, Gland.

de Wit, M.J. (1985) *Minerals and Mining in Antarctica*. Clarendon Press, Oxford.

de Wit, M.J. and Kruger, F.J. (1990) The economic potential of the Dufek complex. In Splettstoesser, J.F. and Dreschhoff, G.A.M. (eds) *Mineral Resources Potential of Antarctica*. American Geophysical Union, Antarctic Research Series 51, Washington DC, 33–52.

de Wit, M.J., Dutch, S., Kligfield, R., Allen, R. and Stern, C. (1977) Deformation, serpentization and emplacement of a dunite complex, Gibbs Island, South Shetland Islands – possible fracture zone tectonics. *Journal of Geology*, 85, 145–162.

Deacon, G. (1984) *The Antarctic Circumpolar Ocean*. Cambridge University Press, Cambridge.

Delille, D. and Vaillant, N. (1990) The influence of crude oil on the growth of subantarctic marine bacteria. *Antarctic Science*, 2, 123–127.

Denton, G.H. (1979) Glacial history of the Byrd–Darwin Glacier area, Transantarctic Mountains. *Antarctic Journal of the United States*, XIV, 57–58.

Denton, G.H. and Hughes, T.J. (1983) Milankovitch theory of ice ages: hypothesis of ice sheet linkage between regional insolation and global climate. *Quaternary Research*, 20, 125–144.

Denton, G.H., Prentice, M.L. and Burckle, L.H. (1991) Cainozoic history of the Antarctic ice sheet. In Tingey, R.J. (ed.) *The Geology of Antarctica*. Clarendon Press, Oxford, 365–433.

Department of the Arts, Sport, the Environment and Territories (1992) *Impact of Climate Change on Antarctica–Australia*. Australian Government Publishing Service, Canberra.

Dickinson, A.B. (1985) South Georgia fisheries: some early records. *Polar Record*, 22, 434–437.

Dickinson, A.B. (1989) The demise of elephant sealing at South Georgia, 1960–68. *Polar Record*, 25, 185–190.

Dickinson, A.B. (1993) Some aspects of Japanese whaling and sealing in South Georgia. In Basberg, B.L., Ringstad, J.E. and Wexelsen, E. (eds) *Whaling and History. Perspectives on the Evolution of the Industry*. Kommandør Chr. Christensens Hvalfangstmuseum Publikasjon 29, Sandefjord, 169–176.

Dingwall, P.R. (1995) *Progress in Conservation of the Subantarctic Islands*. Conservation of the Southern Polar Region No. 2. IUCN, Gland and Cambridge,.

Dingwall, P.R. and Lucas, B. (1992) New Zealand and Antarctica. In *Regional Reviews*. IVth World Congress on National Parks and Protected Areas, Caracas, Venezuela, 10–21 February 1992. IUCN, Gland, 10.1–10.16.

Doake, C.S.M. (1987) Antarctic Ice and Rocks. In Walton, D.W.H. (ed.) *Antarctic Science*. Cambridge University Press, Cambridge, 140–192.

Doake, C.S.M. and Vaughan, D.G. (1991) Rapid disintegration of the Wordie Ice Shelf in response to atmospheric warming. *Nature*, 350, 328–330.

Dodds, K.J. (1994) Geopolitics in the Foreign Office: British representations of Argentina 1945–1961. *Transactions of the Institute of British Geographers*, NS19, 273–290.

Dodds, K.J. (1997) Antarctica and the modern geographical imagination (1918–1960) *Polar Record*, 33, 47–62.

Dodge, C.W. (1977) *Lichen Flora of the Antarctic Continent and Adjacent Islands*. Phoenix Publishing, Canaan, New Hampshire.

Doidge, D.W. and Croxall, J.P. (1985) Diet and energy budget of the Antarctic fur seal, *Arctocephalus gazella*, at South Georgia. In Seigfield, W.R., Condy, P.R. and Laws, R.M. (eds) *Antarctic Nutrient Cycles and Food Webs*. Springer-Verlag, Berlin, 543–550.

Domack, E.W., Jull, A.J. and Nakao, S. (1991) Advance of East Antarctic outlet glaciers during the Hypsithermal: implications for the volume state of the Antarctic ice sheet under global warming. *Geology*, 19, 1059–1062.

Draggan, S. (1992) NSF announces new requirements for waste management and environmental assessment in Antarctica. *Antarctic Journal of the United States*, 27(4), 5–7.

Draggan, S. and Wilkniss, P. (1992) An operating philosophy for the US Antarctic Program. *Marine Pollution Bulletin*, 25, 250–252.

Drake, F. 1995. Stratospheric ozone depletion – an overview of the scientific debate. *Progress in Physical Geography*, 19, 1–17.

Drewry, D.J. (1983) The surface of the Antarctic ice sheet: sheet 2. In Drewry, D.J. (ed.) *Antarctic glaciological and geophysical folio*. Scott Polar Research Institute, Cambridge.

Drewry, D.J. (1988) The challenge of Antarctic science. *Oceanus*, 31(2), 5–10.

Drewry, D.J. (1989) Science in Antarctica – a matter of quality. *Antarctic Science*, 1, 2.

Drewry, D.J. (1991) The response of the Antarctic ice sheet to climatic change. In Harris C.M. and Stonehouse, B. (eds) *Antarctica and Global Climatic Change*. Belhaven Press, London, 90–106.

Drewry, D.J. (1993) The future of Antarctic scientific research. *Polar Record*, 29, 37–44.

Drewry, D.J. (1994) Conflicts of interest in the use of Antarctica. In Hempel, G. (ed.) *Antarctic Science. Global Concerns*. Springer-Verlag, Berlin, 12–30.

Drewry, D.J. and Morris, E.M. (1992) The response of large ice sheets to climatic change. *Philosophical Transactions of the Royal Society of London*, B338, 235–242.

Drewry, D.J., Laws, R.M. and Pyle, J.A. (1992) Antarctica and Environmental Change. *Philosophical*

Transactions of the Royal Society of London, B338(1285), 199–334.

Dudeney, J.R. (1987) The Antarctic Atmosphere. In Walton, D.W.H. (ed.) *Antarctic Science*. Cambridge University Press, Cambridge, 191–247.

Duhamel, G. and Hureau, J.-C. (1990) Changes in fish populations and fisheries around the Kerguelen Islands during the last decade. In Kerry, K.R. and Hempel, G. (eds) *Antarctic Ecosystems: Ecological Change and Conservation*. Springer-Verlag, Berlin, 323–333.

Dunbar, M.J. (1973) Stability and fragility in arctic ecosystems. *Arctic*, 26, 179–185.

Dutkiewicz, L. (1982) Preliminary results of investigations on some periglacial phenomena on King George Island, South Shetland Islands. *Biuletyn Peryglacjalny*, 29, 13–23.

Eastman, J.T. and de Vries, A.L. (1986) Antarctic fishes. *Scientific American*, 254, 106–114.

Edwards, D.M. and Heap, J.A. (1981) Convention on the Conservation of Antarctic Marine Living Resources: a commentary. *Polar Record*, 20, 353–362.

Efremenko, V.N. (1983) Atlas of the fish larvae of the Southern Ocean. *Cybium*, 7, 1–75.

Eklund, C.R. and Beckman, J. (1963) *Antarctica. Polar Research and Discovery during the International Geophysical Year*. Rinehart & Winston Inc., New York.

Elkins, J.W., Thompson, T.M., Swanson, T.H., Butler, J.H., Hall, B.D., Cummings, S.O., Fisher, D.A. and Raffo, A.G. (1993) Decrease in the growth rates of atmospheric chlorofluorocarbons 11 and 12. *Nature*, 364, 780–783.

Elliot, D.H. (1975) Tectonics of Antarctica: a review. *American Journal of Science*, A275, 45–106.

Elliot, D.H. (1985) Physical geography-geological evolution. In Bonner, W.N. and Walton, D.W.H. (eds) *Key Environments: Antarctica*. Pergamon Press, Oxford, 39–61.

Elliot, D.H. (1988) Antarctica: is there any oil and natural gas? *Oceanus*, 31(2), 32–38.

Elliot, D.H. (1991) Triassic–early Cretaceous evolution of Antarctica. In Thomson, M.R.A., Crame, J.A. and Thomson, J.W. (eds.) *Geological Evolution of Antarctica*. Cambridge University Press, Cambridge, 541–548.

Elliot, D.H., Fleck, R.J. and Sutter, J.F. (1985) Potassium–argon age determinations of Ferrar Group rocks, central Transantarctic Mountains. In Turner, M.D. and Splettstoesser, J.F. (eds) *Geology of the Central Transantarctic Mountains*, American Geophysical Union, Antarctic Research Series 36, Washington DC, 197–224.

Elliott, L.M. (1994) *International Environmental Politics. Protecting the Antarctic*. Macmillan, Basingstoke.

Ellis, J. (1991) Antarctica and global climatic change: review of prominent issues. In Harris, C.M. and Stonehouse, B. (eds) *Antarctica and Global Climatic Change*. Belhaven Press, London, 11–20.

Ellis–Evans, J.C. (1996) Microbial diversity and function in Antarctic freshwater ecosystems. *Biodiversity and Conservation*, 5, 1395–1431.

Ellis–Evans, J.C. and Wynn-Williams, D. (1996) A great lake under the ice. *Nature*, 381, 644–46.

El–Sayed, S.Z. (1965) Primary productivity of the Antarctic and SubAntarctic. In Schmidt, W. and Llano, G. (eds.) *Biology of the Antarctic Seas II*. American Geophysical Union, Antarctic Research Series 5, Washington DC, 15–47.

El–Sayed, S.Z. (1978) Primary productivity and estimates of potential yields of the Southern Ocean. In McWhinnie, M.A. (ed.) *Polar Research: to the Present and Future*. Westview Press, Boulder, Colorado, 141–160.

El–Sayed, S.Z. (1985) Plankton of the Antarctic seas. In Bonner, W.N. and Walton, D.W.H. (eds) *Key Environments. Antarctica*. Pergamon Press, Oxford, 135–153.

El–Sayed, S.Z. (1994) *Southern Ocean Ecology: the BIOMASS Perspective*. Cambridge University Press, Cambridge.

El–Sayed, S.Z. and McWhinnie, M.A. (1979) Antarctic krill: protein of the last frontier. *Oceanus*, 22(1), 13–20.

El–Sayed, S.Z, Stephens, F.C., Bidigare, R.R. and Ondrusek, M.E. (1990) Effect of ultraviolet radiation on Antarctic marine phytoplankton. In Kerry, K.R. and Hempel, G. (eds) *Antarctic Ecosystems. Ecological Change and Conservation*. Springer–Verlag, Berlin, 379–385.

Elsom, D. (1992) *Atmospheric Pollution. A Global Problem*, 2nd edn. Blackwell, Oxford.

Elzinga, A. (1993a) Antarctica: the construction of a continent by and for science. In Crawford, E., Shinn, T. and Sörlin, S. (eds) *Denationalizing Science: The Contexts of International Scientific Practice*. Kluwer Academic Publishers, Dordrecht, 73–106.

Elzinga, A. (1993b) *Changing Trends in Antarctic Research*. Kluwer Academic Publishers, Dordrecht.

Elzinga, A. and Bohlin, I. (1989) The politics of science in polar regions. *Ambio*, 18, 71–76.

Engelskjon, T. (1981) Terrestrial vegetation of Bouvetøya: a preliminary account. *Norsk Polarinstitutt Skrifter*, 175, 17–28.

Englehardt, H., Humphrey, N., Kamb, B. and Fahnestock, M. (1990) Physical conditions at the base of a fast moving Antarctic Ice Stream. *Science*, 248, 57–59.

Engler, M., Guichard, A., Le Tavernier, Y. and Regrettier, J-F. (1990) The Dumont d'Urville aerodrome, Terre Adélie, Antarctica. *Cold Regions Science and Technology*, 18, 191–213.

Environmental Policy and Law (1988) Antarctica. Minerals Convention adopted. *Environmental Policy and Law*, 18, 100–102.

Enzenbacher, D.J. (1992a) Tourists in Antarctica: numbers and trends. *Polar Record*, 28, 17–22.

Enzenbacher, D.J. (1992b) Antarctic tourism and environmental concerns. *Marine Pollution Bulletin*, 25, 258–265.

Enzenbacher, D.J. (1993) Tourists in Antarctica: numbers and trends. *Tourism Management*, 14, 142–146.

Enzenbacher, D.J. (1994a) Antarctic tourism: an overview of 1992/93 season activity, recent developments, and emerging issues. *Polar Record*, 30, 105–116.

Enzenbacher, D.J. (1994b) Tourism at Faraday Station. An Antarctic case study. *Annals of Tourism Research*, 21, 303–317.

Enzenbacher, D.J. (1995) The regulation of Antarctic tourism. In Hall, C.M. and Johnston, M.E. (eds) *Polar Tourism: Tourism in the Arctic and Antarctic Regions*. Chichester, Wiley, 179–215.

Eppley, Z.A. (1992) Assessing indirect effects of oil in the presence of natural variation: the problem of reproductive failure in South Polar skuas during the *Bahia Paraiso* oil spill. *Marine Pollution Bulletin*, 25, 307–312.

Eppley, Z.A. and Rubega, M.A. (1989) Indirect effects of an oil spill. *Nature*, 340, 513.

Eppley, Z.A. and Rubega, M.A. (1990) Indirect effects of an oil spill: reproductive failure in a population of South Polar skuas following the *Bahia Paraiso* oil spill in Antarctica. *Marine Ecology Progress Series*, 67, 1–6.

Erickson, A.W. and Hanson, M.B. (1990) Continental estimates of population trends of Antarctic ice seals. In Kerry, K.R. and Hempel, G. (eds) *Antarctic Ecosystems, Ecological Change and Conservation*. Springer-Verlag, Berlin, 253–264.

Evans, A.M. (1993) *Ore Geology and Industrial Minerals*. Blackwell, Oxford.

Evans, A.M. (1995) *Introduction to Mineral Exploration*. Blackwell, Oxford.

Everson, I. (1977) *The Living Resources of the Southern Ocean*. FAO Report GLO/SO/77/1, Rome.

Everson, I. (1978) Antarctic fisheries. *Polar Record*, 19, 233–251.

Everson, I. (1984) Marine interactions. In Laws, R.M. (ed.) *Antarctic Ecology*, Vol 2. Academic Press, London, 783–819.

Everson, I. (1987) Life in a cold environment. In Walton, D.W.H. (ed.) *Antarctic Science*, Cambridge University Press, Cambridge, 125–137.

Everson, I. (1988) Can we satisfactorily estimate variation in krill abundance? In Sahrhage, D. (ed.) *Antarctic Ocean and Resources Variability*. Springer-Verlag, Berlin, 199–208.

Everson, I. (1992) Managing Southern Ocean krill and fish stocks in a changing environment. *Philosophical Transactions of the Royal Society of London*, B338, 311–317.

Everson, I. and Goss, C. (1991) Krill fishing activity in the southwest Atlantic. *Antarctic Science*, 3, 351–358.

Everson, I., Neyelov, A. and Permitin, Y.E. (1992) Bycatch of fish in the South Atlantic krill fishery. *Antarctic Science*, 4, 389–392.

Falk, R. (1991) The Antarctic Treaty System: are there viable alternatives? In Jørgensen-Dahl, A. and Østreng, W. (eds) *The Antarctic Treaty System in World Politics*. Macmillan, Basingstoke, 399–414.

FAO (1995) *The State of World Fisheries and Aquaculture*. Food and Agriculture Organisation of the United Nations, Rome.

Farman, J.C., Gardiner, B.G. and Shanklin, J.D. (1985) Large losses of total ozone in Antarctica reveal seasonal ClO_x/NO_x interaction. *Nature*, 315, 207–210.

Ferrigno, J.E. (1993) Velocity measurements and changes in position of Thwaites Glacier/iceberg tongue from aerial photography, landsat image and NOAA AVHRR data. *Annals of Glaciology*. Vol 17, 239–44.

Ferris, J., Johnson, A and Storey, B. (1988), Form and extent of the Dufek Intrusion, Antarctica, from newly compiled aeromagnetic data. *Earth and Planetary Science Letters*, 154, 185–202.

Fifield, R. (1987) *International Research in the Antarctic*. Oxford University Press, Oxford.

Filson, R.B. (1966) The lichens and mosses of MacRobertson Land. *ANARE Scientific Report Series B(II)*, Botany, 82.

Finkel, R.C., Langway Jr, C.C. and Clausen, H.B. (1986) Changes in precipitation chemistry at Dye 3, Greenland. *Journal of Geophysical Research*, 91, 9849–9855.

Flegal, A.R., Maring, H. and Niemeyer, S. (1993) Anthropogenic lead in Antarctic sea water. *Nature*, 365, 242–244.

Focardi, S. Gaggi, C., Chemello, G. and Bacci, E. (1991) Organochlorine residues in moss and lichen samples from two Antarctic areas. *Polar Record*, 27, 241–244.

Focardi, S., Lari, L. and Marsili, L. (1992) PCB congeners, DDTs and hexachlorobenzene in Antarctic fish from Terra Nova Bay (Ross Sea) *Antarctic Science*, 4, 151–154.

Fogg, G.E. (1977) Aquatic primary production in the Antarctic. *Philosophical Transactions of the Royal Society of London*, B279, 27–38.

Fogg, G.E. (1992) *A History of Antarctic Science*. Cambridge University Press, Cambridge.

Forbes, L.M. (ed.) (1967) *Antarctic*. Annals of the International Geophysical Year 44, Pergamon Press, London.

Forbes, D.L. and Taylor, R.B. (1994) Ice in the shore zone and the geomorphology of cold coasts. *Progress in Physical Geography*, 18, 59–89.

Ford, A.B. (1990) The Dufek intrusion of Antarctica. In Splettstoesser, J.F. and Dreschhoff, G.A.M. (eds) *Mineral Resources Potential of Antarctica*. American Geophysical Union, Antarctic Research Series 51, Washington DC, 15–32.

Ford, A.B. and Himmelberg, G.R. (1991) Geology and crystallisation of the Dufek intrusion. In Tingey, R.J. (ed.) *The Geology of Antarctica*. Clarendon Press, Oxford, 175–214.

Foreign and Commonwealth Office (1997) *List of Protected Areas in Antarctica*. Foreign and Commonwealth Office, London.

Foreman, E.F. (1992) Protecting the Antarctic environment: will a protocol be enough? *American University Journal of International Law and Policy*, 7, 843–879.

Fowbert, J.A. and Smith, R.I.L., (1994) Rapid population increases in native vascular plants in the Argentine Islands, Antarctic Peninsula. *Arctic and Alpine Research*, 26(3), 290–296.

Frakes, L.A. and Moreton, D.L.E. (1990) Manganese nodule provinces of the Southern Ocean. In Splettstoesser, J.F. and Dreschhoff, G.A.M. (eds) *Mineral Resources of Antarctica*. American Geophysical Union, Antarctic Research Series 51, Washington DC, 217–221.

Francioni, F. (1987) Antarctica and the common heritage of mankind. In Francioni, F. and Scovazzi, T. (eds) *International Law for Antarctica*. Giuffrè Editore, Milan, 101–136.

Francioni, F. (1993) The Madrid Protocol on the Protection of the Antarctic Environment. *Texas International Law Journal*, 28, 47–72.

Francioni, F. and Scovazzi, T. (1996) *International Law for Antarctica*. 2nd edn. Kluwer Law International, The Hague.

Francis, J.E. (1991) Paleoclimatic significance of Cretaceous–early Tertiary fossil forests of the Antarctic Peninsula. In Thomson, M.R.A., Crame, J.A. and Thomson, J.W. (eds) *Geological Evolution of Antarctica*. Cambridge University Press, Cambridge, 623–628.

Frank, R.F. (1983) The Convention on the Conservation of Antarctic Marine Living Resources. *Ocean Development and International Law Journal*, 13, 291–345.

Fraser, W.R., Trivelpiece, W.Z., Ainley, D.G. and Trivelpiece, S.G. (1992) Increases in Antarctic penguin populations: reduced competition with whales or a loss of sea ice due to environmental warming? *Polar Biology*, 11, 525–531.

Freckman, D.W. and Virginia, R.A. (1997) Low-diversity Antarctic soil nematode communities: disturbance and response to disturbance. *Ecology*, 78(2), 363–369.

Frederick, J.E. and Alberts, A.D. (1991) Prolonged enhancement in surface ultraviolet radiation during the Antarctic spring of 1990. *Geophysical Research Letters*, 18, 1869–1871.

Frederick, J.E., Diaz, S.B., Smolskaia, I., Esposito, W., Lucas, T. and Booth, C.R. (1994) Ultraviolet solar radiation in the high latitudes of South America. *Photochemistry and Photobiology*, 60, 356–362.

Frederick, J.E. and Snell, H.E. (1988) Ultraviolet radiation levels during the Antarctic spring. *Science*, 241, 438–440.

Frenot, Y., van Vliet–Lanoe, B. and Gloaguen, J.-C. (1995) Particle translocation and initial soil development on a glacier foreland, Kerguelen Islands, Subantarctic. *Arctic and Alpine Research*, 27(2), 107–115.

Friedman, E.I. (1982) Endolithic micro-organisms in the Antarctic cold desert. *Science*, 215, 1045–1053.

Frost, S. (1979) *The Whaling Question*. Friends of the Earth, San Francisco.

Fuchs, V.E. (1982) *Of Ice and Men. The Story of the British Antarctic Survey 1943–73*. Anthony Nelson, Oswestry.

Gambell, R. (1977) Whale conservation. Role of the International Whaling Commission. *Marine Policy*, 1, 301–310.

Gambell, R. (1985) Birds and mammals – Antarctic whales. In Bonner, W.N. and Walton, D.W.H. (eds) *Key Environments. Antarctica*. Pergamon Press, Oxford, 223–241.

Gambell, R. (1990) The International Whaling Commission – quo vadis? *Mammal Review*, 20, 31–43.

Gardam, J.G. (1985) Management regimes for Antarctic marine living resources – an Australian perspective. *Melbourne University Law Review*, 15, 279–312.

Gardiner, B., Jones, A., Roscoe, H., Shanklin, J. and Wynn-Williams, D. (1997) The ozone hole and life beneath it. *Globe*, No. 36, UK Global Environmental Research Office, Swindon, 5–6.

Gardner, H., Knowles, K., Riddle, M., Brouwer, S. and Gleeson, L. (1997) Poultry virus infection in Antarctic penguins. *Nature*, 387, 245.

Garrett, J.N. (1984) The economics of Antarctic oil. In Alexander, L.M. and Hanson, L.C. (eds) *Antarctic Politics and Marine Resources: Critical Choices for the 1980s*. Center for Ocean Management Studies, Kingston, Rhode Island, 185–190.

Gee, C.T. (1989) Permian *Glossopteris* and *Elatocladus* megafossil floras from the English Coast, eastern Ellsworth Land, Antarctica. *Antarctic Science*, 1, 35–44.

Giese, M. (1996) Effects of human activity on Adélie penguin *Pygoscelis adeliae* breeding success. *Biological Conservation*, 75, 157–164.

Giovinetto, M.B. and Bentley, C.R. (1985) Surface balance in drainage systems of Antarctica. *Antarctic Journal of the United States*, 20(4), 6–13.

Giovinetto, M.B., Bentley, C.R. and Bull, C.B. (1989). Choosing between some incompatible regional surface mass balance data sets in Antarctica. *Antarctic Journal of the United States*, 20(1), 7–13.

Gleadow, A.J.W. and Fitzgerald, P.G. (1987) Uplift history and structure of the Transantarctic Mountains: new evidence from fission track dating of basement apatites in the Dry Valleys area, southern Victoria Land. *Earth and Planetary Science Letters*, 822, 1–14.

Gloerson, P. and Campbell, W.J. (1988) Variations in the Arctic, Antarctic and global sea ice covers during 1978–1987 as observed with the Nimbus 7 Scanning Multichannel Microwave Radiometer. *Journal of Geophysical Research*, 93, 10660–10674.

Gloerson, P. and Campbell, W.J. (1991) Recent variations in Arctic and Antarctic sea ice covers. *Nature*, 353, 33–36.

Gloerson, P., Campbell, W.J., Cavalieri, D.J., Comiso, J.C., Parkinson, C.L. and Zwally, H.J. (1992) *Arctic and Antarctic Sea Ice, 1978–1987; Satellite passive–microwave observations and analysis.* (NASA SP–511) Scientific and Technical Information Program, NASA, Washington DC.

Golosov, V.V. (1984) Some biocenological aspects of the exploration of the resources of Antarctic krill. *Polar Geography and Geology*, 8, 63–72.

Goodell, H.G. (1973) The sediments. In *Marine Sediments of the Southern Ocean*. American Geographical Society Antarctic Map Folio Series, 17, 1–9.

Goody, R. (1980) Polar processes and world climate (a brief overview). *Monthly Weather Review*, 108, 1935–1942.

Gordon, A.L. (1988) Spatial and temporal variability within the Southern Ocean. In Sahrhage, D. (ed.) *Antarctic Ocean and Resources Variability*. Springer-Verlag, Berlin, 41–56.

Gordon, J.E. (1987) Radiocarbon dates from Nordenskjöld Glacier, South Georgia, and their-implications for late Holocene glacier chronology. *British Antarctic Survey Bulletin*, 6, 1–5.

Gordon, J.E. and Birnie, R.V. (1986) Production and transfer of subaerially generated rock debris and resulting landforms on South Georgia: an introductory perspective. *British Antarctic Survey Bulletin*, 72, 25–46.

Gordon, J.E. and Hansom, J.D. (1986) Beach forms and changes associated with retreating glacier ice, South Georgia. *Geografiska Annaler*, 68A, 15–24.

Gordon, J.E. and Timmis, R.J. (1992) Glacier fluctuations on South Georgia during the 1970s and early 1980s. *Antarctic Science*, 4, 215–226.

Gordon, J.E., Birnie, R.V. and Timmis, R. (1978) A major rockfall and debris slide on the Lyell Glacier, South Georgia. *Arctic and Alpine Research*, 10, 49–60.

Gore, D. (1990) Examples of ice damming of lakes in the Vestfold Hills, East Antarctica, with implications for landscape development. In Gillieson, D. and Fitzsimmons, S. (eds) *Quaternary Research in Australian Antarctica*. Australian Defence Force Academy, Special Publication No. 3, 37–44.

Görlach, U. and Boutron, C.F. (1992) Variations in heavy metals concentrations in Antarctic snows from 1940 to 1980. *Journal of Chemistry*, 14, 205–222.

Gould, L.M. (1978) The emergence of Antarctica. In McWhinnie, M.A. (ed.) *Polar Research. To the Present and the Future*. Westview Press, Boulder, Colorado, 9–26.

Gow, A.J., Ackley, S.F., Weeks, W.F. and Govoni, J.W. (1982) Physical and structural characteristics of Antarctic sea ice. *Annals of Glaciology*, 3, 113–117.

Grantham, G.J. (1977) *The Utilisation of Krill*. (Report GLO/SO/77/3) FAO, Rome.

Green, G., Skerratt, J.H., Leeming, R. and Nichols, P.D. (1992) Hydrocarbon and coprostanol levels in seawater, sea–ice algae and sediments near Davis Station in Eastern Antarctica: a regional survey and preliminary results for a field fuel spill experiment. *Marine Pollution Bulletin*, 25, 293–302.

Greenpeace (1988) *Antarctic Expedition Report 1987/88*. Greenpeace International, Amsterdam.

Greenpeace (1990) *Antarctic Expedition Report 1989/90*. Greenpeace International, Amsterdam.

Greenpeace (1991) *Antarctic Expedition Report 1990/91*. Greenpeace International, Amsterdam.

Greenpeace (1992) *Antarctic Expedition Report. 1991/92*. Greenpeace International, Amsterdam. 1992.

Greenpeace. (1993) *State of the Ice: An Overview of Human Impacts in Antarctica*. Greenpeace International, Amsterdam.

Greenpeace (1994) *Antarctic Expedition Report 1992/93*. Greenpeace International, Amsterdam.

Gregory, M.R. (1982) Ross Sea hydrocarbon prospects and the IXTOC 1 oil blowout, Campeche Bay, Gulf of Mexico: comparisons and lessons. *New Zealand Antarctic Record*, 4(2), 40–45.

Gregory, M.R. (1987) Plastics and other sea borne litter on the shores of New Zealand's sub-Antarctic islands. *New Zealand Antarctic Record*, 7 (3), 32–47.

Gregory, M.R., Kirk, R.M. and Mabin, M.C.G. (1984a) Shore types of Victoria Land, Ross Dependency, Antarctica. *New Zealand Antarctic Record*, 5(3), 22–40.

Gregory, M.R. Kirk, R.M. and Mabin, M.C.G. (1984b) Pelagic tar, oil plastics and other litter in surface waters of the New Zealand sector of the Southern Ocean, and on Ross Dependency shores. *New Zealand Antarctic Record*, 6(1), 12–28.

Gressitt, J.L. (1970) Subantarctic entomology and biogeography. *Pacific Insects Monographs*, 23, 295–374.

Griffith, T.W. and Anderson, J.B. (1989) Climatic control of sedimentation in bays and fjords of the northern Antarctic Peninsula. *Marine Geology*, 85, 181–204.

Grobelaar, J.U. (1974) Primary production in freshwater bodies of the subantarctic Marion Island. *South African Journal of Antarctic Research*, 4, 40–45.

Gross, M.G. (1977) *Oceanography. A View of the Earth*. Prentice–Hall, New Jersey.

Guinet, C., Jouventin, P. and Georges, J.-Y. 1994. Long term changes of fur seals *Arctocephalus gazella* and *Arctocephalus tropicalis* on subantarctic (Crozet) and subtropical (St Paul and Amsterdam) islands and their possible relationship to El Niño Southern Oscillation. *Antarctic Science*, 6, 473–478.

Guinet, C., Jouventin, P. and Weimerskirch, H. (1992) Population changes, movements of southern elephant seals on Crozet and Kerguelen Archipelagos in the last decades. *Polar Biology*, 12, 349–356.

Gulland, J.A. (1970) The development of the resources of the Antarctic seas. In Holdgate, M.W. (ed.) *Antarctic Ecology*, Vol. 1. Academic Press, London, 217–223.

Gullańd, J.A. (1974) Antarctic whaling. In Gulland, J.A. (ed.) *The Management of Marine Fisheries*. Scientechnia (Publishers) Ltd., Bristol, 10–37.

Gulland, J.A. (1976) Antarctic baleen whales: history and prospects. *Polar Record*, 18, 5–13.

Gulland, J.A. (1987) The Antarctic Treaty System as a resource management mechanism. In Triggs, G.D. (ed.) *The Antarctic Treaty Regime. Law, Environment and Resources*. Cambridge University Press, Cambridge, 116–127.

Gulland, J. (1988a) The management regime for living resources. In Joyner, C.C. and Chopra, S.K. (eds) *The Antarctic Legal System*. Martinus Nijhoff Publishers, Dordrecht, 219–240.

Gulland, J. (1988b) The end of whaling? *New Scientist*, 120(1636), 42–47.

Gulland, J.A. (1990) Commercial whaling – the past, and has it a future? *Mammal Review*, 20, 3–12.

Häder, D.-P. and Worrest, R.C. (1991) Effects of enhanced solar ultraviolet radiation on aquatic ecosystems. *Photochemistry and Photobiology*, 53, 717–725.

Häder, D.-P., Worrest, R.C., Kumar, H.D. and Smith, R.C. (1995) Effects of increased ultraviolet radiation on aquatic ecosystems. *Ambio*, 24, 174–180.

Hall, C.M. and McArthur, S. (1993) Ecotourism in Antarctica and adjacent sub-Antarctic islands: development, impacts, management and prospects for the future. *Tourism Management*, 14, 117–122.

Hall, C.M. and Wouters, M. (1994) Managing tourism in the sub-Antarctic. *Annals of Tourism Research*, 21, 355–374.

Hall, K. and Walton, D.W.H. (1992) Rock weathering, soil development and colonisation under a changing climate. *Philosophical Transactions of the Royal Society of London*, B338, 269.

Hall, K., Curtis, A. and Morewood, C. (1989) Antarctic rock weathering simulations: simulator design, application and use. *Antarctic Science*, 1, 45–50.

Hallam, A. (1976) Continental drift and the fossil record. In Wilson, J.T. (ed.), *Continents Adrift and Continents Aground*. Scientific American, San Fransisco, 186–195.

Hamilton, W. (1964) Diabase sheets differentiated by liquid fractionation, Taylor Glacier Region, southern Victoria Land. In Adie, R.J. (ed.) *Antarctic Geology*. North Holland Publishing Co, Amsterdam, 442–54.

Handmer, J. and Wilder, M. (1993) *Towards a Conservation Strategy for the Australian Antarctic Territory*. Centre for Resource and Environmental Studies, Australian National University, Canberra.

Hanessian, Jr. J., (1965) National interests in Antarctica. In Hatherton, T. (ed.) *Antarctica*. Methuen, London, 3–53.

Hanna, E. (1996) The role of Antarctic sea ice in global climate change. *Progress in Physical Geography*, 20(4), 371–401.

Hannah, L., Lohse, D., Hutchinson, C., Carr, J.L. and Lankerani, A. (1994) A preliminary inventory of human disturbance of world ecosystems. *Ambio*, 23, 246–250.

Hansom, J.D. (1983a) Shore platform development in the South Shetland Islands, Antarctica. *Marine Geology*, 35, 211–229.

Hansom, J.D. (1983b) Ice-formed intertidal boulder pavements in the sub-Antarctic. *Journal of Sedimentary Petrology*, 53, 135–145.

Hansom, J.D. and Kirk, R.M. (1989) Ice in the intertidal zone: examples from Antarctica. In Bird, E.C.F. and Kelletat, D. (eds), *Zonality of Coastal Geomorphology and Ecology*. Proceedings of the Sylt Symposium, *Essener Geographische Arbeiten*, 18, Paderborn, 211–236.

Hardin, G. (1968) The tragedy of the commons. *Science*, 162, 1243–1248.

Hardy, A. (1967) *Great Waters*. Collins, London.

Harley, S.L. (1989) The origins of granulites: a metamorphic perspective. *Geological Magazine*, 126, 215–331.

Haron, M. (1986) Antarctica and the United Nations – the next step? In Wolfrum, R. (ed.) *Antarctic Challenge II. Conflicting Interests, Co-operation, Environmental Protection, Economic Development*. Duncker & Humblot, Berlin, 321–332.

Haron, M. (1988) The issue of Antarctica – a commentary. In Wolfrum, R. (ed.) *Antarctic Challenge III. Conflicting Interests, Co-operation, Environmental Protection, Economic Development*. Duncker & Humblot, Berlin, 271–276.

Haron, M. (1991) Ability of the Antarctic Treaty system to adapt to external challenges. In Jørgensen-Dahl, A. and Østreng, W. (eds) *The Antarctic Treaty System in World Politics*. Macmillan, Basingstoke, 299–308.

Harrington, H.J. (1965) Geology and geomorphology of Antarctica. In van Oye, P. and van Miegham, J. (eds) *Biogeography and Ecology in Antarctica*. Monographiae Biologicae, 15.

Harrington, H.J. and Speden, I.G. (1960) Recent moraines of a lobe of the Taylor Glacier, Victoria Land, Antarctica. *Journal of Glaciology*, 3, 652–653.

Harris, C.M. (1991a) Environmental effects of human activities on King George Island, South Shetland Islands, Antarctica. *Polar Record*, 27, 193–204.

Harris, C.M. (1991b) Environmental management on King George Island, South Shetland Islands, Antarctica. *Polar Record*, 27, 313–324.

Harris, C.M. and Meadows, J. (1992) Environmental management in Antarctica. Instruments and institutions. *Marine Pollution Bulletin*, 25, 239–249.

Harris, C.M. and Stonehouse, B. (1991) *Antarctica and Global Environmental Change*. Belhaven Press, London, 198pp.

Harris, S. (1991) The influence of the United Nations on the Antarctic System: a source of erosion or cohesion? In Jørgensen-Dahl, A. and Østreng, W. (eds) *The Antarctic Treaty System in World Politics*. Macmillan, Basingstoke, 309–327.

Harrowfield, D.L. (1990) Conservation and management of historic sites in the Ross Dependency. In Hay, J.E., Hemmings, A.D. and Thom, N.G. (eds) *Antarctica 150: Scientific Perspectives – Policy Futures*. University of Auckland, New Zealand, 55–66.

Harrowfield, D.L. (1995) *Icy Heritage: The Historic Sites of the Ross Sea Region, Antarctica*. Antarctic Heritage Trust, Christchurch, New Zealand.

Hart, P.D. (1988) The growth of Antarctic tourism. *Oceanus*, 31(2), 93–100.

Hart, T.J. (1942) Phytoplankton periodicity in Antarctic surface waters. *Discovery Reports*, 21, 261–356.

Hatherton, T. (ed.) (1990) *Antarctica: the Ross Sea Region*. DSIR Publishing, Department of Scientific and Industrial Research, Wellington, New Zealand.

Harwood, D.M., Winter, D.M. and Srivastav, A. (1994) Climatic implication for the absence of early Pliocene Antarctic sea-ice: a discussion. *EOS, Transactions American Geophysical Union*, 75(16) suppl. (abstr. 041A–12)

Harwood, J. (1990) Are scientific quotas needed for the assessment of whale stocks? *Mammal Review*, 20, 13–16.

Hatherton, T. (ed.) (1990) *Antarctica: the Ross Sea Region*. DSIR Publishing. Department of Scientific and Industrial Research, Wellington, New Zealand.

Hayes, J.G. (1932) *Antarctica. A Treatise on the Southern Continent*. The Richards Press Ltd, London.

Haynes, V. R. (1995) Alpine valley heads on the Antarctic Peninsula. *Boreas*, 24, 81–94.

Hays, D.E. and Ringis, I. 1973. Seafloor spreading in the Tasman Sea. *Nature*, 243, 454.

Hays, J. D. (1978) A review of the Late Quaternary climatic history of Antarctic Seas. In van Zinderen Bakker, E.M. (ed.) *Antarctic Glacial History and World Palaeoenvironments*. A.A. Balkema, Rotterdam, 57–71.

Headland, R.K. (1984) *The Island of South Georgia*. Cambridge University Press, Cambridge.

Headland, R.K. (1989) *Chronological List of Antarctic Expeditions and Related Historical Events*. Cambridge University Press, Cambridge.

Headland, R.K. (1993a) Geographical discoveries in Antarctica by the whaling industry. In Basberg, B.L., Ringstad, J.E. and Wexelsen, E. (eds) *Whaling and History. Perspectives on the Evolution of the Industry*. Kommandør Chr. Christensens Hvalfangstmuseum Publikasjon 29, Sandefjord, 191–202.

Headland, R.K. (1993b) Winter stations in the Antarctic. *Antarctic*, 13(2), 86–87.

Headland, R.K. (1994a) Historical development of Antarctic tourism. *Annals of Tourism Research*, 21, 269–280.

Headland, R.K. (1994b) First on the Antarctic continent? *Antarctic*, 13(8), 346–348.

Headland, R.K. (1994c) Historic sites and monuments. In Smith, R.I.L. Walton, D.W.H. and Dingwall, P.R. (eds) *Developing the Antarctic Protected Area System*. Conservation of the Southern Polar Region, No. 1. IUCN, Gland and Cambridge, 85–93.

Headland, R.K. (1996) An early Antarctic landing. Captain Cooper's log of the Levant, 1853. *The American Neptune*, 56, 371–381.

Headland, R.K. and Keage, P.L. (1985) Activities on the King George Island Group, South Shetland Islands, Antarctica. *Polar Record*, 22, 475–484.

Headland, R.K. and Keage, P.L. (1995) Antarctic tourist day-flights. *Polar Record*, 31, 347.

Heap, J.A. (1965) Antarctic pack ice. In Hatherton, T. (ed.) *Antarctica*. Methuen, London, 187–196.

Heap, J.A. (1983) Cooperation in the Antarctic: a quarter of a century's experience. In Orrego Vicuña, F. (ed.) *Antarctic Resources Policy. Scientific, Legal and Political Issues*. Cambridge University Press, Cambridge, 103–108.

Heap, J.A. (1988) The role of scientific advice for the decision making process in the Antarctic Treaty System. In Wolfrum, R. (ed.) *Antarctic Challenge III. Conflicting Interests, Cooperation, Environmental Protection, Economic Development*. Duncker & Humblot, Berlin, 21–28.

Heap, J.A. (1989) Science and the environment. *Antarctic Journal of the United States*, 24(4), 12–13.

Heap, J.A. (1990) The political case for the Minerals Convention. In Cook, G. (ed.) *The Future of Antarctica. Exploitation Versus Protection*. Manchester University Press, Manchester, 44–52.

Heap, J.A. (1991a) Antarctic politics and Antarctic science – are they at loggerheads? *Antarctic Science*, 3, 123.

Heap, J.A. (1991b) Has CCAMLR worked? Management policies and ecological needs. In Jørgensen-Dahl, A. and Østreng, W. (eds) *The Antarctic Treaty System in World Politics*. Macmillan, Basingstoke, 43–53.

Heap, J.A. (1994) *Handbook of the Antarctic Treaty System*, 7th edn. US Department of State, Washington DC.

Heap, J.A. and Holdgate, M.W. (1986) The Antarctic Treaty System as an environmental mechanism – an approach to environmental issues. In Polar Research Board, *Antarctic Treaty System. An Assessment*. National Academy Press, Washington DC, 195–210.

Heezen, B.C., Tharp, M. and Bentley, C.R. (1972) *Morphology of the Earth in the Antarctic and Sub-Antarctic. American Geographical Society Antarctic Map Folio Series*, 16, 1–14.

Heilbron, T.D. and Walton, D.W.H. (1984) Plant colonisation of actively sorted stone stripes in the sub-Antarctic. *Arctic and Alpine Research*, 16, 141–172.

Helbling, E.W., Villafañe, V. and Holm-Hansen, O. (1994) Effects of ultraviolet radiation on Antarctic marine phytoplankton photosynthesis with particular attention to the influencing of mixing. In Weiler, C.S. and Penhale, P.A. (eds) *Ultraviolet Radiation in Antarctica: Measurements and Biological Effects*. American Geophysical Union, Antarctic Research Series 62, Washington DC, 207–227.

Helbling, E.W., Villafane, V., Ferrario, M. and Holm-Hansen, O. (1992) Impact of natural ultraviolet radiation on rates of photosynthesis and on specific marine phytoplankton species. *Marine Ecology Progress Series*, 80, 89–100.

Hemmings, A.D. (1990) Human impacts and ecological constraints on skuas. In Kerry, K.R. and Hempel, G. (eds) *Antarctic Ecosystems. Ecological Change and Conservation*. Springer–Verlag, Berlin, 224–230.

Hempel, G. (1994) *Antarctic Science: Global Concerns*. Berlin, Springer-Verlag, 287pp.

Hempel, I. (1985) Variation in geographical distribution and abundance of larvae of Antarctic krill, *Euphausia superba*, in the Southern Atlantic Ocean. In Siegfried, W.R., Condy, P.R. and Laws, R.M. (eds) *Antarctic Nutrient Cycles and Food Webs*. Springer-Verlag, Berlin, 304–307.

Hempel, I. (1988) Antarctic marine research in winter: the Weddell Sea Project 1986. *Polar Record*, 24, 43–48.

Hempel, I. (1991) Life in the Antarctic sea ice zone. *Polar Record*, 27, 249–254.

Hendee, J.C., Stankey, G.H. and Lucas, R.C. (1990) *Wilderness Management*, 2nd edn. North American Press, Colorado.

Herber, B.P. (1991) The common heritage principle: Antarctica and the developing nations. *American Journal of Economics and Sociology*, 50, 391–406.

Herber, B.P. (1992) The economic case for an Antarctic world park in light of recent policy developments. *Polar Record*, 28, 293–300.

Herr, R. (1989) The Antarctic experience: tourism's final frontier. In Vulker, J. and McDonald, G. (eds) *Architecture in the Wild: The Issues of Tourist Development for Remote and Sensitive Environments*. RAIA Education Division, Canberra, 61–67.

Herr, R. (1993) Antarctic tourism: Australia, regulation and the industry. In Handmer, J. and Wilder, M. (eds) *Towards a Conservation Strategy for the Australian Antarctic Territory*. Centre for Resource and Environmental Studies, Australian National University, Canberra, 91–107.

Herr, R. (1996) The changing roles of non-governmental organisations in the Antarctic Treaty System. In Stokke, O.S. and Vidas, D. (eds) *Governing the Antarctic. The Effectiveness and Legitimacy of the Antarctic Treaty System*. Cambridge University Press, Cambridge, 91–110.

Herr, R. and Hall, R. (1989) Science as currency and the currency of science. In Handmer, J. (ed.) *Antarctica: Policies and Policy Development*. Centre for Resource and Environmental Studies, Australian National University, Resource and Environmental Studies No. 1, 13–24.

Herr, R.A., Hall, H.R. and Haward, M.G. (1990) *Antarctica's Future: Continuity or Change?* Australian Institute of International Affairs, Hobart.

Herr, R.A. and Davis, B.W. (1992) Antarctica and non-state actors: the question of legitimacy. *Fridtjof Nansen Institute. International Antarctic Regime Project. IARP Publication Series*, 1992/4.

Heywood, R.B. (1993) Environmentally driven or environmentally benign Antarctic research? In Elzinga, A. (ed.) *Changing Trends in Antarctic Research*. Kluwer, Dordrecht, 129–139.

Hiller, A., Wand, U., Kaempf, H. and Stackebrandt, W. (1988) Occupation of the Antarctic continent by petrels during the past 35,000 years: inferences from a carbon-14 study of stomach oil deposits. *Polar Biology*, 9, 69–77.

Hindell, M.A. and Burton, H.R. (1987) Past and present status of the southern elephant seal (*Mirounga leonina*) at Macquarie Island. *Journal of Zoology, London*, 213, 365–380.

Hindell, M.A. and Burton, H.R. (1988) The history of the elephant seal industry at Macquarie Island and an estimate of the pre–sealing numbers. *Papers and Proceedings of the Royal Society of Tasmania*, 122, 159–176.

Hindell, M.A., Slip, D.J. and Burton, H.R. (1994) Possible causes of the decline of southern elephant seal populations in the southern Pacific and southern Indian Oceans. In Le Boeuf, B.J. and Laws, R.M. (eds) *Elephant Seals: Population Ecology, Behavior and Physiology*. University of California Press, Berkeley, 66–84.

Hodges, C.A. (1995) Mineral resources, environmental issues, and land use. *Science*, 268, 1305–1312.

Hodgson, D.A., Johnston, N.M., Caulkett, A.P. and Jones, V.J. (1997) Palaeolimnology of Antarctic fur seal *Arctocephalus gazella* populations and implications for Antarctic management. *Biological Conservation*, 83, 145–154.

Hoelzel, A.R. and Amos, W. (1988) DNA fingerprinting and 'scientific' whaling. *Nature*, 333, 305.

Hoffman, P.F. (1991) Did the breakout of Laurentia turn Gondwanaland inside out? *Science*, 252, 1409–1412.

Hofman, R.J. (1984) The Convention on the Conservation of Antarctic Marine Living Resources. In Alexander, L.M. and Hanson, L.C. (eds) *Antarctic Politics and Marine Resources: Critical Choices for the 1980s*. Centre for Ocean Management Studies, Kingston, Rhode Island, 113–122.

Holdgate, M.W. (1970) Conservation in the Antarctic. In Holdgate, M.W. (ed.) *Antarctic Ecology*, Vol. 2. Academic Press, London, 924–945.

Holdgate, M.W. (1984) The use and abuse of polar resources. *Polar Record*, 22, 25–48.

Holdgate, M.W. (1987) Regulated development and conservation of Antarctic resources. In Triggs, G.D. (ed.) *The Antarctic Treaty Regime. Law, Environment and Resources*. Cambridge University Press, Cambridge, 128–142.

Holdgate, M.W. (1994) International designations. In Smith, R.I.L., Walton, D.W.H. and Dingwall, P.R. (eds) *Developing the Antarctic Protected Area System*. Conservation of the Southern Polar Region, No.1. IUCN, Gland and Cambridge, 99–104.

Holdgate, M.W. and Tinker, J. (1979) *Oil and Other Minerals in the Antarctic. The Environmental Implications of Possible Mineral Exploration or Exploitation in Antarctica*. SCAR, Cambridge.

Holdgate, M.W. and Wace, N.M. (1961) The influence of man on the floras and faunas of the Southern Ocean. *Polar Record*, 10, 475–493.

Holdgate, M.W., Allen, S.E. and Chambers, M.J.G. (1967) A preliminary investigation of the soils of Signy Island, South Orkney Islands. *British Antarctic Survey Bulletin*, 12, 53–71.

Holdsworth, G. and Bull, C. (1970) The flow law of cold ice; investigations on Meserve Glacier, Antarctica. *International Association of Scientific Hydrology*. Pub. 86, 204–16.

Hole, M.J. (1990) Antarctic volcanoes. *NERC News*, No.14. Natural Environment Research Council, Swindon, 4–6.

Holm-Hansen, O., Helbling, E.W. and Lubin, D. (1993) Ultraviolet radiation in Antarctica: inhibition of primary production. *Photochemistry and Photobiology*, 58, 567–570.

Holt, S. (1985a) Let's all go whaling. *The Ecologist*, 15, 113–124.

Holt, S. (1985b) Whale mining, whale saving. *Marine Policy*, 9, 192–213.

Holt, S. (1993) Harvesting of whales. *Nature*, 361, 391.

Honda, K., Yamamoto, Y. and Tatsukawa, R. (1987) Distribution of heavy metals in Antarctic marine ecosystem. *Proceedings of the NIPR Symposium on Polar Biology*, 1, 184–197.

Hosaka, N. and Nemoto, T. (1986) Size structure of phytoplankton carbon and primary production in the Southern Ocean south of Australia during the summer of 1983–1984. *Memoirs National Institute of Polar Research*, Special Issue, 40, 15–24.

Houghton, J.T., Callander, B.A. and Varney, S.K. (1992) *Climate Change 1992. The Supplementary Report to the IPCC Scientific Assessment*. Cambridge University Press, Cambridge.

Houghton, J.T., Jenkins, G.J. and Ephraums, J.J. (1990) *Climate Change: The IPCC Scientific Assessment*. Cambridge University Press, Cambridge.

Houghton, J.T., Meira Filho, L.G., Callander, B.A., Harris, N., Kattenberg, A. and Maskell, K. (1996) *Climate Change 1995. The Science of Climate Change*. Cambridge University Press, Cambridge.

House of Lords Select Committee on Science and Technology (1996) *Fish Stock Conservation and Management*. Session 1995–96. Second Report. London, HMSO.

House of Representatives Standing Committee on Environment, Recreation and the Arts (1989) *Tourism in Antarctica*. Australian Government Publishing Service, Canberra.

Howard, M. (1989) The Convention on the Conservation of Antarctic Marine Living Resources: a five year review. *International and Comparative Law Quarterly*, 38, 104–149.

Howington, J.P., McFeters, G.A., Barry, J.P. and Smith J.J. (1992) Distribution of the McMurdo Station sewage plume. *Marine Pollution Bulletin*, 25, 324–327.

Hubold, G. (1985) The early life history of the high-Antarctic silverfish, *Pleuragramma antarcticum*. In Siegfried, W.R., Condy, P.R. and Laws, R.M., (eds.) *Antarctic Nutrient Cycles and Food Webs*, Springer-Verlag, Berlin, 445–451.

Hughes, J. (1994) Antarctic historic sites. The tourism implications. *Annals of Tourism Research*, 21, 281–294.

Hughes, J. and Davis, B. (1995) The management of tourism at historic sites and monuments. In Hall, C.M. and Johnston, M.E. (eds) *Polar Tourism: Tourism in the Arctic and Antarctic Regions*. Chichester, Wiley, 235–255.

Hughes, T.J. (1987) Deluge II and the continent of doom: rising sea level and collapsing Antarctic ice sheet. *Boreas*, 16, 89–100.

Hughes, T.J., Denton, G.H. and Fastook, J.L. (1985) Is the Antarctic ice sheet an analogue for Northern Hemisphere palaeo-ice sheets? In Wodenburg, J.J. (ed.) *Models in Geomorphology*. Allen & Unwin, Boston, 25–72.

Hunt Jr, G.L. Heinemann, D. and Everson, I. (1992) Distributions and predator-prey interactions of macaroni penguins, Antarctic fur seals, and Antarctic krill near Bird Island, South Georgia. *Marine Ecology Progress Series*, 86, 15–30.

Huntley, B.J. (1971) Vegetation. In van Zinderen Bakker, E.M., Winterbottom, J.M. and Dyer, R.A. (eds) *Marion and Prince Edward Islands*. Balkema, Cape Town, 98–160.

Hureau, J.-C. (1979) La faune ichthyologique du secteur Indien de l'Ocean Antarctique et estimation du stock de poissons autour des Îles

Kerguelen. *Memoires du Museum National d'Histoire Naturelle*, 43, 235–247.

Hureau, J.-C. (1994) The significance of fish in the marine Antarctic ecosystems. *Polar Biology*, 14, 307–313.

Hureau, J.-C. and Słosarczyk, W. (1990) Exploitation and conservation of Antarctic fishes and recent ichthyological research in the Southern Ocean. In Gon, O. and Heemstra, P.C. (eds) *Fishes of the Southern Ocean*. J.L.B. Smith Institute of Ichthyology, Grahamstown, 52–63.

Huybrechts, P. and Oerlemans, J. (1990) Response of the Antarctic ice sheet to future greenhouse warming. *Climate Dynamics*, 5, 93–102.

Huybrechts, P. (1993) Glaciological modelling of the Late Cenozoic East Antarctic ice sheet: stability or dynamism? *Geografiska Annaler*, 75A(4), 221–238.

Hyland, J., Laur, D., Jones, J., Shrake, J., Cadian, D. and Harris, L. (1994) Effects of an oil spill on the soft–bottom macrofauna of Arthur Harbor, Antarctica compared with long-term natural change. *Antarctic Science*, 6, 37–44.

Ichii, T., Naganobu, M. and Ogishima, T. (1994) An assessment of the impact of the krill fishery on penguins in the South Shetlands. *CCAMLR Science*, 1, 107–128.

Indreeide, T.V. (1990) Antarctic science in a historical perspective: changing roles and functional development. *International Challenges*, 10(4), 36–44.

IUCN (1980) *World Conservation Strategy*. IUCN, Gland, Switzerland.

IUCN (1991) *A Strategy for Antarctic Conservation*. IUCN, Gland, Switzerland.

Ivanhoe, L.F. (1980) Antarctica – operating conditions and petroleum prospects. *Oil and Gas Journal*, 78(52), 212–220.

Ivy-Ochs, S., Slichter, C., Kibuk, P.W., Dittrich-Hannen, B. and Beer, J. (1995) Minimum ^{10}Be ages of early Pliocene for the Table Mountain plateau and the Sirius Group at Mount Fleming, Dry Valleys, Antarctica. *Geology*, 23 (11), 1007–1010.

Jacka, T.H. and Budd, W.F. (1991) Detection of sea ice and temperature changes in the Antarctic and Southern Ocean. In Weller, G., Wilson, J. and Severin, B. (eds) *Role of Polar Regions in Global Change*. Proceedings of the International Conference, University of Alaska, Geophysical Institute and Center for Global Change and Arctic System Research, University of Alaska, Fairbanks 63–70.

Jackson, A. (ed.) (1996) *On the Antarctic Horizon*. Proceedings of the International Symposium on the Future of the Antarctic Treaty System, Ushuaia, Argentina, 20–24 March 1995. Australian Antarctic Foundation, Hobart.

Jackson, G. (1978) *The British Whaling Trade*. Adam and Charles Black, London.

Jacobs, S.S. (1992) Is the Antarctic ice sheet growing? *Nature*, 360, 29–33.

Jacobs, S.S., MacAyeal, D.R. and Ardai, Jr, J.L. (1986) The recent advance of the Ross Ice Shelf, Antarctica. *Journal of Glaciology*, 32, 464–474.

Jacobsson, M. (1994) Asia, Antarctica and the principle of the common heritage of mankind. In Herr, R.A. and Davis, B.W. (eds) *Asia in Antarctica*. Centre for Resource and Environmental Studies, Australian National University, Canberra, 139–157.

James, P.R. and Tingey, R.J. (1983) The Precambrian geological evolution of the East Antarctic metamorphic shield: a review. In Oliver, R.L., James, P.R. and Jago, J.B. (eds) *Antarctic Earth Science*. Cambridge University Press, Cambridge, 5–10.

Janetschek, H. (1970) Environments and ecology of terrestrial arthropods in the High Antarctic. In Holdgate, M.W. (ed.) *Antarctic Ecology*, Vol. 2. Academic Press, London, 871–885.

Jennings, P.G. (1976) Tardigrada from the Antarctic Peninsula and Scotia Ridge region. *Bulletin of the British Antarctic Survey*, 44, 77–95.

Jennings, P.G., (1979) The Signy Island terrestrial reference sites. X. Population dynamics of tardigrada and rotifera. *Bulletin of the British Antarctic Survey*, 47, 89–105.

Johannessen, O.M., Miles, M. and Bjorgo, E. (1995) The Arctic's shrinking sea ice. *Nature*, 376, 126–127.

Johanson, U., Gehrke, C., Björn, L.O., Callaghan, T.V. and Sonesson, M. (1995) The effects of enhanced UV–B radiation on a subarctic heath ecosystem. *Ambio*, 24, 106–111.

John, B.S. (1972) Evidence from the South Shetland Islands towards a glacial history of West Antarctica. In Price, R.J. and Sugden, D.E. (eds) *Polar Geomorphology*. Institute of British Geographers Special Publication 4, 75–92.

Johnson, S.P. (1993) *The Earth Summit: the United Nations Conference on Environment and Development (UNCED)* Graham & Trotman/Martinus Nijhoff, London and Dordrecht.

Johnston, M.E. (1997) Polar tourism regulation strategies: controlling visitors through codes of conduct and legislation. *Polar Record*, 33, 13–20.

Johnston, R.J. (1992) Laws, states and super-states: international law and the environment. *Applied Geography*, 12, 211–228.

Joiris, C.R. and Overloop, W. (1991) PCBs and

organochlorine pesticides in phytoplankton and zooplankton in the Indian sector of the Southern Ocean. *Antarctic Science*, 3, 371–377.

Jones, A.E. and Shanklin, J.D. (1995) Continued decline of total ozone over Halley, Antarctica, since 1985. *Nature*, 376, 409–411.

Jones, A.G.E. (1975) Captain William Smith and the discovery of New South Shetland. *Geographical Journal*, 141, 445–461.

Jones, A.G.E. (1985a) British sealing on New South Shetland 1819–1826. Part 1. *The Great Circle*, 7, 9–22.

Jones, A.G.E. (1985b) British sealing on New South Shetland 1819–1826. Part II. *The Great Circle*, 7, 74–87.

Jones, P.D. (1990) Antarctic temperatures over the present century – a study of the early expedition record. *Journal of Climate*, 3, 1193–1203.

Jones, P.D. (1995) Recent variations in mean temperature and the diurnal temperature range in the Antarctic. *Geophysical Research Letters*, 22(11), 1345–1348.

Jørgensen-Dahl, A. and Østreng, W. 1991. *The Antarctic Treaty System in World Politics*. Macmillan, Basingstoke.

Jouventin P. and Weimerskirch, H. (1990) Long-term changes in seabird and seal populations in the Southern Ocean. In Kerry, K.R. and Hempel, G. (eds) *Antarctic Ecosystems. Ecological Change and Conservation*. Springer–Verlag, Berlin, 208–213.

Jouventin, P. and Weimerskirch, H. (1991) Changes in the population size and demography of southern seabirds: management perspectives. In Perrins, C.M., Lebreton, J.D. and Hirons, G.J.M. (eds) *Bird Population Studies. Relevance to Conservation and Management*. Oxford University Press, Oxford, 297–314.

Jouventin, P., Stahl, J.C., Weimerskirch, H. and Mougin, J.L. (1984) The seabirds of the French subantarctic islands and Adélie Land, their status and conservation. In Croxall, J.P., Evans, P.G.H. and Schreiber, R.W. (eds) *Status and Conservation of the World's Seabirds*. ICBP Technical Publication No. 2. International Council for Bird Preservation, Cambridge, 609–625.

Jouzel, J., Merlivat, L., Pourchet, M. and Lorius, C. (1979) A continuous record of artificial tritium fallout at the South Pole (1954–1978). *Earth and Planetary Science Letters*, 45, 188–200.

Jouzel, J., Lorius, C., Petit, J.R., Genthon, C., Barkov, N.I., Kotlyakov, V.M. and Petrov, V.M. (1987) Vostok ice core: a continuous isotope temperature record over the last climatic cycle (160,000 years). *Nature*, 329, 403–408.

Jouzel, J., Raisbeck, G., Benoist, J.P., Yiou, F., Raynaud, D., Petit, J.R., Barkov, N.I., Korotkevitch, Y.S. and Kotlyakov, V.M. (1989) A comparison of Deep Antarctic ice cores and their implications for climate between 65,000 and 150,000 years ago. *Quaternary Research*, 31, 135–150.

Jouzel, J., Barkov, N.I., Barnola, J.M., Bender, M., Chappellaz, J., Genthon, C., Kotlyakov, V.M., Lipenkov, V., Lorius, C., Petit, J.R., Raynaud, D., Raisbeck, G., Ritz, C., Sowers, T., Stievenard, M., Yiou, F. and Yiou, P. (1993) Extending the Vostok ice-core record of palaeoclimate to the penultimate glacial period. *Nature*, 364, 407–412.

Joyner, C.C. (1986) Protection of the Antarctic environment: rethinking the problems and prospects. *Cornell International Law Journal*, 19, 259–273.

Joyner, C.C. (1989) Antarctic Minerals Convention. *Marine Policy Reports*, 1, 69–85.

Joyner, C.C. (1991) CRAMRA: the ugly duckling of the Antarctic Treaty System. In Jørgensen-Dahl, A. and Østreng, W. (eds) *The Antarctic Treaty System in World Politics*. Macmillan, Basingstoke, 161–185.

Joyner, C.C. (1996a) The legitimacy of CRAMRA. In Stokke, O.S. and Vidas, D. (eds) *Governing the Antarctic. The Effectiveness and Legitimacy of the Antarctic Treaty System*. Cambridge University Press, Cambridge, 246–267.

Joyner, C.C. (1996b) The 1991 Madrid Environmental Protocol. Contributions to marine pollution law. *Marine Policy*, 20, 183–197.

Joyner, C.C. and Chopra, S.K. (1988) *The Antarctic Legal System*. Martinus Nijhoff Publishers, Dordrecht.

Kaczynski, V.M. (1984) Economic aspects of Antarctic fisheries. In Alexander, L.M. and Hanson, L.C. (eds) *Antarctic Politics and Marine Resources: Critical Choices for the 1980s*. Centre for Ocean Management Studies, Kingston, Rhode Island, 141–158.

Kämmerer, J.A. (1992) The Protocol on Environmental Protection to the Antarctic Treaty. *Law and State*, 45, 68–80.

Kapitsa, A.P., Ridley, J.K., Robin, G. de Q., Siegert, M.J. and Zotikov, I.A. (1996) A large deep freshwater lake beneath the ice of central East Antarctica. *Nature*, 381, 684–686.

Kappen, L. (1985) Vegetation and ecology of ice-free areas of northern Victoria Land, Antarctica. I: the lichen vegetation of Birthday Ridge and an inland mountain. *Polar Biology*, 4, 213–225.

Kappen, L. (1993) Lichens. In Friedmann, E.I. (ed.) *Antarctic Microbiology*. New York, Wiley, 433–490.

Kappen, L. and Friedmann, E.I. (1983) Ecophysiology of lichens in the dry valleys of southern Victoria Land. II. CO_2 gas exchange in cryptoendolithic lichens. *Polar Biology*, 1, 227–232.

Karentz, D. (1991) Ecological considerations of Antarctic ozone depletion. *Antarctic Science*, 3, 3–11.

Karentz, D. (1992) Ozone depletion and UV–B radiation in the Antarctic – limitations to ecological assessment. *Marine Pollution Bulletin*, 25, 231–232.

Karentz, D. (1994) Ultraviolet tolerance mechanisms in Antarctic marine organisms. In Weiler, C.S. and Penhale, P.A. (eds) *Ultraviolet Radiation in Antarctica: Measurements and Biological Effects*. American Geophysical Union, Antarctic Research Series 62, Washington DC, 93–110.

Karentz, D. and Gast, T. (1993) Transmission of solar ultraviolet radiation through invertebrate exteriors. *Antarctic Journal of the United States*, 28(5), 113–114.

Karentz, D. and Lutze, L.H. (1990) Evaluation of biologically harmful ultraviolet radiation in Antarctica with a biological dosimeter designed for aquatic environments. *Limnology and Oceanography*, 35, 549–561.

Karentz, D., Cleaver, J.E. and Mitchell, D.L. (1991a) Cell survival characteristics and molecular responses of Antarctic phytoplankton to ultraviolet-B radiation. *Journal of Phycology*, 27, 326–341.

Karentz, D., McEuen, F.S., Land, M.C. and Dunlap, W.C. (1991b) Survey of mycosporine-like amino acid compounds in Antarctic marine organisms: potential protection from ultraviolet exposure. *Marine Biology*, 108, 157–166.

Karev, S.N. (1991) Co-operative momentum? The Antarctic Treaty System and the prevention of conflict. In Jørgensen–Dahl, A. and Østreng, W. (eds) *The Antarctic Treaty System in World Politics*. Macmillan, Basingstoke, 372–378.

Karl, D.M. (1992) The grounding of the *Bahia Paraiso*: microbial ecology of the 1989 Antarctic oil spill. *Microbial Ecology*, 24, 77–89.

Keage, P.L. (1982) The conservation status of Heard Island and the McDonald Islands. University of Tasmania, Environmental Studies, Occasional Paper 13.

Keage, P.L. (1986) Antarctic protected areas: future options. University of Tasmania, Environmental Studies, Occasional Paper 19.

Keage, P.L. and Quilty, P.G. (1988) Future directions for Antarctic environmental studies. *Maritime Studies*, 38, 1–12.

Keage, P.L., Hay, P.R. and Russell, J.A. (1989) Improving Antarctic management plans. *Polar Record*, 25, 309–314.

Kellerman, A. (1990) Catalogue of early life stages of Antarctic notothenioid fishes. *Berichte zur Polarforschung*, 67, 45–136.

Kellogg, D.E. and Kellogg, T.B. (1996) Diatoms in South Pole ice: Implications for eolian contamination of Sirius group deposits. *Geology*, 24 (1), 115–118.

Kennedy, A.D. (1993) Water as a limiting factor in the Antarctic terrestrial environment: a biogeographical synthesis. *Arctic and Alpine Research*, 25, 308–315.

Kennedy, A.D. (1995) Antarctic terrestrial ecosystem response to global environmental change. *Annual Review of Ecology and Systematics*, 26, 683–704.

Kennett, J.P. (1978) Cainozoic evolution of circum-antarctic palaeoceanography. In van Zinderen Bakker, E.M. (ed.) *Antarctic Glacial History and World Palaeoenvironments*. A.A. Balkema, Rotterdam. 41–56.

Kennicutt II, M.C. *et. al.* (1990) Oil spillage in Antarctica. *Environmental Science and Technology*, 24, 620–624.

Kennicutt II, M.C. and Champ, M.A. (1992) Environmental awareness in Antarctica: history, problems and future solutions. *Marine Pollution Bulletin*, 25, 9–21.

Kennicutt II, M.C. and Sweet, S.T. (1992) Hydrocarbon contamination on the Antarctic Peninsula: III. The *Bahia Paraiso* two years after the spill. *Marine Pollution Bulletin*, 25, 303–306.

Kennicutt II, M.C., Sweet, S.C., Fraser, W.R., Stockton, W.L. and Culver, M. (1991a) Grounding of the *Bahia Paraiso* at Arthur Harbor, Antarctica. 1. Distribution and fate of oil spill related hydrocarbons. *Environmental Science and Technology*, 25, 509–518.

Kennincutt II, M.C., Sweet, S.T., Fraser, W.R., Culver, M. and Stockton, W.L. (1991b) The fate of diesel fuel spilled by the *Bahia Paraiso* in Arthur Harbor, Antarctica. *Proceedings 1991 International Oil Spill Conference*, 493–500.

Kennicutt II, M.C., McDonald, T.J., Denoux, G.J. and McDonald S.J. (1992a) Hydrocarbon contamination on the Antarctic Peninsula. I. Arthur Harbor – subtidal sediments. *Marine Pollution Bulletin*, 24, 499–506.

Kennicutt II, M.C., McDonald, T.J., Denoux, G.J.and McDonald S.J. (1992b) Hydrocarbon contamination on the Antarctic Peninsula. II. Arthur Harbor – inter and subtidal limpets (*Nacella concinna*) *Marine Pollution Bulletin*, 24, 506–511.

Kennicut II, M.C., McDonald, S.J., Sericano, J.L., Boothe, P., Oliver, J., Safe, S., Presley, B.J., Liu, H., Wolfe, D., Wade, T.L., Crockett, A. and Bockus, D. (1995) Human contamination of the marine environment – Arthur Harbor and McMurdo Sound, Antarctica. *Environmental Science and Technology*, 29, 1279–1287.

Kerry, E. (1990) Micro-organisms colonising plants and soil subjected to different degrees of human activity, including petroleum contamination, in the Vestfold Hills and MacRobertson Land, Antarctica. *Polar Biology*, 10, 423–430.

Kerry, E. (1993) Bioremediation of experimental petroleum spills on mineral soils in the Vestfold Hills, Antarctica. *Polar Biology*, 13, 163–170.

Keys, J.R. (1984) *Antarctic Marine Environments and Offshore Oil*. Commission for the Environment, Wellington.

Kiernan, V. (1996) US frost chills polar stations. *New Scientist*, 149(2019), 14–15.

Kimball, L. (1984) Environmental issues in the Antarctic minerals negotiations. In Alexander, L.M. and Hanson, L.C. (eds) *Antarctic Politics and Marine Resources: Critical Choices for the 1980s*. Center for Ocean Management Studies, Kingston, Rhode Island, 204–214.

Kimball, L. (1988) The role of non-governmental organisations in Antarctic affairs. In Joyner, C.C. and Chopra, S.K. (eds) *Antarctic Legal Regime*. Martinus Nijhoff Publishers, Dordrecht, 33–63.

Kimball, L. (1989) Antarctica: the challenges that lie ahead. *Ambio*, 18, 77–82.

Kimball, L.A. (1990) Special report on the Antarctic Minerals Convention. In Splettstoesser, J.F. and Dreschhoff, G.A.M. (eds) *Mineral Resources Potential of Antarctica*. American Geophysical Union, Antarctic Research Series 51, Washington DC, 273–310.

Kimball, L.A. (1991) CRAMRA and other environmental regimes in the ATS: how well does it fit? In Jørgensen-Dahl, A. and Østreng, W. (eds) *The Antarctic Treaty System in World Politics*. Macmillan, Basingstoke, 133–143.

King, H.G.R. (1969) *The Antarctic*. The Blandford Press, London.

King, J.C. (1994) Recent climate variability in the vicinity of the Antarctic Peninsula. *Journal of Climatology*, 14, 357–361.

Kirk, R.M. (1972) Antarctic beaches as natural hydraulic models. *Proceedings 7th New Zealand Geography Conference (Hamilton)*, 227–234.

Kirk, R.M. (1990) Raised beaches, late Quaternary sea levels and deglacial sequences on the Victoria Land coast, Antarctica. In Gillieson, D. and Fitzsimons, S. (eds) *Quaternary Research in Australian Antarctica: Future Directions*. Australian Defence Force Academy Special Publication No. 3, Canberra, 85–106.

Kirwan, L.P. (1959) *The White Road. A Survey of Polar Exploration*. Hollis and Carter, London.

Klinck, J.M. and Hoffman, E.E. (1986) Deep-flow variability at Drake Passage. *Journal of Physical Oceanography*, 16(7), 1281–1292.

Knox, G.A. (1984) The key role of krill in the ecosystem of the Southern Ocean with special reference to the Convention on the Conservation of Antarctic Marine Living Resources. *Ocean Management*, 9, 113–156.

Knox, G.A. (1994) *The Biology of the Southern Ocean*. Cambridge University Press, Cambridge.

Kock, K.-H. (1985) Marine habitats – Antarctic fish. In Bonner, W.N. and Walton, D.W.H. (eds) *Key Environments: Antarctica*. Pergamon Press, Oxford, 173–192.

Kock, K.-H. (1991) The state of exploited fish stocks in the Southern Ocean – a review. *Archiv für Fischereiwissenschaft*, 41, 1–66.

Kock, K.-H. (1992) *Antarctic Fish and Fisheries*. Cambridge University Press, Cambridge.

Kock, K.-H. (1994) Fishing and conservation in southern waters. *Polar Record*, 30, 3–22.

Kock, K.-H. and Köster, F.-W. (1990) The state of exploited fish stocks in the Atlantic sector of the Southern Ocean. In Kerry, K.R. and Hempel, G. (eds) *Antarctic Ecosystems. Ecological Change and Conservation*. Springer-Verlag, Berlin, 308–322.

Kock, K.-H., Duhamel, G. and Hureau, J.-C. (1985) Biology and status of exploited Antarctic fish stocks: a review. *BIOMASS Scientific Series*, 6.

Koerner, R.M. (1961) Glaciological observations in Trinity Peninsula, Graham Land, Antarctica. *Journal of Glaciology*, 3, 1063–1074.

Konlechner, J.C. (1985) An investigation of the fate and effects of a paraffin based crude oil in an Antarctic terrestrial ecosystem. *New Zealand Antarctic Record*, 6(3), 40–46.

Koroma, A. (1988) Safeguarding the interests of mankind in the use of Antarctica. In Wolfrum, R. (ed.) *Antarctic Challenge III. Conflicting Interests, Co-operation, Environmental Protection, Economic Development*. Duncker & Humblot, Berlin, 243–252.

Kort, V.G. (1962) The Antarctic Ocean. *Scientific American*, 860, 3–11.

Kriwoken, L.K. and Keage, P.L. (1989) Antarctic environmental politics: protected areas. In Handmer, J. (ed.) *Antarctica: Policies and Policy Development*. Resource and Environmental Studies No. 1. Centre for Resource and Environmental Studies, Australian National University, Canberra, 31–48.

Kriwoken, L.K. (1991) Antarctic environmental planning and management: conclusions from Casey, Australian Antarctic Territory, 1991. *Polar Record*, 27, 1–8.

Kunzmann, A. (1991) Blood physiology and ecological consequences in Weddell Sea fishes. *Berichte zur Polarforschung*, 91, 1–19.

Kureishy, T.W., Sengupta, R., Mesquita, A. and Sanzgiry, S. (1993) Heavy metals in some parts of Antarctica and the Southern Indian Ocean. *Marine Pollution Bulletin*, 26, 651–652.

Kurtz, D.D. and Bromwich, D.H. (1985) A recurring, atmospherically forced polynya in Terra Nova Bay. In Jacobs, S.S. (ed.) *Oceanography of the Antarctic Continental Shelf*. American Geophysical Union, Antarctic Research Series 43. Washington DC, 177–201.

Kyle, P.R., Elliot, D.H. and Sutter, J.F. (1981) Jurassic Ferrar Supergroup theolites from the Transantarctic Mountains, Antarctica, and their relationship to the initial fragmentation of Gondwana. In Creswell, M.M. and Vella, P. (eds) *Gondwana Five*. Balkema, Rotterdam, 283–288.

Lagoni, R. (1984) Convention on the Conservation of Marine Living Resources: a model for the use of a common good. In Wolfrum, R. (ed.) *Antarctic Challenge. Conflicting Interests, Co-operation, Environmental Protection, Economic Development*. Duncker & Humblot, Berlin, 93–108.

Laird, M.G. (1991) The Late Proterozoic–Middle Palaeozoic rocks of Antarctica. In Tingey, R.J. (ed.) *The Geology of Antarctica*. Clarendon Press, Oxford, 74–119.

Laird, M.G., Cooper, R.A. and Jago, J.B. (1977) New data on the lower Palaeozoic sequence of northern Victoria Land and its significance for Australian and Antarctic relations in the Palaeozoic. *Nature*, 265, 107–110.

Laist, D.W. (1987) Overview of the biological effects of lost and discarded plastic debris in the marine environment. *Marine Pollution Bulletin*, 18, 319–326.

Larminie, F.G. (1987) Mineral resources: commercial prospects for Antarctic minerals. In Triggs, G.D. (ed.) *The Antarctic Treaty Regime: Law, Environment and Resources*. Cambridge University Press, Cambridge, 176–181.

Larminie, F.G. (1991) The mineral potential of Antarctica: the state of the art. In Jørgensen-Dahl, A. and Østreng, W. (eds) *The Antarctic Treaty System in World Politics*. Macmillan, Basingstoke, 79–93.

Larsson, P., Järnmark, C. and Södergren, A. (1992) PCBs and chlorinated pesticides in the atmosphere and aquatic organisms of Ross Island, Antarctica. *Marine Pollution Bulletin*, 25, 281–287.

Law, P. (1990) The Antarctic wilderness – a wild idea! In Herr, R.A., Hall, H.R. and Haward, M.G. (eds) *Antarctica's Future: Continuity or Change?* Australian Institute of International Affairs, Hobart, 71–80.

Laws, R.M. (1962) Some effects of whaling on the southern stocks of baleen whales. In Cren, E.D. and Holdgate, M.W. (eds) *The Exploitation of Natural Populations*. Blackwell Scientific Publications, Oxford, 137–158.

Laws, R.M. (1977a) Seals and whales of the Southern Ocean. *Philosophical Transactions of the Royal Society of London*, B279, 81–96.

Laws, R.M. (1977b) The significance of vertebrates in the Antarctic marine ecosystem. In Llano, G.A. (ed.) *Adaptations within Antarctic Ecosystems*. Proceedings of the Third SCAR Symposium on Antarctic Biology. Smithsonian Institution, Washington, DC, 411–438.

Laws, R.M. (1984) Seals. In Laws, R.M. (ed.) *Antarctic Ecology*, Vol. 2. Academic Press, London, 621–715.

Laws, R.M. (1985) The ecology of the Southern Ocean. *American Scientist*, 73, 26–40.

Laws, R.M. (1987) Cooperation or confrontation? Science, the Treaty and the future. In Walton, D.W.H. (ed.) *Antarctic Science*. Cambridge University Press, Cambridge, 249–265.

Laws, R.M. (1989) *Antarctica. The Last Frontier*. Boxtree, London.

Laws, R.M. (1990) Science as an Antarctic resource. In Cook, G. (ed.) *The Future of Antarctica. Exploitation Versus Preservation*. Manchester University Press, Manchester, 8–24.

Laws, R.M. (1991) Unacceptable threats to Antarctic science. *New Scientist*, 129(1762), 4.

Laws, R.M. (1994a) International science and the Antarctic Treaty System. In Hempel, G. (ed.) *Antarctic Science. Global Concerns*. Springer-Verlag, Berlin.

Laws, R.M. (1994b) History and present status of southern elephant seal populations. In Le Boeuf,

B.J. and Laws, R.M. (eds) *Elephant Seals: Population Ecology, Behavior, and Physiology.* University of California Press, Berkeley, 49–65.

Lawver, L.A., Royer, J.-Y., Sandwell, D.T. and Scotese, C.R. (1991) Evolution of the Antarctic continental margins. In Thomson, M.R.A., Crame, J.A. and Thomson, J.W. (eds) *Geological Evolution of Antarctica.* Cambridge University Press, Cambridge, 533–539.

Leader-Williams, N. (1985) The sub-Antarctic islands – introduced mammals. In Bonner, W.N. and Walton, D.W.H. (eds) *Key Environments: Antarctica.* Pergamon Press, Oxford, 318–328.

Leader-Williams, N. (1988) *Reindeer on South Georgia. The Ecology of an Introduced Population.* Cambridge University Press, Cambridge.

Leader-Williams, N., Walton, D.W.H. and Prince, P.A. (1989) Introduced reindeer on South Georgia – a management dilemma. *Biological Conservation,* 47, 1–11.

Ledingham, R. (1993) Reflections of an Antarctic tour guide. *ANARE News,* 73, 6–7.

LeMasurier, W.E. (1972) Volcanic record of Antarctic glacial history: implications with regard to Cenozoic sea levels. In Price, R.J. and Sugden D.E. (eds.) *Polar Geomorphology.* Institute of British Geographers, Special Publication, No. 4, 59–74.

LeMasurier, W.E. and Rex, D.C. (1982) Volcanic record of Cenozoic glacial history in Marie Byrd Land and western Ellsworth Land – revised chronology and evaluation of tectonic factors. In Craddock, C. (ed.) *Antarctic Geoscience.* University of Wisconsin Press, Madison, 725–734.

LeMasurier, W.E. and Rex, D.C. (1991) The Marie Byrd Land volcanic province and its relation to the Cainozoic West Antarctic Rift System. In Tingey, R.J. (ed.) *The Geology of Antarctica.* Clarendon Press, Oxford, 249–284.

LeMasurier, W.E., Harwood, D.M. and Rex, D.C. (1994) Geology of the Mount Murphy Volcano: an 8 Ma history of interaction between a rift volcano and the West Antarctic ice sheet. *Geological Society of America Bulletin,* 106, 265–280.

Lenihan, H.S. (1992) Benthic marine pollution around McMurdo Station, Antarctica: a summary of findings. *Marine Pollution Bulletin,* 25, 318–323.

Lenihan, H.S. and Oliver, J.S. (1995) Anthropogenic and natural disturbances to marine benthic communities in Antarctica. *Ecological Applications,* 5, 311–326.

Lenihan, H.S., Oliver, J.S., Oakden, J.M. and Stephenson, M.D. (1990) Intense and localised benthic marine pollution around McMurdo Station, Antarctica. *Marine Pollution Bulletin,* 21, 422–430.

Lewis, R.S. (1973) Antarctic research and the relevance of science. In Lewis, R.S. and Smith, P.M. (eds) *Frozen Future. A Prophetic Report from Antarctica.* New York, Quadrangle Books, 1–10.

Light, J.J. and Heywood, R.B. (1975) Is the vegetation of continental Antarctica predominantly aquatic? *Nature,* 256, 199–200.

Lindsay, D.C. (1977) Lichens of cold deserts. In Seaward, M.R.D. (ed.) *Lichen Ecology,* Academic Press, London, 183–209.

Lingle, C.S. (1985) A model of a polar ice stream, and future sea level rise due to possible drastic retreat of the West Antarctic Ice Sheet. In *Glaciers, Ice Sheets and Sea Level: Effect of a CO_2-induced climatic change.* United States Department of Energy, Washington DC, 3170–3330.

Lingle, C.S. and Clark, J.A. (1979) Antarctic ice sheet volume at 18,000 yr. B.P. and Holocene sea-level changes at the West Antarctic margin. *Journal of Glaciology,* 24, 213–230.

Lipps, J.H. (1979) Man's impact along the Antarctic Peninsula. In Parker, B.C. (ed.) *Environmental Impact in Antarctica.* Virginia Polytechnic Institute and State University, Blacksburg, Virginia, 333–371.

Lizotte, M.P. and Sullivan, C.W. (1991) Photosynthesis–irradiance relationships in microalgae associated with Antarctic pack ice: evidence for *in situ* activity. *Marine Ecology Progress Series,* 71, 175–184.

Lockyer, C. (1981) Growth and energy budgets of large baleen whales from the Southern Hemisphere. In *Mammals of the Sea.* FAO. Fisheries Series Vol. 3, Rome, 379–487.

Loeb, V., Siegel, V., Holm-Hansen, O., Hewitt, R., Fraser, W., Trivelpiece, W. and Trivelpiece, S. (1997) Effects of sea-ice extent and krill or salp dominance on the Antarctic food web. *Nature,* 387, 897–900.

Loewe, F. (1974) Considerations concerning the winds of Adélie Land. *Zeitschrift für Gletscherkunde und Glazialgeologie.* 10, 189–197.

Loewe, F. (1962) On the mass economy of the interior of the Antarctic ice cap. *Journal of Geophysical Research,* 67, 5171–5177.

Longton, R.E. (1985) Terrestrial habitats – Vegetation. In Bonner, W.N. and Walton, D.W.H. (eds) *Antarctica.* Pergamon Press, Oxford, 73–105.

Longton, R.E. (1988) *The Biology of Polar Bryophytes.* Cambridge, Cambridge University Press.

Longton, R.E. and Holdgate, M.W. (1967) Temperature relationships of Antarctic vegetation.

Philosophical Transactions of the Royal Society of London, B252, 237–250.

Lorius, C. Jouzel, J., Ritz, C., Merlivat, L., Barkov, N.I., Korotkevich, Y.S. and Kotlyakov, V.M. (1985) A 150,000-year climatic record from Antarctic ice. *Nature*, 316, 591–596.

Lovering, J.F. and Prescott, J.R.V. (1979) *Last of Lands . . . Antarctica*. University of Melbourne Press, Victoria.

Luard, E. (1984) Who owns Antarctica? *Foreign Affairs*, 62, 1175–1193.

Lubin, D., Frederick, J.E., Booth, R.C., Lucas, T. and Neuschuler, D. (1989) Measurements of enhanced springtime ultraviolet radiation at Palmer Station, Antarctica. *Geophysical Research Letters*, 16, 783–785.

Lubin, D., Mitchell, B.G., Frederick, J.E., Alberts, A.D., Booth, C.R., Lucas, T. and Neuschuler, D. (1992) A contribution towards understanding the biospherical significance of Antarctic ozone depletion. *Journal of Geophysical Research*, 97(D8), 7817–7828.

Lucas, P.H.C. and Dingwall, P.R. (1987) Protected areas and environmental conservation in Antarctica and the Southern Ocean. In Nelson, J.G., Needham, R. and Norton, L. (eds) *Arctic Heritage. Proceedings of a Symposium August 24–28, 1985, Banff, Alberta Canada*. Association of Canadian Universities for Northern Studies, Ottawa, 219–241.

Lugten, G.L. (1997) The rise and fall of the Patagonian toothfish – food for thought. *Environmental Policy and Law*, 27, 401–407.

Luke, B.G., Johnstone, G.W. and Woehler, E.J. (1989) Organochlorine pesticides, PCBs and mercury in Antarctic and subantarctic seabirds. *Chemosphere*, 19, 2007–2021.

Lyons, D. (1991) Metal mining in polar regions. *Proceedings First 1991 International Offshore and Polar Engineering Conference, Edinburgh, UK, 11–16 August 1991*. 574–581.

Lyons, D. (1993) Environmental impact assessment in Antarctica under the Protocol on Environmental Protection. *Polar Record*, 29, 111–120.

Macdonald, D.I.M and Butterworth, P.J. (1990) The stratigraphy, setting and hydrocarbon potential of the Mesozoic sedimentary basins of the Antarctic Peninsula. In St John, B. (ed.) *Antarctica as an Exploration Frontier – Hydrocarbon Potential, Geology and Hazards*. American Association of Petroleum Geologists, AAPG, Studies in Geology 31, Tulsa, Oklahoma, 101–125.

Macdonald, D.I.M., Barker, P.F., Garrett, S.W., Ineson, J.R., Pirrie, D., Storey, B.C., Whitham, A.G., Kinghorn, R.R.F. and Marshall, J.E.A. (1988) A preliminary assessment of the hydrocarbon potential of the Larsen Basin, Antarctica. *Marine and Petroleum Geology*, 5, 34–53.

MacKenzie, D. (1992) Gone for good. *New Scientist*, 136(1844), 8.

Mackie, P.R., Platt, H.M. and Hardy, R. (1978) Hydrocarbons in the marine environment. II. Distribution of n-alkanes in the fauna and environment of the sub-Antarctic island of South Georgia. *Estuarine and Coastal Marine Science*, 6, 301–313.

Mackintosh, N.A. (1946) The Antarctic Convergence and the distribution of surface temperatures in Antarctic waters. *Discovery Reports*. 23, 177–212.

Mackintosh, N.A. (1965) *The Stocks of Whales*. Fishing News (Books) Ltd, London.

Mackintosh, N.A. (1970) Whales and krill in the twentieth century. In Holdgate, M.W. (ed.) *Antarctic Ecology*, Vol. 1. Academic Press, London, 195–212.

Mackintosh, N.A. (1972) Life cycle of krill in relation to ice and water conditions. *Discovery Reports*, 36, 1–94.

Makarov, R.R., Men'shenina, L.L. and Latogurskiy, V.I. (1994) A current assessment of the Antarctic krill fishery. *Polar Geography and Geology*, 18, 76–94.

Malcolm, H.M., Boyd, I.L., Osborn, D., French, M.C. and Freestone, P. (1994) Trace metals in Antarctic fur seal (*Arctocephalus gazella*) livers from Bird Island, South Georgia. *Marine Pollution Bulletin*, 28, 375–380.

Manabe, S., Spelman, M.J. and Stouffer, R.J. (1992) Transient responses of a coupled ocean–atmosphere model to gradual increases of atmospheric CO_2. Part II: Seasonal response. *Journal of Climate*, 5, 105–126.

Mandelli, E.F. and Burkholder, P.R. (1966) Primary productivity in the Gerlache and Bransfield straits of Antarctica. *Journal of Marine Research*, 24, 15–27.

Manheim Jr, B.S. (1988) *On Thin Ice. The Failure of the National Science Foundation to Protect Antarctica*. Environmental Defense Fund, Washington DC.

Manheim Jr, B.S. (1990) *Paradise Lost? The Need for Environmental Regulation for Tourism in Antarctica*. Environmental Defense Fund, Washington DC.

Manheim Jr, B.S. (1992) The failure of the National Science Foundation to protect Antarctica. *Marine Pollution Bulletin*, 25, 253–254.

Manning, W.J. and Tiedemann, A.V. (1995) Climate change: potential effects of increased atmospheric carbon dioxide (CO_2), ozone (O_3), and ultraviolet-B

(UV-B) radiation on plant diseases. *Environmental Pollution*, 88, 219–245.

Maquieira, C. (1986) Antarctica prior to the Antarctic Treaty: a political and legal perspective. In Polar Research Board, *Antarctic Treaty System. An Assessment*. National Academy Press, Washington DC, 49–54.

Maquieira, C. (1988) The question of Antarctica at the United Nations: the end of consensus? In Wolfrum, R. (ed.) *Antarctic Challenges III. Conflicting Interests, Co-operation, Environmental Protection, Economic Development*. Duncker & Humblot, Berlin, 253–270.

Marchal, A. (1988) Convention for the Conservation of Antarctic Seals: 1988 review of operations. *Polar Record*, 25, 142–143.

Marchant, D.R. and Denton, D.E. (1996) Miocene and Pliocene palaeoclimate of the Dry Valleys region, southern Victoria Land: a geomorphological approach. *Marine Micropaleontology*, 27(1–4), 269–271.

Marchant, D. R., Swisher III, C.C., Lux, D.R., West Jr, D.P. and Denton, G.H. (1993) Pliocene palaeoclimate and East Antarctic Ice-Sheet history from surficial ash deposits. *Science*, 260, 667–670.

Marchant, D.R., Denton, D.E., Bockheim, J.G., Wilson, S.C. and Kerr, A.R. (1994) Quaternary ice-level changes of upper Taylor Glacier, Antarctica: implications for paleoclimate and ice sheet dynamics. *Boreas*, 23, 29–43.

Marchant, D.R., Denton, G.H., Swisher III, C.C. and Potter Jr, N. (1996) Late Cenozoic Antarctic paleoclimate reconstructed from volcanic ashes in the Dry Valleys region of southern Victoria Land. *Geological Society of America, Bulletin*. 108 (20), 181–194.

Marchant, H.J. (1992) Possible impacts of climate change on the organisms of the Antarctic marine ecosystem. In *Impact of Climatic Change on Antarctica*. Australian Government Publishing Service, Canberra.

Marchant, H.J. (1994) Biological impacts of seasonal ozone depletion. In Hempel, G. (ed.) *Antarctic Science. Global Concerns*. Springer-Verlag, Berlin, 95–109.

Markham, C.R. (1921) *The Lands of Silence: A History of Arctic and Antarctic Exploration*. Cambridge University Press, Cambridge.

Marr, J.W.S. (1935) The South Orkney Islands. *Discovery Reports*, 10, 283–382.

Marschall, H-P. (1988) Overwintering strategy of the Antarctic krill under the pack ice of the Weddell sea. *Polar Biology*, 9, 129–135.

Martin, J.H. (1992) Iron as a limiting factor in oceanic productivity. In Falkowski, P.G. and Woodhead, A.D. (eds) *Primary Productivity and Biogeochemical Cycles in the Sea*. Plenum, New York, 123–137.

Martinson, D.G. (1990) Winter Antarctic mixed layer and sea ice evolution, open deep water formation and ventilation. *Journal of Geophysical Research*, 95, 1161–1165.

Masaki, Y. (1978) Yearly change in the biological parameters of the Antarctic sei whale. *Report of the International Whaling Commision*, 28, 421–429.

Masood, E. (1996) Industrial world is unlikely to meet promised carbon cuts. *Nature*, 382, 103.

Massom, R.A. (1988) The biological significance of open water within the sea ice covers of the polar regions. *Endeavour*, 12(1), 21–27.

Mather, K.B. and Miller, G.S. (1967) The problem of the katabatic winds on the coast of Terre Adélie. *Polar Record*, 13, 425–32.

Mawson, D. (1914) Australasian Antarctic Expedition, 1911–1914. *Geographical Journal*, 44, 257–286.

May, J. (1988) *The Greenpeace Book of Antarctica. A New View of the Seventh Continent*. Dorling Kindersley, London.

May, R.M., Beddington, J.R., Clark, C.W., Holt, S.J. and Laws, R.M. (1979) Management of multispecies fisheries. *Science*, 205, 267–277.

Mayes, P.R. (1981) Recent trends in Antarctic temperatures. *Climate Monitor*, 10, 96–100.

Mayewski, P.A., Lyons, W.B., Spencer, M.J., Twickler, M., Dansgaard, W., Koci, B., Davidson, C.I. and Honrath, R.E. (1986) Sulfate and nitrate concentrations from a south Greenland ice core. *Science*, 232, 975–977.

McCann, T.S. and Rothery, P. (1988) Population size and status of the southern elephant seal (*Mirounga leonina*) at South Georgia, 1951–1985. *Polar Biology*, 8, 305–309.

McDonald, S.J., Kennicutt II, M.C. and Brooks, J.M. (1992) Evidence of polycyclic aromatic hydrocarbon (PAH) exposure in fish from the Antarctic Peninsula. *Marine Pollution Bulletin*, 25, 313–317.

McElroy, J.K. (1980) The potential of krill as a commercial catch. In Mitchell, B. and Sandbrook, R. (eds) *The Management of the Southern Ocean*. International Institute for Environment and Development, London, 60–80.

McElroy, J.K. (1984) Antarctic fisheries. History and prospects. *Marine Policy*, 8, 239–258.

McHugh, J.L. (1974) The role and history of the International Whaling Commission. In Schevill, W.E. (ed.) *The Whale Problem. A Status Report*. Harvard University Press, Cambridge, Mass., 305–368.

McIntosh, W.C. and Gamble, J.A. (1991) A subaerial eruptive environment for the Hallett Coast volcanoes. In Thomson, M.R.A., Crame, J.A. and Thomson, J.W. (eds) *Geological Evolution of Antarctica*. Cambridge University Press, Cambridge, 657–662.

McIver, R.D. (1975) Hydrocarbon gases in canned core samples from Leg 28 sites 271, 272 and 273, Ross Sea. In Hayes, D.E. and Frakes, L.A. (eds) *Initial Reports of the Deep Sea Drilling Project*, 28, 815–17.

McLaughlin, W.R.D. (1962) *Call to the South: a Story of British Whaling in Antarctica*. White Lion, London.

McMinn, A., Heijnis, H. and Hodgson, D. (1994) Minimal effects of UVB radiation on Antarctic diatoms over the past 20 years. *Nature*, 370, 547–549.

McNeely, J.A. and Miller, K.R. (1983) IUCN, National Parks, and protected areas: priorities for action. *Environmental Conservation*, 10, 13–21.

Mellor, M. and Swithinbank, C. (1989) Airfields on Antarctic glacier ice. *CRREL Report*, 19–21.

Mercer, J.H. (1978) West Antarctic ice sheet and CO_2 greenhouse effect: a threat of disaster. *Nature*, 271, 321–325.

Mervis, J. (1993a) NSF hangs out wary welcome sign. *Nature*, 361, 106.

Mervis, J. (1993b) Court says US law applies to Antarctic waste cleanup. *Nature*, 361, 482.

Mervis, J. (1995) Pressure on budget triggers review of Antarctic program. *Science*, 270, 1433–1434.

Mervis, J. (1997) U.S. Antarctic panel makes case for replacement station. *Science*, 275, 173.

Mickleburgh, E. (1987) *Beyond the Frozen Sea. Visions of Antarctica*. The Bodley Head, London.

Mill, H.R. (1905) *The Seige of the South Pole*. Alston Rivers Ltd, London.

Miller, D.G.M. (1991) Exploitation of Antarctic marine living resources: a brief history and a possible approach to managing the krill fishery. *South African Journal of Marine Science*, 10, 321–339.

Miller, D.G.M. and Hampton, I. (1989) Biology and ecology of the Antarctic krill (*Euphausia superba* Dana): a review. *BIOMASS Scientific Series*, No. 9.

Miskelly, N. (1994) A comparison of international definitions for reporting mineral resources and reserves. *Minerals Industry International*, July, 28–36.

Mitchell, B. (1977) Resources in Antarctica. Potential for conflict. *Marine Policy*, 1, 91–101.

Mitchell, B. (1983) *Frozen Stakes: The Future of Antarctic Minerals*. International Institute for Environment and Development, London.

Mitchell, B. and Kimball, L. (1979) Conflict over the cold continent. *Foreign Policy*, 35, 124–141.

Mitchell, B. and Sandbrook, R. (1980) *The Management of the Southern Ocean*. International Institute for Environment and Development, London.

Mitchell, J.F.B. and Senior, C.A. (1989) The Antarctic winter: simulations with climatological and reduced sea ice extents. *Quarterly Journal of the Royal Meteorological Society*, 115, 225–246.

Mitterling, P.I. (1959) *America in the Antarctic to 1840*. University of Illinois Press, Urbana.

Moiseev, P.A. (1970) Some aspects of the commercial use of the krill resources of the Antarctic seas. In Holdgate, M.W. (ed.) *Antarctic Ecology*, Vol. 1. Academic Press, London, 213–216.

Molina, M.J. and Rowland, F.S. (1974) Stratospheric ozone delpetion by halocarbons: chemistry and transport. *Nature*, 249, 810.

Monod, J.L., Arnaud, P.M. and Arnoux, A. (1992) The level of pollution of Kerguelen Islands biota by organochlorine compounds during the seventies. *Marine Pollution Bulletin*, 24, 626–629.

Montzka, S.A., Butler, J.H., Myers, R.C., Thompson, T.M., Swanson, T.H., Clarke, A.D., Lock, L.T. and Elkins, J.W. (1996) Decline in the tropospheric abundance of halogen from halocarbons: implications for stratospheric ozone depletion. *Science*, 272, 1318–1322.

Moreno, C.A. (1980) Observations on food and reproduction in *Trematomus bernacchii* (Pisces: Nototheniidae) from Palmer Archipelago, Antarctica. *Copeia*, 1, 171–173.

Morgan, V.I., Goodwin, I.D., Etheridge, D.M. and Wookey, C.W. (1991) Evidence from Antarctic ice cores for recent increases in snow accumulation. *Nature*, 354, 58–60.

Morris, E., King, J., Turner, J., Peel, D. and Doake, C. (1997) Antarctic climate change – an assessment by the British Antartic Survey. *Globe*, UK Global Environmental Research Office, Swindon, No. 36, 3.

Morrison, S.J. (1990) Warmest year on record on the Antarctic Peninsula? *Weather*, 45, 231–232.

Mosley, G. (1984) The natural option: the case for an Antarctic world park. In Harris, S. (ed.) *Australia's Antarctic Policy Options*. Centre for Resource and Environmental Studies, Australian National University, Canberra, Monograph 11, 307–327.

Murphey, B.B. and Hogan, A.W. (1992) Meteorological transport of continental soot to Antarctica? *Geophysical Research Letters*, 19, 33–36.

Murphy, E.J., Clarke, A., Symon, C. and Priddle, J. (1995) Temporal variation in Antarctic sea-ice: Analysis of a long term fast-ice record from the South Orkney Islands. *Deep Sea Research*. 42, 1045–1062.

Murphy, E.J., Morris, D.J., Watkins, J.L. and Priddle, J. (1988) Scales of interaction between Antarctic krill and the environment. In Sahrhage, D. (ed.) *Antarctic Ocean and Resources Variability*. Springer-Verlag, Berlin. 120–123.

Murphy, E.J. and King, J. (1997) Icy message from the Antarctic. *Nature*, 389, 20–21.

Murphy, R.C. (1948) *Logbook for Grace*. Robert Hale, London.

Murray, J. (1886) The exploration of the Antarctic regions. *Scottish Geographical Magazine*, 2, 527–548.

Nagasaki, F. (1990) The case for scientific whaling. *Nature*, 344, 189–190.

National Research Council (1993) *Science and Stewardship in the Antarctic*. National Academy Press, Washington DC.

National Science Foundation. (1989) *A National Science Foundation Strategy for Compliance with Environmental Law in Antarctica*. National Science Foundation, Washington DC.

National Science Foundation (1991) *Final Supplemental Environmental Impact Statement for the United States Antarctic Program*. National Science Foundation, Washington DC.

Neethling, D.C. (1969) Geology of the Ahlmann Ridge, western Queen Maud Land. In Bushnell, V. and Craddock, C. (eds) *Geologic Maps of Antarctica. Antarctic Map Folio Series*, Folio 12, Plate VII. American Geographical Society, New York.

Neftel, A., Beer, J., Oeschger, H., Zürcher, F. and Finkel, R.C. (1985) Sulphate and nitrate concentrations on snow from south Greenland 1895–1978. *Nature*, 314, 611–613.

Nelson, D.M. and Treguer, P. (1992) Role of silicon as a limiting factor to Antarctic diatoms: evidence from kinetic studies in the Ross Sea ice edge. *Marine Ecology Progress Series*, 80, 255–264.

Nemoto, T., Okiyama, M., Iwasaki, N. and Kikuchi, T. (1988) Squid as predators on krill (*Euphausia superba*) and prey for sperm whales in the Southern Ocean. In Sahrhage, D. (ed.) *Antarctic Ocean and Resources Variability*. Springer-Verlag, Berlin, 292–296.

NERC (1989) *Antarctica 2000. NERC Strategy for Antarctic Research*. Swindon, Natural Environment Research Council.

Nichol, S. (1994) Antarctic krill – changing preceptions of its role in the Antarctic ecosystem. In Hempel, G. (ed.) *Antarctic Science Global Concerns*. Springer-Verlag, Berlin, 144–166.

Nicholls, K.W. (1997) Predicted reduction in basal melt rates of an Antarctic ice shelf in a warmer climate. *Nature*, 388, 460–462.

Nichols, R.L. (1966) Geomorphology of Antarctica. In Tedrow, J.C.F. (ed.) *Antarctic Soils and Soil Forming Processes*. American Geophysical Union, Antarctic Research Series, Washington DC, 8, 1–46.

Nichols, R.L. (1968) Coastal geomorphology, McMurdo Sound, Antarctica. *Journal of Glaciology*, 7, 449–478.

Nicholson, I.E. (1986) Antarctic tourism – the need for a legal regime. In Wolfrum, R. (ed.) *Antarctic Challenge II: Conflicting Interests, Cooperation, Environmental Protection, Economic Development*. Duncker & Humblot, Berlin, 191–203.

Nicol, S. (1989) Who's counting on krill? *New Scientist*, 124 (1960), 38–41.

Nicol, S. (1990) The age old problem of krill longevity. *Bioscience*, 40, 833–836.

Nicol, S. (1991) CCAMLR and its approaches to management of the krill fishery. *Polar Record*, 27, 229–236.

Nicol, S. (1992) Management of the krill fishery: was CCAMLR slow to act? *Polar Record*, 28, 155–157.

Nicol, S. (1994) Antarctic krill – changing perceptions of its role in the Antarctic ecosystem. In Hempel, G. (ed.) *Antarctic Science. Global Concerns*. Springer-Verlag, Berlin, 144–166.

Nicol, S. and de la Mare, W. (1993) Ecosystem management and the Antarctic krill. *American Scientist*, 81, 37–47.

Niebauer, H.J. and Alexander, V. (1989) Current perspectives on the role of ice margins and polynyas in high latitude ecosystems. In Rey, L. and Alexander, V. (eds) *Proceedings of the Sixth Conference of the Comité Arctique International, 13–15 May 1985*. E.J. Brill, Leiden, 121–144.

Nowlin, W.D., Whitworth, T. and Pillsbury, R.D. (1977) Structure and transport of the Antarctic Circumpolar Current at Drake Passage from short term measurements. *Journal of Physical Oceanography*, 7, 788–802.

O'Riordan, T. (1995) Managing the global commons. In O'Riordan, T. (ed.) *Environmental Science for Environmental Management*. Longman Scientific & Technical, Harlow, 347–360.

Odum, E.P. (1983) *Basic Ecology*. CBE Publishing, Philadelphia, 613pp.

Oerlemans, J. (1989) A projection of future sea level. *Climatic Change*, 15, 151–74.

Office of Technology Assessment (1989) *Polar Prospects. A Minerals Treaty for Antarctica*. US Government Printing Office, Washington, DC.

Ogrodowczyk, W. (1981) The effects of hydrocarbons on selected hydrobionts in the coastal zone of the Antarctic waters. *Polish Polar Research*, 2, 95–102.

Ohsumi, S. (1972) Examination of the recruitment rate of Antarctic fin whale stocks by the use of mathematical models. *Report of the International Whaling Commission*, 22, 69–90.

Oke, T.R. (1978) *Boundary Layer Climates*. University Paperbacks, Methuen.

Olivero, E.B., Gasparini, Z., Rinaldi, C.A. and Scasso, R. (1991) First record of dinosaurs in Antarctica (Upper Cretaceous, James Ross Island): palaeogeographical implications. In Thomson, M.R.A., Crame, J.A. and Thomson, J.W. (eds) *Geological Evolution of Antarctica*. Cambridge University Press, Cambridge, 617–622.

Open University (1989) *Ocean Circulation*. The Open University in association with Pergamon Press, Oxford.

Oquist, G. (1983) Effects of low temperature on photosynthesis. *Plant, Cell and Environment*, 6, 281–300.

Orheim, O. (1985) Iceberg discharge and the mass balance of Antarctica. In National Research Council, Polar Research Board, *Glaciers, Ice Sheets and Sea Level: the Effects of a CO_2-induced Climatic Change*. Workshop Report, Seattle, Sept. 13–15, 1984 (DOE/ER 60235-1). Department of Energy, Washington DC.

Øritsland, T. (1970) Sealing and seal research in the south-west Atlantic pack ice, September–October 1964. In Holdgate, M.W. (ed.) *Antarctic Ecology*, Vol. 1. Academic Press, London, 367–376.

Orrego Vicuña, F. (1986) Antarctic conflict and international cooperation. In Polar Research Board, *Antarctic Treaty System: An Assessment*. National Academy Press, Washington DC, 55–64.

Orrego Vicuña, F. (1987) The Antarctic Treaty System: a viable alternative for the regulation of resource-orientated activities. In Triggs, G.D. (ed.) *The Antarctic Treaty Regime. Law, Environment and Resources*. Cambridge University Press, Cambridge, 65–76.

Orrego Vicuña, F. (1988) *Antarctic Mineral Exploration: the Emerging Legal Framework*. Cambridge University Press, Cambridge.

Orrego Vicuña, F. (1991) The effectiveness of the decision making machinery of CCAMLR: an assessment. In Jørgensen-Dahl, A. and Østreng, W. (eds) *The Antarctic Treaty System in World Politics*. Macmillan, Basingstoke, 25–42.

Østreng, W. (1989) Polar science and politics: close twins or opposite poles in international cooperation? In Andresen, S. and Østreng, W. (eds) *International Resource Management: The Role of Science and Politics*. Belhaven Press, London, 88–113.

Østreng, W. (1991) The conflict and alignment pattern of Antarctic politics. Is a new order needed? In Jørgensen-Dahl, A. and Østreng, W. (eds) *The Antarctic Treaty System in World Politics*. Macmillan, Basingstoke, 433–450.

Pain, S. (1993) The two faces of the Exxon disaster. *New Scientist*, 138(1874), 11–13.

Palca, J. (1992) Poles apart, science thrives on thin ice. *Science*, 255, 276–278.

Palmer, A.R. (1983) Decade of North American geologic time scale. *Geology*, 11, 503–504.

Pankhurst, R.J., Storey, B.C. and Millar, I.L. (1991) Magmatism related to the break-up of Gondwana. In Thomson, M.R.A., Crame, J.A. and Thomson, J.W. (eds) *Geological Evolution of Antarctica*. Cambridge University Press, Cambridge, 573–580.

Pannatier, S. (1994) Acquisition of consultative status under the Antarctic Treaty. *Polar Record*, 30, 123–129.

Parfit, M. (1993) Reclaiming a lost Antarctic base. *National Geographic*, 183(3), 110–126.

Parker, B.C. (1972) *Proceedings of the Colloquium on Conservation Problems in Antarctica*. Virginia Polytechnic Institute and State University, Blacksburg, Virginia.

Parker, B.C. (1978) *Environmental Impact in Antarctica*. Virginia Polytechnic Institute and State University, Blacksburg, Virginia.

Parker, B.C. and Angino, E.E. (1990) Environmental impacts of exploiting mineral resources and effects of tourism in Antarctica. In Splettstoesser, J.F. and Dreschhoff, G.A.M. (eds) *Mineral Resources Potential of Antarctica*. American Geographical Union, Antarctic Research Series 51, Washington DC, 237–258.

Parker, B.C. and Howard, R.V. (1977) The first environmental impact monitoring and assessment in Antarctica. The Dry Valley Drilling Project. *Biological Conservation*, 12, 163–177.

Parkinson, C.L. (1992) Southern Ocean sea ice distribution and extents. *Philosophical Transactions of the Royal Society of London*, B338, 243–250.

Parsons, A. (1987) *Antarctica: The Next Decade*. Cambridge University Press, Cambridge.

Pascal, M. (1985) Numerical changes in the population of elephant seals (*Mirounga leonina*) in the Kerguelen Archipelago during the past 30 years. In Beddington, J.R., Beverton, R.J.H. and Lavigne, D.M. (eds) *Marine Mammals and Fisheries*. George Allen & Unwin, London, 170–186.

Passacantando, J. and Carothers, A. (1995) Crisis? What crisis? The ozone backlash. *The Ecologist*, 25, 5–7.

Patel, J. (1990) Antarctica – is Britain undermining the world park? *Ecos*, 11, 47–52.

Patel, J. and Mayer, S. (1991) *Antarctica. The Scientists' Case for a World Park*. Greenpeace, London.

Patel, T. (1994) Peace and quiet for penguins. *New Scientist*, 143(1934), 5.

Paterson, W.S.B. (1994) *The Physics of Glaciers*. Pergamon Press, Oxford.

Patterson, R.G. and Seivers, H.A. (1980) The Weddell–Scotia confluence. *Journal of Physical Oceanography*. 10, 1584–1610.

Payne, M.R. (1977) Growth of a fur seal population. *Philosophical Transactions of the Royal Society of London*, B279, 67–79.

Payne, M.R. (1979) Growth in the Antarctic fur seal *Arctocephalus gazella*. *Journal of Zoology*, 187, 1–20.

Payne, A.J., Sugden, D.E. and Clapperton, C.M. (1989) Modelling the growth and decay of the Antarctic Peninsula ice sheet. *Quaternary Research*, 31, 119–134.

Pearce, F. (1989) Methane: the hidden greenhouse gas. *New Scientist*, 6 May, 37–41.

Pearce, F. (1994) Oil leaks as Antarctic talks stall. *New Scientist*, 144(1952), 5.

Pearce, F. (1995) World lays odds on global catastrophe. *New Scientist*, 146(1972), 4.

Pearce, F. (1996) The ones that got away. *New Scientist*, 149(2012), 14–15.

Pearce, F. (1997) Chill winds at the summit. *New Scientist*, 153(2071), 12.

Peel, D.A. (1975) Organochlorine residues in Antarctic snow. *Nature*, 254, 324–325.

Peel D.A. (1989) Trace metals and organic compounds in ice cores. In Oeschger, H. and Langway Jr, C.C. (eds) *The Environmental Record in Glaciers and Ice Sheets*. Wiley, Chichester, 207–223.

Peel, D.A. (1992) Ice core evidence from the Antarctic Peninsula region. In Bradley, R.S. and Jones, P.D. (eds) *Climate since A.D. 1500*. Routledge, London. 549–571.

Peel, D.A. (1997) European project for ice coring in Antarctica (EPICA), *Globe*, No. 36, UK Global Environmental Research Office, Swindon, 7.

Peel, D.A. and Clausen H.B. (1982) Oxygen-isotope and total beta-radioactivity measurements on 10 m ice cores from the Antarctic Peninsula. *Journal of Glaciology*, 28, 43–55.

Peng, T-H. (1992) Possible effects of ozone depletion on the global carbon cycle. *Radiocarbon*, 34, 772–779.

Penhale, P.A. (1989) Research team focuses on environmental impact of oil spill. *Antarctic Journal of the United States*, 24(2), 9–12.

Perovich, D.K. (1993) A theoretical model of ultraviolet light transmission through Antarctic sea ice. *Journal of Geophysical Research*, 98(C12), 22579–22587.

Peterson, M.J. (1980) Antarctica: the last great land rush on Earth. *International Organisation*, 34, 377–403.

Peterson, M.J. (1988) *Managing the Frozen South. The Creation and Evolution of the Antarctic Treaty System*. University of California Press, Berkeley.

Petit, J.R., Basile, I., Leruyuet, A., Raynaud, D, Lorius, C., Jouzel, J., Stievenard, M., Lipenkov, V.Y., Barkov, N.I., Kudryashov, B.B., Davis, M., Saltzman, E. and Kotlyakov, V. (1997) Four climate cycles in Vostok ice core. *Nature*, 387, 359.

Petri, G. and Zauke, G.-P. (1993) Trace metals in crustaceans in the Antarctic Ocean. *Ambio*, 22, 529–535.

Phillips, C. (1992) The management of science in the newly protected Antarctica – the environmental angle. *Circumpolar Journal*, 7(1–2), 139–146.

Phillpot, H.R. (1985) Physical geography – Climate. In Bonner, W.N. and Walton, D.W.H. (eds) *Key Environments: Antarctica*. Pergamon Press, Oxford, 23–38.

Phillpot, H.R. and Zillman, J.W. (1970) The surface temperature inversion over the Antarctic continent. *Journal of Geophysical Research*, 75, 4146–4169.

Picken, G.B. (1985) Marine habitats – Benthos. In Bonner, W.N. and Walton, D.W.H. (eds) *Key Environments: Antarctica*. Pergamon Press, Oxford, 154–172.

Piper, J.D.A. (1982) The Precambrian palaeomagnetic record: the case for the Proterozoic Supercontinent. *Earth and Planetary Science Letters*, 59, 61–89.

Platt, H.M. (1978) Assessment of the macrobenthos in an Antarctic environment following recent pollution abatement. *Marine Pollution Bulletin*, 9, 149–153.

Platt, H.M. and Mackie, P.R. (1979) Analysis of aliphatic and aromatic hydrocarbons in Antarctic marine sediment layers. *Nature*, 280, 576–578.

Polar Record (1992) Recommendations adopted by the XVIth Antarctic Treaty Consultative Meeting

7–18 October 1991, Bonn, Germany. *SCAR Bulletin*, 104, 6–12, (*Polar Record*, 164, 86–92).

Pontecorvo, G. (1982) The economics of the resources of Antarctica. In Charney, J.I. (ed.) The *New Nationalism and the Use of Common Spaces*. Allanhead, Osmun and Co. Publishers, Totowa, New Jersey, 155–166.

Pople, A., Simpson, A., Simpson, R.O. and Cairns, S.C. (1990) An incident of Southern Ocean oil pollution: effects of a spillage of diesel fuel on the rocky shore of Macquarie Island (sub-Antarctic) *Australian Journal of Marine and Freshwater Research*, 41, 603–620.

Post, A. and Larkum, A.W.D. (1993) UV-absorbing pigments, photosynthesis and UV exposure in Antarctica: comparison of terrestrial and marine algae. *Aquatic Botany*, 45, 231–243.

Pourchet, M., Pinglot, F. and Lorius, C. (1983) Some meteorological applications of radioactive fallout measurements in Antarctic snows. *Journal of Geophysical Research*, 88, 6013–6020.

Prézelin, B. B., Boucher, N.P. and Smith, R.C. (1994) Marine primary production under the influence of the Antarctic ozone hole: Icecolors '90. In Weiler, C.S. and Penhale, P.A. (eds) *Ultraviolet radiation in Antarctica: measurements and biological effects*. American Geophysical Union, Antarctic Research Series 62, Washington DC 159–186.

Priddle, J. (1985) Terrestrial habitats – Inland waters. In Bonner, W.N. and Walton, D.W.H. (eds) *Key Environments: Antarctica*. Pergamon Press, Oxford, 118–132.

Priddle, J. and Heywood, R.B. (1980) The evolution of Antarctic lake ecosystems. *Biological Journal of the Linnean Society*, 14, 51–66.

Priddle, J., Croxall, J.P., Everson, I., Heywood, R.B., Murphy, E.J., Prince, P.A. and Sear, C.B. (1988) Large-scale fluctuations in distribution and attendance of krill – a discussion of possible causes. In Sahrhage, D. (ed.) *Antarctic Ocean and Resources Variability*. Springer-Verlag, Berlin, 169–182.

Priddle, J., Nedwell, D., Savidge, G. and Murphy, E. (1997) The Southern Ocean – plant growth and the global oceanic carbon cycle. *Globe*, No. 36, UK Global Environmental Research Office, Swindon, 8–9.

Priddle, J., Smetachek, V. and Bathmann, U. (1992) Antarctic primary production, biogeochemical carbon cycles and climatic change. *Philosophical Transactions of the Royal Society of London*, B338, 289–297.

Priddle, J., Leakey, R.J.G., Archer, S.D. and Murphy, E.J. (1996) Eukaryotic microbiota in the surface waters and sea ice of the Southern Ocean: aspects of physiology, ecology and biodiversity in a 'two-phase' ecosystem. *Biodiversity and Conservation*, 5, 1473–1504.

Pride, D.E., Cox, C.A., Moody, S.V., Conelea, R.R. and Rosen M.A. (1990) Investigation of mineralization in the South Shetland Islands, Gerlache Strait, and Anvers Island, northern Antarctic Peninsula. In Splettstoesser, J.F. and Dreschhoff, G.A.M. (eds) *Mineral Resources Potential of Antarctica*. American Geophysical Union, Antarctic Research Series 51, Washington DC, 69–94.

Priestley, R.E. (1974) *Antarctic Adventure. Scott's Northern Party*. C. Hurst & Co., London.

Prince, P.A. (1985) Population and energetic aspects of the relationship between blackbrowed and grey-headed albatrosses and the Southern Ocean marine environment. In Siegfried, W.R., Condy, P.R. and Laws, R.M. (Eds) *Antarctic Nutrient Cycles and Food Webs*. Springer-Verlag, Berlin, 473–477.

Prince, P.A., Rothery, P., Croxall, J.P. and Wood, A.G. (1994) Population dynamics of black-browed and grey-headed albatrosses *Diomedea melanophris* and *D. chrysostoma* at Bird Island, South Georgia. *Ibis*, 136, 50–71.

Qingsong, Z. and Peterson, J.A. (1984) A geomorphology and Late Quaternary geology of the Vestfold Hills, Antarctica. *Australian National Antarctic Research Expeditions Report*, No. 133. AGPS, Canberra.

Quartermain, L.B. (1967) *South to the Pole. The Early History of the Ross Sea Sector, Antarctica*. Oxford University Press, London.

Quigg, P.W. (1983) *A Pole Apart. The Emerging Issue of Antarctica*. New Press, McGraw-Hill, New York.

Quilty, P.G. (1988) Cooperation in Antarctica in scientific and logistic matters: status and means of improvement. In Wolfrum, R. (ed.) *Antarctic Challenge III. Conflicting Interest, Co-operation, Environmental Protection, Economic Development*. Duncker & Humblot, Berlin, 65–77.

Quilty, P.G. (1990) Antarctica as a continent for science. In Herr, R.A., Hall, H.R. and Haward, M.G. (eds) *Antarctica's Future: Continuity or Change?* Australian Institute of International Affairs, Hobart, 29–37.

Randall, R.M., Randall, B.M. and Bevan, J. (1980) Oil pollution and penguins – is cleaning justified? *Marine Pollution Bulletin*, 11, 234–237.

Ravich, M.G., Fedorov, L.V. and Tarutin, O.A. (1982) Precambrian iron deposits of the Prince Charles Mountains. In Craddock, C. (ed.)

Antarctic Geoscience. University of Wisconsin Press, Madison, 853–858.

Reader's Digest (1985) *Antarctica. Great Stories from the Frozen Continent*. Reader's Digest, London.

Redgwell, C. (1992) Antarctica – wilderness park or Eldorado postponed? *Environmental Politics*, 1, 137–143.

Reed, S.C. and Sletten, R.S. (1989) Waste management practices of the United States Antarctic Program. *Cold Regions Research and Engineering Laboratory*. Special Report, 89–3.

Reich, J. (1980) The development of Antarctic tourism. *Polar Record*, 20, 203–214.

Reid, D.E. and Anderson, J.B. (1990) Hazards to Antarctic exploration and production. In St John, B. (ed.) *Antarctica as an Exploration Frontier – Hydrocarbon Potential, Geology and Hazards*. American Association of Petroleum Geologists, AAPG Studies in Geology 51, Tulsa, Oklahoma, 31–45.

Richards, R. (1992) The commercial exploitation of sea mammals at Îles Crozet and Prince Edward Islands before 1850. *Polar Monographs* 1. Scott Polar Research Institute, Cambridge.

Rigg, K. (1990) Environmentalists' perspectives on the protection of Antarctica. In Cook, G. (ed.) *The Future of Antarctica. Exploitation Versus Protection*. Manchester University Press, Manchester, 68–80.

Rinaldi, C. (1991) SCAR in the ATS: conflict or harmony. In Jørgensen-Dahl, A. and Østreng, W. (eds) *The Antarctic Treaty System in World Politics*. Macmillan, Basingstoke, 153–160.

Risebrough, R.W. (1977) Transfer of organochlorine pollutants to Antarctica. In Llano, G.A. (ed.) *Adaptations within Antarctic Ecosystems*. Proceedings of the Third SCAR Symposium on Antarctic Biology. Smithsonian Institution, Washington DC, 1203–1210.

Risebrough, R.W., de Lappe, B.W. and Younghans-Haug, C. (1990) PCB and PCT contamination in Winter Quarters Bay, Antarctica. *Marine Pollution Bulletin*, 21, 523–529.

Risebrough, R.W., Walker II, W., Schmidt, T.T., de Lappe, B.W. and Connors, C.W. (1976) Transfer of chlorinated biphenyls to Antarctica. *Nature*, 264, 738–739.

Ritchie, N.A. (1990) Archaeological techniques and technology on Ross Island, Antarctica. *Polar Record*, 26, 257–264.

Roberts, B.B (1939) Whale oil and other products of the whaling industry. *Polar Record*, 3, 80–86.

Roberts, B.B. (1977) Conservation in the Antarctic. *Philosophical Transactions of the Royal Society of London*, B279, 97–104.

Roberts, B.B. (1978) International co-operation for Antarctic development: the test for the Antarctic Treaty. *Polar Record*, 19, 107–120.

Roberts, L. (1989) Does the ozone hole threaten Antarctic life? *Science*, 244, 288–289.

Robin, G. de Q. (1983) *The Climatic Record in the Polar Ice Sheets*. Cambridge University Press, Cambridge.

Robin, G. de Q. (1988) The Antarctic ice sheet, its history and response to sea level and climatic changes over the past 100 million years. *Palaeogeography, Palaeoclimatology, Palaeoecology*, 67, 31–50.

Robinson, E., Bodhaine, B.A., Komhyr, W.D., Oltmans, S.J., Steele, L.P., Tans, P. and Thompson, T.M. (1988) Long-term air quality monitoring at the South Pole by the NOAA program Geophysical Monitoring for Climate Change. *Reviews of Geophysics*, 26, 63–80.

Rodhouse, P.G. (1989a) Antarctic cephalopods – a living marine resource? *Ambio*, 18, 56–59.

Rodhouse, P.G. (1989b) Squid. The Southern Ocean's unknown quantity. *Sea Frontiers*, July–August, 35, 206–211.

Rodhouse, P.G. (1990) Cephalopod fauna of the Scotia Sea at South Georgia: potential for commercial exploitation and possible consequences. In Kerry, K.R. and Hempel, G. (eds) *Antarctic Ecosystems. Ecological Change and Conservation*. Springer-Verlag, Berlin, 289–298.

Rodhouse, P.G. and White, M.G. (1995) Cephalopods occupy the ecological niche of epipelagic fish in the Antarctic Polar Frontal Zone. *Biological Bulletin*, 189, 77–80.

Rona, P.A. (1976) Plate tectonics and mineral resources. In Wilson, J.T. (ed.) *Continents Adrift and Continents Aground*. Scientific American, San Francisco, 207–216.

Roots, E.F. (1983) Resource development in polar regions: comments on technology. In Orrego Vicuña, F. (ed.) *Antarctic Resources Policy: Scientific, Legal and Political Issues*. Cambridge University Press, Cambridge, 297–315.

Roots, E.F. (1986) The role of science in the Antarctic Treaty System. In Polar Research Board, *Antarctic Treaty System. An Assessment*. National Academy Press, Washington DC, 169–184.

Rose, G. and McElroy, C.T. (1987) Coal potential of Antarctica. *Resource Report*, 2. Bureau of Mineral Resources, Geology and Geophysics. Australian Government Publishing Service, Canberra.

Rosman, K.J.R., Chisholm, W., Boutron, C.F., Candelone, J.-P. and Patterson, C.C. (1994) Anthropogenic lead isotopes in Antarctica. *Geophysical Research Letters*, 21, 2669–2672.

Ross, J.C. (1847) *A Voyage of Discovery and Research in the Southern and Antarctic Regions during the Years 1839–43*. John Murray, London.

Ross, R.M. and Quetin, L.B. (1986) How productive are Antarctic krill? *Bioscience*, 36, 264–269.

Ross, R.M. and Quetin, L.B. (1988) *Euphausia superba*: a critical review of estimates of annual production. *Comparative Biochemistry and Physiology*, 90B, 499–505.

Rothwell, D.R. (1990a) The Antarctic Treaty System: resource development, environmental protection or disintegration? *Arctic*, 43, 284–291.

Rothwell, D.R. (1990b) A world park for Antarctica? Foundations, developments and the future. *Antarctic and Southern Ocean Law and Policy Occasional Paper*, 3. Institute of Antarctic and Southern Ocean Studies, University of Tasmania, Hobart.

Rothwell, D.R. (1992) The Madrid Protocol and its relationship with the Antarctic Treaty System. *Antarctic and Southern Ocean Law and Policy Occasional Paper*, 5. Institute of Antarctic and Southern Ocean Studies, University of Tasmania, Hobart.

Rothwell, D.R. and Kaye, S. (1994) Law of the Sea and the polar regions. Reconsidering the traditional norms. *Marine Policy*, 18, 41–58.

Rounsevell, D. and Binns, D. (1991) Mass deaths of king penguins (*Aptenodytes patagonica*) at Lusitania Bay, Macquarie Island. *Aurora*, 10(4), 8–9.

Rounsevell, D.E. and Brothers, N.P. (1984) The status and conservation of seabirds at Macquarie Island. In Croxall, J.P., Evans, P.G.H. and Schreiber, R.W. (eds) *Status and Conservation of the World's Seabirds*. ICBP Technical Publication No. 2, International Council for Bird Preservation, Cambridge, 587–592.

Rounsevell, D.E. and Copson, G.R. (1982) Growth rate and recovery of a king penguin, *Aptenodytes patagonicus*, population after exploitation. *Australian Wildlife Research*, 9, 519–525.

Rowedder, U. (1979) Some aspects of the biology of *Electrona antarctica* (Gunther 1878) (family Myctophidae). *Meeresforschung*, 4, 244–251.

Rowland, F.S. (1990) Stratospheric ozone depletion by chlorofluorocarbons. *Ambio*, 19, 281–292.

Rowley, P.D. (1989) 'Antarctica: The Next Decade', by A. Parsons (book review), *EOS*, 70, 4.

Rowley, P.D. and Pride, D.E. (1982) Metallic mineral resources of the Antarctic Peninsula. In Craddock, C. (ed.) *Antarctic Geoscience*. Madison, University of Wisconsin Press, 859–870.

Rowley, P.D., Ford, A.D., Williams, P.L. and Pride, D.E. (1983) Metallogenic provinces of Antarctica. In Oliver, R.L., James, P.R. and Jago, J.B. (eds) *Antarctic Earth Science*. Cambridge University Press, Cambridge, 414–419.

Rowley, P.D., Williams, P.L. and Pride, D.E. (1991) Metallic and non-metallic mineral resources of Antarctica. In Tingey, R.J. (ed.) *The Geology of Antarctica*. Clarendon Press, Oxford, 617–651.

Royds Consulting Ltd (1994) *New Zealand Antarctic Programme Environmental Audit Under the 1991 Madrid Protocol*. Royds Consulting Ltd, Christchurch, New Zealand.

Rubin, M.J. (1982) James Cook's scientific programme in the Southern Ocean, 1772–75. *Polar Record*, 21, 33–49.

Rubin, M.J. and Weyant, W.S. (1965) Antarctic meteorology. In Hatherton, T. (ed.) *Antarctica*. Methuen, London, 375–401.

Rusin, N.P. (1964) *Meteorological and Radiational Regime of Antarctica*. Israel Progam for Scientific Translations Jerusalem.

Rutford, R.H. (1986) *Reports of the SCAR Group of Specialists on Antarctic Environmental Implications of Possible Mineral Exploration and Exploitation (AEIMEE)*. SCAR, Cambridge.

Ruud, J.T. (1956) International regulation of whaling. A critical survey. *Norsk Hvalfangst-Tidende*, 45, 374–387.

Ryan, K.G. and Beaglehole, D. (1994) Ultraviolet radiation and bottom-ice algae: laboratory and field studies from McMurdo Sound, Antarctica. In Weiler, C.S. and Penhale, P.A. (eds) *Ultraviolet Radiation in Antarctica: Measurements and Biological Effects*. American Geophysical Union, Antarctic Research Series 62, Washington DC, 229–242.

Ryan, P.G. (1987) The origin and fate of artefacts stranded on islands in the African sector of the Southern Ocean. *Environmental Conservation*, 14, 341–346.

Rycroft, M.J. (1994) Antarctica: where space meets planet earth. In Hempel, G. (ed.) *Antarctic Science, Global Concerns*. Berlin, Springer-Verlag, 31–44.

Sabella, S.J. (1992) Upon closer inspection. *Marine Pollution Bulletin*, 25, 255–257.

Sahrhage, D. (1985) Fisheries overview. In Alexander, L.M. and Hanson, L.C. (eds) *Antarctic Politics and Marine Resources: Critical Choices for the 1980s*.

Center for Ocean Management Studies, University of Rhode Island, Kingston, Rhode Island, 101–112.

Sahrhage, D. (1989) Antarctic krill fisheries: potential resources and ecological concerns. In Caddy, J.F. (ed.) *Marine Invertebrate Fisheries: Their Assessment and Management*. Wiley, Chichester, 13–23.

Sahurie, E.J. (1992) *The International Law of Antarctica*. New Haven Press, New Haven, Conn., and Martinus Nijhoff Publishers, Dordrecht.

Sansom, J. (1989) Antarctic surface temperature time series. *Journal of Climate*, 2, 1164–1172.

Sanson, L. (1994) An ecotourism case study in sub-Antarctic islands. *Annals of Tourism*, 21, 344–354.

Sarkar, A., Singbal, S.Y.S. and Fondekar, S.P. (1994) Pesticide residues in the sediments from the lakes of Schirmacher Oasis, Antarctica. *Polar Record*, 30, 33–38.

Savin, S.M., Douglas, R.G. and Stehli, F.G. (1975) Tertiary marine palaeotemperatures. *Geological Society of America Bulletin*, 86, 1499–1510.

Sayers, J. (1993) Infrastructure development and environmental management in Antarctica. In Handmer, J. and Wilder, M. (eds) *Towards a Conservation Strategy for the Australian Antarctic Territory*. Centre for Resource and Environmental Studies, Australian National University, Canberra, 109–147.

SCAR (1987) *SCAR Manual 1987*. Scientific Committee on Antarctic Research, Cambridge.

SCAR (1989) *The Role of Antarctica in Global Change. Scientific Priorities for the International Geosphere–Biosphere Programme (IGBP)*. ICSU Press/SCAR, Cambridge.

SCAR (1993a) *The Role of the Antarctic in Global Change. An International Plan for a Regional Research Programme*. Scientific Committee on Antarctic Research, Cambridge.

SCAR (1993b) Protocol on Environmental Protection to the Antarctic Treaty. *SCAR Bulletin* No. 110. (*Polar Record*, 29, 256–275.)

SCAR (1994) *Coastal and shelf ecology of the Antarctic sea-ice zone (CS–EASIZ)*. Scientific Committee on Antarctic Research. Report No. 10, Scott Polar Research Institute, Cambridge, 20pp.

SCAR (1997) Stations of SCAR nations operating in the Antarctic, winter 1997. SCAR Bulletin No. 127. (*Polar Record*, 33)

SCAR Panel of Experts on Waste Disposal (1989) *Waste Disposal in the Antarctic*. Australian Antarctic Division, Kingston, Tasmania.

Schlesinger, M.E. (1986) Atmospheric general circulation model simulations of the modern Antarctic climate. In *Second international conference on Southern Hemisphere meteorology, December 1–5, 1986, New Zealand*. American Meteorology Society, Boston, 111–112.

Schlesinger, M.E. and Mitchell, J.F.B. (1987) Climate model simulation of the equilibrium climatic response to increased carbon dioxide. *Reviews of Geophysics*, 25, 760–798.

Schoeberl, M.R. and Hartmann, D.L. (1991) The dynamics of the stratospheric polar vortex and its relation to springtime ozone depletions. *Science*, 251, 46–52.

Schofield, E. and Rudolf, E.D. (1969) Factors affecting the distribution of Antarctic terrestrial plants. *Antarctic Journal of the United States*, 4(4) 112–113.

Schwerdtfeger, W. (1970) The climate of the Antarctic. In Landsberg, H.E. (ed.) *World Survey of Climatology*, Vol. 14. Elsevier, Amsterdam, 253–355.

Schwerdtfeger, W. (1984) *Weather and Climate of the Antarctic*. Developments in Atmospheric Science No. 15. Elsevier, Amsterdam.

Scott, J.J. (1990) Changes in vegetation on Heard Island 1947–1987. In Kerry, K.R. and Hempel, G. (eds) *Antarctic Ecosystems. Ecological Change and Conservation*. Springer-Verlag, Berlin, 61–76.

Scott, J.J. and Kirkpatrick, J.B. (1994) Effects of human trampling on the sub-Antarctic vegetation of Macquarie Island. *Polar Record*, 30, 207–220.

Scully, R.T. (1984) The Antarctic Treaty System: overview and analysis. In Alexander, L.M. and Hanson, L.C. (eds) *Antarctic Politics and Marine Resources: Critical Choices for the 1980s*. Center for Ocean Management Studies, Kingston, Rhode Island, 3–11.

Scully, R.T. (1986) The evolution of the Antarctic Treaty System – the institutional perspective. In Polar Research Board, *Antarctic Treaty System. An Assessment*. National Academy Press, Washington DC, 391–411.

Scully, R.T. (1991) The Eleventh Antarctic Treaty Special Consultative Meeting. *International Challenges*, 11, 77–90.

Scully, R.T. and Kimball, L.A. (1989) Antarctica: is there life after minerals? *Marine Policy*, 13, 87–98.

Selkirk, P.M. (1992) Climate Change and the Subantarctic. In Department of the Arts, Sport, the Environment and Territories. *Impact of Climate Change on Antarctica – Australia*. Australian Government Publishing Service, Canberra, 43–51.

Sellers, W.D. (1965) *Physical Climatology*. University of Chicago Press, Chicago.

Shabatie, S., Bentley, C.R., Bindschadler, R.A. and MacAyeal, D.R. (1988) Mass-balance studies of Ice Streams A, B and C, West Antarctica, and possible surging behaviour of Ice Stream B. *Annals of Glaciology*, 11, 137–149.

Shabica, S. V. (1971) The general ecology of the Antarctic limpet *Patinigera polaris*. *Antarctic Journal of the United States*, 6(5), 160–162.

Shackleton, E. (1919) *South*. Heineman, London.

Shackleton, Lord (1982) *Falkland Islands Economic Study 1982*. HMSO, London.

Shackleton, N.J. and Kennett, J.P. (1975) Palaeotemperature history of the Cainozoic and the initiation of Antarctic glaciation: oxygen and carbon analyses in DSDP sites 277; 279 and 281. In Kennett, J.P. and Houtz, R. (eds) *Initial Reports of the Deep Sea Drilling Project*, 29, 743–755.

Shapley, D. (1985) *The Seventh Continent. Antarctica in a Resource Age*. Resources for the Future Inc., Washington DC.

Shaughnessy, P.D. and Goldsworthy, S.D. (1990) Population size and breeding season of the Antarctic fur seal *Arctocephalus gazella* at Heard Island – 1987/88. *Marine Mammal Science*, 6, 292–304.

Shaughnessy, P.D., Shaughnessy, G.L. and Fletcher, L. (1988) Recovery of the fur seal population at Macquarie Island. *Papers and Proceedings of the Royal Society of Tasmania*, 122, 177–187.

Shaw, G.E. (1989) Aerosol transport from sources to ice sheets. In Oeschger, H. and Langway Jr, C.C. (eds) *The Environmental Record in Glaciers and Ice Sheets*. Wiley, Chichester, 13–27.

Shears, J. (1994) A review of environmental impact assessments (EIAs) prepared for proposed activities in Antarctica. *XVIII ATCM Information Paper*, 9.

Shears, J.R. (1993) *The British Antarctic Survey Waste Management Handbook*. British Antarctic Survey, Cambridge.

Shears, J.R. (1995) *Initial Environmental Evaluation. Expansion of Rothera Research Station, Rothera Point, Adelaide Island, Antarctica*. British Antarctic Survey, Cambridge.

Shears, J.R. and Hall, J. (1992) Abandoned stations and field huts. The British approach to management. In Melander, O. and Fontana, L.R. (eds) *Proceedings of the Fifth Symposium on Antarctic Logistics and Operations*. Direccion Nacional del Antartico, Buenos Aires, 12–26.

Siegel, V. (1988) A concept of seasonal variation of krill (*Euphausia superba*) distribution and abundance west of the Antarctic Peninsula. In Sahrhrage, D. (ed.) *Antarctic Ocean and Resources Variability*, Springer-Verlag, Berlin, 218–230.

Siegert, M.J., Dowdeswell, J.A., Gorman, M.R. and McIntyre, N.F. (1996) An inventory of Antarctic lakes. *Antarctic Science*, 8(37), 281–286.

Siegfried, W.R., (1985b) Birds and mammals: oceanic birds of the Antarctic. In Bonner, W.N. and Walton, D.W.H. (eds) *Key Environments: Antarctica*. Pergamon Press, Oxford.

Simmonds, I. and Budd, W.F. (1990) A simple parameterisation of ice leads in a General Circulation Model, and the sensitivity of climate to a change in Antarctic ice concentration. *Annals of Glaciology*, 14, 266–269.

Simmonds, I. (1992) Modelling the reaction of Antarctica to climate changes at its periphery. In Department of the Arts, Sport, the Environment and Territories, *Impact of Climate Change on Antarctica – Australia*. Australian Government Publishing Service, Canberra, 16–23.

Simpson, R.D., Smith, S.D.A. and Pople, A.R. (1995) The effects of a spillage of diesel fuel on a rocky shore in the sub-Antarctic region (Macquarie Island). *Marine Pollution Bulletin*, 31, 367–371.

Simpson-Housley, P. (1992) *Antarctica: Exploration, Perception and Metaphor*. Routledge, London.

Sitwell, N. (1993) A safe course through southern waters. *New Scientist*, 140(1895), 45–46.

Sladen, W.J.L. and Leresche, R.E. (1970) New and developing techniques in Antarctic ornithology. In Holdgate, M.W. (ed.) *Antarctic Ecology*, Vol. 1. Academic Press, London, 585–596.

Sladen, W.J.L., Menzie, C.M. and Reichel, W.L. (1966) DDT residues in Adélie penguins and a crabeater seal from Antarctica. *Nature*, 210, 670–673.

Slip, D.J. and Burton IV, H.R. (1991) Accumulation of fishing debris, plastic litter, and other artefacts, on Heard and Macquarie Islands in the Southern Ocean. *Environmental Conservation*, 18, 249–254.

Smetacek, V., Schark, R. and Nothig, E.-M., (1990) Seasonal and regional variation in the pelagial and its relationship to the life history cycle of krill. In Kerry, K.R. and Hempel, G. (eds.) *Antarctic Ecosystems. Ecological Change and Conservation*, Springer-Verlag, Berlin, 103–114

Smith, H.G. (1978) The distribution and ecology of terrestrial protozoa of subantarctic and maritime Antarctic islands. *British Antarctic Survey Scientific Reports*, 95, 1–104.

Smith, J. (1960) Glacier problems in South Georgia. *Journal of Glaciology*, 3, 705–714.

Smith, R.C., Prézelin, B.B., Baker, K.S., Bidigare, R.R., Boucher, N.P., Coley, T., Karentz, D., MacIntyre, S., Matlick, H.A., Menzies, D., Ondrusek, M., Wan, Z. and Waters, K.J. (1992) Ozone depletion: ultraviolet radiation and phytoplankton biology in Antarctic waters. *Science*, 255, 952–959.

Smith, R.I. L., (1984) Terrestrial plant biology of the subantarctic and Antarctic. In Laws, R.M. (ed.) *Antarctic Ecology*, Vol. 1. Academic Press, London, 61–162.

Smith, R.I.L. (1988a) Destruction of Antarctic terrestrial ecosystems by a rapidly increasing fur seal population. *Biological Conservation*, 45, 55–72.

Smith, R.I.L. (1988b) Botanical survey of Deception Island. *British Antarctic Survey Bulletin*, 80, 129–136.

Smith, R.I.L. (1990) Signy Island as a paradigm of biological and environmental change in Antarctic terrestrial ecosystems. In Kerry, K.R. and Hempel, G. (eds) *Antarctic Ecosystems. Ecological Change and Conservation*. Springer-Verlag, Berlin, 32–50.

Smith R.I.L. (1993a) Dry coastal ecosystems of Antarctica. In van der Maarel, E. (ed.) *Ecosystems of the World. 2A. Dry Coastal Ecosystems – Polar Regions and Europe*. Elsevier, London, 51–71.

Smith R.I.L. (1993b) Dry coastal ecosystems on sub-Antarctic islands. In van der Maarel, E. (ed.) *Ecosystems of the World. 2A. Dry Coastal Ecosystems – Polar Regions and Europe*. Elsevier, London, 73–93.

Smith, R.I.L. (1993c) The role of bryophyte propagule banks in primary succession: case study of an Antarctic fellfield soil. In Miles, J. and Walton, D.W.H. (eds) *Primary Succession on Land*. Blackwell, Oxford, 55–78.

Smith, R.I.L. (1994a) Introduction to the Antarctic protected area system. In Smith, R.I.L., Walton, D.W.H. and Dingwall, P.R. (eds) *Developing the Antarctic Protected Area System*. Conservation of the Southern Polar Region No.1. IUCN, Gland and Cambridge, 15–26.

Smith, R.I.L. (1994b) Environmental – geographic basis for the protected area system. In Smith, R.I.L., Walton, D.W.H. and Dingwall, P.R. (eds) *Developing the Antarctic Protected Area System*. Conservation of the Southern Polar Region No.1. IUCN, Gland and Cambridge, 27–36.

Smith, R.I.L. (1995a) Conservation status of South Georgia and the South Sandwich Islands. In Dingwall, P.R. (ed.) *Progress in Conservation of the Subantarctic Islands*. Conservation of the Southern Polar Region No. 2. IUCN, Gland and Cambridge, 3–14.

Smith, R.I.L. (1995b) Human activity and the requirement for environmental impact assessment. In Dingwall, P.R. (ed.) *Progress in Conservation of the Subantarctic Islands*. Conservation of the Southern Polar Region No. 2. IUCN, Gland and Cambridge, 141–146.

Smith, R.I.L. (1996) Introduced plants in Antarctica: potential impacts and conservation issues. *Biological Conservation*, 76, 135–146.

Smith, R.I.L. and Simpson, H.W. (1987) Early nineteenth century sealers' refuges on Livingston Island, South Shetland Islands. *British Antarctic Survey Bulletin*, 74, 49–72.

Smith, R.I.L. and Walton, D.W.H. (1975) South Georgia, Subantarctic. In Rosswall, T. and Heal, O.W. (eds) *Structure and Function of Tundra Ecosystems*. Ecological Bulletins (Stockholm) 20, 399–423.

Smith, R.I.L., Walton, D.W.H. and Dingwall, P.R. (1994) *Developing the Antarctic Protected Area System*. Conservation of the Southern Polar Region No.1. IUCN, Gland and Cambridge.

Smith, V.R. and Smith, R.I.L. (1987) The biota and conservation status of sub-Antarctic islands. *Environment International*, 13, 95–104.

Smith, V.R. and Steenkamp, M. (1990) Climatic change and its ecological implications at a sub-Antarctic island. *Oecologia*, 85, 14–24.

Smith, W.O., Keene, N.K. and Comiso, J.C. (1988) Interannual variability in estimated primary productivity of the Antarctic marginal ice zone. In Sahrhage, D. (ed.) *Antarctic Ocean and Resources Variability*. Springer-Verlag, Berlin, 131–139.

Sollie, F. (1983) Jurisdictional problems in relation to Antarctic mineral resources in political perspective. In Orrego Vicuña, F. (ed.) *Antarctic Resources Policy. Scientific, Legal and Political Issues*. Cambridge University Press, Cambridge, 317–335.

Sollie, F. (1984) The development of the Antarctic Treaty System – trends and issues. In Wolfrum, R. (ed.) *Antarctic Challenge. Conflicting Interests, Co-operation, Environmental Protection, Economic Development*. Duncker & Humblot, Berlin, 17–37.

Solomon, S. (1990) Progress towards a quantitative understanding of Antarctic ozone depletion. *Nature*, 347, 347–354.

Sømme, L. (1985) Terrestrial habitats – Invertebrates. In Bonner, W.N. and Walton, D.W.H. (eds) *Key Environments. Antarctica*. Pergamon Press, Oxford, 106–117.

Speak, P. (1992) William Speirs Bruce and the Scottish National Antarctic Expedition. *Scottish Geographical Magazine*, 108, 138–148.

Spindler, M. and Dieckmann, G. (1994) Ecological significance of the sea ice biota. In Hempel, G. (ed.) *Antarctic Science: Global Concerns*. Springer-Verlag, Berlin, 60–68.

Spivak, J. (1974) Frozen assets? *Wall Street Journal*, February 21, 1.

Splettstoesser, J.F. (1985) Antarctic geology and mineral resources. *Geology Today*, 1(2), 41–45.

Splettstoesser, J.F. and Dreschhoff, G.A.M. (1990) *Mineral Resources Potential of Antarctica*. American Geophysical Union, Antarctic Research Series 51, Washington DC.

Splettstoesser, J.F. and Folks, M.C. (1994) Environmental guidelines for tourism in Antarctica. *Annals of Tourism Research*, 21, 231–244.

Squire, V. (1991) Atmosphere–Ice–Ocean: do we really understand what is going on? In Harris, C.M. and Stonehouse, B. (eds) *Antarctica and Global Climatic Change*. Belhaven Press, London, 82–89.

St John, B. (1986) Antarctica – geology and hydrocarbon potential. In Halbouty, M.T. (ed.) *Future Petroleum Resources of the World*. American Association of Petroleum Geologists, Tulsa, Oklahoma, 55–100.

St John, B. (1990) *Antarctica as an Exploration Frontier – Hydrocarbon Potential, Geology and Hazards*. American Association of Petroleum Geologists, AAPG Studies in Geology 51. Tulsa, Oklahoma.

Staffelbach, T., Neftel, A., Stauffer, B. and Jacob, D. (1991) A record of the atmospheric methane sink from formaldehyde in polar ice cores. *Nature*, 349, 603–605.

Stamnes, K., Jin, Z., Slusser, J., Booth, E. and Lucas, T. (1992) Several-fold enhancement of biologically effective ultraviolet radiation levels at McMurdo Station Antarctica during the 1990 ozone 'hole'. *Geophysical Research Letters*, 19, 1013–1016.

Stark, P. (1994) Climatic warming in the central Antarctic Peninsula area. *Weather*, 49(5), 215–220

Stauffer, B., Fischer, G., Neftel, A. and Oeschger, H. (1985) Increase of atmospheric methane recorded in Antarctic ice core V229. *Science*, 229, 1386–1388.

Steele, W.K. and Hiller, A. (1997) Radiocarbon dates of snow petrel (*Pagodroma nivea*) nest sites in central Dronning Maud Land, Antarctica. *Polar Record*, 33, 29–38.

Stenberg, M., Eriksson, C. and Heintzenberg, J. (in press) Trace substances and firn from the vicinity of two small research stations in Antarctica. *Ambio*.

Stephenson, L. (1993) Managing visitors to Macquarie Island – a model for Antarctica? *ANARE News*, 73, 8–9.

Stevenson, P.J. (1961) Patterned ground in Antarctica. *Journal of Glaciology*, Vol. 3, No. 30, 1163.

Stokke, O.S. (1996) The effectiveness of CCAMLR. In Stokke, O.S. and Vidas, D. (eds) *Governing the Antarctic. The Effectiveness and Legitimacy of the Antarctic Treaty System*. Cambridge University Press, Cambridge, 120–151.

Stolarski, R.S., Kreuger, A.J., Schroebel, M.R., McPeters, R.D., Newman, P.A. and Alpert, J.C. (1986) Nimbus 7 satellite measurements of the springtime Antarctic ozone decrease. *Nature*, 322, 808–811.

Stonehouse, B. (1972) *Animals of the Antarctic. The Ecology of the Far South*. Peter Lowe, Eurobook, London.

Stonehouse, B. (1985) Penguins. In Bonner, W.N. and Walton, D.W.H. (eds) *Key Environments: Antarctica*, Pergamon Press, Oxford, 266–292.

Stonehouse, B. (1989) *Polar Ecology*. Blackie, Glasgow and London.

Stonehouse, B. (1990) A traveller's code for Antarctic visitors. *Polar Record*, 26, 56–58.

Stonehouse, B. (1992a) Monitoring shipborne visitors in Antarctica: a preliminary field study. *Polar Record*, 28, 213–218.

Stonehouse, B. (1992b) IAATO: an association of Antarctic tour operators. *Polar Record*, 28, 322–324.

Stonehouse, B. (1993) Shipborne tourism in Antarctica: Project Antarctic Conservation Studies 1992/93. *Polar Record*, 29, 330–332.

Stonehouse, B. (1994) Ecotourism in Antarctica. In Carter, E. and Lowman, G. (eds) *Ecotourism. A Sustainable Option*? Wiley, Chichester, 195–212.

Stonehouse, B. and Crosbie, K. (1995) Tourist impacts and management in the Antarctic Peninsula area. In Hall, C.M. and Johnston, M.E. (eds) *Polar Tourism: Tourism in the Arctic and Antarctic Regions*. Chichester, Wiley, 217–233.

Storey, B.C. (1993) The changing face of late Precambrian and early Palaeozoic reconstructions. *Journal of the Geological Society of London*. 150, 665–668.

Storey, B.C. (1991) The crustal blocks of West Antarctica within Gondwana: reconstruction and break-up model. In Thomson, M.R.A., Crame, J.A. and Thomson, J.W. (eds) *Geological Evolution of Antarctica*. Cambridge University Press, Cambridge. 587–592.

Storey, B.C. (1995) The role of mantle plumes in continental breakup: case studies from Gondwanaland. *Nature*, 377, 301–308.

Storey, B.C. and Garrett, S.W. (1985) Crustal growth of the Antarctic Peninsula by accretion, magmatism and extension. *Geological Magazine*, 122(1), 5–14.

Stouffer, R.J., Manabe, S., and Bryan, K. (1989) Interhemispheric asymmetry in climate response to a gradual increase of atmospheric CO_2. *Nature*, 342, 660–662.

Stuiver, M., Denton, G.H., Hughes, T.J. and Fastook, J.L. (1981) History of the marine ice sheet in West Antarctica during the last glaciation: a working hypothesis. In Denton, G.H. and Hughes, T.J. (eds) *The Last Great Ice Sheets*. Wiley-Interscience, New York, 319–436.

Stump, E. (1992) The Ross Orogen of the Transantarctic Mountains in the light of the Laurentia–Gondwana split. *Geological Society of America, GSA Today*, 2(2), 25–31.

Subramanian, A., Tanabe, S., Hikada, H. and Tatsukawa, R. (1986) Bioaccumulation of organochlorines (PCBs and p,p'–DDE) in Antarctic Adélie penguins *Pygoscelis adeliae* collected during a breeding season. *Environmental Pollution (Series A)*, 40, 173–189.

Sugden, D.E. (1982) *Arctic and Antarctic*. Blackwell, Oxford.

Sugden, D.E. (1987) The Polar and Glacial World. In Clark, M.J., Gregory, K.J. and Gurnell, A.M. (eds) *Horizons in Physical Geography*. Macmillan, London, 214–231.

Sugden, D.E. (1991) The stepped response of ice sheets to climatic change. In Harris, C.M. and Stonehouse, B. (eds) *Antarctica and Global Climatic Change*. Belhaven Press, London, 107–114.

Sugden, D.E. (1992) Antarctic ice sheets at risk? *Nature*, 359, 775–776.

Sugden, D.E. (1996) The East Antarctic Ice Sheet: unstable ice or unstable ideas? *Transactions of the Institute of British Geographers*, 21, 443–454.

Sugden, D.E. and Clapperton, C.M. (1977) The maximum ice extent on island groups in the Scotia Sea, Antarctica. *Quaternary Research*, 7, 268–282.

Sugden, D.E. and Clapperton, C.M. (1980) West Antarctic ice sheet fluctuations in the Antarctic Peninsula area. *Nature*, 286, 378–381.

Sugden, D.E. and John, B.S. (1976) *Glaciers and Landscape*. Edward Arnold, London.

Sugden, D.E., Denton, G.H. and Marchant, D.R. (1993) The case for a stable East Antarctic Ice Sheet. *Geografiska Annaler*, 75, (4) 151–351.

Sugden, D.E., Denton, G.H. and Marchant, D.R. (1995) Landscape evolution of the Dry Valleys, Transantarctic Mountains: Tectonic implications. *Journal of Geophysical Research*, 100, 9949–9967.

Sullivan, C.W., McClain, C.R., Comiso, J.C. and Smith, W.O. (1988) Phytoplankton standing crops within an Antarctic ice edge assessed by satellite remote sensing. *Journal of Geophysical Research*, 93, 12487–12498.

Sullivan, W. (1961) *Assault on the Unknown. The International Geophysical Year*. Hodder & Stoughton, London.

Sun, S., de la Mare, W. and Nicol, S. (1995) The compound eye as an indicator of age and shrinkage in Antarctic krill. *Antarctic Science*. 7(4), 387–392.

Suter, K. (1991) *Antarctica: Private Property or Public Heritage*? Zed Books, London and New Jersey.

Suttie, E.D. and Wolff, E.W. (1993) The local deposition of heavy metal emissions from point sources in Antarctica. *Atmospheric Environment*, 27A, 1833–1841.

Suzuki, T. and Shibata, N. (1990) The utilisation of Antarctic krill for human food. *Food Reviews International*, 6, 119–147.

Swithinbank, C.W.M. (1988a) Glaciers of the World: Antarctica. *US Geological Survey Professional Paper*, 1386–B.

Swithinbank, C.W.M. (1988b) Antarctic Airways: Antarctica's first commercial airline. *Polar Record*, 24, 313–316.

Swithinbank, C.W.M. (1990) Non-government aircraft in the Antarctic 1989/90. *Polar Record*, 26, 316.

Swithinbank, C.W.M. (1991) Potential airfield sites in Antarctica for wheeled aircraft. *CRREL Special Report*, 91–24.

Swithinbank, C.W.M. (1993a) Airborne tourism in the Antarctic. *Polar Record*, 29, 103–110.

Swithinbank, C.W.M. (1993b) Non-government aircraft in the Antarctic in 1992/93. *Polar Record*, 29, 244–245.

Swithinbank, C.W.M. and Zumberge, J.H. (1965) The ice shelves. In Hatherton, T. (ed.) *Antarctica*. Methuen, London, 199–220.

Szabo, M. (1994) *State of the Ice: An Overview of Human Impacts in Antarctica*. Greenpeace International, Amsterdam.

Szefer, P., Szefer, K., Pempkowiak, J., Skwarzec, B. and Bojanowski, R. (1994) Distribution and coassociations of selected metals in seals of the Antarctic. *Environmental Pollution*, 83, 341–349.

Tanabe, S., Hidaka, H. and Tatsukawa, R. (1983) PCBs and chlorinated hydrocarbon pesticides in Antarctic atmosphere and hydrosphere. *Chemosphere*, 12, 277–288.

Tangley, L. (1988) Who's polluting Antarctica? *Bioscience*, 38, 590–594.

Targett, T. (1981) Trophic ecology and structure of coastal Antarctic fish communities. *Marine Ecology Research Series*, 4, 243–263.

Tauber, G. M. (1960) Characteristics of Antarctic katabatic winds. In *Antarctic Meteorology*. Proc. Symp. Melbourne. 1959. Pergamon Press, Oxford, 52–64.

Taylor, R.H. and Wilson, P.R. (1990) Recent increase and southern expansion of Adélie penguin populations in the Ross Sea, Antarctica, related to climate warming. *New Zealand Journal of Ecology*, 14, 25–29.

Taylor, R.H., Wilson, P.R. and Thomas, B.W. (1990) Status and trends of Adélie penguin populations in the Ross Sea region. *Polar Record*, 26, 293–304.

Testa, J.W., Oehlert, G., Ainley, D.G., Bengtson, J.L., Siniff, D.B., Laws, R.M. and Rounsevell, D. (1991) Temporal variability in Antarctic marine ecosystems: periodic fluctuations in the phocid seals. *Canadian Journal of Zoology*, 48, 631–639.

Thom, G. (1978) Disruption of bedrock by the growth and collapse of ice lenses. *Journal of Glaciology*, 20, 571–575.

Thomas, R.H., Sanderson, T.J.O. and Rose, K.E. (1979) Effect of climatic warming on the West Antarctic ice sheet. *Nature*, 277, 355–358.

Thompson, D.R., Furness, R.W. and Lewis, S.A. (1993) Temporal and spatial variation in mercury concentrations in some albatrosses and petrels from the sub-Antarctic. *Polar Biology*, 13, 239–244.

Thomson, M.R.A. and Swithinbank, C. (1985) The prospects for Antarctic minerals. *New Scientist*, 107(1467), 31–35.

Thomson, M.R.A., Crame, J.A. and Thomson, J.W. (eds) (1991) *Geological Evolution of Antarctica*. Cambridge University Press, Cambridge.

Thomson, M.R.A., Pankhurst, R.J. and Clarkson, P.D. (1983) The Antarctic Peninsula – a late Mesozoic–Cainozoic arc. In Oliver, R.L., James, P.R. and Jago, J.B. (eds) *Antarctic Earth Science*. Cambridge University Press, Cambridge, 289–294.

Thomson, R.B. (1977) Effects of human disturbance on an Adélie penguin rookery and measures of control. In Llano, G.A. (ed.) *Adaptations Within Antarctic Ecosystems. Proceedings of the Third SCAR Symposium on Antarctic Biology*. Smithsonian Institution, Washington DC, 1177–1180.

Thornton, B.S. (1990) A new model convention for Antarctic conservation. *Polar Record*, 26, 329–330.

Tilbrook, P.J. (1970) The terrestrial environment and invertebrate fauna of the maritime Antarctic. In Holdgate, M.W. (ed.) *Antarctic Ecology*, Vol. 2., Academic Press, London, 886–896.

Tilbrook, P.J. (1977) Energy flow through a population of the collembolan *Cryptopygus antarcticus*. In Llano, G.A. (ed.) *Adaptations within Antarctic Ecosystems*. Proceedings of the Third SCAR Symposium on Antarctic Biology. Smithsonion Institution, Washington DC, 935–946.

Tingey, R.J. (1990) Banded iron formations in East Antarctica. In Splettstoesser, J.F. and Dreschhoff, G.A.M. (eds) *Mineral Resources Potential of Antarctica*. American Geophysical Union, Antarctic Research Series 51, Washington DC, 125–131.

Tingey, R.J. (1991) The regional geology of Archean and Proterozoic rocks in Antarctica. In Tingey, R.J. (ed.) *The Geology of Antarctica*. Clarendon Press, Oxford. 1–58.

Tinker, J. (1979) Cold war over Antarctic wealth. *New Scientist*, 83(1173), 867–868.

Tolbert, M.A., Rossi, M.J., Malhotra, R. and Golden, D.M. (1987) Reaction of chlorine nitrate with hydrogen chloride and water at Antarctic stratospheric temperatures. *Science*, 238, 1258.

Tønnessen, J.N. (1970) Norwegian Antarctic whaling, 1905–68: an historical appraisal. *Polar Record*, 15, 283–290.

Tønnessen, J.N. and Johnsen, A.O. (1982) *The History of Modern Whaling*. C. Hurst & Co., London.

Tooley, R.V. (1985) *The Mapping of Australia and Antarctica*, 2nd edn. Holland Press, London.

Townrow, K. (1988) Sealing sites on Macquarie island: an archaeological survey. *Papers and Proceedings of the Royal Society of Tasmania*, 122, 15–25.

Trathan, P.N., Croxall, J.P. and Murphy, E.J. (1996) Dynamics of Antarctic penguin populations in relation to inter–annual variability in sea ice distribution. *Polar Biology*, 16, 321–330.

Treguer, P. (1994) The Southern Ocean: biogeochemical cycles and climate changes. In Hempel, G. (ed.) *Antarctic Science: Global Concerns*. Springer-Verlag, Berlin, 110–125.

Treguer, P. and Jacques, G. (1992) Dynamics of nutrients and phytoplankton and fluxes of carbon, nitrogen and silicon in the Antarctic Ocean: a review. *Polar Biology*, 12, 149–162.

Trenbreth, K.E., Christy, J.R. and Olson, J.G. (1987) Global atmospheric mass, surface pressure and water vapour variations. *Journal of Geophysical Research*, 92, 14815–14826.

Triggs, G. (1984) Australian sovereignty in Antarctica: traditional principles of territorial acquisition versus a 'common heritage'. In Harris, S. (ed.) *Australia's Antarctic Policy Options*, Centre for Resource and Environmental Studies, Australian National University, Canberra, Monograph 11, 29–66.

Triggs, G.D. (1987a) *The Antarctic Treaty Regime. Law, Environment and Resources*. Cambridge University Press, Cambridge.

Triggs, G.D. (1987b) The Antarctic Treaty System: some jurisdictional problems. In Triggs, G.D. (ed.) *The Antarctic Treaty Regime. Law, Environment and Resources*. Cambridge University Press, Cambridge, 88–109.

Triggs, G.D. (1987c) Negotiation of a minerals regime. In Triggs, G.D. (ed.) *The Antarctic Treaty Regime. Law, Environment and Resources*. Cambridge University Press, Cambridge, 182–195.

Triggs, G.D. (1988) Antarctica, the glittering prize. *Marine Pollution Bulletin*, 19, 202–209.

Triggs, G.D. (1990) A comprehensive environmental regime for Antarctica: a new way forward. In Herr, R.A., Hall, H.R. and Haward, M.G. (eds) *Antarctica's Future: Continuity or Change*? Australian Institute of International Affairs, Hobart, 103–118.

Trivelpiece, W.Z., Ainley, D.G., Fraser, W.R. and Trivelpiece, S.C. (1990a) Skua survival. *Nature*, 345, 211.

Trivelpiece, W.Z., Trivelpiece, S.G., Geupel, G.R., Kjelmyr, J. and Volkman, N.J. (1990b) Adélie and chinstrap penguins: their potential as monitors of the Southern Ocean marine ecosystem. In Kerry, K.R. and Hempel, G. (eds) *Antarctic Ecosystems. Ecological Change and Conservation*. Springer-Verlag, Berlin, 191–202.

Trodahl, H.J. and Buckley, R.G. (1989) Ultraviolet levels under sea ice during the Antarctic spring. *Science*, 245, 194–195.

Truswell, E.M. (1991) Antarctica: a history of terrestrial vegetation. In Tingey, R.J. (ed.) *The Geology of Antarctica*. Clarendon Press, Oxford. 499–537.

Tumeo, M.A. and Larsen, M.K. (1994) Movement of fuel spills in the Ross Ice Shelf. *Antarctic Journal of the United States*, 29 (5), 373–374.

Tumeo, M.A. and Wolk, A.E. (1994) Assessment of the presence of oil-degrading microbes at McMurdo Station. *Antarctic Journal of the United States*, 29(5), 375–377.

Usher, M.B. and Edwards, M. (1986a) A biometrical study of the family Tydeidae (Acari, Prostigmata) in the maritime Antarctic, with descriptions of three new taxa. *Journal of Zoology*, 209, 355–383.

Usher, M.B. and Edwards, M. (1986b) The selection of conservation areas in Antarctica: an example using the arthropod fauna of Antarctic islands. *Environmental Conservation*, 13, 115–122.

Usher, M.B., Block, W. and Jumeau, P.J.A.M. (1989) Predation by arthropods in an Antarctic terrestrial community. In Heywood, R.B. (ed.) *University Research in Antarctica*. Proceedings of British Antarctic Survey Antarctic Special Topic Award Scheme Symposium. British Antarctic Survey, Cambridge, 123–130.

van Bennekom, S. (1992) A new regime to protect the Antarctic environment. *Leiden Journal of International Law*, 5, 33–52.

van der Leun, J., Xiaoyan Tang and Tevini, M. (1995) Environmental effects of ozone depletion: 1994 assessment. *Ambio*, 24, 138–142.

van der Veen, C.J. and Oerlemans, J. (eds.) (1987) *Dynamics of the West Antarctic Ice Sheet*. D. Riedel Publishing, Dordrecht.

van Franeker, J.A. and Bell, P.J. (1988) Plastic ingestion by petrels breeding in Antarctica. *Marine Pollution Bulletin*, 19, 672–674.

van Scoy, K. and Coale, K. (1994) Pumping iron in the Pacific. *New Scientist*, 144 (1954), 32–35.

Vaughan, D.G. and Doake, C.S.M. (1996) Recent atmospheric warming and retreat of ice shelves on the Antarctic Peninsula. *Nature*, 379, 328–331.

Vaughan, D.G. and Lachlan-Cope, T. (1995) Recent retreat of ice shelves on the Antarctic Peninsula. *Weather*, 50(11), 374–376.

Vergani, D.F. and Stanganelli, Z.B. (1990) Fluctuations in breeding populations of elephant seals *Mirounga leonina* at Stranger Point, King George Island 1980–1988. In Kerry, K.R. and Hempel, G. (eds) *Antarctic Ecosystems. Ecological Change and Conservation*. Springer-Verlag, Berlin, 241–245.

Verhoeven, J., Sands, P. and Bruce, M. (eds) (1992) *The Antarctic Environment and International Law*. Graham & Trotman, London.

Vernet, M., Brody, A., Holm–Hansen, O. and Mitchell, B.G. (1994) The response of Antarctic phytoplankton to ultraviolet radiation: absorption, photosynthesis, and taxonomic composition. In Weiler, C.S. and Penhale, P.A. (eds) *Ultraviolet Radiation in Antarctica: Measurements and Biological Effects*. American Geophysical Union, Antarctic Research Series, 62, 143–158.

Vidas, D. (1996) The legitimacy of the Antarctic tourism regime. In Stokke, O.S. and Vidas, D. (eds) *Governing the Antarctic. The Effectiveness and Legitimacy of the Antarctic Treaty System*. Cambridge University Press, Cambridge, 294–320.

Vincent, W.F. (1988) *Microbial Ecosystems of Antarctica*. Cambridge University Press, Cambridge.

Vincent, W.F. (1996) *Environmental Management of a Cold Desert Ecosystem: the McMurdo Dry Valleys*. Report of a National Science Foundation Workshop, Santa Fe, New Mexico, 14–17 March 1995. Desert Research Institute, University of Nevada, Reno, Nevada.

Vincent, W.F. and Quesada, A. (1994) Ultraviolet radiation effects on cyanobacteria: implications for Antarctic microbial ecosystems. In Weiler, C.S. and Penhale, P.A. (eds) *Ultraviolet Radiation in Antarctica: Measurements and Biological Effects*. American Geophysical Union, Antarctic Research Series, 62, 111–124.

Voronina, N.M. (1966) Distribution of zooplankton biomass in the Southern Ocean. *Oceanology*, 6, 836–846.

Vosjan, J.H. and Pauptit, E. (1992) Penetration of photosynthetically available light (PAR), UV-A and UV-B in Admiralty Bay, King George Island, Antarctica. *Circumpolar Journal*, 7(1–2), 50–58.

Voss, G.L. (1973) Cephalopod resources of the world. *FAO Fisheries Circular*, 149, 75.

Voytek, M.A. (1989) *Ominous Future Under the Ozone Hole: Assessing Biological Impacts in Antarctica*. Environmental Defense Fund, Washington DC.

Voytek, M.A. (1990) Addressing the biological effects of decreased ozone in the Antarctic environment. *Ambio*, 19(2), 52–61.

Wace, N. (1990) Antarctica: a new tourist destination. *Applied Geography*, 10, 327–41.

Wadhams, P. (1990) The resource potential of Antarctic icebergs. In Splettstoesser, J.F. and Dreschhoff, G.A.M. (eds) *Mineral Resources Potential of Antarctica*. American Geophysical Union, Antarctic Research Series 51, Washington DC, 203–215.

Wadhams, P. (1991) Atmosphere–ice–ocean interactions in the Antarctic. In Harris, C.M. and Stonehouse, B. (eds) *Antarctica and Global Climatic Change*. Belhaven Press, London, 65–81.

Wadhams, P. (1994) The Antarctic Sea Ice Cover. In Hempel, G. (ed.) *Antarctic Science: Global Concerns*. Springer-Verlag, Berlin 45–59.

Wadhams, P. and Crane, D.R., (1991) SPRI participation in the Winter Weddell Gyre Study 1989. *Polar Record*, 27 (160), 29–38.

Wadhams, P. and Davis, N.R. (1997) Climate-related research in the UK on Antarctic sea ice. *Globe*, No. 36, UK Global Environmental Research Office, Swindon, 11–13.

Wadhams, P., Lange, M.A. and Ackley, S.F. (1987) The ice thickness distribution across the Atlantic sector of the Antarctic Ocean in winter. *Journal of Geophysical Research*. 92(C13), 14535–14552.

Walcott, R.I. (1970) Isostatic response to loading of the crust in Canada. *Canadian Journal of Earth Sciences*, 7, 716–727.

Walker, B.C. (1983) The Beacon Supergroup of Northern Victoria Land, Antarctica. In Oliver, R.L., James, P.R. and Jago, J.B. (eds) *Antarctic Earth Science*. Cambridge University Press, Cambridge, 211–214.

Walker, T.R., Reid, K., Arnould, J.P.Y. and Croxall, J.P. (1997) Marine debris surveys at Bird Island, South Georgia 1990–1995. *Marine Pollution Bulletin*, 34, 61–65.

Walton, D.W.H. (1982a) Floral phenology in the South Georgian vascular flora. *British Antarctic Survey Bulletin*, 55, 11–25.

Walton, D.W.H. (1982b) The first South Georgia leases: Compañia Argentina de Pesca and the South Georgia Exploring Company Limited. *Polar Record*, 21, 231–240.

Walton, D.W.H. (1984) The terrestrial environment. In Laws, R.M. (ed.) *Antarctic Ecology*, vol.1. Academic Press, London, 1–60.

Walton, D.W.H. (1985) The subantarctic islands. Bonner, W.N. and Walton, D.W.H. (eds) *Key Environments: Antarctica*. Pergamon Press, Oxford, 293–317.

Walton, D.W.H. (ed.) (1987a) *Antarctic Science*. Cambridge University Press, Cambridge.

Walton, D.W.H. (1987b) Geography, politics and science. In Walton, D.W.H. (ed.) *Antarctic Science*. Cambridge University Press, Cambridge, 6–68.

Walton, D.W.H. (1987c) Antarctic terrestrial ecosystems. *Environment International*, 13, 83–93.

Walton, D.W.H. (1987d) *The Biological Basis for Conservation of Subantarctic Islands*. Report of the Joint SCAR/IUCN Workshop, Paimpont, France, September 1986.

Walton, D.W.H. (1990a) Colonisation of terrestrial habitats: organisms, opportunities and occurrence. In Kerry, K.R. and Hempel, G. (eds) *Antarctic Ecosystems. Ecological Change and Conservation*. Springer-Verlag, Berlin. 51–60.

Walton, D.W.H. (1990b) Waste disposal – expectations and realities. *Antarctic Science*, 2, 103.

Walton, D.W.H. (1991) The Environmental Protocol – scientific advantage or bureaucratic inconvenience? *Antarctic Science*, 3, 349.

Walton, D.W.H. and Bonner, W.N. (1985a) Terrestrial habitats. In Bonner, W.N. and Walton, D.W.H. (eds) *Key Environments: Antarctica*. Pergamon Press, Oxford, 71–72.

Walton, D.W.H. and Bonner, W.N. (1985b) History and exploration in Antarctic biology. In Bonner, W.N. and Walton, D.W.H. (eds) *Key Environments: Antarctica*. Pergamon Press, Oxford 1–20.

Walton, D.W.H. and Morris, E.M. (1990) Science, environment and resources in Antarctica. *Applied Geography*, 10, 265–286.

Walton, D.W.H. and Shears, J. (1994) The need for environmental monitoring in Antarctica: baselines, environmental impact assessments, accidents and footprints. *Journal of Environmental and Analytical Chemistry*, 35, 77–90.

Walton, D.W.H. and Thomson, M.R.A. (1994) Interpreting the Environmental Protocol – a recipe for international confusion? *Antarctic Science*, 6, 431.

Warnke, D.A., Marzo, B. and Hodell, D.A. (1996) Major deglaciation of East Antarctica during the early Late Pliocene? Not likely from a marine perspective. *Marine Micropalaeontology*, 27, 237–251.

Warren, S.G. and Clarke, A.D. (1990) Soot in the atmosphere and snow surface of Antarctica. *Journal of Geophysical Research*, 95, 1811–1816.

Washburn, A.L. (1979) *Geocryology. A Survey of Periglacial Processes and Environments*. Edward Arnold, London.

Washburn, A.L. (1980) Focus on polar research, *Science*, 209, 643–652.

Watkins, B.P. and Cooper, J. (1986) Introduction, present status and control of alien species at the Prince Edward Islands, sub-Antarctic. *South African Journal of Antarctic Research*, 16, 86–94.

Watson, G.E., Angle, J.P. and Harper, P.C. (1975) *Birds of the Antarctic and Sub-Antarctic*. American Geophysical Union, Antarctic Research Series 12, Washington DC.

Watson, R.T., Prather, M.J. and Kurylo, M.H. (1988) *Present State of Knowledge of the Upper Atmosphere 1988: An Assesssment Report*. NASA, Washington DC.

Watts, A.D. (1986a) The Antarctic Treaty as a conflict resolution mechanism. In Polar Research Board, *Antarctic Treaty System: An Assessment*. National Academy Press, Washington DC, 65–75.

Watts, A.D. (1986b) A British perspective. In Millar, T.B. (ed.) *Australia, Britain and Antarctica*. Australian Studies Centre, Institute of Commonwealth Studies, University of London, London, 51–59.

Watts, A.D. (1987) Antarctic mineral resources: negotiations for a mineral resources regime. In Triggs, G.D. (ed.) *The Antarctic Treaty Regime. Law, Environment and Resources*. Cambridge University Press, Cambridge, 164–175.

Watts, A.D. (1992) *International Law and the Antarctic Treaty System*. Grotius Publications Ltd, Cambridge.

Weatherly, J.M. Walsh, J.E. and Zwally, H.J. (1991) Antarctic sea ice variations and seasonal air temperatures. *Journal of Geophysical Research*, 98(C9), 15119–15130.

Webb, D.J., Kilworth, P.P., Coward, A.C. and Thompson, S.R. (1991) *The FRAM Atlas of the Southern Ocean*. Natural Environment Research Council, Swindon.

Webb, P.N. and Harwood, D.M. (1991) Late Cenozoic glacial history of the Ross Embayment, Antarctica. *Quaternary Science Reviews*, 10, 215–233.

Webb, P.N., Harwood, D.M., McKelvey, B.C., Mercer, J.N. and Stott, L.D. (1984) Cenozoic marine sedimentation and ice-volume variation on the East Antarctic craton. *Geology*, 12, 287–291.

Weddell, J. (1825) *A voyage towards the South Pole performed in the years 1822–24*. Longman, Hurst, Orme, Browne and Green, London.

Weeks, W.F. (1980) Iceberg water: an assessment. *Annals of Glaciology*, 1, 5–10.

Weeks, W.F., Ackley, S.F. and Govoni, J. (1989) Sea ice ridging in the Ross Sea, Antarctica, as compared with sites in the Arctic. *Journal of Geophysical Research*, 94(C4), 4984–4988.

Weeks, W.F. and Campbell, W.J. (1973) Icebergs as a fresh water source: an appraisal. *Journal of Glaciology*, 12, 207–232.

Weimerskirch, H. and Jouventin, P. (1987) Population dynamics of the wandering albatross, *Diomedea exulans*, of the Crozet Islands: causes and consequences of the population decline. *Oikos*, 49, 315–322.

Weimerskirch, H., Brothers, N. and Jouventin, P. (1997) Population dynamics of wandering albatross *Diomedea exulans* and Amsterdam albatross *D. amsterdamensis* in the Indian Ocean and their relationship with long-line fisheries: conservation implications. *Biological Conservation*, 79, 257–270.

Weimerskirch, H., Clobert, J. and Jouventin, P. (1987) Survival in five southern albatrosses and its relationship with their life history. *Journal of Animal Ecology*, 56, 1043–1055.

Weissel, J.K. and Hayes, D.E. (1972) *Magnetic anomalies in the southeast Indian Ocean.* American Geophysical Union, Antarctic Research Series 19, Washington DC.

Welch, W.M. (1992) The Antarctic Treaty System: is it adequate to regulate or eliminate the environmental exploitation of the globe's last wilderness? *Houston Journal of International Law*, 14, 597–657.

Weller, G. (1992) Antarctica and the detection of environmental change. *Philosophical Transactions of the Royal Society of London*, B338, 201–208.

Weller, G., Bentley, C.R., Elliot, D.H., Lanzerotti, L.J. and Webber, P.J. (1987) Laboratory Antarctica: research contributions to global problems. *Science*, 238, 1361–1368.

Weller, M. W. (1975) Notes on the formation and life of ponds of the Falkland Islands and South Georgia. *Bulletin of the British Antarctic Survey*, 40, 37–47.

Westermeyer, W. (1982) Resource allocation in Antarctica. *Marine Policy*, 6, 303–325.

Whillans, I.M. and Cassidy, W.A. (1983) Catch a falling star: meteorites and old ice. *Science*, 222, 55–57.

White, G.J. (1995) Microbial ecosystems in Antarctica: is protection necessary? *Antarctic Journal of the United States*, 30, 13–15.

White, W.B. and Peterson, R.G. (1996) An Antarctic circumpolar wave in surface pressure, wind, temperature and sea-ice extent. *Nature*, 380, 699–702.

Whitehead, M.D., Johnstone, G.W. and Burton, H.R. (1990) Annual fluctuations in productivity and breeding success of Adélie penguins and fulmarine petrels in Prydz Bay, East Antarctica. In Kerry, K.R. and Hempel, G. (eds) *Antarctic Ecosystems: Ecological change and Conservation.* Berlin, Springer-Verlag, 213–223.

Whiticar, M.J., Suess, E. and Wehner, H. (1985) Thermogenic hydrocarbons in surface sediments of the Bransfield Strait, Antarctic Peninsula. *Nature*, 314, 87–90.

Wiens, J.A. (1996) Oil, seabirds, and science. The effects of the *Exxon Valdez* oil spill. *BioScience*, 46, 587–597.

Wilder, M. (1993) *Liability on Ice: Environmental Damage and the Antarctic Madrid Protocol.* Centre for Resource and Environmental Studies, Australian National University, Canberra, Working Paper 1993/3.

Wilder, M. (1995) The settlement of disputes under the Protocol on Environmental Protection to the Antarctic Treaty. *Polar Record*, 31, 399–408.

Wilkes, O. and Mann, R. (1978) The story of Nukey Poo. *Bulletin of the Atomic Scientists*, 34(8), 32–36.

Wilkinson, I.S. and Bester, M.N. (1990) Continued population increase in fur seals, *Arctocephalus tropicalis* and *A. gazella*, at the Prince Edward Islands. *South African Journal of Antarctic Research*, 20, 58–63.

Wilkinson, I.S. and Bester, M.N. (1988) Is onshore human activity a factor in the decline of the southern elephant seal? *South African Journal of Antarctic Research*, 18, 14–17.

Wilkniss, P. (1990) Fuel spill clean up in the Antarctic. *Antarctic Journal of the United States*, 25(4), 3–10.

Willan, R.C.R. (1989) Book Review. 'Minerals and Mining in Antarctica – Science and Technology, Economics and Politics' by M.J. de Wit. *Geological Journal*, 24, 229–231.

Willan, R.C.R., Macdonald, D.I.M. and Drewry, D.J. (1990) The mineral resource potential of Antarctica: geological realities. In Cook, G. (ed.) *The Future of Antarctica. Exploitation Versus Preservation.* Manchester University Press, Manchester, 25–43.

Williams, A.J. (1984) Status and conservation of seabirds and some islands in the African sector of the Southern Ocean. In Croxall, J.P., Evans, P.G.H. and Schreiber, R.W. (eds) *Status and Conservation of the World's Seabirds.* International Council for Bird Preservation, ICBP Technical Publication No. 2, Cambridge, 627–635.

Williams, R. (1985) The potential impact of a krill fishery upon pelagic fish in the Prydz Bay area of Antarctica. *Polar Biology*, 5, 1–5.

Wilsher, W.A. and de Wit, M.J. (1990) Toward a quantitative mineral resource assessment of Antarctica from a Gondwana perspective. In Splettstoesser, J.F. and Dreschhoff, G.A.M. (eds) *Mineral Resources Potential of Antarctica.* American Geophysical Union, Antarctic Research Series 51, Washington DC, 1–6.

Wilson, G.S. (1995) The Neogene East Antarctic Ice Sheet: a dynamic or stable feature? *Quaternary Science Reviews*, 14, 101–123.

Wilson, C.A. and Mitchell, J.F.B. (1987) A doubled CO_2 sensitivity experiment with a GCM including a simple ocean. *Journal of Geophysical Research*, 92, 13315–13343.

Wilson J.T. (1976) Continental drift. In Wilson, J.T. (ed.) *Continents Adrift and Continents Aground.* Scientific American, San Francisco, 19–35.

Wilson, K.-J. (1990) Fluctuations in populations of Adélie penguins at Cape Bird, Antarctica. *Polar Record*, 26, 305–308.

Wilson, K.-J., Taylor, R.H. and Barton, K.J. (1990) The impact of man on Adélie penguins at Cape Hallet, Antarctica. In Kerry, K.R. and Hempel, G. (eds) *Antarctic Ecosystems. Ecological Change and Conservation.* Springer-Verlag, Berlin, 183–190.

Wilson, R.P., Culik, B., Danfield, R. and Adelung, D. (1991) People in Antarctica – how much do Adélie penguins *Pygoscelis adeliae* care? *Polar Biology*, 11, 363–370.

Woehler, E.J., Penney, R.L., Creet, S.M. and Burton, H.R. (1994) Impacts of human visitors on breeding success and long-term population trends in Adélie penguins at Casey, Antarctica. *Polar Biology*, 14, 269–274.

Wolff, E.W. (1990) Signals of atmospheric pollution in polar snow and ice. *Antarctic Science*, 2, 189–205.

Wolff, E.W. (1992) The influence of global and local atmospheric pollution on the chemistry of Antarctic snow and ice. *Marine Pollution Bulletin*, 25, 274–280.

Wolff, E.W. and Peel, D.A. (1985) The record of global pollution in polar snow and ice. *Nature*, 313, 535–540.

Wolff, E.W. and Peel, D.A. (1988) Concentrations of cadmium, copper, lead and zinc in snow from near DYE 3 in south Greenland. *Annals of Glaciology*, 10, 193–197.

Wolff, E.W. and Suttie, E.D. (1994) Antarctic snow record of Southern Hemisphere lead pollution. *Geophysical Research Letters*, 21, 781–784.

Wolfrum, R. (1991) *The Convention on the Regulation of Antarctic Mineral Resource Activities.* Springer-Verlag, Berlin.

Woolcott, R.A. (1986) The interaction between the Antarctic Treaty System and the United Nations System. In Polar Research Board, *Antarctic Treaty System. An Assessment.* National Academy Press, Washington DC, 375–390.

World Commission on Environment and Development (1987) *Our Common Future.* Oxford University Press, Oxford.

Worrest, R.C. (1982) Review of literature concerning the impact of UV-B radiation upon marine organisms. In Calkins, J. (ed.) *The Role of Solar Ultraviolet Radiation in Marine Ecosystems.* Plenum Press, New York and London, 429–457.

Wouters, M. and Hall, C.M. (1995a) Managing tourism in the sub-Antarctic islands. In Hall, C.M. and Johnston, M.E. (eds) *Polar Tourism: Tourism in the Arctic and Antarctic Regions.* Wiley, Chichester, 257–276.

Wouters, M. and Hall, C.M. (1995b) Tourism and New Zealand's sub-Antarctic islands. In Hall, C.M. and Johnston, M.E. (eds) *Polar Tourism: Tourism in the Arctic and Antarctic Regions.* Wiley, Chichester, 277–295.

Wright, M.A. and Williams, P.L. (1974) *Mineral Resources of Antarctica.* US Geological Survey Circular 205.

Wright, R., Anderson, J.B. and Fisco, P.O. (1993) Distribution and association of sediment gravity flow deposits and glacial/glacial marine sediments around the continental margin of Antarctica. In Molnia, B.F. (ed.) *Glacial–Marine Sedimentation.* Plenum Press, New York, 233–264.

Wynn-Williams, D.D. (1994) Potential effects of ultraviolet radiation on Antarctic primary terrestrial colonizers: cyanobacteria, algae, and cryptogams. In Weiler, C.S. and Penhale, P.A. (eds) *Ultraviolet Radiation in Antarctica: Measurements and Biological Effects.* American Geophysical Union, Antarctic Research Series 62, Washington DC, 243–257.

Yamamoto, Y., Honda, K. and Tatsukawa, R. (1987) Heavy metal accumulation in Antarctic krill *Euphausia superba. Proceedings of the NIPR Symposium on Polar Biology*, 1, 198–204.

Young, A.R., Björn, L.O., Moan, J. and Nultsch, W. (eds) (1993) *Environmental UV Photobiology.* Plenum, New York.

Young, E.C. (1990) Long-term stability and human impact in Antarctic skuas and Adélie penguins. In Kerry, K.R. and Hempel, G. (eds) *Antarctic Ecosystems. Ecological Change and Conservation.* Springer-Verlag, Berlin, 231–236.

Young, E.C. (1991) Critical ecosystems and nature conservation in Antarctica. In Harris, C.M. and Stonehouse, B. (eds) *Antarctica and Global Climatic Change.* Belhaven Press, London, 117–146.

Young, G.C. (1991) Fossil fishes from Antarctica. In Tingey, R.J. (ed.) *The Geology of Antarctica.* Clarendon Press, Oxford. 538–567.

Young, N.W., Raynaud, D., de Angelis, M., Petit, J.-R. and Lorius, C. (1984) Past changes of the Antarctic ice sheet in Terre Adélie as deduced from ice–core data and ice modelling. *Annals of Glaciology*, 5, 239.

Yuan, X., Cane, M.A. and Martinson, D.G. (1996) Cycling around the South Pole. *Nature*, 380, 673–674.

Zain, A. (1986) The Antarctic Treaty System from the perspective of a state not party to the system. In Polar Research Board, *Antarctic Treaty System. An Assessment*. National Academy Press, Washington DC, 305–313.

Zain, A. (1987) Antarctica: the claims of 'expertise' versus 'interest'. In Triggs, G.D. (ed.) *The Antarctic Treaty Regime. Law, Environment and Resources*. Cambridge University Press, Cambridge, 211–217.

Zemsky, V.A., Berzin, A.A., Mikhaliev, Y.A. and Tormosov, D.D. (1995) Soviet Antarctic pelagic whaling after WWII: review of actual catch data. *Report of the International Whaling Commission*, 45, 131–135.

Zorn, S.A. (1984) Antarctic minerals. A common heritage approach. *Resources Policy*, 10, 2–18.

Zou Keyuan 1993. China's Antarctic policy and the Antarctic Treaty System. *Ocean Development and International Law*, 24, 237–255.

Zumberge, J.H. (1979) *Possible Environmental Effects of Mineral Exploration and Exploitation in Antarctica*. SCAR, Cambridge.

Zumberge, J.H. (1986) The Antarctic Treaty as a scientific mechanism – the Scientific Committee on Antarctic Research and the Antarctic Treaty System. In Polar Research Board, *Antarctic Treaty System. An Assessment*. National Academy Press, Washington DC, 153–168.

Zumberge, J.H. and Kimball, L.A. (1986) Workshop on the Antarctic Treaty System: overview. In Polar Research Board, *Antarctic Treaty System. An Assessment*. National Academy Press, Washington DC, 3–12.

Zwally, H. J. (1991) Breakup of Antarctic ice. *Nature*, 350, 274.

Zwally, H.J. (1994) Detection of change in Antarctica. In Hempel, G. (ed.) *Antarctic Science: Global Concerns*. Springer-Verlag, Berlin 126–143.

Zwally, H.J. and Gloerson, P. (1977) Passive microwave images of the polar regions and research applications. *Polar Record*, 18, 431–450.

Zwally, H.J., Comiso, J.C. and Gordon, A.L. (1985) Antarctic offshore leads and polynyas and oceanographic effects. In Jacobs, S.S. (ed.) *Oceanography of the Antarctic Continental Shelf*. American Geophysical Union, Antarctic Research Series 43, Washington DC, 203–226.

Zwally, H.J., Comiso, J.C., Parkinson, C.L., Campbell, J., Carsey, F.D. and Gloerson, P. (1983) *Antarctic sea ice 1973–1976: satellite passive microwave observations*. (Report SP–459). NASA, Scientific and Technical Information Branch, Washington DC.

Index

Note: Alphabetical arrangement of headings and subheadings is word by word, ignoring words such as 'and', 'in', 'on' and 'under'. Acronyms and abbreviations are treated as words. Page numbers in *italic* refer to diagrams and illustrations.

For Product Safety Concerns and Information please contact our EU representative GPSR@taylorandfrancis.com Taylor & Francis Verlag GmbH, Kaufingerstraße 24, 80331 München, Germany